AA001138

2024 IFIP/IEEE 32nd International Conference on Very Large Scale Integration (VLSI-SoC 2024)

Tanger, Morocco
6-9 October 2024

IEEE Catalog Number: **CFP24LSI-POD**
ISBN: **979-8-3315-3968-9**

**Copyright © 2024 by the Institute of Electrical and Electronics Engineers, Inc.
All Rights Reserved**

Copyright and Reprint Permissions: Abstracting is permitted with credit to the source. Libraries are permitted to photocopy beyond the limit of U.S. copyright law for private use of patrons those articles in this volume that carry a code at the bottom of the first page, provided the per-copy fee indicated in the code is paid through Copyright Clearance Center, 222 Rosewood Drive, Danvers, MA 01923.

For other copying, reprint or republication permission, write to IEEE Copyrights Manager, IEEE Service Center, 445 Hoes Lane, Piscataway, NJ 08854. All rights reserved.

*** This is a print representation of what appears in the IEEE Digital Library. Some format issues inherent in the e-media version may also appear in this print version.

IEEE Catalog Number: CFP24LSI-POD
ISBN (Print-On-Demand):   979-8-3315-3968-9
ISBN (Online):   979-8-3315-3967-2
ISSN: 2324-8432

Additional Copies of This Publication Are Available From:

Curran Associates, Inc
57 Morehouse Lane
Red Hook, NY 12571 USA
Phone: (845) 758-0400
Fax: (845) 758-2633
E-mail: curran@proceedings.com
Web: www.proceedings.com

TABLE OF CONTENTS

A Novel Design Technique for Enhanced Security and New Applications of Ferroelectric-Based Non-Volatile SRAM 1
Lucas Rhetat, Jean-Philippe Noel, Bastien Giraud, Laurent Grenouillet, Julie Laguerre, Cédric Marchand, Ian O'Connor

Capture the Pulse: Impact of FPGA Resource Utilization on EM Fault Injection Attacks Detection 7
Sami El Amraoui, Régis Leveugle, Paolo Maistri

Continuity in Security: Leveraging LLM for Translating Security Properties Across Hardware Designs 13
Bulbul Ahmed, Sujan Kumar Saha, Jingbo Zhou, Sohrab Aftabjahani, Mark Tehranipoor, Farimah Farahmandi

OSHDA: A Containerized CAD Tool for the Design and Analysis of Behavioral FSM Logic Locking 19
Esrat Khan, Shahzad Muzaffar, Lamees M. Al Qassem, Ibrahim M. Elfadel

APPAMM: Memory Management for IPsec Application on Heterogeneous SoCs 25
Ayushi Agarwal, Radhika Dharwadkar, Isaar Ahmad, Krishna Kumar, P. J. Joseph, Sourav Roy, Prokash Ghosh, Preeti Ranjan Panda

Embedded and Real-Time Anomalous Command Classification in Unmanned Ground Vehicle Operations 31
Rafaella Elia, Theocharis Theocharides

Low Power Network-On-Chip Architecture Design Technique 37
Tejas Musale, Arun Ganti, Ankur Gogoi, Kanchan Manna

Minimum Depth Quantum Modular Addition Through Carry-Save Architecture 43
Siyi Wang, Eugene Lim, Xiufan Li, Jerrie Feng, Anupam Chattopadhyay

NEATRouter: A New Method for 2D Global Routing 49
Luis Enrique Murillo Vizcardo, Ricardo Reis

Adaptable FWHW Formal Co-Verification of SoC RISC-V Components 55
Paulette Iskandar, Bryan Olmos, Wolfgang Kunz, Djones Lettnin

Behavioral Simulation of Relative Timed Asynchronous Circuits 61
Sumanth Kolluru, Kenneth S. Stevens

Diagnostic Coverage Estimation for Automotive SoCs Based on Colored Stochastic Petri Nets 67
Ernesto Cristopher Villegas Castillo, Felipe Augusto Da Silva, Michael Glaß

Linear Algebra Approach to Verification of Modular (2^n-1) Multipliers 73
Jiteshri Dasari, Cunxi Yu, Maciej Ciesielski

Adaptive Block-Scaled GeMMs on Vector Processors for DNN Training at the Edge 79
Nitish Satya Murthy, Nathan Laubeuf, Debjyoti Bhattacharjee, Francky Catthoor, Marian Verhelst

A High Throughput, Energy-Efficient Architecture for Variable Precision Computing in DRAM 85
Gian Singh, Ayushi Dube, Sarma Vrudhula

GemIMC: A Configurable HW Architecture for Technology Agnostic IMC Based NN Inference 91
Emilien Taly, Roberto Guizzetti, Pascal Urard, Elena-Ioana Vatajelu

Stochastic Spintronics Device Based Bayesian Networks for Efficient Uncertainty Modeling 97
N/A

AFSRAM-CIM: Adder Free SRAM-Based Digital Computation-In-Memory for BNN 103
Asmae El Arrassi, Mohammad Amin Yaldagard, Xingjian Tao, Taha Shahroodi, Fouwad Mir,
Yashvardhan Biyani, Manil Dev Gomony, Anteneh Gebregiorgis, Rajiv Joshi, Said Hamdioui

DynaCache: A Checkpoint Aware Reconfigurable Cache for Intermittently Powered Computing
Systems 109
Rishabh Mahanta, Hemangee K. Kapoor

In-Memory Mirroring: Cloning Without Reading 115
Simranjeet Singh, Ankit Bende, Chandan Kumar Jha, Vikas Rana, Rolf Drechslert, Sachin
Patkar, Farhad Merchant

Multi-Level FeFET-Based CAM Address Decoder 121
Thomas Makryniotis, Georgi Gaydadjiev, Said Hamdioui, Mottaqiallah Taouil

A Low-Power Linear Phase Interpolation-Based Delay Line in 12nm FinFET Technology 127
Mohammadreza Esmaeilpour, Jan Lappas, Christian Weis, Norbert Wehn

Benchmarking Microfluidic Design Automation Flows 132
Ashton Snelgrove, Skylar Stockham, Pierre-Emmanuel Gaillardon

Understanding Transistor Aging Impact on the Behavior of RRAM Cells 138
Seyed Hossein Hashemi Shadmehri, Supriya Chakraborty, Thiago Santos Copetti, Fabian
Luis Vargas, Letícia Maria Bolzani Poehls

A New Control Law for N-Path Mixer Switches Enhancing Harmonic Rejection 144
Hasan Moussa, Estelle Lauga-Larroze, Laurent Fesquet

A Scalable Hardware Architecture for Efficient Learning of Recurrent Neural Networks at the Edge 148
Yicheng Zhang, Manil Dev Gomony, Henk Corporaal, Federico Corradi

Commercial Evaluation of Zero-Skipping MAC Design for Bit Sparsity Exploitation in DL
Inference 152
Harideep Nair, Prabhu Vellaisamy, Tsung-Han Lin, Perry Wang, Shawn Blanton, John Paul
Shen

Compensating the Load Effect in Quadrature All-Pass Filters 156
U. Esteban Eraso, C. Sánchez-Azqueta, F. Aznar, C. Aldea, S. Celma

Design Co-Processor Based on Partially Homomorphic Encryption Execution Using Open-Source
Tool 160
Mujahid Bilal, M. Kamran Bhatti, Muhammad Kahsif Minhas, Haroon Waris

Exploiting Functional Approximation on Decision-Tree Based Multiple Classifier Systems 164
Mario Barbareschi, Salvatore Barone, Antonio Emmanuele, Nicola Mazzocca

FortBoot: Fortifying Rooted-In-Device-Specific Security Through Secure Booting 168
Sajeed Mohammad, Farimah Farahmandi

FVDCLS: Functional Verification of RISCV Based Dual-Core Lockstep Feature Using Fault Injection Mechanism 172
    Muhammad Kashif Minhas, Haroon Waris, Sajid Baloch

Heterogeneous Approximation of DNN HW Accelerators Based on Channels Vulnerability 176
    Natalia Cherezova, Salvatore Pappalardo, Mahdi Taheri, Mohammad Hasan Ahmadilivani, Bastien Deveautour, Alberto Bosio, Jaan Raik, Maksim Jenihhin

Enhanced Diagnosis of Failing Bits in Memory Built-In Self-Test 180
    Ali Shisha, Balajiraja Ravinarayanan, Knut Mellenthin

Exploring the Role of the Portable Stimulus Standard in Enhancing Security Property Verification 184
    Jaimini Nagar, Thorsten Dworzak, Sebastian Simon, Ulrich Heinkel, Djones Lettnin

High-Density Standard Cell Library for Sequential 3D Integrated Circuits 188
    Arturo Prieto, Joachim Rodrigues

Holistic Framework for Evaluating the Trustworthiness of Integrated Circuits 192
    Mouadh Ayache, Enkele Rama, Saleh Mulhem, Mladen Berekovic, Matthias Korb

Lightweight Active Fences for FPGAs 196
    Anis Fellah-Touta, Lilian Bossuet, Vincent Grosso, Carlos Andres Lara-Nino

MCS-NTT: Multi-Chip System Design for NTT Acceleration 200
    Mohammed Nabeel, Homer Gamil, Johann Knechtel, Michail Maniatakos

MEAN: Mixture-Of-Experts Based Neural Receiver 204
    Bram Van Bolderik, Vlado Menkovski, Sonia Heemstra, Manil Dev Gomony

Resource Management of Automotive Engine Control Units 208
    Istvan Andras Gergely, Sebastian Rausch, Nahla Elaraby, Axel Jantsch

SystemVerilog-SystemC TestBench Architecture for VLSI Chip Design Verification 212
    Mohammad Ismael, Ayman Hroub, Nasib Naser

Time-To-Digital Converter Based Self-Timed Ring Oscillator: An FPGA Implementation 216
    Assia El-Hadbi, Oussama Elissati, Laurent Fesquet

Reliability Assessment of Large DNN Models: Trading off Performance and Accuracy 220
    Junchao Chen, Giuseppe Esposito, Fernando Fernandes Dos Santos, Juan-David Guerrero-Balaguera, Angeliki Kritikakou, Milos Krstic, Robert Limas Sierra, Josie E Rodriguez Condia, Matteo Sonza Reorda, Marcello Traiola, Alessandro Veronesi

QCrop: An IoT Based Framework to Enhance Crop Productivity in Smart Agriculture 230
    Mahdi Shamsa, Laavanya Rachakonda, Saraju P. Mohanty, Elias Kougianos

SanaSolo 2.0: Edge-Based Monitoring and Management of Soil Fertility Using IoT 236
    Laavanya Rachakonda, Samuel Stasiewicz

Transforming Agriculture: A Mini-Review of IoT Innovations and Their Impact 242
    Laavanya Rachakonda

A Novel Current Comparator Enabling Large RRAM Crossbars for BNNs and PUFs 248
    Gokulnath Rajendran, Debajit Basak, Suman Deb, Siyi Wang, Anupam Chattopadhyay

Secure Software/Hardware Hybrid In-Field Testing for System-On-Chip 254
    Saleh Mulhem, Christian Ewert, Andrija Nešković, Amrit Sharma Poudel

The Impact of Logic Synthesis and Technology Mapping on Logic Locking Security.................................. 260
Lilas Alrahis, Mohammed Thari Nabeel, Johann Knechtel, Ozgur Sinanoglu

BlockShield: A TPM-Integrated Blockchain-Based Framework for Shielding Against Deepfakes 266
Venkata K. V. V. Bathalapalli, Aakarshan Kumar, Saraju P. Mohanty, Elias Kougianos,
Venkata P. Yanambaka

Fortified-Edge 5.0: Federated Learning for Secure and Reliable PUF in Authentication Systems 272
Seema G. Aarella, Venkata P. Yanambaka, Saraju P. Mohanty, Elias Kougianos

Performance Analysis of Greedy and Auction-Based Resource Allocation Algorithms in Ubiquitous
Computing Environments... 278
Akshay Nagpal, Vivekananda Jayaram, Manjunatha Sughaturu Krishnappa, Nikhil Jagdish
Bangad, Darshan Mohan Bidkar, Manoj Jayantilal Kathiriya, Seema G Aarella

A Unified Functional Safety EDA Framework for Accurate Diagnostic Coverage Estimation 284
Abhiroop Bhowmik, Subin Babukutty, Mottaqiallah Taouil, Moritz Fieback

3D VNWFET-Based Standard Cell Library Design Flow: From Circuit and Physical Design to
Logic Synthesis 290
Sara Mannaa, Cédric Marchand, Damien Deleruyelle, Bastien Deveautour, Alberto Bosio,
Christoph Lenz, Oskar Baumgartner, Ian O'Connor

Author Index

Foreword

On behalf of the steering committee and the program committee members, we would like to welcome you all to the 32st IFIP/IEEE International Conference on Very Large-Scale Integrated (VLSI-SoC 2024), October 6-9, 2024, in the vibrant city of Tangier, Morocco. Esteemed as a pivotal gathering, this conference brings together the brightest minds, pioneers, and visionaries to share groundbreaking research, innovative concepts, and the exploration of current trends in VLSI and SoC technology. Organized collaboratively by Delft University of Technology (Netherlands) and University of Abdelmalek Saadi University (Morocco), this conference isn't merely an event; it serves as the cornerstone of a comprehensive agenda, hosting specialized workshops, interactive sessions, and a dynamic environment encouraging the exchange of invaluable insights. VLSI-SoC is co-sponsored by IFIP, IEEE, the Council on Electronic Design Automation (CEDA), and the IEEE Circuits and Systems Society (CAS).

The VLSI-SoC 2024 format has one day of tutorials, a three-day technical program, embedded tutorials, and an attractive social program. The conference technical program consists of three plenary keynote speakers, technical paper presentations in two parallel sessions, poster sessions and two plenary panel sessions.

An event of this nature and dimension is the result of the contributions of various individuals and institutions. These include authors' technical contributions, the effort of reviewers on evaluating papers and providing feedback, the dedication of the Local Organization in preparing all the logistics and welcoming facilities, as well as the effort of the session chairs to ensure the correct scheduling of the program and that presenters obtain an interactive discussion with participants. We express our gratitude to all of them. Finally, we thank the support of all public and private institutions that sponsored this event.

We are happy to report that more than 5000 USD of travel grants have been awarded to students from developing countries to help them attend this conference. We welcome these students to the VLSI-SoC community and heartily thank IFIP for funding their travel grants.

We hope that you will find VLSI-SoC 2024 rewarding and exciting. We wish you all an instructive and enjoyable conference in Tangier and hope that you will keep making VLSI-SoC a success by actively participating in it, assisting to its organization, and letting us always know when we can do something better. Thank you all for coming, and welcome at VLSI-SoC in Tangier.

Best wishes to you all for an enriching, memorable VLSI-SoC 2024.

General Co-Chairs

Said Hamdioui   Anass El Haddadi
Delft University of Technology   Abdelmalek Saadi University
The Netherlands   Morocco

Program Committee

Program Chairs

Hussam Amrouch, Technical University of Munich (DE)

Ioana Vatejelu, TIMA Grenoble (FR)

Topic Chairs

T1 - Analog design

Co-Chairs: Salvador Mir - TIMA Laboratory, Martin Andraud - UC Louvain

Topics: modelling, simulation, verification, and prototyping of analog, mixed-signal, RF, sensors, and IoT circuits

T2 - Digital design

Co-Chairs: Joachim Rodrigues - Lunds University, Dimitrios Soudris - ECE-NTUA

Topics: modelling, simulation, verification, and prototyping of circuits and systems, SoC, NoC, reconfigurable, IoT, and low-power architectures

T3 - Design for AI hardware and emerging applications

Co-Chairs: Ian O'Connor - Ecole Centrale de Lyon, Panagiotis Dimitrakis - NCSR "Demokritos"

Topics: unconventional computing, brain inspired computing, computation-in-memory, photonics, quantum, etc.

T4 - EDA tools and methodologies for IC design

Co-Chairs: Katell Morln-Allory - TIMA Laboratory, Tara Ghasempouri - TalTech

T5 - Hardware Dependability

Co-Chairs: Ibrahim Elfadel - Khalifa University, Paolo Bernardi - Politecnico di Torino

Topics: Design for testability, reliability, fault tolerance, security, safety, and variability.

Program Committee Members

Al-Nashash Hasan - American University of Sharjah

Albasha Lutfi - American University of Sharjah

Atef Mohamed - United Arab Emirates University

Baas Bevan - University of California, Davis

BOSIO ALBERTO - Lyon Institute of Nanotechnology

Chattopadhyay Anupam - Nanyang Technological University

Claesen Luc - University Hasselt

El-Araby Nahla - TU Wien, Vienna, Austria and Canadian International College, Cairo, Egypt

Fieback Moritz - Delft University of Technology

Galuzzi Carlo - TU Delft

Gebregiorgis Anteneh - Delft University of Technology

Grimblatt Victor - Synopsys

Guo Xinfei - Shanghai Jiao Tong University

Guthaus Matthew - University of California Santa Cruz

Hamzaoglu Ilker - Ozyegin University

Huebner Michael - Brandenburg University of Technology Cottbus

Jenihhin Maksim - Tallinn University of Technology

Katkoori Srinivas - University of South Florida

Knechtel Johann - New York University

Kravets Victor - IBM

Kumm Martin - University of Applied Sciences, Fulda

Majzoub Sohaib - University of Sharjah

Malcovati Piero - University of Pavia

MARGARIA Tiziana - University of Limerick

Martins Ricardo - Instituto de Telecomunicações / Instituto Superior Técnico - University of Lisbon

Merchant Farhad - Newcastle University

Meribout Mahmoud - Khalifa University of Science & Technology

Miranda Jose - EPFL

MOHANTY Saraju - University of North Texas

Monteiro Jose - INESC-ID, IST ULisboa

Nawaz Kashif - Cryptography Research Centre, Technology Innovation Institute

Poehls Leticia Bolzani - Catholic University of Rio Grande do Sul

Pravadelli Graziano - University of Verona

Rasras Mahmoud - New York University Abu Dhabi

Ruospo Annachiara - Politecnico di Torino

Sa Mihai - Khalifa University of Science and Technology

Saadeh Wala - Western Washington University

Saghir Mazen - American University of Beirut

Shafique Muhammad - New York University Abu Dhabi (NYUAD)

Shamsi Kaveh - University of Texas at Dallas

SILVA-CARDENAS Carlos - Univ. Catolica del Peru

Siozios Kostas - Aristotle University of Thessaloniki

Taouil Mottaqiallah - Delft University of Technology

Theocharides Theocharis - University of Cyprus

Tsai Chun-Jen - National Yang Ming Chiao Tung University

Ugurdag H. F. - Ozyegin University

Viegas Jaime - Khalifa University

VIRAZEL Arnaud - LIRMM

Zakaria Amer - American University of Sharjah

Ziegler Matthew - IBM

<u>Keynote 1 – Secure Heterogeneous Integration and Advanced</u> Packaging: New Attack Surfaces and Grand Challenges Ahead

Mark M. Tehranipoor - University of Florida (US)

Abstract: Heterogeneous integration and advanced packaging have seen resurgence over the past few years. The notion of building a system in package from chiplets is quite attractive, however it comes with challenges of ensuring quality, reliability and security. Providing assurance for each chiplet, establishing a secure chiplet supply chain, ensuring secure and trusted integration, verifying and validating policies, etc are few important challenges that will be discussed in this presentation.

Bio: Mark M. Tehranipoor is currently the Intel Charles E. Young Preeminence Endowed Chair Professor and the Sachio Semmoto Chair of the Department of Electrical and Computer Engineering (ECE) at the University of Florida. His current research projects include: hardware security and trust, supply chain security, IoT security, VLSI design, test and reliability. He has 50+ patents issued/pending, 19 books, and 500+ conference/journal publications. He is a recipient of 18 best paper awards and nominations, as well as the 2008 IEEE Computer Society (CS) Meritorious Service Award, the 2012 IEEE CS Outstanding Contribution, the 2009 NSF CAREER Award, and the 2014 AFOSR MURI award. He received the 2020 University of Florida Innovation of the year as well as teacher/scholar of the year awards. He co-founded the IEEE International Symposium on Hardware-Oriented Security and Trust (HOST), IEEE International Conference on Physical Assurance and Inspection of Electronics (PAINE). He serves on the program committee of more than a dozen leading conferences and workshops. He has also served as Program and General Chair of a number of IEEE and ACM sponsored conferences and workshops (HOST, ITC, DFT, D3T, DBT, NATW, and more). He is currently serving as a founding EIC for Journal on Hardware and Systems Security (HaSS) and served as Associate Editor for TC, JETTA, JOLPE, TODAES, IEEE D&T, TVLSI. He is currently serving as a founding director for Florida Institute for Cybersecurity Research (FICS) and a number of other centers with focus on microelectronics security. Dr. Tehranipoor is the recipient of the Semiconductor Research Corporation (SRC) Aristotle Award, a Fellow of the IEEE, a Fellow of the ACM, a Fellow of the National Academy of Inventors (NAI), a Golden Core Member of IEEE CS, and Member of ACM SIGDA.

<u>Keynote 2 – Long Live Computing Technology</u>

Dr. Rajiv Joshi - IBM Research (US)

Abstract: The explosion of computers and the internet significantly improved the quality of human life. Communication across the globe made the earth as one family. Building blocks of this computing power are fabricated from crucial and key semiconductor technology. This talk covers such building blocks and showcases their impact. Volatile and non-volatile memories (NVM) have proved to be focal points for research over decades. Memories in general are the workhorse of the semiconductor industry. Applications of these spread across many domains such as Artificial Intelligence (AI), servers, high-performance computing, Systems on Chip (SOC), Internet of Things (IoT), quantum computing, etc., and thus are essential components of the computing world. As we march forward the scaling of memories poses a major challenge to achieve functionality, performance, area, power, and yield. To overcome scaling issues the talk will describe alternative techniques and circuits. It will bring out challenges and future directions for various memory applications.

Bio: Dr. Rajiv V. Joshi is an IEEE Fellow, winner of the prestigious IEEE Daniel Noble award, and a key technical lead/Research Scientist at T. J. Watson Research Center, IBM. He received his B. Tech IIT (Bombay, India), M.S (MIT), and Dr. Eng. Sc. (Columbia University). He has successfully led innovations in technology, memories (SRAM, DRAM, and others), and predictive analytic techniques for yield prediction for IBM Server Groups and their products. His statistical techniques are tailored for machine learning and AI which are licensed and commercialized. His memory innovations and work are used in both IBM P and Z servers. His technology innovations set IBM's leadership across the globe. He received 3 Outstanding Technical Achievement (OTAs), 3 highest Corporate Patent Portfolio awards for contributions in interconnect technologies, holds 73 invention plateaus, has over 290 US patents covering front end and back end of the line processes, and structures, volatile and non-volatile memories, Compute in Memory structures, machine learning algorithms, and quantum computing and over 425 international patents. He has authored and co-authored over 235 refereed papers, delivered over 60 invited/keynote talks, and given several Seminars. He received the NY IP Law Association "Inventor of the Year" award in Feb 2020. He is a Mercator Fellow at the University of Siegen, Germany. He received an industrial pioneer award in 2014 from the IEEE Circuits and Systems Society. He received the Best Editor Award from the IEEE TVLSI journal. He was inducted into the New Jersey Inventor Hall of Fame in Aug 2014. He won the Mehboob Khan Award two times from Semiconductor Research Corporation. He won several best paper awards from ISSCC 1992, VMIC 1998, ICCAD 2009, and ISQED 2014. He is a member of the IBM Academy of Technology and a master inventor. He serves on the Board of Governors for IEEE CAS as an industrial liaison. He served on EC for DAC, ISLPED, CICC, ISCAS, AICAS, and APCCAS (2023) committees as well as the AE of TCAS I and TVLSI. He served as a Distinguished Lecturer for IEEE CAS, CEDA, and EDS society. He is an ISQED and World Technology Network fellow and a distinguished alumnus of IIT Bombay. He served on the executive advisory committee for the Center of 3D Ferroelectric and Microelectronics at Penn State. He serves as an IEEE CAS Ambassador to India.

Keynote 3 – AI at the edge: Hype or Hope?

Henk Corporaal - Technical University of Eindhoven (NL)

Abstract: Artificial neural networks (ANNs), especially deep ones, have gained immense popularity. While their potential applications appear boundless, a significant drawback is their heavy reliance on cloud-based servers. This dependence stems from the exponential expansion in the size and computational requirements of these networks. For instance, the latest large language models like ChatGPT-4 boast over a trillion parameters and demand astonishing computational resources, in the order of 10^{25} FLOPS, for training. Achieving this training task in a reasonable timeframe (i.e., less than a month) requires the use of thousands of cloud-based servers. Even a single inference takes about 5.10^{14} FLOPS, making real-time inferencing extremely costly. We believe that there are substantial advantages in bringing intelligence directly to smart sensor devices at the network edge, performing the computation locally, close to the sensing data. However, these devices typically have a sub-Watt or even sub-mWatt power budget, and lack huge memories and compute capabilities. Achieving smart Edge-AI with less reliance on large AI servers necessitates a significant improvement in energy efficiency. This leads to the question: is AI at the edge really feasible, or is it a mindless dream? In this keynote, we address the state of the art (SOTA) in Edge computing and its trends and developments. We discuss what is needed to really bring AI to the Edge, how to bridge this huge energy-efficiency gap. Improvements are needed at all levels of the design stack; perhaps even new neural computing paradigms. We conclude by offering a glimpse into the future, exploring potential breakthroughs on the horizon.

Bio: Henk Corporaal is Professor in Embedded System Architectures at the Eindhoven University of Technology (TU/e) in The Netherlands. He has gained an MSc in Theoretical Physics from the University of Groningen, and a PhD in Electrical Engineering, in the area of Computer Architecture, from Delft University of Technology. His research is on low-power multi-processor, heterogeneous processing architectures, their programmability, and the predictable design of soft- and hard real-time systems. This includes research and design of embedded system architectures, including CGRAs, SIMD, VLIW, and GPUs, on accelerators, the exploitation of all kinds of parallelism, fault-tolerance, approximate computing, architectures for machine and deep learning, optimizations and mapping of deep learning networks, and the (semi-)automated mapping of applications to these architectures. Corporaal has co-authored over 500 journal and conference papers. Furthermore, he invented a new class of VLIW architectures, the Transport Triggered Architectures, which is used in several commercial products, and by many research groups. He initiated and leads the Dutch NWO perspectief program on Efficient Deep Learning (efficientdeeplearning.nl); in this program, many research institutes and over 30 companies participated. He also is the PI of the EU project CONVOLVE (convolve.eu) on seamless design of smart edge processors, with 19 partners. For further details see corporaal.org.

VLSI-SoC Conference Schedule

Time	Day 0 – 6 October 2024	Day 1 – 7 October 2024	Day 2 – 8 October 2024	Day 3 – 9 October 2024
08:00				
08:30		Opening		
09:00	Tutorial 1 - Part 1: Kaushik Roy - Artificial intelligence: from fundamentals to Applications	Keynote 1: Mark M. Tehranipoor - Secure Heterogeneous Integration and Advanced Packaging: New Attack Surfaces and Grand Challenges Ahead	Keynote 2: Rajiv Joshi - Long Live Computing Technology	Keynote 3: Henk Corporaal - AI at the edge: Hype or Hope?
09:30				
10:00	Coffee Break	Coffee Break	Poster Session 1 & Coffee Break	Poster Session 2 & Coffee Break
10:30	Tutorial 1 - Part 2: Kaushik Roy - Artificial intelligence: from fundamentals to Applications	Regular 1: Security / Special Session 1: Reliability Assessment Of Neural Networks: Trading-Off Between Performance and Accuracy	Regular 3: Simulation, Verification and Safety / Regular 4: Design for AI / Embedded Tutorial 1: New Computing Paradigm For Large Language Models (LLMs) / Embedded Tutorial 2: On-Chip Infrastructure For Mission-Mode Monitoring Of Resilient Systems: Towards Silicon Lifecycle Management	Regular 5: Memories / Special Session 2: IoT-Enabled Electronics For Smart Agriculture
11:00				
11:30				
12:00				
12:30	Lunch	Lunch	Lunch	Lunch
13:00				
13:30				
14:00	Tutorial 2 - Part 1: Henk Corporaal - Computers: from Intel 4004 to Neuro Chips	Regular 2: Miscellaneous / Industrial Session	Social Event	Regular 6: Analog, Microfluidics, Aging / Special Session 3: Embedded Hardware Security: Primitives, Architectures, and Test
14:30				
15:00				
15:30	Coffee Break	Coffee Break		Coffee Break
16:00	Tutorial 2 - Part 2: Henk Corporaal - Computers: from Intel 4004 to Neuro Chips	PhD Forum		Panel 2 / Special Session 4: Security by Design
16:30				
17:00				
17:30				Closing
18:00	Welcome reception	Panel 1		
18:30				
19:00				
19:30				
20:00				
20:30				

A Novel Design Technique for Enhanced Security and New Applications of Ferroelectric-based Non-Volatile SRAM

Lucas RHETAT[1], Jean-Philippe NOEL[1], Bastien GIRAUD[1],
Laurent GRENOUILLET[2], Julie LAGUERRE[2], Cédric MARCHAND[3], Ian O'CONNOR[3]

[1]Univ. Grenoble Alpes, CEA LIST
[2]Univ. Grenoble Alpes, CEA LETI
[3]Univ. Lyon, Ecole Centrale de Lyon, INL, CNRS

Abstract—Static Random Access Memories (SRAM) are fast and efficient circuits used as the main working memory of processing units. However, associating these volatile memories with external non-volatile memories leads to energy consumption and area penalties, while leading to security issues. Ferroelectric-based NVSRAMs are one of the most promising ways of combining the high efficiency of SRAMs with non-volatile operations to tackle these challenges. In this work, several design parameters of the bitcell are optimized to ensure error-less data transfer between 6T SRAM internal nodes and 4 ferroelectric capacitors (4C). The presented 6T4C bitcell presents STORE and RECALL energies of 161fJ/bit and 27fJ/bit, respectively, and STORE and RECALL times of 480ns and 245ns, respectively. A high reliability is achieved from -40°C to +85°C for SS, TT and FF fabrication corners. The integration of the four FeCAPs in the bitcell leads to a 46% area overhead, a 94% WRITE time degradation, and a 32% WRITE energy increase. However, an increase of less than 0.5% in both READ time and energy has been observed. A previously developed Fast-Erase system has also been integrated for countering cold-boot attacks. Combining design optimizations and Fast-Erase technique ensures cold-boot attack immunity of the memory and enables error-less RECALL with WRITE operations between STORE and RECALL, leading to new use-cases of NVSRAM circuits.

Index Terms—Non-Volatile SRAM, reliable & secure data, ferroelectric capacitor

I. INTRODUCTION

SRAM are the fastest and most energy-efficient commercial memories. However, they are volatile, which means that data disappear quickly from SRAM after power-off. The simplest solution to avoid volatility could be to use a Battery Backed SRAM (BBSRAM), i.e. maintaining the SRAM supplied permanently using an embedded battery [1]. However, this solution has the major disadvantage of requiring a bulky battery, with limited energy capacity and lifespan. Carrying a battery in the system may also limit its deployment due to the environmentally hazardous substances it contains. If the battery solution is excluded, data can be transferred to an off-chip non-volatile (NV) memory (often Flash memory) before power-off, which is both time-consuming and energy-intensive. Moreover, this can provide an attack window, for example, by probing the bus connections between the two memories. Non-Volatile SRAMs (NVSRAM) overcome theses

problems by finely co-integrating non-volatile storage elements into SRAM bitcell. In this work, NV Ferroelectric CAPacitors (FeCAPs) components that present a promising combination of write speed and energy as well as endurance among others NV memories are used. The resulting circuit can then perform 4 different operations: reading from SRAM (op. READ), writing to SRAM (op. WRITE), copying data from SRAM to FeCAPs (op. STORE) and moving data from FeCAPs to SRAM (op. RECALL) [2]. Using these four operations permits to obtain a fast, efficient and non-volatile memory. NVSRAMs are good candidates and have gained interest for intermittent applications with either (i) unreliable power supplies (e.g. energy harvesting applications) or (ii) strict energy quota where it could be appropriate to power-off parts of the system avoiding stand-by current consumption (particularly relevant in advanced fabrication processes where the I_{on}/I_{off} currents ratio is reduced). These memories are also suitable for secure or critical systems that need to quickly restart after a power supply interruption. NVSRAMs can in particular be integrated in Non-Volatile Processors (NVPs) [3].

On the security side, storing sensitive data in FeCAPs appears to be safer than in Flash [4]. Indeed, it has been shown that a method based on Scanning Electron Microscopy (SEM) in Passive-Voltage-Contrast mode permits to quickly retrieve the whole content of floating-gate based memories [5]. Retrieving data in $HfZrO_2$ (HZO) ferrelectric memories is much more challenging and implies the use of Scanning-Probe Microscopy (SPM) techniques that are far slower and more expensive [6]. In addition, ferroelectric memories are expected to be one of the least sensitive to attacks among emerging non-volatile memories [7].

Still in a security point of view, SRAM memories are prone to cold-boot attacks, enabling attackers to recover data previously stored in the memory just after it has been shut down [8]. To counter this, a fast erasing technique has recently been developed [9]. This system enables data to be erased very quickly (in the nanosecond time scale, reduction of more than 1000x) when a tampering event is detected, at the cost of an additional area of less than 5%.

In terms of reliability, the RECALL operation is sensitive

979-8-3315-3968-9/24 $31.00 © 2024 IEEE

to mismatches between the cross-coupled inverter transistors and that seems to be a main factor limiting the development of these circuits. In [10], [11] the robustness of 6T2C and 6T4C bitcells against threshold voltage mismatches have already been studied, showing that a better RECALL reliability is achieved with 6T4C bitcells with an optimal FeCAPs sizing ratio. However, theses experiments only concerned the case where no WRITE operation was performed between STORE and RECALL.

In this paper, we demonstrate that combining these previous results with the Fast-Erase technique permits a reliable NVSRAM circuit with WRITE operations between STORE and RECALL, leading to new applications of these circuits. For example, an interesting application could be the fast context switching between two processes. An other application could be using this memory to make safe-state snapshots of a system, permitting to recover a previous state after a corruption detection (or an interruption of any kind) [12]. We could also use this NVSRAM as a Physical Unclonable Function (PUF) [13]. These circuits use CMOS process variabilities to associate a unique fingerprint to each circuit. Using an SRAM as a PUF usually means reserving this SRAM to this application only. However, the proposed circuit can be used as a classical SRAM PUF then store the PUF response as a key in the FeCAPs that permits to continue to use the entire SRAM for other applications, keeping the PUF key available at any time. In all of these applications, we benefit from the proximity of the two memories in terms of energy, time and safety.

The remainder of the paper is organized as follow: Section II describes the proposed FeCAP-based NVSRAM architecture and its implementation. Section III presents our simulation setup and the FeCAP model. Section IV presents the simulation results and section V summarizes the paper. The following results come from SPICE simulations on the ST 130nm HCMOS9A design platform.

II. PROPOSED 6T4C NVSRAM ARCHITECTURE

A. 6T2C vs 6T4C NVSRAM bitcells

Figure 1(a) shows the conventional NVSRAM 6T2C bitcell. It is composed of a 6T SRAM bitcell in which 2 FeCAPs have been added, one on each internal node BLTI/BLFI, both connected to a single Plate-Line (PL). By correctly driving PL the data can be copied from the SRAM internal nodes to the FeCAPs (STORE op.) or transferred from the FeCAPs to the SRAM internal nodes (RECALL op., destructive for the FeCAPs data).

Figure 2 shows that data transfer from FeCAPs to SRAM (RECALL operation) can be achieved in several ways. Firstly, V_{DD} can be set from ground to its nominal value (Fig. 2(a)) [14]. However, this method is sensitive to the transistors mismatches. To improve the robustness of the RECALL operation, a second method involves first applying a pulse to the Plate-Line, which inverts one of the two FeCAPs. This will inject

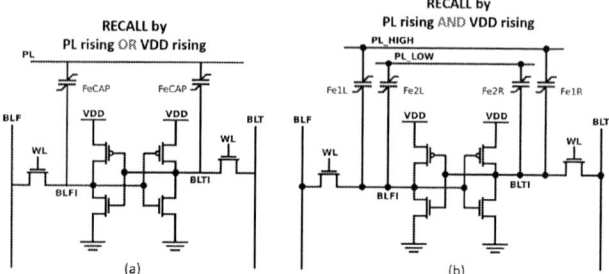

Fig. 1. (a) Conventional 6T2C bitcell implementing one of the two RECALL methods from Fig. 2, (b) 6T4C bitcell with two different Plate-Lines, combining both RECALL methods.

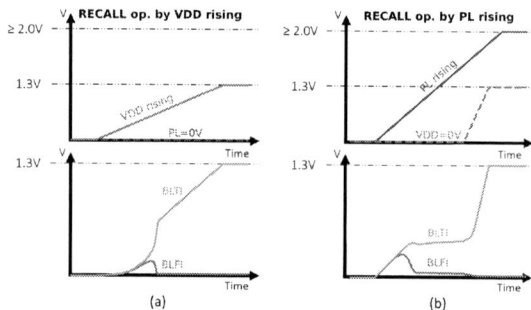

Fig. 2. Simulated waveforms of the two existing RECALL methods: (a) by only pulling up V_{DD} or (b) by applying a pulse on PL before.

a different amount of charge into the SRAM internal nodes, thereby creating a potential difference on these nodes before raising V_{DD} to the nominal value (Fig. 2(b)). Both techniques can be used with 6T2C bitcells. However, these two RECALL methods can be combined by using 6T4C bitcells (Figure 1(b)) and two different Plate-Lines, PL_{HIGH} (rising at RECALL) and PL_{LOW} (staying at 0V at RECALL). By adjusting the sizes of the FeCAPs connected to PL_{HIGH} and PL_{LOW}, it is possible to reach an optimal configuration [11].

B. Fast-Erase technique

When switched off, SRAM bitcell retains its data for some time due to limited leakage current from the internal nodes, up to several minutes if the circuit is cooled. This raises two issues. First, a security issue: it is possible to read the remanent data in SRAM by quickly restoring power to the circuit after shutdown (e.g. cold-boot attack [8]). Secondly, this makes difficult to quickly RECALL the data stored in FeCAPs in case their data are different from the last written data in SRAM. Indeed, the potential difference between BLTI and BLFI corresponding to the remanent data makes an additional bias to counter for a correct sensing of the FeCAPs from the SRAM bitcell. For these reasons, a previously-developed Fast-Erase technique is incorporated into our circuitry [9].

C. Memory architecture and Implementation

Fig. 3 presents the general architecture of the targeted 8-kbit FeCAP-based NVSRAM. The circuit is centered on a

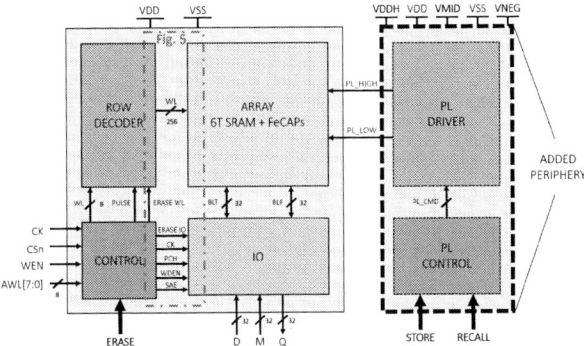

Fig. 3. Global architecture of the proposed 8kbit FeCAP-based 6T4C NVSRAM.

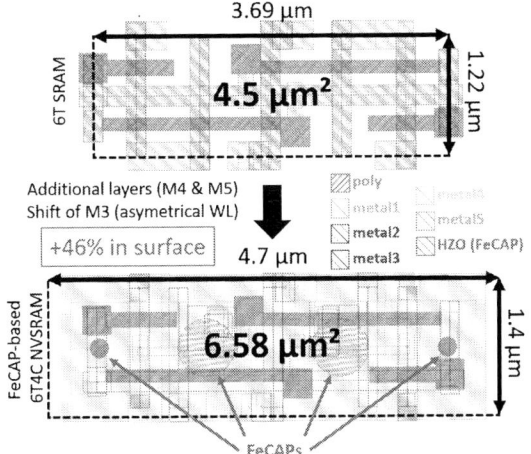

Fig. 4. Layout view of conventional 6T SRAM bitcell vs FeCAP-based 6T4C SRAM bitcell (including the additional back-end levels M4/M5), in 130nm node. The 4 FeCAPs are represented by the red round shapes.

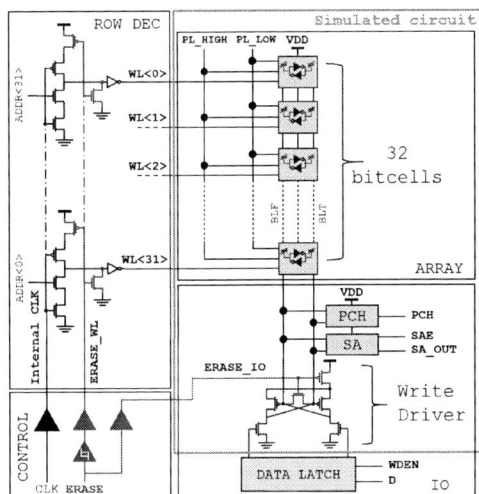

Fig. 5. Schematic view of the simulated circuit and its periphery, including the Fast-Erase circuit [9].

III. TESTBENCH AND SIMULATION SETUP

A. Testbench

The circuit is designed using ST 130nm HCMOS9A design plateform and SPICE simulations are carried out with Synopsys XA. The results being quite sensitive to the precision of the simulations, these were carried out in the highest precision mode. The following error-related results were obtained from Monte-Carlo simulations in different foundry process corners.

Figure 5 presents the simulated circuit and its peripheral circuitry. To keep simulation times acceptable at high precision, the choice was made to restrict the simulation to a 32-bit column only. Our simulated circuit also contains a precharge circuit (PCH), a Write Driver (WD) and a Sense-Amplifier (SA). The Fast-Erase circuit is shown in blue and consists in shorting the two Bit-Lines of each column while activating all the Word-Lines and powering-off the write drivers. In our simulation, the ERASE op. is carried out by directly activating the Word-Lines and the ERASE_IO signal, with $V_{DD} = 0V$.

Fig. 6 shows the overall timing diagram of RECALL and STORE ops. between SRAM READ and WRITE ops. PL_{HIGH} and PL_{LOW} behave the same during the STORE, READ and WRITE operations but differ during the RECALL (see [10], [11]). The Plate-Lines are kept at $V_{MID} = V_{DD}/2$ during normal SRAM operations in order to limit the amplitude of voltage variations across each FeCAP, thereby reducing their aging. The main contribution of our circuit regarding the prior works is the blue part of the diagram: SRAM operations after STORE and the addition of the ERASE signal that equalizes the bitcell internal nodes (see Fig. 8 for details). This Fast-Erase circuit provides immunity against cold-boot attacks and ensure correct RECALL ops. even with READ and WRITE ops. after the last STORE. At the next RECALL, the last data stored in the FeCAPs (which is not necessarily the last written in SRAM; this correspond to a worst case) is going to be

conventional SRAM architecture, comprising a bitcell matrix (ARRAY), a Word-Line decoder (ROW DECODER), write drivers and sense amplifiers (IO) and a circuit managing the timings between the various internal control signals (CONTROL). IO and ROW DECODER blocks are modified to implement the Fast-Erase circuitry, according to [9] (see Figure 5). The PL CONTROL and PL DRIVER blocks are also added to generate control signals for the Plate-Lines.

Figure 4 shows the layout view of the physical implementation of the FeCAPs in the SRAM bitcell. Integrating them in the BEOL level permits to limit the bitcell area overhead. The presented 6T4C bitcell area is roughly increased by 46% with respect to initial 6T SRAM bitcell. FeCAPs are integrated between M4 and M5 levels and the Word-Line is off-center to free enough space to add the M1-M4 vias in the upper part. The Plate-Lines are placed vertically at the M5 level. The layers below M2 have not been modified. A second version of this layout could perhaps be designed to reduce the surface overhead, e.g. by relaxing the BEOL rules or using 3D FeCAPs.

979-8-3315-3968-9/24 $31.00 © 2024 IEEE

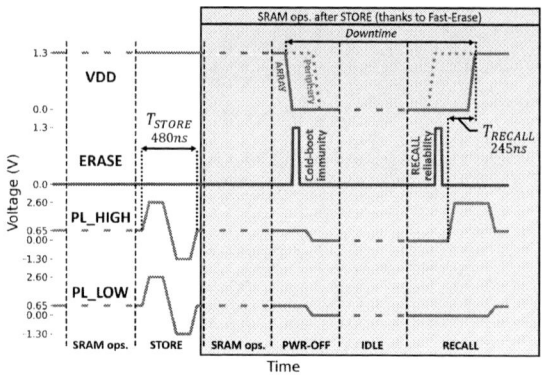

Fig. 6. Chronogram of NVSRAM operations.

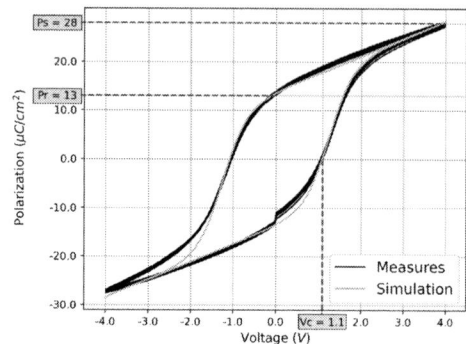

Fig. 7. Comparison between measured (black) and simulated (orange) FeCAP's P-V hysteresis cycle.

transferred to the SRAM. In the RECALL operation, PL and V_{DD} rising edges are shifted by 195ns and their rising times are 30ns and 50ns, respectively. The RECALL is then achieved in 245ns in our simulations but a RECALL time below 100ns can probably be reached if the two rising edges are closed together (not extensively validated). The STORE op. is carried out in a comfortable time of 480ns but this could be reduced.

The nominal supply voltage of the 130nm technology is 1.2V but we choose to supply this circuit at $V_{DD} = 1.3V$ to obtain a sufficient FeCAP programming voltage of $2V_{DD} = 2.6V$. The Plate-Lines vary between $V_{NEG} = -1.3V$ and $V_{DDH} = 2.6V$. These voltages are chosen to be sufficiently high to ensure maximal inversion of the FeCAPs polarizations but not too high to avoid to disturb the state of the first FeCAP while writing the second one (due to the fact that PLs are each common to two FeCAPs polarized in opposite directions).

In the following, the WRITE time is defined as the time between the Word-Line reaches 50% of V_{DD} and the slowest internal node (the rising one) reaches 90% of V_{DD}. The WRITE energy is the instantaneous power integrated on this period. The READ time is measured as the time between the Word-Line reaches 50% of V_{DD} and SA_{OUT} reaches 90% of V_{DD}. The READ energy is the instantaneous power integrated on this period. The STORE energy is defined as the integral of the instantaneous power on the whole operation pattern period and the RECALL energy is the power integrated between PL_{HIGH} reaches 10% of V_{DDH} and the rising internal node reaches 90% of V_{DD}.

B. FeCAP model

For the FeCAPs simulation, a Preisach-based VerilogA model is used [15] [16]. This model simulates both the ferroelectric part of polarization and the paraelectric part. The model also takes into account the capacitive effect between the two electrodes and the leakage current.

To ensure that our model matched the real components, a model calibration phase was carried out. This involved adjusting the model's hysteresis cycle in relation to that measured on real components representative of the technology (average P-V curve on 10nm thick HZO BEOL-integrated scaled FeCAPs) (Fig. 7) [17]. Here are the saturation polarization (P_{sat}), the remanent polarization (P_r) and the coercitive voltage (V_c) of the considered FeCAPs:

$$P_{sat} = 28\mu C/cm^2 \qquad P_r = 13\mu C/cm^2 \qquad V_c = 1.1V$$

IV. SIMULATION RESULTS

A. Fast-Erase benefits

Figure 8 shows the ERASE operation principle, which presents two objectives. First of all, it takes part in the RECALL operation. Indeed, if the system restarts very quickly after a shutdown, it may be difficult to perform a RECALL while the data stored in the FeCAPs is different than the last written data in SRAM. In the example given in Figure 8, the stored data in the FeCAPs is '1' (BLTI=1, BLFI=0) (Fig. 8(a)). The system then performs WRITE ops. in SRAM. Figure 8(b) shows that if the system restarts after a short downtime, the residual charges on BLFI (corresponding to the last written bit in SRAM) prevent the SRAM (that acts like a sense amplifier) from performing a correct RECALL of the stored state '1'. This phenomenon is especially significant at low temperature and slow corners where the leakage currents are weaker. Furthermore, performing WRITE operations after STORE means maintaining PL_{HIGH} and PL_{LOW} at $V_{DD}/2$ to improve the FeCAPs endurance. When the Plate-Lines fall down to 0V after power-off, they can inject charges in the internal nodes. These charges constitute an other voltage bias disturbing the RECALL op. Performing an ERASE pulse just before the RECALL permits to equalize the internal nodes potentials, that greatly improves the reliability of the RECALL in all corners (Fig. 8(c)). For example, for 1000 Monte-Carlo runs of the 32-bit column without ERASE in the -40°C/SS corner, we obtain 30343 RECALL errors for a downtime of $1\mu s$ and still 660 errors for a downtime of $1ms$. However, there are no errors in either case with the ERASE pulse just before the RECALL. Furthermore, the memory is still vulnerable to cold-boot attacks between the power-off and the RECALL op. due to the data remanence, especially when the downtime is long. In order to ensure cold-boot attack immunity, we use another ERASE pulse, just after the power-off (Fig. 8(d)), making cold-boot attacks far more tricky [9].

The duration of these ERASE pulses doesn't matter as long as it is greater than the erasing time of the internal nodes

979-8-3315-3968-9/24 $31.00 © 2024 IEEE

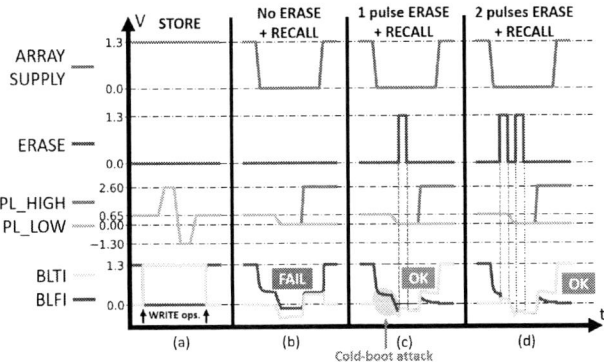

Fig. 8. Comparison of the waveforms of the SRAM internal nodes, with stored data in (a), between RECALL without ERASE pulse (b), with one ERASE pulse (c), with 2 ERASE pulses (d).

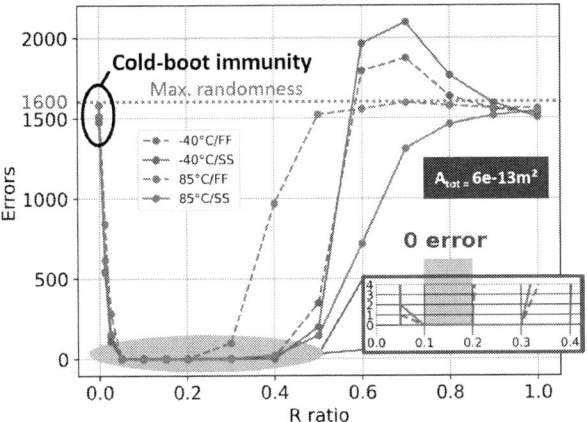

Fig. 9. Number of errors w.r.t. the R ratio (100 Monte-Carlo runs of the 32-bit column for each corner).

(in the order of some nanoseconds, see [9]). The minimal downtime, mainly limited by the RECALL sequence, is on the order of a hundred nanoseconds.

B. FeCAPs sizing optimization

In the following, we define A_{HIGH} and A_{LOW} as the total surface of the two FeCAPs connected to PL_{HIGH} and PL_{LOW}, respectively. We define as well:

$$A_{tot} = A_{HIGH} + A_{LOW} \qquad R = \frac{A_{HIGH}}{A_{HIGH} + A_{LOW}}$$

Figure 9 shows the number of RECALL errors for 100 Monte-Carlo runs in each corner with respect to the R ratio, for $A_{tot} = $ 6e-13m^2. At $R = 0$, the four FeCAPs are connected to PL_{LOW} (staying at 0V at RECALL): the RECALL operation is performed only thanks to the rising edge of V_{DD} (Fig. 2(a)). At $R = 1$, the four FeCAPs are connected to PL_{HIGH} (rising at RECALL): the RECALL is performed by the rising edge of PL_{HIGH} (Fig. 2(b)). All the intermediate values correspond to configurations where the FeCAPs are connected to PL_{HIGH} and PL_{LOW} in the surface ratio R. A low error rate

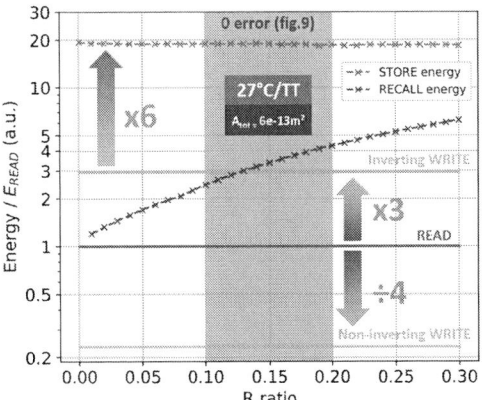

Fig. 10. STORE and RECALL energies w.r.t. the R ratio.

can be achieved with $R \in [0.1; 0.4]$ with an optimum for $R \approx 0.1$, corroborating the results of [11]. From a security point of view, the RECALL error rate must be close to 50% (maximum randomness) for $R = 0$ to ensure cold-boot attack immunity. Indeed, $R = 0$ corresponds to a configuration in which all the FeCAPs are connected to PL_{LOW}, which stays at 0V during RECALL. It is equivalent to a simple power-up, without RECALL, which could be a cold-boot attack.

Figure 10 shows that the RECALL energy grows roughly linearly with increasing R. STORE, WRITE and READ energies and timings are independent on R because PL_{HIGH} and PL_{LOW} behave the same during these operations. This figure shows that it is preferable to minimize R for energy efficiency.

Figure 11 shows that increasing A_{tot} greatly reduces the number of RECALL errors, at the price of increased WRITE energy and timing. The READ energy and timing stay roughly constant (increase very slightly) with increasing A_{tot}. We obtain 0 error with $A_{tot} = $ 4e-13m^2 for 1000 Monte-Carlo simulations of the 32-bit column, in the 85°C/FF corner. However, $A_{tot} = $ 6e-13m^2 can be considered as a better configuration, taking into account a design margin. This corresponds to an increase of roughly 94% in WRITE time compared to the SRAM bitcell without FeCAP. The WRITE energy penalty is 133% for an inverting WRITE and 7% for a non inverting WRITE, both cases with an opposite value stored in the FeCAPs. Assuming that 20% of the data change between two consecutive WRITE ops., the FeCAPs impact on the WRITE energy is reduced to 32% on average.

The combination of $A_{tot} = $ 6e-13m^2 with $R = 0.1$, which is the best case presented here, corresponds to FeCAPs of diameters $D_1 \approx 200nm$ and $D_2 \approx 600nm$, whose can be integrated in the existing SRAM bitcell through an increase of 46% of the initial bitcell area (see Fig. 4). In this configuration, at 85°C in FF corner, storing data in FeCAPs consumes 191fJ/bit or 153fJ/bit, depending if the FeCAPs polarizations are inverted or not, respectively. If we assume that 20% of the SRAM data change between two STORE ops., this operation consumes 161fJ on average. Recalling data in SRAM consumes 27fJ/bit in this corner.

979-8-3315-3968-9/24 $31.00 © 2024 IEEE

TABLE I
NVSRAM TECHNOLOGY COMPARISON TABLE

Reference	This Work	Infineon WhitePaper 2016 [18]	VLSI 2015 [19]	ISSCC 2023 [20]
CMOS bitcell technology	6T 130nm	10T 40nm	7T 90nm	6T 22nm FDSOI
NVM technology	4C BEOL HZO	2T SONOS	1R BEOL HfO_2	2 Charge-Trap Transistors
SRAM ops. after STORE	Yes	No	No	No
Cold-boot immunity	High	Low	Low	Low
Physical attack cost	High	Medium	High	High
Technology scalability	High	Low	High	High
Endurance	High	Medium	Medium	Low
STORE energy	fJ/bit	nJ/bit	fJ/bit	uJ/bit
STORE time	ns	ms	ns	ms
RECALL time	ns	us	ns	Not specified

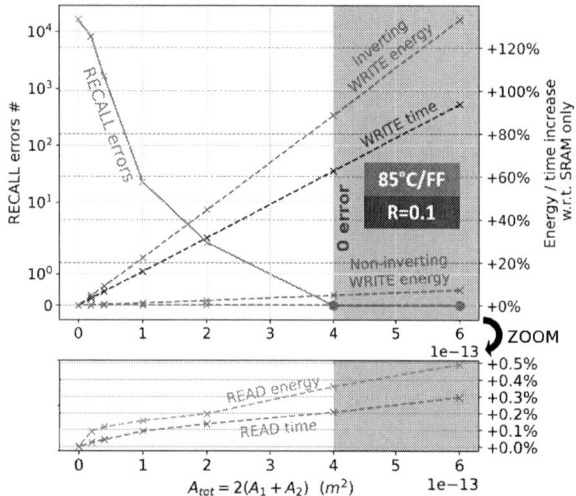

Fig. 11. Number of RECALL errors (for 1000 Monte-Carlo runs) compared to the impact in energy and timing of READ and WRITE ops. w.r.t. the total FeCAPs area A_{tot}.

Table I presents comparisons between our work and several recently presented NVSRAMs, in various technologies. Ferroelectric technology is competitive with other non-volatile technologies in many respects, and our solution offers unique advantages in terms of safety and functionality.

V. CONCLUSION

The proposed ferroelectric-based 6T4C NVSRAM demonstrates the ability to both STORE data in non-volatile part of memory for 161fJ/bit in 480ns and RECALL it to the SRAM for 27fJ/bit in 245ns, with a high reliability. In this work, the combination of previous results in FeCAPs sizing optimization with the recently developed Fast-Erase technique demonstrates the ability to continue using the SRAM after storing data in the FeCAPs and recalling data different from that previously written in the internal nodes of SRAM, without having to wait for a significant amount of time. This solution is also robust against cold-boot attacks and leads to new use-cases of ferroelectric-based NVSRAMs, including fast context switching or safe-state checkpointing. More generally, any application requiring snapshots of a part of its memory at a given time, which is not necessarily the last state before power-off, could benefit from this solution.

REFERENCES

[1] S. Shivendra, *Nonvolatile SRAMs (nvSRAMs) in Gaming Applications*, 2015.

[2] T. Miwa, "NV-SRAM: a nonvolatile SRAM with back-up ferroelectric capacitors," in *CICC*, 2000.

[3] Y. Liu, "4.7 A 65nm ReRAM-enabled nonvolatile processor with 6× reduction in restore time and 4× higher clock frequency using adaptive data retention and self-write-termination nonvolatile logic," in *ISSCC*, 2016.

[4] Fujitsu, *FRAM MCU Key Strengths and Applications*, 2010.

[5] F. Courbon, "Reverse Engineering Flash EEPROM Memories Using Scanning Electron Microscopy," in *Smart Card Research and Advanced Application Conference*, 2016.

[6] S. E. Quadir, "A Survey on Chip to System Reverse Engineering," *J. Emerg. Technol. Comput. Syst.*, 2016.

[7] M. N. I. Khan, "Comprehensive Study of Security and Privacy of Emerging Non-Volatile Memories," *JLPEA*, 2021.

[8] N. A. Anagnostopoulos, "Attacking SRAM PUFs using very-low-temperature data remanence," *Microprocessors and Microsystems*, 2019.

[9] J.-P. Noel, "A Near-Instantaneous and Non-Invasive Erasure Design Technique to Protect Sensitive Data Stored in Secure SRAMs," in *ESSCIRC*, 2021.

[10] S. Masui, "A ferroelectric memory-based secure dynamically programmable gate array," *JSSC*, 2003.

[11] K. Takeuchi, "A Feasibility Study on Ferroelectric Shadow SRAMs Based on Variability-Aware Design Optimization," *JEDS*, 2019.

[12] P.-H. Thevenon, "iMRC: Integrated Monitoring & Recovery Component, a Solution to Guarantee the Security of Embedded Systems," 2022.

[13] D. E. Holcomb, "Power-Up SRAM State as an Identifying Fingerprint and Source of True Random Numbers," *IEEE Transactions on Computers*, 2009.

[14] M. Kobayashi, "Experimental Demonstration of a Nonvolatile SRAM With Ferroelectric HfO2 Capacitor for Normally Off Application," *JEDS*, 2018.

[15] S. L. Miller, "Device modeling of ferroelectric capacitors," *Journal of Appl. Phys.*, 1990.

[16] S. L. Miller, "Modeling ferroelectric capacitor switching with asymmetric nonperiodic input signals and arbitrary initial conditions," *Journal of Appl. Phys.*, 1991.

[17] T. Francois, "Demonstration of BEOL-compatible ferroelectric Hf0.5Zr0.5O2 scaled FeRAM co-integrated with 130nm CMOS for embedded NVM applications," in *IEDM*, 2019.

[18] R. Prakash, *Nonvolatile SRAM (nvSRAM) Basics*, 2016.

[19] A. Lee, "RRAM-based 7T1R nonvolatile SRAM with 2x reduction in store energy and 94x reduction in restore energy for frequent-off instant-on applications," in *VLSIC*, 2015.

[20] S. Nouri, "An 8T eNVSRAM Macro in 22nm FDSOI Standard Logic with Simultaneous Full-Array Data Restore for Secure IoT Devices," in *ISSCC*, 2023.

Capture the Pulse: Impact of FPGA Resource Utilization on EM Fault Injection Attacks Detection

Sami El Amraoui, Régis Leveugle, Paolo Maistri

Univ. Grenoble Alpes, CNRS, Grenoble INP*, TIMA, 38000, Grenoble, France

{sami.el-amraoui,regis.leveugle,paolo.maistri}@univ-grenoble-alpes.fr

Abstract— With the increasing use of Field-Programmable Gate Arrays (FPGAs) in critical applications, safeguarding against malicious attacks becomes necessary. ElectroMagnetic Fault Injection (EMFI) stands out as a potent threat among localized fault attacks with its optimal compromise between cost and effectiveness, without the risk of damaging the target chip. Among potential targets, Ring Oscillators (ROs) are critical components that can be used in secure primitives, as well as detectors against physical attacks. In this paper, we analyze how the use of FPGA resources affects the outcome of EMFI attacks: we experimentally show with single EM pulse injections on three families of Xilinx FPGAs manufactured in 28nm process technology that the harmonic response of a RO heavily depends on its layout and density within the FPGA die. We also highlight the need of considering both EM pulse polarities when evaluating the efficiency of any proposed countermeasures, as this can reveal different sensitive locations on the chip. These findings can be leveraged for designing architectures that address the EMFI threat more effectively.

Keywords— Ring oscillators, EMFI, FPGA, P&R, Harmonic error, Hardware security.

I. INTRODUCTION

In today's digital landscape, ensuring the security of electronic systems is paramount, as they can be targeted by malicious entities seeking to compromise data confidentiality and system integrity. To this end, physical attacks and notably fault attacks have become a prevalent method for circumventing the security mechanisms of embedded devices, with a large arsenal of techniques at the attackers' disposal. These techniques range from disrupting clock signals to inducing sudden variations in temperature, supply voltage or substrate bias, along with injecting parasitic currents using powerful electromagnetic (EM) disturbances [1]. Other sophisticated techniques using laser or X-rays can be more efficient but very expensive and risky for the attackers as they can damage the target device. Among these methods, ElectroMagnetic Fault Injection (EMFI) has gained prominence due to the optimal balance between the attack effectiveness and its low cost and device preparation requirements. In a recent paper [2], an open-source design 'PicoEMP' was introduced whereby an EM pulsed fault injection can be mounted for less than 100$. This makes EMFI attacks more accessible in both academia and industry which will help improve our understanding of its fault mechanisms.

EMFI alters the behavior of a target device by inducing ground bounces or voltage drops as a result of the magnetic flux created by the EM probe [3]. The EM coupling can be performed either through harmonic or pulsed Fault Injection. In the context of our study, we are interested in the second method involving powerful EM pulses.

After recognizing the inherent threat of EMFI, extensive research efforts have been targeted in recent years to fully

---

*Institute of Engineering Univ. Grenoble Alpes

understand and mitigate its impact on diverse devices, ranging from microcontrollers to FPGAs and Application-Specific Integrated Circuits (ASICs). Our focus in this work will be directed towards FPGAs as they represent good candidates for implementing cryptographic algorithms and security primitives such as Physical Unclonable Functions (PUFs) and True Random Number Generators (TRNGs). Since these primitives employ ROs as an essential building block in their design, assessing their response to EMFI is important to improve their robustness.

Within this context, this paper aims to enhance the understanding of the harmonic locking impact on ROs with pulsed EMFI in FPGAs. This effect was introduced in [4] where the authors were able to lock the RO frequency into one of its harmonics through a single properly tuned EM pulse. The RO harmonic response was characterized as a function of its placement in the FPGA chip, the EM pulse width (PW) and amplitude, and the position of the probe relative to the FPGA package. In [5], the investigation sought to understand how multiple pulses and the placement and routing (P&R) constraints (of both the inverters and the input/output (IO) pins) influence the sensitivity of the ring. In this study, we advance from the groundwork of these papers to explore on different modern FPGA platforms how changing the FPGA structure (Spartan7 vs. Artix7) and its type of packaging (Wire-bond vs. Flip-Chip) can affect the harmonic response of the RO. We also elaborate on the effect of P&R constraints to maximize the harmonic response while varying the density and the layout of the RO with the same number of inverters. Furthermore, unlike many papers in the literature that neglect the impact of EM pulse polarity, here we show the fault distribution for both positive and negative polarities to highlight the importance of this parameter when evaluating the efficiency of EMFI across various FPGAs.

The remainder of this paper is structured as follows: Section II introduces the related works on EMFI and the harmonic locking of ROs. Section III details the experimental setup and the methodology. Section IV reports and analyzes the obtained results. Finally, Section V draws conclusions and provides perspectives.

II. BACKGROUND

A. Electromagnetic Fault Injection

EMFI can corrupt the normal operation of an integrated circuit (IC) through parasitic currents that are induced in all wire loops of the IC after a sudden variation of the magnetic or electric field.

Mounting a successful EMFI attack includes multiple steps and identifying the optimal sensitive location on the die is the critical one. To achieve that, one can leverage EM side-channel to find the maximum leakage observed on the target chip. Within this context, a recent paper [6] investigated the design optimization of a hybrid EM probe that can efficiently enable both capturing EM emissions and injecting EM perturbations. This can help reduce the time and complexity of the evaluation that still includes other parameters related to the pulse such as the injection timing, and the width, amplitude, and polarity of the pulse.

Considering all these challenges, ensuring full protection of different devices against EMFI requires accurate fault models describing the mechanism involved in EM-induced faults. Several works [7] [8] [9] have been carried out to explain their nature. They suggested that these faults can be explained either by the Timing or by the Sampling fault model, depending on the clock frequency of the target and the strength of the EM coupling within the circuit. A recent study by Nabhan et al. [10] highlighted that two underlying mechanisms are involved within the timing fault model. The first one relates to the target's clock distribution network which becomes highly susceptible to voltage glitches on the clock tree at low or moderate frequencies. However, at high frequencies associated with small slack, the power distribution network (PDN) is the most sensitive on-chip network as EM perturbation leads to violations in timing constraints. In another paper [11] trying to uncover the complexity of the EM injection and its interaction with circuits at a low physical level, the authors conducted EMFI experiments from the substrate backside of a 180 nm CMOS technology equipped with six crypto cores of 128-bit AES and an on-chip voltage waveform monitor. Their results showed a key finding emphasizing that depending on the polarity of the probe, significant negative voltage drops can be induced at specific locations within the PDN originating from injection positions that are not necessarily close to the impacted point.

Previous works studied the implementation of EMFI countermeasures on FPGAs such as [12] and [13]. However, the optimal number and location of the sensors to guarantee security against EM perturbations with the lowest overhead is still an open question. This primarily stems from the fact that the customization of placement and routing is challenging in FPGA design tools as they are designed by default to guarantee the best performance and resource optimization but do not take into account the circuit's EM emissions or susceptibility issues during the P&R process. To this end, our work will try to give more insights into the effects of P&R constraints on EMFI susceptibility to determine the best strategy for implementing detectors in different FPGA platforms.

B. Harmonic Locking of Ring Oscillators

A RO is harmonically locked when it is forced to oscillate at a harmonic frequency (i.e., a multiple of the fundamental frequency). This phenomenon was achieved with both laser experiments on a custom-designed 40nm Bulk CMOS ring oscillator, and electrical experiments on a RO constructed with discrete components in [14]. They were able to conclude that when one or several Single Event Transients (SETs) with a pulse width smaller than the total loop delay are induced during one oscillation period T of the RO, it deviates from its fundamental frequency and locks to one of its odd harmonics depending on the induced number of extra rising edges. In [4], the authors were able to validate this behavior through

the injection of a single positive EM pulse into a RO composed of 1200 (50 x 24) inverters and implemented in an Artix7 FPGA. In their work, the ring was placed either on the top or bottom clock region and used 1 Look-Up Table (LUT) per Slice. The results showed that the placement of the RO in the chip, and the location and intensity of the injection affect the fault occurrence: when placed on the top clock region, the RO was more vulnerable to harmonic errors. Also, it was shown that in order to force higher harmonics with higher probabilities, one can increase the PW and amplitude of the pulse.

In [5], the characterization included a first analysis of the impact of routing constraints on the harmonic response of the RO. The findings suggested that for the same placement of the ring, favoring short vertical connections (called 'vertical snake') between LUTs results in higher harmonic sensitivity, while the lowest vulnerability was reported with long horizontal connections. Conducting similar tests on Spartan7 and Kintex7 FPGAs enabled us to confirm this conclusion. Therefore in this work, we will adopt the most sensitive routing 'vertical snake' to investigate further parameters: how the negative pulse polarity, the density, and the layout of the RO impact the fault distribution and intensity on different FPGAs.

III. EXPERIMENTAL SETUP & METHODOLOGY

A. EMFI Setup

The setup illustrated in Fig. 1 was used to perform the EMFI experiments. A commercial pulse generator, ChipShouter by NewAE Technology was used to perform EM pulsed fault injection. It generates pulses with amplitudes from 150V up to 500V and variable widths, depending on the probe's diameter and the voltage amplitude. We used two of the EM probes provided with ChipShouter, consisting of a 1mm wire coiled either clockwise (CW) or counter-clockwise (CCW) around a 4mm ferrite core, to induce a positive or negative polarity of the pulse respectively. An XYZ motorized table precisely controls the position of the EM probe on top of the FPGA package in order to perform a full fault cartography. A digital oscilloscope of 200 MHz bandwidth was used to monitor the RO frequency during tests. A PC controls the whole platform through serial ports. The characterization has been achieved on three AMD-Xilinx FPGAs, all manufactured in 28nm process technology: a Spartan7 (xc7s25-1CSGA225) in a Cmod-S7 board; an Artix7 (xc7a100T-1CSG324) within Nexys-A7 board and a Kintex7 (xc7k160T-1FBG676) embedded in a SAKURA-X board. Their characteristics are given in Table I.

Fig. 1. Experimental setup

979-8-3315-3968-9/24 $31.00 © 2024 IEEE

Table I. Characteristics of the targeted FPGAs

FPGA	Packaging	Package Size (mm²)	Die Size (mm²)	Logic Slices
Spartan7	Wire-bond chip-scale	13 x 13	5 x 5.1	3650
Artix7	Wire-bond chip-scale	15 x 15	6.5 x 10	15850
Kintex7	Flip-Chip lidless	27 x 27	9 x 12	25350

It should be noted that each FPGA die covers only a portion of the package. The die size of Kintex7 was measured since it is decapsulated in the board. For Spartan7, this information was revealed through X-ray imaging while it was reported in [15] for Artix7. Both Spartan7 and Artix7 use a wire-bond packaging technology, where the die is attached to the substrate face up and the connections are made with wires. On the other hand, the Kintex7 die is flipped over and placed face down, which eliminates wire bonds and results in lower inductance and better thermal performance. All these differences will be leveraged to explore how they will affect the outcome of EMFI.

B. RO Design

1) RO Architecture

The frequency of a RO depends on various parameters but it can be simplified as in the following formula, where D_i is the propagation delay of each delay element in the RO, D_r is the average delay related to the routing between them, and N is the number of these elements:

$$F = \frac{1}{2 \times (D_i + D_r) \times N}$$

Fig. 2 shows the architecture of our implemented RO with an even number N = 1200 of inverters and a Nand gate used as an activation gate. The control of the RO oscillations was achieved through serial communication.

2) Placement and Routing Constraints

In all targeted FPGAs, the Configurable Logic Block (CLB) tile contains two slices. Each slice includes 4 LUTs and 8 flip-flops. Depending on the density that we want to achieve, our RO was formed by configuring either one or all of the 4 LUTs within each slice as an inverter. This makes the ring more or less compact in the FPGA.

Also, to explore the effect of different RO layouts, we customized the placement and routing of the LUTs in Vivado tool to enable either a horizontal, vertical, or square shape. It should be noted that in all our experiments, we used the same number of inverters (N=1200) for each different P&R implementation of the RO in all three FPGAs.

C. Methodology

After a single EM pulse injection, the RO frequency may change or not, as follows:

- **Unchanged frequency:** In that case, after the attack, the RO still oscillates at the same fundamental frequency. If we disable the RO and it keeps oscillating, we know that the 'Enable' configuration was corrupted, and the FPGA must be reprogrammed for the next test.
- **Harmonic locked frequency:** After the attack, the RO frequency can be locked into one of its odd harmonics (3, 5, 7….). As in the previous case, lack of control over the 'Enable' means that the board must be reprogrammed.
- **Noise signal:** the attack can also force the RO output signal to noise, which means the bitstream was corrupted, and re-setting the Enable signal does not restart the oscillations. Therefore, reprogramming the bitstream is mandatory before the next test.

Based on these effects, a specific methodology, depicted in Fig. 3(a), was adopted to inject and observe a single pulse into the RO. The goal is to detect the occurrence of harmonics while scanning over the FPGA package with the 4mm CW or CCW probe. The probe tip is kept 50μm on top of the package and displaced by steps of 1 mm (due to the probe's size and resolution) from the top left to bottom right corner as depicted in Fig. 3(b). The detailed steps are:

1) Set the initial EM pulse parameters for the test (PW = 100ns and amplitude = 450V). The choice of these values was motivated by the results reported in [4].

2) Place the probe at the initial coordinate (X = 0, Y = 0) above the chip package.

3) Program the FPGA with the bitstream.

4) Enable the RO and trigger the EM pulse injection.

5) Monitor the output RO frequency after injection then disable the RO, to detect the occurrence of harmonic induced frequencies and bitstream corruptions.

6) Repeat 30 times steps 4 and 5 to assess the reproducibility rate for the given (X, Y) coordinate.

7) Move the probe to a new position and repeat from step 3 until the last coordinate to obtain a fault sensitivity map of the FPGA package.

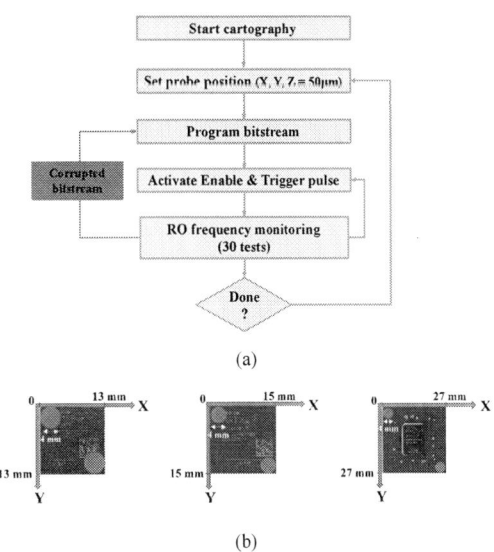

(a)

(b)

Fig. 3. (a) RO faults evaluation flow (b) FPGA Package Scan

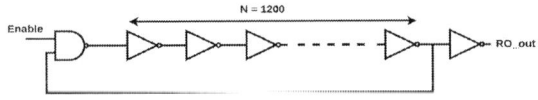

Fig. 2. Architecture of the ring oscillator

IV. EXPERIMENTAL RESULTS

The experimental results reported in this section show the effect of EM pulsed injection with both polarities on the 3 targeted FPGAs along with the impact of different RO layouts on faults.

To improve the readability of the fault sensitivity maps, we assigned a specific color for each effect. It should be noted that the numbers in these maps refer to the ratio between the monitored frequency after EM injection and the fundamental frequency of the targeted RO:

- **White:** No faults (Frequency remained the same and the bitstream was not corrupted).
- **Gradient from Yellow to Red:** shows the highest harmonic error that was monitored within 30 conducted tests.
- **Black:** represents mutes where reprogramming the bitstream was mandatory; the probability of bitstream corruption in this case is 100%.

A. Impact of the FPGA Structure and Packaging

To explore the influence of the FPGA structure and packaging on the fault sensitivity of the RO, we followed the methodology shown in Fig. 3(a) by conducting EM injection campaigns targeting the same RO implemented in the three FPGAs; 1 LUT/Slice and a vertical routing with short connections, as illustrated in Fig. 4, were used. It should be noted that even if the same P&R of the RO was adopted, its frequency slightly varied between them because of the difference in their internal structure and the output pin connection. The RO was originally running at 1250 KHz in Spartan7, at 1075 KHz in Artix7, and at 1417 KHz in Kintex7.

Fig. 5 represents the fault sensitivity maps with the positive polarity of the three FPGAs and shows that changing the FPGA family and packaging, even within the same process technology, can effectively influence the intensity and the location of the induced faults. The maximum impact was achieved in Artix7 with a harmonic error of 43, twice

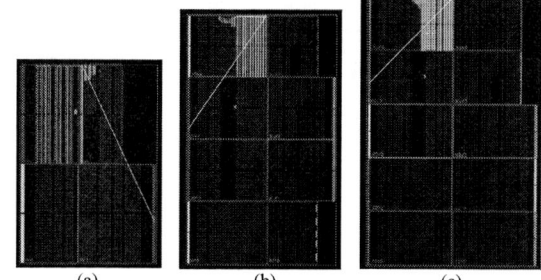

Fig. 4. FPGA floorplan from Vivado showing the implemented RO in (a) Spartan7 (b) Artix7 (c) Kintex7

larger than the impact on Kintex7. Additionally, Spartan7 was the most sensitive to bitstream corruptions, as opposed to Kintex7 with no recorded vulnerability to this type of faults. This shows how Flip-Chip packaging, usually dedicated to high-performance FPGA, can reduce the fault sensitivity compared to wire-bond packaging. On another hand, if we focus on this later technology used for Spartan7 and Artix7, by comparing Fig. 5(a) and Fig. 5(b) we notice some similarities in the location of faults especially in the bottom center coordinates. However, only third harmonic errors were induced for Spartan7, likely because it includes less programmable logic, hence a lower coupling between the probe and its PDN.

B. Impact of the Pulse Polarity

Similar tests were conducted to investigate the effect of changing the pulse polarity and the results are depicted in Fig. 6, showing the new fault locations in the three FPGAs. Upon an initial examination of the three cartographies, one may observe a similar trend compared to the positive polarity with Artix7 being the most vulnerable to harmonics and Spartan7 being the least. However, comparing Fig. 5 and Fig. 6 highlights that the negative polarity reveals complementary fault locations in the FPGA package. This indicates a different susceptibility of the power and the ground network

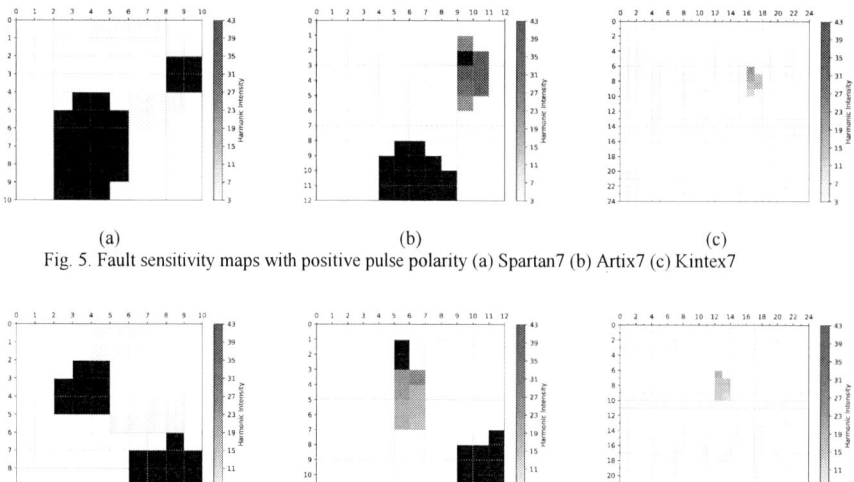

(a)   (b)   (c)

Fig. 5. Fault sensitivity maps with positive pulse polarity (a) Spartan7 (b) Artix7 (c) Kintex7

(a)   (b)   (c)

Fig. 6. Fault sensitivity maps with negative pulse polarity (a) Spartan7 (b) Artix7 (c) Kintex7

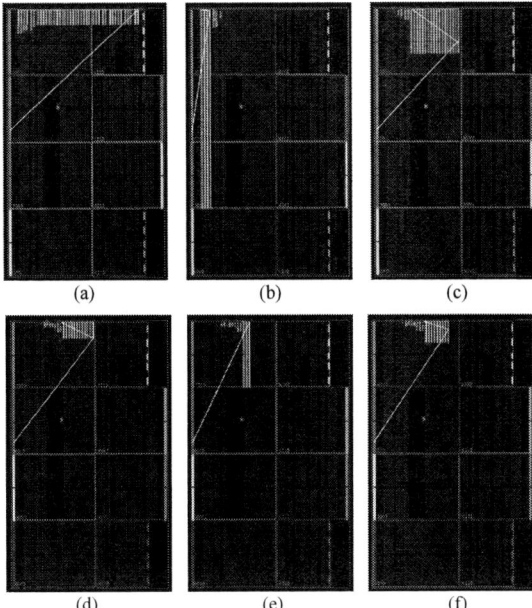

(a)   (b)   (c)

(d)   (e)   (f)

Fig. 7. Different layouts of the RO implemented in Artix7 FPGA using (above) 1LUT/Slice (a) Horizontal (b) Vertical (c) Square and (below) 4LUTs/Slice (d) Horizontal (e) Vertical (f) Square

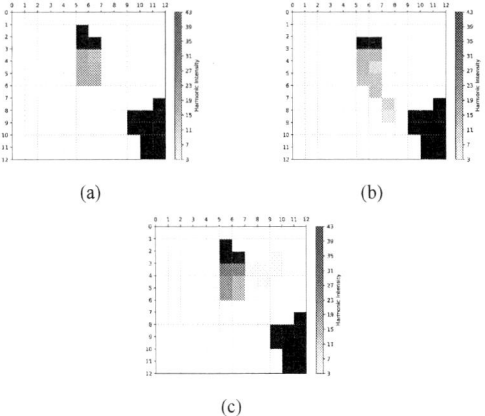

(a)   (b)

(c)

Fig. 8. Fault sensitivity maps with a negative polarity using 1LUT/Slice configuration (a) Horizontal (b) Vertical (c) Square

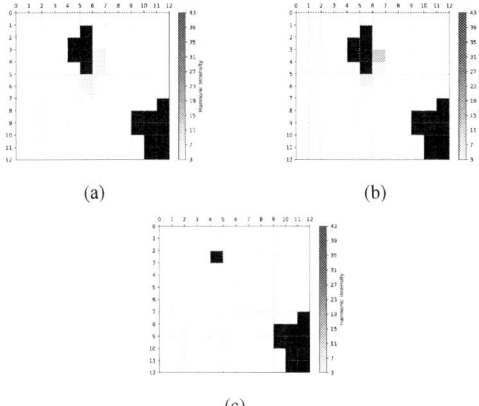

(a)   (b)

(c)

Fig. 9. Fault sensitivity maps with negative polarity using 4LUTs/Slice configuration (a) Horizontal (b) Vertical (c) Square

within FPGAs to the polarity of EMFI, which underlines the importance of this parameter when assessing the robustness of a design against this attack.

C. Impact of the RO Layout

To further investigate the effect of changing the RO layout within the die, we performed new experiments while changing the placement constraints of the LUTs to enable different shapes of the RO. The intended goal is to obtain further insights into the impact that density and connections have on the ring sensitivity. Fig. 7(a) to (c) show the three obtained RO shapes (horizontal, vertical and square) achieved through the use of 1 LUT/Slice. For the sake of conciseness, although tests were conducted on the three FPGAs using positive and negative pulse polarities, we will here present only the results related to the Artix7 FPGA with negative polarity and then compare and discuss the other cases.

Table II lists the frequency of the RO depending on its layout. As shown in this table, changing the layout results in a difference of less than 1% between the output frequencies: this is mainly due to the different use of horizontal and vertical connections.

The results are represented in Fig. 8, showing how the fault sensitivity varies with the layout strategy. The most vulnerable layout was the square one with more sensitive

Table II. Frequency of the RO with different layouts in the FPGAs

FPGA	Density	Layout	Frequency (KHz)
Artix7	1 LUT/Slice	(a)	1086
		(b)	1076
		(c)	1078
	4 LUTs/Slice	(d)	1550
		(e)	1540
		(f)	1546

locations and a harmonic error of 29. This indicates a better coupling between the probe and the FPGA when the RO is more compact within a single clock region, with few horizontal connections between the slices. This is likely due to the optimal density of inverters within the EM pulse. Similar tests on Spartan7 and Kintex7 enabled us to draw the same conclusion for both pulse polarities.

D. Impact of the RO Density

After showing that choices in the RO distribution within the chip lead to consistent differences in the experimental results between the three FPGAs, we aimed to highlight the impact of the RO density on the observed faults. To this end, we implemented similar configurations but using 4LUTs/Slice instead of only one as illustrated in Fig. 7(d) to (f). Their corresponding frequencies are detailed in Table II.

The results reported in Fig. 9 show how minimizing the use of Slices (from 1200 down to 300) had a noticeable impact on the harmonics intensity compared to the results in Fig. 8, especially for the square layout where the harmonic sensitivity was completely suppressed. This was mainly caused by the low spatial resolution of the 4mm probe that cannot couple with elements of a small size, which indicates the likely need for a more advanced setup to induce harmonic errors. The same result was obtained for the two other

979-8-3315-3968-9/24 $31.00 © 2024 IEEE

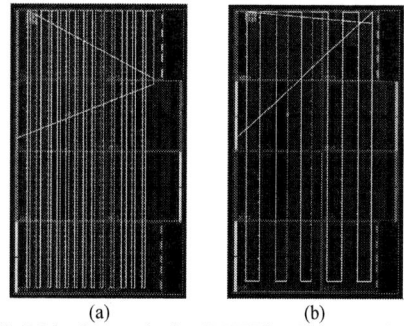

Fig. 10. RO implemented using 1LUT/Slice (a) N=1200 inverters, F=850KHz (b) N=228 inverters, F=4065KHz

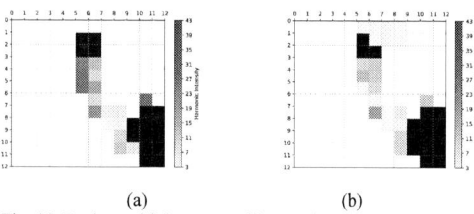

Fig. 11. Fault sensitivity maps with negative polarity (a) N=1200 inverters (b) N=228 inverters

FPGAs, which suggests that if we seek a minimal harmonic response of the RO, then adopting a square layout using 4LUTs/Slice along with horizontal routing as reported in [5] remains the best option. This key finding is important for the robustness of RO-based PUFs and TRNGs against EMFI.

In the opposite direction, we tried to investigate the impact of making the RO layout less compact compared to the previous configurations while distributing the inverters all around the chip using 1LUT/Slice as depicted in Fig. 10(a), with 1200 inverters, and in Fig. 10(b) with only 228 inverters. The fault maps shown in Fig. 11 demonstrate a higher number of impacted locations compared to previous results, even when we reduced the number of LUTs by a factor of 6. This indicates the importance of an optimal utilization of the Programmable Interconnect Points (PIPs) in FPGAs for EMFI detection.

V. CONCLUSION AND FUTURE WORK

In this paper, we discussed the varying EMFI susceptibility of an FPGA while using various layouts of the RO. The evaluation on three FPGAs of different packaging and internal structures gave us valuable insights to consider for designing countermeasures that can effectively mitigate the EMFI threat in critical applications. By carefully analyzing the effect of pulse polarity on the harmonic response of the RO, we showed its significance when evaluating a detection mechanism to achieve an ideal behavior. Based on our analysis, we concluded that adopting a low compactness strategy when implementing the design of EMFI sensors is advantageous for optimal detection.

As future work, we will explore the possibility to leverage the key findings of this paper to propose a new RO-based countermeasure to detect EMFI and different fault attacks.

ACKNOWLEDGMENT

This work is supported by the "France 2030" government investment plan managed by the French National Research Agency (ANR-22-PECY-0004) in the frame of ARSENE project, and co-funded by the Cybersecurity Institute of Grenoble Alpes (ANR-15-IDEX-02).

REFERENCES

[1] A. Barenghi, L. Breveglieri, I. Koren, and D. Naccache, "Fault injection attacks on cryptographic devices: Theory, practice, and countermeasures," *Proceedings of the IEEE*, vol. 100, no. 11.

[2] O'Flynn, Colin. "PicoEMP: A Low-Cost EMFI Platform Compared to BBI and Voltage Fault Injection using TDC & External VCC Measurements." 2023 Workshop on Fault Detection and Tolerance in Cryptography (FDTC) (2023): 60-71.

[3] J.-J. Quisquater and D. Samyde, "Eddy current for magnetic analysis with active sensor," in Proc. Smart Card Programming and Security (E-smart), pages 185–194, 2002.

[4] S. El Amraoui, A. Douadi, R. Leveugle, and P. Maistri, "Harmonic Response of Ring Oscillators under Single ElectroMagnetic Pulsed Fault Injection," in 2024 25th Latin American Test Symposium (LATS). [Online]. Available: https://hal.science/hal-04513585

[5] S. El Amraoui, R. Leveugle, and P. Maistri, "Choose your Path: Control of Ring Oscillators EMFI Susceptibility through FPGA P&R Constraints," in 2024 27th International Symposium on Design & Diagnostics of Electronic Circuits & Systems (DDECS), IEEE, Apr. 2024, pp. 118–123. doi: 10.1109/DDECS60919.2024.10508906.

[6] F. Marrucco, M. Ahmed, B. Bouali, and A. Mady, "EMplifier: Hybrid Electromagnetic Probe for Side Channel and Fault Injection Analysis," in Proceedings of the 10th International Conference on Information Systems Security and Privacy, SCITEPRESS - Science and Technology Publications, Mar. 2024, pp. 815–822.

[7] M. Ghodrati, B. Yuce, S. Gujar, C. Deshpande, L. Nazhandali, and P. Schaumont, "Inducing local timing fault through EM injection," in Proceedings of the 55th Annual Design Automation Conference, New York, NY, USA: ACM, pp. 1–6, Jun. 2018.

[8] A. Dehbaoui, J.-M. Dutertre, B. Robisson, and A. Tria, "Electromagnetic Transient Faults Injection on a Hardware and a Software Implementations of AES," in 2012 Workshop on Fault Diagnosis and Tolerance in Cryptography, IEEE, pp. 7–15, Sep. 2012.

[9] M. Dumont, M. Lisart, and P. Maurine, "Modeling and Simulating Electromagnetic Fault Injection," IEEE Transactions on Computer-Aided Design of Integrated Circuits and Systems, vol. 40, no. 4, pp. 680–693, Apr. 2021.

[10] R. Nabhan, J.-M. Dutertre, J.-B. Rigaud, J.-L. Danger, L. Sauvage, and L. A. Sauvage, "A Tale of Two Models: Discussing the Timing and Sampling EM Fault Injection Models," in FDTC, Sep. 2023.

[11] Hasegawa, R., Monta, K., Wadatsumi, T., Miki, T., Nagata, M. On-Chip Evaluation of Voltage Drops and Fault Occurrence Induced by Si Backside EM Injection. In: Wacquez, R., Homma, N. (eds) Constructive Side-Channel Analysis and Secure Design. COSADE 2024. Lecture Notes in Computer Science, vol 14595. Springer, Cham.

[12] D. El-Baze, J. -B. Rigaud and P. Maurine, "A fully-digital EM pulse detector," in Design, Automation & Test in Europe Conference & Exhibition (DATE), Dresden, Germany, pp. 439-444, Mar. 2016.

[13] S. S. Gujar and L. Nazhandali, "Detecting Electromagnetic Injection Attack on FPGAs Using In-situ Timing Sensors," Journal of Hardware and Systems Security, vol. 4, no. 3, pp. 196–207, Sep. 2020, doi: 10.1007/s41635-020-00096-9.

[14] Y. P. Chen *et al.*, "Single-Event Transient Induced Harmonic Errors in Digitally Controlled Ring Oscillators," *IEEE Trans Nucl Sci*, vol. 61, no. 6, pp. 3163–3170, Dec. 2014.

[15] M. Paquette, B. Marquis, R. Bainbridge, and J. Chapman, "Visualizing Electromagnetic Fault Injection with Timing Sensors," in Proceedings of the 2021 IEEE International Conference on Physical Assurance and Inspection on Electronics, PAINE 2021.

Continuity in Security: Leveraging LLM for Translating Security Properties Across Hardware Designs

Bulbul Ahmed[1], Sujan Kumar Saha[1], Jingbo Zhou[1], Sohrab Aftabjahani[2], Mark Tehranipoor[1], and Farimah Farahmandi[1]

[1]Electrical and Computer Engineering, University of Florida
[2]Intel Corporation

Email: {ahmed.b, sujansaha, jingbozhou}@ufl.edu, sohrab.aftabjahani@intel.com, {tehranipoor, farimah}@ece.ufl.edu

Abstract—Systems on Chips (SoCs) are integral to modern devices, from consumer electronics to critical applications in healthcare, finance, and defense, housing various vital assets. Ensuring comprehensive security verification is crucial to protect these assets from diverse vulnerabilities. However, traditional security verification is time-consuming, and the rapid pace of market-driven design cycles demands new versions within tight time-to-market windows. Conducting exhaustive security verification from scratch for each new design iteration is both challenging and impractical. This paper introduces a novel framework leveraging large language models (LLMs) to translate security properties from legacy designs to new versions at the Register Transfer Level (RTL). By reusing existing verification efforts, this approach significantly reduces verification time while maintaining security continuity. Our methodology not only translates but also extends and expands security properties to detect new vulnerabilities. Experimental results demonstrate substantial improvements in security continuity and vulnerability detection, advancing hardware security verification for evolving SoCs.

Index Terms—Security Property, Property-based Verification, Security Verification Reuse, Security Continuity, Formal Verification.

I. INTRODUCTION

The expanding use of systems on chips (SoCs) in diverse devices, especially those with critical security functions, highlights the need to protect their security assets from various threats, including malicious third-party IP, fault-injection attacks, access control violations, timing side-channel attacks, information leakage, etc. Recent research has uncovered new hardware vulnerabilities [1], [2], but addressing these vulnerabilities often lags behind their discovery, creating a gap between identification and mitigation. Additionally, current computer-aided design (CAD) tools frequently overlook security considerations, inadvertently introducing vulnerabilities into designs [3].

The pressure of limited time-to-market exacerbates security verification challenges, as functional verification is already a time-consuming bottleneck. Besides, the rapid release of new design versions, driven by industry demands for innovation, performance, and competitiveness, complicates maintaining security continuity. Additionally, the lack of comprehensive knowledge about various security threats among design and

verification engineers further complicates thorough security verification.

Security property-based formal verification [4] has become popular in hardware security verification due to its rigorous approach, ensuring robust and reliable designs. However, this process still relies heavily on manual steps such as verification planning, threat modeling, security requirement development, formal property expression, and debugging. Consequently, it is highly time-consuming and costly within the overall design development cycle [5].

A significant challenge in formal verification is the inefficiency in writing security properties, which can be resource-intensive for verification tools. Properties often focus on correctness, neglecting efficiency. For instance, a functionally correct property may still be memory-intensive, hard to debug during verification failures, and dependent on specific SystemVerilog versions, complicating the verification process.

To address these challenges, we propose a novel framework that enables automatic reuse of security verification efforts from legacy designs to new versions, ensuring seamless releases while maintaining security integrity. Additionally, it incorporates techniques for security property extension and expansion to identify new vulnerabilities from feature enhancements and design optimizations. Specifically, our framework offers the following contributions:

- **Security Property Translation**: Our framework, assisted by LLM, automatically translates security properties across hardware designs at the Register Transfer Level (RTL), ensuring security continuity and reduced verification effort.
- **Property Preprocessing**: We propose a Retrieval Augmented Generation (RAG)-based methodology for converting SystemVerilog properties into efficient versions, accelerating formal verification.
- **Variable Mapping**: We develop a RAG-based methodology for variable mapping to facilitate accurate and efficient security property translation.
- **Design bug avoidance**: We develop a methodology to extract timing information for property mapping, ensuring

979-8-3315-3968-9/24 $31.00 © 2024 IEEE

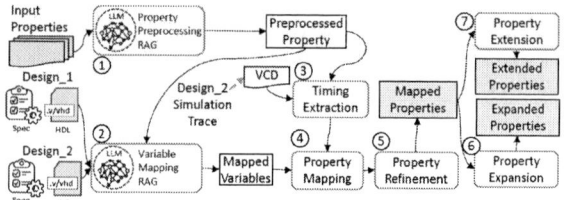

Fig. 1: Framework for Property Mapping, Extension, and Expansion.

security properties do not inadvertently introduce design vulnerabilities.

- **Template-based Property Mapping**: We introduce a technique to generate graph-based property templates, enabling the smooth translation of security properties.

The following sections cover preliminaries, our proposed framework (including property preprocessing, variable mapping, timing extraction), and experimental results demonstrating our approach's efficacy.

II. PRELIMINARIES

In this section, we outline preliminaries on formal verification properties and security properties.

a) Formal Verification Properties: These properties, typically expressed in formal languages like SystemVerilog assertions, ensure hardware designs adhere to specific requirements represented by the properties under all conditions. A typical SystemVerilog property comprises two components: the antecedent, triggering the property, and the consequent, defining the expected outcome. For instance:

property IdleInReset;
```
@(posedge clk) (rst) |-> (state == IDLE)
```
endproperty

Here, the property states that if **rst** is high (antecedent), the system must enter the **IDLE** state (consequent).

b) Security Properties: These are a subset of formal verification properties aimed at ensuring hardware designs are resistant to attacks within a defined threat model. For example:
```
property AccessWhenAuth;
@(posedge clk) (auth_granted == 1) |->
(access_level == HIGH)
endproperty
```
This asserts that if **auth_granted** is high, **access_level** must be set to HIGH, preventing unauthorized access.

III. PROPOSED FRAMEWORK

Figure 1 presents the flow of our proposed framework for security property translation. Each processing element is

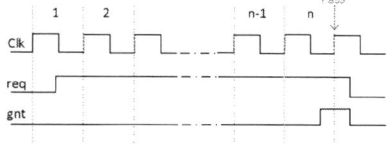

Fig. 2: Example trace for property P1, where a request signal *req* rises and stays high up to and including the clock cycle at which the grant signal *gnt* is asserted.

highlighted with a red border, and their order is indicated by numerical labels. The entire process unfolds through seven pivotal steps: 1) **Property Preprocessing** 2) **Variable Mapping** 3) **Timing Extraction** 4) **Property Mapping** 5) **Property Refinement** 6) **Property Extension** 7) **Property Expansion**.

A. Property Preprocessing

The framework begins by preprocessing the input security property set, making them memory-efficient, debuggable, and independent of the SystemVerilog Reference Manual (SV LRM), which is essential for efficient formal verification. For example, **Property P1** asserts that if there is a *req*, there should eventually be a *gnt*. In the trace shown in Figure 2, *req* rises at cycle 1 and remains high until *gnt* is observed at cycle *n*, validating Property P1.

Property P1: If *req* is asserted, *gnt* will eventually be asserted.
```
property p1;
@(posedge clk) req |-> s_eventually(gnt)
endproperty
```
During Formal Verification (FV), at each positive *clk* edge, the FV tools check if the antecedent (*req*) is true and store a copy of the property in memory until the consequent (*gnt*) becomes true. If *req* stays high for an extended period, like 1000 clock cycles, before *gnt* is asserted, multiple copies of the property are stored in memory. This approach is memory-intensive due to redundant copies. To optimize memory usage, the property formulation can be revised as follows.
```
property p1_1;
@(posedge clk) $rose(req) |->
s_eventually(gnt)
endproperty
```
In this adjustment, replacing req with **$rose**(req) reduces memory usage by storing only one copy of the property, even if *req* remains high for 1000 cycles. Since *req* rises only once during this period, the antecedent becomes true only once. However, a single copy persists until *gnt* asserts after 1000 cycles. Further preprocessing can enhance efficiency as follows.
```
property p1_2;
@(posedge clk) req && !gnt |-> req
endproperty
```
Property **p1_2** addresses the challenges previously discussed. It includes a single-cycle implication ($|->$), ensuring the antecedent and consequent become true within the same clock cycle, thus eliminating the need to store the property. Property **p1_2** enhances **p1** and **p1_1** in the following key areas:

- *Memory Efficiency*: Single-cycle implication removes the need for memory storage.
- *Ease of Debugging*: Simplifies debugging due to its single-cycle nature.
- *SystemVerilog Independence*: **p1_2** is version-independent, enhancing reusability.

B. Property Preprocessing and Variable Mapping using RAG

To enhance the LLM's capabilities for **property preprocessing** and **variable mapping**, we use the Retrieval Aug-

979-8-3315-3968-9/24 $31.00 © 2024 IEEE

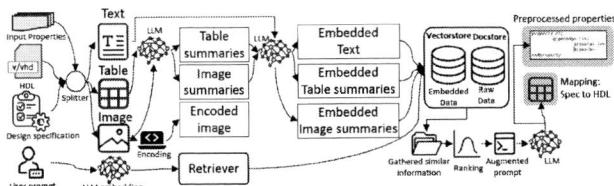

Fig. 3: Retrieval augmented generation for variable mapping and property preprocessing.

mented Generation (RAG) approach. RAG enriches the user prompt with relevant information from the user's dataset before processing by the LLM, ensuring contextually accurate responses. This method achieves precise outcomes with minimal overhead, avoiding the need for extensive LLM fine-tuning. Figure 3 delineates the RAG steps for property preprocessing and variable mapping, highlighting the streamlined process in handling security property translation.

a) RAG-based Property Preprocessing: For the RAG-based property preprocessing, we create a dataset of non-efficient properties and their efficient counterparts in PDF files. This dataset includes original properties in natural language and SystemVerilog, waveform traces of property passing and failure, efficient preprocessed properties, and explanations on their efficiency improvements. Additionally, we customize system instructions via the OpenAI API to ensure the LLM functions as a professional hardware formal verification engineer during preprocessing.

As shown in Figure 3, the dataset is split into text, tables, and images. Summaries of tables and images, which are unstructured data, are generated using LLM models `gpt-3.5-turbo` [6] and `gpt-4-vision-preview` [7], respectively. These summaries and the raw text are embedded using OpenAI tools [8], with the embedded content stored in Vectorstore and the raw content in Docstore. During property preprocessing, user queries containing security properties are embedded via OpenAI embedding and passed to a retriever. The retriever conducts a similarity search against Vectorstore and Docstore, extracting and ranking relevant content. This extracted context augments the user's initial query, which is then fed into the LLM to obtain the preprocessed property.

b) RAG-based Variable Mapping: In this step, variables/signals/registers in a security property for the legacy design are mapped to their counterparts in the newer version using the RAG approach. We utilize both the specification file and the corresponding HDL implementation. Since existing

TABLE I: Extracted Signal Information from Specification Document.

Functional Description	Interconnections
Role: [Brief Description]	**Inputs:** [Related Input Signals]
FSM States: [Applicable States]	**Outputs:** [Related Output Signals]
Conditions for Change: [Change Conditions]	**Interdependent Signals:** [Related Signals]
Reset Behavior	**Usage Context**
Reset Condition: [Reset Condition]	**Module:** [Module Name(s)]
Reset State: [Reset Value/State]	**Process:** [Specific Process]
Additional Notes	**Related Signals**
Special Features: [Special Features]	List
Design Notes: [Design-specific Notes]	[Related Signal Names]

LLM models excel at natural language but are less efficient with HDL coding, we leverage detailed natural language specifications, including block diagrams. Recent LLM models, such as `gpt-4-vision-preview` [7], can effectively analyze text, tables, and images. Using custom system instructions through the *openAI* API, the LLM model acts as an expert specification analyzer, design engineer, and signal mapper. As shown in Figure **??**, the inputs for RAG variable mapping are the specification file and the HDL implementation. Like the RAG steps in property preprocessing, specification documents are split, embedded, and stored in the Vectorstore and Docstore. During property mapping, the user inputs the security property of the legacy design. Then the information about the target variables (shown in Table I) is extracted from the specification file.

The newer design version and its specification undergo processing through RAG, extracting information for all signals, including interfaces, state variables, and registers. This creates one-to-one mappings between HDL variables and their specification information. The specification of the target variable in the legacy design is then compared with variables in the newer design, selecting the most similar as the mapped variable. Leveraging LLM's efficiency in natural language analysis, this process achieves accurate variable mappings. To address scalability, the framework adopts a modular approach, matching modules based on functionality. For instance, a state register's counterpart likely resides within the FSM of the newer design version.

C. Timing Extraction

Once the equivalent variables are identified in the new design, this module establishes temporal relationships among them. We use a hybrid approach combining formal verification and simulation to determine temporal distances. For antecedent peripheral signals, simulation-based methods apply test patterns to make the antecedent true. For internal signals, the hybrid approach constructs a property that negates the antecedent, verifies the property using formal verification, and generates a counterexample that triggers the antecedent. This counterexample aids in creating targeted test patterns, reducing test pattern generation time. The extracted patterns are used to simulate both the original and new design versions, with results recorded in value change dump (VCD) files to extract temporal distances among mapped variables.

D. Property Mapping

Integration of the obtained equivalent variables and their temporal distance is the final step in achieving the translated security property. In this step, a template-based approach is employed. Initially, the original property is converted into a graph-based template. An example of the graph-based template of the following property **a_b_implies_c** is shown in Figure 4.

```
property a_b_implies_c;
@(posedge clk) $rose(a) ## b |->
```

979-8-3315-3968-9/24 $31.00 © 2024 IEEE   15

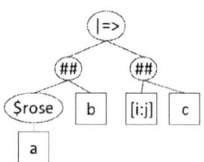

Fig. 4: Example of graph-based property template.

```
## [i:j] $rose(c)
endproperty
```

In the graph representation of the properties, registers, wires, IO ports, and constant values are leaf nodes, while operators are non-leaf nodes. For instance, *a*, *b*, *c*, *i*, and *j* are leaf nodes. Leaf nodes are tagged to indicate if they represent variables or constants, and their association with the antecedent or consequence of the security property. Using variable mapping and temporal distance information from previous steps, node replacement is performed in the graph-based property template, substituting original leaf nodes with their mapped counterparts, resulting in the translated security property

E. Property Refinement

During this phase, mapped properties are refined, potentially incorporating new terms into their antecedents. For instance, an original property A |=> B might be transformed into one of the following formats in the updated design:

1) A_mapped |=> B_mapped
2) A_mapped ∨ C |=> B_mapped
3) A_mapped ∧ C |=> B_mapped

Here, A_mapped and B_mapped correspond to variables in the revised design, akin to A and B in the original design. C is an additional component needed in the antecedent to fully characterize the property in the new design. Extracting C is aided by Goldmine, an open-source assertion mining tool [9], which utilizes simulation traces and RTL implementations. Goldmine also requires information on the clock cycle distance between A_mapped and B_mapped, obtained during timing extraction. For example, in A_mapped |=> B_mapped, the distance is 1. Goldmine then generates a set of possible assertions for B_mapped within this distance. If Goldmine extracts assertions like C |=> B_mapped and D |=> B_mapped, the dependency on either C or D is consolidated as C ∨ D. This unified antecedent, combined with the original one, results in the refined mapped property: A_mapped ∧ (C ∨ D) |=> B_mapped.

F. Property Extension

Property extension, as introduced in [3], adapts security properties to new interfaces in updated designs, detecting vulnerabilities from extended surfaces. For example, consider

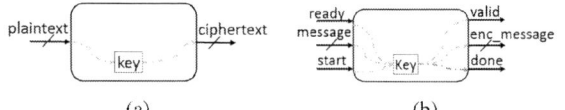

(a)   (b)

Fig. 5: Example interfaces of a Legacy AES and its newer version.

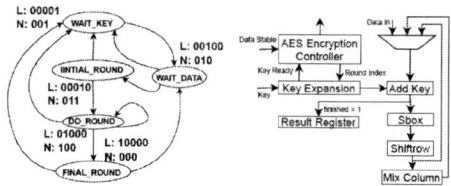

(a)   (b)

Fig. 6: (a) Controller Circuit of AES Encryption Algorithm (b) Data path of AES Encryption Algorithm.

the AES versions in Figure 5. The legacy AES (Figure 5a) has *plaintext* and *ciphertext* interfaces, while the updated version (Figure 5b) adds *start*, *ready*, *valid*, and *done* ports for a valid-ready protocol. Protecting the *key* register from unauthorized access remains crucial. In the legacy design, an information-flow property verifies potential *key* leakage as follows.

```
check_spv -from key -to {plaintext
ciphertext}
```

When mapped to the new AES version, it becomes:

```
check_spv -from key -to {message
enc_message}
```

While this mapped property verifies potential information leakage through mapped counterpart ports, it may miss others, which are additionally introduced in the new version. Property extension extends the mapped property to verify against all possible attack surfaces in the updated design, as shown below:

```
check_spv -from key -to {message
enc_message ready start valid done}
```

This demonstrates property extension tailored for *Information Leakage*, applicable to various other vulnerabilities.

G. Property Expansion

Introduced in [3], property expansion maintains security goals across design versions by creating additional security properties. The following subsection describes security property expansion in terms of the fault-injection attack.

Previous studies [10], [11] show the susceptibility of weakly implemented FSM state encoding to fault-injection attacks. Consider an AES controller circuit, illustrated in Figure 6, which transitioned from one-hot state encoding (**L**) to binary encoding (**N**), reducing the number of state flip-flops to three and minimizing area overhead. The state **FINAL_ROUND** is considered as the protected state due to the risk of prematurely revealing encrypted results if an attacker bypasses

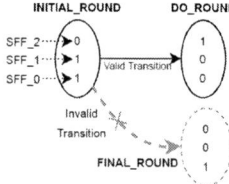

Fig. 7: Fault-injection attack during the transition from INITIAL_ROUND to DO_ROUND to gain unauthorized access to FINAL_ROUND.

979-8-3315-3968-9/24 $31.00 © 2024 IEEE   16

intermediate states and reach this state. In a valid transition from INITIAL_ROUND to DO_ROUND, no direct transition exists to FINAL_ROUND, but during this transition, SFF_2 (0 to 1) and SFF_0 (1 to 0) change states while SFF_1 (0 to 0) remains unchanged (see Figure 7). An attacker can exploit this encoding by manipulating the setup time of SFF_2 and SFF_0 to remain unchanged, accessing the protected state FINAL_ROUND directly from INITIAL_ROUND and bypassing all intermediate states.

Property Expansion for Access Control Verification: This subsection discusses expanding a security property to address access control vulnerabilities due to fault-injection attacks. Lets consider the following property in the legacy AES controller, which verifies an authorized access to the protected state FINAL_ROUND.

Property P2: FINAL_ROUND can only be accessed from DO_ROUND.

While formal verification of Property **P2** in the legacy design ensures resilience, verifying the mapped version in the new design doesn't guarantee the absence of access control violation vulnerabilities. This is because binary encoding introduces additional fault-injection attack opportunities, as discussed earlier. We expand the property using the Fault-Injection (FI) feasibility metric [3], formulating the expanded SystemVerilog property **p3** to verify the absence of access control violation in the enhanced controller (Figure 6) as follows.

```
property p3;
@(posedge clk) !(state == FINAL_ROUND) &&
!($past(state) == FINAL_ROUND) |->
!((!(state[0] ∧ FINAL_ROUND[0]) |
!($past(state[0]) ∧ FINAL_ROUND[0]))
&& (!(state[1] ∧ FINAL_ROUND[1]) |
!($past(state[1]) ∧ FINAL_ROUND[1]))
&& (!(state[2] ∧ FINAL_ROUND[2]) |
!($past(state[2]) ∧ FINAL_ROUND[2])) );
endproperty
```

IV. RESULTS AND DISCUSSIONS

To evaluate our framework, we tested it on AES (5 versions), RSA (3 versions), and SHA256 (5 versions) benchmarks [12], each in legacy and enhanced versions. Enhanced designs included optimizations like binary state encoding, pipelining, additional handshaking signals, and a malicious implant for vulnerability detection in the newer versions. A security verification plan targeted vulnerabilities such as timing side-channels, information flow, malicious implants, and denial of service, resulting in 18 specific security properties as shown in Table II.

AES.P1 ensures that premature encryption results are not stored in the output, preventing data leakage through back-analysis. AES.P2 guarantees constant encryption time to avoid timing side-channels. AES.P3 verifies that there is no encryption key leakage. AES.P4 ensures no unauthorized access to the state register. AES.P5 confirms that the protected **FINAL_ROUND** is accessed only from the authorized

TABLE II: List of properties for AES, RSA, and SHA256 used to evaluate the framework

Design	Property No.	Property Description
AES	AES.P1	The finished signal should be high only at the FINAL_ROUND
	AES.P2	The finished signal should be high 10 cycles after the INITIAL_ROUND
	AES.P3	The key register should not flow to data stable, key ready, finished, and round type sel
	AES.P4	The next FSM register should not be observable or controllable by data stable, key ready, finished, and round type sel
	AES.P5	FINAL_ROUND should only be accessed from DO_ROUND
	AES.P6	DO_ROUND must occur before the FINAL_ROUND
	AES.P7	If the AES encryption starts, it eventually ends
RSA	RSA.P1	The finished signal should be high at the RESULT state
	RSA.P2	RESULT state can only be accessed from SQR state
	RSA.P3	MULT and SQR states must occur before reaching the RESULT state
	RSA.P4	The finished signal should be high 9 cycles after the INIT state
	RSA.P5	The next FSM register should not be observable or controllable by the start and finished signals
	RSA.P6	If the RSA encryption starts, it eventually generates the encrypted data
SHA256	SHA.P1	The done signal should be high only at the st_sha_data_valid state
	SHA.P2	The Data input state must occur before reaching the st_sha_data_valid state
	SHA.P3	The st_sha_data_valid state should only be accessed from st_sha_blk_nxt state
	SHA.P4	The hash_control_st_reg register should not be observable or controllable by the peripheral ports
	SHA.P5	If hashing starts, it eventually produces the hashed output

DO_ROUND state. AES.P6 mandates progression through **DO_ROUND**, preventing the bypassing of scrambling operations. Finally, AES.P7 confirms the availability of the design.

Similar to AES properties, RSA and SHA properties validate encrypted data timing, access control, and operations integrity, ensuring absence of timing side-channels, securing the internal state register, absense of data leakage, collision-free hashing, and design availability. Outcomes of variable and timing mapping, formal verification, and property extension are detailed in Table III, using Cadence JasperGold [13]. However, the framework is not limited to SystemVerilog assertion, and JasperGold.

Variable Mapping Result: In the variable mapping domain, the property translation exhibited a commendable accuracy, with only three out of 18 properties (AES.P4, RSA.P5, SHA256.P4) featuring an incorrect mapping for a single variable. Specifically, the "Next State" register was inaccurately mapped as the "Current State" for these instances. It's noteworthy that despite these discrepancies, the misassigned variables share strikingly similar functionalities within the design context.

Timing Mapping Result: The timing mapping analysis ensured accurate clock-cycle synchronization across AES, RSA, and SHA256 designs. Properties like AES.P2, AES.P7, RSA.P4, RSA.P6, and SHA.P5 exemplify this precision. Notably, the framework adeptly excluded the malicious implant from property mapping, showcasing resilience against inadvertent inclusion of malicious elements.

Design Vulnerability/Bug: Traditional methods [9], [14] risk incorporating design vulnerabilities or bugs into mapped properties. For instance, consider the AES controller in Figure 6. An enhanced version with a malicious implant can exploit a trigger condition, completing operations in just three cycles instead of ten. Our approach employs simulation to preemptively mitigate this risk, delaying discovery of malicious functional-

979-8-3315-3968-9/24 $31.00 © 2024 IEEE

TABLE III: Result of property translation, extension, and expansion with detected vulnerabilities

Design	Prop. No.	# var	# correctly mapped	Temporal	Target Vulnerability	Verification Result (Detected Vulnerability)			
						Original	Mapped	Prop. Ext.	Prop. Exp.
AES (5 version)	P1	3	3		Information Leakage	-	-	-	FSE, FI
	P2	3	3	Yes	Timing SC	-	MI	-	FSE, FI
	P3	5	5	-	InformationLeakage	-	-	IL(HS)	-
	P4	5	4	-	Access Control	-	-	IL	-
	P5	3	3	-	Access Control	-	-	-	FSE, FI
	P6	3	3	-	Access Control	-	-	-	FSE, FI
	P7	2	2	Yes	Denial of Service	-	-	-	DOS
RSA (3 versions)	P1	3	3	-	Information Leakage	-	-	-	FSE, FI
	P2	3	3	-	Access Control	-	-	-	FSE, FI
	P3	4	4	-	Information Leakage	-	-	-	FSE, FI
	P4	3	3	Yes	Timing SC	-	MI	-	FSE, FI
	P5	3	2	-	Access Control InformationLeakage	-	-	IL	-
	P6	2	2	-	Denial of Service	-	-	-	DOS
SHA256 (5 versions)	P1	3	3	-	Information Leakage	-	-	-	FSE, FI
	P2	3	3	-	Information Leakage	-	-	-	FSE, FI
	P3	3	3	-	Access Control	-	-	-	FSE, FI
	P4	15	14	-	Access Control Information Leakage	IL(PIO)	IL(PIO)	IL(HS)	-
	P5	2	2/2	-	Denial of Service	-	-	-	DOS

ity to formal verification. This precaution enhances verification resilience against inadvertent vulnerability inclusion.

Vulnerability Identification: Verification of mapped, extended, and expanded properties using Cadence JasperGold reveals vulnerabilities. The legacy SHA256 design showed information leakage (SHA.P4). Mapped properties identified AES (MI) and RSA (MI, TSC) vulnerabilities. Extended properties uncovered four vulnerabilities (AES.P3, AES.P4, RSA.P5, SHA.P4). Expanded properties revealed 14 vulnerabilities collectively across all designs: AES.P1 (FSE, FI), AES.P2 (FSE, FI), AES.P5 (FSE, FI), AES.P6 (FSE, FI), AES.P7 (DOS), RSA.P1 (FSE, FI), RSA.P2 (FSE, FI), RSA.P3 (FSE, FI), RSA.P4 (FSE, FI), RSA.P6 (DOS), SHA.P1 (FSE, FI), SHA.P2 (FSE, FI), SHA.P3 (FSE, FI), SHA.P5 (DOS).

Figure 8 presents the number of vulnerabilities detected at each stage—Original, Mapped, Extended, and Expanded—for AES, RSA, and SHA256 designs. This analysis demonstrates the efficiency of our framework in identifying security vulnerabilities progressively as follows. 1) *Comprehensive Detection:* The framework detected a significant number of vulnerabilities, with 0 in AES, 0 in RSA, and 1 in SHA256 at the Original stage, which increased to 1 in AES, 1 in RSA, and 1 in SHA256 at the Mapped stage, and further increased substantially at the Extended and Expanded stages. 2) *Progressive Analysis:* The number of vulnerabilities identified increased from the Original to the Expanded stages, highlighting the framework's ability to uncover additional security issues through property extension and expansion. 3) *Adaptive Capability:* The detection of new vulnerabilities in enhanced designs demonstrates the framework's adaptability to evolving hardware features and potential new attack vectors.

V. CONCLUSION

This research introduces an innovative framework that leverages LLM-based security property mapping, extension, and expansion to ensure the security continuity during the evolution of designs from legacy to new versions. The property extension and expansion play a pivotal role in identifying new vulnerabilities that arise due to design optimization and enhancement. In summary, this work significantly contributes to security verification of designs, offering a systematic and effective means for designers to fortify their systems during

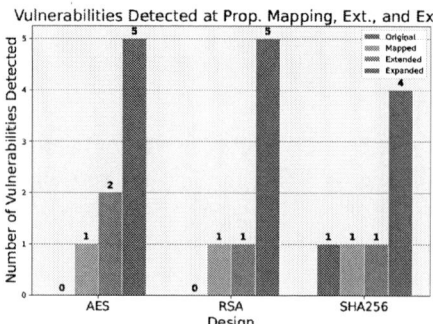

Fig. 8: Number of vulnerabilities detected at each stage of property analysis for AES, RSA, and SHA256 designs.

design updates. Future endeavors will focus on refining and extending our approach to address the evolving diverse set of vulnerabilities.

REFERENCES

[1] S. Bhunia and M. Tehranipoor, *Hardware security: a hands-on learning approach.* Morgan Kaufmann, 2018.

[2] M. Tehranipoor and C. Wang, *Introduction to hardware security and trust.* Springer Science & Business Media, 2011.

[3] B. Ahmed, F. Rahman, N. Hooten, F. Farahmandi, and M. Tehranipoor, "Automap: Automated mapping of security properties between different levels of abstraction in design flow," in *2021 IEEE/ACM International Conference On Computer Aided Design (ICCAD).* IEEE, 2021, pp. 1–9.

[4] N. Farzana, F. Rahman, M. Tehranipoor, and F. Farahmandi, "Soc security verification using property checking," in *2019 IEEE International Test Conference (ITC).* IEEE, 2019, pp. 1–10.

[5] P. Mishra and N. D. Dutt, *Functional verification of programmable embedded architectures: a top-down approach.* Springer Science & Business Media, 2005.

[6] https://platform.openai.com/docs/models/gpt-3-5-turbo.

[7] https://platform.openai.com/docs/models/gpt-4-turbo-and-gpt-4.

[8] https://platform.openai.com/.

[9] https://sites.google.com/view/goldmineillinois/home.

[10] A. Nahiyan, F. Farahmandi, P. Mishra, D. Forte, and M. Tehranipoor, "Security-aware fsm design flow for identifying and mitigating vulnerabilities to fault attacks," *IEEE Transactions on Computer-aided design of integrated circuits and systems,* vol. 38, no. 6, pp. 1003–1016, 2018.

[11] C. Dunbar and G. Qu, "Designing trusted embedded systems from finite state machines," *ACM Transactions on Embedded Computing Systems (TECS),* vol. 13, no. 5s, pp. 1–20, 2014.

[12] https://opencores.org/projects/.

[13] https://www.cadence.com/en_US/home/tools/system-design-and-verification/formal-and-static-verification.html.

[14] R. Zhang and C. Sturton, "Transys: Leveraging common security properties across hardware designs," in *2020 IEEE Symposium on Security and Privacy (SP).* IEEE, 2020, pp. 1713–1727.

979-8-3315-3968-9/24 $31.00 © 2024 IEEE

OSHDA: A Containerized CAD Tool for the Design and Analysis of Behavioral FSM Logic Locking

Esrat Khan*, Shahzad Muzaffar°, Lamees M. Al Qassem*, and Ibrahim (Abe) M. Elfadel*
*Center for Secure Cyber-Physical Systems, Khalifa University, Abu Dhabi, United Arab Emirates
°IMEC, Eindhoven, The Netherlands

Abstract—This paper introduces the Open-source Secure Hardware Design and Analysis (OSHDA) toolchain for the logic locking of finite-state machines (FSMs) at the behavioral level. OSHDA's FSM obfuscation method is based on the recently developed State Permutation Logic Locking (SPeLL) algorithm which obfuscates the behavioral transition graph of the FSM, thus avoiding the use of dummy states and reducing exposure to reverse engineering attacks. In addition to implementing the SPeLL algorithm, the toolchain implements a full logic synthesis flow, including the evaluation of the gate-level SPeLL hardware overhead for both FPGA and ASIC designs. In particular, OSHDA enables the automation of trade-off analysis between the strength of SPeLL security and its hardware overhead. The paper further describes attempted attacks on SPeLL using state-of-the-art de-obfuscation tools and identifies research gaps in behavioral de-obfuscation that must be addressed before one can successfully de-obfuscate SPeLL. OSHDA comes with its own scripting subsystem for augmenting its analysis, adding de-obfuscation methods, and integrating physical design tools. Finally, OSHDA is deployed as a hardware security microservice using the Docker framework.

Index Terms—Hardrware Security, behavioral Logic Locking, Finite-State Machines, Obfuscation, De-obfuscation.

I. INTRODUCTION

In the rapidly evolving landscape of digital circuit design and hardware security, the protection of intellectual property and sensitive information embedded within integrated circuits has become a paramount concern. Third-party involvement in the integrated circuit (IC) design process has increased significantly due to time to market and cost reduction demands [1]. Instead of undertaking complete in-house development of the entire process from chip architecture to fabrication, companies now prefer to outsource key stages of the chip manufacturing process, thus exposing it to the security vulnerabilities of the semiconductor supply chain [2]. These include Intellectual Property (IP) piracy, excessive production, counterfeiting, reverse engineering, and hardware Trojan insertion [3]. Amidst the multitude of design-for-trust methodologies, logic locking is a promising technique that aims to safeguard these assets by introducing deliberate design obfuscations into the logic design. The recent history of logic locking includes successive stages of cat-and-mouse runs between obfuscations and de-obfuscations using a variety of algorithms, most of them geared toward gate-level logic designs [4], [5].

Logic locking could be either key-based or "keyless" [6]. In key-based logic locking, the correct operation of the circuit is guaranteed solely upon providing the secret key value, thereby validating its intended functionality [7]. Typically, the key value is securely loaded from tamper-proof memory (TPM) into the circuit upon powering it up. In keyless logic locking, the confidential data can consist of an implicit sequence of input patterns originating from primary inputs (PIs) that direct the circuit's finite-state machine (FSM) to transition through its intended design states.

Most of the research effort has focused on logic-locking at the gate level for both combinational and sequential circuits. However, behavioral logic locking has emerged as a prominent research domain in recent years [8]. The main goal of this paper is to present a comprehensive hardware security CAD tool chain centered around a recent behavioral logic locking algorithm called State-Permutation Logic Locking (SPeLL) [9]. Unlike recent high-level synthesis approaches [8], this method is systematic, global, and does not require identifying obfuscation points in the behavioral Verilog code. Additionally, it secures the FSM without adding dummy states or fake operations, employing a novel methodology that fortifies the FSM by encrypting its state encodings, leading to a permutation of the original states. Consequently, the conventional static, next-state look-up table of the FSM is transformed into a dynamic, key-dependent lookup table. This confidential permutation serves as the foundation for generating the encryption key. The decryption process restores the initial FSM state arrangement through an inverse permutation of the states in the lookup table [9].

In the current EDA landscape, the absence of a full tool chain capable of obfuscating and de-obfuscating logic designs at the behavioral level presents a significant challenge to the advancement of research in this important area of hardware security. This gap is mainly due to a legacy of existing open source tools the community has been using to implement logic locking algorithms for gate level netlists. In our own experience with existing tools, we have encountered a bewildering heterogeneity in HDL file formats, parser capabilities, gate libraries, output file formats, and of course implemented algorithms for obfuscation and de-obfuscation.

It is against this backdrop that we have developed the Opensource Secure Hardware Design and Analysis (OSHDA), which is our attempt at building an open source, behavioral logic locking tool chain from the ground up. The philosophy of OSHDA is *behavioral HDL in, secured behavioral HDL out* and is meant to ensure that the HDL syntax remains unrestricted both before and after obfuscation. The core of OSHDA is the transition graph obfuscation algorithm, which, unlike existing approaches, does not introduce any fake states or fake transitions to secure behavioural netlists. It should be

979-8-3315-3968-9/24 $31.00 © 2024 IEEE

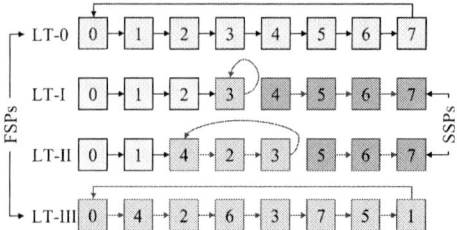

Figure 1. LT-0, LT-I, LT-II, and LT-III FSM lock types.

noted that the selection of state encodings and permutations is executed with strict adherence to a uniform probability distribution, which endows the secured behavioural netlists with a higher degree of resilience against frequency cryptanalysis attacks. OSHDA offers a flexible scripting subsystem that enhances ease of use by allowing users to integrate custom scripts in their preferred languages, such as C or Python. This capability supports specific design and analysis requirements. Using OSHDA, users can effectively engage in logic design using HDL, encrypt behavioral netlists, perform logic synthesis, perform simulations and tests, facilitate gate-level mapping, and perform analysis and optimization processes for both application-specific integrated circuits (ASIC) and field-programmable gate arrays (FPGA). This design flow consists of four main steps: (1) entry of HDL files and logic synthesis; (2) generation of the encrypted file utilizing behavioral logic locking; (3) hardware security testing and evaluation at the logic level; and (4) simulation of the secure digital logic design. The OSHDA tool is packaged and deployed using containerization technology. Containerization involves bundling the CAD tool and its dependencies into a standardized unit called a container, which is easily deployed and runs consistently across different computing environments. This OSHDA container deployment is designed to offer hardware security as a microservice in distributed design environments [10]. The remainder of this paper is organized as follows. In Section II, related research on behavioral logic locking of FSM, state-of-the-art logic locking attack, and deobfuscation tools are outlined. In Section III, details of the OSHDA design and architecture are provided. Test cases of the behavioral logic locking and security analysis are presented in Section IV and Section V, respectively. Section VI explores existing research gaps in the deobfuscation of behavioral logic locking and outlines potential avenues for future advancements in the field of SAT-based attacks on behavioral logic locking. The paper is concluded in Section VII.

II. BACKGROUND

A. Logic Locking

The SPeLL algorithm integrates logic locking at the behavioral level, reducing hardware needs and complicating reverse engineering by obscuring state transitions. The transition obfuscation algorithm assumes attackers have access to the hardware and know the encryption method but not the design details. Correct state transitions require the correct key. Incorrect keys trigger different Lock Types (LT-I, LT-II, and LT-III) and

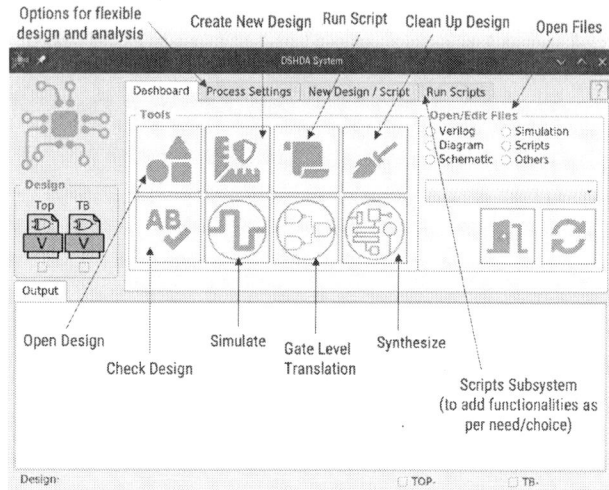

Figure 2. OSHDA user interface (UI).

incorrect transitions, as shown in Figure 1. LT-0 represents the correct transitions. The encryption process of SPeLL involves three steps:

1) Creating an encrypted state table by replacing bits of constant state parameters with their encrypted counterparts.
2) Generating a key insertion pattern, which could be either a Full Key Insertion Pattern (FKIP) for maximum security or a Partial Key Insertion Pattern (PKIP) to reduce hardware overhead.
3) Performing state-bit encryption.

The chosen state permutation dictates the key extraction based on the encoding bits of the permuted states. This method is resistant to frequency analysis and side-channel attacks.

B. Overview of NEOS

The Netlist Encryption and Obfuscation Suite (NEOS) is a state-of-the-art tool developed using a C++ software framework and designed exclusively for the obfuscation and deobfuscation of gate-level netlists [11]. It supports parsing netlist files in Bench format, accommodating various gate types, including AND, NAND, OR, NOR, BUF, NOT, XOR, and XNOR [12]. NEOS employs object-oriented and polymorphic deobfuscation algorithms using the Glucose satisfiability (SAT) solver. Furthermore, NEOS facilitates sequential deobfuscation with model-checking and bounded model-checking (BMC) techniques [13].

NEOS lacks behavioral-level logic-locking encryption, a capability unique to OSHDA for sequential circuits written in Verilog. Deobfuscating the SPeLL algorithm with NEOS has encountered several challenges mainly due to parser limitations. In fact, NEOS is designed for gate-level netlists in Bench format. The limitations of NEOS are further detailed in Section V.

C. Overview of RANE

The Reverse Assessment of Netlist Encryption (RANE) framework is an open source tool that specializes in formal deobfuscation attacks for reverse engineering logic-locked

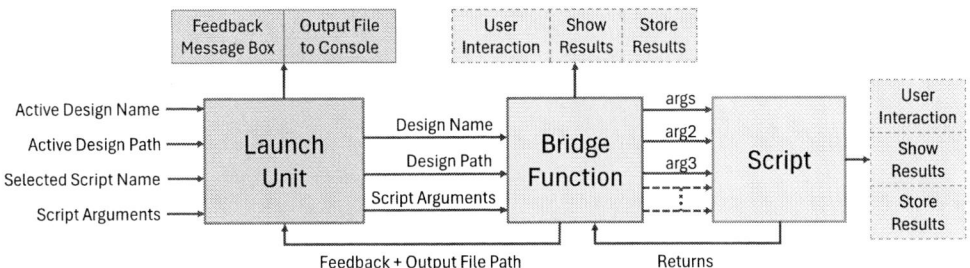

Figure 3. Architecture of scripts subsystem.

circuits. The evaluation of security across larger circuits is demonstrated in four case studies: an oracle-less attack on key-less (implicit key) sequence-based logic locking (HARPOON) [2], an oracle-guided assault on HARPOON [2], an oracle-guided approach targeting sequential logic, and an oracle-guided attack focused on combinational logic locking [7]. Similar to NEOS, RANE lacks behavioral-level logic-locking encryption techniques and cannot decipher behavioral-level encryption due to the absence of the necessary parsers. The limitations of RANE are further detailed in Section V.

A significant gap exists in the availability of an open-source tool that can both secure and deobfuscate behavioral netlists. The OSHDA toolchain addresses the need for securing behavioral netlists. Additionally, current state-of-the-art deobfuscation algorithms are ineffective in breaking the SPeLL encrypted netlists.

III. OSHDA DESIGN AND ARCHITECTURE

A. OSHDA System Philosophy

The containerized OSHDA system is an extensible and adaptable hardware design and analysis toolchain. Figure 2 shows the various features embedded in the OSHDA application. The toolchain validates design and testbench syntax through comprehensive verification, simulates sequential logic circuits, translates designs to gate level, and synthesizes them for FPGA or ASIC deployment. It also automatically generates testbenches and circuit diagrams for encrypted, translated, and synthesized designs. The architecture converts an initial Verilog design into an optimized, secured version with gate primitives and Flip-Flops (FF) for gate-level processing. ASIC synthesis produces a Verilog design with selected gates, muxes, LUTs, Flip-Flops, and derived logic gates like AOI, ANDNOT, and OAI. FPGA synthesis creates a Verilog design tailored to specific FPGA architectures, such as Xilinx, containing LUTs, DFFs, SLRs, and other FPGA elements. Additionally, the obfuscation flow generates secured design features and statistics report.

B. Scripts Subsystem

The OSHDA scripts subsystem allows easy addition and enhancement of functionalities by integrating scripts for handling design, output, logging, and system files. The subsystem consists of three key components as shown in Figure 3: the

TABLE I
ENCRYPTION WHEN $n_k = n_{states}$.

8 States + 8-bit Key + Level 2					
Key-Bits-Insertion Pattern $\longrightarrow$	Bit 0	Bit 1	Bit 2	ZigZag	Triangular
Correct States[a]	Number of Keys out of 256				
-	128	128	128	128	128
0	64	64	64	64	64
1	32	32	32	32	32
2	16	16	16	16	16
3	8	8	8	8	8
4	4	4	4	4	4
5	2	2	2	2	2
6	1	1	1	1	1
CFS[b] 7	1	1	1	1	1
Lock Type	Number of Keys out of 256				
LT-0	1	1	1	1	1
LT-I	240	0	0	80	119
LT-II	0	240	240	168	135
LT-III	15	15	15	7	1

[a]Number of correct states before lock [b]Correct Flow of States

Launch Unit in the main application, the Bridge Function in the bridge script, and the script itself. The Launch Unit activates when a script is initiated, featuring specific input and output arguments. These outputs are then sent to the Bridge Function, which returns feedback and the output file path. This feedback is displayed in a message box, and the output is shown in the console. Each script has a Bridge Function that standardizes the interface with the main application, allows for multiple scripts with unique settings, and prepares arguments for the target script. The default configuration includes three output arguments passed to the script: args for script arguments formatted by the Bridge Function, arg2 indicating the active design's full path, and arg3 representing the path of a predefined output file. Both arg2 and arg3 can be adjusted as needed, and additional arguments can be added if required by the script. The Bridge Function also handles user interactions and helps maintain a clean design directory. When creating new scripts, OSHDA provides a template for Python scripts and a launcher for scripts in other languages, such as C, which may require compilation before execution.

C. Behavioral Logic Locking Integration with OSDHA

The OSHDA toolchain employs the LT-III lock type, shown in Table I, in its logic locking approach to protect FSMs from side-channel attacks by using complex state permutations. This method effectively counters both power and frequency-based attacks by complicating the analysis of power consumption patterns for attackers. Unlike other methods that increase FSM

979-8-3315-3968-9/24 $31.00 © 2024 IEEE

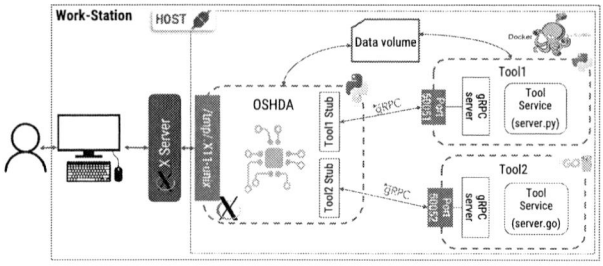

Figure 4. Microservice OSHDA.

TABLE II
HARDWARE RESOURCES REQUIRED FOR FSP IN COUNTER FSMs

Counter Size	S-I-O-T		Original Design	FKIP	PKIP				No. of States in FSM
					3	2	1	0	
4	4-0-2-4	FSP	-	24	-	-	4		24
		Logic Blocks	1	6	-	-	5		6
8	8-0-3-8	FSP	-	40320	-	16			576
		Logic Blocks	2	18	-	9			16
16	16-0-4-16	FSP	-	$> 10^{13}$	256				$> 10^5$
		Logic Blocks	2	104	22				45

TABLE III
SYNTHESIS RESULTS OF LGSYNTH91 BENCHMARK CIRCUITS

Circuit Name	S-I-O-T	Original Design (Logic Blocks)	Locked Design (FKIP) (Logic Blocks)	%↑	Locked Design (PKIP) (Logic Blocks)	%↑
train4	4-2-1-14	6	10	66	134	2133
dk15	4-3-5-32	10	19	90	17	70
beecount	7-3-4-28	35	47	34	57	63
dk14	7-3-5-56	23	38	65	63	174
dk17	8-2-3-32	14	35	150	60	328
ex6	8-5-8-34	36	64	77	85	136
cse	16-7-7-91	71	163	129	145	104
kirkman	16-12-6-370	87	160	84	100	15
sand	32-11-9-184	215	332	54	242	12
s298	218-3-6-1096	1098	2592	136	1270	15

states for security, OSHDA's obfuscation mode introduces deliberate errors in state transitions, making it difficult for attackers to distinguish between protected and unprotected FSMs. Additionally, it safeguards against register overwrite attacks, ensuring that protective mechanisms of the obfuscation mode remain active throughout the entire operation of the FSM, rendering attempts to tamper with registers ineffective.

D. OSHDA Docker Image

In this section, we introduce a cloud-based microservice solution using Docker containers and docker-compose within the OSHDA system framework as shown in Figure 4. Docker images can be versioned for easy rollback or maintaining different OSHDA system versions for troubleshooting or feature testing without affecting performance. Moreover, the application can run natively anywhere with a single command, regardless of the underlying hardware architecture. The microservice implementation of OSHDA is extendable by adding new tools and software programs without impacting the main OSHDA implementation. Achieving this involves containerizing the new tool within a single Docker image and connecting it to the main OSHDA image using gRPC, which is more efficient than HTTP due to its lower network latency and higher throughput [14]. This transition to a lighter-weight virtualization solution (i.e., microservice) helps in better meeting the QoS requirements, lowers costs, enhances availability, reduces virtualization overhead, and increases resource utilization.

IV. USE CASES

A. Full-set State Permutations (FSP) in Counter FSM

Table II illustrates the significant difference in Full-set State Permutations (FSP) complexity and hardware resources between the Full Key Insertion Pattern (FKIP) and Partial Key Insertion Pattern (PKIP). As the size of the counter FSM increases, the number of FSPs increases exponentially under FKIP, while the hardware resources required under PKIP also increase but at a slower rate. This demonstrates the trade-off between FKIP's higher security but greater resource use and PKIP's more efficient but less secure approach.

B. LGSynth91 Benchmark Circuits

Prior work in [7] and [11] implements FSM obfuscation at the gate level using ISCAS89 benchmark circuits to validate and assess security mechanisms. In contrast, OSHDA applies the logic-locking security mechanism at the behavioral level.

While the ISCAS89 circuits provide information on the quantity of D Flip-Flops, they lack the necessary state information for validating and assessing the encryption process at the RTL level. Furthermore, schematic diagrams and functional descriptions for the ISCAS89 circuits are unavailable. Consequently, a direct comparison between OSHDA's obfuscation mode and prior work is not feasible. For the same reason, comparing the unique number of register patterns presented in [15] with the FSPs in the obfuscation implementations is also not possible. Therefore, Table III presents the synthesis results of the LGSynth91 benchmark circuits.

C. Sequential Circuit: Pre and Post-Encryption States

The behavioral netlist obfuscation modifies the FSM state table into a ciphered format by substituting fixed state parameter bits with their encrypted counterparts. This method is shown in Listing 1, with the s27 benchmark circuit. The FKIP approach encrypts all bits within state encodings, thereby generating a boolean matrix for an exhaustive logic lock, whereas the PKIP method opts for selective bit encryption which aims to reduce hardware overhead, but it consequently generates a lesser number of FSPs. During the state bit encryption process, a non-identity permutation is predetermined before key selection to guarantee the obfuscation of FSM transitions. The key itself is extrapolated from the encoding bits of the permuted states. This procedure encompasses the selection of a suitable permutation, linearization, and amalgamation of the state table's binary columns. This ensures that the correct transitions occur with the appropriate key, and conversely, erroneous transitions arise with incorrect keys.

```
1 ...                                                    (a)
2 // Original States
3 parameter STATE_0 = 3'b000;
4 parameter STATE_1 = 3'b001;
5 parameter STATE_2 = 3'b010;
6 parameter STATE_3 = 3'b011;
7 parameter STATE_4 = 3'b100;
8 parameter STATE_5 = 3'b101;
9 ...
```

```
1 ...                                                    (b)
2 // Encrypted States using FKIP
3 wire STATE_0E = {K[12],K[6],K[0]};
4 wire STATE_1E = {K[13],K[7],K[1]};
5 wire STATE_2E = {K[14],K[8],K[2]};
6 wire STATE_3E = {K[15],K[9],K[3]};
7 wire STATE_4E = {K[16],K[10],K[4]};
8 wire STATE_5E = {K[17],K[11],K[5]};
9 ...
```

```
1                                                        (c)
2 module s27_enc(clk, reset, K, In, Out);
3 ...
4 input [17:0] K;
5 ...
6 // States assignments as per decryption pattern
7 assign STATE_0 = STATE_0E;
8 assign STATE_1 = STATE_1E;
9 assign STATE_2 = STATE_2E;
10 assign STATE_3 = STATE_3E;
11 assign STATE_4 = STATE_4E;
12 assign STATE_5 = STATE_5E;
13 ...
14 endmodule
```

Listing 1. s27 Sequential Circuit: (a) Original States (b) Encrypted States using Full Key Insertion Pattern (FKIP) (c) Full Example with Decryption Pattern.

D. Cryptanalysis Process Flow

In the cryptanalysis process shown in Figure 5, a specific key pattern is inserted to simulate the response of an encrypted circuit. This key pattern is carefully chosen to explore the behavior of the circuit under encryption. The selection of keys and the utilization of testbenches are crucial steps that enable a comprehensive exploration of the entire key space.

The simulation generates a Value Change Dump (.vcd) file capturing the circuit's waveform data. This aids in extracting critical features and statistics, such as power consumption and timing, for in-depth analysis. Simultaneously, visualizing

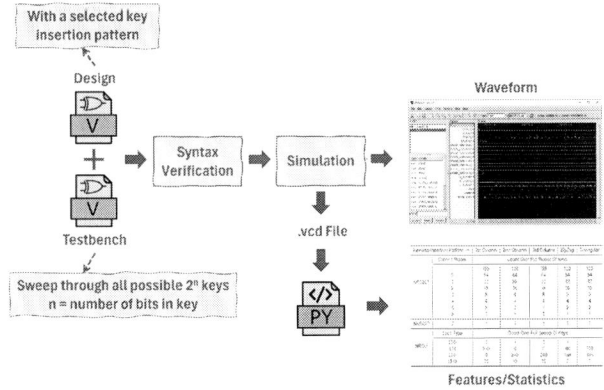

Figure 5. The cryptanalysis process flow for key bit insertion pattern.

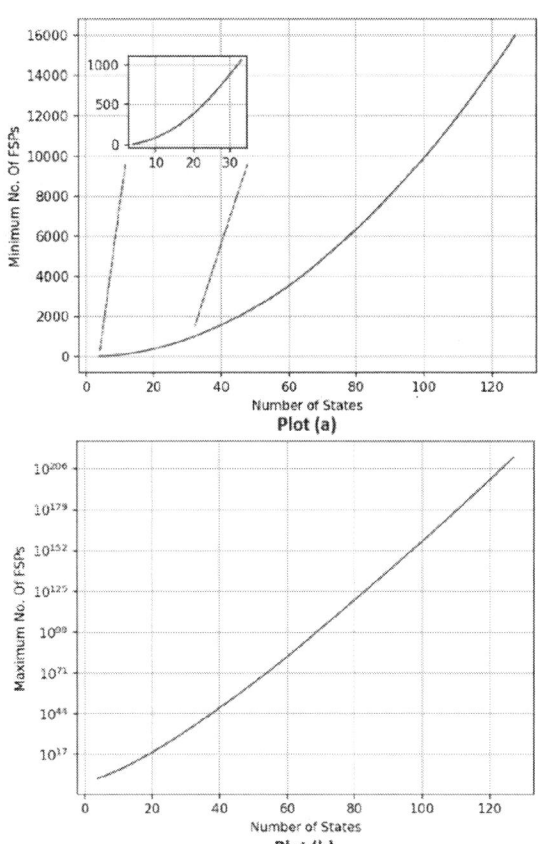

Figure 6. Number of Unique FSPs/Register Patterns (a) Minimum (b) Maximum.

the waveform helps identify vulnerabilities and assess circuit performance under various key conditions.

V. SECURITY ANALYSIS

Figure 6 illustrates the combinatorial increase in obfuscation mode FSPs as the number of states increases. Remarkably, even the minimum number of FSPs shows a significant increase with the number of states when compared with the unique number of register patterns in [15], under the assumption that all the D Flip-Flops in the netlist are state registers. Minimum FSPs occur when specific inputs minimize the states in the FSM transition graph. Identifying these inputs reduces the FSPs for analysis. However, even with reduced FSPs, deobfuscating OSHDA's logic locking remains a significant challenge for attackers.

A. Limitations of NEOS and RANE for Attacking and Deobfuscating the OSHDA Encryption Mode

Table IV provides a feature-based comparison between NEOS, RANE and OSHDA. NEOS supports only the Bench HDL, which is not as widely-used as Verilog, VHDL, or SystemVerilog, thus restricting its analysis range. It is tailored for simpler gate-level netlists, overlooking the complexity of advanced circuit components. NEOS's dependence on the

979-8-3315-3968-9/24 $31.00 © 2024 IEEE

TABLE IV
COMPARATIVE ANALYSIS OF OSHDA TOOLCHAIN WITH STATE-OF-THE-ART NEOS AND RANE TOOLS

Limitations	NEOS [12]	RANE [7]	OSHDA
File Format Support	Bench format	Bench format and Verilog (gate level modelling)	Supports Verilog (gate level, structural, and behavioural modelling)
Circuit Complexity Support	Gate level netlists	Gate level netlists	Behavioral and Gate level netlists
SAT Solver Dependency	Glucose SAT solver	Reverse Assessment of Netlist Encryption attack (SAT based attack)	State-of-the-art SAT attack tools fail to parse behavioral netlists
User Documentation and Support	Unavailable	Unavailable	Comprehensive documentation and user guide available
Advanced Security Scheme Support	Cannot parse Obfuscated Behavioral Netlists	Cannot parse Obfuscated Behavioral Netlists	Parsers available for Obfuscated Behavioral Netlists
Oracle Information Dependency	Oracle guided and oracle less approaches	Oracle guided approach only	Not Dependent
Containerization Support	No support	No support	Hardware security microservice

Glucose SAT solver can limit its deobfuscation effectiveness when other solvers might be more appropriate. Moreover, both tools suffer from a lack of comprehensive documentation and community support, which hampers user engagement and tool utility. They also do not support advanced security schemes like the SPeLL algorithm, potentially weakening their defence against sophisticated obfuscation techniques. RANE's reliance on accurate oracle information for deobfuscation adds another layer of complexity, as such data can be challenging to procure. Finally, the absence of containerization support in both frameworks affects consistent execution and hinders the sharing and reproducibility of research across different computational environments.

VI. DISCUSSION AND FUTURE WORK

Prior research on CAD tools for logic locking mainly used ISCAS89 benchmarks, which are unsuitable for validating OSHDA obfuscation due to their gate-level netlists and lack of state count information. The LGSynth91 benchmarks in Kiss2 format offer comprehensive FSM data, and OSHDA converts this to behavioral Verilog. OSHDA's behavioral-level encryption is robust against SAT attacks, unlike traditional gate-level methods. Future work includes developing a deobfuscation feature that targets behavioral-level encryption in sequential circuits. Such feature should have comprehensive HDL parsers.

VII. CONCLUSIONS

This paper introduces OSHDA, a containerized tool suite for behavioral logic locking. It supports custom scripts in any programming language for sequential logic circuit design and analysis and includes SPeLL for logic obfuscation. OSHDA handles Verilog design, synthesis, encryption, simulation, testing, gate-level mapping, ASIC/FPGA analysis, and design optimization. Built from scratch, it emphasizes logic locking FSMs at the behavioral level.

ACKNOWLEDGEMENT

This research is conducted at Khalifa University with the financial support of the Technology Innovation Institute (TII), Abu Dhabi, UAE, under contract TII/CRP/2036/2020.

REFERENCES

[1] K. Juretus and I. Savidis, "Synthesis of hidden state transitions for sequential logic locking," *IEEE Transactions on Computer-Aided Design of Integrated Circuits and Systems*, vol. 40, no. 1, pp. 11–23, 2021.

[2] R. S. Chakraborty and S. Bhunia, "Harpoon: An obfuscation-based soc design methodology for hardware protection," *IEEE Transactions on Computer-Aided Design of Integrated Circuits and Systems*, vol. 28, no. 10, pp. 1493–1502, 2009.

[3] N. Limaye, E. Kalligeros, N. Karousos, I. G. Karybali, and O. Sinanoglu, "Thwarting all logic locking attacks: Dishonest oracle with truly random logic locking," *IEEE Transactions on Computer-Aided Design of Integrated Circuits and Systems*, vol. 40, no. 9, pp. 1740–1753, 2021.

[4] W. Hu, C.-H. Chang, A. Sengupta, S. Bhunia, R. Kastner, and H. Li, "An overview of hardware security and trust: Threats, countermeasures, and design tools," *IEEE Transactions on Computer-Aided Design of Integrated Circuits and Systems*, vol. 40, no. 6, pp. 1010–1038, 2021.

[5] L. Mankali, S. Patnaik, N. Limaye, J. Knechtel, and O. Sinanoglu, "Vigilant: Vulnerability detection tool against fault-injection attacks for locking techniques," *IEEE Transactions on Computer-Aided Design of Integrated Circuits and Systems*, pp. 1–1, 2023.

[6] M. Zuzak, A. Mondal, and A. Srivastava, "Evaluating the security of logic-locked probabilistic circuits," *IEEE Transactions on Computer-Aided Design of Integrated Circuits and Systems*, vol. 41, no. 7, pp. 2004–2009, 2022.

[7] S. Roshanisefat, H. Mardani Kamali, H. Homayoun, and A. Sasan, "RANE: An open-source formal de-obfuscation attack for reverse engineering of logic encrypted circuits," in *Proceedings of the 2021 on Great Lakes Symposium on VLSI*, ser. GLSVLSI '21. New York, NY, USA: Association for Computing Machinery, 2021, p. 221–228.

[8] C. Pilato, L. Collini, L. Cassano, D. Sciuto, S. Garg, and R. Karri, "Optimizing the use of behavioral locking for high-level synthesis," *IEEE Transactions on Computer-Aided Design of Integrated Circuits and Systems*, vol. 42, no. 2, pp. 462–472, 2023.

[9] S. Muzaffar and I. M. Elfadel, "Logic locking of finite-state machines using transition obfuscation," in *2022 IFIP/IEEE 30th International Conference on Very Large Scale Integration (VLSI-SoC)*, 2022, pp. 1–6.

[10] D. Berardi, S. Giallorenzo, J. Mauro, A. Melis, F. Montesi, and M. Prandini, "Microservice security: a systematic literature review," *PeerJ Computer Science*, vol. 8, 2022.

[11] R. Datta, G. Zhao, K. Basu, and K. Shamsi, "A security analysis of circuit clock obfuscation," *Cryptography*, vol. 6, no. 3, 2022.

[12] K. Shamsi and Y. Jin, "Cad for assurance." [Online]. Available: https://cadforassurance.org/tools/evaluation-of-obfuscation/neos/

[13] Y. Jin and D. Pan, "Quantitative metric and automated toolset for obfuscated logic security evaluation," Air Force Research Laboratory Sensors Directorate, Wright-Patterson Air Force Base, Oh 45433-7320, Air Force Materiel Command United States Air Force, Tech. Rep., Sep 2020.

[14] gRPC, "grpc," https://grpc.io/, Accessed: 2023-09-14.

[15] J. Dofe and Q. Yu, "Novel dynamic state-deflection method for gate-level design obfuscation," *IEEE Transactions on Computer-Aided Design of Integrated Circuits and Systems*, vol. 37, no. 2, pp. 273–285, 2018.

APPAMM: Memory Management for IPsec Application on Heterogeneous SoCs

Ayushi Agarwal*, Radhika Dharwadkar*, Isaar Ahmad[†], Krishna Kumar*[†], P.J. Joseph[†], Sourav Roy[†],
Prokash Ghosh[†] and Preeti Ranjan Panda*
*Indian Institute of Technology Delhi, New Delhi, India
[†]NXP Semiconductors, Noida, India

Abstract—To keep up with the growing computational demands of current-day applications, SoC design has shifted towards heterogeneous architectures with CPUs and domain-specific accelerators. These accelerators demand high on-chip and off-chip memory bandwidth and require efficient management of shared system resources. We characterize Internet Protocol Security (IPsec), a high-throughput application, by collecting the memory traces of this application running on the accelerators and CPU cores of NXP LX2160A SoC. We use this characterization to design a simulation infrastructure for simulating IPsec on possible domain-specific architectural extensions and perform a design-space exploration across various general-purpose memory management policies. We propose APPAMM, an application-specific predictive packet-aware memory management policy using the knowledge of IPsec to improve performance for next-generation SoCs. Using our approach to manage memory for different input packet streams, we report improvements of up to 22x in the packet drop rate and peak throughput.

Index Terms—Internet Protocol Security Accelerators, Dataflow Architectures, Application-Specific Memory Management

I. INTRODUCTION

Domain-specific accelerators (DSA) provide high-performance solutions for heavy computing applications such as machine learning [1], [2], and network security [3]. Complex systems-on-chip (SoCs) increasingly rely on heterogeneous processing components, such as CPUs and DSAs, to meet the performance. Heterogeneous SoCs require high on-chip and off-chip memory bandwidth to meet heavy computational loads. Since the off-chip memory has not kept up with the rising bandwidth demand by applications, an on-chip memory shared by the cores and accelerators provides the necessary bandwidth. However, contention at this memory can cause severe performance bottlenecks. Intelligent on chip memory management leads to significant performance improvement in homogeneous architectures [4], [5]. Memory management is crucial to mitigate contention in heterogeneous SoCs, especially for accelerators with strict Quality-of-Service (QoS) requirements.

Internet Protocol Security (IPsec) is a computationally intensive, high-throughput application that provides network layer security for high-speed secured communication over insecure networks. It addresses security issues such as packet origin authentication via packet header checks and data confidentiality via high-quality encryption and decryption [6], [7]. The application demands heavy compute parallelism to guarantee the required throughput. Hence, it has always been a contender for hardware or software acceleration. However, its impact on the system resources has not been studied. We study the application on a high-performance heterogeneous SoC, NXP LX2160A [8], shown in Figure 1.

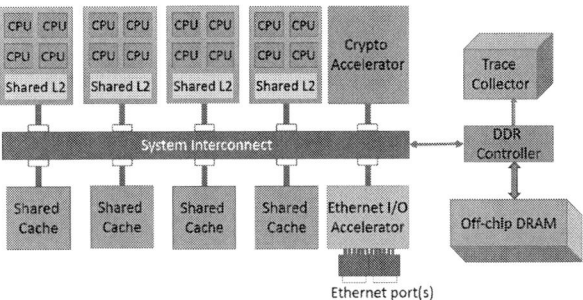

Fig. 1: LX2160A SoC [8] Architecture. The SoC has multiple CPU clusters with an Ethernet I/O and a Crypto Accelerator.

We collect the application's memory traces on the SoC to characterize the memory access patterns and design a simulation infrastructure based on the application's behavior to evaluate the impact on the shared memory resources. We perform a design-space exploration across various memory management policies on multiple architectural configurations to enable better performance and higher throughput on next-generation SoCs. We make the following major contributions:

1) We design a trace collection infrastructure to characterize the IPsec application by analyzing the memory access patterns and packet data flow.
2) We use the application's characteristics to design a simulation infrastructure with architectural flexibility and perform a design-space exploration across on-chip memory management policies to evaluate performance.
3) We propose an application-specific predictive packet-aware memory management policy to improve performance. The policy performs better than general-purpose policies on varied packet streams and can be generalized well to other architectures.

II. RELATED WORK

Modern IPsec implementations in high-performance systems target sustained throughput [8], [9] with minimal computational overhead to achieve higher network bandwidth.

979-8-3315-3968-9/24 $31.00 © 2024 IEEE

Domain-specific hardware or software implementations of IPsec can deliver high throughput. Guilford et al. [9] propose a fast, complete software implementation on multi-core processors. Strongswan [10] and Rockhopper [11] implement the IPsec Encapsulating Security Payload (ESP) protocol in the operating system's user space. Nam et al. [3] and Niu et al. [12] propose a high-performance hardware accelerator to achieve high throughput. Chang et al. [13] design an IPsec accelerator as a co-processor to an ARM9 processor for running cryptographic algorithms. Agrawal et al. [14] analyze the performance achieved by IPsec when offloaded to hardware-based accelerators on SoCs. Thoguluva et al. [15] suggest offloading the cryptographic operations from the SoC's host processor to a programmable security processor.

Most of the above works show that offloading to cryptographic accelerators can meet the throughput requirements. Still, no detailed characterization of the application and the impact of this offloading on the system has been performed. In this work, we characterize the application to study its impact on the SoC's shared memory resources and enable efficient memory management to improve performance.

III. TRACING OF IPSEC APPLICATION ON LX2160A SOC

We discuss the infrastructure used to collect traces for the application running on the SoC and analyze its memory access patterns. The SoC [8], as shown in Figure 1, has four clusters of four ARM Cortex-A72 cores, each with private L1/L2 caches, two custom accelerators for the IPsec application, and a shared banked-cache. The application can achieve high throughput by software-managed offloading of the network packet processing to the Ethernet I/O accelerator and the encryption and decryption tasks to the Cryptographic accelerator on the SoC. A trace collector is connected to the SoC's DDR controller to snoop and collect DDR accesses as traces, which are later stored into a small on-chip memory called packet express buffer instead of DDR to minimize the instrumentation overhead and bandwidth contention between trace packets and the application's data. The traces are read and processed offline upon the application's completion.

Figure 2 shows that the packet dataflow sequence is I/P Stream $\xrightarrow{1}$ I/O Acc. $\xrightarrow{2}$ CPU $\xrightarrow{3}$ Crypto Acc. $\xrightarrow{4}$ CPU $\xrightarrow{5}$ I/O Acc. $\xrightarrow{6}$ O/P Stream. Figure 3 reveals the memory access pattern of a 390B packet stream at a 4 Gbps injection rate. It shows a regular and repeated sequence of memory accesses initiated by computing units plotted over time. Each unit sequence corresponds to a single IPsec packet and aligns with the IPsec dataflow. The collected traces reveal packets with different attributes and memory access behavior. The packets can have various sizes: 86, 150, 250, 390, 162, and 466 bytes, which determines the number of memory accesses and the packet's total memory footprint. However, increasing the packet size only marginally increases the packet processing time (1.22x with a 4.5x increase in size). The packets can be encrypted or decrypted depending on the cryptographic operation required. We found that the memory access pattern has low sensitivity to the packet type. The packet injection rate

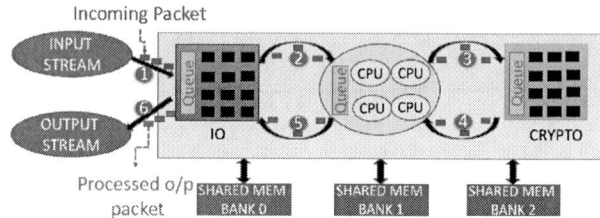

Fig. 2: Abstract Overview of the IPsec dataflow. I/O accelerators, Crypto accelerators, and CPUs have a packet queue and multiple computing resources for parallel computation.

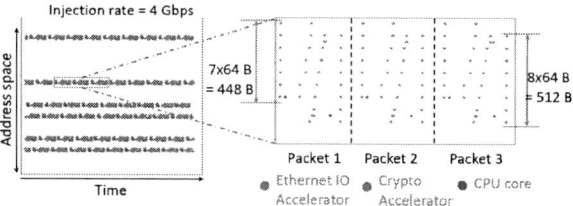

Fig. 3: IPsec traces for 390B packets at 4 Gbps Injection Rate.

is varied between 4-12 Gbps. For a 390B packet, a 3x increase in the injection rate leads to a 3.3x increase in parallelism (facilitated by multiple compute resources).

IV. SIMULATION INFRASTRUCTURE FOR IPSEC

We develop a simulation infrastructure by exploiting the dataflow characteristics to simulate the workload on future projected architectures and facilitate system-level design-space exploration to improve performance on next-generation SoCs.

A. Simulation Model

The IPsec simulation model, shown in Figure 4, has multiple abstraction levels based on its trace characteristics.

1) Dataflow Node: This captures the most elementary task performed on a resource of any processing component. A compute node (C-type) represents a fixed-latency computation task. A memory node (M-type) represents multiple independent accesses to the shared memory.

2) Dataflow Stage: This is a directed sequence of multiple tasks or dataflow nodes scheduled in order of execution on a specific processing component. For any workload, the dataflow stage can capture any custom sequence of tasks.

3) Packet Dataflow Graph (DFG): It is a directed sequence of multiple dataflow stages. The IPsec packet DFG has five dataflow stages, conforming to the dataflow discussed in Section III. These dataflow stages have data dependencies and execute sequentially on different processing components. However, multiple packets' DFGs can execute in parallel on the system and get completed out of order.

B. Architecture Model

The architecture model, as shown in Figure 5, has processing components such as CPUs, accelerators, shared memory, and communication networks.

979-8-3315-3968-9/24 $31.00 © 2024 IEEE

1) Computation Modules: Each computation module, I/O accelerator, CPU, and Crypto accelerator, has an input packet queue and multiple resources for parallel packet processing. A custom packet generator injects packets into the system at a variable injection rate. The packet's five dataflow stages are scheduled sequentially on the respective modules.

2) Shared Memory Model: We model a multi-banked memory with one bank between each pair of computation modules. Each bank has an input queue, associated memory access latency, and queuing delay.

3) Data Network: The computation and memory modules are connected via a bidirectional ring network. The data transfer between a pair of computation and memory modules varies with the number of hops separating them.

4) Control Network: An Auxiliary Address Directory (AAD) is modeled to store the mapping between the memory addresses and banks for dynamic memory management. The control network provides uniform access to the AAD from all the computation modules. On query from a computation module, the AAD returns the mapping for an allocated memory node. An unallocated node is assigned to a bank using a memory management policy, which is then updated in AAD.

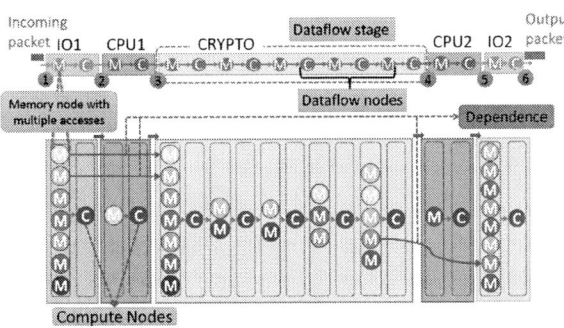

Fig. 4: Packet Dataflow Graph, Stage, and Nodes with an expanded view of M-nodes. Same color M-nodes: Same address.

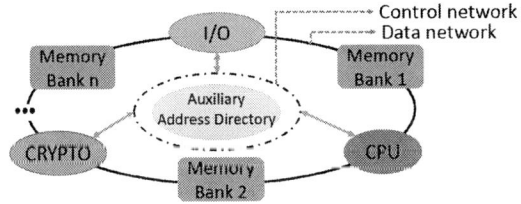

Fig. 5: Architecture Model with Computational and Memory Modules connected over a Bidirectional Ring Network.

C. Comparison against SoC Traces

As mentioned in Section III, an increase in packet size from 86B to 390B increases execution time by 1.22x. We observe a similar trend for execution time in simulation with an error margin of up to 3%. We also reported that a 3x increase in injection rate increased the parallelism by 3.3x. We observe a similar trend in parallelism at multiple injection periods.

V. APPLICATION-SPECIFIC MEMORY MANAGEMENT FOR IPSEC WORKLOAD

We evaluate the performance of traditional memory management policies using our simulation infrastructure on the architecture shown in Figure 5. We leverage IPsec's deterministic memory access patterns to propose an application-specific memory management policy for performance optimization. Table I summarizes the policies explored for performance analysis and optimization.

TABLE I: Design Space Exploration of Memory Management Policies for IPsec workload. STA: Static; DYN: Dynamic; UF: Uniform; NUF: Non-Uniform; GP: General-Purpose; AS: Application-Specific.

Policy Name	STA	DYN	UF	NUF	GP	AS
Fixed Map (FM)	✓		✓	✓	✓	
Round-Robin (RR)	✓		✓		✓	
Address Hash Map (AHM)	✓		✓		✓	
Link Latency-Aware (LLA)		✓		✓	✓	
Memory Queue-Aware (MQA)		✓		✓	✓	
Latency & Queue-Aware (LMQA)		✓		✓	✓	
APPAMM (proposed)		✓		✓		✓

A. Static GP Memory Management Policies

1) Fixed Map (FM): This policy maps all memory nodes from a computation module to a fixed memory bank.

2) Round-Robin (RR): This policy maps unallocated memory nodes to memory banks in a round-robin/circular manner independent of the source computation module.

3) Address Hash Map (AHM): This policy maps memory nodes to memory banks using an address hash function.

B. Proposed Dynamic GP Memory Management Policies

Figure 6 summarizes the proposed general-purpose dynamic memory management policies for IPsec workload.

1) Link Latency-Aware (LLA): This dynamic policy maps an unallocated memory node to the bank closest to the requester/source computation module.

2) Memory Queue-Aware (MQA): This dynamic policy maps an unallocated memory node to the bank with the smallest queue size (least contention), irrespective of the source computation module. Despite balancing inter-bank contention, this policy can lead to higher communication latency due to distant data placement, causing performance deterioration.

3) Latency and Memory Queue-Aware (LMQA): This dynamic policy maps an unallocated memory node to a bank based on its latency from the source computation module and queue size. Between the banks with equal latency from a computation module, the policy selects the bank with a smaller memory queue size for faster turn-around time.

C. APPAMM: Application-specific Predictive Packet-Aware Memory Management

We use the knowledge of the application's memory access patterns to propose APPAMM, an application-specific predictive packet-aware policy for memory management to

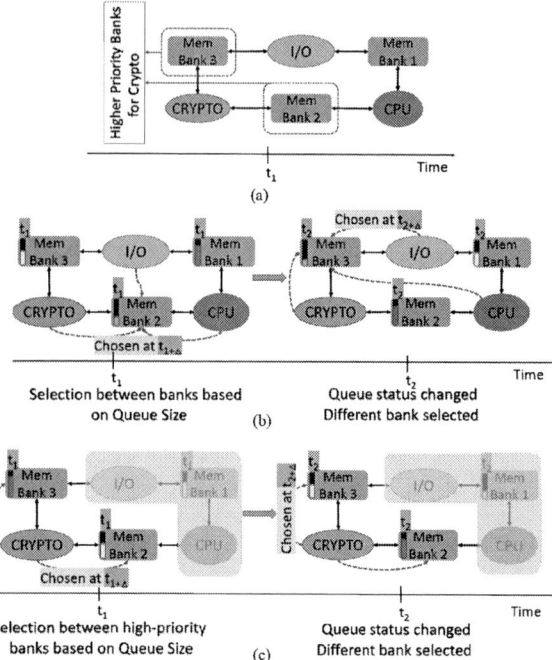

Fig. 6: Proposed Dynamic GP Memory Management Policies for IPsec. (a) LLA, (b) MQA, and (c) LMQA. △: Time taken to query the memory banks' queue sizes.

improve performance. The knowledge of all dependent memory accesses across the packet's dataflow stages provides *packet awareness* to predict the impact on the future state of memory queues and the total packet execution time for efficient memory management.

Equation 1 shows the policy formulation for estimating the cost, in terms of total execution time, of choosing a memory bank k from a computation module i at time t. The equation has the following variables: CM = Number of Computation Modules; B = Number of Memory Banks; $\hat{Q}_{kt}$ = Estimated queue size of bank k at time t; Q_{kt} = Observed queue size of bank k at time t; MAT_{ik} = Total memory access time (memory + network) to access data for computation module i from the bank k; N_{ij} = Number of dependent accesses between computation module i and computation module j for a packet; W = Window length for computing the average queue size; T = Total number of times the queue size has been sampled (every S cycles until time t); $\hat{Cost}_{ik}$ = Predicted cost of choosing bank k to place data from computation module i. At the source computation module i, the policy evaluates two main latency components:

1) Total Memory Access Time (MAT), including queuing delay, witnessed by the packet if a memory node is placed in a particular bank. The first term of Equation 1 shows that this is a function of this bank's access latency and future queue size prediction, as it will process all accesses to this node.

2) Total network latency, the second term of Equation 1, is

the time needed for accessing the chosen bank from all the computation modules according to the number of *dependent* memory accesses to this bank.

$$\hat{Cost}_{ik} = (\hat{Q}_{kt} * MAT_{ik}) + \sum_{j=1}^{CM} N_{ij} * MAT_{jk} \quad (1)$$

$$\forall i \in [1, 2, 3, ..., CM] \quad \forall k \in [1, 2, 3, ..., B]$$

$$\hat{Q}_{kt} = \frac{1}{W} \sum_{y=T-W}^{T} Q_{k(y*S)} \quad (2)$$

The chosen bank for the computation module i, Chosen Bank$_i = k \mid \hat{Cost}_{ik}$ is minimum over $k \in [1, 2, ..., B]$. Figure 7 shows the policy's bank allocation from the I/O module. We compute the cost of placing data from the I/O module into each memory bank and choose the bank with the minimum predicted cost (in this case, Bank 3).

Queue Size Prediction: We experimented with several strategies to predict the memory bank queue size, capturing the effect of all dependent accesses on the queue. Equation 2 proposes a history-based queue size prediction strategy that uses the average queue size in a window large enough to capture the queue state history. Our experiments show that using $W = 10$ and $S = 500$ cycles for the IPsec workload provides the best trade-off between performance achieved and implementation overhead. For $10 < W \leq 100$, the performance improves by upto 0.05% but with significant sampling and storage overhead. Our experiments show that $W = 1$, or Q_{kt}, differs in performance by up to 0.1% compared to the history-based queue size prediction ($W = 10$). Therefore, $\hat{Q}_{kt} = Q_{kt}$ is a good reactive estimate of the future queue state.

Implementation overhead: The policy runs whenever an unallocated memory node is encountered during the packet execution. APPAMM is implemented in each computation module with four 8-bit multipliers and three 10-bit adders for estimating the cost for one bank. Three 10-bit comparators compare the cost across banks to make the final decision. For Synopsys 40nm technology, APPAMM shows an area overhead of 0.044 mm². Additionally, AAD incurs a storage overhead of at most 30KB for all packet streams and sizes. With minimum implementation overhead, APPAMM can easily be integrated into SoC architectures.

VI. EXPERIMENTS AND RESULTS

A. Experimental Setup

We evaluate the performance of the policies using the simulation infrastructure discussed in Section IV for the IPsec workload. The architecture has three computation modules (resources), namely I/O (20), CPU (16), and Crypto (40), connected to three shared memory banks by a bidirectional ring network. The packet generator can generate packets with different parameters discussed in Section III. We model additional packets different from the collected traces by changing the memory access patterns to generalize the packets that can be received from a network. For completeness, we model

979-8-3315-3968-9/24 $31.00 © 2024 IEEE

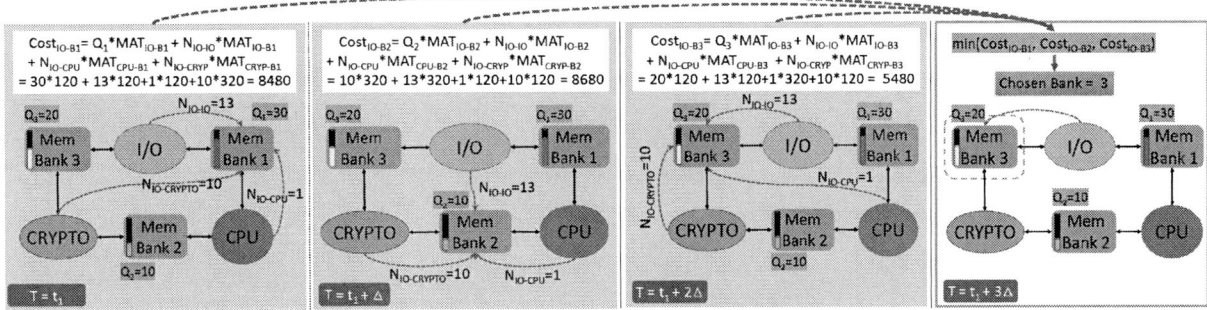

Fig. 7: APPAMM Memory Management Policy; In this example, $i, j \in$ [I/O, CPU, CRYPTO] and $k \in [1, 2, 3]$; $\hat{Q}_{kt} = Q_{kt}$; Memory Access Latency = 20 cycles; 1 hop network turn-around time = 100 cycles; Total MAT for 1 hop = 120 cycles.

status IPsec packets processed only in CPU, PKT14, and PKT15, which are not seen in the SoC traces. Table II summarizes the simulated packets based on their type and size.

TABLE II: Summary of IPsec packets with their Memory Nodes (M_1, M_2, M_3, M_4, M_5) in each Dataflow Stage. Monochromatic entries represent address dependency within a packet. PKTn : Packets modeled accurately from traces.

Packet	Size	Type	IO1	CPU1	CRYPTO	CPU2	IO2
PKT1	390B	enc	7	1	(7,2,2,3,5)	1	(6,2)
PKT2	390B	enc	7	1	2		(6,2)
PKT3	466B	dec	8	3	(8,2,2,3,6)	(2,4)	(1,6)
PKT4	466B	dec	8	2	(8,2,1,2,6)	(2,3)	(1,6)
PKT5	390B	enc	7	5	(3,1,1,2,4)	(3,1,1,4)	(2,1,1,4)
PKT6	390B	enc	7	5	(4,2,2,3,6)	5	(2,6)
PKT7	390B	enc	7	6	(1,2,1,2,6)	(4,3)	(5,3)
PKT8	390B	enc	7	1	(7,2,2,3,5)	1	(3,5)
PKT9	466B	dec	8	1	(8,2,2,3,2)	1	(5,2)
PKT10	86B	enc	2	1	(2,2,1,2)	2	(1,2)
PKT11	150B	enc	3	1	(3,1,7)	1	(2,2)
PKT12	250B	enc	4	1	(2,2,6,2)	1	(3,2)
PKT13	162B	dec	3	1	(3,2,2)	1	2
PKT14	86B	status	2	1	-	1	2
PKT15	390B	status	7	1	-	1	7

We generate large packet streams with 10k-100k packets motivated by real use cases described below. The packet injection period varies around a fixed period ranging from 500–2500 cycles. We show in Section VI-B that APPAMM's performance is insensitive to these values. We report the number of packets dropped at a given mean injection period. We choose one of the traditional GP policies as a baseline for comparison due to the lack of a proper baseline from prior works on IPsec. To evaluate the peak throughput, we consider a 1 GHz operating frequency.

We summarise the performance of the policies in Table III for different packet streams. APPAMM performs consistently better than FM and the best-performing general-purpose policy, such as LMQA. Some observations are consistent for all types of packet streams discussed further in this section. At low injection periods, the difference between the packet injection and packet completion rate increases, causing contention at the I/O module. Hence, the system performs consistently

poorly. APPAMM achieves better *resource utilization*, almost 90%, by efficiently exploiting parallelism across packets.

1) Case 1: Packet Streams for Streaming Applications: Streaming network traffic usually consists of bursts of large packets mixed with status packets for regular handshaking. We randomly mix multiple large packets (PKTs 1-9) with one status packet (PKTs 14,15). Figure 8 shows that at an injection period of 2200 cycles, APPAMM achieves a peak throughput of 1.4 Gbps with a 31x, 7.5x, and 4.3x improvement in drop rate w.r.t. FM, MQA, and LMQA policy, respectively.

2) Case 2: Packet Streams for Browsing Applications: Search engines generate network traffic with alternating bursts of large packets (PKTs 1-9) and small packets (PKTs 10-13). As shown in Table III, at an injection period of 2000 cycles, APPAMM achieves an order-of-magnitude reduction in drop rate w.r.t. FM and 57x and 7.2x reduction w.r.t. MQA and LMQA. This packet stream has smaller packets with fewer total and dependent memory accesses, which complete fast, and release system resources. This lowers system contention, and the overall performance is better than that of Case 1.

3) Case 3: Packet Streams during End-to-End Secured Communications: Most data-sharing and user communication applications provide end-to-end secured communication for data security and privacy. Network traffic on such a channel consists of bursts of encryption and decryption packets (PKTs 1-13). Figure 9 shows that at an injection period of 2200 cycles, APPAMM achieves an order-of-magnitude reduction in drop rate w.r.t. FM policy, 60x and 23x reduction in drop rate w.r.t. MQA and LMQA policy, respectively.

B. Generalizability to other Architectural Configurations

1) Varying Memory Access and Network Delay Ratio: Our experiments assume constant network and memory model delays. Figure 10 shows the performance of different policies on a random packet stream by varying the hop link and memory access latency ratio. The ratio used in all the previous experiments is 5:1 (100:20 cycles). APPAMM performs consistently better than the GP policies for all delay ratios and is therefore insensitive to the absolute delays modeled.

2) Varying Number of Memory Banks and Network Type: We simulate random packet streams on multiple architecture

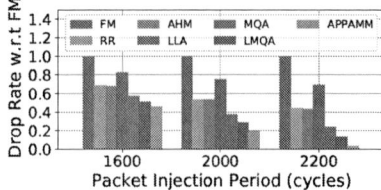
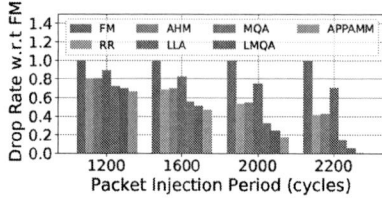
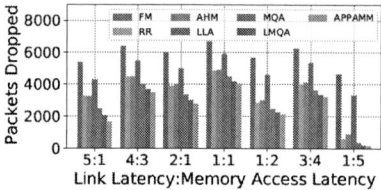

Fig. 8: Case 1: Relative Drop Rates for policies w.r.t. FM.

Fig. 9: Case 3: Relative Drop Rates for policies w.r.t. FM.

Fig. 10: Packets dropped out of n=10000 packets with variable latency ratios.

TABLE III: Performance Comparison of APPAMM with other memory management policies for various packet mixes at multiple Injection Periods (IP). '-' in a cell signifies no packet drops with APPAMM at that injection period.

Pkt. Mix/IP	APPAMM vs. FM				APPAMM vs. LMQA				APPAMM vs. MQA			
	2200	2000	1600	1200	2200	2000	1600	1200	2200	2000	1600	1200
Case 1	**31x**	**5.1x**	**2.1x**	**1.5x**	**4.3x**	**1.5x**	**1.1x**	**1.1x**	**7.5x**	**1.9x**	**1.3x**	**1.1x**
Case 2	-	-	3.5x	1.8x	-	7.2x	1.3x	1.1x	-	57x	1.5x	1.2x
Case 3	-	5.8x	2.2x	1.5x	23x	1.5x	1.1x	1.1x	60x	1.9x	1.2x	1.1x

TABLE IV: Performance Comparison of APPAMM with other policies for random packet streams at multiple Injection Periods (IP) on different architecture configurations to test the generalizability of the simulation model and proposed policies.

Arch./IP	Network Configuration			APPAMM vs. FM					APPAMM vs. LMQA					APPAMM vs. MQA				
	Network Type	Placement	#Banks	2000	1800	1200	1000	800	2000	1800	1200	1000	800	2000	1800	1200	1000	800
Config 1	Bidirec	Sym	3	-	**8.9x**	1.9x	1.6x	1.4x	**7.2x**	**1.7x**	1.1x	1.1x	1.1x	57x	2.4x	1.2x	1.2x	1.1x
Config 2	Unidirec	Sym	3	1.8x	**1.7x**	1.3x	1.2x	1.2x	**1.9x**	1.7x	1.3x	1.2x	1.2x	1.2x	1.2x	1.1x	1.1x	1.1x
Config 3	Bidirec	Sym	9	-	-	-	45x	4.1x	-	-	-	35x	3.4x	-	-	-	7x	1.5x
Config 4	Unidirec	Sym	9	-	**32x**	2.6x	2x	1.7x	-	32x	2.6x	2x	1.7x	10x	3.7x	1.3x	1.2x	1.1x
Config 5	Bidirec	ASym	9	-	-	-	77x	4.3x	-	-	-	61x	3.6x	-	-	-	12x	1.6x
Config 6	Unidirec	ASym	9	-	**22x**	2.6x	2x	1.7x	-	22x	2.6x	2x	1.7x	4.8x	2.6x	1.3x	1.2x	1.1x

configurations by varying the network type and the number of banks between the compute nodes. We vary the number (3, 9 banks) and placement (symmetrical/equal or asymmetrical/unequal) of banks between the compute nodes and the network type (bidirectional or unidirectional ring). Table IV summarizes the improvement in performance achieved by AP-PAMM w.r.t. FM, LMQA, and MQA policy at various packet injection periods on different architectural configurations.

From these experiments, it can be inferred that APPAMM can efficiently manage memory to enable a higher degree of parallelism in the system, thereby improving application throughput. We evaluate the policy on other packet streams for more networking applications and generalize it to 6 memory banks. The details are omitted due to lack of space.

VII. CONCLUSIONS

We characterized the IPsec application by collecting and analyzing the memory access pattern on a commercial SoC platform. We developed a simulation infrastructure based on these characteristics and explored the design space on heterogeneous architectures. The application has deterministic and regular memory access patterns. These helped formulate an application-specific predictive packet-aware policy that significantly improves the performance achieved on next-generation architectures. Such policies can be further enhanced by adding packet-level priorities for efficient arbitration to achieve the required QoS while maintaining high throughput.

VIII. ACKNOWLEDGEMENT

This work was supported by research grant 2017-SD-2738 from Semiconductor Research Corporation under the India Research Program, and NXP Semiconductors, India.

REFERENCES

[1] A. Garofalo et al., "DARKSIDE: A Heterogeneous RISC-V Compute Cluster for Extreme-Edge On-Chip DNN Inference and Training," *IEEE OJSSCS*, vol. 2, pp. 231–243, 2022.
[2] Z. Liu, G. Li, and J. Cheng, "Hardware Acceleration of Fully Quantized BERT for Efficient Natural Language Processing," arXiv 2021.
[3] T. S. Nam et al., "A high-throughput hardware implementation of NAT traversal for IPSEC VPN," *IJCNIS*, vol. 14, no. 1, 2022.
[4] S. Tiwari et al., "REAL: REquest Arbitration in Last Level Caches," *ACM TECS*, vol. 18, no. 6, 2019.
[5] N. R. Holtryd et al., "CBP: Coordinated management of cache partitioning, bandwidth partitioning and prefetch throttling," in *PACT*, 2021.
[6] S. Kent and R. Atkinson. (1998) RFC2402: IP Authentication Header.
[7] D. Harkins and D. Carrel. (1998) RFC2409: The Internet key exchange.
[8] NXP Semiconductors®. (2019) Layerscape® LX2160A Processors.
[9] J. Guilford et al. (2012) Fast Multi-buffer IPsec Implementations on Intel® Architecture Processors. [Online]. Available: https://intel.com/
[10] Strongswan. (2005) Open-source, modular and portable IPsec-based VPN solution.
[11] Rockhopper. (2012) IPsec/IKEv2-based VPN software for Linux.
[12] Y. Niu, L. Wu, and X. Zhang, "An IPSec Accelerator Design for a 10 Gbps In-Line Security Network Processor," *J. Comput.*, 2013.
[13] C. S. Ha et al., "ASIC design of IPSec hardware accelerator for network security," in *IEEE APASIC*, 2004.
[14] H. Agrawal, Y. Dutta, and S. Malik, "Performance Analysis of Offloading IPsec Processing to Hardware-Based Accelerators," in *ISED*, 2012.
[15] J. Thoguluva et al., "Efficient Software Architecture for IPSec Acceleration Using a Programmable Security Processor," in *DATE*, 2008.

Embedded and Real-Time Anomalous Command Classification in Unmanned Ground Vehicle Operations

Rafaella Elia and Theocharis Theocharides
Department of Electrical and Computer Engineering,
KIOS Research and Innovation Center of Excellence,
University of Cyprus, Nicosia, Cyprus

Abstract—Unmanned Ground Vehicles (UGVs) are increasingly used in safety-critical applications, typically controlled under challenging conditions that may cause stress and fatigue on the operator. This can potentially compromise the safety of the mission due to involuntary movements by the operator, resulting in abnormal commands issued on the remote controller of the vehicle. Such movements can be detected by evaluating the mental state of the operator, the operational context of the UGV, and of course real-time movement detection by the operator. To detect such anomalous commands, we propose a three-stage Machine Learning (ML) based approach, which is suitable for embedded implementation providing real-time classification. Firstly, we detect whether there is any movement on the controller and classify the type of the given movement. Next, we classify whether the operator is under incremental stress and finally discern, in relationship to the UGV's operational context extracted by the state of the UGV is considered normal or not. We use a dataset collected through real-world scenarios, to evaluate the proposed approach, and we evaluate the performance of the proposed approach on an embedded platform (Jetson Xavier NX), extracting relevant metrics such as processing time, energy consumption, and classification accuracy. Our findings demonstrate successful high classification rates for the mental state of the operator (96%) and the ground vehicle movements (74%), as well as the associated involuntary command recognition, with low energy ($0.43mJ$) and time ($0.39s$) requirements.

Index Terms—Remotely operated vehicles (ROVs), Unmanned Ground Vehicles (UGVs), Machine-Learning (ML), embedded and real-time systems.

I. INTRODUCTION

Remotely Operated Vehicles (ROVs) find extensive utility across various safety-critical applications, including real-time monitoring, security, surveillance, and search and rescue missions. Additionally, many aerial or ground vehicles often serve as valuable aids equipped with numerous sensors, assisting in decision-making processes or providing support in diverse scenarios. However, in demanding environments such as natural disasters or emergency responses, the operator is expected to control such a vehicle, which increases the stress and fatigue levels. These conditions, along with the high concentration demands, elevate the risk of operator-induced errors that could compromise the outcome of the operation [1]. Consequently, monitoring both the operator and the ROV simultaneously is vital to capture contextual awareness during missions.

The mental state of individuals can be influenced by various factors such as stress, fatigue, environmental conditions, and

Corresponding author: elia.rafaella@ucy.ac.cy

demanding tasks that require high focus and dedication. Continuous monitoring involves analyzing both body movements [2] and biological markers [3]–[5], providing valuable insights for applications in healthcare, sports, and Human-Machine Interaction (HMI). The most promising and non-invasive technique for monitoring mental state is through physiological signals collected using wearable sensors [6]. Physiological signals can offer critical information about the operator's mental and physical state, unlike facial expressions or body movements, which require invasive monitoring procedures [6].

Focusing solely on the human may lead to limited models that fail to capture the operational status of a vehicle during an operation. The operation can be compromised in the case of the vehicle that do not react appropriately, potentially causing damage to itself or its surroundings. To ensure safety during any operation, it is essential to simultaneously monitor the operator and the ROV, identifying any involuntary movements during the mission. Our objective is to develop an approach where we will be able to detect the movement given on the joystick of the vehicle, and the mental state of the operator and detect whether the movement of the vehicle is normal or not. The processing time and energy efficiency aspects were included, as low resource consumption is targeted without compromising accuracy. We aim to optimize recorded data from both the operator and the ROV during missions, reducing dimensionality while preserving high performance and minimizing computational demands. This involves feature extraction, selection, and classifier optimization towards our ML-based approach. To evaluate our approach, a real-life dataset was utilized since many existing studies rely on simulations. The primary focus is on resource-constrained, real-time classification of the operator and ROV states during an operation.

Specifically, the contributions of this work are the following:

- We performed a three-stage classification, by utilizing a dedicated classification algorithm for each group of signals.
- We integrated different pre-processing and dimensionality reduction techniques.
- We applied an oversampling technique in order to overcome the data imbalance observed.
- We evaluated the proposed approach on an embedded device focusing on the classification performance, but at the same time low-power and low processing time aspects were targeted.

The rest of the paper is organized as follows, in Section II we present an overview of the existing similar approaches. In Section III, we describe the real-life dataset utilized and the methodology approach. The methodology includes the description of the procedures followed for pre-processing and the evaluation approach. Further, in Section IV, the evaluation results are presented and discussed. Finally, concluding from the findings of this study, we discuss the conclusions and the future steps, in Section V.

II. BACKGROUND AND RELATED WORK

Numerous studies in the literature focus on monitoring either the operator or the ROV separately, with relatively fewer addressing simultaneous monitoring of both. While many works proposed assistance systems for drivers or critical infrastructure operators by monitoring their mental and physical states, our emphasis lies on efficient and non-invasive techniques for monitoring simultaneously the operator and the ROV, with low-energy and real-time requirements.

Developing a simultaneous monitoring system for anomalous command detection using fused data from both operators and ROVs is crucial, as monitoring solely one entity may lead to a model incapable of capturing the holistic mission state. For instance, in [7], authors designed a vision-based framework for driver activity recognition and prediction in overtaking situations. The proposed framework was evaluated on a real on-road dataset. Cameras were used for driver monitoring and sensing the environment. The signals recorded by the system were the head pose signal, hand location signal, lidar and radar surround features, and foot motion features. Moreover, a real-time approach for detecting cognitive distraction was presented in [8]. In the proposed approach, the driver's eye movements and driving performance data were recorded. The participants had to interact with an In-Vehicle Information System (IVIS) while driving on a simulator. Authors in [9] introduced a haptic driver vehicle steering interface that interacts with the driver through environmentally mediated torque and stiffness changes. The principles of this study were distributed cognition and distributed representation of the task constraints. The aim was for the operator to satisfy a variety of needs, such as managing risk, maintaining contextual awareness, and achieving a satisfactory level of performance. Two types of information were provided to the driver through the steering wheel in the form of steering torques: a. the lateral deviation of the vehicle from a calculated path and b. the distance between the side of the vehicle and the road boundary. The system was tested in the real world with 12 participants and the results indicate higher performance than when the driver was alone.

Another approach proposing the detection and prediction of driver drowsiness in combination with human physiological signals and driving behavior data is described in [10]. The main goal of this study was the detection of drowsiness levels based on physiological and behavioral indicators extracted from Heart Rate (HR), Heart Rate Variability (HRV), Respiration (RSP), head and eyelid movements, and time-to-lane-crossing, speed, steering wheel angle, and position on the lane. Different combinations of the data were tested, with 21 volunteers participating in a car simulator experiment. Two models were designed and tested: a detection and a prediction model, where results from both models were promising and minimized the mean squared error of prediction and detection. A thorough review of personalization approaches in Advanced Driver Assistance Systems (ADAS) and autonomous vehicles systems was presented in [11]. The category of the studies focusing on was learning a user model from the observation of driver behavior and the integration of the driver model with the vehicle control. Driving data were collected using an instrumented vehicle, with the model to have two different controllers: a. high level and b. low level. A driver workload classification through a Neural Network (NN) using physiological signals was proposed in [12]. The data collected and monitored were: a. physiological data, b. driver characteristics, and c. characteristics of the driving conditions. The HR and HRV were the physiological signals acquired from the driver while he/she participated in a driving experiment on a simulator. In data acquisition, 44 subjects participated in two groups: a. a high traffic intensity and b. a low traffic intensity. The proposed classification approach achieved an accuracy of 78%, while the HR was shown to be higher and the HRV lower in the first group, which was associated to increased driver workload.

After all, to the best of our knowledge, limited efforts exist in the literature that combine data from both the vehicle and the operator attempting to concurrently decrease energy consumption and processing time using data monitored in a real-life environment. We aim to achieve resource-constrained and real-time classification of the mental state of the operator, the movement of the vehicle, and the movements of the operator on the vehicle's controller.

III. MATERIALS AND METHODS

A. Real-World Dataset

1) Dataset Description: A real-life dataset was constructed and used for the evaluation of the proposed approach. The dataset consists of data from both the hands of the subject, the surface Electromyography (sEMG) signals and the physiological signals from the operator monitored are the HR, Electrodermal Activity (EDA), Skin Temperature (SKT), and

(a) UGV

(b) UGV with the denoted sensors

Fig. 1: Unmanned Ground Vehicle (UGV) [13].

TABLE I: Pre-processing steps.

Stage of Classification		Pre-processing	Oversampling	Dimensionality Reduction
Normal / Abnormal		RobustScaler	ADASYN	PCA - 5 components
Stressed / Stress-free		-	-	PCA - 2 components
Movement - Right Hand	Level 1	RobustScaler	SMOTE	PCA - 4 components
	Level 2		SMOTE	PCA - 4 components
Movement - Left Hand	Level 1	RobustScaler	ADASYN	PCA - 3 components
	Level 2		ADASYN	PCA - 4 components

Blood Volume Pulse (BVP). Additionally, the data from the Inertial Measurement Unit (IMU) of the Unmanned Ground Vehicle (UGV) are included in this evaluation, consisting of the angular speed (x, y, z), accelerometer (x, y, z), the speed and angle information from the joystick and the Vedder Electronic Speed Controller (VESC) of the UGV. A custom-designed UGV, [13], was utilized and it is shown in Fig. 1.

2) Data Annotation: The data annotation procedure was based on the data acquisition setup and the mental state of the operator. A standardized procedure was followed and presented, in order to prevent any bias. We aim to achieve a three-stage classification based on the following:

1) **Movement / No movement:**
   In the first stage of the proposed classification approach, we aim to detect whether there is a movement or not, and if there is a movement to classify the type of movement performed in each hand. The sEMG signals were labeled by utilizing the values from the controller of the vehicle. The right joystick of the controller gives the value for the right and left movements of the vehicle, which is associated with the angle of the direction of movement, while the left joystick of the controller provides us with the values of speed forward or backward, which is associated with the accelerometer data.

2) **Stressed / Stress-free**
   Following the detection of whether there is a movement or not, we proceed with the second stage, where we aim to classify whether the operator was stressed or not, based on the physiological signals. The physiological signals monitored are the HR, EDA, SKT, and BVP. Based on the fact that each subject has different baseline values from each signal, by observing the videos from the data collection we recognized time-frames that each subject did not perform any movement and was not receiving any instructions from the experiment coordinator. In this way, we identified and calculated the mean and Standard Deviation (STD) values from this window. This results in four baseline mean values and four STD values, for each subject. Afterwards, we compared each sample with the summation of the mean and the STD value, and if the sample value was larger

than this sum then the sample was labeled as stressed, otherwise was labeled as stress-free. Since, the HR, EDA, and SKT signals show an increase as the stress levels increase based on the literature review, they were labeled as we described. However, the BVP decreases as the stress increases, and we labeled, as abnormal the samples that were smaller than the summation of the mean and the SD baseline values. This method provided four labels, one for each signal and then, we proceeded to decide the final label of each sample. Three out of the four signals showed significant differences and these are the HR, EDA, and SKT signals. By using an AND logical statement for the three signals and an OR logical statement for the last signal, the BVP, we concluded with a single final label for each record of data. It is worth mentioning that the baseline values are different for each subject and we followed this procedure in order to avoid inter-subject variability.

3) **Normal / Abnormal:**
   The subjects had to follow the predefined path presented in Fig. 2, where the x and y coordinates are extracted from the path as a baseline. Similar to the physiological signals, we extracted the mean values of the difference observed from the path followed by each subject and the STD values. These values were used for calculating the threshold that we used for the data annotation. The threshold values extracted for the normal mode, where the subject operated at a normal speed, were set to 0.7

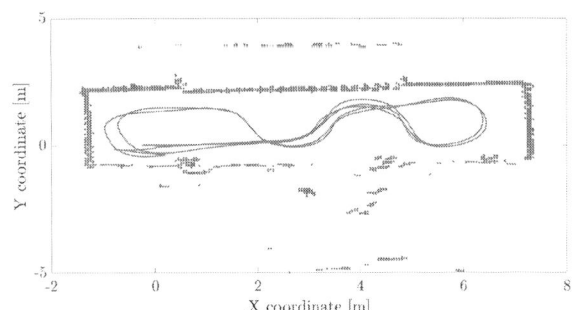

Fig. 2: UGV path to be followed by the participants for data acquisition.

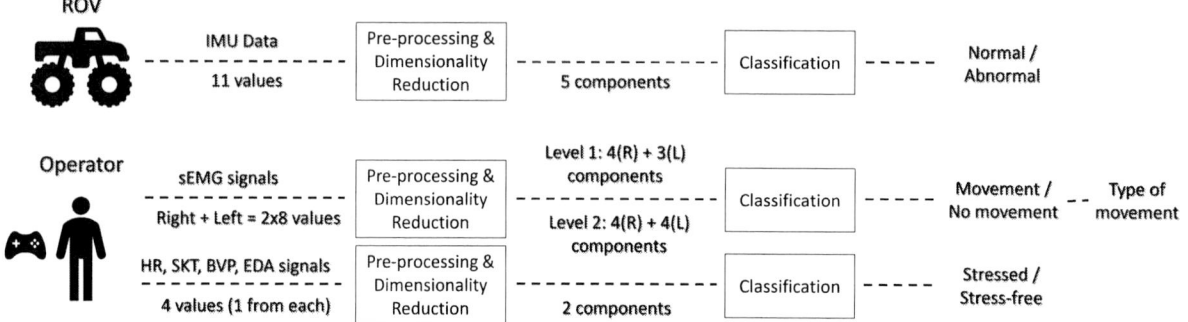

Fig. 3: Three-stage classification, *(R - Right hand, L - Left hand)*.

for the x coordinate and 0.2 for the y coordinate. Moreover, when operating at a higher speed, the threshold values were different and were equal to 0.5 and 0.1 for x and y coordinates respectively. We concluded to these thresholds through analysis of the average values and iterative testing, aiming to achieve a balance between normal and abnormal commands. Basically, the path coordinates that deviate from the baseline values using the thresholds, were marked as abnormal.

B. Methodology

1) Pre-processing: As a pre-processing step, the RobustScaler was applied from sci-kit library in Python [14]. The RobustScaler presents a valuable utility in scenarios involving datasets featuring outliers or non-normal distributions. We applied scaling on the sEMG signals and the data from the UGV only. We experimented with the physiological signals from the operator but we preferred to not apply any scaling or normalization techniques since the results were more promising, without applying any pre-processing method.

2) Oversampling: We noticed a data imbalance between the classes of the movements and the data of the UGV and we decided to utilize an oversampling technique since imbalanced datasets are common in many real-world applications. We employed the ADASYN and SMOTE algorithms from the Imbalanced-learn library in Python [15]. ADASYN sampling is a popular technique used in the field of imbalanced learning, where it generates synthetic samples for the minority class by focusing more on those instances that are difficult to classify, i.e. those instances near the decision boundary [16]. Similarly, SMOTE is a technique used to address the class imbalance problem by generating synthetic samples for the minority class using interpolation between existing minority class instances [17].

The ADASYN was applied to the UGV data, while for the physiological signals, we did not apply any oversampling technique. However, for the sEMG signals, we utilized the SMOTE technique for the right hand and the ADASYN technique for the left hand. We concluded with the use of these techniques for oversampling after observing the number

TABLE II: Parameter selection of classification models.

Stage	Classifier	Parameters
Normal / Abnormal	DT	criterion = entropy max depth = 9 splitter = random min samples split = 5 min samples leaf = 4 random state = 42
Stressed / Stress-free	ADA	no. of estimators = 200 learning rate = 0.0001
Movement Right Hand	*Level 1:* kNN	k = 2 p = 1 weights = uniform
	Level 2: kNN	k = 2 p = 2 weights = uniform
Movement Left Hand	*Level 1:* ANN	Input layer = 32 neurons Hidden layer = 64 neurons, ReLU Batch normalization layer Hidden layer = 128 neurons, ReLU Batch normalization layer Dropout layer = 0.2 Hidden layer = 256 neurons, ReLU Batch normalization layer Hidden layer = 512 neurons, ReLU Batch normalization layer Dropout layer = 0.3 Output layer = 2 neurons, softmax optimizer = Adam learning rate = 0.0001 loss = mean squared error
	Level 2: kNN	k = 2 p = 2 weights = uniform

of samples per class and the results of the evaluation, which were more promising when using these techniques. At this point, I would like to clarify that oversampling techniques were applied only during the training and not the testing stage.

3) Dimensionality Reduction: We integrated the PCA technique for reducing the dimensionality of data instead of the extraction of features. The ICA technique was also evaluated but did not perform as the PCA, and this is the main reason that we decided to proceed with the PCA. The selection of

TABLE III: Classification accuracy (%).

Movement / No Movement				Stressed / Stress-free	Normal / Abnormal
Left Hand		Right Hand		95.86	73.61
Level 1	59.76	Level 1	56.28		
Level 2	83.34	Level 2	63.80		

TABLE IV: Processing time measurements.

Processing Time (s)

Maximum	Average	Minimum
0.9651	0.3916	0.3061

TABLE V: Energy consumption measurements.

Energy Consumption (mJ)

Maximum	Average	Minimum
1.072	0.435	0.340

how many components are extracted from each signal or device was based on the number of extracted features from our previous analyses and evaluations. So, we performed a design space exploration for each signal or group of signals and we concluded with the components presented in Table I.

4) Evaluation Approach: A preliminary evaluation was conducted on a standard laptop, in order to conclude to the top-performing classifiers and combinations of pre-processing steps, presented in Table I. During the evaluation on the personal computer, we performed also a design space exploration on the classification algorithms and their parameters. By experimenting with different models and parameters, we concluded to the algorithms and their parameters presented in Table II.

In this paper, we will focus on presenting the results of the on-board evaluation. A Jetson Xavier NX[1] was utilized for the on-board evaluation. Firstly, we performed the training of all the classification algorithms on the platform and afterwards, the testing stage followed. As a testing set, one subject was left out of the training phase. The goal of this evaluation was to classify sample by sample in order to stimulate the real-time classification. A label was given and presented for each sample which characterizes the state of the operator, the movement given by the operator, and whether the command executed on the vehicle is normal or not. Moreover, the processing time required for pre-processing and classification of each sample was stored in an array. After completing the testing evaluation, we extracted the average value of the processing time required for the pre-processing and classification of a single record of data. The energy requirements were also assessed since we want this mechanism to be deployed on the vehicle in the future and be as lightweight as possible. Since we cannot measure and extract the power consumption required for classifying each sample, we took the measurement displayed on the energy device at the beginning of the evaluation and the maximum value of power required during the evaluation. Following this, we multiplied the difference between these two values by the processing time required to process and classify each of the samples of the testing set. Finally, we divided this energy consumption by the number

[1]NVIDIA Volta, 48 Tensor Cores

of samples included in the testing set, to extract the energy consumption per sample. The classification performance is presented in Table III, while Tables IV and V present the time and energy requirements. The minimum, maximum, and average values were extracted from the processing time and the energy consumption measurements. Regarding classification accuracy, the performance of each stage is presented in Table III.

IV. Evaluation Results

We present the evaluation results in Table III. The physiological signals from the operator achieved the highest accuracy with minimal pre-processing steps applied. However, the classification of the sEMG signals was the most challenging part of this approach since we evaluated numerous classification algorithms but as we observe the results for the first level of classification were lower compared to the second level of classification. Moreover, the sEMG signals did not perform as the rest of the signals of this dataset. In the second level of classification for the sEMG signals, we achieved higher accuracy in both hands. We can conclude, that since we used the joystick values for data labeling in this evaluation, the sEMG signals can be used in the case of ROVs where the data from the joystick are not available. This is due to the fact that low classification performance was observed on both the evaluations performed.

The processing time was measured and stored for each record of data and used afterwards for the following calculations. First of all, we present in Table IV, the maximum and minimum values required for pre-processing and classifying a sample. The processing time required for processing and classifying a single sample, varied from 0.96 to 0.31 s, with an average value equal to 0.39 s (Table IV). As it is shown from the values of the processing time, the real-time requirement is achieved.

Further, the energy consumption was measured by subtracting the initial value of power displayed on the energy device and the maximum value of power required during the evaluation and multiplying this by the total processing time. Finally, to calculate the average energy consumption demands for classifying a record of data, we divided this value by the number of samples. As we expected the energy consumption observed for processing and classifying a single record of data with the proposed approach was proportional to the processing time. The energy required varied from 1.07 to 0.34 mJ, with an average value of 0.43 mJ as it is shown in Table V. Therefore, the battery can run for approximately 217350 hours (or about 9056 days) given the specified energy consumption rate and battery capacity. Note that this was a theoretical calculation assuming constant energy consumption and ignoring factors like battery efficiency and discharge characteristics. The classification demonstrated low energy requirements, confirming it as one of the primary objectives achieved in this work.

V. CONCLUSIONS AND FUTURE WORK

In conclusion, we presented the analysis and evaluation of a real-life dataset that combines data from the operator and a UGV, employing a distinct evaluation approach focused on real-time and low-energy aspects towards a three-stage classification. The pre-processing steps involved data annotation and scaling, which are thoroughly explained. Addressing data imbalance, oversampling techniques such as ADASYN and SMOTE were applied during training, along with dimensionality reduction, using the PCA technique. Results indicated promising classification performance, with physiological signals from the operator outperforming others, albeit with challenges in distinguishing movement efficiently in sEMG components. Real-time classification feasibility was confirmed, with processing time under a second, and low energy consumption, meeting the study's objectives for this evaluation.

As a future step, we would like to assess the exclusion of sEMG signals from our approach, since they did not perform as the rest of the signals. We can investigate the use of the joystick inputs only for movement detection, and not the sEMG signals. Moreover, we would like to explore the implementation of recurrent neural networks (RNNs) or similar models that leverage temporal dependencies to classify signals by incorporating historical data, thereby enhancing the network's ability to interpret and classify time-series data more effectively. Additionally, we would like to evaluate the complexity of the classification algorithms in depth. Finally, a similar evaluation approach can be followed for the case of aerial vehicles.

ACKNOWLEDGMENT

This work was funded by the European Union's Horizon 2020 research and innovation program under grant agreement No. 739551 (KIOS CoE), and from the Republic of Cyprus through the Cyprus Deputy Ministry of Research, Innovation and Digital Policy.

REFERENCES

[1] F. Dell'Agnola, L. Cammoun, and D. Atienza, "Physiological characterization of need for assistance in rescue missions with drones," in *2018 IEEE International Conference on Consumer Electronics (ICCE)*, pp. 1–6, 01 2018.

[2] G. Giannakakis, D. Grigoriadis, K. Giannakaki, O. Simantiraki, A. Roniotis, and M. Tsiknakis, "Review on psychological stress detection using biosignals," *IEEE Transactions on Affective Computing*, 2019.

[3] A. Arza, J. M. Garzón-Rey, J. Lázaro, E. Gil, R. Lopez-Anton, C. de la Camara, P. Laguna, R. Bailon, and J. Aguiló, "Measuring acute stress response through physiological signals: towards a quantitative assessment of stress," *Medical & Biological engineering & Computing*, vol. 57, no. 1, pp. 271–287, 2019.

[4] V. Montesinos, F. Dell'Agnola, A. Arza, A. Aminifar, and D. Atienza, "Multi-Modal Acute Stress Recognition Using Off-the-Shelf Wearable Devices," in *2019 41st Annual International Conference of the IEEE Engineering in Medicine and Biology Society (EMBC)*, pp. 2196–2201, 2019.

[5] U. Côté-Allard, C. L. Fall, A. Drouin, A. Campeau-Lecours, C. Gosselin, K. Glette, F. Laviolette, and B. Gosselin, "Deep Learning for Electromyographic Hand Gesture Signal Classification Using Transfer Learning," *IEEE Transactions on Neural Systems and Rehabilitation Engineering*, 2019.

[6] N. Sharma and T. Gedeon, "Objective measures, sensors and computational techniques for stress recognition and classification: A survey," *Computer Methods and Programs in Biomedicine*, vol. 108, no. 3, pp. 1287–1301, 2012.

[7] E. Ohn-Bar, A. Tawari, S. Martin, and M. M. Trivedi, "Predicting driver maneuvers by learning holistic features," in *2014 IEEE Intelligent Vehicles Symposium Proceedings*, pp. 719–724, June 2014.

[8] Y. Liang, M. L. Reyes, and J. D. Lee, "Real-Time Detection of Driver Cognitive Distraction Using Support Vector Machines," *IEEE Transactions on Intelligent Transportation Systems*, vol. 8, no. 2, pp. 340–350, 2007.

[9] Y. Takada, E. R. Boer, and T. Sawaragi, "Driver assist system for human–machine interaction," *Cognition, Technology & Work*, vol. 19, pp. 819–836, Nov 2017.

[10] C. J. de Naurois, C. Bourdin, A. Stratulat, E. Diaz, and J.-L. Vercher, "Detection and prediction of driver drowsiness using artificial neural network models," *Accident Analysis & Prevention*, vol. 126, pp. 95–104, 2019.

[11] M. Hasenjäger and H. Wersing, "Personalization in advanced driver assistance systems and autonomous vehicles: A review," in *2017 ieee 20th international conference on intelligent transportation systems (itsc)*, pp. 1–7, IEEE, 2017.

[12] R. Hoogendoorn and B. Van Arem, "Driver workload classification through neural network modeling using physiological indicators," in *16th International IEEE Conference on Intelligent Transportation Systems (ITSC 2013)*, pp. 2268–2273, IEEE, 2013.

[13] M. A. Pappas, S. Timotheou, and C. Panayiotou, "Design and experimental evaluation of a model-free controller for autonomous intersection crossing under imperfect localization," in *2023 European Control Conference (ECC)*, pp. 1–6, 2023.

[14] F. Pedregosa, G. Varoquaux, A. Gramfort, V. Michel, B. Thirion, O. Grisel, M. Blondel, P. Prettenhofer, R. Weiss, V. Dubourg, J. Vanderplas, A. Passos, D. Cournapeau, M. Brucher, M. Perrot, and E. Duchesnay, "Scikit-learn: Machine learning in Python," *Journal of Machine Learning Research*, vol. 12, pp. 2825–2830, 2011.

[15] G. Lemaître, F. Nogueira, and C. K. Aridas, "Imbalanced-learn: A python toolbox to tackle the curse of imbalanced datasets in machine learning," *Journal of Machine Learning Research*, vol. 18, no. 17, pp. 1–5, 2017.

[16] H. He, Y. Bai, E. A. Garcia, and S. Li, "ADASYN: Adaptive Synthetic Sampling Approach for imbalanced learning," in *2008 IEEE International Joint Conference on Neural Networks (IEEE World Congress on Computational Intelligence)*, pp. 1322–1328, IEEE, 2008.

[17] N. V. Chawla, K. W. Bowyer, L. O. Hall, and W. P. Kegelmeyer, "SMOTE: Synthetic Minority Over-Sampling Technique," *Journal of Artificial Intelligence Research*, vol. 16, pp. 321–357, 2002.

Low Power Network-on-Chip Architecture Design Technique

Tejas Musale[1], Arun Ganti[2], Ankur Gogoi[3] and Kanchan Manna[4]

[1]Department of Electrical and Electronics Engineering, BITS Pilani, K. K. Birla Goa Campus, India
[2,4]Department of Computer Science and Information Systems, BITS Pilani, K. K. Birla Goa Campus, India
[3]Galgotias University, Uttar Pradesh, India

Email: {[1]f20190409, [2]f20190021, [4]kanchanm}@goa.bits-pilani.ac.in, [3]ankur.gogoi.res@gmail.com

Abstract—In the evolving landscape of many-core architectures, the Network on Chip (NoC) emerges as a pivotal interconnection framework, accommodating the escalating number of cores on a single chip. Despite its widespread adoption, power consumption remains a critical challenge, significantly influenced by factors such as topology, bit toggling, and routing algorithms, with bit-switching power being a predominant concern. In this work, we have proposed an innovative technique in the cores aimed at reducing the power consumption of the NoC. To calculate our method's area, time, and power consumption, we have synthesized it on a Xilinx Virtex-7 FPGA board. We have modified the Noxim open-source simulator to validate our approach using synthetic traffic. Empirical results demonstrate a 29.64% reduction in dynamic power consumption and a 29.09% improvement in switching activity for the average case (Mode 1) traffic scenario.

Index Terms—Network-on-Chip (NoC), Field Programmable Gate Array(FPGA), Switching power, Hamming distance sorting, switch-router architecture, Noxim.

I. INTRODUCTION

The rapid advancement of artificial intelligence (AI) technologies has precipitated an unprecedented demand for more complex and integrated computing cores capable of seamless intercommunication. As Moore's law propels the miniaturization of chip sizes, the resulting intricate architectures and increased logic density pose significant challenges. System-on-chips (SoCs), which feature multiple cores, are particularly affected by the critical issue of core-to-core communication. Cores communicate using shared buses on processors. These buses are prone to scalability issues such as congestion, performance, and power dissipation. In order to address these issues, Network-on-Chip (NoC) architecture provides a scalable, modular, and flexible interconnect fabric that provides high-bandwidth, low-latency data transfer capabilities [1]. The components of a NoC include routers, links, and Network Interfaces (NIs). Network topologies are formed by connecting routers via links. The cores are associated with the routers via NIs. Router fabrics are used to communicate between cores. Due to the scalability, modularity, low power consumption,

This work was supported in part by the Science and Engineering Research Board (SERB), Department of Science and Technology (DST), Government of India, under Grant SRG/2021/001239.

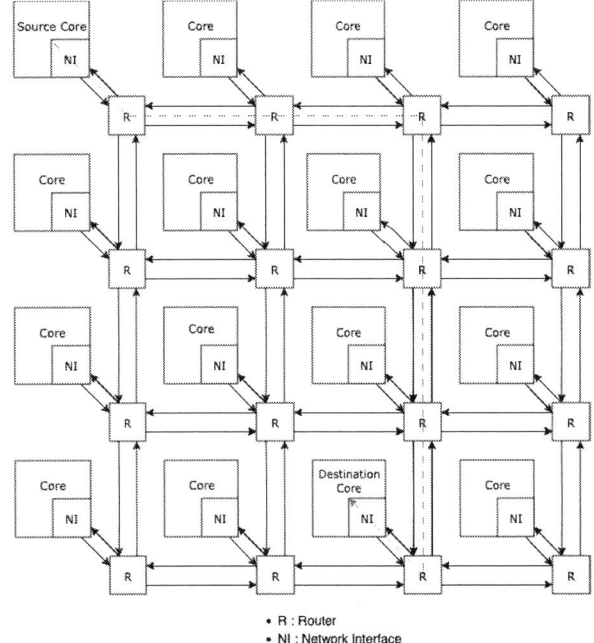

Figure 1: A 4 × 4 mesh-based NoC Architecture

low latency, and high bandwidth, mesh-based NoCs are popular in modern electronic systems. Fig. 1 presents a 4 × 4 mesh-based NoC. Besides, NoC consumes significant power. For example, 30% and 40% of chip power in Intel 80-core Terascale chip [2] and MIT RAW [3] consumed by NoC.

In the quest to minimize the power consumption of the NoC, researchers have proposed several approaches, which are discussed in section 2. In this work, we considered the switching factor, which is determined by the packet's toggle rates, a critical consideration. High toggle rates increase power consumption [4], necessitating strategies to reduce bit toggling at the packet level. The authors in [4] proposed the selective packet interleaving (SIP) technique at the router's output link's different virtual channels to minimize power consumption.

979-8-3315-3968-9/24 $31.00 © 2024 IEEE

Based on the Hamming distance, SIP multiplex flits in the router's output channel so that bits transition between two consecutive flits get reduced. For instance, consider a scenario where a flit 11001100 is transmitted from the router x to the router y. The next available flits are 00110011 and 11011101. By employing the Hamming distance sorting, the next flit chosen for transmission would be 11011101, which requires only 2-bit switches compared to the previous flit, reducing the switching activity and conserving power. The selection of flits is happening in each router, and as a result, the NoC's routers will be slow to process the packet. It works well when more virtual channels (VCs) are present in the routers [4]. These are the gaps present in the SOTA.

Contributions to our work are as follows:

- We have introduced a novel approach at the core or processor level instead of at the router level as in [4] to minimize the bit transition at the packet level, reducing the NoC's power consumption. Our approach does not affect the routers' frequency, whereas the SIP [4] reduces the router's processing speed. Our proposed technique's performance doesn't depend on the number of VCs.

- We have presented an integrated analysis of our proposed approach using an open-source Noxim NoC simulator [5] and Orion [6] for power consumption assessment. We have evaluated the impact of switching factor variations on the NoC's power usage.

- We have synthesized our proposed technique on Field-Programmable Gate Arrays (FPGAs) to find the area, power, and latency information.

The rest of the paper is organized as follows: Section II describes the basics of NoC and reviews related works. Section III explains the proposed methodology. Section IV presents the experimental results and discussion, and conclusions are drawn in Section V.

II. BACKGROUND AND RELATED WORK

The Bus-based communication infrastructures are used to meet the demand for aggressive communication among the ten cores of multi-core architectures. However, when the cores increased, the performance of the bus-based architecture started to decrease which led to the inception of NoC. In NoC, flexible communications are provided by routers and the routers are connected to each other via bidirectional communication links. Each processing element or core is connected to a router via the Network Interface (NI). Instead of fixed communication links, flexible communication is provided in NoC for low latency and dynamic data transmission paths. In the literature, various NoC topologies have been proposed, such as mesh, torus, butterfly, ring, tree, and etc. In our work, we have considered a 2D mesh topology. However, our proposed method can be applicable to other NoC topologies also. The processing elements of a NoC coordinate with each other for concurrent processing of a particular application data, and the application data is transmitted from one core to another using various routing approaches such as XY, Odd-Even, North-Last, etc. In NoC, for convenience of transmission, a packet

is decomposed into smaller units called flits (*flow units*). Three types of flits are available: header, body, and tailer. The header flit determines the transmission path of all the flits of a packet based on the employed routing algorithm. The body flit contains application data, which is to be processed by the processing element, and the tail flit indicates the end of a particular packet.

The power consumption in NoC is dynamic because it is generated by the charging and discharging of the capacitive load on the data-bus wires inside channels and buffers when a transistor changes its state. The overall power is exactly proportional to the frequency of the switching occurrences. During the flit transmission, bit-switching activity occurs in the ports of routers, leading to power dissipation. That is, the higher the flit dissimilarity, the higher the power dissipation in routers. A very few literature have been reported on this issue and are explained in this section. Authors in [7] proposes a highly scalable and portable FPGA-based single cycle, low latency router design for NoC applications. Reducing the input port buffer depth allowed for increased network speed while using low power for data transfer. The communication of several cores of a single chip has become the largest source of power dissipation. To tackle this problem of NoCs, wireless NoCs have been proposed in [8] for multi-core SoCs. The whole architecture has been implemented and integrated over Noxim [5]. The dynamic power consumption of a state-of-art microprocessor is investigated in [9] and further advised power-saving approaches to modify a router, which resulted in an average reduction of 14% of dynamic power. In [10], a unified nanometer-scale bus energy dissipation and thermal model is reported that can be used by designers to track temperature increase and energy dissipation in individual wires during trace-driven or power/performance simulation. Apart from the self-capacitance, the proposed model incorporates the impact of lateral heat transfer between neighboring wires. The authors in [11] proposed an approach that uses routing flexibility to provide a solution for power and performance-aware mapping, resulting in significant energy savings. According to [12], power consumption increases as the hamming distance between adjacent flits increases. Higher hamming distance implies more bit transitions when moving from one flit to another which in turn increases the power consumption due to the charging and the discharging of the capacitive loads. Accordingly, an approach is presented that minimizes the hamming distance between successive elements of the packet, thereby optimizing switching power. In [4], the authors introduced selective packet interleaving (SPI), a flit transmission scheme that reduces dynamic power consumption in NoC links by decreasing the number of bit transitions in links. SPI minimizes the hamming distance between two successive flits and chooses the flit out of all virtual channels, which have a minimum hamming distance with the previously transmitted flit. One drawback of this approach is that sometimes, certain VCs might have to wait for a very long time prior to transmitting their flits. While selecting flits with less bit transition, this approach only considers the flits available at

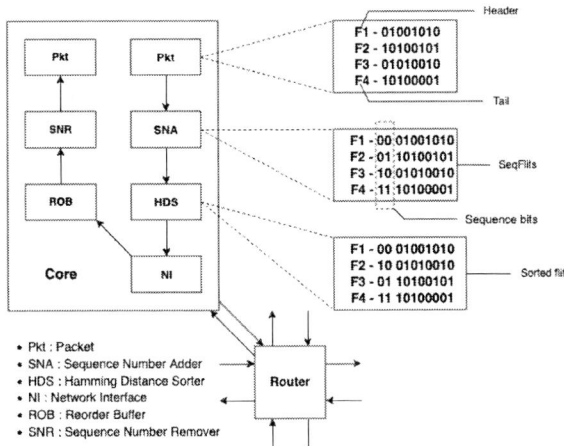

- Pkt : Packet
- SNA : Sequence Number Adder
- HDS : Hamming Distance Sorter
- NI : Network Interface
- ROB : Reorder Buffer
- SNR : Sequence Number Remover

Figure 2: Flow diagram of the proposed technique

the front or head of each buffer and does not consider all the flits available in the link buffers of the NoC router. In [13], the authors conducted a study on reducing power consumption in NoC by minimizing signal transition activity via the use of data encoding methods. Experimental investigations have shown that encoding effectiveness depends on the transition activity patterns. The proposed works modify the basic NoC router architecture to minimize the NoC's power consumption, which can increase the flit processing time. To deal with this problem, we have proposed a technique in the core that can reduce the NoC's power consumption while unchanging the basic router architecture.

III. PROPOSED METHODOLOGY

In response to the escalating power consumption challenges in the NoC architectures, this paper proposes a novel flit processing architecture that efficiently minimizes the bit-switching frequency in the packet. In our approach, the packet generated by the core is split into several flits based on the flit size. It also appends a sequence number with each flit. We have named such an entity as *SeqFlit*. The flit size is taken from the NI, which does not include bits required to identify the flit types. The module then sorts each *SeqFlit* based on the hamming distance with reference to the *header flit*. The header flit contains the source and destination routers' information and 0s for the sequence number field. First, the module finds a *SeqFlit* who's hamming distance is lesser with the header flit. Next, it will find the next *SeqFlit* whose hamming distance will be lesser with the previous *SeqFlit*. Similarly, the rest of the *SeqFlit*s will be arranged for the packet. The sequence number facilitates the reordering of *SeqFlit* into their original sequence at the destination core. The packet's maximum sequence number is fixed, and the bits required for the field is $\log_2(\text{No\_Flits})$.

Our proposed strategy has been presented in Fig. 2. In the diagram, the Pkt unit stores the entire packet as the flits, for example, F1, F2, F3, and F4, and sends it to the Sequence

Number Adder (SNA) module to add the sequences with flits (sequence bits are highlighted as the dotted red box) and generates *SeqFlits*. The Hamming Distance Sorter (HDS) sorts the *SeqFlits* and sends the rearranged packet (set of *SeqFlit*) to the NI. The NI takes the *SeqFlits* one after another and puts them in the flit structure available in the routers. Based on the destination address routing logic, forwards the flits.

The destination router can receive the *SeqFlits* in out-of-order. The *SeqFlits* are sent to NI, and NI stores them in a reorder buffer (ROB) as per the sequence numbers. Next, the entire sequence number is removed from *SeqFlits* in the ROB using the sequence number remover (SNR) module, and the entire packet is stored in another Pkt unit.

The destination core receives the packets in an out-of-order fashion. The sequence numbers in the flits are used to rearrange themselves and are stripped away at the end. Next, the packet is sent to the core.

The HDS module has been presented in the Algorithm 1. The *hamming_distance*() calculates the Hamming distance between two *SeqFlits*. It performs a bitwise XOR operation between consecutive *SeqFlits*, identifying differences bit by bit. The outcome of this operation dictates the arrangement strategy, aiming to position flits in a sequence that minimizes these distances. Crucially, the header and tail *SeqFlit* maintain their positions, while the body *SeqFlit* are reordered based on their Hamming proximity to preceding *SeqFlit*(s). The algorithm has an overall time complexity of $O(N^2)$, where N is the number of flits in SeqFlits. While there is potential to optimize this time complexity further, which we propose as part of future work, the current implementation should suffice for the purposes of this study.

Algorithm 1 Rearrangement of *SeqFlit*s

Result: Rearrange *SeqFlit*s based on Hamming distance

/*HD: Hamming Distance*/
for $i \leftarrow 0$ **to** $N-1$ **do**
  $min\_dist \leftarrow \text{HD}(SeqFlits[i], SeqFlits[i+1])$;
  $min\_idx \leftarrow i+1$
  for $j \leftarrow i+2$ **to** $N-1$ **do**
    $dist \leftarrow \text{HD}(SeqFlits[i], SeqFlits[j])$
    if $dist < min\_dist$ **then**
      $min\_idx \leftarrow j$
      $min\_dist \leftarrow dist$
    end
  end
  if $min\_idx \neq i+1$ **then**
    $\text{swap}(SeqFlits, i+1, min\_idx)$
  end
end

IV. EXPERIMENTS RESULTS AND DISCUSSION

We have implemented our proposed approach in the open-source Noxim simulator [5]. For the experiment purpose, we have used the following simulation parameters:

In this experiment, we have generated random synthetic traffic. We have generated random numbers between minimum

TABLE I: Noxim Settings

Parameters	Values
Topology	Mesh (4×4)
Buffer depth	4
VCs	1
Flit size	32 bits
Routing	Dimension-order (XY)
Selection logic	Random
Warmup time	10,000 clk cycles
Simulation time	200,000 clk cycles
Traffic pattern	Random

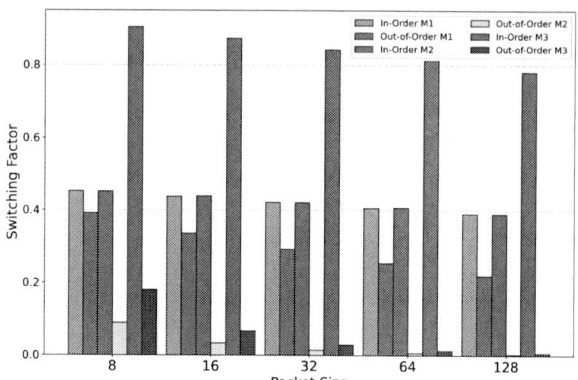

Figure 3: Switching factor variation

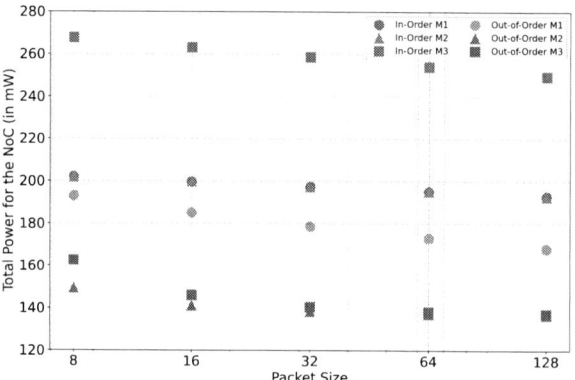

Figure 4: Total Power variation

and maximum values and used those values as body flits. The minimum value is 0, and the maximum is $2^k - 1$, where k is the number of bits required for body flits; it doesn't include the bits required for the sequence number. We have generated three modes of random traffic: Mode 1, 2, and 3.

1) Mode 1 (average case): Randomly generated body flits between minimum and maximum values.
2) Mode 2 (not average case): Alternative transmission of two randomly selected numbers between minimum and maximum values as the body flits.
3) Mode 3 (not average case): Sending the maximum and minimum numbers of the range alternatively as the body flits.

A. Switching Factor and Power Consumption Analysis

We have generated synthetic traffic for the three modes, sent them "In-Order" and "Out-of-Order" forms and measured the switching activity for each router's input ports. The bit switching factor for each packet was calculated using the Noxim simulator, with packet sizes ranging from 8 to 132 flits and flit of size 32 bits. Figure 3 illustrates the comparison of the switching factor for all three modes. The "In-Order" transmission refers to the scenario without applying our proposed technique, whereas the "Out-of-Order" transmission applies our proposed technique to optimize the Hamming distance between consecutive flits. Significant improvements have been observed across all modes, with average improvements in the switching factor of 29.09%, 92.95%, and 92.96% for Modes 1, 2, and 3, respectively. These results indicate a definitive improvement even in the random generation case and more drastic improvements in the other two modes. The reported power consumption doesn't include the power consumed by our proposed technique as it is done at the core or processor, not in the network or NoC.

The total power consumed by the NoC has been calculated using the ORION3.0 simulator [6], which employs 65 nm technology. The total power, P_{total}, is given by:

$$P_{\text{total}} = P_{\text{internal}} + P_{\text{switching}} + P_{\text{leakage}} + P_{\text{link}}$$

where $P_{\text{switching}}$ and P_{link} are the dynamic power components determined by the switching factor, and P_{leakage} and P_{internal} represent the static power components of the NoC. The link power depends on the number of links available in the NoC topology. Figures 4 and 5 show the variations in total power and switching power for varying packet sizes across all modes. Our simulations demonstrate average improvements of 29.64%, 93.28%, and 93.28% in switching power for Modes 1, 2, and 3, respectively. Overall, the random generation case of the NoC saves an average of 17.8 mW in power across varying packet sizes. We observe significant improvements in both the switching factor and total power consumption, underscoring the efficacy of our approach in enhancing the efficiency of NoC operations. The experimental results confirm that the benefits of our proposed technique increase with the packet size, leading to significant power savings and enhanced efficiency in NoC operations. However, we have used additional bits for sequence numbers.

Our proposed technique does not impact other NoC parameters such as network latency and throughput, as the implementation is at the core or processor level rather than the NoC itself. However, it can increase the processing time at the core, leading to a potential increase in overall latency, including the core's processing time, which may subsequently reduce throughput. Despite this, the technique offers significant benefits in applications requiring low power NoCs, such as high-performance computing environments, wireless communication systems (such as smartphones and

979-8-3315-3968-9/24 $31.00 © 2024 IEEE   40

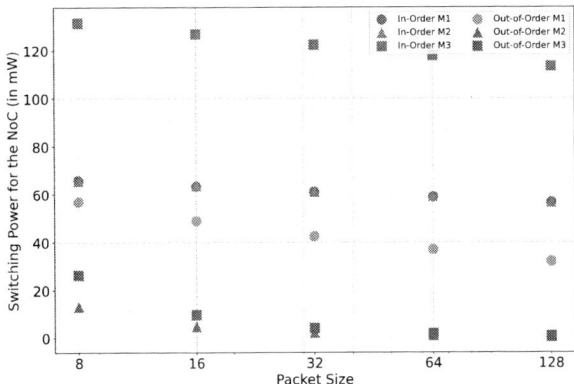

Figure 5: Switching Power variation

IoT devices), embedded systems in medical and industrial applications, and AI/ML on edge or wearable devices [14]. These applications would benefit significantly from our design due to their need for energy efficiency, extended operational life, improved performance, and reliable operation in power-constrained environments.

B. FPGA Implementation and Performance Evaluation of Hamming Distance Sorter IP

We have implemented the proposed method using Verilog HDL and executed on a Virtex-7 FPGA board using the Xilinx Vivado Design Suite for synthesis, placement, and routing. This process aimed to assess the resource utilization and the maximum clock frequency achievable by the IP. As indicated in Table II, resource utilization was minimal, with Look-Up Tables (LUTs) and Flip-Flops (FFs) showing very low usage, which demonstrates an efficient use of the FPGA's logic resources. Specifically, the IP utilized only 673 LUTs and 632 FFs, corresponding to just 0.05% and 0.02% of the total available resources, respectively. The maximum clock frequency achieved by the IP was 447 MHz, underscoring the IP's ability to operate at high speeds, a critical feature for the cores in a multi-core system. This IP, which can be integrated alongside the core, highlights its potential for efficient incorporation into large NoCs that require rapid flit reordering capabilities.

TABLE II: Resource Utilization

Resource	Usage	Available	Utilization %
LUT	673	1303680	0.05
FF	632	2607360	0.02
IO	530	624	84.94
BUFG	2	1008	0.2

A qualitative comparison with [4] shows that our approach consumes less power. The method in [4] is buffer-dependent, and its power efficiency improves with larger buffer sizes. In contrast, our approach computes the next flit selection within the core, eliminating the need for additional buffers and

thereby reducing power consumption. Unlike [4], which can face starvation due to changes in packet transmission order, our method reorders flits within a packet without altering the order of packet transmission. This ensures that all packets are transmitted in their generation order, avoiding starvation and the need for counters in each virtual channel. Consequently, our approach is more area-efficient, as it eliminates the area overhead associated with the counters used in [4].

V. CONCLUSION AND FUTURE WORK

In this paper, we presented a flit sorting technique before transmission to achieve energy-efficient NoCs. The flit sorting is performed by optimizing the Hamming distance between successive flits, which significantly reduces the switching activity. Our technique, which uses a simple and very low complexity circuit addition to the core, has demonstrated significant improvements in dynamic power consumption. Specifically, we observed a 29.09% improvement in switching activity and a 29.64% improvement in dynamic power consumption in the average case traffic scenario (Mode 1). Furthermore, in scenarios where alternate body flits are the same, which is common in real workloads, our technique achieved remarkable average power consumption and switching activity improvements of 92.9% and 92.8%, respectively. The benefits of our technique grow with increasing packet size. For future work, we plan to conduct simulations using the Gem5 simulator with real workloads to further validate our approach. Additionally, we look to explore other encoding and power optimization techniques in integration with our sorting approach to address the low power challenges introduced by modern multi-core systems, particularly the dynamic power consumption due to the switching activity of interconnects. In addition, a more ambitious study could investigate the use of other sorting techniques for flits.

REFERENCES

[1] S. Kumar, A. Jantsch, J.-P. Soininen, M. Forsell, M. Millberg, J. Oberg, K. Tiensyrja, and A. Hemani, "A network on chip architecture and design methodology," in *Proceedings IEEE Computer Society Annual Symposium on VLSI. New Paradigms for VLSI Systems Design. ISVLSI 2002*, 2002, pp. 117–124.

[2] Y. Hoskote, S. Vangal, A. Singh, N. Borkar, and S. Borkar, "A 5-ghz mesh interconnect for a teraflops processor," *IEEE Micro*, vol. 27, no. 5, pp. 51–61, 2007.

[3] M. B. Taylor, W. Lee, J. Miller, D. Wentzlaff, I. Bratt, B. Greenwald, H. Hoffmann, P. Johnson, J. Kim, J. Psota, A. Saraf, N. Shnidman, V. Strumpen, M. Frank, S. Amarasinghe, and A. Agarwal, "Evaluation of the raw microprocessor: An exposed-wire-delay architecture for ilp and streams," in *Proceedings of the 31st Annual International Symposium on Computer Architecture*, ser. ISCA '04. IEEE Computer Society, 2004, p. 2.

[4] A. Berman, R. Ginosar, and I. Keidar, "Order is power: Selective packet interleaving for energy efficient networks-on-chip," in *2010 18th IEEE/IFIP International Conference on VLSI and System-on-Chip*, 2010, pp. 37–42.

[5] V. Catania, A. Mineo, S. Monteleone, M. Palesi, and D. Patti, "Noxim: An open, extensible and cycle-accurate network on chip simulator," in *2015 IEEE 26th International Conference on Application-specific Systems, Architectures and Processors (ASAP)*, 2015, pp. 162–163.

[6] A. B. Kahng, B. Lin, and S. Nath, "Orion3.0: A comprehensive noc router estimation tool," *IEEE Embedded Systems Letters*, vol. 7, no. 2, pp. 41–45, 2015.

979-8-3315-3968-9/24 $31.00 © 2024 IEEE

[7] P. Gupta, A. Akoglu, K. Melde, and J. Roveda, "Fpga based single cycle, reconfigurable router for noc applications," in *2013 IEEE International Symposium on Circuits and Systems (ISCAS)*, 2013, pp. 2428–2431.

[8] S. Mnejja, Y. Aydi, and M. Abid, "Exploring hybrid noc architecture for chip multiprocessor," in *2018 30th International Conference on Microelectronics (ICM)*, 2018, pp. 307–310.

[9] N. Magen, A. Kolodny, U. Weiser, and N. Shamir, "Interconnect-power dissipation in a microprocessor," in *Proceedings of the 2004 International Workshop on System Level Interconnect Prediction*, ser. SLIP '04. New York, NY, USA: Association for Computing Machinery, 2004, p. 7–13. [Online]. Available: https://doi.org/10.1145/966747.966750

[10] K. Sundaresan and N. Mahapatra, "Accurate energy dissipation and thermal modeling for nanometer-scale buses," in *11th International Symposium on High-Performance Computer Architecture*, 2005, pp. 51–60.

[11] J. Hu and R. Marculescu, "Exploiting the routing flexibility for energy/performance aware mapping of regular noc architectures," in *2003 Design, Automation and Test in Europe Conference and Exhibition*, 2003, pp. 688–693.

[12] A. Agrawal, "An architectural power model for networks on chip," 2023.

[13] J. C. Palma, L. S. Indrusiak, F. G. Moraes, A. G. Ortiz, M. Glesner, and R. A. Reis, "Inserting data encoding techniques into noc-based systems," in *IEEE Computer Society Annual Symposium on VLSI (ISVLSI '07)*, 2007, pp. 299–304.

[14] E. Ofori-Attah and M. O. Agyeman, "A survey of low power noc design techniques," in *Proceedings of the 2nd International Workshop on Advanced Interconnect Solutions and Technologies for Emerging Computing Systems*, ser. AISTECS '17. New York, NY, USA: Association for Computing Machinery, 2017, p. 22–27. [Online]. Available: https://doi.org/10.1145/3073763.3073767

Minimum Depth Quantum Modular Addition through Carry-Save Architecture

Siyi Wang[1], Eugene Lim[1], Xiufan Li[2], Jerrie Feng[1], and Anupam Chattopadhyay[1]

[1]Nanyang Technological University, 50 Nanyang Avenue, Singapore 639798.

[2]Centre for Quantum Technologies, National University of Singapore, 3 Science Drive 2, Singapore 117543.

siyi002@e.ntu.edu.sg

Abstract—Shor's factorization algorithm, as one of the most significant achievements in quantum computing, exhibits an exponential speedup compared to the corresponding classical algorithm. In Shor's factorization algorithm, modular exponentiation is one of the most computationally intensive components, which relies on the modular addition building block.

This work aims to explore novel designs for enhancing the efficiency of quantum modular addition. In particular, we introduce a novel quantum modular addition framework based on carry-save architecture, which facilitates the conversion of multiple 2-addend quantum operations within modular addition into a single 3-addend operation, thereby reducing the computational depth. Compared to the most efficient existing quantum modular addition, our design has achieved an impressive result - a reduction in Toffoli Depth by up to 33.33%, while maintaining comparable Toffoli Count and Qubit Count. This research underscores the potential of carry-save architecture as a promising technique for accelerating quantum modular arithmetic as well as advancing the development of quantum computing in general.

Index Terms—Quantum Computing, Quantum Arithmetic, Toffoli-depth Optimization, Carry-save Adder, Modular Addition.

I. INTRODUCTION

Shor's algorithm [1] has firmly established itself as an approach in quantum algorithms to efficiently factor large integers into their prime factors. By leveraging on its two key operations, modular exponentiation and its component modular addition, Shor's algorithm can work exponentially faster than classical algorithms. Therefore numerous efforts have been made to push the boundaries on modular exponentiation, as it's one of the most computationally intensive components in Shor's algorithm.

Recent works have achieved many improvements in reducing the bottleneck of modular exponentiation by enhancing the design of its main component, modular addition, through efficient implementations of Clifford+T gates. However, it still remains challenging to develop modular addition frameworks with a lower gate depth and gate count.

In this work, we propose a novel framework for quantum modular addition that relies on the Clifford+T gate set to leverage classical carry-save adders (CSAs). We consolidate multiple 2-addend quantum operations within modular addition into a single 3-addend operation, greatly reducing the overall depth of the quantum circuit.

Our work addresses a hot-in-discussion arithmetic problem: quantum modular addition, expressed as $(a+b) \bmod N$. Here,

N represents an n-bit binary number, while a and b represent n-bit binary numbers less than N. This operation usually involves checking if $a + b > N$, and if so, subtracting N to ensure the result is within the range $[0, N-1]$. In our specific design, we implement the modular addition by involving a three-operand addition $a + b + (-N)$ in the framework.

This paper has the following contributions:

- Highlight the potential of carry-save architecture by enhancing the performance of quantum arithmetic designs.
- Reduce the depth of quantum modular addition by proposing a novel framework based on carry-save architecture.
- Accelerate Shor's algorithm through improving modular addition efficiency.

II. BACKGROUND

Current research on relevant designs can be categorized into the following two main sorts:

A. Clifford+T-based Modular Addition Framework

Typical architectures realize modular addition with gates in the Clifford+T set. We denote the elementary adder block and subtractor block by *Add* and *Sub* separately. The circuits input quantum states that correspond to the binary representation of the numbers and output $|S\rangle$ as the result of modular addition. Here we enumerate four types of popularly cited modular addition frameworks in literature.

- *VBE Modular Addition Framework.*
  Quantum modular adder based on ripple-carry adders was proposed by Vedral, Barenco, and Ekert in 1996, which we refer to as VBE in this paper [2]. As shown in Figure 1, VBE framework contains 5 adders and subtractors with reversible CNOT gates and Toffoli gates. After that, Cucarro et, al. [3] proposed an improved modular adder with more efficient full adder blocks. However, this framework only solves the modulo calculation when N is the power of two, causing a lack of extendability to general problems. So we omit the Cucarro modular design in this paper.
- *Beckman Modular Addition Framework.*
  Figure 2 depicts Beckman's quantum modular framework made from adder-based comparators, full adders, and subtractors [4]. To standardize the performance estimation, we transform all adder-based operations, especially

979-8-3315-3968-9/24 $31.00 © 2024 IEEE

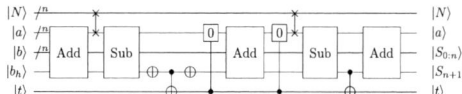

Fig. 1. VBE Modular Addition Framework [2].

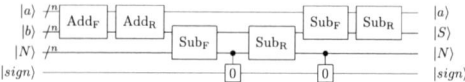

Fig. 4. Rines-Chuang Modular Addition Framework[9].

the comparators, into quantum adders and subtractors. Inspired by Beckman's work, Takahashi and Kunihiro propose their modular addition framework composed of 2 comparators and 1 adder or subtractor [5]. But their implementation of comparators is based on Draper's Fourier basis addition [6] which implies a large number of T gates. Haner et, al. [7] utilize recursive adder and Carry gates to build the comparators. The intuition is straightforward such that, by comparing the values of b and $N-a$, we can determine whether to add a or subtract $N-a$ and revert the indicator to the zero state.

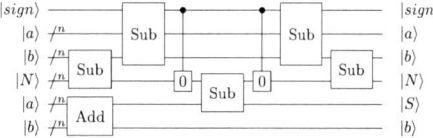

Fig. 2. Beckman Modular Addition Framework [4].

- **Van Meter Modular Addition Framework.**
  In 2005, Van Meter and Itoh constructed a more efficient modular addition framework by removing the last 2 adder blocks in VBE that were used to undo the carry overflow [8], as is shown in Figure 3. Controlled U_a and $U_{-a,N}$ represent the data loading oracles that set the register with respect to the value of the control qubit. The limitation of this framework is that it requires multiple calls of data loading oracle which complicates circuit operations.

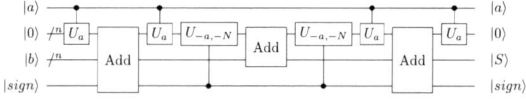

Fig. 3. Van Meter Modular Addition Framework [8].

- **Rines-Chuang Modular Addition Framework.**
  In 2018, Rine et al [9] enhanced Beckman's design [4] by splitting each integer adder into forward and reverse subsections. The implementation of Rines-Chuang's modular addition framework is depicted in Figure 4. It is constructed with 1 integer adder and 2 integer subtractors. Every forward and reverse block-pairs are denoted with the subscripts F and R respectively. Intermediate results of the forward block are utilized to control its subsequent reverse block in the second integer adder.

B. QFT-based Modular Addition Framework

Apart from the circuit designs with Clifford+T gates, modular addition frameworks are also developed in Fourier ba-

sis [10], [11], [12]. These algorithms utilize quantum Fourier transformation (QFT) and its inverse as subroutines [13], [14]. However, the complexity associated with QFT arithmetic contributes to its limited adoption in experimental studies. Specifically, the controlled rotation gates with arbitrary angles are practically infeasible to operate in the current stage of hardware than the Toffoli gate. Besides, even the state-of-the-art approximation method to build a QFT block requires $\mathcal{O}(n \log n)$ T gates, complicating the experimental realizations [15]. Hence, we only focus on quantum designs based on the Clifford+T gate set in this work.

III. CARRY-SAVE ARCHITECTURE

We introduce the carry-save architecture and consider its key implementations.

A. Carry-Save Structure

Various types of adders are available in classical arithmetic, such as ripple carry adders, carry-lookahead adders, and etc. Among these designs, the carry-save adder (CSA) [16] emerges as a high-speed, low-cost design with low latency, making it ideal for efficiently calculating multi-operand additions.

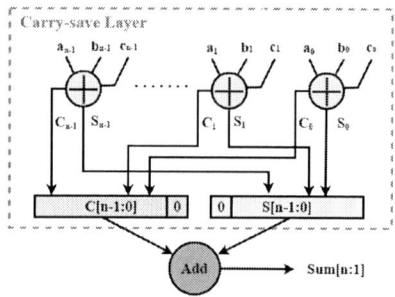

Fig. 5. A depiction of the Carry-Save architecture for multi-operand addition, illustrating the case where the number of operands is 3.

The carry-save architecture is summarized in Figure 5. It achieves the addition of three n-bit inputs by first calculating the partial sum S and shifted carry-out vector C outputs using a Carry-save layer. Subsequently, the partial sum and shifted carry-out vector can be added using any n-bit adder, such as a ripple carry adder or carry look-ahead adder, to produce the sum of the multi-operand addition.

B. Quantum Implementation

The quantum implementation of carry-save architecture relies on two primary components: the Carry-Save layer comprising a series of full adders operating in parallel, and a quantum full adder. Extensive research has already been done on the quantum full adder design [17][18][19][20]. As such, our work will focus on the Carry-Save layer.

$$S = a \oplus b \oplus c \tag{1}$$

$$C = ab \oplus ac \oplus bc \tag{2}$$

We are particularly interested in the shifted carry-out vector C and partial sum S of a 1-bit CSA that can be computed using Formulas 1 and 2. Particularly, it is worth noting that for hardware implementation, the quantum XOR logic is generally cheaper than OR logic. We note that the function of this basic unit aligns with that of a quantum full adder. Thus by leveraging the quantum full adder circuit illustrated in Figure 6, the quantum Carry-save layer can be constructed.

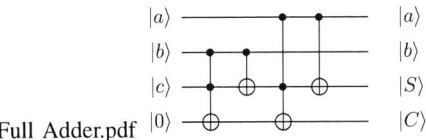

Fig. 6. Quantum Full Adder.

C. Advantages in Efficiency

Previous research [2][4][8], suggests that quantum implementations of addition involving three numbers using only 2-addend additions will lead to multiple sequential addition operations. The resulting circuit experiences the full propagation delay of both full adders, giving a depth of $O(2 \cdot TD_{Add})$, where TD_{Add} denotes the Toffoli Depth of a specific quantum adder for n-bit addition.

In comparison, our quantum implementation uses a Carry-save layer and one full adder. The quantum CSA layer only has a Toffoli Depth 2. Its resulting CSA-based quantum circuit for 3-addend addition will possess a Toffoli Depth of only $O(2 + TD_{Add})$, which significantly improves the sequential addition implementation.

IV. OUR DESIGN

Our proposed quantum carry-save based modular addition framework is summarized in Figure 9. It contains six operations. Below, we describe some of the novel components and structure of our framework.

A. Evaluation Metrics

To evaluate the importance of different components in our framework, we employ three metrics commonly used in many works [21][22] in quantum computing. These metrics are widely accepted benchmarks for comparison and can be conveniently expressed at the T-gate level:

- **Toffoli Depth(TD)**: By counting the number of computational layers containing Toffoli gates, this metric quantifies the time complexity of the quantum circuit.
- **Toffoli Count(TC)**: This metric is utilized to evaluate the gate complexity and resource consumption of quantum circuits, directly calculating the cumulative number of Toffoli gates.

- **Qubit Count(QC)**: This metric reveals the qubit requirement of the quantum circuit.

Decomposing the Toffoli gates in the quantum circuits using Toffoli decomposition methods can be considered if a comprehensive evaluation at the T-gate level is necessary. Correspondingly, TD, TC, and QC can be used to evaluate the overall performance.

B. Main Components

The following are the five main components of our design:

- **CSA Module** represents the Carry-save Architecture utilized in our work. As demonstrated in Section III, the CSA module consists of a series of full adders operating in parallel, as depicted in Figure 7.

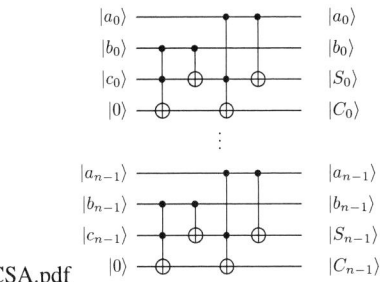

Fig. 7. Quantum Carry-save Architecture.

- **Add Module** represents a quantum adder. In quantum arithmetic, there are various types of adders, such as ripple carry adders[2], [3], carry-lookahead adders[23], [18], [17], [24], etc. In this work, we adopt prominent works detailed in Section V.
- **Sub Module** denotes a quantum subtractor, which is a direct application of the quantum adder.

$$a - b = (a' + b)' \tag{3}$$

As shown in Equation 3, the process involves bit-wise complementation of a, followed by its addition to b, yielding $a' + b$. Subsequently, the bits of a and the output qubits are complemented. Given that bit-wise complementation can be achieved using quantum NOT gates, the cost of a quantum subtractor is close to that of the corresponding adder.

- **Set 0 Module** offers same as the VBE Modular framework [2], the set 0 operation can be implemented using a series of CNOT gates.
- **Uncompute S Module** is used to uncompute the partial sum S, which can be achieved through a series of CNOT gates, as detailed in Figure 8.

C. Overall Structure

- **Step 1**: The carry-save layer is performed to calculate the partial sum, S, and the shifted carry, C. For n-bit modular addition, all inputs and outputs in this step have a length of n qubits. In terms of the detailed costs, the TC is $2n$, the QC is $4n$, and the TD is 2.

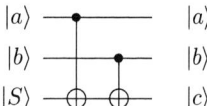

Fig. 8. Quantum Uncompute S Module.

- **Step 2**: A signed addition is applied to S and C to calculate the sum of $(a + b - N)$, with the initial sign set to 0. The associated costs include a TC of TC_{Add}, a TD of TD_{Add}, and there's an additional ancilla qubit required, totaling $1 + Anc_{\text{Add}}$.
- **Step 3**: Utilize Set 0 Module, which is constructed only with CNOT gates. The sign qubit serves as the control qubit for this module. If the sign is in the state $|0\rangle$, indicating $(a+b-N) \geq 0$, the Set 0 gate toggles $|-N\rangle$ to the state $|0\rangle$. Otherwise, $|-N\rangle$ passes through the module unchanged. This step incurs no requirement for TC, TD, and additional ancilla.
- **Step 4**: Perform a subtraction if the sign qubit is $|0\rangle$, computing $(a+b-N)-0$. Otherwise, $(a+b-N)-(-N)$ is calculated to restore the original value of $a+b$. As for the cost, TC is TC_{Add}, TD is TD_{Add}, with no additional ancilla required.
- **Step 5**: The previous Set 0 Module is uncomputed by applying all the CNOT gates in reverse order. For this step, there is no additional TC, TD, or ancilla required.
- **Step 6**: After applying the Uncompute S Module in the final step, all qubits except the final result of $(a + b)$ mod N are uncomputed. Here, no additional TC, TD, or ancilla are required.

$$
\text{TD} = \overbrace{2}^{\text{Step 1}} + \overbrace{TD_{Add}}^{\text{Step 2}} + \overbrace{0}^{\text{Step 3}} + \overbrace{TD_{Add}}^{\text{Step 4}} + \overbrace{0}^{\text{Step 5}} + \overbrace{0}^{\text{Step 6}} \quad (4)
$$
$$
= 2 + 2 \times TD_{Add}
$$

$$
\text{TC} = \overbrace{2n}^{\text{Step 1}} + \overbrace{TC_{add}}^{\text{Step 2}} + \overbrace{0}^{\text{Step 3}} + \overbrace{TC_{add}}^{\text{Step 4}} + \overbrace{0}^{\text{Step 5}} + \overbrace{0}^{\text{Step 6}} \quad (5)
$$
$$
= 2n + 2 \times TC_{Add}
$$

$$
\text{QC} = \overbrace{4n}^{\text{Step 1}} + \overbrace{1 + Anc_{Add}}^{\text{Step 2}} + \overbrace{0}^{\text{Step 3}} + \overbrace{0}^{\text{Step 4}} + \overbrace{0}^{\text{Step 5}} + \overbrace{0}^{\text{Step 6}} \quad (6)
$$
$$
= 4n + 1 + Anc_{Add}
$$

We present a comprehensive construction for a carry save-based quantum modular addition framework in these six steps. The detailed costs are illustrated in Formulas 4, 5, and 6, where TC_{Add} represents the Toffoli Count, TD_{Add} represents the Toffoli Depth, and Anc_{Add} denotes the ancilla count of the n-bit quantum Add building block.

V. RESULTS AND DISCUSSION

A. Experiment 1: Comparison with Prominent Quantum Modular Addition Frameworks.

After reviewing Paper [24], we choose three adders in Table I as our adder blocks since they exhibit optimal TC,

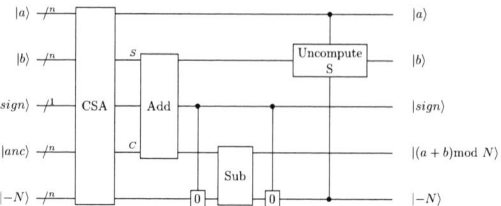

Fig. 9. Our Modular Addition Framework

TD, QC among all other quantum adders, respectively. Table II clearly demonstrates that the effectiveness of a proposed quantum modular addition framework largely relies on the Add module, representing a quantum adder. A performance comparison between our proposed quantum modular addition framework and other relevant prominent works is summarized in Figure 10, where we note the following:

TABLE I
EFFICIENT QUANTUM ADDERS UTILIZED IN EXPERIMENT 1.

Adder	TD	TC	QC
Gayathri[25]	n	n	$3n + 1$
Takahashi[26]	$2n - 1$	$2n - 1$	$2n + 1$
Optimal TD [24]	$\log n + 1$	$\frac{n \log n}{2}$	$n + n \log n + \lceil \log n \rceil + 2$

- The framework structure determines the modular adder efficiency. Our structure consistently demonstrates the lowest TD compared to all others, as shown in Figure 10(a). In terms of TC, the Rines-Chuang framework with Gayathri Add block achieve the lowest TC cost, with our design closely following as the second lowest, as depicted in Figure 10(b). However, when using a low TC Add block, the Rines-Chuang framework gains more significant advantages. Our design exhibits a QC that falls slightly below the Beckman framework, as shown in Figure 10(c). When the ancilla of the Add block is high, our design shows similar QC to the Rines-Chuang framework.
- The choice of Add Block significantly impacts the overall performance of modular adders. Within the same modular framework, utilizing Gayathri Adder achieves the lowest TC, while employing Optimal TD Adder results in the lowest TD. Besides, using Takahashi Adder allows for achieving the lowest QC.
- Our Optimal TD Add block design minimizes notable time requirements by achieving the lowest TD. Comparing our Gayathri Add block variation, which scores the second lowest TC and shows its advantage in reducing Toffoli resource usage. On the other hand, our design incorporating the Takahashi Add block achieves the fourth lowest QC.

Our designs exhibit commendable efficiency across various cost metrics in comparison to existing efficient quantum modular addition framework. We achieved up to 33.33% reduction in TD, while keeping TC and QC at comparable levels.

TABLE II
PERFORMANCE ANALYSIS OF DIFFERENT QUANTUM MODULAR ADDITION FRAMEWORKS.

Modular Addition Framework	Year	Toffoli Depth	Toffoli Count	Qubit Count
VBE [2]	1996	$5 \times TD_{Add}$	$5 \times TC_{Add}$	$3n + 2 + Anc_{Add}$
Beckman [4]	1996	$5 \times TD_{Add}$	$6 \times TC_{Add}$	$5n + 1 + 4 \times Anc_{Add}$
Van Meter [8]	2005	$3 \times TD_{Add}$	$3 \times TC_{Add}$	$3n + 1 + Anc_{Add}$
Rines-Chuang [9]	2018	$3 \times TD_{Add}$	$3 \times TC_{Add}$	$3n + 1 + 2 \times Anc_{Add}$
Our Design	-	$2 + 2 \times TD_{Add}$	$2n + 2 \times TC_{Add}$	$4n + 1 + Anc_{Add}$

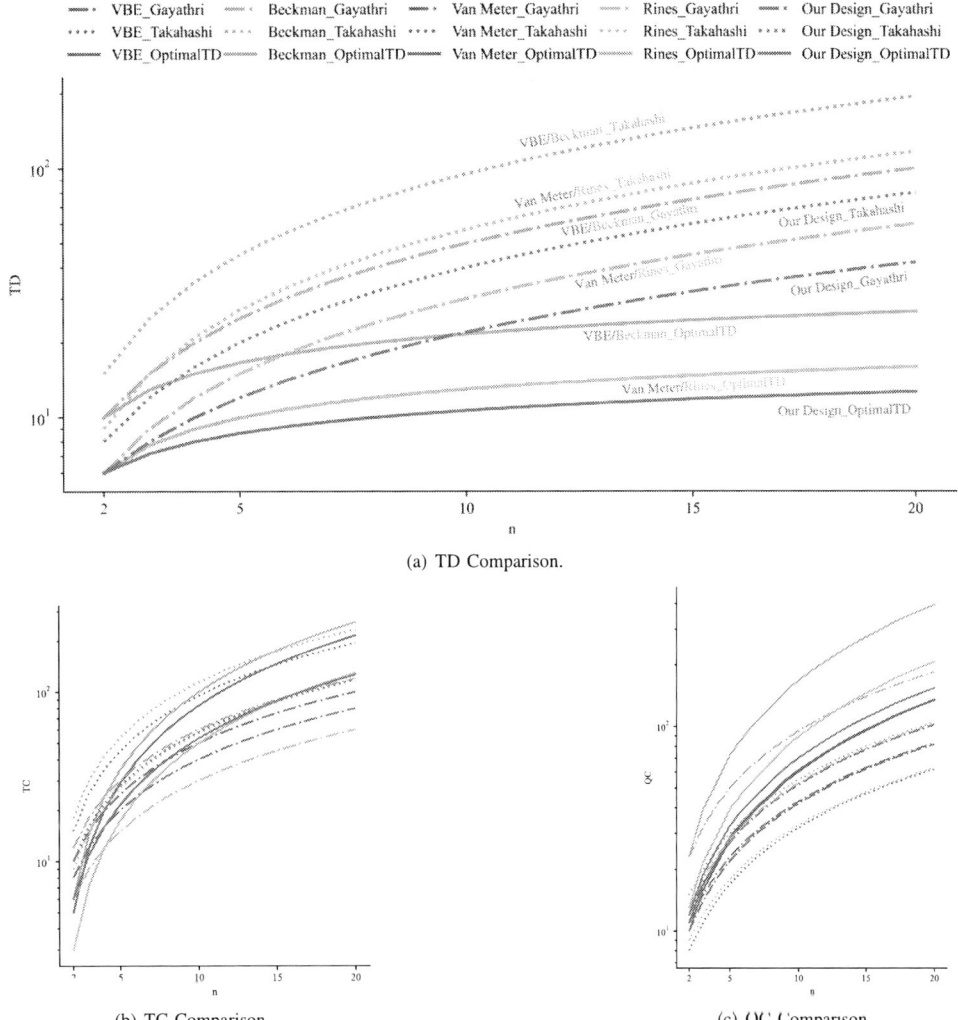

Fig. 10. Comparative Cost Analysis of the Proposed Quantum Modular Adders. Here, n denotes the qubit-width of the modular addition.

B. Experiment 2: Influence on Quantum Modular Exponentiation

We leverage the VBE modular exponentiation framework [2] to analyze the influence of our proposed modular addition framework on quantum modular exponentiation circuits. We start with constructing an n-bit controlled multiplier to compute $ax \bmod N$, where x is represented as an n-bit binary number. This task is accomplished by using repeated modular

N adders. The corresponding TD, TC and QC cost of this circuit are detailed in Formulas 7, 8, and 9 respectively. Additionally, it's important to note that the cost of the Set State Module within this circuit depends on the specific scenario. Under the best case, TC and TD cost are 0, whereas under less favorable circumstances, TC and TD cost are n.

$$TD_{C-MODMul} = 2n \times SETState + n \times TD_{MODAdd} \quad (7)$$

979-8-3315-3968-9/24 $31.00 © 2024 IEEE   47

$$TC_{C-MODMul} = 2n \times SETState + n \times TC_{MODAdd} \quad (8)$$

$$QC_{C-MODMul} = 1 + n + QC_{MODAdd} \quad (9)$$

Building upon the n-bit controlled modular multiplier, the VBE modular exponentiation can be constructed to calculate $a^x \bmod N$, where x is an n-bit binary number. The Formulas 10, 11, and 12 illustrate the number of calls to our modular adder block required to perform an n-bit modular exponentiation.

$$\begin{aligned} TD_{MODExp} &= 2n \times TD_{C-MODMul} \\ &= 4n^2 SETState + 2n^2 TD_{MODAdd} \end{aligned} \quad (10)$$

$$\begin{aligned} TC_{MODExp} &= 2n \times TC_{C-MODMul} \\ &= 4n^2 SETState + 2n^2 TC_{MODAdd} \end{aligned} \quad (11)$$

$$\begin{aligned} QC_{MODExp} &= n + QC_{C-MODMul} \\ &= 1 + 2n + QC_{MODAdd} \end{aligned} \quad (12)$$

By employing our proposed structure, a 60% reduction in addition operations in both VBE quantum modular multiplication and VBE quantum modular exponentiation can be achieved.

VI. Conclusion

In this paper, we focused on exploring innovative designs to enhance the efficiency of quantum modular addition. We presented a novel quantum modular addition framework based on carry-save architecture, reducing the computational intensity associated with modular addition. When compared against existing quantum modular addition framework, our framework can reduce up to 33.33% in TD while maintaining comparable TC and QC. Upon acceptance of this paper, the Qiskit code for the proposed framework will be made available to facilitate future benchmarking.

We are excited about future promising directions based on this research. There is still a lack of further exploration on related quantum modular arithmetic designs, such as modular multiplication and modular exponentiation. Another research goal of ours is to assess the performance and scalability of this proposed modular addition framework within large-scale quantum algorithms, such as the Shor's algorithm.

Acknowledgement

This research is supported by the National Research Foundation (NRF), Singapore under its Quantum Engineering Programme Initiative. This research is partly supported by MoE Tier 1 grant on Quantum Cryptanalysis.

References

[1] P. W. Shor, "Polynomial-time algorithms for prime factorization and discrete logarithms on a quantum computer," *SIAM Journal on Computing*, vol. 26, no. 5, p. 1484–1509, Oct. 1997. [Online]. Available: http://dx.doi.org/10.1137/S0097539795293172

[2] V. Vedral, A. Barenco, and A. Ekert, "Quantum networks for elementary arithmetic operations," *Physical Review A*, vol. 54, no. 1, p. 147–153, Jul. 1996. [Online]. Available: http://dx.doi.org/10.1103/PhysRevA.54.147

[3] S. A. Cuccaro, T. G. Draper, S. A. Kutin, and D. P. Moulton, "A new quantum ripple-carry addition circuit," 2004. [Online]. Available: https://arxiv.org/abs/quant-ph/0410184

[4] D. Beckman, A. N. Chari, S. Devabhaktuni, and J. Preskill, "Efficient networks for quantum factoring," *Physical Review A*, vol. 54, no. 2, p. 1034–1063, Aug. 1996. [Online]. Available: http://dx.doi.org/10.1103/PhysRevA.54.1034

[5] Y. Takahashi and N. Kunihiro, "A quantum circuit for shor's factoring algorithm using 2n + 2 qubits," *Quantum Info. Comput.*, vol. 6, no. 2, p. 184–192, mar 2006.

[6] T. G. Draper, "Addition on a quantum computer," 2000. [Online]. Available: https://arxiv.org/abs/quant-ph/0008033

[7] T. Häner, M. Roetteler, and K. M. Svore, "Factoring using 2n + 2 qubits with toffoli based modular multiplication," *Quantum Info. Comput.*, vol. 17, no. 7–8, p. 673–684, jun 2017.

[8] R. Van Meter and K. M. Itoh, "Fast quantum modular exponentiation," *Physical Review A*, vol. 71, no. 5, May 2005. [Online]. Available: http://dx.doi.org/10.1103/PhysRevA.71.052320

[9] R. Rines and I. Chuang, "High performance quantum modular multipliers," 2018. [Online]. Available: https://arxiv.org/abs/1801.01081

[10] S. Beauregard, "Circuit for shor's algorithm using 2n+3 qubits," 2002. [Online]. Available: https://arxiv.org/abs/quant-ph/0205095

[11] A. G. Fowler, S. J. Devitt, and L. C. L. Hollenberg, "Implementation of shor's algorithm on a linear nearest neighbour qubit array," 2004. [Online]. Available: https://arxiv.org/abs/quant-ph/0402196

[12] A. Pavlidis and D. Gizopoulos, "Fast quantum modular exponentiation architecture for shor's factorization algorithm," 2012. [Online]. Available: https://arxiv.org/abs/1207.0511

[13] P. Atchade-Adelomou and S. Gonzalez, "Efficient quantum modular arithmetics for the isq era," 2023. [Online]. Available: https://arxiv.org/abs/2311.08555

[14] Y. Yuan, C. Wang, B. Wang, Z.-Y. Chen, M.-H. Dou, Y.-C. Wu, and G.-P. Guo, "An improved qft-based quantum comparator and extended modular arithmetic using one ancilla qubit," *New Journal of Physics*, vol. 25, no. 10, p. 103011, Oct. 2023. [Online]. Available: http://dx.doi.org/10.1088/1367-2630/acfd52

[15] Y. Nam, Y. Su, and D. Maslov, "Approximate quantum fourier transform with o(n log(n)) t gates," *npj Quantum Information*, vol. 6, no. 1, Mar. 2020. [Online]. Available: http://dx.doi.org/10.1038/s41534-020-0257-5

[16] J. Earle, "Latched carry-save adder," *IBM Technical Disclosure Bulletin*, vol. 7, no. 10, pp. 909–910, 1965.

[17] S. Wang and A. Chattopadhyay, "Reducing depth of quantum adder using ling structure," in *2023 IFIP/IEEE 31st International Conference on Very Large Scale Integration (VLSI-SoC)*. IEEE, 2023, pp. 1–6.

[18] S. Wang, A. Baksi, and A. Chattopadhyay, "A higher radix architecture for quantum carry-lookahead adder," *Scientific Reports*, vol. 13, no. 1, Sep. 2023. [Online]. Available: http://dx.doi.org/10.1038/s41598-023-41122-4

[19] S. Wang, E. Lim, and A. Chattopadhyay, "Boosting the efficiency of quantum divider through effective design space exploration," in *IEEE International Symposium on Circuits and Systems (ISCAS)*, 2024.

[20] S. Wang, X. Li, W. J. B. Lee, S. Deb, E. Lim, and A. Chattopadhyay, "A comprehensive study of quantum arithmetic circuits," *arXiv preprint arXiv:2406.03867*, 2024.

[21] S. Lim, H. Kim, K. Jang, S. Wang, A. Baksi, A. Chattopadhyay, and H. Seo, "Optimized quantum circuit implementation of payoff function," in *2023 IFIP/IEEE 31st International Conference on Very Large Scale Integration (VLSI-SoC)*. IEEE, 2023, pp. 1–6.

[22] W. Zi, S. Wang, H. Kim, X. Sun, A. Chattopadhyay, and P. Rebentrost, "Efficient quantum circuits for machine learning activation functions including constant t-depth relu," *arXiv preprint arXiv:2404.06059*, 2024.

[23] T. G. Draper, S. A. Kutin, E. M. Rains, and K. M. Svore, "A logarithmic-depth quantum carry-lookahead adder," *Quantum Info. Comput.*, vol. 6, no. 4, p. 351–369, jul 2006.

[24] S. Wang, S. Deb, A. Mondal, and A. Chattopadhyay, "Optimal toffoli-depth quantum adder," in *arXiv*, 2024. [Online]. Available: https://arxiv.org/abs/2405.02523

[25] G. S S, R. Kumar, D. Samiappan, B. K. Kaushik, and M. Haghparast, "T-count optimized wallace tree integer multiplier for quantum computing," *International Journal of Theoretical Physics*, vol. 60, pp. 1–13, 08 2021.

[26] Y. Takahashi, S. Tani, and N. Kunihiro, "Quantum addition circuits and unbounded fan-out," *Quantum Information and Computation*, vol. 10, 10 2009.

NEATRouter: A New Method for 2D Global Routing

Luis Enrique Murillo Vizcardo
PPGC-Instituto de Informática
Universidade Federal do Rio Grande do Sul
Porto Alegre, Brazil
luis.vizcardo@inf.ufrgs.br

Ricardo Reis
PPGC/PGMicro-Instituto de Informática
Universidade Federal do Rio Grande do Sul
Porto Alegre, Brazil
reis@inf.ufrgs.br

Abstract—**More and more components are integrated into a single chip. This advance makes difficult the task of placement and connecting these components, since more components means more connections to be performed. To address the place and route problem, there are automatic design techniques which allow to find near optimal solutions. In this work, we present NEATRouter as an alternative algorithm to optimize 2D routing of 2-pin nets. Our method focuses on complementing or replacing the MazeRouter algorithm, which is used in several state-of-the-art global routing algorithms. NEATRouter uses the Neuroevolution of Augmenting Topologies (NEAT) algorithm to generate neural networks capable of finding the shortest and most resource-efficient path for a 2-pin net. The results of our experiments suggest that this method can successfully compete with traditional approaches such as MazeRouter, generating quality routing in terms of wirelength and congestion management.**

Index Terms—**VLSI, Global Routing, NEAT algorithm, Genetic Algorithms, Physical Design, EDA, Microelectronics**

I. Introduction

With the advancement of the semiconductor industry, the chip integration density has improved. More and more components can be integrated in a single chip. In terms of chip size and capacity, the routing problem should be able to address a circuit chip scale of tenths of thousands of large modules, and millions of small modules, and requires that routing should be completed within a reasonable time [2]. Innovation in EDA tools is fundamental. In the IC design flow, the routing task plays a crucial role, since it is responsible for establishing the connections between the pins of an integrated circuit. Routing is a complex task as integrated circuits are composed of millions of components in very small areas. For this reason, the routing stage is divided into two substages: global routing and detailed routing. The global routing stage is responsible for performing the initial routing of all the nets of an integrated circuit, reducing both wirelength and congestion. This first routing is known as guides, which then serve as input into the second stage. In detailed routing, guides are used as a guide to perform the final routing, observing the design rules. In this way, the definitive routing is obtained for each net of the integrated circuit.

We propose NEATRouter as a two-dimensional global routing algorithm for 2-pin nets, with the objective of generating high-quality routes in terms of wirelength and congestion.

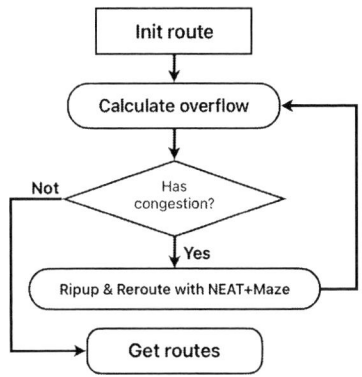

Fig. 1: NEATRouter algorithm workflow.

Our algorithm focuses on complementing or replacing the MazeRouter algorithm, which is used in several state-of-the-art global routing algorithms. This approach is based on the Neuroevolution of Augmenting Topologies (NEAT) algorithm [1], a variant of genetic algorithms designed to generate neural networks capable of optimizing specific problems. We adapt the NEAT algorithm to generate neural networks that optimize two-dimensional routing of simple 2-pin nets.

The NEATRouter workflow (Fig. 1) begins with an initial routing of all nets followed by an iterative process of rip-up and reroute which is used to eliminate congestion. The neural networks will determine the set of movements necessary for routing a net, with the goal of reducing congestion and wirelength throughout the routing area. In case a net cannot be routed, the MazeRouter algorithm will be used for its routing, thus ensuring the routing of all nets in the design. After eliminating congestion or after reaching a defined limit of rip-up and reroute iterations, the algorithm will return to 2D routing of the nets. In summary, our proposal combines the power of NEAT algorithm and MazeRouter algorithm to effectively address global routing of 2-pin nets, prioritizing routing quality and runtime efficiency.

For the experiments, we implemented NEATRouter in the Python3 programming language, and we used the available NEAT-python library [12]. The results of our experiments

979-8-3315-3968-9/24 $31.00 © 2024 IEEE

demonstrate that NEATRouter has a potential to improve resource utilization in the routing area. In some test cases, our method even improves the wirelength, although this improvement is related to a significant increase in runtime compared to the MazeRouter.

The rest of the paper is organized as follows. In Section II we provide essential information on the key concepts and methodologies relevant to our research. In Section III, we review some state-of-the-art algorithms for global circuit routing as well as a summary of works that use ML techniques to solve the global routing task. In section IV we detail the implementation of our NEATRouter algorithm. In Section V we present the experiments and results together with a comparison with the MazeRouter algorithm. Finally, section VI shows the conclusions and future work.

II. FUNDAMENTAL CONCEPTS

A. Global Routing

Routing is one of the top ten problems that the current physical design needs to solve [4]. In the global routing stage the objective is to route the pin nets using the smaller wirelength and to minimize the final slack. As the current IC industry continues to advance, global routing task is a very difficult one. One reason is that the nets are composed of many pins and also have shared routing resources, which they must negotiate to obtain the best solution and reduce the congestion. The quality of the global routing (GR) solution directly affects chip area, speed, manufacturability, power consumption and the number of iterations required to complete the design cycle. Hence this step plays an important role in determining circuit performance [5].

B. Maze Router

Maze routing introduced by Lee's [6], is a heuristic method which is responsible for finding the least expensive path between two positions. This method is commonly used in various global routing algorithms such as FastRouter [14], CUGR [15], among others. In addition, Johann and Reis [7] use the bidirectional A* algorithm which speeds up convergence and improves performance.

C. Genetic Algorithm

Genetic Algorithm (GA) is a programming technique based on the biological evolution [9]. The GA work by generating an initial population of mostly random values which will be their first candidates. In each generation the population is subjected to a process of selection, reproduction, crossing and mutation to create new candidates. The process is similar to natural selection, where the fittest candidates have a greater probability of surviving and transmitting their genes to subsequent generations.

D. Artificial Neural Networks (ANN)

Artificial neural networks are generalizations of mathematical models of biological nervous systems [10]. The ANN consists of a group of nodes called artificial neurons which are interconnected and organized into layers. These nodes are responsible for processing information by passing signals through connections from the input layer to the output layer.

E. NEAT algorithm

Neuroevolution of Augmenting Topologies (NEAT) [1] is a variation of GA which is used to generate ANN. Each individual in the population stores the information of an ANN, which includes a list of its nodes and their connections. NEAT operations not only allow the weights of the ANN to be optimized, but also allow its structure to be modified as necessary. The crossover and mutation operations are performed as follows:

- *Mutation:* It can change both connections weight and ANN structure. Connection mutation in which there is a probability of changing the weight of an existing connection. Structural mutation, which can be done in two ways, add a new connection between two disconnected nodes and add a new node, by disabling an existing connection, to connect its two nodes to the new node with two new connections.
- *Crossover:* Each connection of the ANN stores an innovation number which indicates at what time this connection was created. Then, the crossover between two ANN will be carried out by joining the connections with the same innovation number. If a connection does not exist in the other neural network, it will be inherited, if it belongs to the parent with more fitness.
- *Speciation method:* That helps protect the topology of the new generations by dividing them into groups with similar topologies and making them compete with each other. So, the generated topologies have the opportunity to innovate before competing with individuals from other groups.

The NEAT algorithm has proven to be a useful tool for the optimization of the ANN; its applications are found in fields such as robotics, video games and the optimization of control systems.

III. RELATED WORKS

In FastRouter [14], the authors proposed an algorithm that addresses global routing. Initially, a construction of congestion-driven and via-sensitive Steiner topologies is performed for each net in the design, followed by segment shifting techniques. Then, the tree structures are decomposed into 2-pin nets and pattern routing is executed using L and Z shapes to perform the first routing. Subsequently, the virtual capacity of the design is initiated based on the first routing. This will be useful to guide the iterative stage of rip-up and reroute and address the congestion problem. During the rip-up and reroute stage, it is used two techniques: 3-bend routing and multi-source multi-sink maze routing, with the aim of avoiding the generation of congestion and minimizing the use of vias. Finally, after obtaining a 2D routing, they converted it to a 3D routing using the spiral layer assignment algorithm. FastRouter managed to reduce the wirelength and runtime compared to

979-8-3315-3968-9/24 $31.00 © 2024 IEEE

the algorithms classified in the top positions of the ISPD08 contest [16]. They considered the possibility of improving the maze routing stage to achieve a balance between reducing congestion and maintaining a small wirelength.

CUGR [15], is an algorithm that solves the global routing problem directly in a 3D area. To achieve this, the authors proposed an algorithm that decomposes the routing problem into three phases: initial routing, 3D maze routing, and generation of route guides. They compared the results obtained with the two algorithms that occupied the first places in the ICCAD19 contest [17]. They managed to match the wirelength results and the via score of these algorithms, considerably improving the final runtime.

In NCTU-GR 2.0 [13], a heuristic maze-type routing algorithm is proposed that aims to improve the wirelength estimation, reduce the execution time and minimize the final wirelength. NCTU-GR 2.0 uses Rectilinear Steiner Minimum Tree (RSMT) to generate routing trees containing paths with short wirelengths. A multi-threaded version of the same collision-aware global routing algorithm is presented. It is observed that the algorithm manages to reduce both the wirelength and the execution time compared to other algorithms.

In [3], the author applied the NEAT algorithm to address the problem of nets routing in integrated circuits. In his research he adapted the NEAT algorithm to route 2-pin nets and compared his obtained results with industry standard algorithms. To carry out his experiments, he generated two-dimensional circuits represented by two-dimensional grids, where each position has a value of 0 if it is not routable and 1 if it is routable. The input values for the NEAT neural networks include all the values of the two-dimensional grid ($N \times N$), in addition to the position of each agent and the position of its destination point. The output nodes are defined in an array that represents the preferred movements that a net can make during its routing. In its evaluation method, he defined the fitness function to reduce the wirelength and the number of vias. He used the same function to compare his results with the results of the standard A* and BFS algorithms. The conclusion reached by the author is that the routing generated by the NEAT algorithm do not compete with the industry standard algorithms A* and BFS since in some test cases the final routing ends with disconnected nets or with detours that impact the wirelength. He also highlights that a better definition of the fitness function could equal or even improve net routing, since neural networks are able to take into account more parameters compared to standard algorithms.

In [11], the authors applied Deep Reinforcement Learning (DRL) to address the global nets routing problem in integrated circuits. They used a Deep Q-network (DQN) as their main Reinforcement Learning (RL) algorithm to route the nets simultaneously. They defined 12 variables as input to the Deep Q-network, the first 3 representing the current position of the agent in x, y and z coordinates. The next 3 variables reflect the distance from the current position to the target in the x, y and z directions. The latter variables encode the capability information of all edges that the agent can traverse in its next

step. The output of the network is encoded into 6 values, which represent the preferred direction for next movement during routing of a net. For the experiments, they generated test cases with various characteristics, which were routed with both their proposed RL algorithm and the standard A* algorithm. The results of both algorithms were evaluated based on the total congestion and the total wirelength of the routing. They concluded that the results produced by the DQN algorithm are superior to those obtained by the A* router, highlighting the importance of fine tuning the variables of the DQN algorithm.

IV. IMPLEMENTATION

In this section, we explain the configurations used for the training phase of the NEAT algorithm. We also explain the methodology used to apply the trained NEAT neural networks to the global routing problem. Our algorithm was implemented using the Python3 programming language. We used the NEAT-python library [12], which has an implementation of the NEAT algorithm.

A. NEAT Configuration

For the configuration of the NEAT algorithm, we use the ReLU function as the activation function for the NEAT neural networks. We set a probability of 50% to add a new connection and 40% to add a new node. Additionally, nodes in the initial population will be disconnected, allowing NEAT neural networks to form their own structures from scratch.

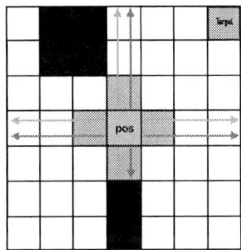

Fig. 2: The gray, green, blue, and black gcells represent, respectively, the source pin position, the destination pin position, the neighbors of the current position, and the gcells that have no available resources. The green arrows represent the number of gcells in the 4 directions, from which the minimum resource value will be taken, and the red arrows represent gcells in the 4 directions, from which the maximum historical value will be taken.

B. Input nodes

The values assigned to the input nodes are presented in a list that covers the minimum amount of resources available on the four sides (north, south, east and west) of the current position. In addition, the measurement of the distance from the four neighbors of the gcell in the current position to the desired goal is incorporated. This measurement is complemented with the maximum history value between the aforementioned neighbors of the gcell at the current position (Fig. 2). This inclusion of detailed information provides the algorithm with a comprehensive view of available resources and distance to the target, allowing for more informed decisions in the routing process.

C. Output nodes

During the routing process, output nodes play a critical role in generating a complete list of four values, each representing a possible movement, either north, south, east or west. The final choice of the next move is determined by identifying the maximum value in this list. By assigning numerical values to each possible move, the algorithm can effectively weigh its options and select the most promising direction, contributing to a more dynamic and adaptive routing mechanism.

D. Fitness function

The fitness function plays a crucial role in the evaluation of each individual, with the best of a generation being defined by the maximum fitness value achieved. To determine the fitness of a NEAT neural network, we carry out a comprehensive definition that involves the evaluation of specific rewards.

- The *connection reward* is used to score how close a net is to be connected, the reward is defined as follows:

$$rew_{con} = 1.0 - \frac{dist(pos, pos_{target})}{height + width} \quad (1)$$

where $dist$ is the Manhattan distance between two positions, pos is the current position of net, pos_{target} is the target pin position of the net, $height$ is the height of the grid and $width$ is the width of the grid.

- The wirelength reward is used to score the final routing wirelength of a net, the reward is calculated in relation to the Half Perimeter Wirelength value:

$$rew_{WL} = \frac{HPWL_{net}}{WL_{net}} \quad (2)$$

where WL_{net} is the wirelength of net route and $HPWL_{net}$ is the half perimeter wirelength of net.

- The history reward is used to score the total amount of history used to route a net. Then it is defined by the following equation:

$$rew_h = 1.0 - \frac{usedHistory_{net}}{totalHistory} \quad (3)$$

Where $usedHistory_{net}$ is the sum of the history used by the route of the net and $totalHistory$ is the total sum of the history in the grid.

- Finally, to calculate the total fitness we analyze two cases: when the routing fails to connect the net to its pin node, the total fitness is defined as:

$$Fitness_{total} = HW \times (\theta \times rew_{con}) \quad (4)$$

- If, on the other hand, the final routing manages to connect the net to its target pin, then the fitness of the individual is defined as follows:

$$Fitness_{total} = HW \times (\theta \times rew_{con} + \alpha \times rew_{WL} + \beta \times rew_h) \quad (5)$$

Where HW is defined as $height \times width$, rew_{con} is defined in equation 1, rew_{WL} is defined in equation 2, rew_h is defined in equation 3 and the values of θ, α, and β are coefficients that help balance the training of the NEAT neural network.

E. Global Routing with NEAT algorithm

The NEATRouter flow is shown in figure 1, where it begins with initial routing using the previously trained NEAT neural network. If a net cannot be routed, the MazeRouter algorithm is used as an alternative to ensure the routing of all the nets. After routing all the nets, a congestion calculation is performed which will detect the number of gcells that have excess of used resources. If there is congestion, then an iterative rip-up and reroute stage will be carried out to try to eliminate the overflow of the gcells. At the end of this process, the 2D routing of all the nets will be returned.

The figure 3 shows the steps of the rip-up and reroute stage.

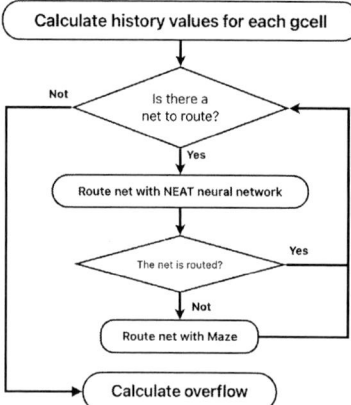

Fig. 3: Flow of the Rip-up and Reroute stage of NEATRouter algorithm.

Initially, it is calculated the history of each gcell, introducing an additional cost to gcells that experienced congestion in the last routing. This is intended to encourage the exploration of alternative routes by nets, ensuring that only those nets that need it use those specific gcells. Next, it is disconnect each net to reroute, thus freeing the used resources. The algorithm will route each net using first a NEAT neural network previously trained to handle 2-pin nets, obtaining the routing environment information as input data. In the case that the NEAT neural network is unable to route the net, the MazeRouter algorithm will be used to ensure valid routing. After all nets are routed, the algorithm will re-evaluate congestion across the entire grid. If any gcell shows congestion, the same steps will be repeated. The maximum number of iterations is set to 50, in case of not achieving congestion-free routing within this limit, the algorithm will provide congestion routing.

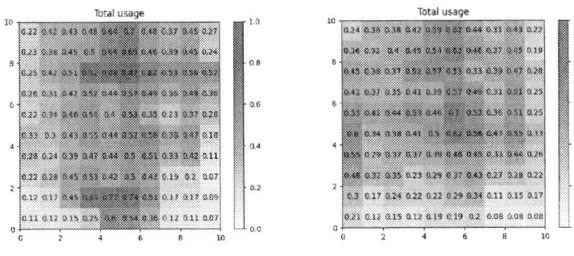

(a) MazeRouter congestion   (b) NEATRouter congestion

Fig. 4: Comparison of congested gcells using MazeRouter and NEATRouter for Test 5.

V. EXPERIMENTS AND RESULTS

A. Experiments

To test the algorithm, we generate test cases randomly. We try to ensure that routing each test with MazeRouter requires mandatory use of iterations to remove congestion. The routing area of all the generated tests is composed of 5 layers which will be 10×10 in size and will have alternating routing directions starting with the bottom layer with horizontal direction. During the training of the NEAT neural network, 1000 generations were carried out, and the best individuals from the 100th, 250th, 500th and 1000th generations were selected to carry out the experiments.

B. Results

To analyze our results, we perform a comparison of our NEATRouter algorithm with the MazeRouter algorithm. We select the final wirelength, and the total congestion after performing the routing in each of the generated tests. In addition, we also consider as secondary metrics the final runtime of each algorithm as well as the number of rip-up and reroute iterations that were necessary to remove congestion. Table I shows the results obtained for the test cases, the results show that the NEATRouter algorithm manages to obtain the same results as the MazeRouter algorithm, generating routes with the same wirelength in most of the tests. We also obtain positive results in the resource used metric (Avg. used), showing that NEATRouter manages to perform a more balanced routing without generating many congestion areas compared to the MazeRouter algorithm. Another interesting metric to analyze is the number of iterations necessary to route the nets without overflow gcells. MazeRouter, by using the NEAT neural networks that learned from experience to route nets, manages to converge on the desired result in fewer number of iterations, compared to just using the MazeRouter algorithm. In contrast, the results also show that MazeRouter has a better runtime compared to NEATRouter, this may be because it needs to obtain and process the input information in each movement, making the routing process slower.

To obtain a deeper understanding, we selected the test 5 to closely examine its results and deepen our understanding of the algorithm's behavior. The choice of this test is based on

NEATRouter's ability to more efficiently find an optimal route, without congestion, which significantly reduces the need to perform additional iterations. This case will give us a more detailed view of how the implementation of NEATRouter can positively impact routing effectiveness, allowing us to explore in depth the advantages it offers in terms of speed and efficiency. The Figure 4 shows the comparison of the congestion generated for Test 5 using MazeRouter and NEATRouter as can it be seen when applying NEATRouter to route some nets. It is possible to reduce the use of resources in some congestion areas, thus helping the routing of other nets being more efficient. To understand how the best NEAT neural networks

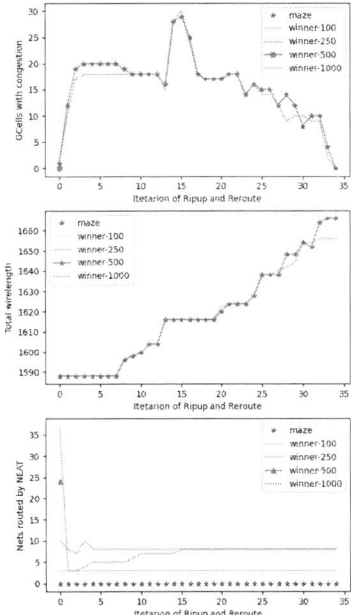

Fig. 5: Comparison of MazeRouter with the best NEAT neural networks generated in each generation for Test 5. The X-axis represents the number of rip-up and reroute iterations, while the Y-axis represents congestion, wirelength, and the number of nets routed by NEATRouter, respectively.

perform on the routing problem, we analyze the congestion, wirelength and number of successful nets per NEAT neural network vary in each iteration of rip-up and reroute. As figure 5 shows, the best individual of 100th generation is able to route around 3 nets in each iteration, matching the results of MazeRouter. The best 250th generation individual can route between 8-10 nets in each iteration, contributing to a more effective reduction of wirelength and congestion compared to MazeRouter. In 500th generation, the best individual manages to route 24 nets on the first attempt, addressing congested areas and eliminating the need for additional iterations. However, in 1000th generation, although this individual can route a greater number of nets, he worsens the routing of other nets, resulting in the need for additional iterations to address congestion.

TABLE I: Comparison of the results using MazeRouter algorithm and NEATRouter. The WL is the wirelength used to route all nets, # iter is the number of iterations that the algorithm performed to remove congestion, Avg. used is the average used resources of a gcell and runtime is the time that takes the algorithm to do the total routing.

Tests	# Nets	MazeRouter				NEATRouter			
		# iter	WL	Avg. used	Runtime(s)	# iter	WL	Avg. used	Runtime(s)
Test 1	200	19	1519	57.4	**0.88**	19	1519	57.4	9.595
Test 2	200	5	1581	**58.9**	0.428	3	1581	61.5	2.067
Test 3	200	5	**1615**	61.7	0.444	1	1617	59.7	0.995
Test 4	200	2	1551	56.5	0.312	1	1551	55.3	0.523
Test 5	200	34	1666	61.4	1.574	0	**1588**	**56.2**	**0.559**
Test 6	500	**8**	3771	61.9	**1.035**	11	3771	**59.6**	8.197
Test 7	500	8	**3764**	60.0	**1.01**	8	3766	**59.8**	14.343
Test 8	500	3	3886	**61.7**	**0.56**	2	3886	61.8	1.67
Test 9	500	12	3756	62.8	**1.342**	10	3756	62.4	17.568
Test 10	500	3	**3879**	60.3	**0.538**	0	3881	**57.4**	1.776
Test 11	1000	3	7730	62.5	**0.901**	1	7730	**59.9**	2.241

VI. CONCLUSIONS

We present NEATRouter as a new method to perform two-dimensional global routing of 2-pin nets. NEATRouter is implemented using the NEAT algorithm, which is a genetic algorithm used to optimize neural networks to perform a specific task. The results show that the proposed method is efficient, facilitating the search for the shortest paths for a net, which impacts the final wirelength and reduces the use of resources in the routing area. We performed a series of experiments to evaluate our method in different scenarios and compare it with another method called MazeRouter, which is a reference considering several global routing algorithms and can be complemented/replaced by our method. Based on the results, we can conclude that our method manages to find more optimal paths, reducing the number of iterations necessary to eliminate congestion. Furthermore, we observe that the resource distribution across the entire circuit tends to be more balanced compared to the MazeRouter. As the execution time of the algorithm can become high with a large number of nets, our proposed algorithm previously requires the use of an algorithm that divides the total nets into sets. These sets will be routed by NEATRouter and then subjected to global routing using another algorithm. As routing is a major problem, any gain we can obtain is important, the congestion reduction is a main contribution of our proposal. As future work, our study suggests that the use of neural networks can improve the quality of routing, so the application of machine learning (ML) techniques have significant potential to achieve better results. For example, the Reinforcement Learning (RL) technique, which learns by reward stimulation to perform various tasks, could be useful for deciding movements during global routing. Another interesting point to investigate is the appropriate choice of input information for neural networks, as this can lead to better movement decisions.

VII. ACKNOWLEDGMENTS

This work was financed in part by CAPES, Brazil- Finance Code 001, CNPq, and FAPERGS.

REFERENCES

[1] Kenneth O. Stanley and R. Miikkulainen, "Efficient Evolution of Neural Network Topologies", Proceedings of the 2002 Congress on Evolutionary Computation (CEC '02), IEEE, 2002. DOI: 10.1109/CEC.2002.1004508.

[2] X. Chen, G. Liu, N. Xiong, Y. Su, and G. Chen "A Survey of Swarm Intelligence Techniques in VLSI Routing Problems". IEEE Access (Volume: 8), 2020. DOI: 10.1109/ACCESS.2020.2971574.

[3] J. Rinnarv, P. Brink. "Machine learning - neuroevolution for designing chip circuits/pathfinding". Master of Science in Engineering - Computer Science and Technology, URN: urn:nbn:se:kth:diva-210178, 2017.

[4] J. Parkhurst, N. Sherwani, S. Maturi, D. Ahrams, and E. Chiprout, "SRC physical design top ten problem", Proc. Int. Symp. Phys. Design (ISPD), 1999. DOI: 10.1145/299996.300022.

[5] H. Tang, G. Liu, X. Chen and N. Xiong, "A Survey on Steiner Tree Construction and Global Routing for VLSI Design", IEEE Access (Volume: 8), 2020. DOI:10.1109/ACCESS.2020.2986138.

[6] C. Y. Lee, "An algorithm for path connections and its applications," IEEE Trans. Electron. Comput., vol. EC-10, no. 3, pp. 346–365, Sep. 1961. DOI: 10.1109/TEC.1961.5219222.

[7] M. Johann and R. Reis, "Net by net routing with a new path search algorithm", Proc. 13th Symp. Integr. Circuits Syst. Design, 2000, pp. 144–149. DOI: 10.1109/SBCCI.2000.876022.

[8] Jordan, M. I.; Mitchell, T. M. "Machine learning: Trends, perspectives, and prospects". Science Magazine, v. 349, p. 255–260, July 2015. DOI: 10.1126/science.aaa8415.

[9] Mitchell, Melanie. "An Introduction to Genetic Algorithms". MIT Press, 1996. DOI: 10.7551/mitpress/3927.001.0001.

[10] Abraham, "A. Artificial neural networks". Handbook of Measuring System Design. John Wiley and Sons, Ltd, 2005. chp. 129. ISBN: 9780471497394.

[11] Liao, H. et al. "A deep Reinforcement Learning approach for Global Routing". Jornal of Mechanic Design, v. 142, 2019. DOI: 10.1115/1.4045044.

[12] A. McIntyre, M. Kallada, C. G. Miguel and C. F. da Silva, "NEAT - python", https://github.com/CodeReclaimers/neat-python.

[13] Liu, W. et al. "Nctu-gr 2.0: Multithreaded collision-aware global routing with bounded-length maze routing". IEEE Transactions on Computer-Aided Design of Integrated Circuits and Systems, IEEE, v. 32, 2013. DOI: 10.1109/TCAD.2012.2235124.

[14] Pan, M. et al. "FastRoute: An efficient and high-quality global router". Hindawi Publishing Corporation, 2012. DOI: 10.1155/2012/608362.

[15] Liu, J. et al. "CUGR: Detailed-Routability-Driven 3D Global Routing with Probabilistic Resource Model". ACM/IEEE Design Automation Conference (DAC), ACM/IEEE, 2020. DOI: 10.1109/DAC18072.2020.9218646.

[16] Nam, G.-J.; Sze, C.; Yildiz, M. "The ispd global routing benchmark suite". In: Proceedings of the 2008 International Symposium on Physical Design. New York, NY, USA: Association for Computing Machinery, 2008. (ISPD '08), p. 156–159. DOI: 10.1145/1353629.1353663.

[17] Schlichtmann, U. et al. "Overview of 2019 cad contest at iccad", IEEE/ACM International Conference on Computer-Aided Design (IC-CAD). 2019. p. 1–2. DOI: 10.1109/ICCAD45719.2019.8942133.

Adaptable FWHW Formal Co-Verification of SoC RISC-V Components

Paulette Iskandar*†, Bryan Olmos*†, Wolfgang Kunz†, Djones Lettnin*

*Infineon Technologies AG †Rheinland-Pfälzische Technische Universität Kaiserslautern-Landau

Abstract—**The increasing shift towards the RISC-V open-source instruction set architecture requires the development of new design techniques. In recent years, it has been demonstrated that RISC-V designs can be generated in a modular and scalable manner by utilizing metamodeling techniques. However, verifying these designs presents a significant challenge, because the verification must consider both Register Transfer Level (RTL) components and firmware components such as drivers. Furthermore, the interaction between firmware and hardware components is susceptible to various issues, including incorrect transaction sequences, synchronization problems, encoding mismatches, and reserved values. Traditionally, verifying the interaction between hardware and firmware requires simulation/emulation tools and verification engineers with expertise in both firmware and hardware. To overcome these challenges, this paper introduces an automated formal verification approach for FWHW Co-verification of peripherals such as timers and interrupt controllers, and their respective drivers in generated RISC-V designs. This verification process employs formal verification methods. This methodology enables the detection of bugs in both hardware and firmware because it consists of the verification of individual components that can be reused later in the integration process. By implementing this methodology in the early design stages, developers can identify and address potential issues more efficiently and avoid later corrections.**

Index Terms—**Formal Verification, Firmware Verification, Co-verification, SoC Verification**

I. INTRODUCTION

The use of RISC-V designs has increased during the last few years, especially in fields like automotive. This is due to their open-source nature and removal of proprietary constraints [1]. Previous studies have demonstrated that the system-on-chip components for RISC-V cores can be generated in seconds or minutes using methodologies like Model-Driven Architecture (MDA) [2]. This generation includes peripheral components such as timers or interrupt controllers (IC), which are instantiated and handled using drivers written in C or C++. However, verifying the interaction between the peripheral and the software components is challenging because the analysis and simulations must consider all the components of the System-on-Chip and all their possible values and states. One alternative is formal verification, which has been well-accepted in the industry for verifying HW designs. However, the use of formal methods for the verification of FW and HW interactions is not common. This is mainly due to four reasons. First, formal verification is not friendly for all designs because it is limited by the state explosion problem and designs such as FPUs, GPUs, and complex bus protocols require expert verification engineers [3]. Thus, the problem becomes more complex if additional elements are added to the verification environment. Second, most of the approaches require an intermediate model conversion of the Hardware or Software designs to allow a common syntax. These intermediate models are usually verified through a model checker [4]. However, due to these intermediate conversions, these approaches are difficult to debug and identify whether the problem comes from the FW or the HW. Third, the difference in the inherent nature of FW and HW makes it difficult to have a cycle-accurate verification —the HW is synchronous and the FW is asynchronous. Fourth, it is not clear what properties need to be verified, how they should be written or in which language. Only in [5], a general classification into transaction-level properties and component-level properties is proposed.

To tackle these challenges and improve the verification of our in-house RISC-V designs, we present an automated formal verification setup for the FWHW Formal Co-verification of SoC RISC-V components. These hardware components were verified beforehand. However, it was necessary to assume some constraints according to their current state, for example, it was assumed that they were initialized with the right sequence of transactions and were in a reachable state. In our approach, we trigger the sequence of transactions instead of assuming the state.

The verification setup in this paper is designed for the verification tool OneSpin [6], which supports model generation from components in Hardware Description Languages (HDL) such as SystemC and SystemVerilog in its front end. However, the methodology can be implemented for other tools that support multiple HDL or the direct use of C++ and it can also be applied in simulation. In this work, we analyze the components using formal verification. Additionally, we focus on transaction-level properties written in SystemVerilog Assertions (SVA) and propose their classification into 5 types depending on the interaction of the FW and HW with the HAL (Hardware Abstraction Layer) and GPIO (General Purpose Input/Output), as shown in section IV-B. Some of these types can be generated based on the design, as an example in this work, we generate all the type IV properties following the Model Driven Architecture (see: subsection II-A).

To prove our setup, we verified the designs of our in-house RISC-V code generator, which was previously implemented. This generator is capable of producing various types of cores with different capabilities, such as customized RISC-V Core Local Interrupt Controllers (CLIC) or timer counter peripherals, depending on the implementation and the targeted

979-8-3315-3968-9/24 $31.00 © 2024 IEEE

application. Additionally, the drivers and the firmware code to access the hardware layer are also generated.

The main contributions of this paper are:

- Proposing an innovative methodology for FWHW Co-verification, which can be used for simulation or formal verification.
- Automation of the verification for FWHW Co-verification.
- Generation and formal verification of properties for FWHW Co-verification of SoC RISC-V components. The results show that the verification elements can be generated in seconds based on the specification, for example, a RISC-V Timer including 531 lines of C code and 11560 lines of RTL code needs 592 lines of additional code for the application of the methodology, which we were able to generate. With this approach, 141 properties were verified in 7,5 s and 4 bugs were found related to the priorities between the processor and the peripheral.

This paper is organized as follows: In Section II, the main concepts of MDA, the features of the tool and the related work are discussed. Section III introduces the methodology. The automatic setup and generation of properties are presented in Section IV. Section V shows the evaluation and results. Conclusions and future work are presented in Section VI.

II. BACKGROUND

A. Model Driven Architecture

Model-Driven Architecture (MDA) is a software development framework that utilizes models to describe a system at various levels of abstraction [7]. It is based on models with independent implementation details. In the case of hardware description, the authors of [8] propose the use of three models: Model of Things (MoT), Model of Properties (MoP) and Model of View (MoV). MoT is the model containing the specification, for example, the information about the drivers and the peripherals under verification; MoP contains the property classes, for example, the 5 types of properties of subsection IV-B can be defined here; MoV contains the final format of these properties depending of the target application and platform, in this work the properties are generated for SVA and following the rules of the methodology.

B. Tool and SystemC front end

The tool OneSpin has been selected to verify the proposed methodology using assertion-based verification (ABV) due to its capability to include SystemC designs. Both C language assertions and SVA are accommodated, allowing the utilization of well-established specifications for assertions. Additionally, it facilitates the integration of C/C++ code with RTL, hence, enabling transactional flow between SW and HW. Some additional beneficial features include formal AutoChecks e.g. initialization checks, truncation checks, and toggle checks [6].

C. Related Work

In [4], the HW-CBMC tool translates Verilog/VHDL designs into an intermediate model for analyzing each clock cycle by instrumenting the C code with its own syntax. Our approach directly works with Verilog/VHDL and SystemC/C designs, utilizing SVA properties to simplify setup and debugging. In [9], the tool CoVer translates embedded software and hardware into a C model. In our approach, we do not translate the RTL design. Firmware and hardware interaction is tested in [10] using simulation for concurrent designs using concolic-generated patterns. Our approach focuses on formal verification considering the concurrency of the FW and the hardware. In [11], a formal HW/SW co-verification method is introduced, which includes the processor in the design under verification (DUV) and requires program compilation before analysis. We directly work with C code, allowing verification of individual drivers and peripherals without the need for a processor. Additionally, other approaches involve converting assembler code into a computational model known as program netlist (PN), as discussed in various works [12]–[14].

III. METHODOLOGY

A. Firmware and Hardware Connection

One of the goals of this paper is the verification of SoC RISC-V components which contain thousands of lines of code. However, the explanation of the methodology in these examples requires many details. For this reason, we illustrate the methodology with a small example. However, it is important to remark that the same signals and signal connections can be used for other designs. Let's consider a C driver and an RTL design implementing a 32-bit addition, where the driver needs to set up the values, and then the RTL module computes the operation and returns the result to the driver. As shown in Figure 1, the DUT (Design Under Test) contains the following main files:

1) driver.c: FW driver which sets up the values for the addition and reads the results. It contains 2 C functions *adder_start(op1,op2)* and *adder_get()*; the first function triggers the calculation of the operands *op1* and *op2*; the second function reads the calculated value.
2) hal.h: defines the relative addresses of the hardware abstraction layer (HAL)
3) adder.vhd: RTL implementation of the adder.

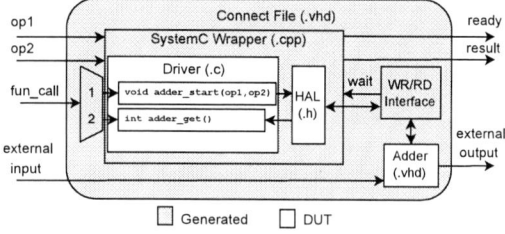

Fig. 1. Example of Top Module for FWHW Co-verification

The verification of this DUT faces 2 main challenges. The first challenge is related to the integration of a C code and the hardware description languages (HDL), particularly in managing timing dependencies —C code typically does not have explicit timing dependencies and focuses on the functionality of the program and HDL work at the level of individual clock cycles. The second challenge is related to the CPU architecture. In a normal situation, the software execution depends on clock speed, instruction set, data size, execution pipelines, cache and memory hierarchy, parallelism, branch prediction, microarchitecture, overhead and latency [15]. Additionally, the adder will return the result in a specific clock cycle —and it depends on the number of bits and the adder design such as Ripple Carry Adder (RCA) or Carry Lookahead Adder (CLA). For this reason, it is needed to abstract all the intermediate operations of the CPU. It is important to remark that in this paper we focused on the application of the methodology to the peripherals of our in-house RISC-V generator. However, since the methodology is abstracting the CPU architecture, it can enable a hierarchical verification of Embedded Systems independent of the CPU architecture.

This abstraction is done by adding an element called "Wrapper" implemented in SystemC, as shown in Figure 1. This wrapper is the interface between the C code and the RTL design, for example, it calls the C functions and waits until the completion of them. In this example, this is done by the declaration of two SystemC functions. These functions are *read_reg_fun* and *write_reg_fun*, which read and write the values of the registers abstracting the bus interface of the processor. In a SoC, this bus interface is more complex, e.g. ARM's Advanced Microcontroller Bus Architecture (AMBA) specification includes features for address mapping and translation [16]. In this paper, the translation is done by the so-called "WR/RD interface". Moreover, all elements are mapped to the element "Connect File", which is the top module designed in HDL. As shown in Figure 1, the main signals of the Connect File are:

1) *op1* and *op2*: operands for the addition.
2) *fun_call*: the function *adder_start(op1,op2)* is called when the input signal is 1 and receives the operands to trigger the calculation. If the input signal is 2, it invokes *adder_get()* to read the result.
3) *result*: output signal returning the transaction result.
4) *ready*: output signal, it indicates that the result was read successfully.
5) *wait*: internal signal waiting for the result of the adder.
6) *external input/output*: signal from other HW modules or GPIO.

Note that the selected C function depends on the value of the input signal *fun_call*. In the RISC-V module designs of the next section, all the functions in the driver file are read and a value is assigned automatically to each function in the driver (See: Subsection IV-A). Afterwards, this value is associated with the memory addresses of the hardware registers within the memory-mapped I/O space defined in the HAL file (hal.h). This is done for the automation of the generation of properties.

B. Timing Behavior

The methodology is based on triggering read-and-write transactions using *fun_call*, which then calls the driver functions. The waveform for the adder example is shown in Figure 2. In the clock cycle t, the function *adder_start(op1,op2)* is called with the operands values 0x98 and 0x78 by setting $fun\_call = 1$. In this example, the driver function multiplexes the 16-bit values of op1 and op2 into a 32-bit value, but it can also be a single transaction for each value. In the clock cycle $t + 1$, the wait signal is activated until the adder ends its calculation in the clock cycle $t + n$. In the next clock cycle, the function *adder_get()* is called by setting $fun\_call = 2$ and the result is displayed in the clock cycle $t + n + 2$. Note that the C functions are executed in one clock cycle, i.e., the tool considers them as a single-cycle combinatorial implementation.

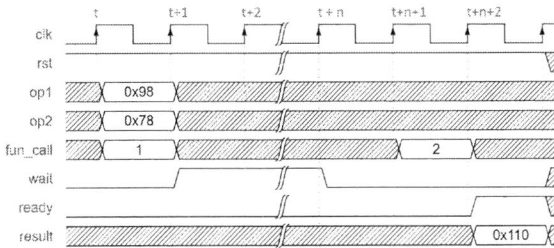

Fig. 2. Example of Timing Diagram

IV. AUTOMATIC SETUP AND GENERATION OF PROPERTIES

The previous methodology can be set up manually. However, it requires a lot of effort if we consider hundreds of driver functions and hundreds of files generated by our in-house RISC-V generator. For this reason, the next following steps are followed for the automation of the approach.

A. Initial Setup Generation

Figure 3 shows the procedure to generate the verification elements such as *connect.sv*, *wrapper.cpp* and *wrapper.h*. The first script will extract the name of the HW entity and the input and output port names. This would result in a nested dictionary in a .inc format. The second script parses the *driver.c* file. This is performed using the Ctags Python library, which returns all function declarations in the driver. The script would then adjust this data to be in a dictionary.inc format. The information in this dictionary is necessary to generate the SystemC function calls. The third script parses the register interface's header file. This file contains the address offsets for each HW register and their corresponding bitfields.

B. Classification of SVA Properties

The setup and connection (explained in Section III-A) can be used for simulation and formal verification tools that support SystemC and other HDL languages, such as

979-8-3315-3968-9/24 $31.00 © 2024 IEEE

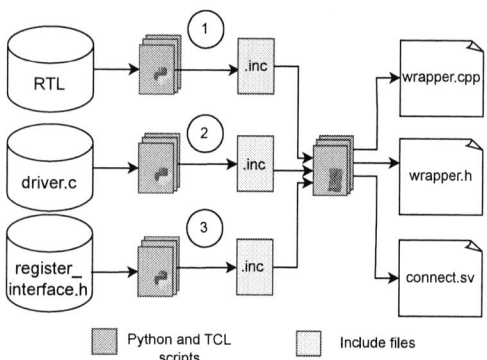

Fig. 3. Generation of verification elements

VHDL. The proposed methodology was performed to verify SVA properties using formal verification. In this paper, these properties are classified into five main categories depending on their interaction with the registers:

1) *Type I* (Register modified by FW/HW interaction): This property verifies that after the FW triggers a series of transactions requesting an operation of the HW, the HW executes the operation and returns the right value. In the adder example, the FW sends the operands *op_1* and *op_2* by triggering *fun_call = 1*, then the HW computes the value and the result is read by triggering *fun_call = 2*. It can be verified using the property adder_result of Listing 1.

2) *Type II* (Register modified by the interaction of an external HW value): This property verifies that HW gets the right value of an external signal such as GPIO or other module connected to the DUT and that it reflects changes to the registers than can be monitored through the FW. As shown in the property ExtRes_ACTVAL of Listing 2 of the Timer; in clock-cycle (CC) 1, an FW transaction is initiated with *fun_call == 7'd31* to read the value of the MAXVAL register. Two cycles later, *fun_result* is assumed to not be zero, indicating that the MAXVAL has a positive value to start counting from. In CC 3, the Timer's configuration register is written by triggering the FW with *fun_call == 7'd30*. The configuration value is provided with *default_interface_data_in == 4'b0011*. Here, the Timer is configured to read the external GPIO on the rising edge, as the edge sensitivity is configurable in this design. In CC 4, the *tim0_extres* input is assumed to be 1, meaning that the external GPIO signal is asserted from the processor. When the external Timer reset is activated, the ACTVAL register should be reset, in this case copying the MAXVAL content to itself. In CC 5, another FW transaction is invoked with *fun_call == 7'd1* to read the ACTVAL register. Two cycles later, the ACTVAL register value should be equal to that of the MAXVAL register, previously read by the FW.

3) *Type III* (HW external output signal modified by FW transaction): For example, the property OvfInt_output

of Listing 3. This property verifies that if the ACTVAL reaches 0 and the overflow feature is enabled by setting the correct bitfield in the configuration register and there is not a timer overflow reset (*tim0_ovfintres*), then the *tim0_ovfint* output will be set.

4) *Type IV* (Transaction check): a simple property to check that the FW writes the correct value into the registers and bitfields of the peripheral, as shown in the property write_ACTVAL_register_FW of Listing 4. This property is useful for the teams working in the HAL.

5) *Type V* (Interface Properties): Properties related to the WR/RD interface verification.

```
property adder_result;
    (!wait && fun_call == 1) ##3 wait
    ##1 (!wait && fun_call == 2)
    |-> ##1 result == $past(op_1,6)
        + $past(op_2,6) && ready;
endproperty
```
Listing 1. SVA Property: adder_result

```
property ExtRes_ACTVAL;
    !tim0_extres
    //Read MAXVAL
    ##1 fun_ready && fun_call == 7'd31
    ##1 !fun_ready ##1 fun_result != 32'b0
    && !tim0_extres
    //Configure timer
    && fun_ready && fun_call == 7'd30
    && default_interface_data_in == 4'b0011
    ##1 tim0_extres
    //Read ACTVAL
    ##1 fun_ready && fun_call == 7'd1
    |-> ##1 read_s ##1 fun_result ==
        $past(fun_result,4);
endproperty
```
Listing 2. SVA Property: ExtRes_ACTVAL

```
property OvfInt_output;
    //Check if MAXVAL is not zero
    ##1 fun_ready && fun_call == 7'd31
    ##1 !fun_ready ##1 fun_result != 32'b0
    //Configure timer
    && default_interface_data_in == 32'
        b10001111 && fun_call == 7'd30
    //Assume ACTVAL is zero
    ##2 fun_call == 7'd1 ##1 !tim0_ovfintres
    ##1 fun_result == 32'b0
    |-> tim0_ovfint;
endproperty
```
Listing 3. SVA Property: OvfInt_output

```
property write_ACTVAL_register_FW;
    fun_call = =3 && FW_value != 0 #3 fun_call
        == 4
    |-> ##2 result == $past(FW_value,5);
endproperty
```
Listing 4. SVA Property: write_ACTVAL_register_FW

C. Generation of Properties

Type IV properties were generated using the Model Driven Architecture. The property generation flow is depicted in

Figure 4. The other properties can be generated in the same way following three steps. First, the information of XML specification and *fun_calls* of the Wrapper are extracted automatically and associated with the Metamodel of Figure 5 —MoT Layer. This generates a MoT instance that is the input for the MoP Layer. In this layer, the properties are generated based on Python Classes. Third, the properties are generated in the final syntax in this case SVA —they can also be generated for ITL (Interval Language) —OneSpin's proprietary assertion language.

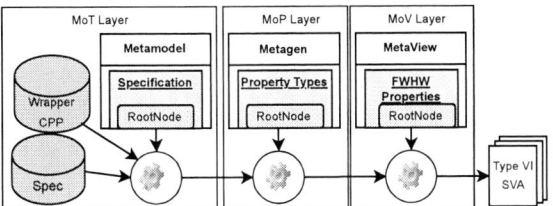

Fig. 4. MDA - Property Generation Flow

Figure 5 shows the metamodel used for the automation. The main elements are: *Component*: contains the information about the DUV such as code location. *Register HAL*: contains the name of the register, the address in the HAL and the width of the register e.g. 32 bit. *Bitfield*: contains information about the bitfields of the registers.

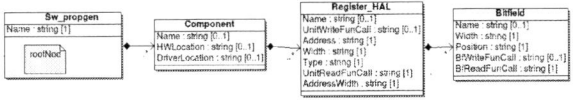

Fig. 5. Simplified Metamodel

V. EVALUATION AND RESULTS

The proposed methodology was applied for the verification of the next components.

A. Timer Controller for RISC-V SoC

A general-purpose 32-bit Timer/Counter peripheral module that can decrease or increase a counter register on different input signal events and provide an output signal for other devices. It has the capacity to configure different values from the FW side using around 100 C functions. It is connected to a Control and Status Register (CSR), where the values such as the actual value (ACTVAL) of the timer can be monitored. The main requirements are: 1) The timer counts down starting from the maximum value (MAXVAL) if the timer is enabled, and the reset input is disabled. 2) The timer can be reset from the GPIO module, that is, the MAXVAL is copied into the ACTVAL. 3) The ACTVAL can be configured by the FW or the HW. The FW should have a higher priority over the HW.

The first two requirements can be verified using a hardware-only approach such as FPV(Formal Property Verification). However, they also can be verified by reading the ACTVAL

from the CSR using a property Type I, for example, it is possible to verify that the FW restarts the timer through the CSR and after the reset the ACTVAL matches the MAXVAL. The third requirement demands a FW/HW Co-verification approach. A bug was detected related to the FW priority, as shown in Figure 6. In the clock cycle t, the function to modify the ACTVAL is called by setting $fun\_call == 3$, and the input value for the function is $0x40$. In the clock cycle $t+2$, it is expected that ACTVAL takes the value of the CSR modified by the FW interaction. However, it ignores the FW request. In the clock cycle $t+3$, the function to read the ACTVAL is called by setting $fun\_call == 4$. In the clock cycle $t+5$, the result is $0x7C$ instead of the expected value $0x40$. After debugging, this error was due to a corner case in the priority block during the FW transactions.

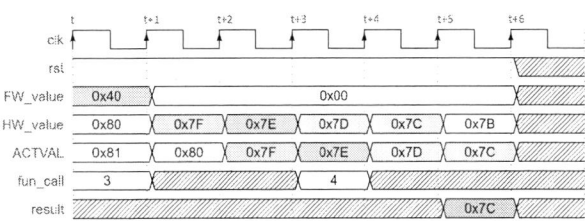

Fig. 6. Counterexample of FW priority - Property Type I

As shown in Table I, 592 lines of code were generated for the automation of the methodology. This was done in a few seconds. Additionally, 42 Type IV properties were generated. The verification runtime is 7,5s (see: Table II) and 4 properties failed exposing a bug in the design (FWHW priority bug). The rest of the properties hold (full proof).

TABLE I
GENERATED LINES OF CODE (LoC)

RISC-V Peripheral	Design (LoC)		Generated Elements (LoC)		
	Driver	RTL	Wrapper	Connect File	WR/RD Interface
Timer	531	11560	416	135	41
Interrupt Controller	657	3490	293	119	41

TABLE II
VERIFICATION RESULTS

Verification Results	Timer			Interrupt Controller		
Property	Number	Time(s)	Fail	Number	Time(s)	Fail
Type 1	6	2	0	3	5	0
Type 2	56	1,5	0	2	10	0
Type 3	36	2	0	3	2,3	0
Type 4	42	1	4	17	3,6	0
Type 5	1	1	0	1	1	0
Total	141	7,5	4	26	21,9	0

B. Interrupt Controller for RISC-V SoC

The interrupt controller is a peripheral that manages the various interrupt sources in a system and ensures that they are properly handled by the CPU. The overview of the verified

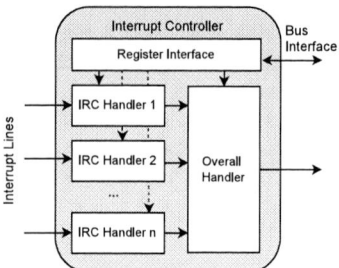

Fig. 7. Overview of Interrupt Controller

RTL design is shown in Figure 7. It consists of three main blocks; first, the IRC Handler is responsible for pre-processing the incoming interrupts. Second; the CSR is the interface between the FW and HW. It provides a structured way to configure signals of the peripheral. The FW can activate or deactivate the IRC handler and modify its features, for example, specify if an interrupt is masked/unmasked or set its priority [17]. The HW defines the states of the IRC Handler such as pending, active, requested or paused. Third, the Overall Interrupt Handler is mainly responsible for managing the priority of the IRC handlers. The main requirements related to the current methodology are:

1) The FW enables the interrupts by setting the CSR to 1.
2) In the case of Dynamic Priority, the FW sets the priorities of the interrupt sources. If not, the HW sets the priorities according to the default values specified in the model of design.
3) The FW unmasks the incoming interrupt sources, if the maskable attribute is set to true in the model of design.
4) IRC handler receives an interrupt request (IRQ).
5) If there are no other pending interrupt requests from other sources, the intr_CPU output is asserted indicating that an IRQ is ready to be serviced.

As shown in Table I, a total of 453 lines of code were generated for the automation of the methodology. Additionally, 17 properties Type IV properties were generated. The verification runtime is 21,9s (see: Table II). All properties hold (full proof).

VI. CONCLUSION

We provided an automated methodology for the formal FWHW Co-Verification of our in-house RISC-V generator and showed how properties can be generated using the MDA. 167 properties were verified in 29,4s, from them, 4 properties failed. It helped to identify a new bug due to the interaction of FW and HW components. Due to the automation, the methodology can be easily adapted for the verification of other timers and ICs for different RISC-V designs increasing the reliability of our products. Moreover, we presented a use case for our in-house RISC-V generator, however, we found a way to abstract the processor enabling a modular verification of Embedded Systems. Finally, it was developed mainly using Python, so the presented approach can be reused by other companies or institutions.

ACKNOWLEDGMENT

This work has been developed in the project VE-VIDES (project label 16ME0243K) which is partly funded within the Research Programme ICT 2020 by the German Federal Ministry of Education and Research (BMBF). Furthermore, we thank Jörg Bormann of Siemens EDA for his support in setting up the initial methodology in OneSpin. Finally, we thank our colleagues Ares Tahiraga, Paritosh Sinha, Johannes Grinschgl and Shuhang Zhang for their feedback and support.

REFERENCES

[1] "RISC-V," https://riscv.org/, accessed: 26-04-2024.
[2] A. Tahiraga, "Implementation of an RTL Generator for RISC-V Privileged Architecture," Master's thesis, Technical University of Kaiserslautern, 2023.
[3] K. Devarajegowda, L. Servadei, Z. Han, M. Werner, and W. Ecker, "Formal Verification Methodology in an Industrial Setup," in *2019 22nd Euromicro Conference on Digital System Design (DSD)*, 2019, pp. 610–614.
[4] D.-A. Lee, J.-H. Lee, and J. Yoo, "Verification Process of Behavioral Consistency between Design and Implementation programs of pSET using HW-CBMC," 2011.
[5] R. Mukherjee, M. Purandare, R. Polig, and D. Kroening, "Formal Techniques for Effective Co-verification of Hardware/Software Co-designs," in *Proceedings of the 54th Annual Design Automation Conference 2017*. Austin TX USA: ACM, Jun. 2017, p. 1–6. [Online]. Available: https://dl.acm.org/doi/10.1145/3061639.3062253
[6] "OneSpin Solutions," https://www.onespin.com/, accessed: 26-04-2024.
[7] S. J. Mellor, K. V. Scott, A. Uhl, and D. Weise, "Model-Driven Architecture," in *OOIS Workshops*, 2002. [Online]. Available: https://api.semanticscholar.org/CorpusID:14083108
[8] W. Ecker and J. Schreiner, "Introducing Model-of-Things (MoT) and Model-of-Design (MoD) for simpler and more efficient hardware generators," in *2016 IFIP/IEEE International Conference on Very Large Scale Integration (VLSI-SoC)*, 2016, pp. 1–6.
[9] K. Liu, W. Kong, G. Hou, and A. Fukuda, "A Survey of Formal Techniques for Hardware/Software Co-verification," in *2018 7th International Congress on Advanced Applied Informatics (IIAI-AAI)*. Yonago, Japan: IEEE, Jul. 2018, p. 125–128. [Online]. Available: https://ieeexplore.ieee.org/document/8693095/
[10] S. Ahn and S. Malik, "Automated firmware testing using firmware-hardware interaction patterns," in *2014 International Conference on Hardware/Software Codesign and System Synthesis (CODES+ISSS)*, 2014, pp. 1–10.
[11] M. D. Nguyen, M. Wedler, D. Stoffel, and W. Kunz, "Formal hardware/software co-verification by interval property checking with abstraction," in *Proceedings of the 48th Design Automation Conference*. San Diego California: ACM, Jun. 2011, p. 510–515. [Online]. Available: https://dl.acm.org/doi/10.1145/2024724.2024843
[12] M. Schwarz, C. Villarraga, D. Stoffel, and W. Kunz, "Cycle-accurate software modeling for RTL verification of embedded systems," in *2017 IEEE 20th International Symposium on Design and Diagnostics of Electronic Circuits & Systems (DDECS)*, 2017, pp. 103–108.
[13] C. Villarraga, B. Schmidt, J. Bormann, C. Bartsch, D. Stoffel, and W. Kunz, "An equivalence checker for hardware-dependent embedded system software," in *2013 Eleventh ACM/IEEE International Conference on Formal Methods and Models for Codesign (MEMOCODE 2013)*, 2013, pp. 119–128.
[14] B. Schmidt, C. Villarraga, J. Bormann, D. Stoffel, M. Wedler, and W. Kunz, "A computational model for sat-based verification of hardware-dependent low-level embedded system software," in *2013 18th Asia and South Pacific Design Automation Conference (ASP-DAC)*, 2013, pp. 711–716.
[15] D. A. Patterson and J. L. Hennessy, "Computer organization and design: The hardware/software interface," 2005.
[16] "AMBA Specification," https://www.arm.com/architecture/system-architectures/amba/amba-specifications. accessed: 26-04-2024.
[17] N. M. Mutzel, "Implementing Interrupts in HW and SW," 2018.

Behavioral Simulation of Relative Timed Asynchronous Circuits

Sumanth Kolluru, Kenneth S. Stevens
Electrical and Computer Engineering, University of Utah

Abstract—**Relative Timed design represents timing constraints in an integrated circuit as mathematical equations. This differs from current state of the art integrated circuit design methodologies and electronic design automation tools which employ a finite state machine methodology by using a clock to implement circuit timing. This paper presents a behavioral model for relative timing constraints that enables simulation and evaluation of relative timed circuits written in behavioral Verilog. The model enables the source behavioral Verilog to be directly synthesized into relative timed circuits. The model provides support for rapid design exploration, architectural timing experiments, design validation, and multi-mode behavioral and structural simulations. Experiments show a $50\times$ speedup over current Verilog designs that require a structural implementation.**

I. INTRODUCTION

The correct behavior of an integrated circuit (IC) depends on two related but independent domains: logic and timing. Failures in either of these domains will result in an inoperative chip.

Traditional IC design flows implement chip timing as a finite state machine methodology where the next state updates on the edge of a clock. This work focuses on an alternative timing methodology called relative timing (RT) that implements timing as the mathematical equation shown in Eqn. 1 [1]. This equation states that from a common timing reference, called a point of divergence or pod, the maximum delay to one point of convergence (poc_0) is always less than the minimum delay to another point of convergence (poc_1). This ensures that the poc_0 event always occurs before the poc_1 event with minimum separation specified by the margin m. This effectively reduces the reachability graph that defines possible circuit behavior.

$$pod \mapsto poc_0 + m \prec poc_1 \qquad (1)$$

If relative timing is coupled with handshaking, modular design methods can be implemented. For example, a single stage pipeline is shown in Figure 1. The handshake employs request and acknowledge signaling between LC modules. The LC modules implement a sequential handshake protocol that controls data transfer between register banks. Relative timing ensures that the delay from pod to poc_0 (the red line) is always less than the delay from pod to poc_1 (the blue line). RT constraints guarantee that these pipelined circuits operate correctly in the timing domain, while the handshake protocol implemented in the LC blocks guarantee that the circuit operates correctly in the logic domain. By combining sequential handshake protocols with relative timing constraints, arbitrary

pipelines, including fan-in and fan-out can be implemented and verified to operate correctly in the logic and timing domains.

Verification of a relative timed handshake design proceeds as follows. Handshake protocols are specified with a formal language (such as a Petri Net) and synthesized to a standard cell library [2], [3]. This gate level sequential circuit is formally verified using model checking applying an unbounded delay model against the specification. This verification generates a set of RT constraints that must hold that allow the circuit to conform to the specification correctly [4]. Modular handshake designs are implemented using these LC modules which are formally proven correct in both the logic and timing domains. These designs are now combined with combinational functions and composed in pipelines to implement complex functions. The circuit is fully verified and operational when the combinational functions are verified as well ans the handshake control and timing [5], [6]. For example, the pipeline in Figure 1 implements the x^3 function.

Synthesis flows have been developed to create relative timed circuits whose internal delays meet their timing constraints [7]. Handshake RT designs are synthesized by placing a maximum delay constraint between the pod and poc_0 that is less than a minimum delay constraint between the pod and poc_1 [8], [9]. These behavioral circuits can be simulated and validated to ensure correct behavior and that the relative timing constraints have been met. Timing verification tools such as PrimeTime are employed to ensure these constraints hold in the final implementation. The RT modules are characterized to be compatible with clocked CAD flows, using Synopsys Design Constraints (SDC) commands like `set_size_only`, `set_dont_touch`, `set_data_check`, `set_max_delay`, `set_min_delay` and `set_disable_timing`. These constraints help preserve the functional correctness of the asynchronous design during the synthesis step.

This paper provides behavioral models for relative timing constraints (RTCs), allowing designers to simulate the designs with full functional correctness pre-synthesis (without the SDC). This approach streamlines development and supports mixed-mode simulation, which is particularly beneficial for unit testing and comprehensive design verification. By integrating RT models into a simulator testbench, this method allows for the automatic generation of timing constraints from RTCs, relieving designers from the intricate details of timing implementation. This process not only enhances efficiency but also offers a versatile and efficient tool for

979-8-3315-3968-9/24 $31.00 © 2024 IEEE

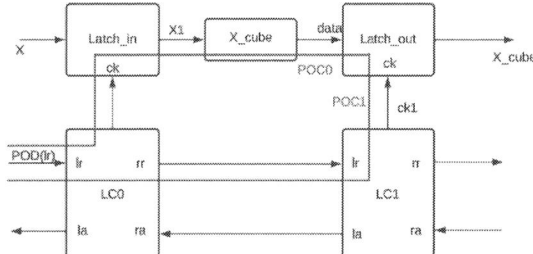

Figure 1: Pipelined RT implementation of x^3

mixed-mode simulations. Behavioral simulations are markedly quicker than structural simulations, facilitating a more efficient design exploration and behavioral validation process [10], [11], [12], [13]. Further, simulation times are prohibitive for large structural designs [14], [15].

To evaluate benefits and demonstrate feasibility of the model, the equations are applied to routers in a network-on-chip. Simulations are performed on an 8×8 array of 64 routers, and comparisons are made between a behavioral, structural, and mixed mode instance of the array.

II. Background and Related Work

Commercial and open source EDA tools are used to implement relative timed design, including synthesis, place and route (PnR), and verification flows. However, additional models, and at times custom tools, are required. (Sequential synthesis and automatic verification and generation of RT constraints use custom tools). Behavior and timing verification has been fully supported for some time for synthesized and PnR designs. This work creates models that enable behavioral simulation.

Numerous designs have been implemented using RT design flows showing advantages in power, performance, and area compared to clocked flows [16], [17]. However, these designs have only been able to use post synthesis or PnR designs for simulation. The Verilog models introduced in this work enable simulation and validation of behavioral relative timed designs for the first time.

RT equations have been proven complete to implement any timed circuit application [18]. This proof of generality of RT constraints requires the specification of multi-cycle instances. However, designs normally only require single cycle constraints on the pod and pocs of Eqn. 1. Verilog models for RT constraints presented in Sec. IV are only shown for single cycle relationships for clarity. However, these models can be extended to the arbitrary indexed transitions that are required by the proof of generality.

A pipelined design flow based on formal timing and logic verification was introduced in Sec. I. This flow partitions the design into data path logic that consists of combinational logic and registers, and a control path that consists of formally verified sequential controllers. The general RT design flow is not restricted to this approach. Arbitrary design constructions

and user generated RT constraints are also supported by this flow.

The Verilog models in this paper can also be used for performance verification (PV) as well as functional validation of an RT design. Performance verification of a RT handshake design is more complex than that of a clocked finite state machine [19]. Each request acknowledge handshake pair implements a silicon oscillator, that serves as a "local clock". Some designs have millions of these silicon oscillators. Most designs have architectural feedback that span numerous oscillators. These dependencies create complex timing relationships [20]. If no margin (m from Eqn. 1) is specified in each RT constraint, a fast "zero delay" simulation can be implemented. To perform behavioral performance verification, a minimum requirement is to specify a margin m for each RTC that results in a handshake frequency that matches the performance targets for each oscillator associated with every pipeline stage. This is not shown in this paper for to help maintain clarity.

III. Approach

The Verilog for Figure 1 is shown in Figure 2. This Verilog will be used as an example for building the RT simulation models. This example assumes four cycle return-to-zero handshake protocols where data is valid on the rising edge of request. Latches are assumed to be pulse clocked.

```
module Xcube (lr, la, rr, ra, x, X_cube);
  input lr, ra; output la, rr;
  input [31:0] x; output [31:0] X_cube;
  latch Latch_in (.D(x), .clk(ck0), .q(x1));
  ctrl LC0 (.lr(lr), .la(la), .rr(rr0), .ra(la0), .ck(ck0));
  assign data = x**3;
  latch Latch_out (.D(data), .clk(ck1), .q(X_cube));
  ctrl LC1 (.lr(rr0), .la(la0), .rr(rr), .ra(ra), .ck(ck1));
endmodule // Xcube
```

Figure 2: Verilog for circuit in Figure 1.

In this example the output latch is pulse clocked after the process X_cube produces its result. For correct functionality, data must arrive at Latch_out before clock ck1 stores the value, with a margin of s. This is expressed by the RT constraint of Eqn. 2. These delay paths are highlighted in red and blue in Figure 1. The path highlighted in red from pod to poc_0 is the maximum delay path and the path highlighted in blue from pod to poc_1 is the minimum delay path. The common causal event or point of divergence of this design is LC0/lr, and the two signals racing in time are data and ck1.

$$\text{LC0/lr}\uparrow \mapsto \text{data} + s \prec \text{ck1}\downarrow \qquad (2)$$

Correct signal ordering is enforced in simulation using RT equations implemented with an event based model that detects the arrival of ck1 relative to data for every causal transition on LC0/lr. The proposed model utilizes Verilog's force, wait, and release statements to ensure the sequential occurrence of poc_0 (data) before poc_1 (ck1) as shown in Figure 3. The force statement plays a crucial role by fixing the net at a specified logic level, maintaining this value irrespective of

979-8-3315-3968-9/24 $31.00 © 2024 IEEE

any other logic-driven changes to the net until the `release` statement is executed.

```
always @(posedge LC0.lr) begin
    force ck1 = ck1;
    wait(data);
    release ck1;
end
```

Figure 3: Verilog behavioral RT constraint for Eqn. 2

The model in Figure 3 first checks for transitions on pod. In scenarios involving concurrent system operations, the model can freeze the value of poc_1 upon the occurrence of pod. The state of poc_1 is now maintained until the poc_0 event takes place. This model narrows the simulation's reachability, thereby ensuring the precedence of poc_0 over poc_1. The poc_0 event is then detected using the Verilog `wait` statement. At this point poc_1 is allowed to proceed based on the logic and timing of the design. This technique is instrumental in establishing the event order as dictated by a relative timing constraint within a behavioral design context.

This basic *force-wait-release* mechanism is the foundation of the behavioral simulation model for an RT constraint. This method offers a robust implementation of a mathematical RT constraint. In the context of Figure 1, when a rising transition on LC0/lr happens, transitions on ck1 are delayed until data arrives at Latch_out.

Based on the implementation and behavior of a circuit and its sequential protocols, multiple RT constraints may be required to work in concert to correctly prune the behavioral reachability of a design. These include constraints to enforce minimum clock pulse width and transparency window control for the latches among others. The singular constraint of Eqn. 2 used in this section is not sufficient for Figure 1 to operate correctly in the time domain. For instance, Eqn 2 alone does not guarantee that ck1 is low when LC0/lr rises. A verification flow is therefore recommended to ensure a necessary and sufficient set of constraints.

IV. Key Requirements

The purpose of an RT constraint is to prune the reachability of a circuit based on the ordering of events that is imposed based circuit timing. Multiple constraints often interact to produce the correctly pruned reachability graph (such as both setup and hold constraints). This produces a number of important but subtle requirements when pruning reachability in Verilog with `force` and `wait` statements. Likewise, the amount of concurrency in a design can add complexity to design. This includes complexity in terms of bit timing of a wide function as well as the amount of concurrency between individual signals in a sequential design.

To illustrate these relationships a hypothetical constraint set is introduced in Figure 4 to outline the key requirements and challenges that must be considered in the development of the Verilog RTC models. The hypothetical constraints rtc0, rtc1 and rtc2, rtc3 have a common poc_1 (y_+ and lr−

respectively). Constraint rtc3 has a datapath with a 32-bit endpoint (din[31 : 0]).

```
rtc0:   lr+ ↦ rr+ ≺ y_+
rtc1:   lr+ ↦ la+ ≺ y_+
rtc2:   la+ ↦ y_+ ≺ lr-
rtc3:   lr+ ↦ din[31:0] ≺ lr-
```

Figure 4: Example Constraint Set

Each RT constraint is mapped to a Verilog code segment such as shown in Figure 3. These are included in the Verilog testbench. In the interest of brevity and clarity, the Verilog code snippets related to the following requirements represent segments of the full mapping codebase. Portions of the code which include various necessary checks and flags have been replaced with comments or omitted entirely to maintain readability and focus on the critical components.

Requirement 1. *Causality:* The pod is assumed to be causal to both poc_0 and poc_1 in rt constraint pod $\mapsto poc_0 + m \prec poc_1$. For example, in RTC a+ ↦ b- ≺ c+, a rising edge on a will cause a falling transition on b and a rising transition on c.

Requirement 2. *Trigger Condition:* A poc signal is *triggered* when the logic level of the poc is in the opposite boolean state than the condition specified by the RT constraint (e.g. a logic low value for poc+ and logic high for poc-). If the poc is not triggered when the pod event occurs, a transition on pod must occur to place it in a triggered condition.

This assumption implies that the causal path from pod to poc is not multicycle, which can be extended in future work.

Requirement 3. *Breaking combinational cycles.* Verilog simulators are not required to preserve signal causality when a design contains combinational cycles. Since causality is specified in Requirement 1, combinational cycles must be broken into different simulation steps.

This is performed by adding delay to the feedback signal in a combinational cycle specified behaviorally. For example, assume a C-element, which is a state-holding circuit that can be implemented with combinational feedback. The following snippet adds one unit of delay that breaks the combinational cycle into different simulation steps.

```
assign #1 c = (a & b) | (a & c) | (b & c);
```

Requirement 4. *Reachability Pruning:* Concurrency is reduced in the design by ensuring the poc events are sequential with Verilog `force` and `release` statements. Signal poc_1 is prevented from occurring until the event on poc_0 has occurred. This forms the basis for the Verilog statements that enforces a timing constraint in the system.

This requirement can be implemented as follows. This form of reachability pruning requires that poc_1 is triggered. Timing endpoint poc_1 is forced to remain in the triggered state until poc_0 fires, at which point it is released and can follow the logical and standard timing behavior of the circuit.

```
always @(pod) begin
```

979-8-3315-3968-9/24 $31.00 © 2024 IEEE

```
    force poc1 = poc1; // force to current value
    wait(data);
    release ck1;
end
```

Requirement 5. poc *sharing:* The semantics of the Verilog `release` statement will immediately return the specified net to the control of the logic behavior of the circuit regardless to the number of times it has been forced. Thus any set of RT constraints that share the same poc_1 net must work in concert to ensure the net is properly released. Any RTC with a shared poc_1 may not release the poc_1 signal until all of restraining poc_0 signals have fired in active RTCs where the pod event has occurred.

The constraint set example shown in Figure 4 has shared poc_1 signals y_+ in constraints rtc0 and rct1. Since they also share the same pod signal lr+, signal y_+ may not assert until after both rr and la rise. A counter is added for each poc_1 that exists in multiple RTCs. It is incremented each time a pod event occurs in one of these RTCs, and decremented when the poc_0 event fires. If the counter reaches zero then the `force` statement can be released. A code snippet for the model for rtc0 is shown. A similar model will be implemented for rtc1.

```
int poc1_y_UP = 0;
always @(posedge lr) begin
    poc1_y_UP = poc1_y_UP + 1;
    force y_ = y_; // force to current value
    wait(rr);
    poc1_y_UP = poc1_y_UP - 1;
    if (poc1_y_UP == 0) release poc1;
end
```

Requirement 6. *Continuous Assignments:* The semantics of the Verilog `force` statement are limited in scope as it operates only on continuous assignments. This requires that `reg` variables must be assigned to a continuous statement to have their value forced.

One primary goal of this work is to allow the source Verilog to be synthesized as well as simulated. To implement this requirement the source code must be modified. Fortunately this can be done in such a way that still allows the code to be synthesized. Assume that net y_ from rtc0 is declared a `reg`. A new wire is created and assigned the `reg` value. This wire can be forced and released. With this change the poc_1 signal y_ is assigned to y_RTC_CA. Net y_RTC_CA must replace all instances of net y_ in the source code, and is the signal that is forced and released to prune reachability of the design.

```
reg y_; wire y_RTC_CA;
assign y_RTC_CA = y_;
```

Requirement 7. *Multibit timing endpoints:* Data validity of a multi-bit signals such as a data bus must be identified with a single boolean value.

Identifying data validity in a behavioral simulation using zero-delay functions is usually readily performed. For example, data will always be stable per Requirement 3 when handshake signals are asserted in a zero delay simulation

of a bundled data design style when handshake controllers with combinational loops are employed. However, the bundled data timing requirement including setup and hold times must be correctly modeled in structural and mixed mode designs. This issue can be addressed by modern hardware description languages that implement a framework that determines data validity [21].

The constraint rtc3 shown in Figure 4 has a multibit data bus at the first point of convergence. A boolean flag must be generated indicating data is valid, shown as dinvalidp in the code snippet below. The specific approach for determining when to assert dinvalidp depends on the implementation.

```
reg dinvalidp = 0;
...
assign dinvalidp = 1; // when din is stable
...
always @(posedge lr) begin
    force lr = lr;
    wait(dinvalidp);
    dinvalidp = 0;
    release lr;
end
```

Requirement 8. *Margins of separation:* Minimum margins of separation between poc0 and poc1 are implemented by adding a delay between the `wait` and `release` statements.

Minimum separation between events is important for pulse shaping, to ensure correct setup and hold times in a design, and for using behavioral simulations for performance verification. The RT constraint pod $\mapsto poc_0 + m \prec poc_1$ indicates that poc_1 is triggered after poc_0 plus a margin m. The example below shows how this is modeled. Performance verification of a constraint is implemented when m is set to match the desired phase delay of silicon oscillator.

```
always @(pod) begin
    force poc1 = 0;
    wait(poc0);
    #m;           //introduce margin or delay
    release poc1;
end
```

Requirement 9. *Mixed mode simulation:* Delay tracing of relative timed paths is required to ensure RT constraints hold in a mixed mode relative timed design.

This work enables simulations of mixed designs which are part structural and part behavioral. The behavioral design can be either "zero delay" or implement delays for performance verification.

Structural design that have been synthesized with RT constraints will simulate without requiring any models developed in this work. However, handshake designs have signal paths at the interfaces that are outside of the synthesized logic. For example, a pipeline controller at the interface of a synthesis block will have request acknowledgment handshakes with adjacent modules [22]. Relative timing constraints where the timing path goes outside of the synthesized module can be violated, particularly when interfacing with zero delay behavioral models.

The RT constraint path in Figure 5 illustrates this point.

979-8-3315-3968-9/24 $31.00 © 2024 IEEE

The red max delay path is internal to the circuit, whereas the blue min delay path is outside the circuit. In a zero delay behavioral model, the blue min delay path can be less than the red max delay path, violating an RT constraint. Thus all RT constraints which have paths that pass outside of a synthesized module must be properly modeled with delays appropriate for performance verification and verified for correct RT behavior.

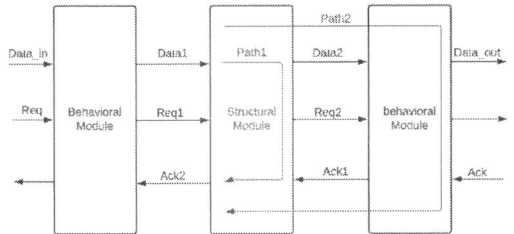

Figure 5: Internal and external paths in a mixed mode design

Requirement 10. *Pulse Logic:* Pulses are generated when the same net is specified for poc_0 and poc_1. To model minimum pulse width the margin of separation m must be specified in the Verilog model.

Pulse based logic is particularly challenging for traditional design flows, but is easily modeled with relative timing. Consider the pulse generator example as shown in Figure 6 which generates a high pulse on the rising transition of in. The RT constraint of Eqn. 3 can be used to define the correct timed behavior of this pulse.

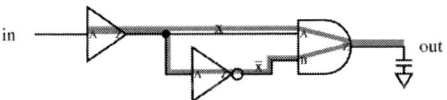

Figure 6: Pulse generator circuit.

$$\text{in+} \mapsto \text{out+} + m \prec \text{out-} \tag{3}$$

This can be modeled as follows.

```
always @(posedge in) begin
    wait(out);
    force out = out;
    #m;
    release(out);
end.
```

V. RESULTS

Simulation models for the rules presented in Sec. IV are developed for a network-on-chip (NoC) [23]. The NoC's architecture comprises 64 routers arranged in an 8×8 grid, configured in a wrapped torus topology. It utilizes dimension-ordered source routing with single-flit packets. The sequential controllers implement the communication protocols and control latch clocking similar to the LC blocks in Figure 1.

As a considerably extensive design, the NoC incorporates hundreds of timing constraints across its 64 submodules, ensuring precise operation. A complete set of RTCs are generated for the sequential controllers with formal verification. These constraints are largely mapped to behavioral simulation constraints by hand in order to validate this approach. The Verilog models generated for the different RT constraint sets are added to simulation testbenches.

Simulations are performed in QuestaSim using SPICE modeled NoC link delays and sdf back annotated timing for the synthesized designs. The base testbench employs 303 lines of Verilog code. The testbench code interfaces with a C++ NoC simulator that generates various work loads and delivery patterns that are offered to the NoC. Each simulation run verifies correct message delivery, records latency statistics, and measures activity factors for power evaluation.

The comparison between four designs is shown in Table I. All of the four designs passed the testbench with no errors. The second column lists the size of the testbench. All the additional lines of testbench code beyond 303 are comprised of the RT Verilog models developed from the requirements enumerated in Sec. IV. The third column lists the size of the NoC under test. The fourth column lists the maximum memory used during simulation, and the last column lists the runtime of the testbench in seconds.

The structural design was synthesized and timing closed in a 65nm technology node using the RT constraints to enforce the signal ordering specified in the RTCs. The behavioral and behavioral min designs implement the NoC with behavioral Verilog. In the mixed design, one of the routers is structural, and the other 63 are behavioral.

Zero delay behavioral simulation coupled with Requirement 3 (breaking combinational cycles) will prune many of the error states from a Verilog simulation of a handshake design. Therefore the behavioral min design used a subset of the RT constraints for simulation. The behavioral and mixed designs implement the full set of RT constraints. The size of the RT Verilog models for the NoC comprises 19,126 lines of code, or an average of approximately 300 lines of Verilog code per router. The reduced model reduces the size of the Verilog RT models by 21%, and contains an average of approximately 235 lines of Verilog code per router.

The synthesized structural design contains over 11× more lines of Verilog code than the behavioral or mixed mode versions, as synthesis has mapped the design to standard cell library gates. No behavioral timing constraints are required for the behavioral simulation of the structural design.

The process memory for the structural design simulation was approximately 12× larger than the behavioral design and 9× bigger than the mixed design. The simulation time for the structural design is over 50× longer than the behavioral designs.

The mixed mode design allows one to focus on a particular design block, given its full synthesized or PnR delays, while the rest of the design is behavioral. This is advantageous when performing unit testing. Mixed mode simulation can hence

979-8-3315-3968-9/24 $31.00 © 2024 IEEE

Table I: NoC Results

Design	Lines of code(Testbench)	Lines of code(Design)	Simulation process memory (KB)	Simulation time (s)
Structural Design	303	95,207	3,100,000	1800
Behavioral Design	19,429	8,563	239,788	34
Mixed Design	19,361	19,469	325,952	180
Behavioral Min	15,332	8,563	230,487	32

help in reducing the time for the overall design debugging and provide a way for easier integration in the RT design space. The proposed work also paves the way for design modularity in asynchronous circuit design. In this mode, the runtime memory is reduced by 89% and simulation time by 90% over the fully structural design.

VI. CONCLUSION

A methodology is presented for mapping relative timing constraints to Verilog models. This enables the ability to behaviorally simulate designs with arbitrary timing expressed as relative timing constraints. The Verilog models are explicitly designed to be added to a testbench. Only small modifications to the behavioral code are required to implement the RT simulation models. The modifications do not prevent the behavioral code from being directly used for synthesis and place and route. The mapping approach is rule based and can be automated through CAD development. This approach enables rapid design and architectural exploration and evaluation of circuit designs that employ relative timing constraints.

To prove feasibility and correctness, four testbenches with simulation models were developed and tested on three versions of a 64 node toroidal network on chip design. All designs passed the regression tests. This experiment showed the advantages in memory and run time of a behavioral design over a synthesized structural design, with a 92% and 98% reduction in simulation memory and runtime respectively. This demonstrated significant advantages and feasibility of this approach at simulation time. A mixed mode design with one structural block and 63 behavioral blocks showed a 89% and 90% reduction in simulation memory and runtime, while providing detailed gate-level delay data of the structural router.

REFERENCES

[1] K. S. Stevens, R. Ginosar, and S. Rotem, "Relative Timing," *IEEE Transactions on Very Large Scale Integration (VLSI) Systems*, vol. 1, no. 11, pp. 129–140, Feb. 2003.

[2] J. Cortadella, M. Kishinevsky, A. Kondratyev, L. Lavagno, and A. Yakovlev, "Petrify: A Tool for Manipulating Concurrent Specifications and Synthesis of Asynchronous Controllers," in *XI Conference on Design of Integrated Circuits and Systems*, Barcelona, November 1996.

[3] K. Y. Yun and D. L. Dill, "Automatic Synthesis of Extended Burst-Mode Circuits: Part I (Specification and Hazard-Free Implementation)," *IEEE Transactions on Computer-Aided Design*, vol. 18, no. 2, pp. 101–117, Feb 1999.

[4] Y. Xu and K. S. Stevens, "Automatic Synthesis of Computation Interference Constraints for Relative Timing," in *26th International Conference on Computer Design*. IEEE, Oct. 2009, pp. 16–22.

[5] M. Bozga, H. Jianmin, O. Maler, and S. Yovine, "Verification of Asynchronous Circuits using Timed Automata," *Electronic Notes in Theoretical Computer Science*, vol. 65, no. 6, pp. 47–59, 2002.

[6] K. Desai, K. S. Stevens, and J. O'Leary, "Symbolic Verification of Timed Asynchronous Hardware Protocols," in *Annual Symposium on VLSI (ISVLSI)*. IEEE Computer Society, Aug 2013, pp. 147–152.

[7] K. S. Stevens, Y. Xu, and V. Vij, "Characterization of Asynchronous Templates for Integration into Clocked CAD Flows," in *15th International Symposium on Asynchronous Circuits and Systems*. IEEE, May 2009, pp. 151–161.

[8] E. Quist, P. Beerel, and K. S. Stevens, "Enhanced SDC Support for Relative Timing Designs," in *Digital Automation Conference*. IEEE/ACM, July 2009, user Track Poster.

[9] G. Gimenez, A. Cherkaoui, G. Cogniard, and L. Fesquet, "Static Timing Analysis of Asynchronous Bundled-Data Circuits," in *24th International Symposium on Asynchronous Circuits and Systems*. IEEE, May 2018, pp. 110–118.

[10] M. Renaudin and A. Fonkoua, "Tiempo Asynchronous Circuits System Verilog Modeling Language," in *18th International Symposium on Asynchronous Circuits and Systems*. IEEE, Oct 2009, pp. 105–112.

[11] A. Saifhashemi and P. A. Beerel, "High Level Modeling of Channel-Based Asynchronous Circuits Using Verilog," in *Communicating Process Architectures*, J. Broenink, H. Roebbers, J. Sunter, P. Welch, and D. Woods, Eds. IOS Press, 2005, pp. 275–287.

[12] E. Esimai and M. Roncken, "Flexible Compilation and Refinement of Asynchronous Circuits," in *28th International Symposium on Asynchronous Circuits and Systems*. IEEE, July 2023, pp. 109–119.

[13] A. Saifhashemi and P. A. Beerel, "SystemVerilogCSP: Modeling Digital Asynchronous Circuits Using SystemVerilog Interfaces," *Communicating Processor Architectures*, vol. 68, pp. 287–302, 2011.

[14] Y. Zhang, H. Cheng, D. Chen, H. Fu, S. Agarwal, M. Lin, and P. A. Beerel, "Challenges in Building An Open-source Flow from RTL to Bundled-Data Design," in *24th International Symposium on Asynchronous Circuits and Systems*. IEEE, May 2018, pp. 26–27.

[15] C. Chau, W. A. Hunt Jr., M. Roncken, and I. Sutherland, "A Framework for Asynchronous Circuit Modeling and Verification in ACL2," in *Hardware and Software: Verification and Testing*, ser. Lecture Notes in Computer Science, vol. 10629, Nov 2017, pp. 3–18.

[16] K. S. Stevens, S. Rotem, R. Ginosar, P. Beerel, C. J. Myers, K. Y. Yun, R. Kol, C. Dike, and M. Roncken, "An Asynchronous Instruction Length Decoder," *IEEE Journal of Solid State Circuits*, vol. 36, no. 2, pp. 217–228, Feb. 2001.

[17] Y. Chen, X. Zhang, Y. Lian, R. Manohar, and Y. Tsividis, "A Continuous-Time Digital IIR Filter With Signal-Derived Timing and Fully Agile Power Consumption," *IEEE Journal of Solid State Circuits*, vol. 53, no. 2, pp. 418–430, Feb 2018.

[18] R. Manohar and Y. Moses, "Timed Signalling Processes," in *28th International Symposium on Asynchronous Circuits and Systems*. IEEE, July 2023, pp. 10–19.

[19] C. E. Molnar, I. W. Jones, W. S. Coates, J. K. Lexau, S. M. Fairbanks, and I. E. Sutherland, "Two FIFO Ring Performance Experiments," *Proceedings of the IEEE*, vol. 87, no. 2, pp. 297–307, February 1999.

[20] G. Gill and M. Singh, "Bottleneck Analysis and Alleviation in Pipelined Systems: A Fast Hierarchical Approach," in *15th International Symposium on Asynchronous Circuits and Systems*. IEEE, May 2009, pp. 195–205.

[21] T. Bourgeat, C. Pit-Claudel, A. Chlipala, and Arvind, "The Essence of Bluespec: A Core Language for Rule-Based Hardware Design," in *41st International Conference on Programming Language Design and Implementation*. ACM, June 2020, pp. 243–257.

[22] C.-F. Law, B.-H. Gwee, and J. S. Chang, "Modeling and Synthesis of Asynchronous Pipelines," *IEEE Transactions on Very Large Scale Integration (VLSI) Systems*, vol. 19, no. 4, pp. 682–695, April 2011.

[23] V. Nori, B. Chauviere, M. J. Wibbels, and K. S. Stevens, "A Novel Asynchronous Network-On-Chip Based on Source Asynchronous Signaling," in *28th International Symposium on Asynchronous Circuits and Systems*. IEEE, July 2023, pp. 71–77.

Diagnostic Coverage Estimation for Automotive SoCs based on Colored Stochastic Petri Nets

Ernesto Cristopher Villegas Castillo ©*†, Felipe Augusto da Silva ©*, and Michael Glaß©†

*Cadence Design Systems, Munich, Germany

†Institute of Embedded Systems/Real-Time Systems, Ulm University, Ulm, Germany

E-mail: {*ernesto, dasilva*}*@cadence.com, michael.glass@uni-ulm.de*

Abstract—Safety-critical systems can cause catastrophic effects when particular failures occur during their operation. These systems, used in diverse domains, including automotive SoCs, incorporate Safety Mechanisms (SMs) to enhance their safety performance and meet certification standards (e.g., ISO26262). A critical measure of safety performance is the Diagnostic Coverage (DC) of SMs, determined through extensive and expensive Gate-Level Fault Injection (FI) campaigns, as recommended by ISO26262. To address this challenge, designers need early DC estimation methods to efficiently develop more reliable SMs by reducing simulation times, redesign stages, and computational resources. Our previous work proposed an interactive simulation framework based on Colored Generalized Stochastic Petri Nets (CGSPN) for Fault Coverage (FC) estimation. This work incorporates the requirements of automotive safety standards to predict the efficiency of SMs. We propose a methodology for early-stage estimation of the DC, enabling efficient SM development and its Design Space Exploration (DSE), and the discovery of Failure Modes through CGSPN simulations. To the best of our knowledge, the proposed work is the first DC estimation approach based on high-level models such as CGSPN. The methodology was verified in an automotive SoC, showing an average estimation accuracy of 97.2% and a 175x speed-up for a Software Test Library (STL) compared to results obtained through an exhaustive RTL FI campaign.

Index Terms—Functional Safety, FMEDA, Diagnostic Coverage, Fault Injection, Safety Mechanisms

I. INTRODUCTION

Automotive SoC designs must comply with safety standards such as ISO26262 [1] to ensure they incorporate effective measures for reducing the risk of catastrophic events during operation. The ISO26262 recommends evaluating the SoC's safety performance through Failure Mode Effect Diagnostic Analysis (FMEDA), which involves creating a table describing the SoC's Failure Modes (FMs) along with their respective failure rate (λ), measured in Failure in Time units (FIT, indicating one failure per million hours). To mitigate the impact of these FMs, the designer must integrate Safety Mechanisms (SMs). Their effectiveness is measured by Diagnostic Coverage (DC), which is generally obtained through simulation-based Fault Injection (FI) applied to the SoC Gate-Level (GL) model as recommended by ISO26262. However, conducting an FI campaign on a GL model, even when applying sampling methods [2] to reduce the number of injected faults, can lead to complex and time-consuming computational tasks. This complexity may result in delays when new vulnerabilities are discovered during later design stages, implicating redesign

cycles and architectural redefinitions and ultimately extending the time-to-market of the SoC.

Previous works have tackled the challenges of long fault injection campaigns by proposing structural and formal methods. In [3], [4], the authors proposed structural and formal methodologies to detect safe faults in SoC RTL models. Their approaches aim to reduce the number of FI simulations while obtaining more reliable DC values. Additionally, [5] introduced a Flip-Flop weighting technique to predict DC values during RTL design phases before the availability of the SoC netlist. Although these methods provide more reliable DC values of the selected SMs and reduce the number of FI simulations, they still require the RTL models, which are not applicable at early design stages. Instead, high-level models are a potential solution for designers to perform Design Space Exploration (DSE) and obtain safer SoC architectures while avoiding previously mentioned problems. Other related work focuses on other SoC dependability properties outside the automotive domain. In [6], the authors proposed a cross-layer multi-objective DSE algorithm based on Bayesian networks to support the development of soft error-resilient electronic systems. In [7], the authors proposed a vulnerability model approach based on a few FI simulations to identify vulnerable model parameters in Deep Neural Networks (DNN), considering a trade-off between accuracy and required computational resources. Additionally, [8] estimates the criticality of all the DNN parameters and their bits to provide information for cost-effective DNN protection. These methodologies efficiently support identifying critical parameters of complex architectures as DNNs. On the other hand, safety performance metrics such as DC require more accurate estimations to perform DSE of SMs at early design stages. To the best of our knowledge, no such estimation approach exists for safety performance metrics such as DC.

This work presents an efficient methodology for estimating DC using a high-level modeling approach based on Colored Generalized Stochastic Petri Nets (CGSPNs). We propose significant contributions demonstrating that our approach effectively addresses the challenge of estimating safety performance metrics in the automotive domain. These innovations include our CGSPN modeling pattern for early DC estimation, supporting the efficient development of SMs, and identifying the SoC's FMs through rapid CGSPN simulations.

979-8-3315-3968-9/24 $31.00 © 2024 IEEE

We utilize the Software Test Libraries (STLs) of an industrial complexity automotive test case called AutoSoC [9] to evaluate and validate the proposed methodology. We compare our results with an exhaustive RTL FI campaign of the OpenRISC 1000 mor1kx CPU. Our findings show that by considering the proposed framework using the CGSPN strategy, we can achieve results equivalent to the traditional RTL-based FI experiments (e.g., with an accuracy degradation of just 2.1%). Additionally, we can reduce the overall execution time of the FI campaigns by up to 43%. In particular, our CGSPN approach provides a speed-up of up to 175 times.

This paper is structured as follows: Section II presents the background; Section III, the proposed approach; Section IV discusses the test case; Section V presents the results; and Section VI outlines the conclusions.

II. BACKGROUND

A. High-level SoC fault propagation model based on CGSPN

CGSPN is a well-known graph modeling approach based on discrete event-triggering (firings) [10]. Colors are incorporated into the Generalized Stochastic Petri Nets (GSPN) to identify the tokens and define weight functions in the arcs. In [11], we leveraged the CGSPN features and created a modeling pattern for faulty SoC architectures. The simulation of these models allows the user to inject faults on each component, observe their propagation, and determine the final effect on the SoC-delivered service.

Figure 1 depicts this modeling pattern. It consists of a *Fault Dispatcher* containing the number of faults (blue tokens) to be injected into the system. The FI is modeled using timed transitions (white rectangles) being fired at the component's failure rate λ. The injected blue token reaches the component $C0$, which is processing data (dark tokens). When a dark and a blue token are present in the same component's place, and immediate transition (dark rectangles) are fired, the tokens are taken, and a red token (faulty data) is transferred to the interconnected component (place $C1$). Depending on the immediate transition rate values, red tokens can be transferred to the places representing the final states of the injected faults: *Observed* or *Unobserved*.

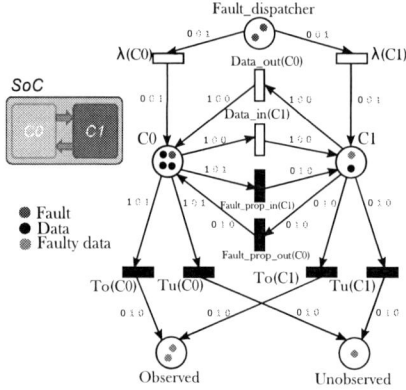

Fig. 1: CGSPN model of an SoC with two components

B. Functional Safety

It is essential to understand functional safety fundamentals well to appreciate the work's significance we are proposing entirely. Part 5 of ISO26262 [1] contains all the definitions related to hardware development. To clarify, we will introduce some basic definitions used in this work.

1) Failure Mode Effect and Diagnostic Analysis (FMEDA): It is an analytical methodology that recognizes and evaluates the effects of FMs' SoC components so that the failure effect can be minimized or eliminated by selected SMs. One variation, Architectural FMEDA, is used in early design phases to identify the SoC's vulnerabilities, explore different SMs, and determine compliance with ISO26262.

2) Diagnostic Coverage (DC): It is a performance property of SMs that detects dangerous faults to a certain extent in a specific functionality. ISO26262 recommends using simulation-based FI at GL SoC model using two types of monitors, also known as strobes: functional and checker. During the safety verification, engineers set the functional strobes at the SoC-delivered service, and the checkers at the SM indicator signals. After running the FI campaign, injected faults are classified by the FI tool according to their effect on the selected strobes. Table I shows the fault's classification according to their observability at the defined strobes. We will use the following acronyms to identify the faults according to their observability at the strobes: *Dangerous_detected* (DD), *Dangerous_undetected* (DU), *Unobserved_detected* (UD), and *Unobserved_undetected* (UU).

TABLE I: Fault annotattion

		Checker	
		Undetected	Detected
Functional	Unobserved	UU	UD
	Dangerous	DU	DD

The DC value is expressed in percentage and calculated using Equation (1).

$$\mathbf{DC}\% = \frac{\text{DD}}{\text{DD} + \text{DU}} \tag{1}$$

ISO26262 recommends measuring the SM's DC using FI campaigns on the GL models. Engineers can functionally verify SMs using RTL models or high-level models based on virtual prototypes. Equation (2) represents the residual failure rate (λ_{res}) obtained after incorporating one SM with DC value to a specific SoC FM with a failure rate λ. It means the higher the DC, the lower the fault probability of causing catastrophic events during SoC operation.

$$\lambda_{res} = \lambda * (1 - \mathbf{DC}\%) \tag{2}$$

3) Failure Modes (FMs): It describes a cause/effect case based on the ISO26262. This description helps the engineers to map all possible failure causes into the SoC. The diagram in Figure 2 illustrates how the FMEDA methodology maps the FMs. The developer assigns relative weight distribution, expressed in percentage, to $FM0$, $FM1$, $FM2$, and $FM3$ for the design sub-part SP. Subsequently, various SPs comprise a part P of the SoC.

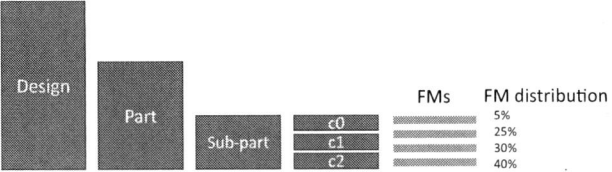

Fig. 2: Failure Mode mapping

Sub-components $c0$, $c1$, and $c2$ constitute a design sub-part SP. Therefore, the failure rate of the FM is the sum of the failure rates of the sub-components $\lambda(c)$, which cause the system failure.

III. Proposed approach

We propose an approach based on Colored Generalized Stochastic Petri Nets (CGSPN) to support the estimation of the DC value of any SM at the early stages of the RTL design and verification process, in addition to the Fault Coverage (FC) metric presented in [11]. Such an approach impacts the functional safety process since we demonstrate that engineers can efficiently apply our CGSPN approach to more relevant use cases, such as automotive safety performance.

This work's contributions are the following:

A. CGSPN modeling pattern extension

Figure 3 shows the CGSPN model, including the fault states already presented in Table I. It also displays an FI setup of an illustrative SoC composed of two interconnected components, $C0$ and $C1$, with an SM incorporated on $C0$. The engineer sets the functional strobes at the SoC output delivered service, while the checker strobes at SM's indicator signals. The engineer performs two independent CGSPN simulations to calculate the DC value of $C0$ and $C1$, enabling the respective timed transitions $\lambda(c)$. In that way, we can simulate one component failing and observe the fault propagation through the other component. We selected the PIPE2 tool [12] as the CGSPN simulator.

B. SM development support based on DC estimation

Instead of aiming for an FC estimation, this work targets the DC value of a selected group of components of an SoC architecture that interests the designer. This contribution enables fault propagation analysis through components and the overall effect on the SoC and its SMs.

Assuming we dispose of an existing SM library of SMs, we can avoid running an entire FI campaign and update the immediate transition rates to the final state places of each

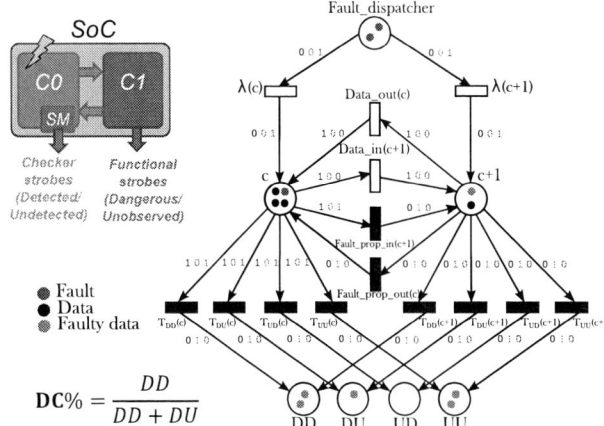

Fig. 3: CGSPN approach supporting DC estimation

component. The rates $T_{DD}(c), T_{DU}(c), T_{UD}(c)$ and $T_{UU}(c)$ represent the probabilities of faulty data (red token) reaching one of these final states. Assuming the DC value of the applied SM is 90%, the $T_{DD}(c)$ value is set to this value, while the $T_{UU}(c)$ is reduced to 10%.

Furthermore, our approach can facilitate the development of a new SM by optimizing the FI campaign injecting faults at the locations where the faults propagate to the SoC output service instead of the indicator signals of the SMs. This effect means the designer can extract the DU fault list from the demo run and reuse it for a reduced FI campaign.

Figure 4 shows our extended SM development approach, outlining the steps to develop a new SM.

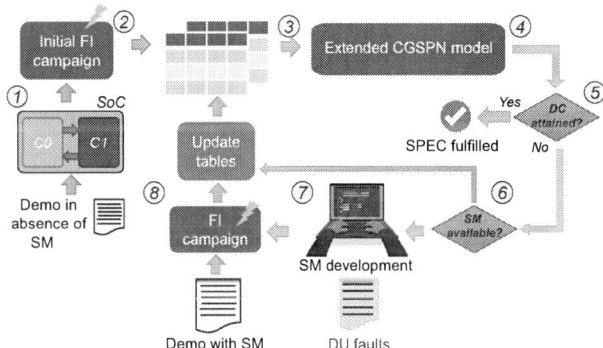

Fig. 4: Extended CGSPN methodology

Our proposed methodology consists of the following steps:

1) The user sets the functional strobes at the interconnections between components $C0$ and $C1$ and at the SoC output service signals to track fault propagation. The user should select a low-complexity application (e.g., *hello_world*) without SMs. Since SM indicator signals are not available at this stage, we consider *Observed* and *Unobserved* faults at the functional strobes as DU and UU faults, respectively.

979-8-3315-3968-9/24 $31.00 © 2024 IEEE   69

2) The user runs an initial FI campaign on the SoC's RTL model to obtain the FI reports.

3) The user takes the FI reports to create probability tables based on the number of faults propagated through the interconnections and faults, altering the SoC output service against the total number of faults injected into the respective component. The user takes the calculated probability values to set immediate transition rates of the CGSPN model. Timed transitions representing the component's failure rates ($\lambda(c)$) are obtained using the target technology information and its area estimation, represented by the number of gates and Flip-Flops.

4) The user uses the PIPE2 tool to run the CGSPN simulations. The user should run each component simulation until a stable value of the estimated DC is obtained.

5) The user should verify whether the DC values calculated from the CGSPN simulation are appropriate to guarantee the desired safety level of the SoC specification.

6) If the user does not achieve the required safety conditions at this stage, they should improve the SM efficiency to cover more faults or include an available SM whose DC is high enough to protect the system

7) The user refines the SM to increase the number of *DD* faults and hence its DC value.

8) Afterward, the user reruns the FI campaign of the SoC with the incorporated SM, only injecting the faults from the DU list obtained in the last FI run. Then, the user takes the new FI campaign reports to update the fault probability tables. Finally, the user repeats the process from Step 3 until the safety requirements are achieved.

C. Failure mode identification

We enable FM identification by simulating a CGSPN model whose fault propagation probabilities are calculated using the mean of the faulty component's probability tables obtained from the FI campaign. For instance, in Figure 5, three different SoC FMs are depicted, with the SoC components and their respective $\lambda(c)$ on the left-hand side of the Table. Every FM considers a set of faulty components.

Users can test various FMs by enabling the timed transitions of their components of interest, avoiding time-consuming and resource-consuming faulty simulations. When the user discovers FMs through CGSPN simulation, they can be validated using a faulty simulation with multiple faults injected in different component locations of the SoC. In future works, the user will validate discovered FMs through FI simulation.

IV. TESTCASE

In our methodology evaluation, we use the AutoSoC open-source platform [9], a complex industrial platform containing hardware and software components for automotive applications. AutoSoC is based on the OpenRISC 1000 mor1kx CPU and includes various features such as dual-lock steps, Error Correction Code (ECC), and STLs of the CPU and other peripherals. We tested our approach using the mor1kx CPU STL, which is not constrained to this type of SM.

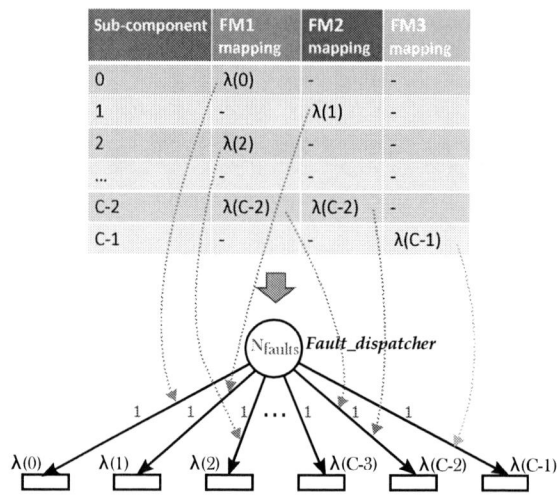

Fig. 5: FMs mapped to the CGSPN model for fault dispatching

To evaluate the DC, we focus on the following components: *fetch, decode, decode_execute, execute, execute_control, control*, and *load_store_unit (lsu)*. We do not consider other mor1kx CPU components as the STL does not have meaningful coverage on them, and their complexity would extend the FI campaign time without contributing to our approach. Additionally, memory blocks within *fetch* and *lsu* components are excluded from the analysis, as memory reliability can be addressed through SM, such as ECC, which has a DC value of 99%.

AutoSoC provides the necessary setup scripts for running the RTL FI campaign with a commercial fault simulator tool. This provided setup enables the execution of faulty simulation processes in parallel, with one simulation per process. We configured the FI campaign to be run in 100 parallel processes, reflecting the infrastructure commonly used in the automotive SoC industry. We only considered permanent faults (i.e., stuck at 0/1) for the FI campaign. Transient faults will be explored in the future.

V. RESULTS

A. RTL FI campaigns

We conducted an exhaustive FI campaign using the selected components with the demo application, *Hello_world.exe*. A total of 21,778 faults were injected during this stage, and it took some days to generate the final FI report. Following the demo FI campaign, the DC values did not meet the expected level due to the absence of SM in the AutoSoC configuration. Consequently, we extracted the FI campaign report for the demo application to identify *DU* and *UU* faults. These initial campaign results were utilized to develop the demo CGSPN models, as outlined in Section III.

The obtained DC values were low in the demo, so we used a new application called *stlDeterSafe.exe*, which includes an SM, such as the CPU STLs, to increase the DC values of the selected pipeline components. The CPU STLs are part of the AutoSoC repository and were not developed as part of this work. We reused the *DU* faults from the demo FI campaign report to run a new FI campaign since these faults were already observed at the CPU output service. The new FI campaign took a few more days than the demo FI campaign. We did not consider *UD* and *UU* faults since *UD* faults were zero, and *UU* faults are dormant and are not covered by the STL than *DU* faults after incorporating the CPU STLs. The new FI report enhanced the demo CGSPN models by updating the transition rates, as explained previously in Section III.

B. CGSPN simulation

We simulated the demo and the enhanced CGSPN models for each pipeline component using PIPE2 [12]. We ran each CGSPN model for five minutes on average, and the estimated DC value stabilized. Initially, users can manually run the demo CGSPN and adjust the transition rate values to explore SMs with known DC values. This feature allows users to skip running a new reduced FI campaign of SoC with SMs and manually create the enhanced CGSPN models.

In addition, CGSPN simulation enables the users to identify the components that are more likely to fail, allowing them to focus their efforts on improving safety by including various SMs until the desired DC is achieved.

The number of blue tokens in the *Fault dispatcher* equals the number of possible faults for the respective component. The number of black tokens at the *fetch* place was the same as that of blue tokens. The black tokens represent the application being processed on the mor1kx CPU.

The efficiency of the PIPE2 engine in simulating CGSPN models is a testament to its reliability. With just one CPU process, it can simulate a CGSPN model, significantly reducing the number of processes needed to run an RTL FI campaign in our test case. This optimization not only saves computational resources but also allows us to run other tasks simultaneously with the PIPE2 engine process, instilling confidence in the reliability of our simulation process.

Furthermore, the power of CGSPN simulation in identifying difficult-to-detect FMs is genuinely enlightening. By combining various failures on multiple components, we can reveal FMs that are typically elusive in FI simulations. The GL FI simulation can then confirm these FM cases, empowering the user to include them in the FMEDA. We validate an FM example through CGSPN simulation, which describes failures at the *decode*, *decode_execute*, and *execute_ctrl* components. After simulating the FM, we obtained a DC value of 71.7%, demonstrating the high mitigation effect of the CPU STL on the FM. We also discovered other FM with failures at the *fetch* and *control*. The CGSPN simulation estimates a DC value of 23.1%, serving as a crucial warning about the safety of our SoC architecture.

C. Methodology speed-up

Our approach also enhances the development process for SMs by reducing the size of the FI campaign, injecting only

the *DU* faults from the FI campaign of the last development cycle. Regarding the CPU STLs validation, we required 43.2% fewer faults than an exhaustive RTL FI campaign for one development loop, as shown in Figure 6. The chart also depicts the percentage of reduced injected faults for each selected component.

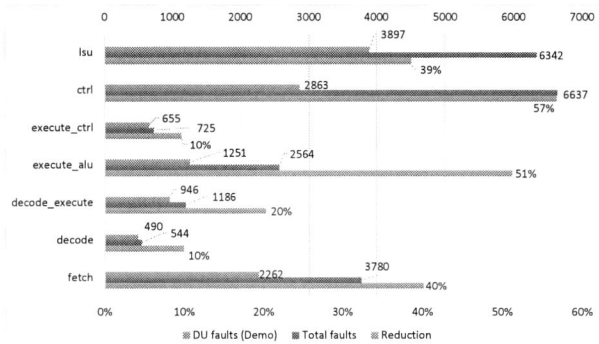

Fig. 6: Reduced number of faults required for the STL FI campaign

As the PIPE2 engine only requires one single process and we use 100 parallel processes to run the FI campaigns, the total speed-up would be 175x. Figure 7 depicts the speed-up process of our approach. In the calculation, we dismissed the simulation time for both the demo and enhanced CGSPN models, as the CGSPN simulations took around 70 minutes compared to the required number of days.

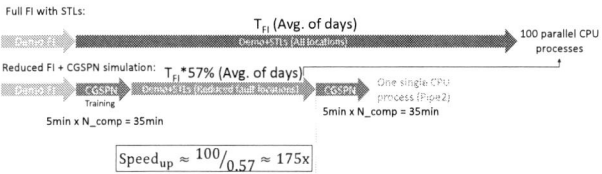

Fig. 7: CGSPN speed-up

D. Estimation Accuracy

After simulating the CGSPN models, we compared our estimated DC results with the ones obtained from the exhaustive FI campaign of the mor1kx CPU, including the STLs. In Figure 8, the blue bars represent the DC values of each component obtained from the exhaustive RTL FI campaign, the orange bars represent our estimated DC values obtained from the CGSPN simulation, and the gray bars the absolute estimation error.

Our approach's average absolute estimation error is 2.8% (i.e., 97.2% of average accuracy) for the selected CPU's pipeline components, demonstrating its practicality. The highest obtained value was 4.9% for the *execute_ctrl* component, followed by 4.5% for the *execute_alu* component. Conversely, the lowest obtained absolute estimation error value was 0.1%

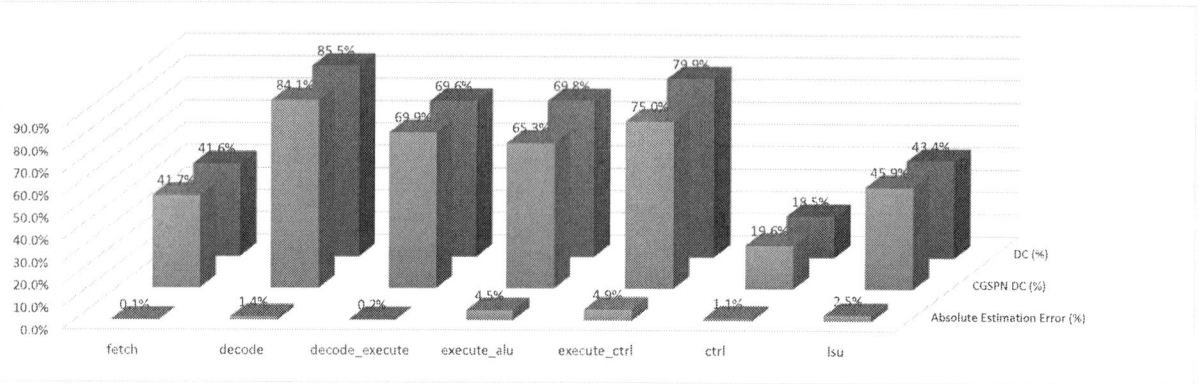

Fig. 8: DC estimation using CGSPN models

for *fetch* component. Even considering the worst-case estimation error (e.g., 4.9%), it is reassuringly low enough to be confidently applied to DSE during RTL design phases, demonstrating the real-world applicability of our approach.

We anticipated that estimation errors for each component would vary based on the complexity of the component, as shown in our previous work [11]. However, the difference between the estimated results and the values from the exhaustive FI campaign is influenced by the following reasons: disregarding the *UU* faults from the demo FI campaign to be reused by our approach, inclusion of STLs in the application that alters the fault propagation behavior through pipeline components, and conversion of some *DU* faults from the demo to *UU* faults in the new run.

VI. CONCLUSIONS

Safety-critical SoC development requires extensive verification processes based on FI simulations to ensure that the final design meets safety certification standards such as ISO26262. More reliable estimation approaches must be developed and implemented to enable early-stage estimations that allow DSE. The main contributions of this work are as follows: (i) An extension of a high-level modeling approach based on CGSPN for estimation of safety performance metrics, such as DC, with a low prediction error, (ii) an efficient development approach of SMs, like STLs, based on the DC estimation requiring less FI simulations, (iii) a new safety simulation framework for discovering new FMs that users can later incorporate on the FMEDA. The research problems stated in this work were addressed through our contributions and validated with experimental results: 1) Accurate estimation of DC of SMs, with a 97.2% accuracy compared to an exhaustive RTL FI campaign, 2) Acceleration of the SM's development process, requiring fewer computational resources (e.g. 175x)., 3) Enablement of SMs DSE during RTL design phases at the cost of low estimation error.

To the best of our knowledge, this work represents the first step towards predicting safety metrics for Automotive SoCs using high-level modeling approaches such as CGSPN.

ACKNOWLEDGEMENT

This work was partially supported by the German Federal Ministry of Education and Research (BMBF) under research grant No. 16ME0278.

REFERENCES

[1] ISO, "26262 road vehicles-function safety-part 5: Product development at the hardware level," *International Standardization Organization Std*, 2018.

[2] A. Ruospo *et al.*, "Assessing convolutional neural networks reliability through statistical fault injections," in *2023 Design, Automation & Test in Europe Conference & Exhibition (DATE)*. IEEE, 2023.

[3] F. A. da Silva *et al.*, "Determined-safe faults identification: A step towards iso26262 hardware compliant designs," in *2020 IEEE European Test Symposium (ETS)*. IEEE, May 2020.

[4] A. C. Bagbaba *et al.*, "Automated identification of application-dependent safe faults in automotive systems-on-a-chips," *Electronics*, vol. 11, no. 3, p. 319, Jan. 2022.

[5] F. A. da Silva *et al.*, "Flip flop weighting: A technique for estimation of safety metrics in automotive designs," in *2021 IEEE 27th International Symposium on On-Line Testing and Robust System Design (IOLTS)*. IEEE, Jun. 2021.

[6] A. Savino *et al.*, "ReDO: Cross-layer multi-objective design-exploration framework for efficient soft error resilient systems," *IEEE Transactions on Computers*, vol. 67, no. 10, pp. 1462–1477, oct 2018.

[7] Y. Zhang *et al.*, "Vulnerability estimation of DNN model parameters with few fault injections," *IEICE Transactions on Fundamentals of Electronics, Communications and Computer Sciences*, vol. E106.A, no. 3, pp. 523–531, mar 2023.

[8] M. Traiola *et al.*, "hardnning: a machine-learning-based framework for fault tolerance assessment and protection of dnns," in *2023 IEEE European Test Symposium (ETS)*. IEEE, May 2023.

[9] F. A. da Silva *et al.*, "Special session: Autosoc - a suite of open-source automotive soc benchmarks," in *2020 IEEE 38th VLSI Test Symposium (VTS)*, 2020, pp. 1–9.

[10] E. Roubtsova, "Chapter two - advances in behavior modeling," ser. Advances in Computers, A. M. Memon, Ed. Elsevier, 2015, vol. 97, pp. 49–109.

[11] E. C. Villegas *et al.*, "An efficient approach for stls development of automotive socs using colored petri nets," in *2024 27th International Symposium on Design and Diagnostics of Electronic Circuits and Systems (DDECS)*, 2024, pp. 1–6.

[12] N. J. Dingle *et al.*, "Pipe2: a tool for the performance evaluation of generalised stochastic petri nets," *SIGMETRICS Perform. Evaluation Rev.*, vol. 36, pp. 34–39, 2009. [Online]. Available: https://api.semanticscholar.org/CorpusID:3265173

Linear Algebra Approach to Verification of Modular (2^n-1) Multipliers

Jiteshri Dasari, Cunxi Yu, Maciej Ciesielski,
University of Massachusetts, Amherst, MA, USA
jdasari@umass.edu, cunxiyu@umd.edu, ciesiel@umass.edu

Abstract—This paper describes an original approach to formal verification of a special class of modular multipliers, namely modulo ($2^n - 1$) multipliers, critical components of cryptographic and error correction circuits. The proposed method completely avoids the expensive SAT, symbolic computer algebra, and rewriting techniques, typically used in formal verification of arithmetic circuits. Instead, recognizing a regular structure of such multipliers, constructed as an array of adders, the problem is modeled as a system of linear equations. Each adder is represented by a linear equation with an appropriate and easy to compute weight; the resulting linear system is solved by eliminating the intermediate signals, exposing the direct relation between the primary inputs and outputs. The results obtained for large ($2^n - 1$) modular multiplier circuits show several orders of magnitude improvement in CPU time compared to those in the published literature.

I. INTRODUCTION

Multipliers are some of the most important blocks of digital systems and are widely used in digital signal processing, machine learning, artificial intelligence, and more. A special class of multipliers are modular multipliers, essential components of communication systems that rely on complex hardware for cryptography and error correction. Verifying such hardware is paramount to guarantee security of digital systems.

A large body of work is available on verification of standard multipliers [1][2], dividers [3][4] and other arithmetic circuits. The state of the art in formal verification of such circuits is based largely on Symbolic Computer Algebra (SCA) approach and Groebner basis theory. In these methods, an arithmetic circuit is represented in algebraic domain, in which the input and output operands as well as the internal logic gates are modeled as pseudo-Boolean polynomials. The verification is based on transforming the output polynomial (output signature) into an input polynomial (input signature), and checking if the resulting polynomial matches the expected arithmetic function of the circuit. This process is popularly known as *algebraic* or *backward rewriting* [5]. Instead of rewriting the output polynomial to the inputs, some researchers work with a *specification polynomial*, a difference between the two signatures, and use backward rewriting to prove that it reduces to zero.

However, algebraic rewriting is plagued by a prohibitively large number of multi-variable *vanishing monomials* - terms that are generated during rewriting, but which reduce to zero as the polynomials move towards the primary inputs. A number of original techniques have been devised to avoid the explosive growth of such monomials. A notable example is that of [2], which claims to eliminate up to 38 million vanishing monomials in large multipliers to avoid memory explosion.

Another approach to formal verification, particularly popular in industry, is Theorem Proving. Theorem provers are interactive, inductive systems for proving that an implementation satisfies the specification using mathematical reasoning. The proof system is based on a large and strongly problem-specific set of axioms and inference rules, such as simplification, term rewriting, and induction. These systems require intimate knowledge of the design domain, extensive user guidance and expertise for efficient use. The success of the proof strongly depends on the choice and the order in which the rules are applied, with no guarantee for a conclusive termination. Theorem provers have been used in multiplier verification in the Amulet2 system [6] and in [7] based on ACL2 prover.

Modular arithmetic is frequently used in computer algebra, error detection and correction, and cryptography, where modular multipliers form basic implementation blocks of those systems. They are also essential components of homomorphic computing that performs computation on an encrypted data. Modular multipliers can be constructed with a parallel architecture [8] or as Montgomery modular digit-serial multiplication [9]. In general, design and structure of modular multipliers is more complex than those of the standard ones and creates a formidable verification challenge. An example of the verification of a Montgomery multiplier using theorem proving is given in [10], but it does not apply to the type of modular multipliers considered here. Other than that, work on modular multiplier verification is almost non-existent.

A notable exception in this area is the work of [11], which considers a class of ($2^n - 1$) and ($2^n + 1$) modular multipliers, where n is the number of bits of the operands. Despite the fact that these multipliers are characterized by a regular structure, their verification still poses a serious challenge. Their paper describes a verification approach using SCA-based rewriting and SAT-based procedure for removal of vanishing monomials. To deal with the modular nature of the circuit they developed a coefficient correction technique that is applied after each rewriting step to avoid memory explosion. This is followed by a specialized SAT-based technique to locally remove the vanishing monomials. A final step verifies that the result Z satisfies the modular condition, i.e. that $Z < m$ for modulus $m = 2^n - 1$ or $2^n + 1$, using standard SAT.

While original and well implemented, this technique is

unnecessarily computationally expensive, requiring about two hours of CPU time to verify a 512-bit modular multiplier. In contrast, the work described in this paper, recognizing a regular array-based architecture, can solve the problems on the same benchmark set several orders of magnitude faster. Our paper concentrates on a modulo $(2^n - 1)$ multiplier and develops a simpler technique that can verify large multipliers in single minutes of CPU time.

II. BACKGROUND

A. Modular Multipliers

A *modular multiplier* computes a product of two integer numbers modulo m, noted $Z = (A \times B) \bmod m$, with the inputs and outputs satisfying conditions: $A, B < m,\ Z < m$.

Modular multiplication can be implemented by generating product of $A \times B$, followed by modular reduction mod m. Alternatively, it can be obtained by integrating the modular reduction into the multiplier structure. The latter approach offers a more efficient hardware, both in terms of hardware area and delay.

Figure 1 shows a typical structure of a modular $(2^n - 1)$ multiplier, an $n \times n$-bit multiplier for bit width $n = 3$ and modulus $m = 7$. The structure is very similar to that of a regular multiplier, composed of the following parts: a partial product generation; partial product accumulation, typically using carry save adders; and the final stage of carry propagation and computation of the final result. The only difference between the two types of multipliers is how the adders in the most significant bit (MSB) position pass the carry-out to the next layer. While in a standard multiplier the MSB carry-out is passed to the next level in the same bit position (column $n - 1$), in the modular multiplier the MSB carry-out signal is wired to an adder in the least significant bit (LSB) position at column 0. This can be seen clearly in Figure 1.

B. SCA/Rewriting Approach

In a standard SCA approach, backward rewriting is applied to single-output logic gates, represented by nonlinear algebraic expressions. The following expressions are used to perform the rewriting over basic logic gates:

$$
\begin{aligned}
AND &: f = a \cdot b \\
OR &: f = a + b - ab \\
XOR &: f = a + b - 2ab
\end{aligned}
\tag{1}
$$

During rewriting, an output of each logic gate is handled separately, typically followed in a breadth-first fashion, from primary outputs to primary inputs. For example, the carry-out of a half-adder (HA) is rewritten as $C = a \cdot b$ and the sum output as $S = a + b - 2ab$; then each of the inputs, a, b, is rewritten further using the expression of the gate they correspond to, etc., creating more and more complex polynomials. Rewriting a full adder (FA) with $C = Maj3(a, b, c)$ and $S = XOR3(a, b, c)$ produces even more complex expressions. It is precisely this rewriting of nonlinear expressions that

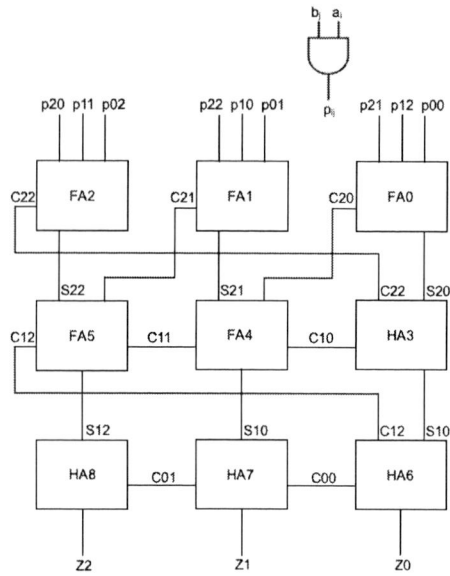

Fig. 1: Modular 3-bit mod 7 multiplier $A \times B$; inputs $p_{ij} = a_i b_j$ are partial products of A and B.

causes exponential growth in the number of polynomials, including vanishing monomials, that can get out of control.

Some rewriting system try to combine the carry-out C and sum S terms of the adders, using the following characteristic function of the half-adder (HA) and full-adder (FA) modules,

$$
HA : a + b = 2C + S \quad FA : a + b + c = 2C + S \tag{2}
$$

and rewrite an entire expression $2C + S$ across the adder. However, rewriting of a modular multiplier is more complicated, because some polynomials may not have the expected form $\{2kC + kS\}$ of an adder, for some signals associated with bit position log_k. As an example, consider the case of verifying a modulo 17 multiplier, given in [11], and look at the rewriting of the FA at bit position 3 (i.e., with $k = 8$). At some point, the polynomial that needs to be rewritten contains a sub-expression $\{-c_3 + 8s_3\}$, which cannot be directly rewritten into the inputs of the FA. In the standard multiplier each of the terms (c_3 and s_3) would have to be rewritten separately, offering no possibility of immediate reduction. However, in computation performed modulo 17, the term $-c_3$ can be converted to $\{-c_3 + 17c_3\} = 16c_3$, resulting in the expression $\{16c_3 + 8s_3\}$, which can be used for a clean rewriting across the adder.

To address this issue, the authors of [11] solve the following problem: given an expression $\{pC + kS\}$, check if there is a constant p such that $p = 2k + m \cdot q$, for some integer q, where m is the value of the modulus (here $m = 17$). If such a constant exists, then pC is replaced by $2k \cdot C$ and the term $\{2k \cdot c_3 + k \cdot s_3\}$ can be rewritten with the sum of inputs of the FA. This must be done for each occurrence of such a case. In all other cases, the individual elements of $\{pC + kS\}$ have

979-8-3315-3968-9/24 $31.00 © 2024 IEEE

to be rewritten separately, causing proliferation of monomials (possibly vanishing later).

In contrast, our approach does not encounter such a problem, as it does not apply backward rewriting and instead uses a much simpler approach based on Linear Algebra.

III. LINEAR ALGEBRA APPROACH

The proposed verification technique takes advantage of the regular structure of the modular multiplier and the natural separation between the partial product terms and the computation of sums. In this approach, the multiplier is partitioned into two parts:

1) A non-linear front-end part composed of partial products $p_{ij} = a_i b_j$, where a_i, b_i are individual bits of the operands A, B; and.

2) A linear array part composed of an array of interconnected HA and FA modules.

Inputs to the array come from the outputs of the non-linear part (symbolically shown in Figure 1 with a single AND gate). The output is encoded in n bits, $Z = \sum_{k=0}^{n-1} z_k$. The goal is to show that the polynomial representing output Z corresponds (i.e., is numerically equivalent) to the polynomial $A \times B$ modulo $(2^n - 1)$ for all values of primary inputs A, B.

The proposed verification process is illustrated with the modular 3-bit modulo 7 multiplier shown in Figure 1. Partial products p_{ij} are generated in the front-end from the bits of A, B using AND gates as $p_{ij} = a_i b_j$. Using those partial products, one can compute the *reference signature* of the multiplier, reduced modulo 7 by removing all terms with coefficients equal to 7.

$$(A \times B) \bmod 7 =$$
$$(4a_2 + 2a_1 + a_0)(4b_2 + 2b_1 + b_0) \bmod 7 =$$
$$[16p_{22} + 8(p_{12} + p_{21}) + 4(p_{11} + p_{02} + p_{20})$$
$$+ 2(p_{01} + p_{10}) + p_{00}] \bmod 7 =$$
$$4(p_{11} + p_{02} + p_{20}) + 2(p_{22} + p_{01} + p_{10}) + (p_{00} + p_{12} + p_{21})$$
$$\bmod 7$$
$$(3)$$

The expression in the last line is the expected *reference signature* of this modular multiplier. The terms in red are those whose coefficient have changed by $\bmod 7$ operation ($16p_{22} \to 2p_{22}$ and $8p_{12} + 8p_{21} \to p_{12} + p_{21}$). That is, in a functionally correct mod 7 multiplier the relation between the primary inputs and outputs should be as follows:

$$4z_2 + 2z_1 + z_0 =$$
$$4(p_{11} + p_{02} + p_{20}) + 2(p_{22} + p_{01} + p_{10}) + (p_{00} + p_{12} + p_{21})$$
$$\bmod 7$$
$$(4)$$

The difference between these two expressions is called *Specification Polynomial (SP)*. In our approach each half-adder (HA) and full-adder (FA) module is represented by a linear expression. Each expression is multiplied by the weight of the corresponding bit position (column) - to be determined by the procedure described next - and all such expressions are combined together to form a system of linear equations. The resulting system is then solved by adding all the equations.

In a functionally correct circuit, the internal signals should be eliminated, exposing the desired relation between the outputs and the input partial product terms.

The resulting expression is analyzed to see if it contains polynomials only in the input and output variables. The input portion of the expression (the *input signature*) is compared to the expected functional specification of the modular multiplier, expressed in partial products, to check if it is correct. If the final expression contains polynomials in any of the internal variables, it is examined to see if it is reducible mod m, in which case it can be discarded. Otherwise the multiplier is declared faulty, with the residual expression providing the proof of a bug.

IV. IMPLEMENTATION

The proposed verification procedure starts with a gate-level description of the multiplier in an AIG (and-inverter-graph) or BLIF (Berkeley logic intermediate form) format and is composed of the following steps:

1) Performing "reverse engineering" by extracting an array of interconnected adders.

2) Separating the design into the partial product generation part and the array of adders.

3) Creating equation for each adder module, using its characteristic HA or FA equation (2).

4) Computing the weight of each equation and representing them as a system of linear equations.

5) Solving the linear system and reducing the residual expression with internal variables modulo m. The remaining part, if any, is an indication of a bug.

6) Comparing the resulting linear expression with the expected reference signature of the multiplier.

7) Checking if the modular condition $Z < m$ is satisfied.

A. Extracting Network of Adders

Extraction of the set of interconnected adders is accomplished with the ABC tool [12] using command *&show -a*, which produces a list of interconnected adders and product terms. Figure 2 shows the result of such an extraction.

B. Creating a System of Linear Equations

The resulting file is parsed to create equations for each HA and FA, with the partial product terms considered as primary inputs to the system. The following are the equations generated for our 3-bit mod 7 multiplier example. An FA is represented by equation $a + b + c = 2C + S$, and a HA by $a + b = 2C + S$, where a, b, c are inputs and C, S outputs of the adder.

```
p00+p12+p21 -2*c20-s20 = 0
p01+p10+p22 -2*c21-s21 = 0
p02+p11+p20 -2*c22-s22 = 0
s20+c22 -2*c10-s10 = 0
c10+c20+s21 -2*c11-s11 = 0
c11+c21+s22 -2*c12-s12 = 0
s10+c12 -2*c00-z0 = 0
s11+c00 -2*c01-z1 = 0
s12+c01 -z2 = 0
```

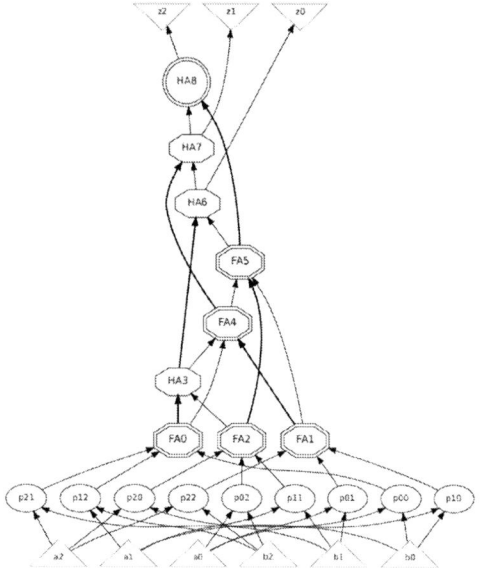

Fig. 2: A mult3-mod7 multiplier - structural ABC view.

It should be emphasized that the array structure and the resulting linear equations are not provided by the original circuit, which is given in an encoded AIG form, but instead are extracted using ABC.

The next step of the procedure is to find the multiplicative constant α_i for each equation i, in an attempt to eliminate the internal variables, while matching the input and output signatures. For example, in the case of our 3-bit mod 7 multiplier, the output signature $Z = 4z_2 + 2z_1 + z_0$ immediately dictates the weights α_i of the last three equations to be 1, 2, and 4. Similarly, the input signature expressed by:

$$4(p_{11} + p_{02} + p_{20}) + 2(p_{22} + p_{01} + p_{10}) + (p_{00} + p_{12} + p_{21}) \quad (5)$$

imposes weight 4, 2, and 1 on the first three equations, involving the respected partial product term. The composition of the input product terms (the input signature) does not need to be matched, and the weights can be computed by a separate procedure, described below.

A similar approach has been proposed for conventional (standard) multipliers in the original work of [13]. The goal there was to find integer coefficients α_i for each equation i that would reduce the weighted sum of the equations to an expression containing only the primary inputs (partial product terms p_{ij}) and output variables (z_i). The intention was to eliminate the internal variables; subsequently, the presence of a *residual expression*, containing any of the internal signals, would be an indication of a bug in the circuit. This approach, however, does not directly apply to modular multipliers, as some of the internal signals (variables) in the final sum may appear with the modulus constant m, and thus are reducible to zero. These signals can be considered analogous to the "vanishing monomials" which plague the SCA and algebraic

rewriting that systems like [2] and [11] so bravely fight off.

C. Computing the Weights

In our approach, instead of solving the linear system for α, we compute these coefficients based on the topology of the network of interconnected adders. Figure 2 shows such a network obtained by the ABC adder extraction tool. Starting with the least significant bit z_0 of the network, we can trace the Sum outputs of the adders leading to the primary inputs (product terms), namely

$$z_o \rightarrow HA_6 \rightarrow HA_3 \rightarrow FA_0 \quad (6)$$

and similarly for other bits of Z. All the equations associated with these adders will have coefficient $\alpha = 1$, which correspond to output z_0 in column 0 (LSB) of the array in Figure 1. Similarly, the equations associated with adders HA_7, FA_4, FA_1 will have coefficient $\alpha = 2$, corresponding to z_1 in column 1; and those associated with adders HA_8, FA_5, FA_2 will have coefficient $\alpha = 4$, corresponding to z_2 in column 2. In summary, the computed coefficients α correspond to the respective columns of the array, even if the array structure is not directly available from the input file and needs to be computed.

D. Solving the Linear System

Once the coefficients α have been determined, the system of linear equations is transformed by multiplying each equation by the corresponding coefficient, as shown below.

```
1*[p00+p12+p21 -2*c20-s20] = 0
2*[p01+p10+p22 -2*c21-s21] = 0
4*[p02+p11+p20 -2*c22-s22] = 0
1*[s20+c22 -2*c10-s10] = 0
2*[c10+c20+s21 -2*c11-s11] = 0
4*[c11+c21+s22 -2*c12-s12] = 0
1*[s10+c12 -2*c00-z0] = 0
2*[s11+c00 -2*c01-z1] = 0
4*[s12+c01 -z2] = 0
-------------------------------------------
SUM:
4(p11+p02+p20)+2(p22+p01+p10)+(p00+p12+p21)
 +7c12+7c22 -4z2-2z1-z0=0
```

E. Analyzing the Results

The equations are then added together, resulting in an expression that links the inputs $\{pp_i\}$ and outputs $\{z_i\}$. For our mod7 example, this results in the following equation:

$$4(p_{11} + p_{02} + p_{20}) + 2(p_{22} + p_{01} + p_{10}) + (p_{00} + p_{12} + p_{21})$$
$$+ 7c_{12} + 7c_{22} - 4z_2 - 2z_1 - z_0 = 0$$

One can notice that, in addition to the expression involving the product terms (input signature) and the outputs (output signature), this equation includes an expression associated with the internal variables, namely $7c_{12} + 7c_{22}$. This expression however reduces to zero under modulus 7.

The result of solving the above system of linear equations (adding all equations together) can be illustrated graphically

979-8-3315-3968-9/24 $31.00 © 2024 IEEE

using a canonical TED representation [14] in Figure 3. Note the correct coefficients of the individual partial products and those with coefficient 7 that reduce modulo 7 to 0.

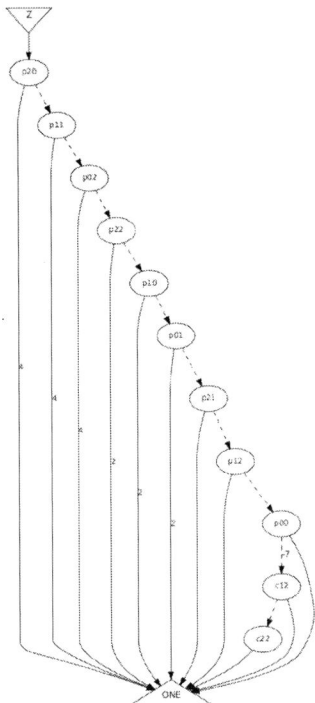

Fig. 3: TED view of the mult3-mod7 multiplier.

F. Checking the Output Condition

The final step of the verification is to check the modular condition for the output, $Z < m = 2^n - 1$. As already suggested and done by [11], the satisfiability of this condition can be easily checked using a SAT solver by checking if the CNF for $Z \geq m$ combined with input conditions $A, B < m$ is unSAT. One should notice that for modulus $m = 2^n - 1$ the sum of the terms of output Z with coefficients 2^k for $k = 1, .., 2^{n-1}$ evaluates to value less or equal to $2^n - 1$. The only case when this condition is violated is when all terms $z_i = 1$. Thus, alternatively, one can use SAT solver to check if $\Pi_i z_i \neq 1$. Both of these methods are solvable in surprisingly short amount of time, taking less than a second of CPU time, as also reported in [11].

V. Experimental Results

The verification method described in this paper was implemented as program written in Python with interfaces to the ABC tool [15]. The input to the program is a file representing a gate-level modular circuit in AIG or BLIF format. ABC tool was used to extract adders and create a list of adders and AND gates using the *&show -a* command. The resulting hierarchical netlist file, visualized in Figure 2, was parsed to create linear equations for the multipliers with partial product terms as inputs. A python program was written to compute the weights of the equations and to create a system of linear equations. A python library was used to generate and solve the system and the results were reduced by removing mod $2^n - 1$ terms and compared to the expected specification polynomial.

The program was tested on a benchmark set of $(2^n - 1)$ modular multipliers in AIG format, ranging from $n =4$ to 512 bit-widths, courtesy of authors of [11]. The experiments were run on a M1 MacBook Pro computer using a single core and 8 GB of RAM. The results shown in Table I give the size of the circuits, the CPU time of the adder extraction, generating and solving the linear equations, and the total CPU time to generate the results. As shown in the table, our approach offers a significant (up to 125×) improvement compared to those published in [11], the only one available for comparison [1]; neither SAT nor combinational equivalence checking (CEC) of ABC were able to solve the problem for multipliers with operand sizes greater than 16 bits.

VI. Summary and Conclusions

The paper presents a novel approach to formal verification of modular $2^n - 1$ multipliers. It partitions the design into two components: 1) a Boolean/nonlinear part composed of partial products, modeled in Boolean or algebraic domain; and 2) an array of adders, modeled as a system of linear equations. In case of the modular $(2^n - 1)$ multiplier the nonlinear part is simple, composed of a single layer of AND gates. In other cases, such other modular multipliers or Booth multipliers, the nonlinear section can be more complex. For example, a modular $2^n + 1$ multiplier has several layers of logic gates, but as long as they are separable from the linear array and their outputs serve as inputs to the linear part, the proposed approach applies equally well.

The approach described here avoids the numerous problems faced by CSA- and SAT-based approaches (dealing with excessive number of polynomials and vanishing monomials), by avoiding having components with non-linear equations. Instead it groups some of the single-output components or logic gates (e.g., Majority-3, XOR, AND) into multiple-output adders (FA, HA), characterized by linear equations. Verification of the internal array of the multiplier is modeled as a system of linear equations, which can be computed very fast. An important and critical part of the approach is efficient computation of weights, which replaces the expensive process of modulo reduction and vanishing monomial elimination encountered in backward rewriting.

Advantages of this approach are simple linear formulation and fast execution. In principle this method can be extended to other arithmetic designs that i) can be partitioned into linear and nonlinear portions; and ii) contain multiple-output modules that can be represented as linear equations. Many arithmetic circuits fall in this category and are composed of linear and nonlinear part that can be treated separately.

---

[1]It should be noted that a portion of the difference can be contributed to the performance difference between their processor i7-8565U and ours, M1.

TABLE I: Verification results for $2^n - 1$ modular multiplier (in CPU seconds)

Size	Nodes	Extracting adders	Computing weights	Solving linear sys.	**Total time**	Verif. time [11]
4x4	160	< 0.01	< 0.01	< 0.01	< 0.01	< 0.01
8x8	744	< 0.01	< 0.01	< 0.01	< 0.01	0.01
16x16	3,096	< 0.01	< 0.01	0.01	0.02	0.05
32x32	12,344	0.02	0.01	0.05	0.08	0.21
64x64	49,784	0.07	0.09	0.19	0.35	1.40
128x128	197,880	0.29	0.62	0.81	1.71	15.16
256x256	787,960	1.15	4.81	3.44	9.40	384.39
512x512	3,152,888	4.58	38.26	12.78	55.62	6,948.99

The main limitation of the method is that it only applies to designs with a fixed, regular structure, composed of modules that can be modeled with linear equations. It takes advantage of the structural regularity of the circuit and may not work for structures that deviate from it. Specifically, it may not be directly applied to multipliers with arbitrary value of modulus m, unless the adder array structure similar to that considered in this paper is given. By comparison, the traditional rewriting approach of [11] is more general and applies to designs that do not possess such structure - but at a cost of high computational complexity. On the other hand, with current complexity of designs reaching levels of intractability, the priority should be given to designs that can assure their verification, embracing the concept of "design for verification", even if at the cost of the area/delay sub-optimality. This is very much in line with the idea of "polynomial time verification", advocated in [16].

Future work will focus on extending the linear algebra technique described here to $(2^n + 1)$ modular multipliers. These circuits, while characterized by similar linear array structure, exhibit a more complex nonlinear part that includes several layers of logic AND and OR gates providing inputs to the linear portion and the inversion of the carry out signal between the rows. Our initial attempts to solve this problem were unsuccessful due to a lack of reliable reverse-engineering techniques to produce such an array structure. We believe that the complication comes mostly from the inversion of the carry out signal between the MSB cell of a given row of adders and the LSB cell of the next row. Our detailed simulation confirm this suspicion. Knowing the exact structure instead of deriving it from the lower level Verilog, AIG or Blif netlists, would solve this problem. An additional challenge is to determine the reference signature (expressed in terms of partial products p_{ij}), similar to the one developed for this modular design, taking that increased complexity into account.

ACKNOWLEDGEMENT

This work has been supported by a grant from the National Science Foundation, Award No. CCF-2006465.

REFERENCES

[1] M. Ciesielski, T. Su, A. Yasin, and C. Yu, "Understanding Algebraic Rewriting for Arithmetic Circuit Verification: a Bit-Flow Model," *IEEE TCAD*, vol. 39, no. 6, pp. 1346–1357, 2019.

[2] A. Mahzoon, D. Große, and R. Drechsler, "REVSCA-2.0: SCA-based formal verification of non-trivial multipliers using reverse engineering and local vanishing removal," *IEEE Transactions on Computer-Aided Design of Integrated Circuits and Systems*, 2021.

[3] C. Scholl, A. Konrad, A. Mahzoon, D. Große, and R. Drechsler, "Verifying dividers using symbolic computer algebra and don't care optimization," in *2021 Design, Automation & Test in Europe Conference & Exhibition (DATE)*. IEEE, 2021, pp. 1110–1115.

[4] J. Dasari and M. Ciesielski, "Efficient formal verification and debugging of arithmetic divider circuits," in *2023 IEEE/ACM International Conference on Computer Aided Design (ICCAD)*, 2023, pp. 1–9.

[5] C. Yu, W. Brown, D. Liu, A. Rossi, and M. Ciesielski, "Formal verification of arithmetic circuits by function extraction," *IEEE Transactions on Computer-Aided Design of Integrated Circuits and Systems*, vol. 35, no. 12, pp. 2131–2142, 2016.

[6] D. Kaufmann and A. Biere, "AMulet 2.0 for verifying multiplier circuits." in *TACAS (2)*, 2021, pp. 357–364.

[7] M. Temel, A. Slobodova, and W. A. Hunt, "Automated and scalable verification of integer multipliers," in *International Conference on Computer Aided Verification*. Springer, 2020, pp. 485–507.

[8] R. Zimmermann, "Efficient vlsi implementation of modulo (2/sup n//spl plusmn/1) addition and multiplication," in *Proceedings 14th IEEE symposium on computer arithmetic (Cat. No. 99CB36336)*. IEEE, 1999, pp. 158–167.

[9] S. Fatemi, M. Zare, A. F. Khavari, and M. Maymandi-Nejad, "Efficient implementation of digit-serial montgomery modular multiplier architecture," *IET Circuits, Devices & Systems*, vol. 13, no. 7, pp. 942–949, 2019.

[10] C. Walther, "Formally verified montgomery multiplication," in *Computer Aided Verification: 30th International Conference, CAV 2018, Held as Part of the Federated Logic Conference, FloC 2018, Oxford, UK, July 14-17, 2018, Proceedings, Part II 30*. Springer, 2018, pp. 505–522.

[11] A. Mahzoon, D. Große, C. Scholl, A. Konrad, and R. Drechsler, "Formal verification of modular multipliers using symbolic computer algebra and boolean satisfiability," in *Proceedings of the 59th ACM/IEEE Design Automation Conference*, 2022, pp. 1183–1188.

[12] R. Brayton and A. Mishchenko, "ABC: An Academic Industrial-Strength Verification Tool," in *Proc. Intl. Conf. on Computer-Aided Verification*, 2010, pp. 24–40.

[13] M. A. Basith, T. Ahmad, A. Rossi, and M. Ciesielski, "Algebraic Approach to Arithmetic Design Verification," in *Formal Methods in CAD*. FMCAD, 2011, pp. 67–71.

[14] M. Ciesielski, P. Kalla, and S. Askar, "Taylor Expansion Diagrams: A Canonical Representation for Verification of Data Flow Designs," *IEEE Trans. on Computers*, vol. 55, no. 9, pp. 1188–1201, Sept. 2006.

[15] A. Mishchenko *et al.*, "ABC: A system for sequential synthesis and verification," *URL http://www. eecs. berkeley. edu/~alanmi/abc*, 2007.

[16] R. Drechsler and A. Mahzoon, "Preserving design hierarchy information for polynomial formal verification," in *2022 IFIP/IEEE 30th International Conference on Very Large Scale Integration (VLSI-SoC)*. IEEE, 2022, pp. 1–7.

Adaptive block-scaled GeMMs on vector processors for DNN training at the edge

Nitish Satya Murthy[12], Nathan Laubeuf[1], Debjyoti Bhattacharjee[1], Francky Catthoor[12] and Marian Verhelst[12]

[1]imec, Kapeldreef 75, 3001 Leuven, Belgium, [2]ESAT - KU Leuven, Kasteelpark Arenberg 10, 3001 Leuven, Belgium
Email: {nitish.satyamurthy, nathan.laubeuf, debjyoti.bhattacharjee, francky.catthoor, marian.verhelst}@imec.be

Abstract—Reduced precision datatypes have become essential to the efficient training and deployment of Deep Neural Networks (DNNs). A recent development in the field has been the emergence of block-scaled datatypes: tensor representation formats derived from floating-point, that share a common exponent across multiple elements. While these formats are being broadly adopted and optimised for by DNN-specific inference accelerators, the potential benefits for training workloads on general-purpose (GP) vector processors has yet to be thoroughly explored. This work proposes a benchmarked implementation of block-scaled general matrix multiplications (GeMM) for DNN training at the edge using commercially available vector instruction sets (ARM SVE). Using this implementation, we highlight an accuracy-speed trade-off involving the shape of shared exponent blocks — vectors or squares. We exploit this result to optimize the training of fully connected networks by dynamically adapting the shared exponent block shapes during training. This strategy yields on average around 1.95× faster training with 2× lower memory footprint compared to standard IEEE 32-bit floating point (FP32), while achieving similar accuracy.

Index Terms—DNN training, Vector processors, Block-scaled datatypes, ARM SVE ISA

I. INTRODUCTION

Recent concerns regarding privacy, reliability, and latency have led to a growing interest in techniques to efficiently train DNNs at the edge [1]. This becomes even more essential for tasks that depend on real-time or online learning, for effective functionality. Continuous control robotics tasks encounter such challenges, where the robots are mostly battery-powered and need to operate in dynamic environments [2]. Desirable robotic control in such scenarios demands efficient DNN training through the robot's environment interactions in the field [3].

Some of the most notable efforts towards efficient DNN training at the edge have focused on using reduced precision datatypes to accommodate the limited memory and compute resources typically available to edge processors. Despite the success of these efforts at deploying DNNs in their inference regime using low precision fixed-point integers [4], the training of DNNs under these conditions remains a significant challenge. Faced with more stringent precision and dynamic range constraints, the execution of backpropagation – DNNs' principal training algorithm – has motivated the exploration of more complex data representations [5], [6].

This research received funding from the Flemish Government (AI Research Program).

Among such representations, block-scaled datatypes are gathering increasing attention. By associating a common scale to a block of numbers, block scaled datatypes are able to combine the deployment efficiency of fixed point datatypes with the dynamic range of floating-point types [7]–[9]. However, most reported implementations of block-scaled datatypes have so far required specific hardware support [7], [10], [11].

Performing block-scaled quantization requires the selection of which and how many elements share a common exponent or scale. To maximize computational efficiency when using block-scaled DNNs, the block quantization axis should align with the accumulation axis in GeMM, as that enables efficient fixed-point accumulations within each block. DNN training typically involves transpose and quantization operators in the backward pass, in contrast to only forward passes in inference. If these operators are non-commutative, block-scaled DNN training will result in extra block transformation overheads [12], [13], which we analyze later in section III-A.

To achieve high accuracy within the latency and energy constraints of edge deployment, it is crucial to make the training process adaptable. Adaptive training for block scaled datatypes has been considered in the past for custom hardware, by means of varying bit precision [10] and blocksizes [11]. However, these methods cannot be adapted directly for deployment on general-purpose (GP) vector processors.

To address these research challenges, we make the following contributions in the paper:

- We propose square block shapes as opposed to conventional vector shapes for block-scaled datatypes and implement GeMMs using them on ARM vector processor.
- We develop an adaptive block-scaled DNN training approach, by initially performing "coarse-commutative" training followed by "fine-adaptation" to allow accuracy-training time trade-off.
- Our proposed method obtained around 1.95× speedup with 2× lower memory footprint for fully-connected NN (FCNN) using adaptive training over baseline FP32 models.

The rest of the paper is organized as follows: Section II discusses the background and related works. Section III explains our methods. Sections IV and V describe our experimental setup and results respectively. Lastly, we conclude in section VI.

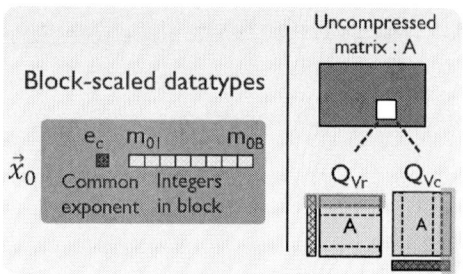

Fig. 1: Block-scaled datatypes with vector (row/column) are demonstrated.

Instruction	Description
svld1	Loading data into vector registers.
svld1rq	Loading 128-bits data replicated along the total vector length.
svld1ub	Extended load to 32-bit vector registers.
svdup	Duplication of a scalar value along total vector length.
svst1b	Storing truncated 8-bit values into memory.
svdot_lane	Indexed 4-way int8 dot product accumulating in 32-bit integer outputs, with lane index specifying which 8-bit integer group is selected for first vector operand.
svasr / *svlsl*	Right / Left shift lanes of vector register by shift values in another vector register operand.
svmaxv	Return maximum value of the vector register.

TABLE I: ARM Vector Instructions used for the implementation of block-scaled GeMM kernels.

II. BACKGROUND AND RELATED WORK

A. Block-scaled datatypes for DNN training

Block-scaled datatypes: Extensive research has been done on number representations for efficient DNN executions [14]. Most popular datatypes are floating point (FP) and fixed-point/integer (FXP) types. FP numbers generally have good dynamic range over FXP, however require more complex HW computation units. Block-scaled formats combine the 2 datatypes, by sharing a common exponent among multiple FXP numbers in a block as depicted in Fig. 1. Block-scaled datatypes are recently gaining traction due to their deployment ease and desirable HW-SW trade-offs [7], [9], [12]. Block-scaled dot-product computation with operands $\vec{x_0}, \vec{x_1}$ with block size B, common scales e_0, e_1, integers $m_{0,i}, m_{1,i}$ for $i = (0, 1, \ldots, B-1)$ is as shown in Eqs. (1)-(2). The equations also showcase efficient FXP computation within a block, and enhanced dynamic range through common exponents.

$$\vec{x_j} = 2^{e_j}[m_{j,0}, m_{j,1}, \ldots, m_{j,B-1}] \text{ for } j = (0,1) \quad (1)$$

$$\vec{x_0} \cdot \vec{x_1}^T = 2^{e_0 + e_1} \sum_{i=0}^{B-1}(m_{0,i} * m_{1,i}) \quad (2)$$

There are multiple variants of block-scaled quantizations depending on the number representation of each block element being sub-byte [7], minifloat [12], [15]; multi-level scaling [9]; static requantizations [16], etc. Most of the previous approaches involved custom HW implementations. Few works targeted GP vector processors, but only evaluated tensor-level

exponents [17] [18]. In contrast, we focus on optimizing block-scaled GeMM kernels for DNN training through flexible block shapes and sizes utilizing only the available vector processor instructions (described in section II-B) in commercial ISAs [19].

Quantized DNN training: Considering a DNN with layer L involving forward propagation of uncompressed activations $A^{[L]}$ and weights $W^{[L]}$, the following equations represent the computations involved during block-scaled training.

$$A^{[L+1]} = Q_{Vr}(A^{[L]}) \times Q_{Vc}(W^{[L]}) \quad (3)$$

$$G^{[L]} = Q_{Vr}(Tr(A^{[L]})) \times Q_{Vc}(E^{[L+1]}) \quad (4)$$

$$E^{[L]} = Q_{Vr}(E^{[L+1]}) \times Q_{Vc}(Tr(W^{[L]})) \quad (5)$$

where, Q_{Vr} and Q_{Vc} are block-scaled quantizers like in Fig. 1 for vectors in the row and column direction respectively (quantization direction along GeMM accumulation axis), Tr is a transpose operator, L is the layer index, $\times$ denotes matrix multiplication operator. Eq. (3) is the forward pass which is calculated end-to-end for all indices L, and all intermediate uncompressed $A^{[L]}$ are stored for later usage. Eq. (4) calculates the gradients $G^{[L]}$ for their respective weights $W^{[L]}$ using all stored $A^{[L]}$. Eq. (5) represents the error propagation proceeding in the opposite direction of Eq. (3). Note that the quantizer nodes enable the GeMM operations to be computed in efficient block-scaled computations.

B. ARM Vector Instructions

Vector extensions (such as SVE and SVE2) offer architectural enhancements designed to improve the performance of compute-intensive applications, such as AI workloads due to their SIMD compute capabilities [20]. These extensions introduce a scalable vector length (ranging from 128 to 2048 bits) that allows flexible and efficient parallel processing. Block-scaled computing formats allow representing wide dynamic ranges with fewer bits and offer the possibility to improve the compute efficiency. The mapping of the block-scaled types on vector instructions presents multiple challenges, since the instruction set extension does not natively support these formats [19]. Furthermore, auto-vectorization of the block scaled computation kernels cannot be performed by an off the shelf compiler effectively. Careful mapping is required to allow full utilization of the available vector lanes. Furthermore, the conversion and alignment of data formats needs to be carefully performed to minimize the overheads associated with management of the data movement between the processor and the memory subsystem. The support for integer operands by these extensions can be effectively leveraged while computing in the block-scaled domain, as we demonstrate in the following sections. Some of the instructions used for the mapping are summarily shown in Table I.

979-8-3315-3968-9/24 $31.00 © 2024 IEEE

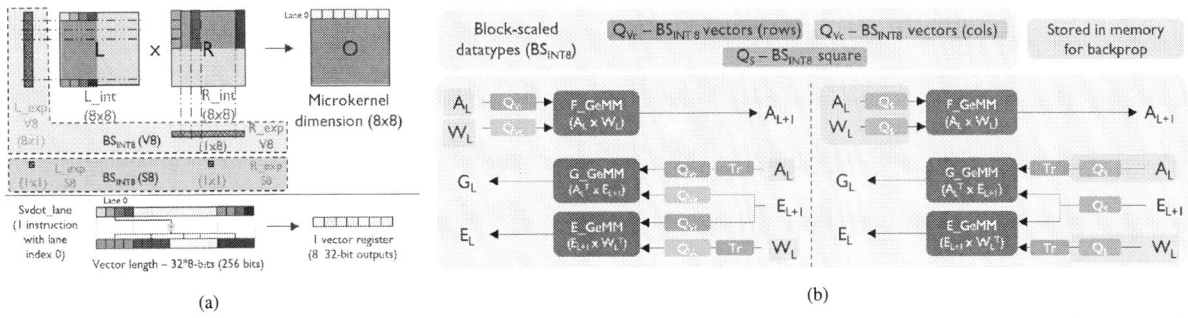

Fig. 2: (a) Illustration of matrix multiplication microkernel (8x8) with integer parts and exponent parts (V8 and S8 configurations). Note that each number and the common exponents are 8-bit integers. An example of lane-based dot product using ARM SVE instruction on a 256-bit vector processor is also illustrated. (b) The optimizations in DNN training computation graph after reversing the orders of quantization and transpose operators in backward pass for square shapes.

III. BLOCK-SCALED INTEGER (BS_{INT8}) GEMMS FOR DNN TRAINING

A. Optimizing block shapes for DNN training

Fig. 2a illustrates an 8x8 microkernel calculation part of a bigger GeMM kernel. With BS_{INT8} quantization, the matrix multiplication involves a series of parallel integer operations on the mantissa's and a single shared exponent operation shown previously in Eq. (2).

The integer matrix-multiplication part is mainly computed using lane-based dot product instructions that output 32-bit integers. With a vector block-scaling approach, based on the alignments of block-quantization and accumulation axes, the left (right) matrix should involve row-wise (column-wise) blocks resulting in a column (row) exponent vector.

On the other hand, square blocks will have a single scalar exponent per block tile. We propose utilizing square blocks instead of conventional vector blocks, due to their advantage of being commutative in quantization-transpose operations as illustrated in Fig. 3. It details how the operation orders of performing transpose and block-scaled quantization leads to different outputs for vector shapes (patch 1), while producing same outputs for square shapes (patch 2). This allows us to reverse these operation orders for squares in BW passes, leading to optimization opportunities as shown in the training computation graphs in Fig. 2b.

Fig. 2b demonstrates that block-scaled activations ($Q_S(A^{[l]})$) and weights ($Q_S(W^{[L]})$) can be stored with their common exponent for later usage in the BW passes

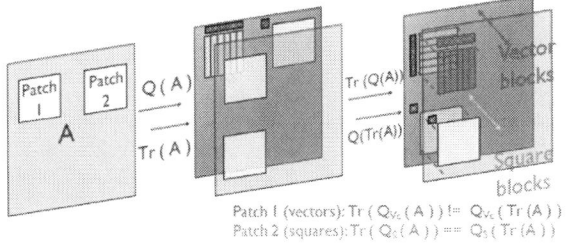

Fig. 3: Investigating commutative property of transpose and quantization operators for different block-scaled quantizations.

for square blocks. This leads to lower memory footprint and data transfer costs. Moreover, this circumvents having to perform extra quantizations before GeMMs by storing already block-quantized values. Vector block-scaled values, on the other hand, would after transpose, lose their common exponent per element of the block, and hence have to be stored in non-compressed formats.

Notations: In the rest of the paper, we represent our block-scaled datatypes by $BS_{INT8}(V/S)$ for vector/square block shapes respectively. Block size is denoted by a number next to the shape. Ex.: $BS_{INT8}(V8)$ denotes vector blocks sharing an 8-bit exponent among 8 elements of INT8 numbers. $BS_{INT8}(S8)$ forms a square block of 8x8, sharing an exponent among 64 elements.

B. Vectorized BS_{INT8} GeMM kernels for efficient computation on vector processors

While the previous subsection justified our square blocks and its impact on the DNN training complexity, there is additional impact on the GeMM computations themselves. Figure 4 shows the pseudo-code of our BS_{INT8} vectorized kernel implementations using ARM SVE vector instructions, along with the difference in instruction counts due to block shapes. Our block-scaled GeMMs involve an integer calculation part, and separate exponent handling computations (highlighted in blue) for aligning exponents of different block partial accumulations.

The reduced GeMM instruction count for square versus vector shapes mainly arises from exponent handling overheads due to their different number of exponents for same matrix dimensions. The block size along accumulation axis directly relates to the alignment overhead needed for performing block-scaled GeMM kernels. It is important to note that the number of elements sharing a common exponent is not identical among the two shapes, rather the number of elements in the accumulation direction is kept similar instead. This leads to square shapes having higher compression ratios than vectors, leading to also lower instruction counts, however, at the cost of higher approximation errors as will be seen in section V-A.

```
L, R: Left/right integer. Dimensions: L-(MxK), R-(KxN)
B: Block size (8),
Lexp, Rexp: exponent matrices for L and R resp.

                                              Instruction
                                              (count and type)
Lp = pack(L)
Rp = pack(R)                                  N_exp    V8  8x
for (x=0; x<M; x+=8):                                  S8  1x
  for (y=0; y<N; y+=8):
    Vaccs[0:7] = 0;                           1- vector load (R)
    Ecur[0:Nexp] = Lexp[0:Nexp] + Rexp[0]; }  1- scalar load (L)
    for (b=0; b<K/B; b+=1):                    1- Compute
      for (z=0; z<B; z+=4):
        Lv1 = svld1rq(Lp[x : x+16, y]);        3-Load
        Lv2 = svld1rq(Lp[x+16 : x+32, y]);     (vector)
        Rv = svld1(Rp[z : z+4, y : y + 8]);

        Vaccs[0:4] = svdot_lane(Lv1, Rv, index=[0:4]); } 8-Computes
        Vaccs[4:7] = svdot_lane(Lv2, Rv, index=[4:7]); } (vector)

    Enew[0:Nexp] = Lexp[x*K/B + (b+1)] + Rexp[(b+1)*K/B + y]; }
                                              1- vector load (R)
    Sh = | Enew - Ecur |;         3 - Exponent 1- scalar load (L)
    Pred = svcmpge(Enew, Ecur);   handling     1- Compute

    svasr/svlsl(Pred, Vaccs[0:7], Sh[0:7]); } 16 compute

  Requantized output store of Vaccs


  ┌──────────────────────────────────────────────────┐
  │ Instr. counts excluding packing/requantization:   │
  │ (M/8) * (N/B) * ( (1+2*N_exp) + (K/B*B/4*(3+8)) + (K/B-1)*(1+5*N_exp+16) ) │
  │                                                    │
  │ V8 (N_exp=8) : M*N/64*(17+11/4*K+57*(K/B-1))      │
  │ S8 (N_exp=1) : M*N/64*(3+11/4*K+22*(K/B-1))       │
  └──────────────────────────────────────────────────┘
```

Fig. 4: Pseudo-code for our GeMM kernel using BS_{INT8} formats with square and vector shapes, also highlighting the instruction counts for either scenario.

C. Adaptive DNN training through vectorized BS_{INT8} GeMMs (vector and square)

We propose an adaptive training approach here to obtain an optimized solution that properly trades off HW and SW metrics as illustrated in Fig. 5. The previous subsections described the differences in HW metrics (instr. counts, memory footprint) due to different block-shaped kernels. However, these block shapes which have different compression ratios would also lead to different training qualities (SW metrics - loss/accuracy) due to different quantization errors.

Algorithm 1 details our adaptive approach to change block shape during training process. A quantization configuration is set to first perform $BS_{INT8}(S8)$ training and then switch to $BS_{INT8}(V8)$. We define a hyperparameter α that is used to split the total training process amongst the 2 configurations. This follows the principle of performing a coarse training search that is more computationally efficient and later performing fine adaptation which is slower, but more accurate. As shown in section V-C, performing this multi-stage training helps in reaching optimized solutions based on accuracy-training speed trade-offs.

IV. EXPERIMENTAL SETUP

Workload: We analyze FCNN models which are trained to predict the system dynamics of continuous control robotics tasks for benchmarking our approach. As mentioned in section I, these tasks require real-time or online learning based on the robot interactions with the environment at the edge. Hence, efficient embedded runtime training techniques would aid in optimizing task performance with limited resource constraints.

Algorithm 1 Adaptive BS_{INT8} DNN training

1: Conf_list: [BS_{INT8} (S8), BS_{INT8} (V8)], Max epochs: E_{max}, Epoch ratios with α: $E_r = [\alpha, 1-\alpha]$, DNN model (weights W): M_W.

2: **for** id in [0, 1] **do**
3:     Model quantization set to Conf_list[id]
4:     **if** $id==0$ **then**
5:         M_W random weights initialization
6:     **else**
7:         $M_W[id] = M_W[id\text{-}1]$ (Initialize with prev weights)
8:     **end if**
9:     **for** t = 1 to $E_r[id] \cdot E_{max}$ **do**
10:         Train DNN on M_W with training data
11:         Monitor average loss, L = []
12:     **end for**
13: **end for**

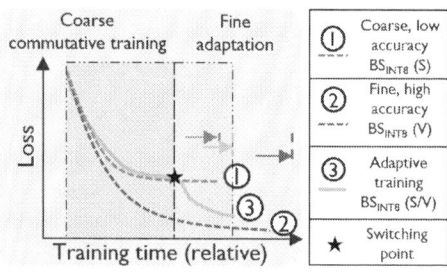

Fig. 5: Adaptive training for accuracy-speed trade-offs.

These robotics tasks involve FCNNs with state-action pair inputs – current robot state values and actions that control the robot. It outputs a probability distribution (mean and standard deviation) for the next state values capturing the uncertainty in its predictions, which is utilized for effectively controlling the robot based on user specifications [3]. This work focuses on analyzing different training performances by simulating our BS_{INT8} quantization configurations in PyTorch using quantization libraries [21]. The cartpole models use 3 layer FCNN with 500 hidden neurons (balance a pole in upright position). The pusher and reacher models use 3 layers with 200 neurons (push an object and reach a goal using a robotic arm). The halfcheetah model uses 4 layers and 200 neurons (should move/run fast). ReLU activations, an Adam optimizer with initial learning rate of 0.001 and a Gaussian loss function are used in all scenarios. This is mainly used as a demonstration task, while our technique is applicable to any generic DNN training task which involve GeMM operations.

Target HW details: We target general-purpose vector processors, in particular ARM v8.4a+ architectures with SVE extension because of its wide-spread usage in the edge and mobile community. Our experiments are run with the 256 bits SVE vector length. We benchmark our kernel runs through instruction/cycle counts on a gem5 [22] MinorCPU model with 64kB L1 cache and 1MB L2 cache memories.

Kernel code configurations: All matrices of GeMM operands are assumed to be stored in row-major formats. Furthermore, in BS_{INT8} configurations, the exponent and integer matrices are stored separately. The code implementation is written in C using vector intrinsic functions from ARM

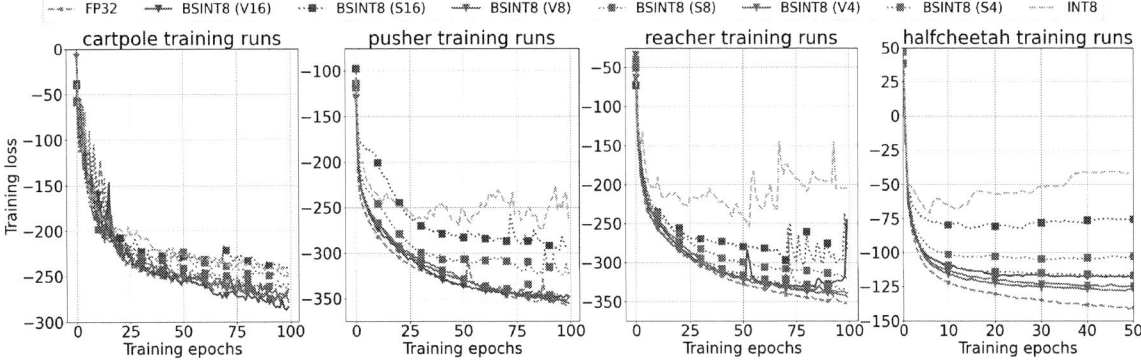

Fig. 6: Impacts of block-scaled datatypes (shape / sizes) on DNN training quality (Baseline models – FP32 and INT8 (tensor-level scale)).

C Language Extensions (ACLE) for high efficiency. ARM GCC cross-compiler 12.3 was used on x86-64 host platforms targeting aarch64 platforms.

V. RESULTS AND ANALYSIS

A. DNN training curves due to block-scaled datatypes (vector and square)

We analyze here the impacts of block shapes (S/V) and sizes (4/8/16) on the training quality for different datatypes. As observed from Fig. 6, the quality is worse for square configurations due to higher block elements that share a common exponent. The quality also reduces for increasing block sizes (highest for INT8 with tensor-level scales), which is more prominent in the relatively more complex task - halfcheetah. Hence, although the square blocks are more suited for training computationally, it is still prone to error. This provides us with crucial design parameter choices that can be tuned based on the desired trade-off user requirements.

B. Benchmarking BS_{INT8} vector and square GeMM kernels

While previously, we looked at the training quality impacts with $BS_{INT8}(S/V)$ formats, the GeMM kernel deployment was not analyzed. We compare the instruction/cycle counts for performing GeMMs in the different phases of backpropagation and analyze the instruction/cycle counts for different datatypes (baseline - FP32 and INT8 formats). In Fig. 7, we can firstly observe the split-up of instruction and cycle counts for the cartpole model. Each operation phase (FGeMM for FWD, GGeMM and EGeMM for BWD gradient and error computations as in Eqs. (3), (4) and (5) respectively) require similar counts as they involve GeMMs of similar dimensions. Additionally, we observe increasing speedups as we go from FP32, BS_{INT8} (V8), BS_{INT8} (S8) to non-blocked INT8, while also having increasing quantization errors. This speedup increase from V8 to S8 is mainly due to the reduced overheads of exponent handling as seen earlier in the pseudo-code in Fig. 4. The figure also shows the compression ratios leading to reduced memory footprints and data transfer costs, while handling intermediate activations during backpropagation. Note that BS_{INT8} (V8) has only 2× compression, as we need to store exponents for each element as seen in earlier sections.

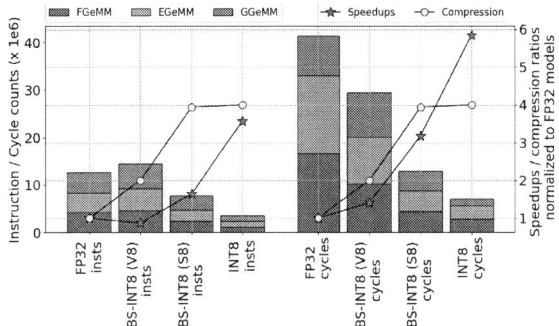

Fig. 7: Vectorized GeMM kernels execution profiling for cartpole FCNN model with different datatypes depicting instruction/cycle counts, speedup and memory compression ratios. Cycle counts are a better system-level metric and would hence be used hereafter for relative speedup comparisons.

Lastly, these flexible deployment choices arise from vector processor programmability while not needing any extra custom HW, and yet providing different options for optimized HW-SW trade-offs.

C. Adaptive DNN training through vectorized BS_{INT8} GeMMs (vector and square)

We observed in the previous subsections the trade-off of $BS_{INT8}(V8)$ vs $BS_{INT8}(S8)$ in terms of HW instruction/cycle count metrics and SW training loss metrics. In this subsection, we showcase and analyze the results of this trade-off as seen in Fig. 8. Compared to baseline FP32 models, our non-adaptive vector block models achieve 0.9 – 1.4× faster training times, while the square block models are 2.1 – 3.2× faster across different tasks. At the extreme end lies baseline INT8 models with tensor-level exponents, which are fast but suffer significantly in quality as also observed in the convergence noise of loss curves in Fig. 6. Figure 8 also shows the results of our adaptive approach that follows Algorithm 1 through a tunable parameter α to decide the ratio of training epochs performed in $BS_{INT8}(S)$ vs $BS_{INT8}(V)$. This enables us to reach more optimized solutions than non-adaptive approaches. For $\alpha = 0.95$, we observe very close speedups to $BS_{INT8}(S)$, while ensuring no significant accuracy loss for most tasks.

979-8-3315-3968-9/24 $31.00 © 2024 IEEE   83

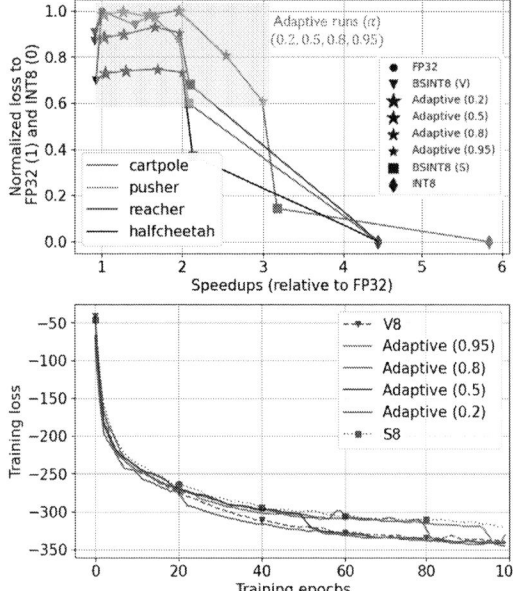

Fig. 8: HW-SW trade-offs through our various BS_{INT8} training configurations (in comparison to baseline FP32 and INT8 models). Training curves for different adaptive training ratios is also showed for the reacher task here.

VI. CONCLUSION

DNN training at the edge induces many challenges due to resource constraints. These are addressed by efficient number representations such as block-scaled datatypes. We propose the use of square shape blocks to exploit the commutative property of quantization and transpose operators, which allows more optimized DNN training. We implement efficient square and vector shape block-scaled ($BS_{INT8}(S/V)$) GeMM kernels on vector processors by only using commercially available vector instruction sets (ARM SVE). Furthermore, we propose the adaptive combination of different shapes during training process to effectively trade-off HW metrics such as cycle counts and memory footprint, with SW training loss metrics. Benchmarking on fully-connected networks, our training approach can yield around $1.95\times$ faster training with $2\times$ lower memory footprint, compared to conventional FP32 models while reaching similar training performances. In the future, we plan to evaluate our adaptive training approach on different network architectures, varied on-device learning scenarios, and other diverse hardware platforms.

REFERENCES

[1] C.-H. Lu, Y.-C. Wu, and C.-H. Yang, "A 2.25 tops/w fully-integrated deep cnn learning processor with on-chip training," in *2019 IEEE Asian Solid-State Circuits Conference (A-SSCC)*, pp. 65–68, IEEE, 2019.

[2] S. M. Neuman, B. Plancher, B. P. Duisterhof, S. Krishnan, C. Banbury, M. Mazumder, S. Prakash, J. Jabbour, A. Faust, G. C. de Croon, *et al.*, "Tiny robot learning: Challenges and directions for machine learning in resource-constrained robots," in *2022 IEEE 4th International Conference on Artificial Intelligence Circuits and Systems (AICAS)*, pp. 296–299, IEEE, 2022.

[3] K. Chua, R. Calandra, R. McAllister, and S. Levine, "Deep reinforcement learning in a handful of trials using probabilistic dynamics models," *Advances in neural information processing systems*, vol. 31, 2018.

[4] R. Krishnamoorthi, "Quantizing deep convolutional networks for efficient inference: A whitepaper," *arXiv preprint arXiv:1806.08342*, 2018.

[5] P. Micikevicius, S. Narang, J. Alben, G. Diamos, E. Elsen, D. Garcia, B. Ginsburg, M. Houston, O. Kuchaiev, G. Venkatesh, and H. Wu, "Mixed precision training," in *International Conference on Learning Representations*, 2018.

[6] G. Alsuhli, V. Sakellariou, H. Saleh, M. Al-Qutayri, B. Mohammad, and T. Stouraitis, *Number Systems for Deep Neural Network Architectures*. Springer, 2023.

[7] B. Darvish Rouhani, D. Lo, R. Zhao, M. Liu, J. Fowers, K. Ovtcharov, A. Vinogradsky, S. Massengill, L. Yang, R. Bittner, *et al.*, "Pushing the limits of narrow precision inferencing at cloud scale with microsoft floating point," *Advances in neural information processing systems*, vol. 33, pp. 10271–10281, 2020.

[8] G. Yang, T. Zhang, P. Kirichenko, J. Bai, A. G. Wilson, and C. De Sa, "Swalp: Stochastic weight averaging in low precision training," in *International Conference on Machine Learning*, pp. 7015–7024, PMLR, 2019.

[9] B. Darvish Rouhani, R. Zhao, V. Elango, R. Shafipour, M. Hall, M. Mesmakhosroshahi, A. More, L. Melnick, M. Golub, G. Varatkar, *et al.*, "With shared microexponents, a little shifting goes a long way," in *Proceedings of the 50th Annual International Symposium on Computer Architecture*, pp. 1–13, 2023.

[10] S. Q. Zhang, B. McDanel, and H. Kung, "Fast: DNN training under variable precision block floating point with stochastic rounding," in *2022 IEEE International Symposium on High-Performance Computer Architecture (HPCA)*, pp. 846–860, IEEE, 2022.

[11] S. Lee, J. Choi, S. Noh, J. Koo, and J. Kung, "DBPS: Dynamic Block Size and Precision Scaling for Efficient DNN Training Supported by RISC-V ISA Extensions," in *2023 60th ACM/IEEE Design Automation Conference (DAC)*, pp. 1–6, IEEE, 2023.

[12] B. D. Rouhani, R. Zhao, A. More, M. Hall, A. Khodamoradi, S. Deng, D. Choudhary, M. Cornea, E. Dellinger, K. Denolf, *et al.*, "Microscaling data formats for deep learning," *arXiv preprint arXiv:2310.10537*, 2023.

[13] M. G. d. Nascimento, V. A. Prisacariu, R. Fawcett, and M. Langhammer, "Hyperblock floating point: Generalised quantization scheme for gradient and inference computation," in *Proceedings of the IEEE/CVF Winter Conference on Applications of Computer Vision*, pp. 6364–6373, 2023.

[14] T. Liang, J. Glossner, L. Wang, S. Shi, and X. Zhang, "Pruning and quantization for deep neural network acceleration: A survey," *Neurocomputing*, vol. 461, pp. 370–403, 2021.

[15] S. Fox, S. Rasoulinezhad, J. Faraone, P. Leong, *et al.*, "A block minifloat representation for training deep neural networks," in *International Conference on Learning Representations*, 2020.

[16] H. Fan, G. Wang, M. Ferianc, X. Niu, and W. Luk, "Static block floating-point quantization for convolutional neural networks on fpga," in *2019 International Conference on Field-Programmable Technology (ICFPT)*, pp. 28–35, IEEE, 2019.

[17] D. Das, N. Mellempudi, D. Mudigere, D. Kalamkar, S. Avancha, K. Banerjee, S. Sridharan, K. Vaidyanathan, B. Kaul, E. Georganas, *et al.*, "Mixed precision training of convolutional neural networks using integer operations," in *International Conference on Learning Representations*, 2018.

[18] B. de Bruin, Z. Zivkovic, and H. Corporaal, "Quantization of deep neural networks for accumulator-constrained processors," *Microprocessors and microsystems*, vol. 72, p. 102872, 2020.

[19] N. Satya Murthy, F. Catthoor, and M. Verhelst, "Optimization of block-scaled integer gemms for efficient dnn deployment on scalable in-order vector processors," *Journal of Systems Architecture*, p. 103236, 2024.

[20] N. Stephens, S. Biles, M. Boettcher, J. Eapen, M. Eyole, G. Gabrielli, M. Horsnell, G. Magklis, A. Martinez, N. Premillieu, *et al.*, "The arm scalable vector extension," *IEEE micro*, vol. 37, no. 2, pp. 26–39, 2017.

[21] T. Zhang, Z. Lin, G. Yang, and C. De Sa, "Qpytorch: A low-precision arithmetic simulation framework," in *2019 Fifth Workshop on Energy Efficient Machine Learning and Cognitive Computing-NeurIPS Edition (EMC2-NIPS)*, pp. 10–13, IEEE, 2019.

[22] J. Lowe-Power, A. M. Ahmad, A. Akram, M. Alian, R. Amslinger, M. Andreozzi, A. Armejach, N. Asmussen, B. Beckmann, S. Bharadwaj, *et al.*, "The gem5 simulator: Version 20.0+," *arXiv preprint arXiv:2007.03152*, 2020.

979-8-3315-3968-9/24 $31.00 © 2024 IEEE

A High Throughput, Energy-Efficient Architecture for Variable Precision Computing in DRAM

Gian Singh
Arizona State University
Tempe, AZ, USA
gsingh58@asu.edu

Ayushi Dube
Arizona State University
Tempe, AZ, USA
adube9@asu.edu

Sarma Vrudhula
Arizona State University
Tempe, AZ, USA
svrudhul@asu.edu

Abstract—DRAM-based near-memory architectures are recognized for their ability to deliver substantial energy efficiency and throughput to execute data-intensive tasks. However, the inherent limitations regarding area, power, and timing within DRAM allow the integration of only primitive processing elements with limited operations and application support. This paper introduces a near-memory processing architecture based on DRAM featuring a novel computing unit termed the neuron processing element (NPE). NPEs are capable of performing multiple arithmetic, logical, and predicate operations. With a well-defined instruction set, the NPEs can be programmed to support standard data formats for floating point and fixed point precision used in different AI/ML and signal processing applications. They can be dynamically reconfigured to switch operations during run-time without increasing overall latency or power consumption. The NPEs have a small area and power footprint compared to conventional MAC units and other functionally equivalent implementations, making them suitable for integration with DRAM without compromising its organization or timing constraints. Furthermore, this paper shows a substantial improvement in latency and energy consumption compared to prior in-memory architectures and demonstrates the efficacy of the proposed architecture for the acceleration of neural network inference.

Index Terms—Processing-in-Memory, Low-power, Deep Neural Networks, DRAM, Memory Wall, Energy Efficiency.

I. INTRODUCTION

Machine learning (ML) is fast becoming a dominant paradigm of computing in almost every domain. ML algorithms are realized as parametric *function graphs* (i.e., deep neural networks–DNN) in which nodes represent the composition of inner products and non-linear functions, and connections represent function composition. The major internet companies like Google, Meta, Microsoft, and others deploy DNNs with hundreds of billions of parameters, performing trillions of large dimensional matrix operations. Thus, DNNs are often both memory and compute-bound. Consequently, they require massive amounts of memory and large server farms with thousands of high-performance GPU processors. The electricity usage of such server farms is approaching that of whole industries and some nation states, and for such systems to be sustainable, at least one to two orders of magnitude improvements in energy-efficiency are required [1]–[3].

Existing CPU/GPU processors based on the traditional von Neumann processor architecture in which the computation units

This work was supported in part by the NSF I/UCRC IDEAS center and from the NSF grant #2324945.

and main memory (i.e. DRAM) are separated are wholly inadequate [4] for large-scale memory and compute-intensive applications such as ML because the energy required to data transfer between the CPU and DRAM is nearly two orders of magnitude greater than that required to perform computations on a CPU [5]. Processing-in-memory (PIM), in which the computation units are integrated with the DRAM, can eliminate the energy consumption of the data transfers and achieve the required improvement in energy efficiency.

While the concepts of underlying PIM are not new, the recent proliferation of ML has led to a rapid development of PIM architectures. Both SRAM and DRAM have been proposed for the design of PIM architectures. SRAM implementations either utilize analog multiply and accumulate (MAC) with small ADCs (quantization < 4 bits) [6], or bit-serial digital computation [7]. SRAM-based compute-in-memory (SRAM-CIM) is usually implemented in the cache of CPU/GPU of small size (< 100 MB), requiring at least one-time data transfer from an external larger capacity memory such as a DRAM. Thus, for large models, data transfers on the memory channel dominate the energy consumption of the system.

A DRAM-based PIM architecture eliminates all the data movement on the memory channel. A DRAM, with its much larger capacity (> 10 GB), can provide various amounts of parallelism depending on where the compute elements are placed. The closer they are to the memory array, the more available parallelism can be exploited.

In-array architectures modify the memory array itself [8]–[10]. This exploits the maximum available parallelism by performing bit-wise operations on entire rows of banks. Simple logic operations are performed by exploiting the charge-sharing characteristics of the transistors that comprise the memory cells or by modifying the sense amplifiers of the memory banks. These require changes to the memory access protocol and timing, and incurs high delay when performing arithmetic operations. Tools like DRAM bender [11] are used to implement and test In-array PIM architectures on commercial DRAM chips.

Near-array processing places the compute elements just outside the memory array after the bit line sense amplifiers (BLSA) [12], [13]. Here, the compute elements can receive the maximum number of bits (e.g., 8K to 16K) from a memory array. Unfortunately, due to stringent constraints on how much memory capacity can be sacrificed, only compute elements of

979-8-3315-3968-9/24 $31.00 © 2024 IEEE

limited functionality (e.g. a full adder as in [12]) can be used.

Near-bank processing places compute elements further away from the memory array after the local I/O circuits. Here the bandwidth is reduced to 64 bits for a DDR memory or 256 bits in the case of 3D-HBM (High Bandwidth Memory) [4], while still sacrificing nearly 50% of the memory capacity. Finally, compute elements can be placed entirely outside the memory chip at the RANK level, as in [14]. This is essentially a CPU and a small DRAM memory on a separate chip. All of these approaches present trade-offs between memory capacity and the amount of parallelism. The near-array design [13] places LUTs and the near-bank design [4], with reduced bandwidth, places 16-bit floating-point MACs–both incurring 50% reduction in memory capacity.

A. Main Contributions

In this paper, we present a PIM design that addresses some of the basic disadvantages of all existing PIM designs. As explained above, the fundamental challenge is to introduce compute elements that have high *compute density*, very low area, and low latency near the memory array so as to access the maximum number of bits in parallel. The low area should allow many compute elements to be deployed near the array to operate in a SIMD mode so as to maximize throughput. With conventional CMOS logic, the requirements of high compute density, low area and latency are conflicting.

We propose a novel solution to this problem by using a new circuit element, referred to as a *configurable neuron* (CN), that can realize complex functions within an ultra-small area. For instance, the carry-out of a 5-input carry-lookahead adder can be realized in an area the size of a single D-flipflop. A small network of CNs forms the basic compute unit that can be configured, without any area or delay penalty, to compute arithmetic, bit-wise logic, and comparison functions. The area and power of a CN is 50% of a functionally equivalent, equi-delay version using conventional logic. A combination of the compute unit and additional logic forms a novel processing element, called NPE (neuron processing element), that can be introduced into a DRAM at the output of each memory array to maximally exploit all the available parallelism.

The proposed PIM architecture supports multiple data formats (integer and floating point) and multiple bit-precision (4, 8, 12, 16, 32 bits) as the instruction set of the NPEs supports all logic, arithmetic, and comparison operations. The proposed architecture achieves on an average $1.78\times$ improvement in throughput against [13], and $2.64\times$ against [4], and $9.27\times$ improvement in energy efficiency as compared to [13]. [4] and [13] are state-of-the-art PIM architectures.

II. PROPOSED PIM ARCHITECTURE

A. Top-Level Architecture

Fig. 1(a) shows the top-level architecture of the proposed PIM design. It consists of a High Bandwidth Memory (HBM) cube, which has multiple DRAM chips and a base logic layer connected using the through silicon vias (TSVs) to form a 3D integrated memory with high density. Each DRAM layer

TABLE I: Comparison of different Processing Elements/cores used in PIM architectures.

PE Type	Area (20nm)	#PEs/Bank	Relative Throughput /Bank
MAC INT 16 [4] (w/ 48-bit Acc.)	61.77	16	1.00
MAC INT 8 [4] (w/ 48-bit Acc)	27.80	32	2.00
MAC INT 8 [4] (w/ 32-bit Acc)	21.62	32	2.00
MAC FP 16 [4]	81.54	16	1.00
MAC BFLOAT 16 [4]	71.04	16	1.00
NPE-INT 16	1	1024	3.70
NPE-INT 8	1	1024	11.11
NPE-BFLOAT 16	1	1024	4.49

in the HBM consists of a collection of 2D memory arrays called banks, as shown in Fig. 1(b). The main innovation of this paper is the design of the Neuron Processing Elements (NPEs) (shown in Fig. 1(c)), which can be interfaced directly to the bit-line sense amplifier arrays of the DRAM banks without interfering with the timing constraints or access protocols of the memory as shown in Fig. 1(b). The array of NPEs, *NPE-Array*, connected to the BLSA operates in a SIMD fashion to enable massive compute parallelism inside the DRAM array. The NPE is the basic computing element that can be configured using the control signals generated by an instruction decoder in the logic layer of the HBM. These signals are broadcast to all the NPEs to operate in a SIMD fashion. NPEs perform different operations without any delay penalty and without having to include separate units for different functions. The instruction set of the NPE consists of multi-bit carry-lookahead operation, multi-bit logic, and comparison operations, which provide an ability to map any higher-level arithmetic, linear, or non-linear functions (multiplication, pooling, ReLU, etc.) in different data formats (integer and floating point) to the NPE. The NPE-Array enables the parallel execution of the matrix-matrix (MM) and matrix-vector (MV) multiplications which are common to many workloads (DNNs, LLMs, etc.) that drive much of the applications today. The integration of the NPE-Array to the memory banks can be easily adapted to all the 2D (DDR, GDDR, LPDDR, etc.) and 3D (HMC and HBM) DRAM organizations, making such a PIM architecture scalable from edge devices to high-end servers. This paper uses the HBM organization to demonstrate the efficacy of the NPEs for DRAM-based PIM architectures.

Why is NPE suitable for BLSA integration? The DRAM provides maximum memory parallelism (8K bits) at the array level or at the output of bit-line sense amplifiers (BLSAs). The major constraint in adding logic near BLSA is the high area overhead of the compute elements. Therefore, prior architectures have been able to place only primitive gates near BLSA [12]. Such architectures have high latency for multi-bit arithmetic operations, which makes them unsuitable for many data-intensive applications such as DNNs. On the other hand, several PIM architectures have used multiply and accumulate (MAC) units of different precision to accelerate ML applications inside DRAM [4], [15]. These architectures, however, place the MAC units outside the bank I/O, thus operating on a much lower data width (64 or 256 bits). Furthermore, these

979-8-3315-3968-9/24 $31.00 © 2024 IEEE

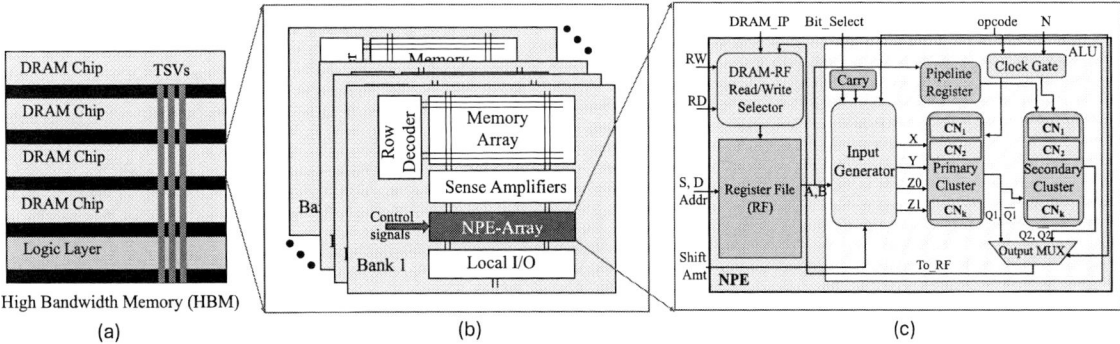

Fig. 1: (a) High Bandwidth Memory (HBM) with multiple DRAM chips; (b) Bank-level description of the proposed PIM architecture showing the proposed NPE array interfaced with a bank; (c) The proposed NPE microarchitecture.

architectures require more hardware for other operations of the DNNs such as pooling and ReLU. McDRAM [15] showed that adding 8-bit MAC units near BLSA in LPDDR4 leads to an area overhead of 120%. Thus, such an architecture is not feasible.

The unique design of the NPE, as explained later in this section, provides the flexibility to perform multi-bit operations in a very low area footprint. Table I provides an area and throughput/bank comparison of NPE and different MAC units [4]. The MAC units are placed outside the DRAM bank with parallel access of 256 bits, and NPEs are placed near BLSA with parallel access of 16K bits [15]. The area advantage of the NPE is clearly evident, and moreover, the NPE can be configured to perform MAC operations of different bit-width, unlike conventional MAC units.

B. Neuron Processing Element (NPE)

Fig. 1(c) shows the major components of the proposed NPE microarchitecture. At the microarchitecture level, the NPE resembles an execution unit of a conventional microprocessor. The NPE is designed to issue a single instruction in every cycle. The three main components of the NPE include **DRAM-RF Read/Write Selector** (referred to as the Selector), **Register File (RF)** with associated circuitry, and the **Arithmetic Logic Unit (ALU)**. The design of ALU is what distinguishes the NPE from the conventional CMOS implementations.

The main computing elements in the ALU are the *Primary Cluster (PC)* and the *Secondary Cluster (SC)*. These clusters consist of k *Configurable Neurons (CNs)* which can perform k-bit operations (for some k) in a single clock cycle. The CNs are the basic compute elements that perform the operations on each bit of the operands. The main operations of the cluster are k-bit carry look ahead, k-bit sum, k-bit comparison, and k-bit logic operations. The cluster can implement all these functions just by using k-CNs. No additional hardware is required for any of the functions. The function on the cluster is selected by simply generating the appropriate set of inputs to the neurons. This leads to a very compact design of the primary and secondary clusters. For instance, for $k=5$ as used in this paper, in 40nm technology, the CN-based primary cluster is **4×** smaller than an equivalent CMOS implementation.

The other components of the ALU comprise an Input Generator (referred to as the Generator), a Clock Gate (CG) module, a Carry register, and an output multiplexer (Output_MUX).

The Instruction Set Architecture (ISA): The ISA of the proposed NPE design is formally defined in Fig. 2. It consists of the 11 functions represented by the 4 bits of the opcode field of the instruction encoding as shown in Fig 2. The instructions of the NPE include arithmetic (ADD (addition)), COMP (comparison) and logical (AND, OR, NOT, XOR, XNOR, LADD (addition after logical left shifting second operand), RADD (addition after logical right shifting second operand), RCAR (reset the carry register to zero) and MAND (AND of all bits of first operand with a single bit)).

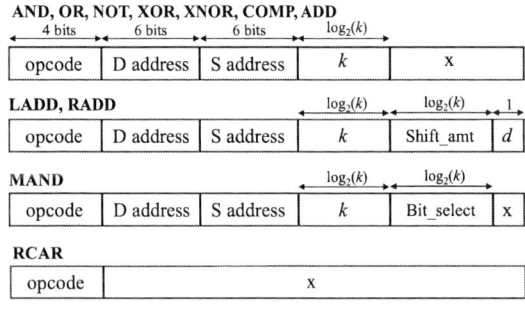

Fig. 2: ISA for the NPE (D and S denote the destination and source register respectively; k is the bit-width of the operands; d is a bit denoting left or right direction; x denotes unused bit(s).

NPE Operation: The primary inputs to the NPE are control signals (*RW and RD*), input data (*DRAM_IP*), and the opcode. The **Selector** interfaces the data from DRAM to the register file (RF) in the NPE. The NPE operates in two modes: *functional mode* and *buffer mode*. In the buffer mode, the NPE either reads data from the DRAM and writes to the RF or writes data back to the DRAM from the RF. In this mode, the ALU of the NPE is clock-gated to save power consumption. In functional mode, the write port of the RF is connected to data coming from the ALU to write the output of the operation performed by the ALU. The control signals *RW and RD* are used to select the mode of the operation of the NPE.

979-8-3315-3968-9/24 $31.00 © 2024 IEEE

In the ALU, the Input Generator is responsible for formatting the ALU operands A and B based on the opcode such that they can be directly interfaced with the individual inputs of the CNs in the clusters. The function performed by a CN is concisely described by Equation 1 below.

$$Q(p, Z_0, X, Z_1, Y) = Z_0 + \sum_{j=0}^{p-1} 2^j X_j \geq Z_1 + \sum_{j=0}^{p-1} 2^j Y_j \quad (1)$$

For k-bit ALU operands A and B, the primary and secondary clusters each consist of k CNs and each with $2k + 2$ inputs. Therefore, the total number of configuration bits per cluster is $2k^2 + 2k$ (X (k^2 bits), Z0 (k bits), Y (k^2 bits) and Z1 (k bits)) to evaluate the Q function as defined in Equation 1. These configuration bits for different functions are listed in Table II.

TABLE II: Arguments to a Q function for k-bit operations.

operation	p	Z0	X	Z1	Y
AND	1	0	A_k	1	$\sim B_k$
OR	1	0	A_k	0	$\sim B_k$
NOT	1	0	A_k	1	0
COMP	k	$COMP_{out}$	A[k-1:0]	1	B[k-1:0]
ADD	k-1, 2	$Carry_{in}, A_k$	A[k-1:0], B_k	1, 0	$\sim$B[k-1:0], {$Carry_k, \sim Carry_{k-1}$}
XOR	2	A_k	B_k	1	{AND_k, 0}
XNOR	2	A_k	B_k	1	{AND_k, 0}
LADD	k-1, 2	$Carry_{in}, A_k$	A[k-1:0], B_k	1, 0	$\sim$B[k-1:0], {$Carry_k, \sim Carry_{k-1}$}
RADD	k-1, 2	$Carry_{in}, A_k$	A[k-1:0], B_k	1, 0	$\sim$B[k-1:0], {$Carry_k, \sim Carry_{k-1}$}
MAND	1	0	A_k	1	$\sim B_k$
RCAR	0	0	0	0	0

This explains how different functions can be computed by simply setting the appropriate bits. All functions except for SUM and XOR/XNOR need only the primary cluster (PC) for execution. A carry-look-ahead adder is implemented for SUM operation wherein the PC computes the carry bits C_k. The output of PC, $Q1$, and $\overline{Q1}$ are supplied to the secondary cluster (SC) and the Output MUX. The SC evaluates the SUM bits where each of the k bits is computed using one out of k CNs. The outputs to the SC are $Q2$ and $\overline{Q2}$. For the XOR/XNOR operation, the PC evaluates the AND operation between operands A and B, which is used as an intermediate operation to compute XOR/XNOR. A clock gate module operates only the necessary clusters and/or CNs. The output multiplexer (output_MUX) selects the required output among $Q1$, $\overline{Q1}$, $Q2$, and $\overline{Q2}$ as per the opcode.

Values of the inputs to the Q function depend on the operation to be performed, as shown in Table II. For example, consider an AND operation between two 1-bit operands A and B, which can be calculated using $Q(1,0,A,1,\overline{B})$. By substituting the appropriate values into Equation 1, results in $0+A \geq 1+\overline{B}$, which in turn can be rewritten as $A + B \geq 2$. Other k-bit logic operations are similarly defined. They are computed in one cycle using a neuron cluster in an NPE. For $XOR(A_{[i]}, B_{[i]})$, where $i \leq k$, there is a two-level cluster network and therefore requires two cycles.

For N-bit operands ($N > k$), addition, comparison, logic, **multiplication** in **integer or floating point** format can be decomposed into a sequence of k-bit operations, executed sequentially on a single NPE. In this paper, $k = 5$.

The implementation of floating-point operations requires the additional step of *Normalization*. The normalization procedure

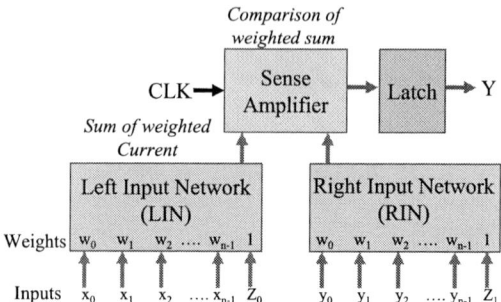

Fig. 3: Configurable Neuron (CN): a mixed-signal circuit implementation based on Threshold Logic (TL) [18].

is dependent on the specific output value of the operation on the input operands which can break the SIMD operation of the NPEs. This paper uses the Normalization algorithm presented in [16] for the SIMD operation. This Normalization procedure depends only upon the bit-width of the operands; thus, the same operations are performed for all the NPEs irrespective of the value of the input operands. The algorithm is mapped onto the ISA of the proposed NPE to be performed at the end of each floating point operation.

In summary, the NPE can: (1) implement multiple functions on the same hardware structure, (2) instantaneously switch between the various functions by activating specific inputs of the CN, and (3) utilize the exact number of CNs required in a cluster depending upon the operand bit-width. All of these factors result in extremely low power and a very small area footprint of the NPE.

C. Circuit Implementation of Q (Configurable Neuron)

The Q function in the Equation 1 is realized by a *mixed-signal* circuit called Configurable Neuron (CN) as shown in Fig. 3. The inputs x_i and Z_0 are mapped to the left input network (LIN) of the structure, where each x_i has an associated weight w_i. The w associated with Z_0 is 1. Similarly, the inputs y_i and Z_1 are mapped to the right input network (RIN). For the NPE design $w_i = 2^i$. When the clock is enabled, LIN and RIN evaluate the weighted sum of the inputs in the form of a cumulative current and connect to the Sense Amplifier (SA) as two differential signals, as shown in Fig. 3. The SA evaluates to 1 (0) if the LIN current is greater (lesser) than the RIN and stores the result in the SR latch. A detailed discussion on the threshold functions is provided in [17]. **Note:** The weights w_i in a CN are implemented by programmable resistors, whose resistance values are set after fabrication. The details of the CN's circuit architecture and the realization of the w_i are beyond the scope of this paper. The circuit level details of the CN design are presented in [18]. It is not the main contribution of this paper.

The advantages of a single CN over a functionally equivalent CMOS are substantial [18]. At the individual cell level, a 5-input CN results in improvements in area, power, and delay of [80%, 60%, 40%] respectively, over the performance-optimized, functionally equivalent CMOS circuit.

979-8-3315-3968-9/24 $31.00 © 2024 IEEE   88

III. EXPERIMENTS AND RESULTS

A. Design and evaluation methodology

The NPE is designed in Verilog and synthesized in 40nm technology using the TSMC standard cell library of the CMOS and the configurable neuron data obtained from the work [18]. The frequency of the NPE is 300 MHz, which is equal to the internal HBM frequency [4]. The area, power, and timing information of the NPEs extracted using the industry standard CAD tools and the HBM are listed in Table III. A behavior-level simulator is designed using the NPE power and timing data along with DRAMPower [19] simulator to characterize the latency and the energy consumption of the proposed and baseline PIM architectures FlutPIM [13] and FIMDRAM [4] for different workloads.

Hardware Configuration: The proposed NPEs can be integrated with memory banks in both 2D-DIMM and 3D-HBM memory. In this paper, the parameters of the HBM are used according to Samsung's HBM2-based PIM industrial product [4]. The PIM-HBM2 consists of 8 DRAM dies stacked using the TSVs. The memory bus frequency of the PIM-HBM2 is 1.2 GHz, and the operating frequency of the DRAM is $4\times$ slower than the bus frequency (= 300 MHz). In Samsung's PIM-HBM2, half of the DRAM dies consist of compute elements, and the capacity of each DRAM die with compute elements is equal to 4 Gb. Therefore, this paper simulates a 4 Gb DRAM die with the bank organization and the timing parameters as listed in Table III. The area of each DRAM die is 84.4 mm^2 at 20nm technology [4] and therefore, the area overhead of the proposed design with 16384 NPEs/chip is only **10.6%**. All the presented results of the proposed and the baseline architecture are based on the workload mapping and simulation on a single DRAM chip.

TABLE III: Configuration of the platforms used.

Attributes	FIMDRAM [4]	FlutPIM [13]	This Work
PE Area (20 nm) in mm^2	0.045	0.021	0.00055 (0.0026 in 40 nm)
PE Power (mW)	N.A.	14	0.051
PE Operating Freq	300 MHz	1.69 GHz	300 MHz
PIM HBM Freq	1.2 GHz	1.2 GHz	1.2 GHz
PIM HBM Die Capacity	4 Gb	4 Gb	4 Gb
#PEs/bank	16	32	1024
#Banks/die	16	16	16
MAC Latency (Tinyfloat-12)	N.A.	19 Cycles	67 Cycles
MAC Latency (INT-8)	N.A.	6 Cycles	33 Cycles
MAC Latency (FP-16)	2 Cycles	N.A.	96 Cycles
Data-width/bank	256 bits	4096 bits	8192 bits
PIM HBM timing	tRAS = 29, tRP = 14, tRCD = 16, tCCD_S = 4, tCCD_L = 2, tWR = 16, tRC = 45, tRRD = 2		

B. Workloads

This paper evaluates the performance and energy efficiency of the proposed PIM architecture on CNN inference workloads. CNN inference is a popular workload for PIM architectures as

it requires high computing and data parallelism, which is available in PIM designs. Furthermore, the CNNs are executed at different data precision in both integer (INT) and floating-point (FP) data formats. Hence, the CNN workloads are ideal for testing the efficacy of the proposed PIM architecture, which can be programmed to support multiple data formats. This paper evaluates popular CNN architectures such as ALEXNET [20], RESNET-18, RESNET-50 [21], VGG-16, and VGG-19 [22] on the Imagenet dataset. Fig. 4 shows how the weight matrix (shown in different colors on top of the bank) and input vector V are mapped to the DRAM bank in the proposed architecture for parallel computation of the vector-matrix product, a common operation in CNN inference.

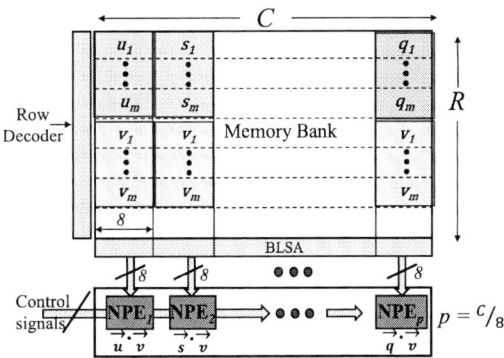

Fig. 4: Mapping of a matrix of weights and activation vectors to a DRAM bank for parallel computation.

C. Results and Discussion

Support For Variable Precision: The instruction set of the NPE consists of basic logic functions, carry-lookahead addition, and comparison operations. Hence, a micro-program consisting of NPE instructions can be designed to implement different arithmetic and logic operations involving various data formats. The proposed architecture's throughput (Images/s) and energy efficiency (Images/J) are evaluated for all workloads at popular integer formats for inference such as INT-4 and INT-8. Popular floating point formats for machine learning, including Tinyfloat (12 bits) and brain floating point (BFLOAT-16), are also evaluated. Fig. 5 shows the throughput and Fig. 6 shows the energy efficiency of the CNN workloads for different data formats. It is observed that as the size (number of parameters) of the CNN architecture increases, both the throughput and energy efficiency of the proposed architecture decrease. The same trend can be observed with respect to the data format and precision and higher precision involves computing more NPE instruction per operation.

High Throughout and Energy Efficiency: Table IV compares throughput and the energy efficiency of the proposed PIM architecture with another LUT-based PIM architecture FlutPIM [13]. The throughput of the proposed architecture is also compared with Samsung FIMDRAM [4]. The processing elements of the FlutPIM consist of multiple lookup tables that

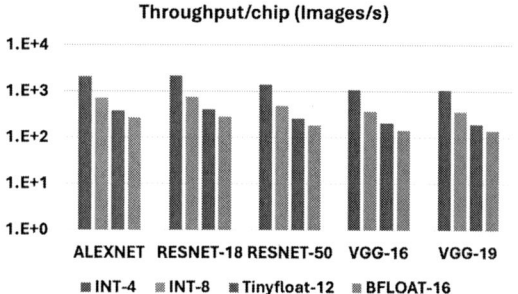

Fig. 5: Demonstrating variable precision. Throughput of the proposed PIM architecture for computing different CNNs at different data formats.

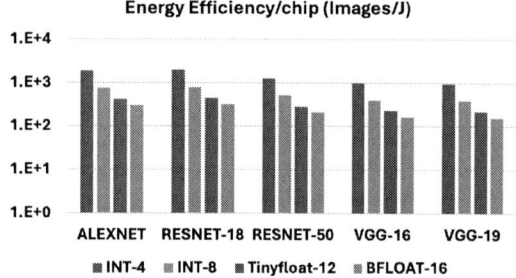

Fig. 6: Demonstrating variable precision. Energy efficiency of the proposed PIM architecture for computing different CNNs at different data formats.

implement 8-bit functions and, therefore, are about **38×** larger than the NPEs used in the proposed PIM architecture when scaled to 20 nm process node. To limit the area overhead of the PEs, only 32 of the FlutPIM PEs are used per bank as opposed to 1024 NPEs. This results in a larger compute bandwidth and compute parallelism per bank for the proposed PIM architecture. As a result, the proposed PIM architecture is able to achieve improvements in throughput over FlutPIM in spite of NPE being slower as compared to FlutPIM PE[1].

The NPEs are compact and do not overprovision hardware to achieve flexibility; hence, they achieve about **274×** lower power than a FlutPIM PE. This results in an order of magnitude improvement in the energy efficiency of the proposed PIM architecture as compared to FlutPIM, as shown in Table IV.

IV. CONCLUSION

This paper presents a PIM architecture that integrates novel neuron processing elements (NPEs) into a conventional DRAM to utilize the maximum available parallelism inside the memory. The aim of this work is to design a general-purpose computing element with a very small area and power footprint that is, therefore, non-invasive to the DRAM design and makes the PIM architecture easily adaptable. This paper shows the NPE

[1]The FlutPIM uses a clock frequency of 1.69 GHz for its PE which is greater than the HBM frequency (300 MHz). This is not a feasible design according to Samsung Industrial PIM product [4]. Any components inside the HBM have to operate at the HBM frequency, or else it will lead to timing violations.

TABLE IV: Throughput of the proposed architecture normalized to FlutPIM [13] and Samsung FIMDRAM [4] and Energy-Efficiency normalized to FlutPIM [13].

| | Throughput Ratio | | | Energy-Efficiency Ratio | |
| | FlutPIM = 1 | | FIMDRAM = 1 | FlutPIM = 1 | |
	INT-8	Tinyfloat-12	FP-16	INT-8	Tinyfloat-12
ALEXNET	1.56	2.01	2.55	6.94	11.41
RESNET18	1.56	2.01	2.59	6.97	11.75
RESNET50	1.56	2.01	2.66	6.96	11.43
VGG16	1.56	2.01	2.69	6.95	11.71
VGG19	1.56	2.01	2.69	6.95	11.71

can perform both integer and floating point operations and can be integrated into BLSA to extract maximum memory parallelism. When compared to an existing PIM architecture with floating point support, the proposed PIM architecture achieves about 1.56× to 2.69× higher throughput and about 6.94× to 11.75× higher energy efficiency.

REFERENCES

[1] C. Lai et al. AI is harming our planet: addressing AI's staggering energy cost. https://numenta.com/blog/2022/05/24/ai-is-harming-our-planet, 2022.

[2] E. Strubell et al. Energy and Policy Considerations for Deep Learning in NLP, 2019.

[3] C. Wu et al. Sustainable AI: Environmental Implications, Challenges and Opportunities, 2021.

[4] S. Lee et al. Hardware architecture and software stack for pim based on commercial dram technology : Industrial product. In *ACM/IEEE ISCA*, 2021.

[5] Mark Horowitz. Computing's energy problem (and what we can do about it). In *IEEE ISSCC*, 2014.

[6] S. Yin et al. Vesti: Energy-Efficient In-Memory Computing Accelerator for Deep Neural Networks. *TVLSI'20*.

[7] H. Kim et al. Colonnade: A Reconfigurable SRAM-Based Digital Bit-Serial Compute-In-Memory Macro for Processing Neural Networks. *IEEE JSSC*, 2021.

[8] V. Seshadri et al. Ambit: in-memory accelerator for bulk bitwise operations using commodity DRAM technology. In *MICRO'17*.

[9] Q. Deng et al. DrAcc: a DRAM based accelerator for accurate CNN inference. In *DAC'18*.

[10] İ. Yüksel et al. Functionally-Complete Boolean Logic in Real DRAM Chips: Experimental Characterization and Analysis. In *IEEE HPCA*, 2024.

[11] A.and others Olgun. DRAM Bender: An Extensible and Versatile FPGA-Based Infrastructure to Easily Test State-of-the-Art DRAM Chips. *IEEE TCAD*, 2023.

[12] S. Li et al. DRISA: a DRAM-based Reconfigurable In-Situ Accelerator. In *IEEE/ACM MICRO'17*.

[13] P.R. Sutradhar et al. FlutPIM: A Look-up Table-based Processing in Memory Architecture with Floating-point Computation Support for Deep Learning Applications. In *ACM GLSVLSI*, 2023.

[14] J. Gomez-Luna et al. Benchmarking a New Paradigm: Experimental Analysis and Characterization of a Real Processing-in-Memory System. *IEEE Access*, 2022.

[15] H. Shin et al. McDRAM: Low Latency and Energy-Efficient Matrix Computations in DRAM. *IEEE TCAD*, 2018.

[16] O. Leitersdorf et al. AritPIM: High-Throughput In-Memory Arithmetic. *IEEE Transactions on Emerging Topics in Computing*, 2023.

[17] S. Muroga. *Threshold logic and its applications*. Wiley-Interscience, New York, 1971.

[18] A. Wagle et al. A Novel ASIC Design Flow Using Weight-Tunable Binary Neurons as Standard Cells. *IEEE TCAS I: Regular Papers*, 2022.

[19] K. Chandrasekar et al. DRAMPower: Open-source DRAM Power and Energy Estimation Tool,. http://www.drampower.info/.

[20] A. Krizhevsky et al. ImageNet Classification with Deep Convolutional Neural Networks. In *NeurIPS*, 2012.

[21] K. He et al. Deep residual learning for image recognition. *CoRR*, 2015.

[22] S. Simonyan and A. Zisserman. Very deep convolutional networks for large-scale image recognition. *CoRR*, 2015.

979-8-3315-3968-9/24 $31.00 © 2024 IEEE

GemIMC: A Configurable HW Architecture for Technology Agnostic IMC based NN Inference

Emilien Taly*[†], Roberto Guizzetti*, Pascal Urard* Elena-Ioana Vatajelu[†]
*STMicroelectronics, Crolles, France
[†]Univ. Grenoble Alpes, CNRS, Grenoble INP, TIMA, F-38000 Grenoble, France

Abstract—This paper presents GemIMC, a High Level Synthesis (HLS) based configurable digital unit architecture for accelerating Neural Networks (NN) at the edge using In Memory Computing (IMC). The proposed architecture is capable of supporting any type of memory technology, be it CMOS or resitive, with digital or analog storage capabilities. GemIMC aims to facilitate design space exploration among different IMC-related parameters and provide a top architecture for prototyping. By taking as input the IMC-tile parameters such as storage type (analog/digital), size, latency, power and process variability, GemIMC provides a an estimation of the full system's latency, area, and power consumption, taking into account the full digital control. The Tensorflow-to-GemIMC environment flow allows for direct evaluation of the ineference accuracy for any type of NN, considering the variability associated with analog computation (when needed). The proposed work provides a flexible and efficient solution for IMC-based NN inference on the edge, along with a methodology for writing the IMC tile model to ensure compatibility with the top architecture.

I. INTRODUCTION

In recent years, Deep-Edge computing has become essential in various systems by bringing computations close to the sensors. This approach saves a significant amount of energy by running only a wake-up system continuously in the background [1] and it typically includes a Neural Network (NN) accelerator to perform computations and make decisions. The Deep-Edge environment is one of frugal resources, therefore, the notions of energy and surface efficiency are critical. To comply with energy constraints more and more architectures based on In Memory Computing (IMC) for NN computations have emerged. The current research focuses on the development of the IMC-tile. An IMC-tile is a essentially a memory block that has computation capabilities; it contains the memory array and the periphery (address decoders and read/write drivers) specifically designed to enable Multiplications ACcumulations (MACs) directly in memory. Even though they are all designed to perform the same computation, their operation mode, performance and power efficiency depend strongly on the type of memory employed, such as Static Random Access Memory (SRAM) or Non Volatile Memory (NVM). Moreover, at the architecture level, the digital control which encompasses the tile, is also diverse and dependent on factors such as computing parallelism, NN model mapping, or weights- and data-encoding [2]–[10]. The state of the art presents numerous solutions of NN inference accelerated by IMC but they are either build from bottom-up, which means that the proposed architectures are custom designed for a specific memory technology and

operation mode, or top-level only where the IMC-tile is seen as a black box which outputs the MAC result. Besides many other works are only focusing on the design and optimization of the IMC-tile without considering the architecture-level implications, despite the fact that the architecture view can play a crucial role in evaluating performance based on the intended application. In addition, several frameworks exist such as NeuroSim [11] [12], which provide performance estimations of specific IMC-tile accelerating NN inference without considering a top-down view, while the framework proposed in [13] addresses this limitation but it is limited to only two types of SRAM architecture. Therefore, further research is needed to explore the potential benefits of a comprehensive approach that considers both the memory device specificity and the architecture level design.

This paper introduces GemIMC, a technology agnostic and configurable hardware (HW) architecture to control an IMC-tile for NN inference. The architecture is based on Gemini, a Neural Processing Unit (NPU) for NN inference introduced in [14] and [15]. Specifically, we consider small NN models and map the entire model onto the tile without weights reloading. To evaluate a IMC-tile, we propose a flow that takes the behavioral model of the IMC-tile as input, including power consumption estimation and analog variability for Analog IMC (AIMC) and outputs the characteristics of the resulting NN implementation such as inference accuracy, and estimation of the area, latency and power consumption. The proposed flow also supports Digital tile (DIMC).

The objective of this study is to understand the impact of various IMC-tile parameters on digital HW cost and latency for a given NN model. Additionally, we investigate the different in term of parallelism and NN models mapping for our architecture. The proposed configurable HW architecture is designed in C++ using High-Level Synthesis (HLS) and can be directly integrated with one or many tiles at the Register Transfer Level (RTL). This paper is organized as follows. Section II presents an overview of the current state-of-the-art in IMC-tile design and existing frameworks for their evaluation. Section III describes the various parameters of the proposed architecture, their respective impacts on the design and the dataflow for only single tile. Section IV explains the evaluation metrics of the proposed architecture such as latency and digital HW cost function of architectural and NN parameters. Section V outlines the methodology for writing the IMC-tile model and presents the results for one NN model. Section VI concludes the paper.

979-8-3315-3968-9/24 $31.00 © 2024 IEEE

II. RELATED WORK

The types of memories used for IMC range from SRAM [2]–[5] to NVM with PCM [6] [7] or RRAM [8]–[10] cells. This technology versatility leads to a large variety of IMC-tile designs, each taking advantage of specific technology characteristics - such as type of memory point (charge-based or resistance-based), type of data storage (analog or digital) and ability to access multiple memory cells simultaneously - and implementing different strategies for performing the MAC operations. The points of difference between existing IMC-tiles can be classified into seven categories: input encoding (binary [5] [7] [9] [10] or on multiple bits [2], [3] [4] [6] [8], physical computation (current-based or charge-based [5]), memory cell (SRAM or NVM with single [8] or multi level storage [6] [7] [9] [10]), implementation of analog-to-digital converters (ADC), parallelism (maximum number of WordLines (WL) and BitLines (BL) that can be activated at the same time), weight mapping scheme (the order in which the 3D filter weights will be unrolled on a column, and the way in which the weight bits will be mapped), and data/weights precision.

All these device-related differences between IMC-tiles and the large spectrum of operation modes lead to unique functionality of the digital control. This complexity also makes it difficult to compare the effects of using different IMC solutions at system level or even to comprehensively estimate the performance and accuracy of an NN implemented using IMC. To alleviate these issues, frameworks have been developed in support to IMC-tile evaluation in the NN inference context, such as Neurosim [11]. This C++ environment supports SRAM or NVM tile (digital or analog), and accepts different levels of abstraction, ranging from a cell-to array-level. DNN+Neurosim [12] adds a circuit-level view with consideration of the digital control architecture of the IMC-tile. It also includes an algorithm-level view with a wrapper to interface NeuroSim and Tensorflow (TF) for the software part of the NN.

The limitations of these frameworks lie in their configurability. For input encoding, only binary is supported. In terms of parallelism, the assumption is that all WLs can be and are activated (even though some architectures [2] [7] [8] [9] activate only a portion for calculations). This has an impact on the digital HW part. Additionally, the user must provide a tile size equal to the NN size, which can be limiting in cases where a fixed tile size needs to be tested on different NN sizes or if the technology does not allow for the fabrication of large tiles (due to variability or other constraints). Reference [13] focuses more specifically on the impact of the digital control on the NN inference. However, they only focus on SRAM-based IMC-tile with two types of architecture, supporting voltage or pulse encoding. Parallelism is not taken into account, and only one type of weight mapping is considered.

In this paper we propose GemIMC which goes beyond the state of the art solutions by being flexible enough to allow the use of any IMC-tile design and operation mode, the parallelism at different levels and it includes the digital HW for realistic cost and performance evaluations.

III. ARCHITECTURE DESCRIPTION

The GemIMC environment consists of a Python wrapper, the technology-agnostic GemIMC core in C++, and a metrics calculation block, as shown in Figure 1. The Python wrapper describes the NN model and performs the training taking into account TF quantization using Quantization Aware Training (QAT). The GemIMC core includes the HLS of the digital HW architecture which to controls the IMC-tile and performs the calculations that are not IMC compatible. The IMC tile model provides the MACs result at each cycle. Metrics such as inference accuracy, latency, power, and surface are then calculated.

Moreover, the IMC performance and usability is evaluated on a the single-tile architecture. The latter allows for higher parallelism and it could be the optimal choice depending on the latency requirements of the application.

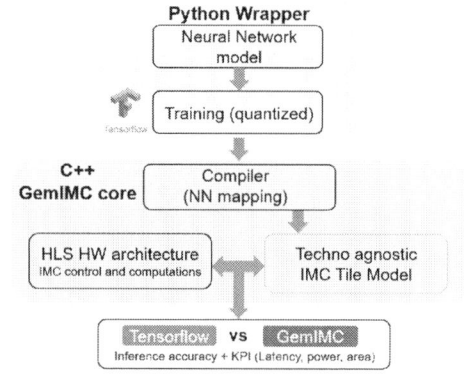

Fig. 1: General overview of GemIMC

A. Configurable parameters

The implemented architecture is user-configurable according to the parameters list shown in Table 1. Where the first column contains the parameter notation, the second column indicates to which type of IMC this parameter is applicable: AIMC (A), DIMC (D), or both (A/D), while the third column briefly describes the parameter. These parameters can be divided into 2 categories: (1) parameters that define the IMC-tile and (2) parameters related to the parallelism.

The parameters that define the IMC-tile are: the tile size (Tx, Ty), the number of states a memory cell can store: Level Cells (LC), the data precision (Dn), the weights precision (Wn), the number of data bits which can be processed simultaneously (DP), the number of Activated WordLines (AWL) and the number of Activated BitLines (ABL), the precision of ADC and of DAC (ADCn, DACn).

The LC parameter allows to support Single or Multi-Level Cell (SLC, MLC). It should be noted that the parameter is represented in terms of the bits number per cell. To enable encoding of a weight precision (Wn) on cells with a different number of levels, the value of LC can vary across the same tile (as described, for instance, in [7]). The DP parameter supports any type of input data encoding. Regarding the ADCs, their

979-8-3315-3968-9/24 $31.00 © 2024 IEEE

precision may be lower than the maximum output range given by :

$$ADC_{n_{max}} = \lceil \log_2(2^{LC-1} \cdot 2^{DP-1} \cdot AWL + 1) \rceil$$

Their number may also be lower than ABL depending on the IMC-tile architecture. The maximum number of ADCs per tile is equal to ABL.

The parameters related to the parallelism are: MPAR, MACPAR, and BITPAR. MPAR parallelism refers to the calculation of pixels on multiple filters simultaneously. MACPAR parallelism refers to the calculation of multiple MACs operations of the same convolution simultaneously. BITPAR parallelism refers to the calculation of multiple data bits simultaneously. The maximum value of BITPAR is equal to the data precision (Dn).

The different parallelisms are illustrated in Fig. 2(a) on a convolution layer, while Fig. 2(b) shows the parameters for AIMC- and DIMC-tile as well as the differences between them. AIMC is based on analog computation through the activation of multiple WLs with a multiplication that can be on multiple bits. DIMC is based on deterministic computation through the activation of a single WL with binary multiplication and accumulation via an adder tree in the tile. Generally, for AIMC, the filter coefficients are placed on the different columns and the 3D filters are unrolled on the rows. Conversely, for DIMC, the filters are placed on the different rows and the 3D filters are unrolled on the columns. The parameters ADCn/DACn, LC, DP, and AWL are specific to AIMC. For a single-tile AIMC, the parameters AWL and DP are equivalent to MACPAR and BITPAR respectively. The filter parallelism, noted MPAR, is define by :

$$MPAR = \frac{ABL}{Cell_{per_{Wn}}} \quad with \quad Cell_{per_{Wn}} = \lceil \frac{Wn}{LC} \rceil$$

For DIMC, the parameters AWL and DP are equal to 1, while ABL defines the MACPAR parallelism on a single tile.

The rest of this chapter includes a detailed description of the single-tile architecture.

TABLE I: List and description of architectural parameters

Name	Memory	Description
Tx, Ty	A/D	Tile size (columns, row)
LC	A	Number of weights bits encoded on one cell
Dn, Wn	A/D	Data and Weights precision
DP	A	Number of data bits processed at same time
AWL	A	Activated Wordline per tile
ABL	A/D	Activated Bitline per tile
ADCn, DACn	A	ADC and DAC precision
MPAR	A/D	Filter parallelism
MACPAR	A/D	MAC parallelism
BITPAR	A/D	Data bits parallelism

B. Single-tile

Figure 3 shows a block diagram of the single-tile architecture. It consists of two SRAM blocks (Param SRAM and Data SRAM), the IMC-tile, the control and the digital post-tile computation. The Param SRAM stores the parameters of each layer of the NN as well as the biases of the filters. The

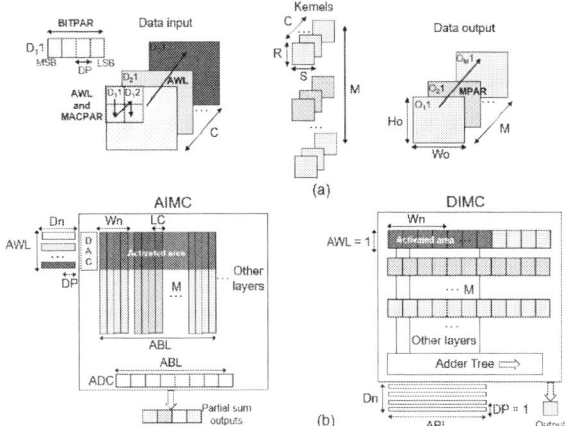

Fig. 2: Parameters for AIMC and DIMC : (a) a convolution layer with parallelism description. R, S, and C represent the 3D filter, and M the filters number. $D_1 1$ and $O_1 1$ respectively represent the first input data and output data pixel. The arrows indicate the directions of parallelism, (b) Difference between AIMC and DIMC with the activated area. depending on the parameters

Data SRAM stores the input image of the NN as well as the intermediate computation data.

The AWL parameter defines the sizing of a memory word and the number of data processed at same time. For convolution and maxpool layers, the AWL parallelism prioritize the channels. For depthwise layers, where there is only one input channel (C) the parallelism is done by rows and the filter is traversed column by column. The number of consecutive words in a channel block is equal to $\frac{MPAR}{AWL}$. This placement allows for easy management of the dataflow and ensures that the necessary data is available at each cycle.

The AWL data read is then placed in registers for processing. Depending on the DP parameter, the row controller, consisting of multiplexers, will place the current DP bits in the correct location in the Ty-sized bus. They will be placed according to the current sequencing of the NN, which is dependent on the NN mapping. The column controller will generate a mask (0/1) of Tx bits to activate the necessary columns according to ABL and the NN mapping. Figure 4 illustrates the row and column controller module. Four inputs data channels (AWL=4) are read from the DATA SRAM. Multiple stages of multiplexers are employed to select the appropriate DP bits from each input data based on the bits to be processed. Subsequently, additional multiplexer stages direct the bits into the row controller registers at the correct positions depending on the IMC tile memory area to be activated. This activation is determined by the inference sequencing. Any remaining registers are set to zero. At the output of the architecture, the number of wires matches the number of rows in the IMC tile (Ty), facilitating direct connection to the WLs or the DAC inputs. Decoding is performed externally to the IMC tile. This method is highly efficient for serial computation (DP=1) or temporal PWM modulation of input data due to the low DAC cost. Requiring one DAC per row, which would be inefficient

979-8-3315-3968-9/24 $31.00 © 2024 IEEE

for voltage modulation. Similarly, columns are activated in the same manner using column controller registers. The number of registers equal the number of columns in the IMC tile. These registers are set to 1 or 0 based on the filters to be activated according to the NN sequencing. The NN mapping in the IMC-tile is performed by the compiler to optimize IMC-tile space. The layers can be placed on the same column or on different columns.

The digital post-tile component consists of several blocks, including constant shifters and an adder for summing the different bits of the weights placed on the Wn column. This block can be removed if the sum is performed in the IMC-tile, as in [5]. Additionally, there is a shifter block followed by an accumulator to manage the Dn bits, as well as a number of MACs greater than AWL. A quantization block is included to return to a real value between each layer, based on the TF quantization. Finally, there is the activation function component, where only ReLU is supported. The number of digital post-tile and quantization blocks depends on parallelism and will be discussed in the next subsection.

The storing stage block manages the writing of intermediate images to the data SRAM. The maxpool stage block manages the case of maxpool layers where the IMC-tile is bypassed. The registers of the post-tile blocks are reused to perform comparison operations. The architecture supports convolutional, depthwise, dense, and maxpool layers.

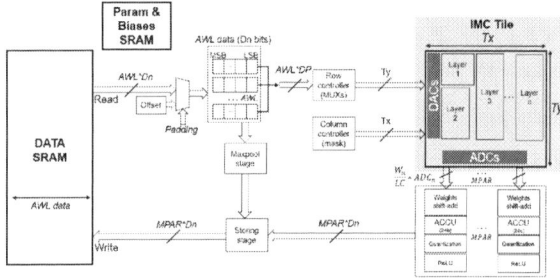

Fig. 3: Block diagram of single-tile architecture

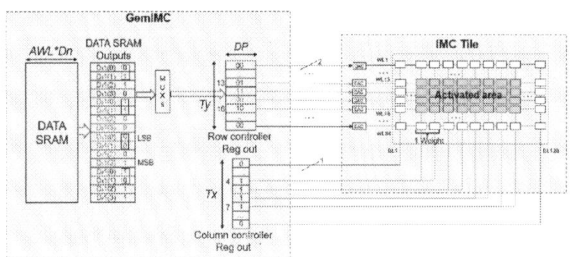

Fig. 4: Operating principle of the IMC Tile control with the row and column controllers registers and the wires directly connected to the WLs and BLs of the IMC Tile

Figure 5 illustrates the scheduling of the architecture, where different functions are pipelined to enable one processing of the IMC-tile per cycle. The beginning of a layer involves reading the parameters into param SRAM to initialize the

control registers. The bias registers are then updated in several cycles, depending on the number of filters and param SRAM width. During the last cycle of the bias update, a read of the data SRAM is performed, and the IMC control ports (rows and columns) are updated. The same data is processed for $\frac{D_n}{DP}$ cycles before starting the second wave of data. At the start of a new convolution, the quantization of the outputs of the previous convolution's accumulator is done over several cycles before writing the outputs to memory.

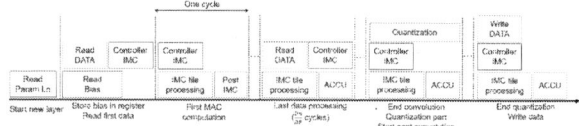

Fig. 5: Scheduling of the architecture for convolution layer starting from the beginning of the layer up to the writing of one wave of results

IV. ARCHITECTURE EVALUATION

The digital HW architecture is evaluated based on the configuration parameters described in Section III and is dependent on the NN. The latency is accurately determined, while the digital HW cost is determined by analyzing the number and precision of the main digital blocks. An extrapolation can be performed to obtain the power and area for a given technology.

A. Latency

The ideal latency for a convolution is defined by Equation 1. It is the ratio of the number of MACs required to perform a convolution and the product of all parallelisms. The BITPAR parallelism is considered to be well-chosen with 100% efficiency, which is lower than $BITPAR_{max}$ (see III.b). For a depthwise layer and maxpool, C is equal to 1 for the number of required MACs. For a dense layer, it is the product of the input and output neurons.

$$Lat_{conv_{ideal}} = \frac{R \cdot S \cdot C \cdot W_o \cdot H_o \cdot M}{MPAR \cdot AWL \cdot MACPAR} \cdot \left\lceil \frac{Dn}{DP \cdot BITPAR} \right\rceil \quad (1)$$

Equation 2 provides the real latency of the architecture based on different parallelisms. The latency is based on the number of cycles of the digital HW part. It is assumed that the next DP bits are provided to the IMC-tile at each cycle, and the actual number of cycles may vary depending on the encoding type, such as PWM or pulse count. The ceil of each parallelism indicates the loss of efficiency, i.e., the times when all parallelisms are not used at 100%. For instance, when C=9 and AWL=8, the equivalent latency is C=18.

$$Lat_{conv_{reel}} = R \cdot S \cdot W_o \cdot H_o \cdot \left\lceil \frac{C}{AWL} \right\rceil \cdot \left\lceil \frac{M}{MPAR} \right\rceil \cdot \left\lceil \frac{Dn}{BITPAR} \right\rceil \quad (2)$$

Equation 3 shows the loss of efficiency, with the ratio of the actual latency of the architecture to the total number of MACs in the layer. The objective is to maximize efficiency for each layer of a given NN, and to find the parallelisms that optimize efficiency with a fixed architecture. A reconfigurable architecture is the best-case scenario to have the adequate parallelisms for each layer, but at the cost of the maximum digital HW of each parallelism. The reconfigurability aspect is not discussed in this paper.

$$Efficiency_{conv}(\%) = \frac{Lat_{conv_{ideal}}}{Lat_{conv_{reel}}} \cdot 100 \quad (3)$$

B. Digital HW cost

Figure 6 provides a detailed view of the HW implementation. For the post-tile computation phase, the weights shift-add component is responsible to aggregate the results into a single weight value derived from multiple columns. The shifter adjusts the result based on the specific bits being processed within the data input. Finally, the accumulation unit aggregates these results. In the quantization section, the final convolution result is multiplied by the scaling factor over 16 bits, which is executed in four cycles. The row controller section illustrates the hierarchical levels of multiplexers used to initially select the DP bits from among the Dn bits. Subsequently, it addresses the Ty registers appropriately, ensuring that each data point can be directed to any register as required.

Table II provides the operand precision for the input/output modules of post-tile, quantization, and IMC tile control for a single-tile. The number of post-tile and quantization modules is equal to MPAR. The adder weights sub-module's precision is determined by the number of adder inputs (i), each with a different precision. The constant shifter preceding the adder varies for each weight bit position, ranging from MSB to LSB. This sub-module is not present if the operation is performed directly in the IMC-tile (Cf. II).

The register ACCU size of 24 bits for a single-tile was chosen arbitrarily. It is dependent on Wn and Dn: $Reg_{ACCU_{1tile}} = D_n \cdot W_n + 8$.

related to uncertainties. The calculation core of the IMC-tile array provides the results of the IMC-tile to the digital HW at each cycle.

In this study, we will exclusively evaluate the digital architecture component to compare it with Gemini, an internal NPU at STMicroelectronics. This evaluation aims to determine the hardware cost of the digital part within an IMC-based architecture. For a fair comparison, the cost of the IMC Tile should also be considered and can be included during the evaluation process. The evaluation was performed on a VGG-like neural network model, detailed in Figure 8. The results presented in Table III are derived from a gate-level simulation based on an inference of the VGG-like model. The equivalent number of PEs corresponds to the maximum number of MAC operations that can be performed in one cycle with the associated hardware. For GemIMC, the architectural parameters were set as follows: AWL=32, MPAR=32, DP=2, LC=2, Tx=128, and Ty=256. The equivalent number of PEs is calculated as $\frac{MPAR \times AWL}{DP}$.

The number of cycles required for inference with a VGG-like model does not follow a straightforward 2:1 ratio between the two architectures. This discrepancy is due to GemIMC's specific efficiency of 75% with this neural network. Layers 1 through 4 lack a sufficient number of channels and filters relative to the parallelization. The area cost is mainly driven by the SRAM DATA cost, which varies based on the targeted application. The standard cell portion of GemIMC is relatively small, being 4.5 times smaller than that of the NPU. The dynamic power cost of the digital part in an IMC-based architecture is notable, being 2.5 times lower than that of the NPU, excluding the IMC tile power consumption.

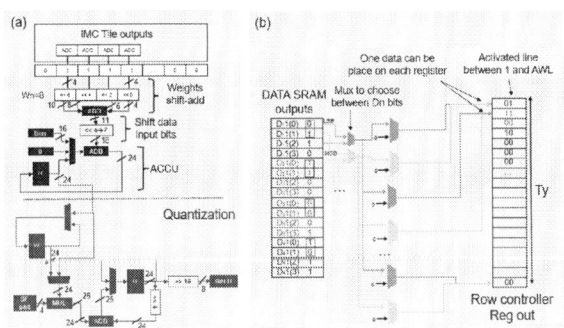

Fig. 6: HW details : (a) Post-tile and quantization part (b) Row controller

V. METHODOLOGY AND RESULTS

To evaluate the digital HW architecture with an IMC model, a methodology to write the IMC model is provided in Figure 7. The digital HW provides control for the rows and columns to be activated at each cycle. An architectural parameters file in the form of defines is used to configure the environment, and the compiler fills arrays with the mappings of weights on the IMC-tile. The IMC model must include a method for mapping numerical weight values onto the physical cell arrays, as well as parameters such as current, operating voltage, and equations

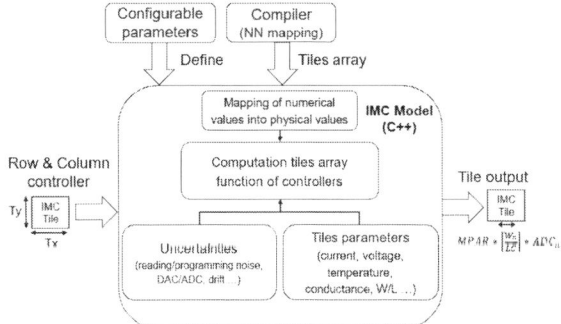

Fig. 7: Block scheme of IMC model

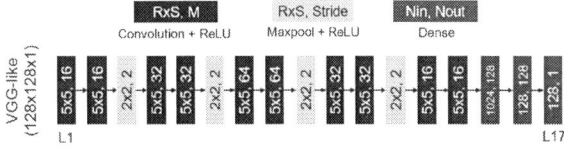

Fig. 8: Details of VGG-like NN model for model evaluation

979-8-3315-3968-9/24 $31.00 © 2024 IEEE   95

TABLE II: Digital HW operand precision function of architectural parameters

Module	Sub-module	Operand Precision (bits)	Number
Post-tile	Adder weights	$Add_{Weights_{in(i)}} = ADC_n + W_n - (i \cdot \lceil \frac{W_n}{\lceil \frac{W_n}{LC} \rceil} \rceil)$ with $1 \leq i \leq \lceil \frac{W_n}{LC} \rceil$ $Add_{Weights_{out}} = ADC_n + W_n$	MPAR
	Register ACCU	24	MPAR
	Adder ACCU	$Add_{ACCU_{in1}} = Adder_{Weights_{out}} + D_n$ $Add_{ACCU_{in2}} = 24$ $Add_{ACCU_{out}} = 24$	MPAR
	Register bias	16	MPAR
Quantization	Multiplier SF	$In = 24 \times 4$ and $Out = 28$	MPAR
	Register SF	16	MPAR
	Register	24	2×MPAR
IMC tile controller	Mux row	2->1	AWL×Ty×DP
	Mux column	2->1	Tx
	Register row	DP	Ty
	Register column	1	Tx

TABLE III: Comparative results between Gemini NPU and GemIMC architecture

18nm 0.68V 25° @200MHz	PEs equivalent	SRAMs (KB)	Cycles (VGG-like)	Area (mm²)	Dynamic power (µW/MHz)
Gemini NPU	128	Weights: 128 Data: 128	4M	Logic : 0.1 SRAM : 0.55	Logic : 100 SRAM : 37
GemIMC (without IMC Tile)	256	Data: 128	3M	Logic : 0.023 SRAM : 0.265	Logic : 33 SRAM : 19.5

VI. CONCLUSIONS

In this paper, a configurable digital HW architecture for technology-agnostic IMC-based neural network inference at the edge has been proposed. This configurability takes into account the IMC-tile parameters and provides a HW architecture that can adapt to any IMC technology and the physics of the MAC operation, it can be adapted to any type of NN and can incorporate parallelism at different levels. In addition to providing an RTL that can be directly integrated, it enables an evaluation of the entire system in terms of latency, power, and area by taking as input a variabilty-aware IMC model (to guarantee a thorowgh evaluation of AIMC solutions). The paper also presents a methodology to evaluate the resulting architecture and highlights the impact of different parallelisms on latency and digital HW cost.

REFERENCES

[1] Weiwei Shan et al. "a 510-nw wake-up keyword-spotting chip using serial-fft-based mfcc and binarized depthwise separable cnn in 28-nm cmos". *IEEE Journal of Solid-State Circuits*, 56(1):151–164, 2021.

[2] Qibang Zang et al. 4b/4b/8b precision charge-domain 8t-sram based cim for cnn processing. *2023 IEEE 5th International Conference on Artificial Intelligence Circuits and Systems (AICAS)*, pages 1–5, 2023.

[3] Kaili Zhang et al. A novel 9t1c-sram compute-in-memory macro with count-less pulse-width modulation input and adc-less charge-integration-count output. *IEEE Transactions on Circuits and Systems I*, 2023.

[4] Heng Zhang et al. Ssm-cim: An efficient cim macro featuring single-step multi-bit mac computation for cnn edge inference. *IEEE Transactions on Circuits and Systems I: Regular Papers*, pages 1–12, 2023.

[5] Yiming Chen et al. Samba: Single-adc multi-bit accumulation compute-in-memory using nonlinearity- compensated fully parallel analog adder tree. *IEEE Transactions on Circuits and Systems I:*, 70(7), 2023.

[6] Riduan Khaddam-Aljameh et al. Hermes-core—a 1.59-tops/mm2 pcm on 14-nm cmos in-memory compute core using 300-ps/lsb linearized cco-based adcs. *IEEE Journal of Solid-State Circuits*, 57(4), 2022.

[7] Win-San Khwa et al. A 40-nm, 2m-cell, 8b-precision, hybrid slc-mlc pcm computing-in-memory macro with 20.5 - 65.0tops/w for tiny-al edge devices. *2022 IEEE International Solid- State Circuits Conference*.

[8] Hongwu Jiang, Shanshi Huang, Wantong Li, and Shimeng Yu. Enna: An efficient neural network accelerator design based on adc-free compute-in-memory subarrays. *IEEE Tran. on Circuits and Systems I*, 70(1).

[9] Jong-Hyeok Yoon et al. A 40-nm 118.44-tops/w voltage-sensing compute-in-memory rram macro with write verification and multi-bit encoding. *IEEE Journal of Solid-State Circuits*, 57(3):845–857, 2022.

[10] Dingbang Liu et al. An energy-efficient mixed-bit cnn accelerator with column parallel readout for reram-based in-memory computing. *IEEE Journal on Emerging and Selected Topics in Circuits and Systems*, 12(4):821–834, 2022.

[11] Pai-Yu Chen, Xiaochen Peng, and Shimeng Yu. Neurosim: A circuit-level macro model for benchmarking neuro-inspired architectures in online learning. *IEEE Transactions on Computer-Aided Design of Integrated Circuits and Systems*, 37(12):3067–3080, 2018.

[12] Xiaochen Peng et al. Dnn+neurosim: An end-to-end benchmarking framework for compute-in-memory accelerators with versatile device technologies. *2019 IEEE International Electron Devices Meeting (IEDM)*, pages 32.5.1–32.5.4, 2019.

[13] Yimin Wang, Zhuo Zou, and Lirong Zheng. Design framework for sram-based computing-in-memory edge cnn accelerators. *2021 IEEE International Symposium on Circuits and Systems (ISCAS)*, 2021.

[14] Ali Oudrhiri et al. Performance modeling and estimation of a configurable output stationary neural network accelerator. *hal-04168803*, 2023.

[15] Nathan Bain et al. Quantization modes for neural network inference: Asic implementation trade-offs. *2023 International Joint Conference on Neural Networks (IJCNN)*, pages 01–08, 2023.

Stochastic Spintronics Device Based Bayesian Networks for Efficient Uncertainty Modeling

Abstract—The conventional computing paradigm for the Bayesian network(BN) is inefficient due to resource-intensive floating-point operations. Spintronics-based systems, offer high parallelism, enabling faster and more energy-efficient Bayesian Networks. The work presents a systematic approach to transforming a Bayesian Network into an electronic circuit, where stochastic nodes mimic real-world variables. The relationships between these variables can be deduced through electrical measurements on the circuit nodes. The electronic circuit is built using Spin-Orbit Torque Random Access Memory (SOTRAM) technology, employing MTJ constructed from thermally unstable nanomagnets with adjusted free layer thickness. This results in an energy-efficient and faster implementation of the BN. This shift to Spintronics-based approaches opens new possibilities for high-performance energy-efficient probabilistic models, particularly suited for edge computing and real-time applications. The novel work has 40% energy efficiency and high inference speed compared to other direct implementations of Bayesian networks using spintronic counterparts. The proposed design has significant potential as they open up the possibility of developing hardware that directly emulates the computational units necessary for real-time applications.

Index Terms—Bayesian networks, stochastic spintronic devices, spin orbit torque, Magnetic Tunnel Junction

I. INTRODUCTION

Spintronics devices are well-suited for edge computing due to their unique properties that enable efficient and low-power data processing. Edge computing scenarios involve handling large volumes of data from sensors and IoT devices, requiring real-time processing capabilities [1]. By configuring the magnetic tunnel junction (MTJ) characteristics of spintronics devices can be fine-tuned to meet the specific needs of various applications, making them versatile and adaptable solutions for a wide range of computing tasks [2]. The MTJ consists of two ferromagnetic layers: a free layer and a pinned layer (Refer Fig 1). These layers are separated by an non-magnetic layer. The magnetization of the free layer can switch between parallel and anti-parallel orientations relative to the pinned layer, resulting in distinct resistance states(RP and RAP). By utilizing the discrete resistance state of MTJ, a novel hardware computing framework called Probabilistic Spin Logic (PSL) has been introduced (fundamental unit known as Probabilistic Bits (p-bits)) [3]. P-bits can be interconnected to solve a wide array of problems, ranging from optimization tasks to inference processes [4].

This work explores the use of this p-bit to construct a p-circuit that can mimic the functionality of a Bayesian network (BN). BN are widely used in AI-related sectors for probabilistic inference and causal reasoning, representing probabilistic dependencies and causal relationships between variables [5]. The BN is defined using conditional probability tables (CPTs) to describe how parent nodes influence the behavior of each child node.

This novel work improves upon conventional SOT-based BN designs by taking advantage of the random switching behavior inherent to MTJ. The key innovation lies in precisely controlling

the switching probability of the MTJ by judiciously modifying the free layer thickness, thereby modulating the thermal stability factor Δ. This strategic reduction in Δ lowers the energy barrier for magnetization reversal, enabling faster and more random switching dynamics within the MTJ. Remarkably, this random switching phenomenon adheres to the Poisson distribution, where events occur discretely and independently at a constant average rate over a fixed time interval. This stochastic behavior is indispensable for generating high-quality random numbers and executing probabilistic computations with unparalleled efficiency in BN. By unlocking the true potential of MTJ randomness, this novel approach paves the way for a paradigm shift in the design and implementation of next-generation probabilistic computing architectures. The proposed model SOT device is thoroughly studied, examining critical switching current and magnetoresistance. Utilizing this ΔSOT, BN nodes (referred to as ΔSOT-BN) are designed, demonstrating improved performance in switching speed without the need for additional circuitry. The design is assessed using the 45nm technology node in Cadence Virtuoso. The proposed BN system achieves reduced power consumption and high inference speed by leveraging stochastic computing with spintronic devices. The key contributions of this work are as follows:

- To analyze and identify the optimal free layer thickness for addressing random switching of SOT-MTJ.
- Characterize the Stochastic Behavior of MTJ to determine the switching probabilities for the BN by different values of the free layer thickness.
- Validate the functionality and performance of the BN system using practical data and real-world scenarios.
- Perform extensive simulations and tests to evaluate the performance of the designed BN system. Analyze the inference speed, power consumption, and accuracy of the system to ensure that it meets the desired efficiency and speed requirements.

II. PRELIMINARIES

A. Overview of Bayesian Network

BN, built on probabilistic directed acyclic graphs (DAGs), provide an interpretable framework to model complex relationships and uncertainties. Unlike neural networks, BN offer insights from all nodes, making them valuable for prediction and inference tasks, especially when expert knowledge is crucial. Conventional computing platforms struggle with these intricate calculations, impacting real-time applications and energy-efficient solutions, particularly as BN find diverse applications necessitating efficient resolutions. Literature on hardware implementations of BN has predominantly utilized CMOS technology, with FPGA-based approaches extensively explored [6]. Novel digital circuit abstractions and probabilistic CMOS hardware [7] has also been proposed. However, despite these advancements, the energy, time, and space inefficiencies intrinsic to floating-point operations remain a significant

---

979-8-3315-3968-9/24 $31.00 © 2024 IEEE

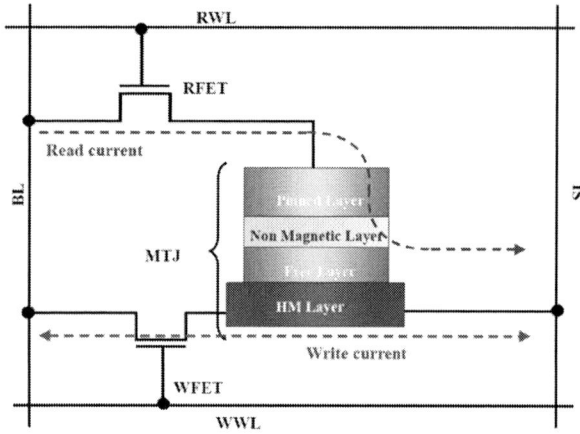

Fig. 1: Bit cell Schematic view of Spin-Orbit Torque Random Access Memory

problem. The key issue with conventional approaches is their inability to efficiently handle the computationally intensive operations and random number generation required for Bayesian inference. Spintronics-based BN emerge as a promising solution, harnessing the inherent parallelism of Bayesian inference to address the limitations of conventional implementations, promising enhanced computational efficiency and potentially revolutionizing Bayesian inference.

The implementation of BN using spintronics has been explored through various innovative approaches, each leveraging the unique characteristics of spintronic devices to enable efficient probabilistic computing. One notable method harnesses the stochastic behavior of MTJ, which exhibits random switching between resistance states, a crucial aspect for Bayesian inference Another intriguing approach involves representing BN with probabilistic circuits constructed using p-bits [8]. Techniques like Neural Sampling Machine (NSM) and Drop Connect exploit the inherent noise in MTJs for probabilistic inference in BN [9]. Strain-switched MTJs have been employed as non-volatile devices in Bayesian hardware implementations [10], while reconfigurable architectures with Stochastic Bayesian Nodes (SBN) accelerate structure learning [11]. Additionally, Stochastic Bitstream Generator Blocks, utilizing MTJs, generate stochastic bitstreams for Bayesian inference systems [12]. In [13], an MTJ array design was proposed for efficient BN implementation. Each of these approaches exploits the stochastic behavior of spintronic devices to enable accurate and energy-efficient probabilistic computing in BN. Despite the promise of energy-efficient computing in BN using spintronics, existing literature and research may not comprehensively address the energy efficiency considerations for large-scale applications. Achieving optimal energy efficiency in spintronics-based BN, especially for complex and high-dimensional problems, remains an ongoing challenge that warrants further investigation and innovation.

III. MODELING SOTRAM IN TELEGRAPHIC SWITCHING REGIME FOR ENERGY EFFICIENT BAYESIAN NETWORK

The MTJ with its free and pinned layers separated by a non-magnetic oxide layer exhibits distinct resistance states (Parallel and Antiparallel) based on the orientation of the free layer's magnetization (Refer Fig1). In stochastic SOTRAM design, a current is applied to the MTJ placed on a non-magnetic heavy metal (HM) layer, inducing magnetization switching through the Spin Hall effect. The three-terminal SOTRAM offers greater endurance compared to conventional two-terminal STTRAM devices. The stochastic characteristics of MTJs are leveraged in BN to set up nodes with defined probabilities. This behavior is enhanced when the MTJ operates in the "superparamagnetic regime," making it particularly advantageous for applications involving BN.

A. Integration of Spintronics Devices within Bayesian Network Nodes

The work introduces an innovative spintronic bit cell to effectively and efficiently incorporate Gaussian-distributed values within Bayesian neural networks. The study delves into the direct hardware integration of BN for tasks involving probability assessment and inference. In this framework, every node within the BN is translated into a stochastic device featuring distinct probabilities for its binary states. These probabilities are influenced by input signals derived from the states of parent nodes, which are regulated by the connection weights linking them. The encoding of the Conditional Probability Table (CPT) is achieved through these connection weights. This hardware-based rendition of the BN simplifies the process of estimating event probabilities by sampling the output of the corresponding stochastic device. Additionally, inference about the potential cause of a specific event is facilitated through an examination of the joint distribution of the two stochastic devices connected to the "event" node and the pertinent "cause" node. Each node configures an MTJ device with a specific probability by adjusting its physical configuration to realize this implementation.

Designing spintronics devices for BN involves meticulous consideration of essential criteria to ensure optimal performance and smooth integration. These criteria are vital for leveraging spintronic technology to enhance the efficiency and capabilities of BN implementations. The key design factors include Stochastic nature, Energy efficiency, Tunability, and Inference speed. Spintronics devices must exhibit inherent stochastic behavior to function as probabilistic elements, capturing the essence of Bayesian inference for an accurate representation of conditional probabilities. The probability of an MTJ's switching behavior can be regulated through parameters such as the switch current and the thermal stability factor (Δ). The relationship between switching probability and Δ is encapsulated in the subsequent equation [14]:

$$P_{SW}\left(I_{SW}, t\right) = 1 - \exp -\frac{t}{\tau_0} \exp\left[-\Delta\left(1 - 3\frac{I_{SW}}{I_{CO}}\right)^2\right] \quad (1)$$

In this equation, τ_0 represents the attempt time, I_{CO} denotes the critical switching current at 0 Kelvin, I_{SW} signifies the applied current magnitude, and t corresponds to the width of the current pulse applied to the MTJ. This implies that manipulating Δ and I_{SW} enables the configuration of switching probability tailored to specific applications.

The Δ of the SOT device is defined by the subsequent equation [15]:

$$\Delta = \frac{\mu_0 M_s H_{\text{Keff}} t_F \pi r^2}{2k_B T} \quad (2)$$

Here, μ_0 denotes vacuum permeability, k_B signifies the Boltzmann constant, T represents temperature, M_s stands for saturation magnetization, H_{Keff} characterizes effective perpendicular anisotropy field, and t_F corresponds to the free layer's thickness [14]. Equation 2 underscores the direct influence of free layer thickness on Δ. Moreover, the equation

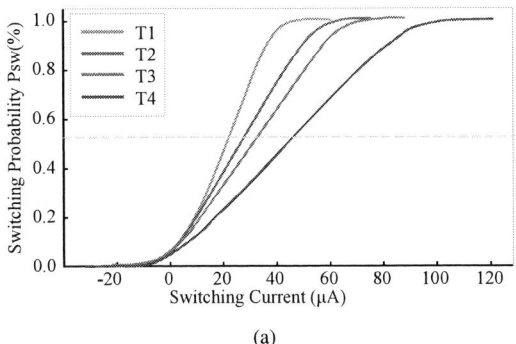
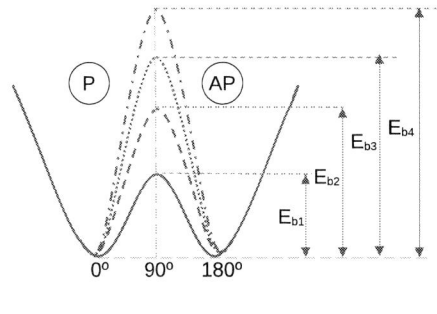

(a)   (b)

Fig. 2: (a) Switching probability in SOT as a function of applied switching current for different free layer thickness T1, T2, T3, and T4 respectively, obtained using Monte Carlo simulations, (b)Representation of energy barrier for different free layer thicknesses,E_{bT4}, E_{bT3}, E_{bT2}, E_{bT1} for T4, T3, T2, T1, respectively

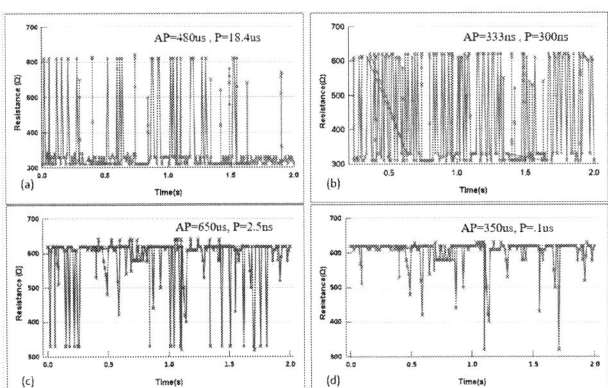

Fig. 3: Four different telegraphic switching produced by the MTJ under varying free layer thickness (a) T1, (b) T2,(c)T3, and (d)T4 respectively.

$\Delta = E_b/k_BT$ with E_b represents the energy barrier between an MTJ's two stable states [16].

This work unveils a remarkable capability to precisely tune the energy barrier between the Parallel (P) and Anti-Parallel (AP) states by adjusting the free layer thickness, enabling controlled stochastic switching in MTJs. The energy barrier primarily stems from exchange interactions between magnetic layers and the shape anisotropy of the free layer, which aligns the magnetic moment along the easy axis, typically perpendicular to the layer plane (Fig2(b)) Reducing the free layer thickness ($T1 < T2 < T3 < T4$, where T1: 0.8-1 nm, T2: 1.1-1.2 nm, T3: 1.3-1.5 nm, T4: 1.5-2 nm) lowers the energy barrier, simplifying the switching process. These Low Barrier Magnets have been rigorously investigated through Monte Carlo simulations using Cadence, covering the aforementioned thickness range with key parameters adopted from [14]. The simulations vividly illustrate that reducing the free layer thickness enhances the switching probability while concurrently decreasing the corresponding switching current (Fig2), paving the way for efficient and tunable stochastic switching, pivotal for Bayesian node implementation

Reducing the free layer thickness increases MTJ stochasticity due to decreased thermal stability. When the energy barrier drops below 1 kbT, the MTJ enters a Telegraphic regime, causing random switching between resistance states due to thermal fluctuations. Telegraphic signals can be manipulated using bias currents/fields and are characterized by time spent in different states. The random spiking signals produced by MTJs

are assessed in terms of the percentage of time spent in the anti-parallel (AP) magnetization state, referred to as the "AP rate" known as dwell time. Additionally, the average dwell times in both AP-state and P-state pulses can be controlled separately. These findings indicate that neural spiking signals produced by MTJs can be decoded based on both spike rate and spike count. The range of t_F in T3 and T4 corresponds to MTJs suitable for non-volatile storage applications.

The stochastic telegraphic switching behavior of an MTJ, manifesting as unpredictable resistance fluctuations between high and low states, unlocks its potential for Bayesian node implementation. Fig3(a)-(d) showcases the resistance changes over time for various free layer thicknesses (t_F) within the T1, T2, T3, and T4 ranges, observed with a critical switching current and an applied magnetic field of H=0kOe at 1 Mbits per second sampling rate. While resistance oscillates between approximately 650Ω (high) and 340Ω (low), the impact of t_F on switching behavior is evident. Low-resistance states dominate in T1 (Fig3(a)), high-resistance states prevail in T3 and T4 (Fig3(c, d)), and equal likelihoods for R_P and R_{AP} emerge in T2 (Fig3(b)). Notably, MTJs with T2 range t_F exhibit the desired stochastic telegraphic switching, offering low-energy barrier Δ_{SOT} MTJs controllable using spin currents. Reducing t_F shortens dwell times, enhancing entropy rates and randomness crucial for Bayesian node tuning and energy efficiency, with AP dwell times ranging from 37.3 to 366 s (0.8-2 nm thickness) and P dwell times spanning 1.51 to 15.4 s. The reduced free layer thickness MTJs offer an exceptional solution for implementing reliable, efficient, and scalable spintronics-based BN. They ensure consistent and precise probabilistic calculations, with adjustable parameters for optimized network behavior. Their low current switching requirements render them highly energy-efficient, while enabling rapid state transitions for real-time computations. Seamlessly integrating with existing architectures and accommodating larger networks, these MTJs comprehensively fulfill the essential criteria, emerging as a viable and efficient option for high-performance BN implementation.

IV. IMPLEMENT BAYESIAN NETWORK BASED ON ΔSOT

BN excel in managing missing data and revealing causal relationships. Their probabilistic nature integrates prior knowledge and data seamlessly, preventing overfitting and yielding coherent models. Fig4 shows a simple BN with four binary variables - Cloudy (C), Humidity (H), Windspeed (W), and Rainy (R). The relationships between variables are quantified

979-8-3315-3968-9/24 $31.00 © 2024 IEEE   99

using CPT associated with transitions to nodes from their parent nodes. Bayesian inference estimates the probability of hidden causes based on observed evidence. For example, given the observation of rain, we can calculate the posterior probability of possible causes, such as high wind speed or high humidity, using Bayes' rule. By configuring the MTJ's characteristics based on each node's conditional probability, the BN's variables can be effectively represented using MTJs to enable efficient and hardware-efficient probabilistic inference. Instead of traditional floating-point calculations, this method counts pulses from each variable within a large time window to estimate the likelihood of inference operations.

After exploring the MTJ's switching behavior across various free layer thicknesses, the T2 range is chosen for Bayesian applications due to its equal probabilities in both 0 and 1 states. In this context, the estimation of inference operation probabilities involves pulse counting from each variable within a sufficiently large time window. The core concept revolves around representing the BN's variables using a Poisson pulse train generator, converting probability information into pulse frequency. This strategy harnesses the controlled stochastic switching characteristic within the T2 range SOT to generate Poisson spikes. Achieving this involves interfacing the device with a reference resistor, generating a Poisson spike or pulse train. The count of spikes in the time window encodes information about the frequency and magnitude of the incoming spike train.

The operational process involves a write/read/reset cycle. A specific stochastic element, shown in Fig1, receives input from another stochastic element, corresponding to nodes in the BN. During writing, the causal stochastic element generates a write current through the heavy metal, determined by a current source adjusted according to the CPT and free layer thickness. This current source activates at a frequency matching the incoming spike train. After writing, a read cycle checks device switching, and resetting if needed. The output train's frequency directly corresponds to switching probability, controlled by the current source's magnitude, adjusted by the CPT and pulse train frequency from the causal element. The pulse train frequency from the causal stochastic element signifies event probability, while the current source driven by it encodes conditional probability of the receiving element's occurrence.

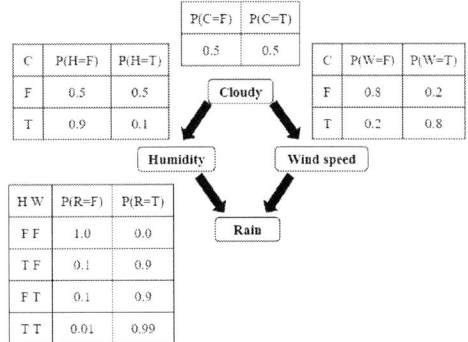

Fig. 4: To demonstrate the independence between variables, a basic Bayesian Network comprising four variables is constructed, along with its corresponding Conditional Probability Table (CPT).

V. EXPERIMENT RESULTS AND DISCUSSION

An integrated simulation framework combining device and circuit components was developed to assess the performance of the proposed BN using ΔSOT. The magnetization switching characteristics of MTJ were simulated using *MuMax3* simulator, while the nonequilibrium Green's function (NEGF) and Landau–Lifshitz–Gilbert (LLG)-based transport simulation framework were employed to model the MTJ resistance variation with Free layer thickness. The resulting device characteristics from MuMax3 and then cadence virtuoso simulator to evaluate the circuit's functionality.

Fig5 provides a comprehensive overview of the BN implementation using the proposed device-circuit configuration. Each variable in the BN corresponds to a single stochastic device equipped with necessary peripherals. The stochastic switching in each device generates a telegraphic spike train that represents the next stage of the variable. The switching frequency information is then conveyed to the subsequent variable as pulses through the interface circuitries. The last variable in the network has four AND gates, enabling the generation of a multiplication output between two telegraphic spike trains. We can infer the probabilities of each variable, such as P(Humidity), P(Windspeed), and P(Rain), which might be challenging to obtain directly from the CPTs. However, this can be achieved by counting the number of output pulses corresponding to each variable over an extended duration.

For instance, if 220 spike pulses are observed at the Vs+ output of the 'Humidity' node during 500 write cycles, it implies that the probability of 'Humidity' is estimated to be 44%. Similarly, for 'Wind speed', the estimated probability will be 56%. This approach enables efficient and accurate inference in the BN by utilizing the telegraphic switching behavior of the MTJ-based devices and the counting of output pulses for probability estimation.

The timing waveform illustrates the calculation of the probability of occurrence for different events, and the average number of spikes generated in each case from 500 sample points closely aligns with the actual analytical values(Refer Fig6). The data presented in the format A/B denotes that A represents the analytical solution, while B corresponds to the computed probability obtained from the 500 sample points of each output. Furthermore, the probability distribution function (PDF) of the two conditional events (W/R) and (H/R) converges towards the mean value as the number of inference samples increases(Refer the Table I). This convergence is evident in the graph, indicating that as more sample points are considered, the computed probabilities approach the actual mean values derived from the analytical solution.

To evaluate the network's performance under realistic conditions, an analysis was conducted with the addition of random Gaussian transient noise to the input charge current provided through the current source and heavy metal layer. Notably, the network effectively utilized a current range of approximately 30-40uA from the switching probability characteristics of the magnetic stack, contributing to the proper functioning of the BN. This current range represents a significant reduction of about 33.3% compared to previously state-of-the-art methods [17], resulting in a 40% decrease in energy consumption. While the variance of the probability distribution function (PDF) increased slightly due to noise, this effect can be mitigated by increasing the number of sample points used for

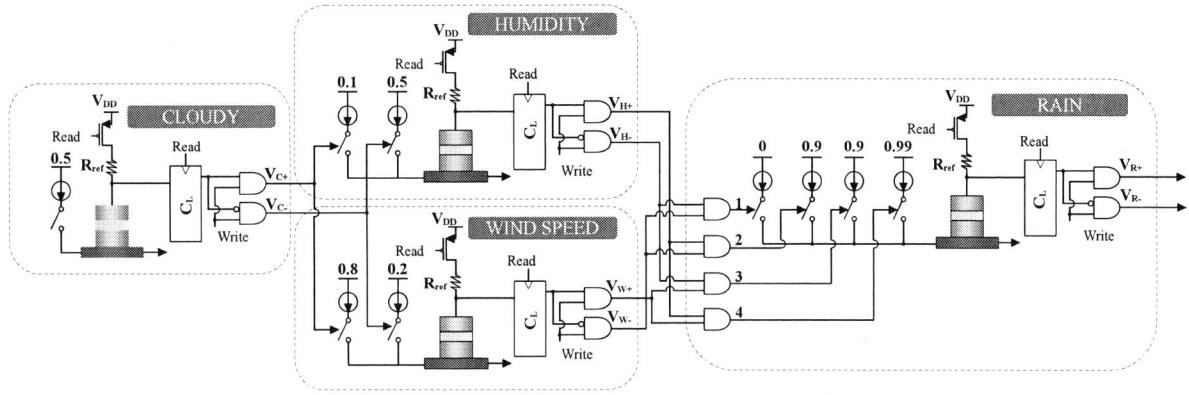

Fig. 5: The proposed device-circuit configuration is utilized to implement a Bayesian Network with four variables.

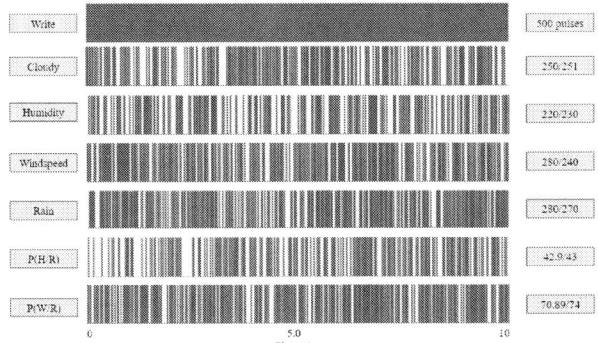

Fig. 6: Timing waveforms were utilized to compute the probabilities of different events occurring, with the average count of spikes generated in each scenario over 500 sample points closely mirroring the actual analytical results

TABLE I: The probability distribution function (PDF) of the two conditional events (H/R) and (W/R) tends to converge towards the mean value as the number of inference samples increases,

Infer	Analytic	# of Samples		
		1,000	5,000	10,000
P(H/R)	0.429	0.4321 +0.0219	0.4294 0.0168	0.4294 +0.0092
P(W/R)	0.7089	0.700 +0.0230	0.6998 +0.0173	0.6994 +0.0093

inference. The proposed work is proof-of-concept experiments showcasing probabilistic SOT-induced magnetization reversal. Such stochastic devices offer a direct mapping to the computing elements required for Bayesian inference. Furthermore, the potential for estimating more intricate inferences is expanded through the incorporation of supplementary arithmetic components like multiplication and division between two Poisson pulses.

Comparative analysis: This study introduces a novel strategy for the hardware implementation of BN by integrating an s-bit generator with a 2D memtransistor-based 2x1 multiplexer (MUX). The BN architecture employs 29 memtransistors and requires about 1.2 nJ/bit of energy for accurate computations [18]. In contrast, the MTJ-based probabilistic matrix-vector multiplication (MVM) engine significantly enhances energy efficiency by amalgamating multi-bit random number genera-

tion (RNG) with multiply-accumulate (MAC) operations within the same array, which relies on separate and energy-intensive true random number generator (RNG) circuits in advanced CMOS designs. For instance, the energy cost of true RNG in such CMOS circuits is approximately 1.6 pJ/bit or 6.4 pJ for generating a 4-bit random value [13]. In contrast, the proposed approach showcases remarkable energy efficiency, with an energy requirement below 100fJ. This novel design marks a notable advancement over existing methodologies, holding significant potential for substantial energy savings. Furthermore, this study observes that MTJ devices exhibit an autocorrelation time related to magnetoresistance fluctuation of around 2 to 5 ns (dependent on free layer thickness within the T1 and T2 ranges). Consequently, this leads to an increased interference speed in the implemented approach.

VI. CONCLUSION

The proposed work marks a significant advancement in achieving energy-efficient and faster processing of real-world probabilistic challenges. The integration of inherently stochastic spintronic devices with BN introduces novel capabilities for probabilistic graphical modeling across various domains. The BN is intricately built using SOTRAM-MTJs with precisely tailored free layer thickness to enable controlled manipulation of thermally unstable nanomagnets. This transformative shift to Spintronics-based approaches paves the way for creating high-performance, energy-conscious probabilistic models with a specific focus on applications such as edge computing and real-time tasks. This pioneering approach showcases a remarkable 40% increase in energy efficiency compared to other spintronic BN implementations. These achievements lay the groundwork for extending similar functionalities to neural networks, which is a potential avenue for future exploration.

REFERENCES

[1] D. Pocher, Lathrop, "Empirical modeling of superparamagnetic magnetic tunnel junctions with application to probabilistic computing," *Bulletin of the American Physical Society*, 2023.

[2] S. Chowdhury and Grimaldi, "A full-stack view of probabilistic computing with p-bits: devices, architectures and algorithms," *IEEE Journal on Exploratory Solid-State Computational Devices and Circuits*, 2023.

[3] B. Sutton, R. Faria, L. A. Ghantasala, R. Jaiswal, K. Y. Camsari, and S. Datta, "Autonomous probabilistic coprocessing with petaflips per second," *IEEE Access*, vol. 8, pp. 157238–157252, 2020.

979-8-3315-3968-9/24 $31.00 © 2024 IEEE

[4] P. Alisha and T. S. Warrier, "True random number generator based on voltage-gated spintronic structure," in *2023(VLSID)*. IEEE, 2023, pp. 377–382.

[5] S. Hong, "Solving inference problems of bayesian networks by probabilistic computing," *AIP Advances*, vol. 13, no. 7, 2023.

[6] A. Saidi and Othman, "Fpga-based implementation of classification techniques: A survey," *Integration*, vol. 81, pp. 280–299, 2021.

[7] Y. Lin and Zhang, "Uncertainty quantification via a memristor bayesian deep neural network for risk-sensitive reinforcement learning," *Nature Machine Intelligence*, pp. 1–10, 2023.

[8] R. Faria, "Autonomous probabilistic hardware for unconventional computing," Ph.D. dissertation, Purdue University, 2020.

[9] S. Dutta and Detorakis, "Neural sampling machine with stochastic synapse allows brain-like learning and inference," *Nature Communications*, vol. 13, no. 1, p. 2571, 2022.

[10] S. Khasanvis, "Architecting for causal intelligence at nanoscale," *Computer*, vol. 48, no. 12, pp. 54–64, 2015.

[11] X. Jia and Yang, "Spintronics based stochastic computing for efficient bayesian inference system," in *2018 23rd Asia and South Pacific Design Automation Conference (ASP-DAC)*. IEEE, 2018, pp. 580–585.

[12] X. Jia, H. Gu, Y. Liu, J. Yang, X. Wang, W. Pan, Y. Zhang, S. Cotofana, and W. Zhao, "An energy-efficient bayesian neural network implementation using stochastic computing method," *IEEE Transactions on Neural Networks and Learning Systems*, 2023.

[13] S. Liu, T. P. Xiao, J. Kwon, B. J. Debusschere, S. Agarwal, J. A. C. Incorvia, and C. H. Bennett, "Bayesian neural networks using magnetic tunnel junction-based probabilistic in-memory computing," *Frontiers in Nanotechnology*, vol. 4, p. 1021943, 2022.

[14] K. Zhang, D. Zhang, C. Wang, L. Zeng, Y. Wang, and W. Zhao, "Compact modeling and analysis of voltage-gated spin-orbit torque magnetic tunnel junction," *IEEE Access*, vol. 8, pp. 50 792–50 800, 2020.

[15] Y. Suzuki, A. A. Tulapurkar, and C. Chappert, "Spin-injection phenomena and applications," in *Nanomagnetism and Spintronics*. Elsevier, 2009, pp. 93–153.

[16] L. Pan, L. Huang, M. Zhong, X.-W. Jiang, H.-X. Deng, J. Li, J.-B. Xia, and Z. Wei, "Large tunneling magnetoresistance in magnetic tunneling junctions based on two-dimensional crx 3 (x= br, i) monolayers," *Nanoscale*, vol. 10, no. 47, pp. 22 196–22 202, 2018.

[17] Y. Shim, S. Chen, A. Sengupta, and K. Roy, "Stochastic spin-orbit torque devices as elements for bayesian inference," *Scientific reports*, vol. 7, no. 1, p. 14101, 2017.

[18] Y. Zheng, H. Ravichandran, T. F. Schranghamer, N. Trainor, J. M. Redwing, and S. Das, "Hardware implementation of bayesian network based on two-dimensional memtransistors," *Nature communications*, vol. 13, no. 1, p. 5578, 2022.

AFSRAM-CIM: Adder Free SRAM-Based Digital Computation-in-Memory for BNN

Asmae El arrassi*, Mohammad Amin Yaldagard*, Xingjian Tao†, Taha Shahroodi*, Fouwad Mir*,
Yashvardhan Biyani *,Manil Dev Gomony†, Anteneh Gebregiorgis*, Rajiv Joshi ‡, Said Hamdioui*

*Department of Quantum and Computer Engineering, Delft University of Technology, Delft, The Netherlands
Email: {a.elarrassi, m.a.yaldagard, t.shahroodi, f.j.mir, y.biyani, a.b.gebregiorgis, S.Hamdioui} @tudelft.nl
†Eindhoven University of Technology, Eindhoven, The Netherlands
Email: {x.tao1@student, m.gomony@} tue.nl
‡Thomas J. Watson Research Center IBM, New York, USA Email: rvjoshi@us.ibm.com

Abstract—Binary Neural Networks (BNNs) have demonstrated significant advantages in reducing computation and memory costs, all while maintaining acceptable accuracy on various image detection tasks. Thus, BNNs have the potential to support practical cognitive tasks on resource-constrained platforms, such as edge computing devices. To realize this, SRAM-based digital Computation-in-Memory (CIM) has gained growing attention as it overcomes the analog CIM architecture bottlenecks such as limited computing accuracy due to process variation, non-linearity, power and area-hungry Analog-to-Digital Converters (ADCs), etc. However, digital CIM architectures are highly dominated by power-hungry adder-trees, which can nullify the benefits of SRAM-based digital CIM. To address this issue, this paper proposes an adder free SRAM-based digital CIM, AFSRAM-CIM, for BNN acceleration. The proposed CIM architecture utilizes a multi-functional 10-T SRAM cell-based crossbar array and a new energy-efficient approach to perform the popcount operation. Simulation results using the MNIST dataset show that the proposed architecture maintains the state-of-the-art inference accuracy of 99.21% with only 11.86 fJ energy per operation. Moreover, AFSRAM-CIM achieves over $3\times$ energy and $\approx17\times$ area savings when compared to the conventional digital CIM approaches.

Index Terms—Computation-in-Memory, SRAM, Fully-digital, BNN, MAC

I. INTRODUCTION

Smart applications have become a crucial part of human daily life. These applications are gaining growing interest for their potential to deliver high accuracy and performance in multiple domains such as computer vision applications [1]. However, due to various architectural challenges and technological limitations, the existing Von-Neumann architecture cannot deliver the computing efficiency required by resource-constrained platforms such as edge devices [2], [3]. Hence, designing energy-efficient architectures and simplified network models is important to deploy AI on edge devices. As a result, emerging CIM architectures are widely studied for their potential to eliminate the excessive time and energy spent on moving massive amounts of data between the memory and processing unit [4], [5]. Moreover, BNNs have shown significant improvements in reducing the complexity of neural network models [6] by aggressively quantizing the parameters to 1-bit precision while maintaining reasonable accuracy [7].

Therefore, there is a clear need to exploit the huge potential of energy-efficient CIM architectures for BNN acceleration.

CIM integrates computing and storage together and provides an efficient implementation of Vector-Matrix Multiplication (VMM), which is the key operation in BNN [8]. Therefore, CIM provides a huge potential for energy-efficient BNN implementation on edge devices [9]. CIM can be realized in an analog domain using emerging/CMOS technologies [10]–[13] or in a digital domain using SRAM cells as its core building block [14], [15]. Digital CIM has an edge over its analog counterpart as it avoids the need for energy and area-hungry ADCs, which represents a vital block for analog CIM [16]. Moreover, CMOS technology maturity and EDA tool support availability make it favorable for fabrication and near future industry-scale adaptation of digital CIM [16]. Several works have presented solutions for SRAM-based digital CIM to perform Multiply-and-Accumulate (MAC) operations [8], [14]–[17]. However, state-of-the-art approaches [8], [14]–[17] require additional circuits to perform multiplication and adder-trees to perform accumulation which dominates the energy and area consumption of the array. As a result, digital SRAM-based CIM computing overheads and challenges need to be minimized to further enhance and optimize CIM architecture for edge applications.

In this work, we propose an energy-efficient adder SRAM-based CIM architecture, AFSRAM-CIM, for BNN accelerators. For this purpose, a multi-functional 10-T SRAM cell is designed to support multiple logic operations such as (XOR, XNOR, OR, etc...). The logic operations are performed within the cell-specific to the targeted application. We propose a new approach to perform accumulation that eliminates the use of the power-hungry adder-trees used in state-of-the-art solutions [9], [14]. The proposed AFSRAM-CIM is evaluated with BNN using the MNIST dataset. Simulation results of post-layout extraction show that the proposed AFSRAM-CIM is highly energy-efficient with a consumption of 11.86 fJ per operation while maintaining state-of-the-art inference accuracy of 99.21%. Thus, AFSRAM-CIM realizes more than $3\times$ and $\approx17\times$ energy and area savings over the conventional adder-tree based digital CIM approaches. The main contributions of the paper are summarized as follows:

979-8-3315-3968-9/24 $31.00 © 2024 IEEE

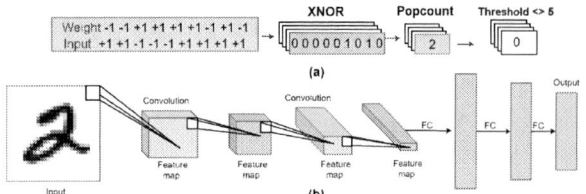

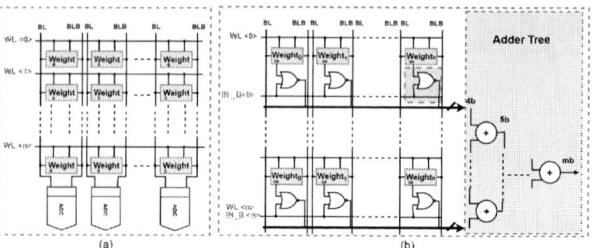

Fig. 1. an illustration of a) BNN inference binary MAC operation and b) Binary Convolutional Neural Network (CNN) architecture.

Fig. 2. An illustration of a) analog CIM architecture [24] and b) all digital CIM architecture [14].

- A multi-functional SRAM cell design to perform multiple logic operations within the cell;
- An adder-tree free SRAM-based digital CIM architecture to perform binary MAC operation for BNN applications;
- Validation of the proposed architecture shows high energy efficiency of 11.86 fJ per operation with more than $3\times$ energy efficiency improvement compared to state-of-the-art approaches;

The remainder of the paper is organized as follows, Section II presents the basic concepts. Section III presents the proposed architecture. Section IV presents experimental results and discussion. Finally, Section V concludes the paper.

II. PRELIMINARIES

A. Binary Neural Network

BNN is a type of artificial neural network where weights and activation are binarized. BNN reduces memory and computation requirements and offers great potential to improve energy efficiency [18]. Despite the extreme quantization to 1-bit weights and activations, BNN still delivers good inference accuracy for several applications such as object detection and classification [19]. The main operation for BNN is binary MAC operation. The binary MAC operation can be expressed as follows:

$$BinMAC(In, W) = \textbf{popcount}(In \textbf{ XNOR } W) \quad (1)$$

Where In is the input/activation vector and W is the weight vector. Fig. 1(a) illustrates the MAC operation for BNN inference. The signed weights and inputs/activations can be represented by two values "+1" or "-1" [6]. However, previous works [20] explored further simplification of the computation by encoding the values to "1" or "0". The first phase of the MAC operation consists of the XNOR boolean operation between the inputs/activation and the weights. The popcount operation is then performed. The outputs of the popcount represent the accumulation of the output of the XNOR operation. An activation function is then applied to the output of the popcount. The sign activation function can be expressed as follows:

$$sign(x) = \begin{cases} 1 & \text{if } x \geq Threshold \\ 0 & \text{if } x < Threshold \end{cases} \quad (2)$$

Where $Threshold$ can be expressed as follows:

$$Threshold = \frac{Th + N}{2} \quad (3)$$

Where Th is the activation threshold (ex. $Th = 1$) and N is the input vector size (ex. N = 9). Fig. 1(b) illustrates an example of a CNN architecture for binary classification. The binary MAC operation reduces the computation costs and memory storage of BNNs.

B. SRAM-based CIM architectures

CIM architectures have the potential to overcome the Von-Neumann challenges by integrating computation and storage in the same physical location [21]. CIM can be realized using different memory technologies such as memristors [22], SRAMs [21], and DRAMs [23]. SRAM-based CIM can be realized in the analog [24] or digital [9] domain. In the analog domain, the first operands (e.g., weight values) are stored in the SRAM cells while the second operands (e.g., activation inputs) are provided through the wordlines (WLs). Then, each column performs the MAC operation by multiplying the input operands and the operands stored in the SRAM cells, the output current is accumulated through the bitlines (BLs) and forms a dot-product according to Kirchhoff's law [3]. The output current through the Bitline (BL) is then fed to an ADC to be converted to its digital value. Fig 2.(a) shows an SRAM-based analog CIM macro with SRAM crossbar array to perform the MAC operation and ADC to convert the MAC output current to digital values. Several works have studied SRAM-based analog CIM. The work in [25] proposes a 10-T SRAM cell to store 1-bit of the weight. The MAC operation is performed in the charge domain, and each column has a dedicated ADC. This approach has addressed the write disturb limitation by decoupling the write path and the read path. However, the power-hungry ADCs used for each column decrease the energy efficiency of the array. Analog-based CIM offers advantages such as high parallelism [9], [10]. However, there are certain design challenges associated with analog-based CIM, including process variation, ADCs overhead, and computing non-linearity, which can limit its scalability [16].

To address these limitations SRAM-based digital CIM eliminates the need for energy and area excessive ADCs [9]. MAC operation can be realized in the digital domain by storing the first operand (eg. weight) in the SRAM-based CIM array. The second operand is provided through the WLs. In each column, the two operands are provided to multipliers to perform multiplication. The multiplication outputs are then fed to an adder-tree to perform the accumulation. Fig. 2(b) illustrates an example of an SRAM-based digital CIM for MAC operation [14]. The work in [9] has proposed an

979-8-3315-3968-9/24 $31.00 © 2024 IEEE

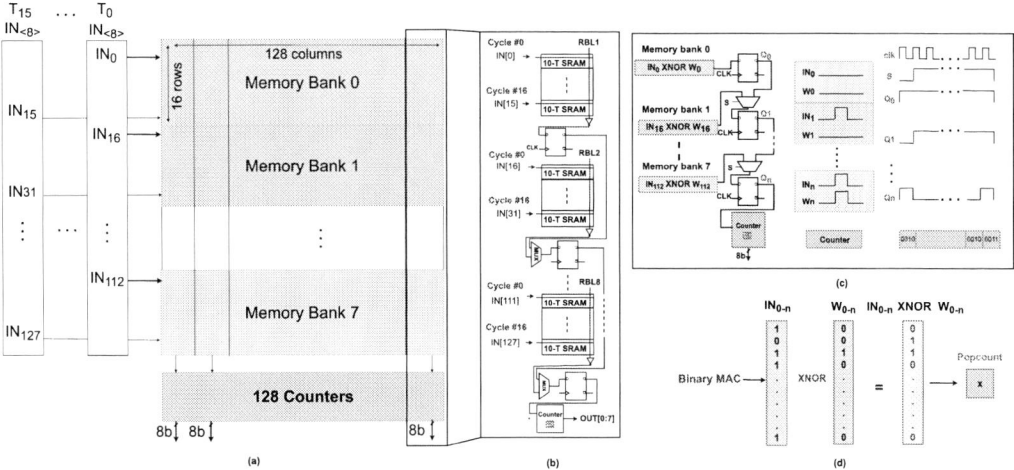

Fig. 3. Illustration of a) the overview of the proposed CIM architecture, b) A column and periphery structure to perform binary MAC operation,c) the Binary MAC operation implementation, and d) A BNN inference binary MAC operation.

SRAM-based CIM architecture for BNN applications. In this approach [9], the weights and activations are stored in the SRAM array. The XNOR operation is performed by activating two rows of the array. For each column, two sense amplifiers (SAs) are added to read the output of the XNOR operation. Furthermore, the outputs of each column are accumulated using an adder-tree. Additionally, the work in [14] has presented a new adder-tree structure to perform accumulation. In this approach, each 4-bit column has a dedicated adder tree which results in a high area consumption. SRAM-based digital CIM offers advantages such as high precision, energy efficiency, and high scalability. However, these approaches suffer from the overhead of the adder-tree units that dominate the area and energy consumption of the array. To address these limitations, we propose a new energy-efficient SRAM-based digital CIM architecture, AFSRAM-CIM, for BNN accelerators.

III. AFSRAM-CIM ARCHITECTURE

The proposed AFSRAM-CIM architecture is realized as follows. First, a multi-functional 10-T SRAM bit cell is designed to perform different logic operations. Then, an adder-tree free CIM macro is designed using the 10-T SRAM cell and popcount logic as its building blocks. In this section, first the design of the 10-T SRAM bit cell is discussed. Then, the proposed CIM macro is presented followed by the discussion of the BNN mapping to the CIM macro.

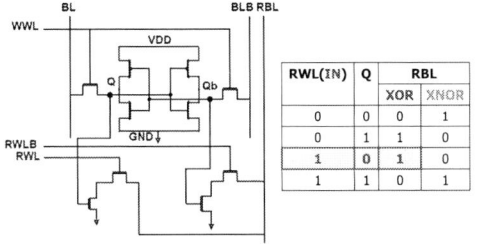

Fig. 4. Illustration of a) the 10-T SRAM cell and b) the XOR operation performed within the memory cell.

A. 10-T SRAM cell design

In this work, we propose a 10-T SRAM cell design that supports multiple logic operations within the cell-specific to the targeted application. Fig. 4(a) illustrates a schematic of the proposed 10-T SRAM cell. The write operation is performed by driving the Write Wordline (WWL) and BL to write 1 (or 0). To perform the read operation, The Read Bitline (RBL) is first precharged to V_{DD} and the Read Wordline (RWL) is then activated. The RBL is discharged to 0 when the content of the cell Q = 1 and remains charged when Q = 0. The output of the RBL is then inverted for each column using an inverter as shown in Fig. 3(b). Additionally, the 10-T SRAM cell can support multiple logic operations. The XOR/XNOR operation can be performed within the cell by storing the first operand in the cell and driving the second operand to the RWL/RWLB and its invert to RWLB/RWL. The RBL is precharged to V_{DD} initially when the RWL or RWLB is activated and the content of the cell is Q = 1 or Qb = 1, respectively, the RBL discharges while it remains charged in the other cases. Fig. 4 illustrates how XOR and XNOR operations can be performed within the memory cell.

The SRAM cell supports other logic operations such as NAND and OR operation depending on how the inputs are driven. NAND and OR operations can be performed by disabling RWLB/RWL and driving the IN/INB to RWL/RWLB. The RBL is discharged when (IN = 1 and W = 1) or (IN = 0 and W = 0) to perform NAND or OR operation, respectively. The 10-T SRAM cell has decoupled read-and-write paths which eliminate the read-disturb limitation present in the conventional 6-T SRAM. The cell design is used as the core building block of the CIM macro.

B. 10-T SRAM-based CIM macro

The proposed CIM macro is composed of 10-T SRAM crossbar array, shared flip-flops, shared multiplexers and counters. The crossbar array consists of multiple subarrays that are arranged in a bank structure. Each bank is organized into

979-8-3315-3968-9/24 $31.00 © 2024 IEEE

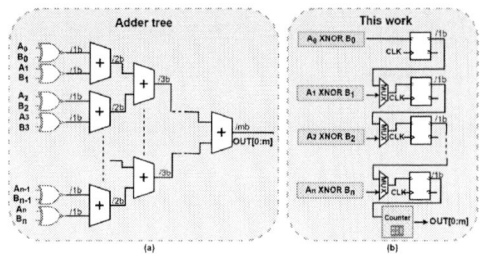

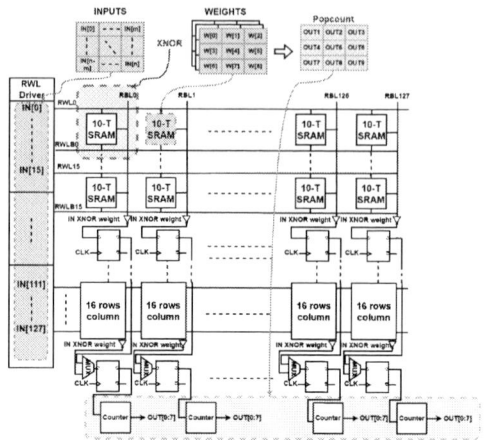

Fig. 5. a) Conventional adder-tree versus b) proposed adder-tree free architecture.

16 rows and 128 columns. For each column in a bank, the rows share the RBLs, BLs, and BLBs while for each row, all the columns share the WWLs, RWLs, and RWLBs as shown in Fig. 3(a). The columns in the first bank are connected to dedicated flip-flops that receive as input the output of their respective RBLs (digitalized with the column inverters). However, the columns in the remaining banks are connected to a multiplexer (MUX) and a flip-flop. The MUXs receive an input signal from the column RBLs and a signal from the flip-flops outputs of the preceding bank. The MUX outputs are then stored in their respective flip-flops. Finally, the flip-flop outputs of the last bank are connected to the counters dedicated for each column as shown in Fig. 3(b).

The proposed AFSRAM-CIM architecture increases the parallelism and reduces the RBLs, BLs, and BLBs charging and discharging delays. The digital counter offers high area efficiency compared to the adder tree approach. Fig. 5 illustrates a comparison between an adder-tree approach and the proposed adder-tree free architecture.

C. BNN mapping

1) BNN implementation of the proposed SRAM-based CIM: The BNN architecture adopted in this work is the LeNet-5 network topology [26]. Fig. 6 illustrates the mapping of the BNN architecture on the proposed CIM design. To map the convolutional layers each filter weight vector is stored in the same column and distributed on different banks while, for the fully connected layer, the weights connected to the same output neuron are stored in one column of the SRAM-based crossbar array. We activate in parallel one row from each bank where the input is provided through the RWLs. The memory bank structure can allow maximum use of the memory storage and parallelism in the array for different layer parameters.

2) Binary MAC operation implementation: To implement binary MAC operation, We perform the XOR operations within the SRAM cell between the weights stored in each cell and the inputs provided through the RWLs. The XOR results are inverted in the RBLs to get the intended XNOR results as shown in equ. 1. We activate n RWLs in parallel, where n is the number of banks in the SRAM-based crossbar array.

To perform the popcount operation, the RBLs of each bank drive the output of the XNOR operation of one SRAM cell at a time. Next, The first bank RBLs are connected to flip-flops that store the signal and provide it to the next bank. The next bank consists of MUXs that receive input signals from

Fig. 6. BNN mapping scheme to the proposed CIM architecture.

the RBLs of the same memory bank and signals from the flip-flops output of the preceding bank. The MUXs are connected to flip-flops. During the first clock cycle, the MUXs provide as output the XNOR outputs provided by the connected RBLs. During the remaining cycle time, These MUXs provide as output the outputs of the preceding flip-flops. The last flip-flops provide inputs to digital counters connected to each column. These flip-flops deliver the RBLs output signal of each bank sequentially. Therefore, in each clock cycle the counter receives the XNOR operation outputs from all the array rows sequentially as shown in Fig. 3(c). The binary MAC operation implementation using digital counters is illustrated in Fig. 3(c)-(d).

To implement large kernel sizes of the convolutional layer, multiple arrays can be used to store the weights, and the resulting partial sum from the digital counters can be accumulated. AFSRAM-CIM architecture offers high accuracy due to the digital logic units used for computation that give an exact and accurate output similar to software implementation. additionally, The low area overhead of the proposed adder-tree free architecture enables accumulation units to be assigned to every column, which offers high parallelism with low overhead.

IV. EXPERIMENT RESULTS AND DISCUSSION

A. Experiment setup

The simulation setup used in this work is presented in Table. I. The proposed AFSRAM-CIM architecture is simulated with SPICE using TSMC 40 nm CMOS technology. The energy, area, and latency results are extracted using post-layout simulations. It is worth noting that the results were reported in a worst-case scenario with high sparsity (IN = 1, weights 50%). The network is trained offline and then implemented in hardware with VHDL. We performed synthesis using digital design flow for area evaluation.

In this work, We have trained a BNN architecture based on the LeNet-5 network topology [27] as shown in Table. I. The architecture delivers high performance and accuracy in a compact topology [26]. For the MNIST dataset, the size of

TABLE I
SIMULATION SETUP.

Technology	40 nm
Supply voltage (V)	1 V
Temperature	27 °C
SRAM cell	10-T
Unit macro size	16 kb (128x128b)
BNN topology	Lenet-5
Dataset	MNIST
Conv1	(5,5)×20
Conv2	(5,5)×50
Fully-connected layer 1	800×500
Fully-connected layer 2	500×10

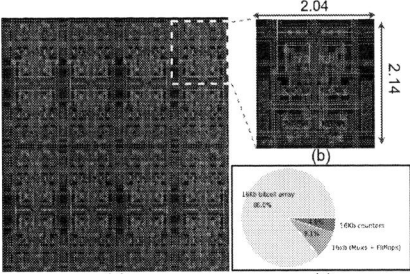

Fig. 7. a) crossbar array layout, b) bitcell layout and c) Post-layout area breakdown of different units.

the input matrix is (28,28). In the first convolutional layer, we apply a filter of dimension (5x5) with 20 channels followed by a pooling layer performing max pooling with a pool size of (2x2). In the second convolutional layer, the applied filter size is (5x5) with 50 channels and a pooling layer with a pool size of (2x2). The network contains two fully connected layers with a size of (800,500) and (500,10). We have trained the network using the BCIM framework [27] using a feedforward model and backpropagation algorithm. The trained network is mapped to implement the inference on the proposed AFSRAM-CIM architecture. the network parameters are presented in Table. I.

B. Energy and area results

To perform energy and area evaluation, a 128×128 SRAM-based crossbar array was simulated in SPICE using the TSMC 40 nm technology. Results show that the proposed AFSRAM-CIM implementation performs MAC operation in an energy-efficient manner with an energy consumption of 11.86 fJ per operation. An operation is defined as one binary multiplication and accumulates operation performed in the array. The recorded energy efficiency and peak throughput are 157 TOPS/W and 496 GOPS, respectively.

Fig. 7 illustrates the array and Bitcell layout and a comparison of the area consumption of different units. The Bitcell occupies an area of 3.407 μm^2. The SRAM 16 kb array occupies 86% area. The counters and the additional units to perform the popcount operation occupy 14% of the total area of the array as shown in Fig. 7(c).

C. Accuracy results

To evaluate the application accuracy, we trained the BNN network with LeNet-5 topology using BCIM framework [27]. The MNIST dataset is used for training. We have mapped the trained weights to the array for inference. The weights are preloaded before performing the MAC operation through the write ports to the SRAM cell. To ensure the maximum use of parallelism and memory storage the weights of the same filter were stored in the same column and the same row of each memory bank. The inputs were provided in parallel through the read ports as described in Section. III-A. The AFSRAM-CIM architecture has achieved the ideal accuracy of 99.28%.

D. State-of-the-art comparison

Table. II shows the performance comparison of the proposed AFSRAM-CIM architecture and state-of-the-art SRAM-based

digital CIM neural network implementations. The work in [16] has presented an SRAM-based digital CIM architecture where each bitcell consists of a 6-T SRAM, full adder, two MUXs, and an XNOR gate. The multiplication and partial sum are performed within the array. However, the additional computing units in the bitcell increase the area consumption, resulting in a low-density array. Additionally, the work in [14] has presented an SRAM-based digital CIM architecture. In this work, additional circuitries and adder-trees are used to perform multiplication and accumulation, respectively. However, the high number of adder-trees makes it area inefficient.

In order to compare our solution with the state-of-the-art SRAM-based digital CIM presented in [8], we simulated the CIM architecture using 28 nm technology. The power consumption comaprison of AFSRAM-CIM and the digital CIM [8] is presented in Fig. 8. The reported average power shown in Fig. 8 represents wide range of weight sparsities (percentage of binary weights with the value of "1") in order to compare the worst-case and best-case scenarios. Fig. 8 demonstrates that the proposed AFSRAM-CIM architecture consumes $\approx 4\times$ less power compared to the approach in [8]. As illustrated in the figure low-weight sparsity (less binary "1" weight values) reduces the average power consumption. The 50% weight sparsity shows the highest average power consumption due to the high switching activity of the XNOR results. Moreover, it is worth mentioning that the reported reduction in power consumption is without considering the power reduction benefit of 28 nm implementation over AFSRAM-CIM's 40 nm. Thus, for similar technology nodes, the power saving of AFSRAM-CIM will be even higher than what is shown in Fig. 8.

For area evaluation, we simulated a 128-bit adder-tree and the proposed adder-tree free approach including all the additional units (digital counter, flip-flops, and MUXs) using the

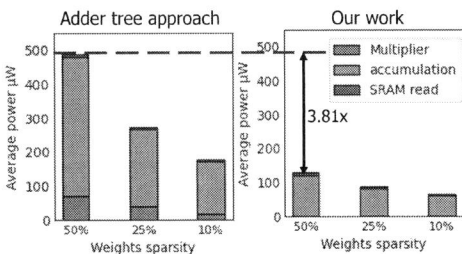

Fig. 8. Power consumption comparison of (right) the proposed architecture and (left) an adder-tree based approach [8] for 32 × 32 array.

979-8-3315-3968-9/24 $31.00 © 2024 IEEE

TABLE II
State-of-the-art comparison of the SRAM-based digital CIM macros.

	ICTA'22 [17]	ISSCC'21 [14]	VLSIS'22 [15]	JSSC'21 [16]	ISSCC'22 [8]	This Work
Technology (nm)	28	22	12	65	5	40
Supply Voltage (V)	0.9-0.8	0.72	0.72	0.6-0.8	0.5-0.9	1
Input/Output Precision (bit)	4-to-4	(1-8)-to-4	(1-8)-to-(4-16)	1-to-16	4-to-4	1-to-1
GOPS/mm^2	N/a	16000	N/A	6750 (1b)	221000 (4b)	7638 (1b)
Throughput (GOPS)	2632 (4b)	N/A	1343 (4b)	576 (1b)	N/A	496.48 (1b)
Energy efficiency (TOPS/W)	61 (4b)	89 (4b)	121 (4b)	117.3 (1b)	254 (4b)	157.1 (1b)
Energy per operation (pJ/op)	N/A	N/A	N/A	0.017 (1b)	N/A	0.0118
Bitcell density (kb)	64	64	8	64	64	16
Bitcell area (μm^2)	N/A	0.379	N/A	10.53	0.075	3.407
Bitcell array area (mm^2)	0.19	0.202	0.0323	0.19	0.0133	0.0659

same 40 nm technology node. The results show that the pop-count units of the proposed AFSRAM-CIM architecture for 128-bit binary MAC operations occupy $\approx 17\times$ less area than a conventional adder-tree structure. Furthermore, the proposed solution occupies 74.91 μm^2, while the conventional adder-tree approach occupies 1246 μm^2 for 128-bit accumulation. Therefore, AFSRAM-CIM achieves high energy efficiency while occupying a relatively low area with an area efficiency of 7638 GOPS/mm^2.

V. Conclusion

In this work, we proposed an energy-efficient BNN implementation using SRAM-based digital CIM. The proposed AFSRAM-CIM architecture minimizes the high energy overhead of the accumulation units by presenting an adder-tree free SRAM-based digital CIM architecture to perform binary MAC operation. Simulation results demonstrated that the proposed AFSRAM-CIM architecture is highly energy-efficient with an energy consumption of 11.86 fJ per operation while maintaining state-of-the-art accuracy of 99.21%. This work achieved over $3\times$ energy and $\approx 17\times$ area savings when compared to the conventional adder-tree approach.

Acknowledgements

This work is funded by CONVOLVE (Grant No. 101070374).

References

[1] G. Wang and J. Gong, "Facial expression recognition based on improved lenet-5 cnn," in *CCDC*, 2019.

[2] S. Hamdioui et al., "Memristor based computation-in-memory architecture for data-intensive applications," in *DATE*, 2015.

[3] G. W. Burr, A. Sebastian, T. Ando et al., "Ohm's law+ kirchhoff's current law= better ai: Neural-network processing done in memory with analog circuits will save energy," *IEEE Spectrum*, 2021.

[4] H. A. Du Nguyen, J. Yu, L. Xie, M. Taouil, S. Hamdioui, and D. Fey, "Memristive devices for computing: Beyond cmos and beyond von neumann," in *VLSI-SoC*, 2017.

[5] M. Hu, C. E. Graves, C. Li, Y. Li, Ge et al., "Memristor-based analog computation and neural network classification with a dot product engine," *Advanced Materials*, 2018.

[6] M. Courbariaux, I. Hubara, D. Soudry et al., "Binarized neural networks: Training deep neural networks with weights and activations constrained to+ 1 or-1," *arXiv*, 2016.

[7] M. Rastegari, V. Ordonez et al., "Xnor-net: Imagenet classification using binary convolutional neural networks," in *ECCV*, 2016.

[8] H. Fujiwara, H. Mori, Zhao et al., "A 5-nm 254-tops/w 221-tops/mm 2 fully-digital computing-in-memory macro supporting wide-range dynamic-voltage-frequency scaling and simultaneous mac and write operations," in *ISSCC*, 2022.

[9] A. Agrawal, A. Jaiswal, D. Roy, B. Han et al., "Xcel-ram: Accelerating binary neural networks in high-throughput sram compute arrays," *ISCAS-I*, 2019.

[10] Q. Dong, M. E. Sinangil, B. Erbagci et al., "15.3 a 351tops/w and 372.4 gops compute-in-memory sram macro in 7nm finfet cmos for machine-learning applications," in *ISSCC*, 2020.

[11] R. Kozma, R. E. Pino, and G. E. Pazienza, *Advances in neuromorphic memristor science and applications*, 2012.

[12] S. Yin, Z. Jiang, J.-S. Seo, and M. Seok, "Xnor-sram: In-memory computing sram macro for binary/ternary deep neural networks," *IEEE Journal of Solid-State Circuits*, 2020.

[13] H. Benmeziane et al., "Analognas: A neural network design framework for accurate inference with analog in-memory computing," in *EDGE*, 2023.

[14] Y.-D. Chih, P.-H. Lee, Fujiwara et al., "16.4 an 89tops/w and 16.3 tops/mm 2 all-digital sram-based full-precision compute-in memory macro in 22nm for machine-learning edge applications," in *ISSCC*, 2021.

[15] C.-F. Lee, C.-H. Lu, C.-E. Lee, H. Mori et al., "A 12nm 121-tops/w 41.6-tops/mm2 all digital full precision sram-based compute-in-memory with configurable bit-width for ai edge applications," in *ISVLSI*, 2022.

[16] H. Kim, T. Yoo, T. T.-H. Kim, and B. Kim, "Colonnade: A reconfigurable sram-based digital bit-serial compute-in-memory macro for processing neural networks," *IEEE Journal of Solid-State Circuits*, 2021.

[17] D. Wanq, Z. Li, C. Chang, W. He et al., "All-digital full-precision in-sram computing with reduction tree for energy-efficient mac operations," in *ICTA*, 2022.

[18] S. Liang, S. Yin, L. Liu, W. Luk, and S. Wei, "Fp-bnn: Binarized neural network on fpga," *Neurocomputing*, 2018.

[19] R. Liu, X. Peng, X. Sun, W.-S. Khwa et al., "Parallelizing sram arrays with customized bit-cell for binary neural networks," in *DAC*, 2018.

[20] X. Sun, X. Peng, P.-Y. Chen et al., "Fully parallel rram synaptic array for implementing binary neural network with (+ 1,- 1) weights and (+ 1, 0) neurons," in *ASP-DAC*, 2018.

[21] Z. Lin, Z. Tong, J. Zhang et al., "A review on sram-based computing in-memory: Circuits, functions, and applications," *Journal of Semiconductors*, 2022.

[22] A. Gebregiorgis, A. Singh, A. Yousefzadeh et al., "Tutorial on memristor-based computing for smart edge applications," *Memories-Materials, Devices, Circuits and Systems*, 2023.

[23] D.-Y. Lim, I.-J. Jung, D.-H. Kim et al., "Computing-in-memory using 1t1c embedded dram cell with micro sense amplifier for enhancing throughput," in *ICCE-Asia*, 2022.

[24] J.-S. Kim, J.-W. Lee et al., "10t sram computing-in-memory macros for binary and multibit mac operation of dnn edge processors," *IEEE Access*, 2021.

[25] A. Biswas and A. P. Chandrakasan, "Conv-sram: An energy-efficient sram with in-memory dot-product computation for low-power convolutional neural networks," *IEEE Journal of Solid-State Circuits*, 2018.

[26] E. Kussul and T. Baidyk, "Improved method of handwritten digit recognition tested on mnist database," *Image and Vision Computing*, 2004.

[27] M. Zahedi, T. Shahroodi, S. Wong, and S. Hamdioui, "Bcim: Efficient implementation of binary neural network based on computation in memory," *arXiv*, 2022.

979-8-3315-3968-9/24 $31.00 © 2024 IEEE

DynaCache: A checkpoint aware reconfigurable cache for Intermittently powered Computing Systems

Rishabh Mahanta and Hemangee K. Kapoor

Dept. of Computer Science and Engineering
Indian Institute of Technology Guwahati, India
{rishabh.mahanta, hemangee}@iitg.ac.in

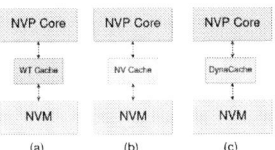

Fig. 1: Cache designs. (a) write-through caches (b) Non-volatile caches (c) Our proposed system

Abstract—Energy harvesting devices are rapidly evolving to rival battery-backed technologies. Batteries have a shorter lifetime and need maintenance compared to capacitors. Moreover, the usage of batteries comes with undeniable environmental costs. Distinct challenges have emerged such as performance enhancement, crash consistency of data, energy management, etc.
To tackle these challenges we propose DynaCache: A checkpoint-aware reconfigurable cache for Intermittent powered computing systems. It is an intermittent computing architecture consisting of a reconfigurable L1 cache that dynamically transforms its write policy to strike a favorable trade-off between performance, data consistency, and forward progress of a program. Our proposal achieves a speedup in performance of 1.85x and 1.5x compared to non-volatile write-back cache and volatile write-through cache respectively. It also achieves a 99% reduction in dirty data compared to a volatile write-back cache guaranteeing a near absolute data consistency in a scenario of a power failure.

Index Terms—Intermittent computing, Energy harvesting, Caches, Non-Volatile memory, Embedded systems

1. INTRODUCTION

The quest for improving the size and latency of computing devices has led mankind to take computation at the edge of the source of its fuel. However, not all sources of energy are stable and consistent with time. Involving batteries results in the overhead of maintenance and recycling in the long run. To mitigate the shortcomings of batteries, energy is cultivated by exploiting irregular bursts of energy through energy harvesting devices. Thus, the research community has referred to such battery-less systems as intermittent computing (ImC) systems ImC systems leverage a small capacitor that buffers the harvested energy. Such systems employ non-volatile memory (NVM) that lets the system offload the on-chip volatile data as a checkpoint and later restore the same back to the volatile memory. Most ImCs are devoid of caches and instead have

to rely on the NVM residing in the system. Exclusion of the cache is due to the frequent power failures making the cache inconsistent resulting in a phenomenon known as crash consistency problem [1], [2]. Directly accessing the NVM results in a large overhead in terms of latency and energy where the capacitor buffer drains quickly due to the repeated NVM accesses. It is well known that caches help to improve performance by providing quicker access to frequently used data. An SRAM cache shows a promising potential to lower energy costs and improve the performance of ImC systems. However, implementing cache in ImC systems comes with the challenge of crash consistency. Incoherent checkpointing of the cache can lead to an incoherent program state post-backup which can result in an overall corruption of the program output. Thus, backing up of cache contents should be given a crucial priority which results in cache contents getting persisted to the NVM. Caches need to be implemented with write policies that suit the application. (i) Write-through policy is suitable for maintaining consistency as writes are always synchronized with the main memory in this policy. However, the Write-through policy turns out to be costly in terms of latency and energy due to constant writes to NVM. (ii) The write-back policy brings out the speed benefits of a cache in complete totality. However, the lingering issue of data consistency cannot be overlooked, especially when it comes to ImC devices running on harvested energy. Un-persisted data lost in a power failure is lost forever and becomes irrecoverable.

In this paper, we propose DynaCache (Dynamic Cache), a dynamic cache architecture for intermittent computing systems. It comprises of one-level volatile SRAM cache with the PCM-based NVM acting as the main memory. The objective of DynaCache is to reduce the checkpointing overheads incurred by write-back caches. It re-configures itself to behave as a write-through cache after a certain threshold of cycles, thus significantly reducing the amount of dirty lines to backup. Our proposed work can combine the speed of a write-back cache and also provide near-perfect crash consistency of a write-through cache all in a volatile SRAM cache.

979-8-3315-3968-9/24 $31.00 © 2024 IEEE

II. RELATED WORK

An Intermittent computing system majorly consists of a volatile and a non-volatile memory component. Converting volatile structures to non-volatile may eliminate some of the memory inconsistency issues. However, fully non-volatile architectures and main memories have two drawbacks. First, efficiency suffers, because the relatively low-latency, low-energy volatile memory accesses become relatively high-latency, high-energy non-volatile memory accesses. Second, some states are supposed to be atomic and become ineffectual if saved during a power failure requiring a re-initialization by executing a code again. For example, a MEMS(Micro-electromechanical systems) sensor must perform an initialization routine before it can be sampled. This is an example where a fully non-volatile architecture is at best a partial solution to the problem of preserving progress across power failures.

A. Caches

Caches are available in all sorts of varieties like completely volatile SRAM cache [3], completely non-volatile STT-RAM cache [4], hybrid SRAM-STTRAM cache. [5]. STT-RAM caches are one of the most promising alternatives to SRAM caches as the former is faster compared to other Non-volatile memory technologies and also preserves consistency. Unfortunately, the write latency is still almost 5X [6] to that of SRAM and each write operation consumes an amount of energy that still cannot be overlooked given the constrained energy budget. In addition to that, STT-RAM is a costlier version of NVM compared to SRAM.

Proper checkpointing approaches [7]–[10] ensure atomicity and consistency at the cost of writing back the dirty data. Non-volatile cache(NVCache) [11], [12] is designed as a full non-volatile cache system architecture. However, a traditional SRAM cache unarguably outperforms the NVCache in terms of latency and energy consumption. Since NVCache does not need to be checkpointed, it guarantees consistency.

B. Hybrid Architectures

Hybrid caches bring together the speed benefits of SRAM with the capacity and non-volatility features of STT-RAM. This allows for energy-efficient checkpointing across power cycles as the STT-RAM portion is used also for caching the data as well as for taking backup. Badri et al. [13] propose an STT-RAM-based hybrid architecture that can save the process state by optimally utilizing a fixed amount of energy and using effective cache management policies. Another technique PROWL [?] is proposed which takes advantage of skewed associative cache [14] where each way of a cache has a unique mapping function. It results in the enhancement of the hit ratio.

C. Challenges of Hybrid Architectures

Commercial microcontrollers come with in-order pipelined CPUs. Installing a hybrid cache on the chip will lead to non-volatile writes command the critical latency of execution, resulting in higher cycles per instruction than simulated. Moreover, the cost of additional hardware such as dirty-block

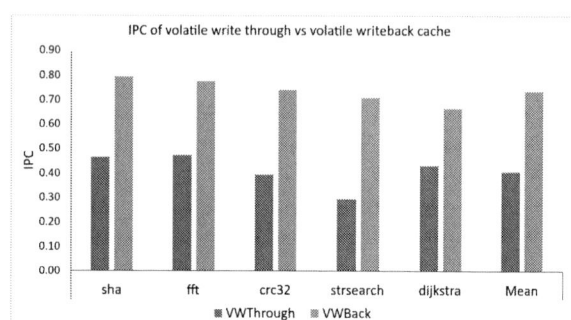

Fig. 2: IPC of volatile write-through (VWThrough) vs volatile write-back (VWBack)

counters for keeping track of dirty lines and write-back queues for lazy write-backs to the NVM needs to be factored in. Our approach is a low-cost solution where only a volatile SRAM cache is used effectively to provide speed with optimum consistency.

III. PROPOSED METHOD

A. Motivation

The prime motivation of this work lies in the working of a conventional write-through cache where each write on the cache is synchronously updated to the main memory. This behavior is particularly cooperative for an ImC system where power failure can occur at any moment. Also, data from the volatile portion of memory can be lost without warning. We conducted extensive experiments on a subset of benchmarks from the Mibench [15] benchmark suite and found that the IPC of the write-through cache is many folds lower than the IPC of the write-back cache. This is because, the processor stalls until the write completes wrt to the cache and NVM. This experimental evidence from Figure 2 suggests that the data consistency of an application will come at a huge cost of latency which is difficult to disregard. We note that from figure 2, write-through IPC is 45.2% lower compared to a volatile write-back cache system. Moreover, in certain applications where the recency of data is given prime importance in the calculation of results, the forward progress of the same cannot be ignored. In such cases, achieving maximum forward progress is as significant as maintaining data consistency.

This is where the need to achieve a trade-off between latency and data consistency comes into play. On top of that, if components such as non-volatile caches are used, then the energy factor also comes into play as non-volatile writes consume a significant amount of energy which will certainly put more than a mild dent in the energy budget of the whole system. We experimented with write-back caches on the MiBench benchmark suite and found the results quite encouraging. We can see from figure3 that dirty lines are minuscule compared to total writes to the NVM.

To sum up the motivation, we will propose a solution that will fit in an optimized spot between latency and energy

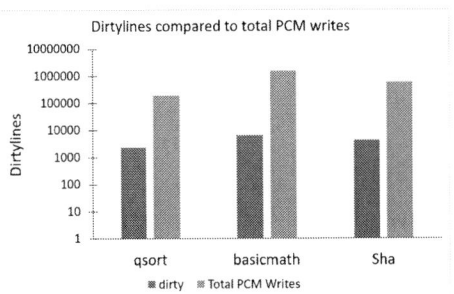

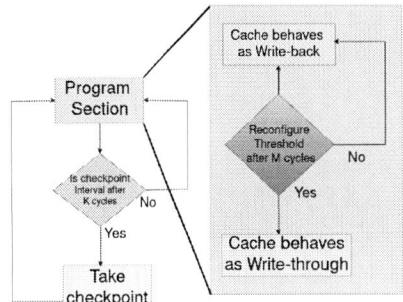

Fig. 5: DynaCache system flow

Fig. 3: Cumulative dirty lines at each checkpoint, compared to total PCM (NVM memory) writes in write-back caches over the complete execution

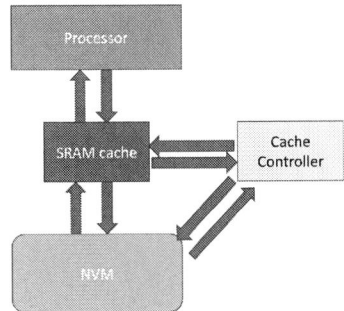

Fig. 4: Abstract Architecture of DynaCache

while ensuring correctness and minimal hardware and software requirements.

B. Proposed Policy: DynaCache

Figure 4 describes the overall architecture of an intermittent computing system. The power and energy component of the system is not described in the figure as it is not the prime focus of our discussion but our proposed architecture works on a general energy harvesting system.

The processor consists of a non-volatile processor(NVP). NVP has non-volatile flip-flops embedded in the core along with volatile registers to facilitate register backup in the scenario of a power failure. The SRAM cache acts as the L1 cache which is the primary data access instrument of the processor. The cache controller coordinates the accesses of the main memory by the cache for fetching and storing the data. Backup of the data to the NVM is done when the voltage monitor signals the drop of voltage below the threshold specified for the safe operation of the energy harvesting system. Data, specifically from the cache is backed up to the NVM main memory. As application response time is crucial, spending time for backup must be reduced as much as possible. In addition, to reduce the power consumption and area overheads, we do not assume the presence of write buffers in the system.

Our proposal begins the application with a cache in the write-back mode. Closer to the instant of taking a backup it

changes mode to write-through. The idea behind changing to write-through policy is to reduce the number of dirty lines to be backed up. Once the regular backup (i.e. checkpoint) is taken, the cache moves back to write-back mode. This way the cache is in write-back for a longer duration (to cater to better performance) and in the write-through mode for a lesser duration (to reduce impact on execution time due to the write-through mode). However, the write-through mode helps reduce the number of dirty lines that accumulate in the write-back mode. It significantly reduces the load of backing up the dirty lines which results in lesser consumption of energy during backup as well as time duration for a checkpoint. This leads to a lower probability of encountering stale data in case of time-critical operations such as sampling of data by a sensor. The number of cycles after which the write-policy is altered is termed in this paper as *Reconfigure-threshold*.

The system applies just-in-time(JIT) [10], [16]–[18] checkpointing where a checkpointing procedure is executed whenever the voltage monitor signals a drop in operating voltage. A JIT operation demands to reserve a portion of energy so that data can be backed up to the NVM. This is the reason why it is extremely beneficial to reduce the number of dirty lines to be written back. It will also enable the ImC to extract more from the capacitor for making forward progress in the application. Although we apply JIT checkpointing in our proposed mechanism, it is important to note that our policy will be able to work on any different checkpointing scheme. Figure-5 shows the overall flow of the mechanism.

Decision of Reconfiguration threshold:

- Let there be K cycles between two checkpoints. K cycles represent the checkpointing interval. We divide K into two parts: one for write-back mode and the second for write-through mode.

$$K = M + N$$

- M= No. of cycles for which the cache behaves according to a write-back policy.
- N= No. of cycles for which the cache behaves as a write-through cache after being on writeback mode. Thus, we call M-cycles the Reconfigure-threshold (RT).
- For the remaining N-cycles, the cache follows the write-

979-8-3315-3968-9/24 $31.00 © 2024 IEEE   111

TABLE I: system configuration

Component	Description
CPU core	1core,500MHz
L1 cache	64-byte block size, 4-way associative cache, 16KB dcache 16KB Icache, LRU replacement policy
Main memory	128MB PCM
Others	Clock Period=2ns, SRAM Read=1 cycle, SRAM Write=2 cycles STT-RAM Read=2 cycles, STT-RAM Write=10 cycle PCM Read=35 cycle, PCM Write=100 cycles
Benchmarks	Mibench suite- (a)memory-intensive- Bitcnt, strsearch, susan, sha (b)compute-intensive- qsort, basicmath, FFT, crc32, djikstra

TABLE II: Parameters of SRAM, STTRAM Caches, and PCM main memory

Parameters	16KB SRAM	16KB STTRAM	128MB PCM
Read latency	0.792 ns	1.994 ns	204.584 ns
Read Energy	0.006 nJ	0.081 nJ	1.553 nJ
Write Latency	0.772 ns	10.52 ns	199.93 ns
Write Energy	0.002 nJ	0.217 nJ	6.9365 nJ

IV. EXPERIMENTAL EVALUATION

The proposed system is evaluated using Gem5 [19] simulator and MiBench [15] benchmark suite for embedded systems. Table I shows the parameters used for the implementation of our system. Note that ImC systems are mostly simple single-core systems with low-capacity caches & memory. Dynamic energy values and latency values for a single read-write operation to SRAM, STTRAM, and PCM memory are calculated using the NVSim [20] tool. It is provided in the Table II.

We modeled two major baseline designs to compare with our proposed architecture and one minor baseline design for comparing the checkpointing overhead. The list of policies is given below:

1) **VWThrough** (Volatile Write Through) cache architecture: This architecture uses a volatile SRAM L1 cache which follows a write-through policy throughout the execution of a program.

2) **NVCache [21]** (Non-volatile write-back Cache) architecture: This architecture uses a non-volatile L1 STT-RAM cache which follows a write-back policy throughout the execution of a program.

3) **VWBackCache [22]** (volatile write-back Cache) architecture: This architecture uses a volatile L1 SRAM cache which write-back policy throughout the execution of a program. This baseline is used only for comparing the checkpointing overhead.

4) **DynaCache**: Our proposed policy of dynamic cache reconfiguration which uses a volatile L1 SRAM cache.

To simulate the frequent power failures, we interrupt the simulator every 2 million cycles. We have selected this interval because 2 million cycles usually take between 25-30ms to execute. So, we can infer that there is a power failure every 30ms. Existing work by Badri et al. [13], Xie et al. [5] assumed that a power failure occurs between every 25ms to 500ms. Thus, we would like to claim that our results are more conservative. Since we assume a checkpoint interval of 2M cycles we use RT = M = 1.8M cycle. Thus N becomes 0.2M cycles.

through policy. The cache takes a compulsory JIT checkpoint after the K cycles.

Let e_{wb} be energy consumed per cycle during write-back mode. Let e_{wt} be energy consumed per cycle during write-through mode. Let total energy consumed during program execution of K cycles in a pure write-through mode be TE_{wt}. Where,

$$TE_{wt} = K \times e_{wt} \tag{1}$$

Let total energy consumed in our system be TE_{dyna} where M is the RT. So,

$$TE_{dyna} = M \times e_{wb} + (K - M) \times e_{wt}$$

Simplifying,

$$TE_{dyna} = K \times e_{wt} - M \times (e_{wt} - e_{wb}) \tag{2}$$

Also,

$$e_{wb} < e_{wt} \tag{3}$$

Substituting equation 1 in 2, we get,

$$TE_{dyna} = TE_{wt} - M \times (e_{wt} - e_{wb}) \tag{4}$$

Therefore, we can infer that

$$M \propto \frac{1}{TE_{dyna}} \tag{5}$$

The reconfiguration threshold, ie. M is a crucial design point because it directly affects the accumulated dirty lines and also the performance. If M is very small, then we will accumulate fewer dirty lines because most of them will get written back during the (longer) write-through mode of N cycles. In the case of a large value of M, there will be superior performance but we may have a little more overhead of backup compared to a smaller value of M. The value of M is chosen based on empirical evaluations. We shall discuss the sensitivity analysis in section IV-B

A. Results and Analysis

(1) Performance: We performed experiments on various benchmarks of the MiBench benchmark suite and found the results quite encouraging. DynaCache has outperformed both NVCache and VWThrough by achieving a speedup of 1.85X and 1.5X respectively. The results are shown in Figure-8. The volatile cache was successfully leveraged by DynaCache

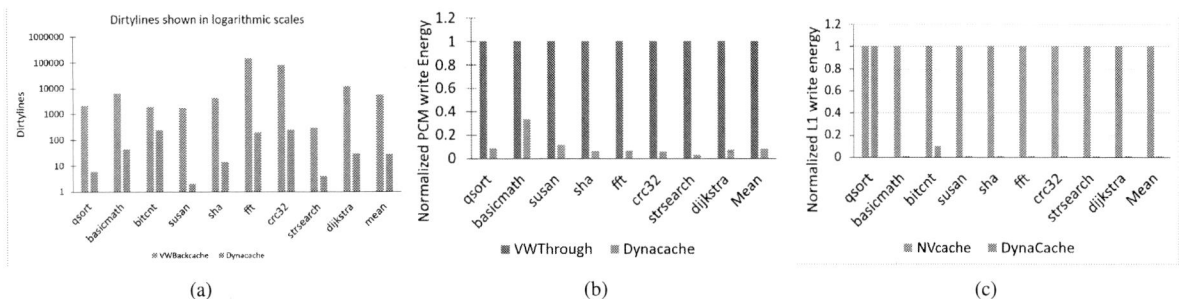

Fig. 6: (a) Number of dirty lines to be backed up during checkpointing(lower is better) (b) PCM write energy comparison w.r.t VWThrough cache design, (c) L1 write energy comparison w.r.t NVcache design

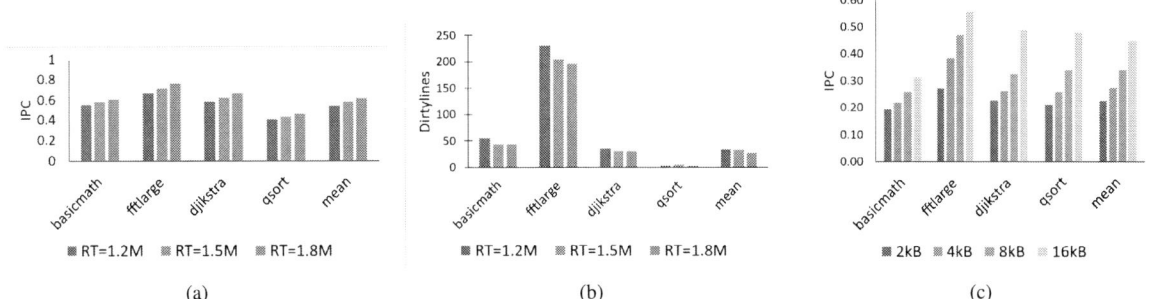

Fig. 7: (a)Sensitivity analysis w.r.t IPC for determining Reconfigure Threshold(RT) calculated in millions of cycles (Higher is better) (b)Sensitivity analysis w.r.t dirtylines for determining Reconfigure Threshold(RT) calculated in millions of cycles (c)Sensitivity analysis w.r.t cache size

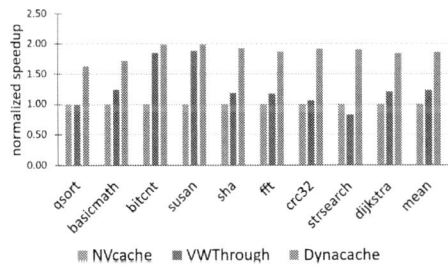

Fig. 8: Normalized speedup of Dynacache (higher is better)

which has enhanced the instruction per cycle (IPC) due to the faster read and write operations to SRAM. During the execution of time-critical operations where data freshness is crucial, the performance of the system becomes the deciding factor to determine whether the data integrity is intact or not.

(2) Checkpoint Overhead due to dirty lines: As a portion of the data in the cache is dirty, there arises a necessity to back up that dirty data. This necessity creates a checkpoint overhead due to which the system is liable to reserve a portion of the energy reservoir as a buffer. That buffer in the worst case can be a sizeable value that can't be neglected. Dynacache

despite being volatile ameliorates the challenging phenomenon of checkpoint overhead. The transformation of the cache from being a write-back cache to a write-through one cleanses the dirty lines as write hits on a dirty cache line will transform the block to a clean block in the write-through mode. Dirty write hits will be written to the NVM which will result in the reduction of dirty lines in the cache. The effect on the checkpoint overhead by our proposed method can be seen in Figure-6a. This is in contrast with the VWBackCache design where dirty lines are colossally higher than our proposed method. There is almost a 99% reduction in the number of dirty lines. Figure-6a shows the log-transformed values to eliminate the visual asymmetry in the data. The residual dirty lines in DynaCache at the end of K cycles are the lines that did not incur any writes after the reconfiguration threshold was triggered. Note that NVCache is not compared in Figure-6a as dirty lines are not needed to back up in such devices.

(3) Energy consumption: Experiments were also performed on various benchmarks to unearth the retrenchment in energy consumption. Figure-6b describes the comparison of Dynacache against VWThrough cache design in terms of total PCM write energy. We found that Dynacache consumes a diminutive **0.08X** times the energy of VWThrough cache design which culminates in a reduction of 92%. Figure-6c shows how L1 write energy is reduced by the usage of our proposed system in comparison to NVcache which uses an STT-RAM cache. We found that on average our proposed system consumes a minuscule **0.011x** times the energy of NVcache. A massive reduction of 98.9% compared to NVcache.

B. Sensitivity Analysis for Reconfiguration threshold

We performed a sensitivity analysis to determine the optimal value of the reconfigured threshold. First, to keep performance in consideration we need to start the transformation of the write policy as late as possible to negotiate the increased write latency due to repeated writes in the NVM. The relation of reconfiguring the threshold concerning IPC is shown in figure-7a. It is clear that a higher threshold is more rewarding in terms of performance. Second, the value had to be such that the number of dirty lines is minimized while also keeping writes to the NVM lower. Figure-7b shows the dirty lines at various reconfigure thresholds. The reduction in dirty lines saturates around 1.5M cycles; but we select a reconfigure threshold of 1.8M cycles to improve the performance and keep the writes to NVM lower.

C. Sensitivity Analysis for L1 cache

We also performed a sensitivity analysis for the L1 cache keeping the reconfigure threshold at 1.8M cycles as shown in figure 7c. Higher cache sizes display a noticeable performance gap due to fewer misses and lower accesses to main memory. However, due to space and energy constraints we cannot keep expanding the cache sizes.

V. CONCLUSION

In this paper, we propose an architecture for intermittent computing systems(ImC) harboring an NVM-based main memory with an SRAM cache and non-volatile processor. The cache has a re-configurable write policy which transforms itself from having a write-back policy to a write-through one based on a threshold linked to number of cycles executed during runtime of the system before experiencing a power failure.

We have compared our mechanism with a non-volatile writeback cache and a volatile write-through cache. The proposed architecture outperforms the compared systems significantly in terms of performance as well as energy. Our system achieves an average speedup of 1.85x. Apart from this, it also achieves 98.9% reduction in power consumption compared to a fully non-volatile architecture.

REFERENCES

[1] M. A. De Kruijf, "Compiler construction of idempotent regions and applications in architecture design," Ph.D. dissertation, USA, 2012, aAI3523614.

[2] Q. Liu, J. Izraelevitz, S. K. Lee, M. L. Scott, S. H. Noh, and C. Jung, "ido: Compiler-directed failure atomicity for nonvolatile memory," in *2018 51st Annual IEEE/ACM International Symposium on Microarchitecture (MICRO)*, 2018, pp. 258–270.

[3] N. Bellas, I. Hajj, and C. Polychronopoulos, "An analytical, transistor-level energy model for sram-based caches," in *1999 IEEE International Symposium on Circuits and Systems (ISCAS)*, vol. 6. IEEE, 1999, pp. 198–201.

[4] E. Chen, D. Lottis, A. Driskill-Smith, D. Druist, V. Nikitin, S. Watts, X. Tang, and D. Apalkov, "Non-volatile spin-transfer torque ram (stt-ram)," in *68th Device Research Conference*, 2010, pp. 249–252.

[5] M. Xie, C. Pan, Y. Zhang, J. Hu, Y. Liu, and C. J. Xue, "A novel stt-ram-based hybrid cache for intermittently powered processors in iot devices," *IEEE Micro*, vol. 39, no. 1, pp. 24–32, 2019.

[6] M. Imani, S. Patil, and T. Rosing, "Low power data-aware stt-ram based hybrid cache architecture," in *2016 International Symposium on Quality Electronic Design (ISQED)*, 2016, pp. 88–94.

[7] K. Maeng and B. Lucia, "Adaptive dynamic checkpointing for safe efficient intermittent computing," in *Proceedings of the 13th USENIX Conference on Operating Systems Design and Implementation*, ser. OSDI'18. USA: USENIX Association, 2018, p. 129–144.

[8] Y. Choi, N. Chang, and T. Kim, "Dc–dc converter-aware power management for low-power embedded systems," *IEEE Transactions on Computer-Aided Design of Integrated Circuits and Systems*, vol. 26, no. 8, pp. 1367–1381, 2007.

[9] A. Mirhoseini, E. M. Songhori, and F. Koushanfar, "Idetic: A high-level synthesis approach for enabling long computations on transiently-powered asics," in *2013 IEEE International Conference on Pervasive Computing and Communications (PerCom)*, 2013, pp. 216–224.

[10] Y. Wang, Y. Liu, S. Li, D. Zhang, B. Zhao, M.-F. Chiang, Y. Yan, B. Sai, and H. Yang, "A 3us wake-up time nonvolatile processor based on ferroelectric flip-flops," in *2012 Proceedings of the ESSCIRC (ESSCIRC)*, 2012, pp. 149–152.

[11] M. Baker, S. Asami, E. Deprit, J. Ouseterhout, and M. Seltzer, "Non-volatile memory for fast, reliable file systems," *SIGPLAN Not.*, vol. 27, no. 9, p. 10–22, sep 1992.

[12] C. Herdt and C. Paz de Araujo, "Analysis, measurement, and simulation of dynamic write inhibit in an nvsram cell," *IEEE Transactions on Electron Devices*, vol. 39, no. 5, pp. 1191–1196, 1992.

[13] S. NVcachel, M. Saini, and N. Goel, "An efficient nvm-based architecture for intermittent computing under energy constraints," *IEEE Transactions on Very Large Scale Integration (VLSI) Systems*, vol. 31, no. 6, pp. 725–737, 2023.

[14] F. Bodin and A. Seznec, "Skewed associativity improves program performance and enhances predictability," *IEEE Transactions on Computers*, vol. 46, no. 5, pp. 530–544, 1997.

[15] M. Guthaus, J. Ringenberg, D. Ernst, T. Austin, T. Mudge, and R. Brown, "Mibench: A free, commercially representative embedded benchmark suite," in *Proceedings of the Fourth Annual IEEE International Workshop on Workload Characterization. WWC-4 (Cat. No.01EX538)*, 2001, pp. 3–14.

[16] D. Balsamo, A. S. Weddell, G. V. Merrett, B. M. Al-Hashimi, D. Brunelli, and L. Benini, "Hibernus: Sustaining computation during intermittent supply for energy-harvesting systems," *IEEE Embedded Systems Letters*, vol. 7, no. 1, pp. 15–18, 2015.

[17] D. Balsamo, A. S. Weddell, A. Das, A. R. Arreola, D. Brunelli, B. M. Al-Hashimi, G. V. Merrett, and L. Benini, "Hibernus++: A self-calibrating and adaptive system for transiently-powered embedded devices," *IEEE Transactions on Computer-Aided Design of Integrated Circuits and Systems*, vol. 35, no. 12, pp. 1968–1980, 2016.

[18] Q. Liu and C. Jung, "Lightweight hardware support for transparent consistency-aware checkpointing in intermittent energy-harvesting systems," in *2016 5th Non-Volatile Memory Systems and Applications Symposium (NVMSA)*, 2016, pp. 1–6.

[19] N. Binkert, B. Beckmann, G. Black, S. K. Reinhardt, A. Saidi, A. Basu, J. Hestness, D. R. Hower, T. Krishna, S. Sardashti, R. Sen, K. Sewell, M. Shoaib, N. Vaish, M. D. Hill, and D. A. Wood, "The gem5 simulator," *SIGARCH Comput. Archit. News*, vol. 39, no. 2, p. 1–7, aug 2011.

[20] X. Dong, C. Xu, Y. Xie, and N. P. Jouppi, "Nvsim: A circuit-level performance, energy, and area model for emerging nonvolatile memory," *IEEE Transactions on Computer-Aided Design of Integrated Circuits and Systems*, vol. 31, no. 7, pp. 994–1007, 2012.

[21] K. Ma, Y. Zheng, S. Li, K. Swaminathan, X. Li, Y. Liu, J. Sampson, Y. Xie, and V. Narayanan, "Architecture exploration for ambient energy harvesting nonvolatile processors," in *2015 IEEE 21st International Symposium on High Performance Computer Architecture (HPCA)*, 2015, pp. 526–537.

[22] H. Li, Y. Liu, Q. Zhao, Y. Gu, X. Sheng, G. Sun, C. Zhang, M.-F. Chang, R. Luo, and H. Yang, "An energy efficient backup scheme with low inrush current for nonvolatile sram in energy harvesting sensor nodes," in *2015 Design, Automation & Test in Europe Conference & Exhibition (DATE)*, 2015, pp. 7–12.

979-8-3315-3968-9/24 $31.00 © 2024 IEEE

In-Memory Mirroring: Cloning Without Reading

Simranjeet Singh[*], Ankit Bende[‡], Chandan Kumar Jha[§], Vikas Rana[‡],
Rolf Drechsler[§¶], Sachin Patkar[*], Farhad Merchant[†]

[*]IIT Bombay, India, [‡]Forschungszentrum Jülich GmbH, Germany,
[§]University of Bremen, Germany, [¶]DFKI GmbH, Germany, [†]Newcastle University, UK
{simranjeet, patkar}@ee.iitb.ac.in, {a.bende, v.rana}@fz-juelich.de,
{chajha, drechsler}@uni-bremen.de, farhad.merchant@newcastle.ac.uk

Abstract—**In-memory computing (IMC) has gained significant attention recently as it attempts to reduce the impact of memory bottlenecks. Numerous schemes for digital IMC are presented in the literature, focusing on logic operations. Often, an application's description has data dependencies that must be resolved. Contemporary IMC architectures perform read followed by write operations for this purpose, which results in performance and energy penalties. To solve this fundamental problem, this paper presents in-memory mirroring (IMM). IMM eliminates the need for read and write-back steps, thus avoiding energy and performance penalties. Instead, we perform data movement within memory, involving row-wise and column-wise data transfers. Additionally, the IMM scheme enables parallel cloning of entire row (word) with a complexity of $\mathcal{O}(1)$. Moreover, we analyzed the energy consumption of the proposed technique on an RRAM crossbar with an experimentally validated JART VCM v1b model. The IMM increases energy efficiency and shows $2\times$ performance improvement compared to conventional data movement methods.**

Index Terms—**RRAM, cloning, data dependency, energy efficiency, performance**

I. INTRODUCTION

The gap between the processing unit and memory leads to speed limitations known as the memory wall. This challenge is addressed by processing data within the memory and has emerged as a solution to alleviate memory bottleneck issues [1]. One solution is to design the logic-in-memory (LiM). The fundamental approach in LiM involves storing input logical states in memory cells, with the computed output remaining in memory as a logical state. Various memory technologies, including resistive random access memory (RRAM), phase change memory (PCM), and spin-transfer torque magnetic random access memory (STT-RAM), have been utilized to implement LiM.

Among these technologies, RRAM stands out as a contender for LiM, where logical states are represented by the resistive state of the device, enabling computation within the memory [2]. Several schemes of LiM using RRAM devices have been demonstrated, such as MAGIC [3], IMPLY [4], FELIX [5], and Majority [6], and so on [7]. Fig. 1(a) shows the schematic to implement the MAGIC OR gate in the crossbar given in Fig. 1(b). Furthermore, experimental demonstrations of some schemes have been conducted, validating the schemes [8]–[10].

Various architectures have been proposed to map the logic on the crossbar to compute serially and in parallel. Considering the area and latency constraint, parallel mapping has been proposed, such as CONTRA [11] and SIMPLE [12]. These schemes allow the mapping of the logic function in parallel,

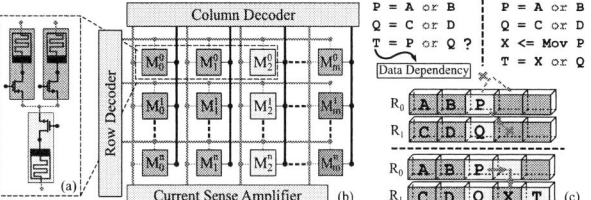

Fig. 1: (a) schematic for implementing the OR operation within the crossbar architecture, utilizing three memristors. Two memristors are designated as inputs, while the third is output storage. (b) The crossbar architecture demonstrates the parallel execution of the same operation. (c) The challenge of data dependency and proposes a solution by transferring data between devices within rows and/or columns.

where both the rows are columns considered. However, data dependencies arise in many cases, particularly when the operations are performed in multiple columns. Fig. 1(c) presents a scenario where three operations need to be performed, with operation T dependent on data from P and Q (left code snippet). Currently, these dependencies are handled by copying the data from the current cell to the required cell, which either requires a read and write-back cycle [12] or two complement operations [13], thereby resulting in increasing the overall energy and latency of the computation.

This paper proposes an in-memory mirroring (IMM) technique to mitigate data dependency by facilitating data cloning within the memory, thereby eliminating the need for energy-intensive read and write-back operations. Unlike previous approaches, inspired by the RowClone methods for DRAM [14], IMM operates directly within the crossbar memory and does not require a read cycle, improving the overall computation latency. Furthermore, IMM introduces the capability for parallel cloning of entire rows with a one-cycle latency, further enhancing its efficiency and scalability. To the best of our knowledge, our paper represents the first pioneering demonstration of IMM using an experimentally validated device model. The following are the contributions of this paper:

- Concept of cloning: bit cloning and word cloning.
- Simulation analysis and validation using JART VCM v1b [15] SPICE model aims to demonstrate the cloning data in rows, columns, and in parallel.
- Finally, energy and latency calculations during the cloning operation are conducted, and the results are compared with those found in the literature.

979-8-3315-3968-9/24 $31.00 © 2024 IEEE

II. Background and Related Work

A. Memristive Devices

Memristive devices have emerged as a significant advancement in non-volatile memory technology. Initially proposed as a concept by Leon Chua in 1971 [16], memristive devices have gained prominence due to their unique ability to store data by modulating resistance states [17]. One such device is RRAM, where resistance modulation is achieved by applying a voltage across these device terminals. In response, the resistance of the devices changes based on the magnitude and direction of the current flow. Typically, these memristive devices can be interconnected to form a crossbar structure. However, issues related to forming and sneak-path currents can arise when individual memristive devices are connected without a CMOS transistor in series. To mitigate these concerns, memristive devices are fabricated with a CMOS transistor in series, resulting in what is known as a 1T1R cell.

A memristive 1T1R cell possesses at least two distinct states: high resistive state (HRS) and low resistive state (LRS) that are mapped to Boolean logic '0' and '1' for LiM implementation. To simulate the characteristics of RRAM cells, several models have been introduced in the literature for characterization at the SPICE level [18]. Among all the models, JART VCM v1b is particularly noteworthy as an open-source model that is based on experimental data from devices fabricated at Forschungszentrum Jülich GmbH, Germany [15]. Moreover, the logic gate using the MAGIC design style has been experimentally validated, resembling the device from the JART VCM 1b model [9]. So, in this study, JART VCM v1b has been used to conceptualize and conduct simulation analyses.

B. Related Work

Previous attempts to handle the energy and latency during data dependency for LiM have been focused on synthesizing the logic function to reduce the number of copy operations [12]. However, the problem remains the same: computation still requires some copy operations that are expensive in energy and latency. Another approach that has been recently used is performing the two-time complement operation to copy within the memory [13]. Even though this scheme still does not require a read cycle, the copy operation needs two cycles, leading to latency and energy inefficiency. Another approach for copying using the IMPLY logic style has been demonstrated with passive devices, requiring an extra R_s resistor and an additional isolation voltage. Adding R_s and addressing the sneak path problem in the passive crossbar makes this method impractical in real crossbar implementations [19].

Although previous literature provides limited evidence of copy operations, typically performed using four devices in series [20] or a passive devices, a comprehensive analysis of this approach is lacking. To the best of our knowledge, our paper represents the first pioneering demonstration of IMM using an experimentally validated device model.

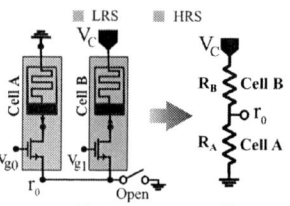

Fig. 2: (a) RRAM schematic for cloning. (b) the approximate equivalent circuit.

TABLE I: Parameters (JART)

Params	Value
LRS (R_{On})	$\approx 3.5 - 4.5 K\Omega$
HRS (R_{Off})	$\approx 65 - 70 K\Omega$
V_{Set}	1V
V_{Reset}	>2.0V
V_{Read}	0.5V
V_C	1.5V
V_g	2.5V

III. Operating Principle of Cloning

The IMM process utilizes just two memristors for cloning; one holds the data to be cloned, while the second is the target for cloning. Initially, all devices are considered to be in the HRS state. Fig. 2(a) shows the schematic for one-bit cloning, where two cells named Cell A and Cell B have been used. Cell A contains the data to be cloned into Cell B, which begins in the HRS state. An equivalent representation of Fig. 2(a) is shown in Fig 2(b) using the simple resistors. The resistance of Cell A is marked as R_A; similarly, the resistance of Cell B is marked as R_B. Two possible values of R_A and R_B could be either HRS or LRS.

In the cloning operation, a positive voltage V_C is applied across Cell B, while Cell A is connected to the ground. Throughout this process, the gates of both transistors are fully open, and devices sharing the common line are kept open (r_0). This establishes a pathway between Cells A and B, where R_A and R_B are connected in series. The parameters list used for the analysis is shown in Table I. This creates a potential drop at r_0, which can be calculated according to the Equation 1. The cloning voltage (V_C) is chosen to be slightly greater than the Set threshold voltage (V_{Set}) of the memristive cell. In this case, 1.5V as V_C has been chosen for a successful cloning operation.

$$V_{r0} = \frac{R_A}{(R_A + R_B)} \times V_C \qquad (1)$$

Based on the values of R_A and R_B, there are two scenarios to consider: (1) when the input data (R_A) is logic '0' (HRS), and (2) when the input data (R_A) is in LRS or logic '1'. The IMM technique allows for a change of the state of R_B according to the state of R_A.

A. Case 1: Cloning Logic '1'

In this case, Cell A contains logic '1', which needs to be cloned in Cell B and is initialized to logic '0' (HRS). When the V_C is applied, the V_{r0}, according to Equation 1, becomes very less because the $R_A << R_B$. As the $(V_C - V_{r0}) \approx V_C$ is greater than the Set voltage ($> V_{Set}$), it generates a sufficient voltage across Cell B, enabling it to transition from HRS to LRS. Cell B will switch from HRS to LRS in accordance with the data in Cell A, eventually cloning the Cell A data into Cell B. Due to the high reset-to-set ratio (≈ 2) in this model, Cell A will retain its original state.

979-8-3315-3968-9/24 $31.00 © 2024 IEEE

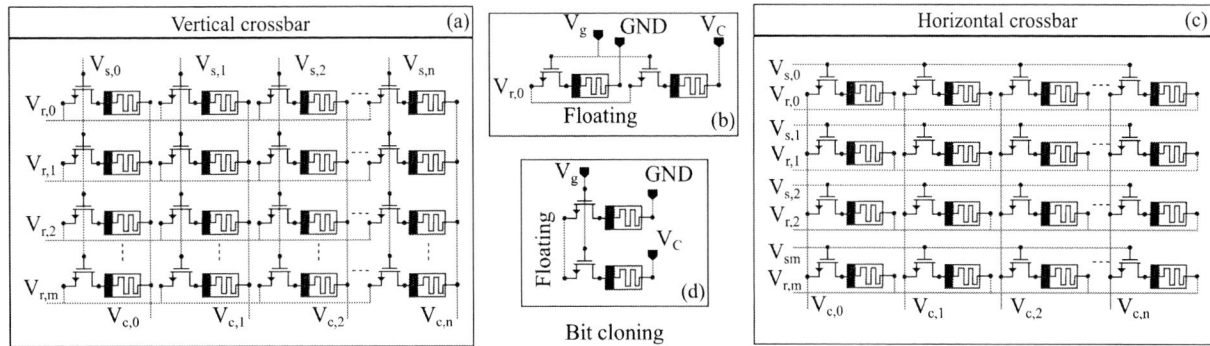

Fig. 3: Different configurations of crossbar architectures. In (a), the vertical crossbar layout is presented, wherein the gates of transistors are connected vertically. (b) Bit-cloning in the vertical crossbar. (c) Horizontal crossbar, characterized by horizontally connected transistor gates. (d) Bit-cloning in horizontal crossbar

B. Case 2: Cloning Logic '0'

In this scenario, Cell A is in the logic '0', and the target cell (Cell B) is already in the HRS state. According to Equation 1, the voltage across the row will be evenly distributed ($R_A = R_B$), and voltage at r_0 will be $\approx V_C/2$. The voltage across Cell B, $(V_C - V_{r0}) \approx V_C/2$ is not sufficient to switch the state ($< V_{Set}$). Thus, Cell B will maintain its original state of HRS, which is the same as the state of Cell A.

IV. The Proposed Methodology

RRAM often adopts a crossbar structure, which is known for its ability to facilitate dense memory. In its simplest form, the array consists of horizontal and vertical lines, where each RRAM cell is connected at each junction. There are multiple ways to connect the RRAM cell at the junction. Fig. 3 shows the vertical and horizontal crossbar structure for $m \times n$ (rows $\times$ columns) size. In a vertical crossbar structure, as shown in 3(a), the vertical lines are connected to the electrode of the device, and horizontal lines are connected to the transistor source. The gate of all the transistors is connected horizontally. On the other hand, in the horizontal crossbar, as shown in Fig. 3(d), horizontal lines are connected to the electrode of the RRAM cell while the vertical lines are connected to a source of the transistors. Moreover, the gate of transistors is shorted horizontally. An appropriate voltage at horizontal and vertical lines is applied to write (0 or 1) and read an individual cell in the crossbar. In Fig. 3, $V_{r,x}$ and $V_{c,y}$ represent the voltage at row and column lines, respectively, where $|0 \leq x < m|$ and $|0 \leq y < n|$. $V_{s,z}$ is the gate switch voltage, which is given as $|0 \leq z < n|$ for vertical and $|0 \leq z < m|$ for horizontal crossbar.

In this work, the data is copied from one cell to another without performing the read operation called cloning or mirroring. Initialization of both input and output follows a standard memory write operation procedure, and the data cloning process aligns closely with this write operation method. To integrate a cloning operation within a crossbar array, two requirements must be fulfilled: the structure of the crossbar and the connections of the memristive cells configured within the array as shown in 3. Additionally, the logical state of the memristive cell should be represented as resistance. Fig. 3

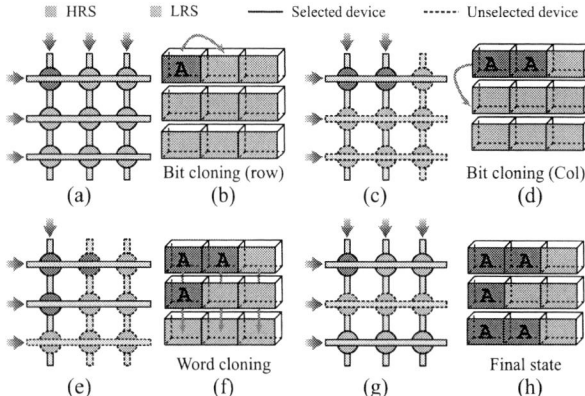

Fig. 4: (a) 3x3 crossbar array structure sketch. The gray arrow exemplifies the voltage source. The colored circle at the junction represents different memristors' states. (b) Method of bit cloning in the same row, where the data is marked as "A," which will be moved according to the representation of the green arrow. (c) Row-wise bit operations and the dotted blocks show the unselected rows and columns. (d) Column-wise bit operation where only one-bit value will be moved is marked in the green arrow. (e) Selected voltage source and devices to perform the column-wise bit cloning. (f) Representation of full column movement where the first complete row word will move to the third row in parallel. (g) Selection of the required cell to perform word cloning. (h) final state after performing all operations from (a) to (g) in a sequence

shows the memristive crossbar structure with the necessary connections. While the crossbar architecture allows for various other connection configurations, this study focuses on demonstrating cloning within the structures depicted in both Fig. 3(a) and Fig. 3(d).

The vertical crossbar structure has a connection that is similar to the connection depicted in Fig. 2(a) for bit cloning. The V_C is applied to the vertical lines connected to the electrodes of the memristors. For instance, to execute the clone operation on the first two devices in a row, V_C is applied at $V_{c,1}$ while ground is applied at $V_{c,0}$. Gate voltages $V_{s,0}$ and $V_{s,1}$ are applied to open both transistors (>2V in

979-8-3315-3968-9/24 $31.00 © 2024 IEEE   117

this case). The first row ($V_{r,0}$) remains floating, facilitating cloning as illustrated in Fig. 2(b), effectively transferring data from the first device to the second. Similarly, voltages can be applied to rows, columns, and gates of a horizontal crossbar, as depicted in Fig. 3(d), to execute the cloning operation. Fig. 3(c) illustrates the implementation of the bit cloning operation in the horizontal crossbar. It is noteworthy that the operation conducted on the row-wise vertical crossbar resembles the operation in the horizontal crossbar for the column due to crossbar connections. The voltage sequence in various configurations enables data to be cloned bit-wise and column-wise, and even the entire word can be cloned to another word. Subsequently, we delve into the specifics of each operation. It is important to note that we exclusively focus on the vertical crossbar as similar operations can be performed on the horizontal crossbar.

A. Bit Cloning

In bit cloning, a single bit of data is moved either within the same row or column, although it can be transferred to any location within the same row or column. Row-wise and column-wise operations require distinct voltage schemes. A 3×3 crossbar structure elucidates the cloning methodology shown in Fig. 4. The device in operation is highlighted using a solid line, while the dashed line indicates the unselected devices for that specific operation. Fig. 4(a) presents the states of devices in a 3x3 crossbar, where 'A' represents the LRS cell while all other cells are in HRS.

1) Row-Wise

In row-wise bit operation, 'A' has to be moved within the same row, as depicted in Fig. 4(b). The voltage applied to the device follows the configuration shown in Fig. 3(b), selecting the first two devices connected in $r, 0$. The gates of selected devices are connected to $V_{s,(0,1)} = V_g$ while other devices are deselected by applying 0V to their gates. During the cloning phase, V_C is applied at $V_{c,1}$ and $V_{c,0}$ is connected to GND. The $r, 0$ line remains floating, while other row lines are connected to a voltage, $V_{r,x} = V_C/2$, which is due to shared $V_{s,z}$, where $1 \leq x < m$ and $0 \leq z < 2$. This results in proper device selection, which is shown in Fig. 4(c), and clone 'A' to the desired cell. It's important to note that V_C is applied across the output device during the row-wise cloning phase.

2) Column-Wise

Similar to row-wise bit cloning, column-wise bit cloning allows data to be cloned within the column, requiring a distinct voltage configuration in the opposite direction compared to row-wise bit cloning. Fig. 4(d) displays the current state of devices after bit-wise cloning and the subsequent operation with the green arrow. Fig. 4(e) shows the selected and unselected cells for operation. In column-wise cloning, a single gate line is shared among the devices in the column, with the commonly shared line being a row line rather than a column line, as in row-wise bit cloning. The required voltages are applied to the column side. To select the cell as shown in Fig. 4(d), $V_{s,0}$ is connected to V_g, while all other gates are connected to 0V. During the cloning phase, $V_{r,0}$ is connected

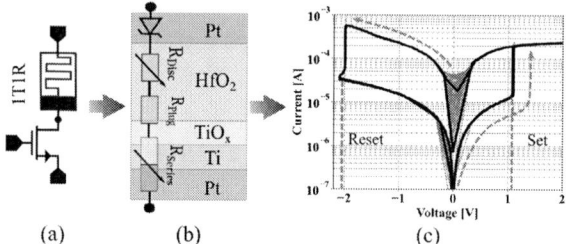

Fig. 5: (a) 1T1R cell schematic, (b) material stacks of memristor, (c) I-V characteristics for 100 cycles

to V_C, and $V_{r,1}$ is connected to the ground, while all other row voltages are connected to $V_C/2$. The column-wise operation requires V_C at the input device, which creates a positive voltage drop at the device electrode, resulting in state change for the output device. The applied voltage terminal is opposite to the operation during row-wise cloning, where the V_C is applied at the output device.

B. Word Cloning

As in a vertical crossbar, the gates are shared vertically (refer Fig. 3(a)); it's possible to do the operation in parallel by opening the gate of all the lines connected vertically, allowing for the complete row to be cloned to another row. As depicted in Fig. 4(f), the example involves moving the first row to the last row (marked by the green arrow). The first Row contains two 'A' cells. For word cloning operation, $V_{s,z}$ is connected to V_g, where $0 \leq z < n$, and $V_{r,0}$ is connected to V_C, $V_{r,1}$ is connected to ground, while $V_{c,x}$ remains floating, where $0 \leq x < n$. The unselected rows are connected to $V_C/2$. Fig. 4(g) depicts the selected and unselected devices after applying the specified voltages across the row and column lines. This allows the complete row to be moved to another row, as each clone operation can be performed in parallel with a complexity of one cycle only. The operation is similar to the column-wise operation, where all column operations happen simultaneously. Fig. 4(h) illustrates the final state after all operations.

All operations are depicted with respect to the vertical crossbar structure. Nevertheless, similar operations can also be conducted on the horizontal crossbar. Additionally, due to the crossbar structure, the operations interchange between row-wise and column-wise. Furthermore, in the vertical crossbar, word cloning conducted row-wise will transition to column-wise word cloning. The voltage required for the basic clone operation will remain the same.

V. RESULTS

This section unveils the outcomes derived from SPICE level simulation using the Cadence spectre. The JART VCM v1b model has been used as an RRAM and the *gpdk 45nm* technology node for CMOS integration. This section also provides insights into the energy consumption associated with cloning operations.

A. 1T1R RRAM Switching Characteristics

The 1T1R cell design comprises a single memristor and one NMOS transistor, forming the 1T1R structure. Fig. 5(a)

979-8-3315-3968-9/24 $31.00 © 2024 IEEE

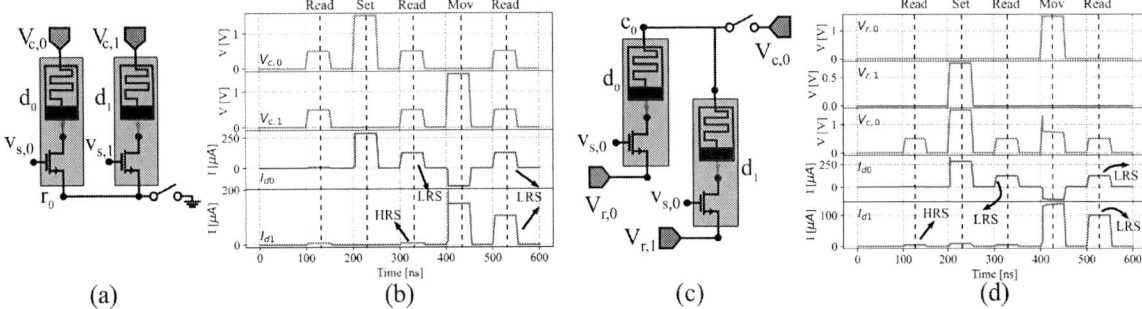

Fig. 6: (a) Schematic for bit-wise cloning in the same row. In the schematic, we have four voltage sources, $V_{c,0}$, $V_{c,1}$, $V_{r,0}$, $V_{r,1}$, and currents in the memristors are marked in colors. (b) The simulation results' waveform shows each cycle marked on the top of the graphs. The last Read cycle shows the state of both devices. (c) Column-wise bit operation. The voltage sources are at r_0 and r_1 along with c_0 instead of c_0 and c_1. (d) Waveform during the column-wise bit operation.

illustrates the cell design alongside the equivalent stack of the memristor depicted in Fig. 5(b). The memristor model utilized in this investigation is experimentally validated and consists of a Pt/Ti/TiO$_x$/HfO$_2$/Pt material stack known for its favorable electroforming voltage and thermal stability. The I-V characteristics of the 1T1R cell are depicted in Fig. 5(c), which shows Set and Reset voltages marked with the red lines, with the Set voltage approximately 1V, and the Reset voltage approximately 2V. Any voltage between the Set and Reset voltages can be employed for the Read operation, with 0.5V utilized in this study to read the device state. Throughout the I-V characteristics, the gate of the transistors is maintained at 2.5V.

B. Cloning Implementation

The schematic depicting row-wise cloning in the vertical crossbar structure is presented in Fig. 6(a). Initially, all devices are set to HRS by default. To validate the cloning operation, d_0 is initialized to LRS/HRS by applying the necessary voltage at $V_{c,0}$. Fig. 6(b) has the waveform applied to $V_{c,0}$ and $V_{c,1}$, along with the current flowing through devices d_0 and d_1, denoted as I_{d0} and I_{d1}, respectively. A Read pulse of 0.5V is initially applied across both devices to ascertain their current state. Subsequently, a Set voltage is applied to $V_{c,0}$ to transition device d_0 to LRS, as indicated in the subsequent read cycle. The 'Mov' pulse marked in Fig. 6(b) signifies the clone operation pulse. During the clone operation, $V_{c,0} = 0V/GND$ and $V_{c,1} = V_C$ are applied, while r_0 remains floating. The final read operation verifies the state of both devices, confirming that both are in LRS, thus validating the clone operation.

Fig. 6(c) shows the simulated schematic for column-wise bit cloning. Similar to row-wise cloning, all devices are initially in HRS. However, the voltage scheme differs due to the crossbar connection. After Set and Read pulses, the clone operation is executed by applying $V_{r,0} = V_C$ while $V_{r,1} = 0V$. The column line $V_{c,0}$ remains floating during the operation. The final Read cycle confirms the successful cloning of d_0 data to d_1, as shown in Fig. 6(d).

Finally, word cloning on a 2x2 crossbar structure is depicted in Fig. 7. Similar to column-wise bit cloning, multiple columns share the same gate and row lines in the vertical crossbar,

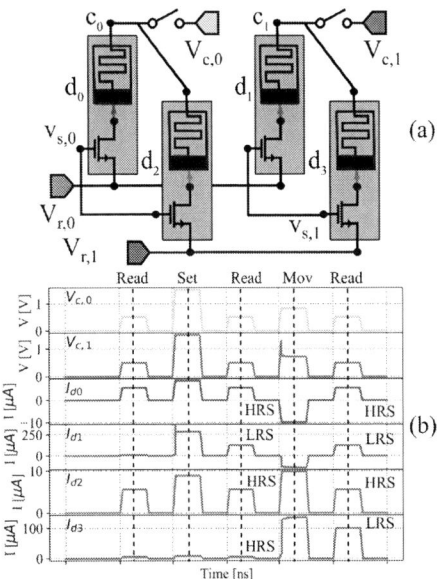

Fig. 7: (a) Schematic for word cloning, showing 2x2 crossbar, where the data in d_0 and d_1 will be cloned to d_2 and d_3, respectively. The current flowing from each device is marked in colors. (b) Waveform for the cloning "10"

enabling parallel data cloning. As shown in Fig. 7(a), the same $V_{r,0}$ is applied to devices d_0 and d_2, while similarly, $V_{r,1}$ is applied to devices d_1 and d_3. Since $V_{s,0}$ and $V_{s,1}$ are shared for devices connected in r_0 and r_1, respectively, both column lines c_0 and c_1 are kept floating during the 'Mov' cycle. Fig. 7(b) shows the waveforms for performing word cloning. Devices d_0 and d_1 are programmed to logic '1' and logic '0', respectively. The state of d_0 and d_1 after the clone cycle (Mov) can be observed in the final Read operation, reflecting the transfer of d_0 and d_1 states to d_2 and d_3, respectively.

TABLE II: Energy consumption for cloning

Device operations			Cloning		
Operation	Voltage V	Energy (pJ)	Operation	Voltage V	Energy (pJ)
Reset	2.25	15.54	Bit (1)	1.5	9.52
Set	1.5	20.17	Bit (0)	1.5	0.71
Read (0 & 1)	0.5	3.1	Word (2-bit)	1.5	11.28

C. Energy and Latency Calculations

Comparing energy consumption and latency of existing methods with the IMM approach is essential for assessing efficiency. Table II outlines energy usage during device operations such as Set, Reset, Read, and cloning, classified as bit and word cloning. The energy is obtained by multiplying the voltage waveforms with the sensed current and then integrating the product over the measurement time as per $\int_0^t v(t) \times i(t)\, dt$, where t is the pulse time. In existing computing, copying operations typically require two cycles: read and writeback, consuming approximately 18 pJ and 23 pJ for copying logic '0' and logic '1', respectively, in the worst case. In terms of latency, the literature shows the two cycles for copying; in [11], one for reading and another for writeback, and in [10], two NOT operations to copy. In contrast, the proposed IMM scheme enables data cloning within the memory in one cycle, requiring around 10 (9.52 + 0.71) pJ for bit cloning. An average 2-bit word cloning consumes only 11.28 pJ of energy. The proposed scheme reduces overall energy consumption by the sum of the energy expended during a read operation and the energy consumed by the peripheral during that operation. This reduction is particularly significant in large-scale applications. Additionally, it achieves a $2\times$ improvement in latency.

VI. Discussion

This study demonstrates a technique for cloning data within memory without reading. Table II shows that bit-wise ('0' & '1') cloning consumes approximately 10.2 pJ, as the output cell is in a HRS, limiting switching current and reducing energy consumption. For 2-bit word cloning, the average energy is dominated by the "11" combination at around 22 pJ, while "00" consumes 0.7 pJ, and "01 & 10" consumes 11.11 pJ. These energy values are based on the operational voltage, which can be adjusted for optimization. However, it is assumed that all devices are in an HRS, similar to the logic operation implementation. If the cell has been previously utilized, it must be switched to HRS before the cloning operation, necessitating an additional cycle. This procedure is similar to handling a reused cell and initializing the output cell to HRS. For output devices in use or unknown states, dependencies can be verified at the compiler level and initialized with the output device for logic operations, like regular memory write operations. Since the voltage requirements are compatible with other logic operations, the area needed for the control circuit for cloning will remain the same.

Conclusion

This paper introduced the concept of IMM in RRAM memory crossbars, enabling data cloning without reading the state within a single cycle. IMM facilitates efficient data cloning within the memory, eliminating the necessity for read and write-back cycles and resulting in substantial energy savings. Moreover, the scheme supports bit-wise data movement in both columns and rows while enabling parallel cloning of entire columns with the complexity of a single cycle. The effectiveness of the IMM concept has been demonstrated through SPICE-level simulation analysis with an experimentally validated RRAM model. Furthermore, we have thor-

oughly examined the energy consumption and latency associated with our proposed technique. The IMM scheme exhibits energy efficiency and is $2\times$ faster than the existing method of copying data in the RRAM crossbar. Moving forward, we plan to validate the proposed scheme experimentally using a fabricated RRAM crossbar.

Acknowledgments

This work was supported in part by the Federal Ministry of Education and Research (BMBF, Germany) in the project NEU-ROTEC II under Project 16ME0398K, Project 16ME0399, German Research Foundation (DFG) within the Project PLiM (DR 287/35-2) and through Dr. Suhas Pai Donation Fund at IIT Bombay.

References

[1] A. Sebastian *et al.*, "Memory devices and applications for in-memory computing," *Nature nanotechnology*, vol. 15, no. 7, pp. 529–544, 2020.

[2] R. Waser *et al.*, "Redox-based resistive switching memories – nanoionic mechanisms, prospects, and challenges," *Advanced Materials*, vol. 21, pp. 2632–2663, 07 2009.

[3] S. Kvatinsky *et al.*, "MAGIC—Memristor-Aided Logic," *IEEE TCAS-II*, vol. 61, no. 11, pp. 895–899, Nov. 2014.

[4] S. Kvatinsky *et al.*, "Memristor-based material implication (IMPLY) logic: Design principles and methodologies," *IEEE TVLSI*, vol. 22, no. 10, pp. 2054–2066, 2013.

[5] S. Gupta *et al.*, "Felix: Fast and energy-efficient logic in memory," in *2018 IEEE/ACM ICCAD*. IEEE, 2018, pp. 1–7.

[6] A. Deb *et al.*, "Automated Equivalence Checking Method for Majority based In-Memory Computing on ReRAM Crossbars," in *2023 ASP-DAC*, Jan. 2023, pp. 19–25.

[7] S. Singh *et al.*, "Should We Even Optimize for Execution Energy? Rethinking Mapping for MAGIC Design Style," *IEEE Embedded Systems Letters*, vol. 15, no. 4, pp. 230–233, 2023.

[8] B. Hoffer *et al.*, "Experimental demonstration of memristor-aided logic (MAGIC) using valence change memory (VCM)," *IEEE TED*, vol. 67, no. 8, pp. 3115–3122, 2020.

[9] A. Bende *et al.*, "Experimental Validation of Memristor-Aided Logic Using 1T1R TaOx RRAM Crossbar Array," in *VLSID*, 2024, pp. 565–570.

[10] H. Padberg *et al.*, "Experimental Demonstration of Non-Stateful In-Memory Logic With 1T1R OxRAM Valence Change Mechanism Memristors," *IEEE TCAS II*, vol. 71, no. 1, pp. 395–399, 2024.

[11] D. Bhattacharjee *et al.*, "Contra: Area-constrained technology mapping framework for memristive memory processing unit," in *2020 IEEE/ACM ICCAD*, 2020, pp. 1–9.

[12] R. Ben Hur *et al.*, "Simple magic: Synthesis and in-memory Mapping of logic execution for memristor-aided logic," in *2017 IEEE/ACM ICCAD*, 2017, pp. 225–232.

[13] B. Perach *et al.*, "Understanding Bulk-Bitwise Processing In-Memory Through Database Analytics," *IEEE TETC*, vol. 12, no. 1, p. 7–22, Jan. 2024.

[14] V. Seshadri *et al.*, "RowClone: Fast and energy-efficient in-DRAM bulk data copy and initialization," in *2013 46th Annual IEEE/ACM MICRO*, 2013, pp. 185–197.

[15] C. Bengel *et al.*, "Variability-aware modeling of filamentary oxide-based bipolar resistive switching cells using SPICE level compact models," *IEEE TSCAS I*, vol. 67, no. 12, pp. 4618–4630, 2020.

[16] L. Chua, "Memristor-The missing circuit elemet," *IEEE Transactions on Circuit Theory*, vol. 18, no. 5, pp. 507–519, 1971.

[17] D. Strukov *et al.*, "The missing memristor found," *Nature*, vol. 453, pp. 80–3, 06 2008.

[18] F. Staudigl *et al.*, "A Survey of Neuromorphic Computing-in-Memory: Architectures, Simulators, and Security," *DATE*, vol. 39, no. 2, pp. 90–99, 2022.

[19] L. Xie *et al.*, "Fast boolean logic mapped on memristor crossbar," in *2015 33rd IEEE International Conference on Computer Design (ICCD)*, 2015, pp. 335–342.

[20] L. Luo *et al.*, "Reconfigurable stateful logic circuit with cu/cui/pt memristors for in-memory computing," *IEEE TVLSI*, vol. PP, pp. 1–13, 05 2024.

Multi-Level FeFET-Based CAM Address Decoder

Thomas Makryniotis[1], Georgi Gaydadjiev[1,2], Said Hamdioui[1] and Mottaqiallah Taouil[1]

[1]Delft University of Technology, the Netherlands
[2]Rijksuniversiteit Groningen, the Netherlands

Abstract—Address decoders are an integral part of random access memories. They are typically implemented using fast logic optimised for low latency. The latter, however, are difficult to test, while their repair is considered to be impossible. In this work we propose a highly scalable and testable address decoder solution, based on Content-Addressable Memories build with ferroelectric transistors (FeFET). Our solution has a transistor count close to the state of the art, while it outperforms it in terms of latency. Due to its regular 2D structure, our proposal's testability is comparable to that of memory arrays. Moreover, adding a few spare rows will enable end-of-production repair, in the presence of manufacturing defects. By additionally increasing the number of address bits stored in a single FeFET CAM cell, potential area reductions of 30% - compared to the traditional dynamic NAND decoders - can be achieved.

Index Terms—FeFET, decoders, ferroelectric, memory, testing

I. Introduction

Low latency address decoding remains a challenge in high-performance memory designs ranging from cache memories up to large main memory arrays [1]. This typically results in highly customised logic implementations for the targeted memory array that are fast but difficult to test [2]. Detecting address decoder faults is challenging since the tests have to "recover" the effects from the expected values obtained from different locations in the memory array [3, 4]. Moreover, detecting linked address decoder faults requires long march tests [2]. Implementing testable and ideally, repairable address decoders will not only help reducing end of production test times but will also improve yield.

Ferroelectric FET (FeFET) technology is emerging in many application domains ranging from non-volatile storage, up to reconfigurable hardware. Some examples of promising FeFET based applications are emulating atomic neuromorphic operations [5], hyper dimensional encoding [6], multiply-accumulate operation crossbars [7] and energy/area efficient FPGA fabrics [8]. Clearly, both Academia and Industry consider using FeFET devices in promising solutions for various limitations faced by contemporary computing technology.

In this paper we propose a novel, low latency address decoder built using an array of two-FeFET Content Adressable Memory (CAM) cells. Each FeFET CAM cell stores at least two bits in order to reduce the overall array size. Our address decoder was simulated in 14nm CMOS technology and outperforms the state of the art in latency, by at least a factor of

This publication is part of the project NL-ECO: Netherlands Initiative for Energy-Efficient Computing (with project number NWA. 1389.20.140) of the NWA research programme Research Along Routes by Consortia which is financed by the Dutch Research Council (NWO).

1.7x. At the same time, its area stays on par with the baseline solutions. When more than two bits per CAM cell are used, additional area gains can be achieved. Moreover, due to its memory-like regular organization, our decoder is easy to test and by adding few spare rows end-of-production-line repair can be facilitated.

The main contributions of this paper are:

- a fast FeFET CAM based address decoder design;
- a careful investigation of the possible partial open defects in the proposed two-transistor CAM cells;
- a set of march tests able to detect the above defects.

The remainder of the paper is organized as follows. Section II introduces the necessary background. Section III describes the proposed solution, while Section IV presents our experimental results and the comparison against state of the art. Finally, Section V concludes the paper and provides some directions for future work.

II. Background and Related Work

First, we introduce the Ferroelectric FETs and the two main compact models used by the research community. Next, we provide the necessary background on Multi-bit Content Addressable Memories (MCAMs) and describe the most relevant proposals using FeFETs. At the end of this section we describe related work on CAMs usage in TLBs.

A. Ferroelectric FETs

Ferroelectric Field-Effect Transistors (FeFETs) are created by integrating a layer of a ferroelectric material (FE layer) into the gate stack of a regular MOSFET. This layer introduces an additional capacitance, coupled with the gate capacitance of the MOSFET device, while its polarization can be manipulated by an appropriate voltage pulse applied on the gate of the transistor. The implications of such a device are obvious; by controlling the polarization (and as a consequence, the voltage) of the FE layer, it is possible to store binary values in a FeFET and read them afterwards, in a non-destructive way, as I_{ON} and I_{OFF} will be associated with a logic '1' or '0'. Since the polarization of the FE layer within a FeFET is retained, the device behaves as a non-volatile storage element.

FeFETs offer a way to realize a MOSFET with a "programmable" threshold voltage. An invaluable property is the partial polarization switching of the FE layer; by utilizing that, a FeFET storing different voltage levels is possible. The partial polarization of the FE layer can be achieved by controlling the duration and/or the amplitude of the programming pulses.

979-8-3315-3968-9/24 $31.00 © 2024 IEEE

In this work we are using constant-width, variable-amplitude pulses for both programming and reading.

FeFET devices are modelled using two main approaches: the Landau-Khalatnikoff (L-K) model which is based on time-dependent equations that describe the relationship between the polarization (P) and the electric field (E) [9] and the Preisach model, built upon the fact that a FE thin-film consists of multiple independent domains, with a distribution of coercive fields [10], [11]. Although the L-K model is useful in specific cases, it also has a few drawbacks. The first being that it assumes a single domain FE material, while in practice it is usually poly-crystalline with multiple domains. The second limitation is that the V_{gs} applied to the FeFET can either keep the device polarization or switch it, as reported in [12]. It is impossible to turn the transistor completely off and maintain polarization, which means that the model cannot reproduce the basic non-volatile operation of FeFET [12]. In our work, we make use of a Preisach compact model, that can well mimic the aforementioned behaviour [11].

B. Multi-bit Content Addressable Memories

Content-addressable memories (CAMs) function as associative memories, comparing input data with stored data and yielding addresses for matching data [13]. Their parallel processing capability across multiple data rows renders CAM-based searches notably faster compared to other alternatives. CAMs find particular appeal in applications requiring rapid and energy-efficient search operations, such as network packet forwarding [14] and wide database searches [15]. Ternary CAMs (TCAMs) enjoy widespread adoption and are being researched for, among others, neuromorphic computing [16], while Analog CAMs (ACAMs) have been used as TCAM alternatives with increased data density, reduced operational energy and area for in-memory processing circuits [17].

An ACAM cell that stores multiple, narrow, non-overlapping ranges, can be viewed as a high-density digital CAM referred here as Multi-bit Content Addressable Memories (MCAM), with each range signifying a distinct state. The primary disparity between MCAMs and ACAMs lies in their search capabilities: every MCAM cell exclusively seeks within a restricted set of input values, each corresponding to a state, while ACAMs search across an infinite range of inputs. In MCAMs, the stored ranges maintain a one-to-one correlation with the inputs. Therefore, if there are four narrow, non-overlapping ranges and specific inputs, the ACAM would manifest as a 4-state or 2-bit MCAM. Consequently, MCAM can be viewed as a discretised and robust case of ACAM [18].

C. FeFET-based MCAM designs

Almost all of the recent efforts aiming to realize MCAMs are based on emerging memory technologies such as FeFETs and ReRAM [16], [17], [18], while in a few cases, floating-gate FLASH cells are being proposed [19]. When it comes to FeFETs there are three main designs: the 4T-2FeFET [12], the 2FeFET [12], [20] and the 2FeFET-1T [21]. The above designs promise a much better energy efficiency as compared

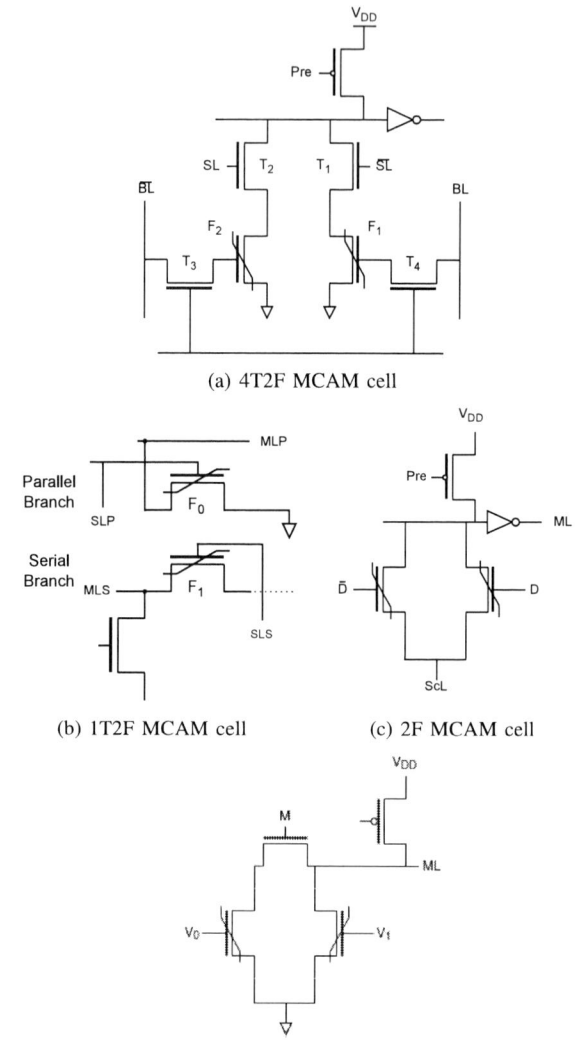

(a) 4T2F MCAM cell

(b) 1T2F MCAM cell

(c) 2F MCAM cell

(d) TC-MEM CAM cell

Fig. 1: Four different designs of FeFET-based CAM cells.

to their memristor counterparts and marginally better than the 16T CMOS implementations [12], while they all require significantly less area (see Figure 1 for the different FeFET CAM cells). In addition to these designs, a few others have been proposed; the TC-MEM design proposed in [22] extends the 2-FeFET cell so it can be accessed either by content or by address (Figure 1-d), while the authors of [23] introduced a complementary FeFET CAM design utilizing a pair of p- and n- FeFET devices and reporting significantly reduced search latency and failure probability. Another novel design was also proposed in [24], which uses two n-FeFET devices connected in serial topology; the authors reported promising results with regards to search energy, compared to other designs.

In this work we use the most compact design, which consists of only two FeFETs.

979-8-3315-3968-9/24 $31.00 © 2024 IEEE

D. CAM-based Address Decoders

Hardwired dynamic NAND address decoders are widely used in random access memories to decode parts of the address. The function of a dynamic NAND decoder is quite simple: each row of the decoder corresponds to a row of the decoder's truth table and a certain number of transistors are hardwired to either the address signals, or their inverted ones. Based on which transistors are ON or OFF, the appropriate line of the address decoder is activated. Practically, the same can be achieved by using a content-addressable memory which has all the available addresses pre-programmed: an address is used as a query and if a CAM row contains it, it signals a Match Line (ML). An implementation was suggested for the first time in [25] where the proposed CAM array was based on ReRAM. This design was discussed in a more elaborate way in [26] where a few applications were also proposed such as fully associative TLBs and virtually addressable memory.

III. PROPOSED FCAM-BASED SOLUTION

A. Concept

Our work combines carefully chosen elements and ideas from the aforementioned works. The FeFET-based CAM cell provides a viable alternative to the originally proposed ReRAM devices. The idea of replacing a logic based address decoder with a CAM array is appealing not only because of the potential performance advantages but also from a testability perspective.

Testing a decoder implemented with CMOS logic can be challenging, while repairing such a structure could turn virtually impossible [3]. Replacing the decoder with a CAM array immediately shifts the address decoder testing challenges to the realm of memory array testing.

Despite the focus on testing, we could not overlook other possible advantages, such as area and power consumption. Previous works suggest a single-level resistive CAM cell; in this work we leverage the benefits of the multi-bit CAM cells in order to reduce the overall decoder area. Combining this with the virtually effortless integration with existing CMOS processes, the intrinsic advantage of FeFETs against ReRAM on search energy [16] and the comparable search delay, the FCAM address decoder becomes an appealing solution.

B. Structure of the array

As we have already mentioned, we chose to base our design on the ultra-compact 2FeFET MCAM cell (see Figure 1c). This compact design however, comes at the cost of increased complexity for the writing and reading scheme and parasitics from neighbouring cells. In order to deal with these challenges, we implemented a programming scheme similar to those proposed in [18], [27] and [28]. The programming scheme is based on analog-inverse voltages applied on each of the FeFETs. The map of the different states is shown in Figure 3.

A simple 2×3 FCAM array is depicted in Figure 2 as a motivational example of the array organization. Bitlines are shared among cells in the same column, while match lines and source lines (ML and ScL) are shared by all cells in the same

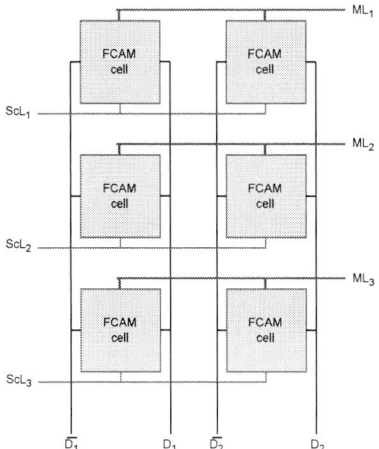

Fig. 2: A 2×3 FCAM array.

row. All match lines are precharged and retain their charges in case of a match. However, if a mismatch occurs, at least one of the FeFETs in one of the CAM cells from that row will open, leading to a discharge of the corresponding ML resulting in a '0' output.

C. Program/Read Scheme

The Program/Read scheme idea is straight-forward; the *right* FeFET of the cell is programmed to the threshold voltage on the *right side* of the state to be stored. Then, the *left* FeFET is programmed to the analog-inverse of the threshold voltage on the *left side* of the state. The analog-inverse of a signal is defined as the one that has the same distance from the centre line as the original, hence, they are symmetrical. Despite this scheme working well for a single cell, applying it to an array introduces certain challenges. On each column, the gates of the FeFETs are connected to the same D and $\bar{D}$ bitlines. This creates problems during subsequent program (or erase) of a cell, since the write voltage will be applied to every other cell on the same column. To mitigate this, it is necessary to implement an inhibition bias scheme (or protection voltage) with which a $V_W/2$ voltage is applied on the ScL lines of the cells not being programmed. The same inhibition bias scheme has been used in the aforementioned works and is derived by the results reported at [28].

D. Testing and Fault Tolerance

One of the main advantages of the proposed FCAM address decoder is its testability. In order to demonstrate this, we introduced partial open defects inside a FCAM cell and evaluated both its functional margins and a simple cell failure detection approaches. We simulated the different defects in isolation, by introducing a variable resistor in four different locations inside the FCAM cell. Figure 4 depicts the sites where the resistors emulating partial open defects were placed. Figure 5 shows the ranges within which the FCAM cell operates correctly

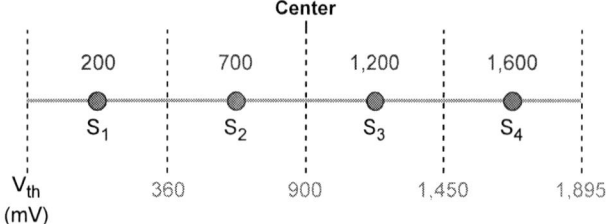

Fig. 3: The four levels of the proposed MCAM cell. The blue dots are read voltages (V_R) while the dotted lines represent the threshold voltages.

Defect site	Tests
R_{defA}	$\{(W_{S2}, R_{S1})\}$
R_{defB}	$\{(W_{S3}, R_{S4})\}$
R_{defC}	$\{(W_{S3}, R_{S4})\}$
R_{defD}	$\{(W_{S2}, R_{S1})\}$

TABLE I: March tests used to detect faulty FCAM cells.

(the light areas on the left), and thus a partially open defect will not disturb the correct cell operation. Moreover, it also shows the areas where a defect will cause all of the reading operations to fail (the dark areas on the right). The dashed area in the middle shows the range where a defect may or may not cause incorrect operation depending on the specific read operation used. Obviously we have to look for the 'most sensitive' read operation that will ensure detection of defects in the two concatenated faulty areas on the right side. Please note the FCAM design tolerance to partially open defects. Values from zero up to few kilo Ohms will not have any impact on its correct operation.

Detecting faulty FCAM cells is easy requiring a short march test, consisting of a single, two-operation, march element per defect. A fail in any of these tests can immediately flag the cell under test as defective and corrective actions, e.g., row repair when spare rows are available, can be taken. Table I lists the tests for every defect considered in our study. The first operation writes a specific state to the cell, e.g., W_{S3}, and the second operation queries another state, e.g., R_{S1}. The reason for attempting to read back a different state than the one written, is that partial open faults are affecting the cell's ability to properly discharge the matching line (ML) in case of a mismatch, thus leading to "false-match" errors. By querying a state different from the one written, we ensure that a match indicates the cell is definitely defective.

IV. EVALUATION AND METHODOLOGY

We conducted the simulations of our 5-to-32 FCAM address decoder with Cadence Spectre, using the model described in [11]. We were provided two versions of the model, one based on the PTM 14nm FinFET process and the other based on the PTM 45nm planar process. We used the 14nm process for both our FCAM cells and the sensing circuitry. The FeFET dimensions were set to $W \times L$ of 75×30 nm, respectively.

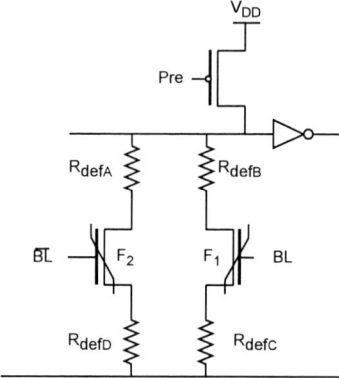

Fig. 4: The sites where resistive defects in a cell are inserted (high resistor values mimic partially open defects).

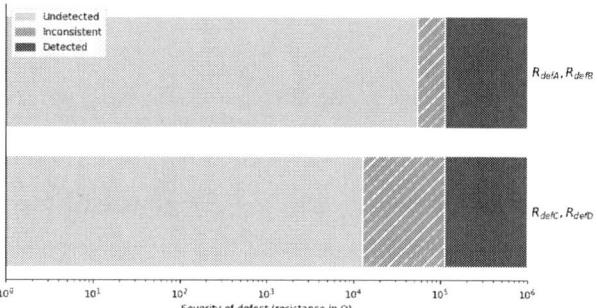

Fig. 5: Detectability ranges of resistive defects located on the (R_{defA}-R_{defD}) sites.

The width reported here is the effective width W_{eff} which is calculated based on the dimensions of the Fin according to the equation $W_{eff} = 2 \times Fin_{Height} + Fin_{Width}$. The sensing circuitry consists of two inverters connected in series, with each of their transistors sized 80×30 nm. The supply voltage used in our circuit simulation was 0.8V.

The FeFET programming pulses had constant duration of 200ns, while the read pulses used were 2.5 ns. Evaluating the decoder delay included determining the total capacitance of the decoder row. For a 5-to-32 decoder, assuming an average of three cells per decoder row, together with the peripheral circuitry (the pre-charge PMOS and the two inverters used for sensing the output signal), as well as the capacitance of the transmission line itself, we estimated the total capacitance as $35fF$. Under the above conditions, our simulations resulted in a propagation delay of approximately 870 ps.

Compared to related work as depicted in Table II [29] [30] our implementation is $1.68\times$ faster than the fastest, hybrid logic implementation and up to $2.6\times$ faster than the slowest static implementation. With regards to area, our 2-bit/cell solution performs slightly better than the static AND logic based CMOS decoder. However, as it was shown before [18], an 8-state, 3-bit FCAM cell is feasible; by exploiting this

979-8-3315-3968-9/24 $31.00 © 2024 IEEE

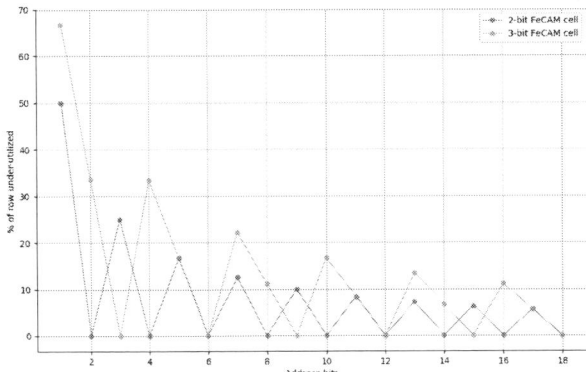

Fig. 6: Under-utilization of 2- and 3-bit FCAM cells with regards to the address length.

Design (5-to-32 decoder)	Transistor Count	Delay (ns)
Static CMOS (AND Logic)	308	2.25
Static CMOS (NOR Logic)	220	2.16
Pseudo-nMOS Logic	176	1.69
Mixed Logic [30]	154	1.58
Hybrid Logic	132	1.46
FeFET CAM (2-bits / cell)	224	0.87

TABLE II: Comparison of the transistor count and delay of different implementations.

denser design we can further reduce the number of transistors down to 160 for the considered 5-to-32 address decoder. This will result in a significant improvement over most of the designs shown in Table II.

A downside of using a multi-bit approach for address decoding is that the address must be aligned to the number of the bits each cell is able to store. When this is not the case, under-utilization of the hardware occurs, which might put the proposed CAM solution at a disadvantage against other solutions. This reality is reflected in Figure 6, where the percentage of row under-utilization is shown, in relation to the address length. For shorter addresses (≤ 8 bits) the under-utilization can be significant; as the address width increases, any under-utilised cells have lesser impact. On the plot we are considering 2-bit and 3-bit cells; the 2-bit FCAM decoder is more efficiently used with even-sized addresses while the 3-bit when the address width is a multiple of 3.

V. DISCUSSION, CONCLUSION AND FUTURE WORK

In this paper we proposed a novel address decoding circuit built upon FeFET based Content Addressable Memory cells. We, (i) demonstrated a FeFET CAM based address decoder design with superior latency, (ii) conducted a thorough investigation of the partial open defects in the proposed CAM cells and (iii) defined a set of march tests able to quickly detect the aforementioned defects. Our simulation results showed an improved delay against state of the art designs, by at least $1.68\times$, while the transistor count remained on par with those

implementations. The proposed address decoder structure can be tested for defects, using simple march tests consisting only of a single two-operations element.

Research in ferroelectrics has yielded promising results in recent years, especially when compared to other technologies such as ReRAM and traditional CMOS. Specifically for CAM designs, the authors of [27] have reported $3.5\times$ to $3200\times$ reduced write energy compared to CMOS and ReRAM designs, respectively, while the reductions on the search EDP were between $4.1\times$ and $2.8\times$, respectively. The authors of [24] reported a further reduced number for the search energy, by more than 60%, using their in-series design. Considering the several advancements in materials and their compatibility with the existing CMOS manufacturing processes, FeFETs are the most promising candidate for CAM cells from all the emerging device technologies.

However, there are still important aspects being worth of further research such as the robustness of multi-bit cell with regard to process variation, assessing the reliability limitations compared to other technologies and evaluating other, device specific defects. These topics will be explored as continuation of the current research.

Future work will also include completing the design and the simulation of the decoder version based on 3-bit FCAM cells, while exploring its energy consumption and comparing it to other state of the art decoders is a natural part of this.

VI. ACKNOWLEDGEMENTS

We are grateful to Dr. Kai Ni, from University of Notre Dame, for providing us the FeFET compact model used in our simulations.

REFERENCES

[1] Leonid Yavits et al. "Dual Mode Logic Address Decoder". In: *2020 IEEE International Symposium on Circuits and Systems (ISCAS)*. 2020, pp. 1–5. DOI: 10.1109/ISCAS45731.2020.9180587.

[2] A. J. van de Goor et al. "March LA: A test for linked memory faults". In: *Proceedings of the 1997 European Design and Test Conference, pp. 627-632, Mar. 1997, issn.* Mar. 1997, pp. 1066–1409.

[3] A. J. van de Goor. *Testing Semiconductor Memories. Theory and Practice.* J. Wiley & Sons, 1991.

[4] A. J. van de Goor and Georgi Gaydadjiev. "March U: A test for unlinked memory faults". In: *IEE Proceedings on Circuits, Devices and Systems, IEE Proceedings* 144 (June 1997), pp. 155–160. DOI: 10.1049/ip-cds:19971147.

[5] S. Beyer et al. "FeFET: A versatile CMOS compatible device with game-changing potential". In: *2020 IEEE International Memory Workshop (IMW), Dresden, Germany: IEEE.* May 2020, pp. 1–4. DOI: 10.1109/IMW48823.2020.9108150.

979-8-3315-3968-9/24 $31.00 © 2024 IEEE

[6] Q. Huang et al. "FeFET Based In-Memory Hyperdimensional Encoding Design". In: *IEEE Trans. Comput.-Aided Des. Integr. Circuits Syst.* (2023), pp. 1–1. DOI: 10.1109/TCAD.2023.3253766.

[7] S. De et al. "Demonstration of Multiply-Accumulate Operation With 28 nm FeFET Crossbar Array". In: *IEEE Electron Device Lett.* 43.12 (Dec. 2022), pp. 2081–2084. DOI: 10.1109/LED.2022.3216558.

[8] X. Chen et al. "Power and Area Efficient FPGA Building Blocks Based on Ferroelectric FETs". In: *IEEE Trans. Circuits Syst. I* 66.5 (May 2019), pp. 1780–1793. DOI: 10.1109/TCSI.2018.2874880.

[9] A. Aziz et al. "Physics-based circuit compatible spice model for ferroelectric transistors". In: *IEEE Electron Device Letters* 37 (2016), pp. 805–808.

[10] K. Ni et al. "Critical role of interlayer in hf 0.5 zr 0.5 o 2 ferroelectric fet nonvolatile memory performance". In: *IEEE Transactions on Electron Devices* 65 (2018), pp. 2461–2469.

[11] K. Ni et al. "A Circuit Compatible Accurate Compact Model for Ferroelectric-FETs". In: *2018 IEEE Symposium on VLSI Technology, Honolulu, HI, USA.* 2018, pp. 131–132.

[12] X. Yin et al. "Ferroelectric FET Based TCAM Designs for Energy Efficient Computing". In: *2019 IEEE Computer Society Annual Symposium on VLSI (ISVLSI), Miami, FL, USA.* 2019, pp. 437–442.

[13] K. Pagiamtzis et al. "Content-addressable memory (cam) circuits and architectures: A tutorial and survey". In: *JSSC* (2006).

[14] H. J. Chao. "Next generation routers". In: *Proceedings of the IEEE* (2002).

[15] J. P. Wade et al. "A ternary content addressable engine". In: *JSSC* (1989).

[16] K. Ni et al. "Ferroelectric ternary content-addressable memory for one-shot learning". In: *Nature Electronics* (2019).

[17] C. Li, C. E. Graves, X. Sheng, et al. "Analog content-addressable memories with memristors". In: *Nat Commun* 11 (2020), p. 1638.

[18] A. Kazemi et al. "FeFET Multi-Bit Content-Addressable Memories for In-Memory Nearest Neighbor Search". In: *IEEE Transactions on Computers* 71.10 (Oct. 1, 2022), pp. 2565–2576.

[19] A. Kazemi et al. "A Flash-Based Multi-Bit Content-Addressable Memory with Euclidean Squared Distance". In: *2021 IEEE/ACM International Symposium on Low Power Electronics and Design (ISLPED), Boston, MA, USA.* 2021, pp. 1–6.

[20] X. Yin et al. "FeCAM: A Universal Compact Digital and Analog Content Addressable Memory Using Ferroelectric". In: *IEEE Transactions on Electron Devices* 67.7 (July 2020), pp. 2785–2792.

[21] C. Li et al. "A Scalable Design of Multi-Bit Ferroelectric Content Addressable Memory for Data-Centric Computing". In: *2020 IEEE International Electron Devices Meeting (IEDM), San Francisco, CA, USA, 2020, pp. 29.3..3.4.* 2020, pp. 1–29.

[22] Cédric Marchand et al. "A FeFET-Based Hybrid Memory Accessible by Content and by Address". In: *IEEE Journal on Exploratory Solid-State Computational Devices and Circuits* 8.1 (2022), pp. 19–26. DOI: 10.1109/JXCDC.2022.3168057.

[23] O. Bekdache et al. "Scalable Complementary FeFET CAM Design". In: *2023 IEEE International Symposium on Circuits and Systems (ISCAS).* 2023, pp. 1–5. DOI: 10.1109/ISCAS46773.2023.10181788.

[24] Chengji Jin et al. "A Multi-Bit CAM Design With Ultra-High Density and Energy Efficiency Based on FeFET NAND". In: *IEEE Electron Device Letters* 44.7 (2023), pp. 1104–1107. DOI: 10.1109/LED.2023.3277845.

[25] L. Yavits, U. Weiser, and R. Ginosar. "Resistive Address Decoder". In: *IEEE Computer Architecture Letters* 16.2 (July 2017), pp. 141–144.

[26] L. Yavits, R. Kaplan, and R. Ginosar. "Enabling Full Associativity with Memristive Address Decoder". In: *IEEE Micro* 38.5 (Sept. 2018), pp. 32–40.

[27] X. Yin et al. "An Ultra-Dense 2FeFET TCAM Design Based on a Multi-Domain FeFET Model". In: *IEEE Transactions on Circuits and Systems* 66.9 (Sept. 2019), pp. 1577–1581.

[28] K. Ni et al. "Write Disturb in Ferroelectric FETs and Its Implication for 1T-FeFET AND Memory Arrays". In: *IEEE Electron Device Letters* 39.11 (Nov. 2018), pp. 1656–1659.

[29] M. Madhusudhan Reddy, K. V. Challa, and B. Srinivasa Raja. "An Energy Efficient Static Address Decoder for High-Speed Memory Applications". In: *2022 7th International Conference on Communication and Electronics Systems (ICCES), Coimbatore, India.* 2022, pp. 50–53.

[30] Dimitrios Balobas and Nikos Konofaos. "Design of low-power high-performance 2-4 and 4-16 mixed-logic line decoders". In: *IEEE transactions on circuits and systems-II: express briefs* 64 (2017), pp. 176–180.

A Low-Power Linear Phase Interpolation-Based Delay Line in 12nm FinFET Technology

Mohammadreza Esmaeilpour*, Jan Lappas*, Christian Weis*, Norbert Wehn*

*University of Kaiserslautern-Landau (RPTU), Germany

{m.esmaeilpour, jan.lappas, christian.weis, norbert.wehn}@rptu.de

Abstract—A novel low-power high-linear phase interpolation-based delay line in 12nm FinFET technology is detailed in this paper. The proposed delay line exhibits 50% improvement in terms of power consumption compared to the previous work. In addition, the presented architecture to the best of our knowledge is the most efficient delay line for fine tuning in advanced technology nodes due to the low complexity and complete controllability over resolution, delay range and target frequency. The analysis in this paper indicates that the input slew rate plays an indispensable role in the linearity of the delay line. Consequently, two identical resistors are added to the input of the phase interpolator unit to decrease the slew rate. This approach significantly improves the linearity over a wide frequency range. The proposed delay line dissipates 0.56 mW from a 0.8 V supply voltage and 5 GHz operating frequency.

Index Terms—Digitally Controlled Delay Line (DCDL), Phase interpolator, Fine tuning, Delay element, Delay-locked loop, Memory interface.

I. INTRODUCTION

Variable delay lines are widely employed in a range of VLSI applications such as clock generation [1], delay locked loops (DLLs) [2]–[4], ring oscillators and time-to-digital converters (TDCs) [5]. In addition, the increasing demand for high performance and low power DRAM interfaces requires power efficient and high-resolution delay lines to maximize the timing margin through accurate phase shift during read training operation [6]. Moreover, high linearity and wide frequency range are further important design metrics for delay lines which should be considered in the aforementioned applications.

Several delay line implementations have been proposed for different purposes. The most common approach is binary-weighted capacitors which are implemented by using a cascade of two inverters and digitally controlled capacitors connected to the output of the first inverter. Since metal-oxide-metal (MOM) capacitors occupy significant silicon area and also additional transistors are required to control each capacitor, an array of MOSCAPs is employed to provide tunable delay [4], [7], [8]. However, to achieve an acceptable resolution as well as wide delay range in advanced technology nodes, in which the parasitic capacitors are small, the MOSCAPs should be large enough to provide enough capacitance [9]. Furthermore, the first stage inverter needs to be sufficiently wide to charge the capacitors at the desired operating frequency. Due to the mentioned design constraints, this approach consumes high power and silicon area. Another common implementation is current-starved inverters (CSI) [3], in which NMOS and PMOS networks are utilized to control the current and change the delay accordingly. Although CSI topology benefits from high resolution and good jitter performance, the delay range is extremely limited. They also suffer from poor linearity and high sensitivity to process variations [7]. Thyristor-based delay element [10]

is another conventional design which is controlled by current sources. It demonstrates a high power supply rejection ratio (PSRR) while suffering from complex biasing circuits and static power consumption. In addition, the analog nature of this type of delay elements makes it difficult to migrate over state-of-the-art technology nodes.

Phase Interpolators (PI) is a promising alternative to be used as a high linear and power efficient delay line. The phase interpolation is done by integrating two input clocks with different phases and generating an output with an in-between phase [11]. Wide frequency range and constant resolution are the main advantages of using phase interpolators. The conventional phase interpolator design is based on dumping two weighted currents triggered by different phase clocks into a shared load [12]. This technique, however, not only dissipates high power but also faces severe non-linearity due to PVT variations. In [13] an all-digital phase interpolator which consists of an array of connected output multiplexers is proposed. Despite low complexity, our analysis indicates that uncontrolled slew rate of input signals causes high fluctuations in differential non-linearity (DNL) and integral non-linearity (INL) in delay steps.

In this paper, a new wide range, high-linear and low-power phase interpolation-based delay line in 12nm FinFET technology is presented. The proposed architecture can be utilized in high performance memory interfaces and all digital delay-locked loops for fine tuning due to the wide frequency range and high resolution in advanced technology nodes. Compared to other topologies, the proposed delay line demonstrates noticeable improvement in terms of power consumption due to the reduced hardware complexity. In order to control the slew rate of input clock signals, a pair of resistors are connected to the inputs of phase interpolator block, which noticeably improves the linearity over a wide frequency range. Since the number of control bits, input slew rate and the time difference between the input signals significantly affect the linearity of the delay line, detailed design considerations and trade-offs will be discussed in the following sections. The remaining of this brief is organized as follows. Section II introduces the architecture and circuitry of the proposed delay line. Post-layout simulation results and detailed design constraints are discussed in Section III. Finally, Section IV concludes this work.

II. CIRCUIT DESCRIPTION

The overall architecture of the proposed phase interpolation-based delay line is illustrated in Fig. 1. It consists of 15 units of PI cell, a replica delay cell to produce desired time difference between the input signals and a 4-bit binary to thermometer converter (BTC). To improve the linearity over a wide frequency range, two identical resistors are added to the

979-8-3315-3968-9/24 $31.00 © 2024 IEEE

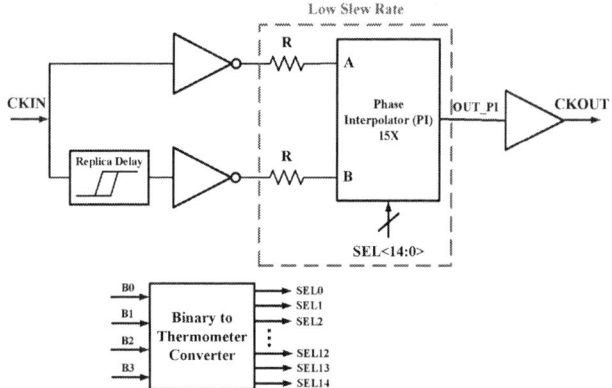

Fig. 1: Architecture of the proposed phase interpolation-based delay line.

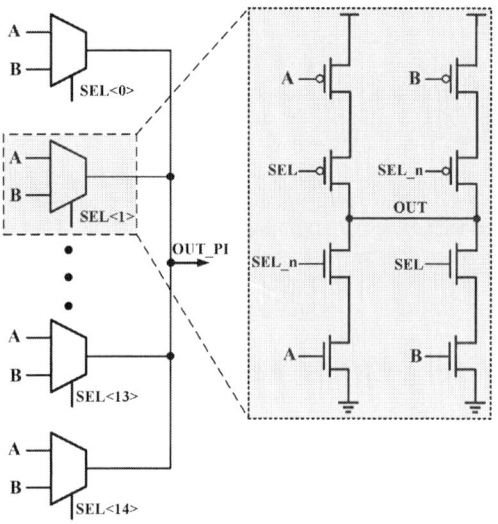

Fig. 2: Phase interpolation unit.

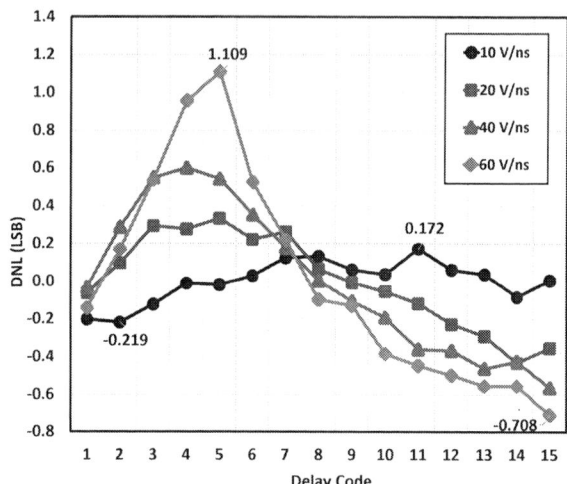

(a)

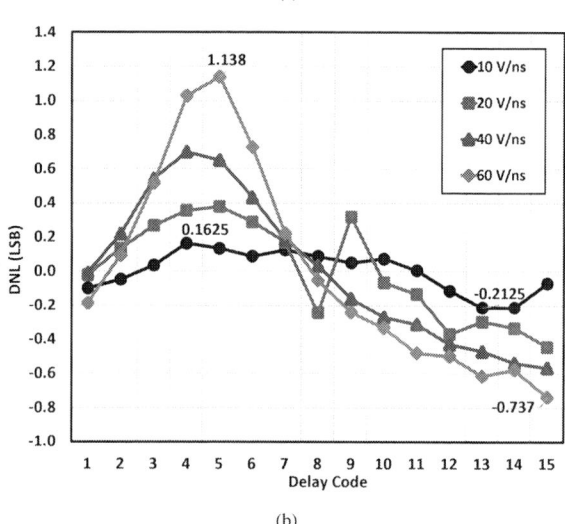

(b)

Fig. 3: Effect of input slew rate on DNL at , (a) 2.5 GHz, (b) 5 GHz.

input of the PI block. These resistors decrease the slew rate of the input signals, as a result of which enough time is provided for the phase interpolator block to merge input signals. The BTC is utilized to generate control signals for the PI units to perform the interpolation process.

In order to achieve high linearity over a wide frequency range, it is crucial to maintain a low slew rate in the inter-polation region. Thus, the type of PI unit is also crucial for sustaining low slew rate. The PI unit consists of an array of 2:1 multiplexers which are controlled by 15 thermometer codes as shown in Fig. 2. By increasing the control bit, the phase of CKOUT shifts from A to B and provides delay steps for different bit streams. The interpolation is accomplished directly at the output of the multiplexers, maintaining the slew rate of OUT_PI at its lowest possible value to ensure high linearity while adjusting the control bits.

Using pass transistor logic (PTL) based multiplexers is an alternative solution to achieve low slew rate at the output node.

However, our analysis shows that due to the direct coupling between input and output of the PTL logic, the input will be affected during the interpolation process at the output. Thus, the linearity of the delay line deteriorates due to the continuous changes in input and output.

III. POST-LAYOUT SIMULATION RESULTS AND DISCUSSIONS

A. Slew rate

As mentioned in the previous sections, slew rate of the input signals is the most important factor that affects the linearity of the phase interpolation-based delay line. Fig. 3, illustrates the effect of slew rate on DNL values at 2.5 GHz and 5 GHz. As can be seen in Fig. 3, decreasing the slew rate can effectively improve the linearity of the phase interpolator regardless of the frequency. Consequently, a pair of resistors are added to

979-8-3315-3968-9/24 $31.00 © 2024 IEEE

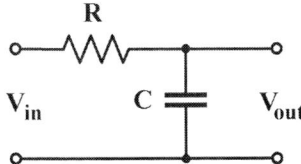

Fig. 4: Equivalent RC circuit of the phase interpolation unit.

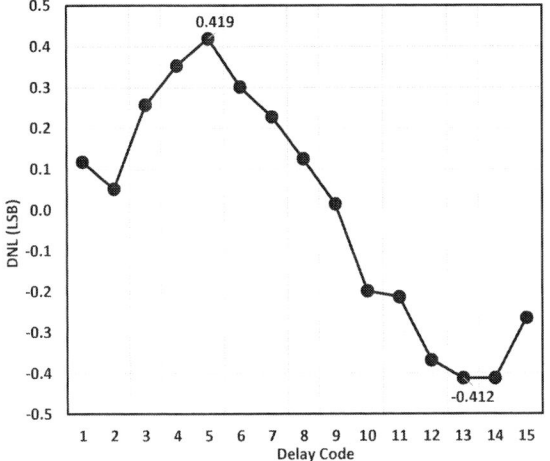

Fig. 5: DNL variations at 10 GHz.

the input of the phase interpolator unit to control the slew rate based on the target frequency and number of control bits.

The phase interpolation unit in Fig.1 can be simplified as an RC network illustrated in Fig.4, where the input signal is an ideal voltage step of height V_0. The output voltage V_{out} can be expressed as [14]

$$V_{out} = V_0[1 - exp(-t/\tau)] \qquad (1)$$

in which $\tau = R \times C$ and C is the equivalent parasitic capacitance of the multiplexers. From Eq.(1), the slew rate can be derived as [14]

$$SR = \frac{dV_{out}}{dt} = \frac{V_0}{\tau} exp(\frac{-t}{\tau}) \qquad (2)$$

As shown in Eq.2, the slew rate value is proportional to height of input voltage step V_0 and $1/\tau$. Considering fixed values for V_0 and C, the slew rate can be controlled via changing the input resistor to achieve the optimum performance of the phase interpolation unit.

On the other hand, increasing the frequency of the input signal along with decreasing the slew rate can cause duty cycle distortion at the output node due to the inability to reach a rail-to-rail signal. The simulation results indicate that an input slew rate within the range of 10 V/ns is the optimum value for the aforementioned frequency range to achieve both good linearity and maintain 50% duty cycle at the output node. However, to achieve a proper function in terms of linearity and output duty cycle for higher frequencies, the input slew rate of the proposed delay line should be fixed to higher values. For example, Fig.5

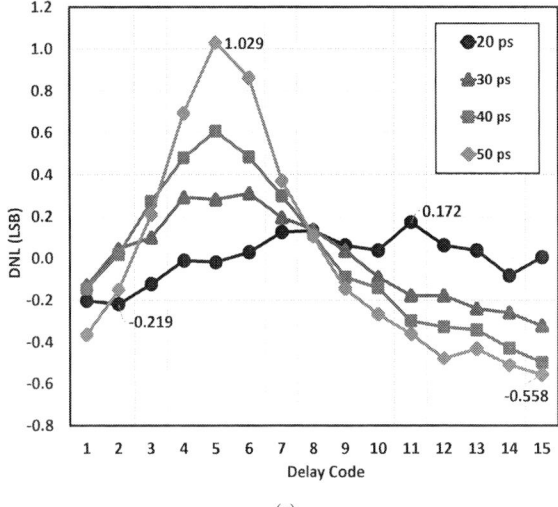

(a)

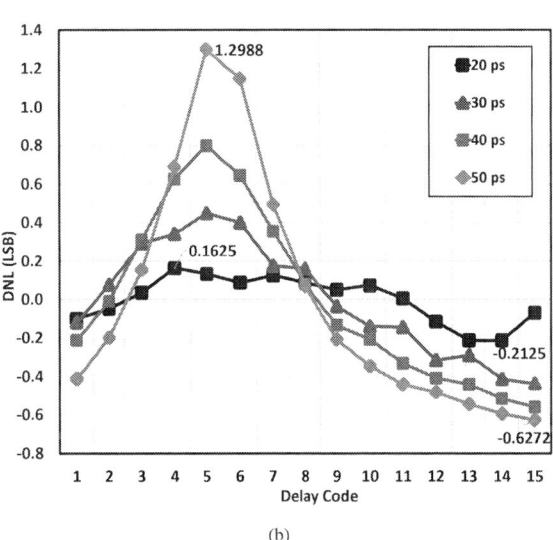

(b)

Fig. 6: Effect of time difference on DNL at , (a) 2.5 GHz, (b) 5 GHz

illustrates the DNL values of the proposed delay line at 10 GHz. In this case, the input slew rate is set to 20 V/ns to achieve acceptable linearity and output duty cycle.

B. Time Difference

Another important metric that significantly affects the linearity of the phase interpolation-based delay line is the time difference between the inputs which is supposed to quantize by the PI units and provide delay steps. Fig. 6 illustrates the effect of input delay difference on DNL values at 2.5 GHz and 5 GHz. Increasing the input delay difference with the same number of control bits resulting in higher resolution which degrades the linearity of the proposed delay line. As shown in Fig. 6, maintaining a resolution in the range of 1 ps to 2 ps provides lower fluctuations in DNL values. Consequently, the proposed delay line in this work is recommended to be utilized for fine tuning.

979-8-3315-3968-9/24 $31.00 © 2024 IEEE   129

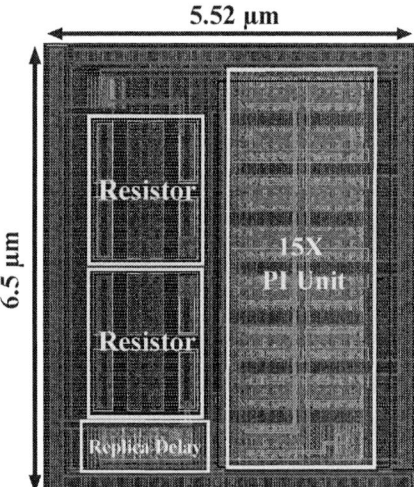

Fig. 7: Layout view of the proposed delay line.

In order to increase further the input time difference, the number of control bits should also be increased proportionally to achieve the mentioned resolution range. However, increasing the number of bits causes drastic growth in circuit complexity and power consumption. Thus, an appropriate choice between the number of control bits and the time difference between inputs can effectively improve the linearity and power dissipation.

C. Power Consumption and Comparison

The proposed phase interpolation-based delay line is designed and symmetrically laid out in 12nm FinFET technology. The distinctive feature of the proposed delay line compared to previous studies is the complete control over delay resolution and slew rate which effectively enhances the linearity especially in advanced technology nodes. Furthermore, less hardware complexity and lower power dissipation makes this structure a viable choice for low-power applications. The 4-bit structure with an input time difference of 20 ps consumes only 0.56 mW at 5 GHz and a 0.8 V supply voltage.

Considering the significance of good matching in the linearity of the delay lines, the proposed architecture with a delay difference of 20 ps and slew rate of 10 V/ns was symmetrically laid out as shown in Fig. 7. It occupies 5.52 μm×6.5 μm of area.

In order to evaluate the performance of the proposed phase interpolation-based delay line under the process, voltage and temperature (PVT) variations, the DNL values for typical, best and worst corners were calculated based on the post-layout simulations at 5 GHz as indicated in Fig. 8. The DNL values varies between -0.3 LSB and 0.2 LSB at all PVT corners demonstrating the feasibility of the proposed delay line due to its linearity and monotonic characteristic.

The performance and characteristics of the proposed delay line is summarized and compared with previous studies as shown in Table I. The proposed architecture in this brief provides a high resolution and linear delay steps while it has a remarkable low power usage. Since the mentioned studies in Table I were implemented in different technologies and operating frequencies, for a fair comparison, a figure-of-merit (FOM) for power consumption is utilized and defined as [15]

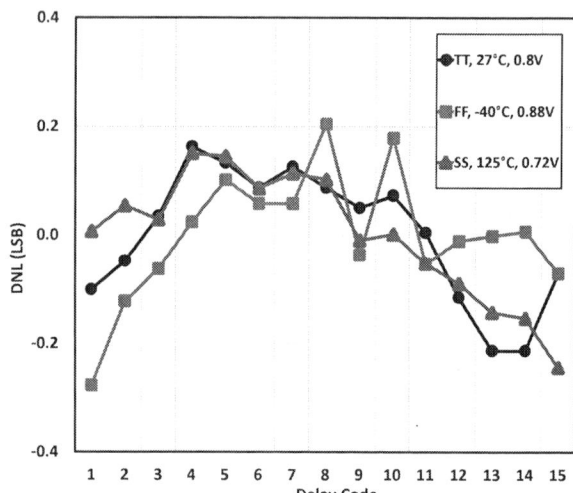

Fig. 8: DNL variations at PVT corners.

TABLE I: PERFORMANCE SUMMARY AND COMPARISON.

Parameter	[7]	[16]	[11]	[12]	This Work
Technology (nm)	65	130	28	65	12 FinFET
Supply Voltage (V)	1.2	1.2	0.9	1.2	0.8
Frequency (GHz)	2.5	1.2	6	8	5.0
Resolution (ps)	0.82	1.95	-	-	1.3
Power (mW)	1.25	11.25	2.2	13	0.56
FOM_{power}	0.347	6.5	0.452	1.16	**0.175**

$$FOM_{power} = \frac{Power Consumption(mW)}{Operating Frequency(GHz) \times VDD^2(V^2)} \quad (3)$$

IV. CONCLUSION

A novel low-power and high linear phase interpolation-based delay line is presented in this brief. The proposed architecture can be a perfect choice for fine tuning due to the less hardware complexity and low power consumption. In addition, the full controllability over delay range, resolution and operating frequency allows the designers to utilize this delay line for different applications especially in advanced technology nodes. The presented analysis in this paper demonstrates that input slew rate and time difference are the most important metrics in the linearity of the phase interpolation-based delay line. To control the input slew rate, a pair of identical resistors are added to the input of the phase interpolation unit which noticeably improves the linearity of the delay line. The proposed delay line was designed and simulated in 12nm FinFET technology and consumes 0.56 mW at 5 GHz operating frequency and 0.8 V supply voltage.

ACKNOWLEDGEMENT

This work receives EuroHPC-JU funding under grant no. 101034126, with support from the Horizon2020 programme.

REFERENCES

[1] H. H. Cheong and S. Kim, "A power-efficient and fast-locking digital quadrature clock generator with ping-pong phase detection," in *2021 IEEE International Symposium on Circuits and Systems (ISCAS)*, 2021, pp. 1–4.

[2] H. Park, J. Sim, Y. Choi, J. Choi, Y. Kwon, and C. Kim, "A 2.4–8 ghz phase rotator delay-locked loop using cascading structure for direct input–output phase detection," *IEEE Transactions on Circuits and Systems II: Express Briefs*, vol. 69, no. 3, pp. 794–798, 2022.

[3] R.-J. Yang and S.-I. Liu, "A 2.5 ghz all-digital delay-locked loop in 0.13 μm cmos technology," *IEEE Journal of Solid-State Circuits*, vol. 42, no. 11, pp. 2338–2347, 2007.

[4] M. E. Quchani and M. Maymandi-Nejad, "Design of a low-power linear sar-based all-digital delay-locked loop," in *2019 27th Iranian Conference on Electrical Engineering (ICEE)*, 2019, pp. 118–124.

[5] C.-C. Chen, C.-L. Chen, W. Fang, and Y.-C. Chu, "All-digital cmos time-to-digital converter with temperature-measuring capability," *IEEE Transactions on Very Large Scale Integration (VLSI) Systems*, vol. 28, no. 9, pp. 2079–2083, 2020.

[6] C.-W. Tsai, Y.-T. Chiu, Y.-H. Tu, and K.-H. Cheng, "A wide-range all-digital delay-locked loop for ddr1–ddr5 applications," *IEEE Transactions on Very Large Scale Integration (VLSI) Systems*, vol. 29, no. 10, pp. 1720–1729, 2021.

[7] N. Angeli and K. Hofmann, "A low-power and area-efficient digitally controlled shunt-capacitor delay element for high-resolution delay lines," in *2018 25th IEEE International Conference on Electronics, Circuits and Systems (ICECS)*, 2018, pp. 717–720.

[8] J. Song and W.-S. Choi, "A highly linear digitally controlled delay line with reduced duty cycle distortion," in *2022 19th International SoC Design Conference (ISOCC)*, 2022, pp. 398–399.

[9] J. I. Morales, F. Chierchie, P. S. Mandolesi, and E. E. Paolini, "Design and evaluation of an all-digital programmable delay line in 130-nm cmos," in *2019 XVIII Workshop on Information Processing and Control (RPIC)*, 2019, pp. 209–213.

[10] M. Esmaeilzadeh, Y. Audet, M. Ali, and M. Sawan, "A wide-range low-power thyristor-based delay element with improved temperature sensitivity," *IEEE Transactions on Circuits and Systems II: Express Briefs*, vol. 70, no. 7, pp. 2370–2374, 2023.

[11] A. AbdelHadi, M. Allam, and S. Ibrahim, "A high-linearity 6-ghz phase interpolator in 28-nm cmos technology," in *2017 Japan-Africa Conference on Electronics, Communications and Computers (JAC-ECC)*, 2017, pp. 41–44.

[12] G. Wu, D. Huang, J. Li, P. Gui, T. Liu, S. Guo, R. Wang, Y. Fan, S. Chakraborty, and M. Morgan, "A 1–16 gb/s all-digital clock and data recovery with a wideband high-linearity phase interpolator," *IEEE Transactions on Very Large Scale Integration (VLSI) Systems*, vol. 24, no. 7, pp. 2511–2520, 2016.

[13] J. Sim, H. Park, Y. Kwon, S. Kim, and C. Kim, "A 1-3.2 ghz 0.6 mw/ghz duty-cycle-corrector using bangbang duty-cycle-detector," in *2021 IEEE International Symposium on Circuits and Systems (ISCAS)*, 2021, pp. 1–4.

[14] B. Razavi, *Design of analog CMOS integrated circuit*, 1st ed. McGraw Hill, 2001.

[15] E. Bayram, A. F. Aref, M. Saeed, and R. Negra, "1.5–3.3 ghz, 0.0077 mm2, 7 mw all-digital delay-locked loop with dead-zone free phase detector in 0.13 μm cmos," *IEEE Transactions on Circuits and Systems I: Regular Papers*, vol. 65, no. 1, pp. 39–50, 2018.

[16] K. Ryu, D.-H. Jung, and S.-O. Jung, "All-digital process-variation-calibrated timing generator for ate with 1.95-ps resolution and a maximum 1.2-ghz test rate," in *2013 Proceedings of the ESSCIRC (ESSCIRC)*, 2013, pp. 41–44.

979-8-3315-3968-9/24 $31.00 © 2024 IEEE

Benchmarking Microfluidic Design Automation Flows

Ashton Snelgrove, Skylar Stockham, Pierre-Emmanuel Gaillardon
Electrical and Computer Engineering
University of Utah
Salt Lake City, Utah, USA
pierre-emmanuel.gaillardon@utah.edu

Abstract—In this paper, we propose a methodology for measuring figures of merit relevant to microfluidics practitioners. We also present a benchmark suite for microfluidics design automation (MFDA). The suite is composed of generated circuits designed to challenge tools for specific figures of merit identified in the measurement methodology. We survey the MFDA literature to evaluate the state-of-the-art for benchmark and measurement methodologies. We include in the benchmark distribution a complete set of transcriptions of the major benchmarks used in the MFDA literature, including designs previously unavailable. Benchmarks are distributed in the major hardware description languages along with reported measurements

Index Terms—Microfluidics, Electronic Design Automation, 3D Printing, Placement and Routing

I. INTRODUCTION

Microfluidics concerns the precise manipulation of small amounts of fluids at the nanoliter scale. This is used to create so-called "lab-on-chips" which consolidate chemical or biological assays into small automated devices. The goal of microfluidic design automation (MFDA) is to introduce computer tools and algorithms to automate aspects of the design process. [1]

The MFDA community is faced with several related challenges. The community is lacking an available set of comprehensive, shared, and open-source benchmarks. Previous works identified this problem and began a collection of open-source benchmarks[2][3] However, this still remains a limited set that does not bring together all of the major benchmarks currently in use in the literature, and has not seen adoption. Current benchmarks do not reflect the scale currently achieved in electronic design. A set of benchmarks is needed that will challenge MFDA at scale.

The community is also lacking a clear consensus for appropriate figures of merit, in particular metrics aligned with the needs of microfluidics practitioners[4]. A pressing concern is the adoption of new manufacturing techniques such as 3D printing which challenge existing assumptions and requirements.

To address these issues, this work proposes:

This material is based upon work supported by the National Science Foundation under Grant No. 2245494. Any opinions, findings, and conclusions or recommendations expressed in this material are those of the author(s) and do not necessarily reflect the views of the National Science Foundation.

1) A complete set of transcriptions of the major benchmarks in use in the MFDA literature.
2) A methodology for measuring figures of merit relevant to microfluidics designers.
3) A set of synthetic benchmarks designed to challenge tools on specific figures of merit.
4) Initial reported measurements for each benchmarks.

Our proposed figures of merit fall into three categories: practitioner needs, manufacturability, and optimization. Practitioners are particularly interested in tool runtime and meeting design constraints[4]. Manufacturing concerns include area/volume and component density. Optimization concerns are related to the output of synthesis algorithms, including control line minimization and component counts. Our methodology also seeks to address the question of comparing traditional planar manufacturing and 3D printing.

Our proposed benchmark suite provides a range of synthetic benchmarks designed to challenge specific metrics in the placement and routing (PNR) stages of design automation.

The remainder of our paper is organized as follows. Section II evaluates previous measurement methodologies in the literature. Section III proposes a methodology for measuring and evaluating MFDA tools. Section IV surveys the common benchmarks used in the literature. Section V proposes a set of synthetic benchmarks. Section VI concludes the paper.

II. CURRENT USE OF METRICS AND FIGURES OF MERIT

A. Survey of commonly reported metrics in the literature

We surveyed 39 papers which were referenced from two literature reviews, and evaluated each for metrics reported and benchmarks used. Papers were taken from Table 1 in Sanka et al. [3] and from Tables 1 and 2 in Huang et al. [1]. We noted which metrics and which benchmarks were used in each paper. All papers were targeting planar devices implemented in PDMS.

Table I summarizes the results for metrics used. The most commonly used metric was runtime of algorithm or tool, with 87% reporting. 79% reported which type of compute resources were used: usually the processor model, available memory, and operating system. Programming language used for implementation was usually given. Tools using an external solver usually specified which was used (such as an integer

Metric	# papers	% papers
Component count	26	67%
Number of valve switching states	7	18%
Valve or control line count	21	54%
Number of flow/control intersections	9	23%
Total linear channel length	20	51%
Delay or latency	10	26%
Area	16	41%
Algorithm specific synthesis metrics	12	31%
Provides design rules or dimensions	11	28%
Runtime (seconds)	34	87%
Specifies compute resources used	31	79%
Success/failure rate of algorithm	4	10%

TABLE I

SURVEY OF METRICS USED IN 39 PAPERS REFERENCED FROM [1][3]

linear programming solver). Runtime is easily quantifiable, and non-specific to any particulars of algorithm or benchmark. However, while most papers reported descriptions of the environment, comparing run time between tools is difficult due to no clear scaling between different computing resources. Huang et al.[1] assigned a general "fast, medium, slow" rating to each algorithm in their review.

The next most commonly specified metric was some type of operational component count, with 67% reporting. 50% of papers which specified a component count also specified a count of each types of components (such as mixers, heaters, etc.). This metric is difficult to compare due to potentially conflicting meanings. Some papers are simply specified the size of the benchmark, such as those optimizing placement and routing without component minimization. Other papers were specifically attempting to do resource allocation and scheduling, and minimizing the components used in the schedule.

54% of papers gave a count of valves or control lines. Intersections between flow and control layers was reported in 23%. Both of these represent synthesis targets for the flow/control layer interaction, for control line minimization and planarity layout optimization respectively. Control input ports in particular represent a resource bottle neck.

Size information was present in many of the papers. 51% of papers specified a total linear channel length. 26% specified a delay or latency metric of a synthesized schedule. 41% specified an area measurement of the final device. However, only 28% specified details of design rules, component or channel dimensions, or device specifics. This is a potential blind spot in comparing dimensions metrics with possible unknown dimensions.

Finally, 31% gave other metrics specific to the algorithm. Papers sometimes provided their own reference algorithm implementation to compare against, stating uniqueness of their target optimization.

B. The move to 3D printing

All of the papers surveyed represent design flows targeting planar PDMS manufacturing. Metrics such as minimizing control/flow layer intersection for planarity are imposed by the underlying manufacturing technology. New manufacturing techniques stemming from advances in 3D printing challenge

existing limitations, but also introducing new constraints[5]. Volume in planar devices is defined by the two-dimensional area with a fixed depth of the feature. Three-dimensional features can utilize varying depths to reduce feature surface area and footprint. Multi-layer routing, typical of electronic manufacturing, removes the routing restrictions imposed by a single pair of control and flow layers.

Some challenges remain the same between technologies. Input/output line utilization remains a problem, perhaps even worse in 3D printed devices due to the reduced surface area of a three-dimensional device and the fixed width and height of the print area in a stereolithography (SLA) printer. Innovations in connection miniaturization are promising for increasing the density of external connections [6]. Control line and valve count minimization also remains a concern. Recent work proposes transistor like microfluidic structures, bringing the potential for on-chip control logic[7], but some amount of external control will continue to be required. Minimizing total channel length will remain a consistent concern.

C. Metrics of relevance to designers

McDaniel et al[4] interviewed approximately 100 lab-on-chip designers to determine what factors are relevant to their work. Of primary concern was correct device function, meeting design constraints, reliability, legibility, and testability. They are critically interested in maintaining control of the design process. Designers recognized a need for fast algorithms and supplemental tools which boost productivity and address specific pain points of their workflows. Slow algorithms would not be adopted, and designers showed no trust or interest in full turn-key solutions. Metrics associated with EDA such as channel length and skew, were considered of little utility. Transport delay has little bearing when assay times are measured in minutes or hours. Minimizing area is not a significant cost concern[4]. Reducing reagent and sample volumes and waste is often cited as a benefit of lab-on-chips[1], which places interest on minimization of channel lengths. However, channel length is not considered relevant by device designers[4].

The adoption of 3D printing introduces new demands on tools. Given the desire for fast algorithms, it is likely faster prototyping turnaround times would also be desirable. For SLA 3D printing, print time is determined by device depth[5], which could be a target metric for minimization. SLA printing works within a fixed print area, which places a manufacturing constraint on area that may not be previously present when working with planar technologies. SLA printed devices require post-processing flushing of uncured resin trapped in the device during printing. Channel lengths are a major contributor to flushing time.

As is the case with EDA designers, area or timing optimizations are not a problem right up until constraints are violated. MFDA tools currently do not have a format for design constraint specification, which will be required for future work. Synthesis algorithms require a level of trust that users do not appear to be willing to give[4]. Future development of tools for verification may improve this situation.

III. PROPOSED MEASUREMENT METHODOLOGY

We identify three groups of metrics: universal metrics that affect user workflow, metrics for synthesis and optimization, and metrics for physical design.

Optimization is only of practical use when it is in service of meeting design constraints. When comparing algorithms we often target absolute value improvement on a metric. Absolute improvement reflects on the capability of a tool, but does not necessarily translate to improvements in outcomes to meet design constraints. Measurements are also often taken without first doing predictions of outcome. We propose that benchmarks also include a set of design constraints of varying levels of strictness that challenge tools in multiple conditions.

Testing and quality assurance metrics such as validation time are also relevant metrics, but are not within the scope of this work. Finally, the implicit metric for any tool is a need for maintaining correctness. This needs to be explicitly noted here, as design verification is not currently an explored problem in MFDA. Verification is beyond the scope of this work, but must be addressed in the future.

A. Universal metrics

Metrics applicable to all scenarios. These metrics are critical to two identified user priorities, fast runtime and meeting design constraints [4].

- Compute environment. Results should be placed in the context. Papers should report CPU model and speed, total system memory, programming language, and operating system.
- Runtime; in seconds. Despite the difficulty in comparing time between compute environments, runtime remains a critical metric. Ideally, papers should report runtimes for both proposed tools and the tools being compared against, all run in the same environment. Runtimes should be reported for each flow step, so that comparisons can be made for individual steps if some steps were not performed.
- Failure to generate a solution, meet user design constraints, or pass manufacturer's technology design rules; binary success/failure. If an algorithm fails to complete or cannot provide a solution, this should be noted. If algorithms are non-deterministic, percent failure rate should be given.

B. Synthesis metrics

Metrics applicable to tools doing high level device synthesis. Expected tool input is a high level description of device operation. Expected tool output is a netlist. Goals for synthesis are optimization.

- Component count. Synthesis algorithms performing allocation and mapping for schedule. Specifying the types of allocated component is useful.

- Valve count. Synthesis algorithms performing control line minimization.
- Input/output port count. Fluid transport requirements and control line minimization.

C. Physical design metrics

Metrics applicable to physical device generation. Expected tool input is a netlist. Expected tool output is manufacturing plans. Goals for physical design is manufacturability. Benchmarks are defined with the following set of characteristics:

- Component, valve, and IO counts. Reported values used to describe scale of benchmark problem.
- Minimal channel feature size; in micrometers. Establishing bounds on the manufacturing technology to adjust for differences in sizes.

The figures of merit themselves are reflective of the physical characteristics of the manufactured device.

- Simple dimensions height, width, and depth; in either micrometers or multiple of minimum feature size. For SLA printing, depth is the primary factor impacting print time.
- Spacial utilization; percent of device size. For two-dimensional devices, this a percentage of area occupied by components and channels. For three-dimensional devices, this is percentage of volume.
- Effective planar area; in either micrometers or multiples of feature size. For two-dimensional devices, this is the simple area, height × width. For three-dimensional devices, multiply the simple area by the number of routing layers to get an approximate equivalent planar area. If routing is not done in a layered fashion, replace number of layers with depth divided by the minimum feature size.
- Total and maximum channel length; in either micrometers or multiple of feature size. For SLA printing, this metric is also related to post-print flush time.

IV. CURRENT STATE OF THE ART FOR BENCHMARKS

A. Commonly used benchmarks

Table II lists the frequency of benchmarks used in the surveyed papers. No benchmark was used in a majority of MFDA papers surveyed. Three sets of benchmarks appear most often in the papers surveyed. The first is a set originating from Duke University [8]. It consists of three designs derived from real-life assays: polymerase chain reaction (PCR), in-vitro diagnostic (IVD), and colorimetric protein assay (CPA).

The second is a set originating from Technical University of Denmark (DTU)[9]. It consists of five synthetic designs of increasing size from 10 to 50 components.

A third set consists of devices with similar structure derived from published fluidic devices[10]–[14], first appearing in papers from Technical University of Munich (TUM)[15].

Of these benchmarks the PCR benchmark is referenced in 44% of papers surveyed. 26% of papers used benchmarks from two sets, 38% used benchmarks from only one set, and 36% used no benchmarks from any set. 56% of papers used a benchmark unique to that paper.

979-8-3315-3968-9/24 $31.00 © 2024 IEEE

Benchmark	# papers	% papers
PCR[8]	17	44%
IVD[8]	14	36%
CPA[8]	12	31%
DTU Synthetic 1-5 [9]	12	31%
Kinase[11]	4	10%
MNAcid Process [12]	4	10%
MRNA[10]	6	15%
ChIP[13]	6	15%
HIV1[14]	3	8%
Other paper specific benchmark	22	56%

TABLE II

SURVEY OF BENCHMARKS USED IN 39 PAPERS REFERENCED BY [3][1]

Benchmarks specific to papers were typically random graphs or synthetic benchmarks designed for the paper. No reproducible details of the structure of these benchmarks was given.

B. Benchmark analysis

These benchmarks suffer from several drawbacks. The Duke and DTU sets are given as abstract sequence graphs of fluidic operations. This is an appropriate input for a high level synthesis algorithm, such as allocating the sequence onto shared resources and doing scheduling. However, no standard synthesized netlist is available for these designs to use in evaluating stages such as placement and routing. The distributed format of these benchmarks is an image showing the sequence graph. No hardware description language (HDL) is known to the authors which has been used to encode the graph - the image is the canonical definition. Both of these sets are frequently cited at the original URLs - both of which are no longer directly available, nor are they publicly archived. Recent papers increasingly cite older papers that use the benchmarks rather than the original URLs.

The TUM set is limited by its focus on benchmarks that represent a single class of design. The benchmarks are all of designs utilizing a circular reaction/mixer structure. The benchmarks themselves are taken from peer reviewed papers with clear device images and operational descriptions which can be transcribed. The example netlists given on the Cloud Columba website [15] utilizes a custom HDL, which specifies two possible components, the circular reactor and a reservoir. However, the examples are not an exact match to the devices presented in the source papers. Simplifications were made to the device layout, particularly in combining control and pump lines. Some simplifications appear to change the function of the design from the procedure described in the original paper.

A set of benchmarks *Parchmint* has been proposed as a common standard[2]. Parchmint provides netlists in a consistent JSON format. JSON does have an advantage of being easily read by libraries in any programming language. The benchmarks can also be found in the MINT HDL netlist format. The suite consist of benchmarks aggregated from those used in the authors' previous works. This consist largely of benchmarks originally designed for the Fluigi design flow [16]. Parchmint also proposes a synthesized netlist of some of the DTU synthetic benchmarks. However it is unclear which of the

seven designs presented in Parchmint is associated with which of the five designs in the original DTU and what modifications where made. Several additional benchmarks were transcribed from device images presented in assay design papers, including one[14] also used in the TUM set. The transcriptions do not appear to reflect the exact structure presented in the original paper, but have simplifications or modifications. At time of writing, only one paper was found to cite and use the Parchmint benchmarks, also written by the primary author of the benchmark suite[17].

C. Necessary characteristics

The literature survey identified several important characteristics:

1) Applicability. The Duke and DTU sets are operation sequence graphs, appropriate as input to a high level synthesis flow. The TUM and Parchmint sets are netlists, appropriate as input to placement/routing and manufacturing flows. The two types are not targeting the same goals.

2) Availability. The Duke and DTU sets were referenced by online location and became unavailable when the original website was no longer hosted. A copy of the Duke set has since been shared online[8], and the DTU sets were reprinted in a dissertation[9]. Benchmarks transcribed from fluidic devices presented in published papers could be retranscribed as long as the papers remain accessible.

3) Reproducibility. A majority (56%) of surveyed papers used benchmarks unique to the work with no way to reproduce the benchmark.

4) Consistency. Transcribed benchmarks from the TUM and Parchmint sets did not match the exact structure presented in the original device image.

5) Accuracy. Benchmarks need to capture design constraints. Transcriptions should accurately match dimensions in the original work. The MINT and Columba HDLs as well as the Parchmint JSON structure include dimensional information for channels and components, necessary for accurate representation. None of these formats are capable of capturing the temporal information inherent to the operation sequence graphs from DTU and Duke. Because valve control logic largely lies external to the bounds of the chip, this information must also be captured.

6) Interoperability. Research groups developed benchmarks that would be consistently used within their own work, but rarely used in work by other groups. Project specific HDLs inhibit access without tools to convert between formats.

7) Utility. Benchmarks should demonstrate the key figures of merit. While inspired by real assays, the Duke and DTU benchmarks are only partially representative of actual assays. The TUM benchmarks do not provide a sufficient variety to demonstrate a range of capabilities. The use of a limited set of benchmarks could lead

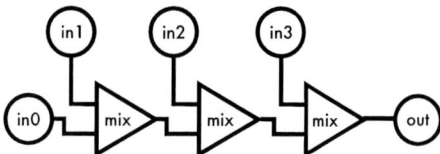

Fig. 1. Example of an n-stage chain of mixers. Benchmark challenges linearly scaling the number of components and ports.

to overfitting algorithms to the benchmarks rather than actual use cases.

8) Complexity. While the PCR benchmark is the most frequently used, it consists of only 7 cascaded mixing operations. This is valuable as a baseline, but does not demonstrate the complexity of design that MFDA tools are promising to handle[1].

V. PROPOSED BENCHMARK METHODOLOGY

We now propose our physical design benchmark suite. The suite consists ten benchmarks designed to challenge figures of merit in the automation of physical design. A summary of the new proposed benchmarks and their features is presented in Table III

The repository also includes a transcription of the Duke and DTU sequence graphs. Transcriptions from the original designs are included for benchmarks discussed in section IV and the Parchmint benchmarks. Designs are rendered when possible in the Columba HDL[15], MINT HDL, Parchmint[2], and Verilog.

The benchmarks are distributed as a set of netlists through a GitHub repository.

A. Functional descriptions

The *complete* and *complete-bipartite* benchmarks are complete graphs - all vertices are connected with an edge to every other vertices. Each vertex is a port, and each edge connects a pass through connection directly between ports.

The *chain* benchmark is a sequence of small chambers connected in series. One input port sources the first element in the chain, and one port sinks the last element.

The *mixer-chain* is a sequence of two-to-one mixers connected in series (Figure 1). One input to the mixer comes from the previous stage, and the other input comes from an input port. One input port sources the first element in the chain, and one port sinks the last element.

The *binary-tree* is a full balanced binary tree of depth n. Each node is a two-to-one mixer. There is one output port, and 2^n input ports. Each input port is connected to one input of a mixer.

The *gradient-generator* implements an i input o output gradient generator (Figure 2. Each input port is connected to a serpentine channel. Serpentine channels are place in layers. Each layer has one more layer serpentine than the previous layer. Each serpentine in a layers is connected by a shared channel. Each output is connected to an serpentine connected to the final shared channel.

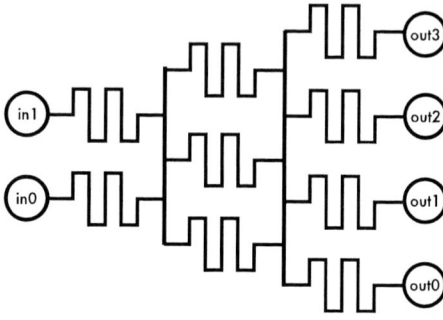

Fig. 2. Example of a 3-stage 2-input gradient generator. Benchmark designed to challenge geometric growth of area utilization.

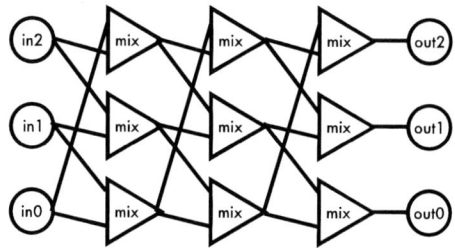

Fig. 3. Example of a 3-wide 3-long braid. Benchmark is designed to challenge linear scaling with mild congestion.

The *braid* implements repeated layers of gates with interwoven fanout (Figure 3). Each layer consists of w nodes with fan-in 2 and fan-out 2. Each node i, fans out into the next layer nodes i and $i + 1 \mod w$. The braid is fed by w source ports and w sink ports.

The *multiplexer* implements a full balanced 2^n-to-one multiplexer tree. A core two-to-one multiplexer is made of two valves fed from two input ports, connected to a shared output channel. The larger multiplexer is constructed with a tree of core multiplexers. Each layer of the tree has two control lines shared among all multiplexers in the layer.

The *crossbar* consists of two sets of channels fed from ports, a vertical and horizontal set. Each horizontal channel is connected through a valve to each vertical channel. Connections can be made between a vertical port and a horizontal port by activating the intersecting valve.

B. Design methodology

The largest benchmarks currently in use are small compared to contemporary expectations for EDA. We identified a need to test tools at different design scales and for different design constraints. Benchmarks were selected with different scaling patterns.

Structures were chosen to challenge specific patterns and cases for metrics. The *complete* and *complete-bipartite* graphs were chosen to challenge the densest possible routing. A mix of planar and non-planar graphs were selected to challenge different levels of edges crossing. Trees are commonly found

979-8-3315-3968-9/24 $31.00 © 2024 IEEE

Name	Vertices	Edges	IO	Challenge
Complete bipartite $K_{n,m}$	0	$2mn$	$n+m$	dense routing, high fanout degrees, high intersections, symmetric
Complete graph K_n	0	n^2	n	dense routing, high fanout degrees, high intersections, symmetric
Chain n	n	$n+2$	2	long planar paths, low IO
n-stage mixer chain	n	$2n$	$n+2$	long planar paths, linear component count, high IO/component connection.
Binary tree n	2^n-1	2^n-1	2^n+1	symmetric planar connections, exponential component count
Gradient Generator i,o	$\sum_{k=i}^{o} k$	$o-i$	$i+o$	symmetric planar connections, exponential component count
Braid w,l	wl	$2wl$	$2w$	linear component count, non-planarity, low intersection
Multiplexer n	2^n-1	2^n-1	2^n+2n+1	planar, linear scale control line, exponential scale component count
Crossbar m,n	mn	$4mn$	$m+n+mn$	non-planar, geometric scale control line and component count, symmetric
Nucleic Acid Processor [12] n	$16n+6$	$14n+25$	$24+n$	typical circular reactor circuit, n parallel reactors

TABLE III

PROPOSED SYNTHETIC BENCHMARKS FOR PHYSICAL DESIGN.

structures, with the *multiplexer* and *gradient generator* representative of practical usage.

Benchmarks are intended to be reproducible from the description in this paper. Common recognizable graph types were used where applicable. Random graphs were avoided - beyond the availability and reproducibility problems, designs are generated in structured ways by human designers, not randomly. The proposed benchmarks do suffer from artificial symmetry and regularity due to the synthetic origin. Additional assay transcriptions are needed, but beyond the scope of this work.

The nucleic acid processor benchmark[12] is representative of the class of design. As specified in the original source, the design already has all control lines shared between parallel stages, with no simplification needed. Scaling the design by increasing parallel stages will not change the structure presented in the source.

VI. CONCLUSION

MFDA is still a growing discipline, and our survey of the literature shows the use of benchmarks and figures of merit to be fragmented and inconsistent. Previous attempts to create a shared benchmark suite or consistent figures of merit have so far been unsuccessful.

In this work we have proposed a methodology for measuring figures of merit for MFDA tools, with particular attention to the needs of microfluidics designers and emerging manufacturing technologies. Our proposed benchmark suite aims to address the issue of benchmarking design scaling in physical design. We have also included transcriptions of the major benchmarks used in the literature. Future work is needed, particularly in the area of accurate transcription of a broad range of published assays.

We hope that our efforts here improve the accessibility of the current benchmarks in use to the community. Benchmarks in the major HDLs and comparative results for major MFDA tools are distributed through GitHub at https://github.com/utah-MFDA/mfda_benchmarks

REFERENCES

[1] X. Huang, T.-Y. Ho, *et al.*, "Computer-aided design techniques for flow-based microfluidic lab-on-a-chip systems," *ACM Computing Surveys*, vol. 54, no. 5, pp. 1–29, 2022. DOI: 10.1145/3450504.

[2] B. Crites, R. Sanka, *et al.*, "Parchmint: A microfluidics benchmark suite," in *2018 IEEE International Symposium on Workload Characterization (IISWC)*, 2018, pp. 78–79.

[3] R. Sanka, B. Crites, *et al.*, "Specification, integration, and benchmarking of continuous flow microfluidic devices: Invited paper," in *2019 IEEE/ACM International Conference on Computer-Aided Design (ICCAD)*, 2019, pp. 1–8. DOI: 10.1109/ICCAD45719.2019. 8942171.

[4] J. McDaniel, W. H. Grover, and P. Brisk, "The case for semi-automated design of microfluidic very large scale integration (mVLSI) chips," in *Design, Automation & Test in Europe Conference & Exhibition (DATE), 2017*, 2017, pp. 1793–1798. DOI: 10.23919/DATE.2017.7927283.

[5] L. A. Pradela Filho, T. R. L. C. Paixão, *et al.*, "Leveraging the third dimension in microfluidic devices using 3d printing: No longer just scratching the surface," *Analytical and Bioanalytical Chemistry*, 2023. DOI: 10.1007/s00216-023-04862-w.

[6] H. Gong, A. T. Woolley, and G. P. Nordin, "3d printed high density, reversible, chip-to-chip microfluidic interconnects," *Lab on a Chip*, vol. 18, no. 4, pp. 639–647, 2018. DOI: 10.1039/C7LC01113J.

[7] K. A. Gopinathan, A. Mishra, *et al.*, "A microfluidic transistor for automatic control of liquids," *Nature*, vol. 622, no. 7984, pp. 735–741, 2023. DOI: 10.1038/s41586-023-06517-3.

[8] "Benchmarks — microfluidics." (), [Online]. Available: https://microfluidics.cs.ucr.edu/dmfb_static_simulator/benchmarks.

[9] W. H. Minhass, "System-level modeling and synthesis techniques for flow-based microfluidic very large scale integration biochips," Ph.D. dissertation, 2012.

[10] J. S. Marcus, W. F. Anderson, and S. R. Quake, "Microfluidic single-cell mRNA isolation and analysis," *Analytical Chemistry*, vol. 78, no. 9, pp. 3084–3089, 2006. DOI: 10.1021/ac0519460.

[11] C. Fang, Y. Wang, *et al.*, "Integrated microfluidic and imaging platform for a kinase activity radioassay to analyze minute patient cancer samples," *Cancer Research*, vol. 70, no. 21, pp. 8299–8308, 2010. DOI: 10.1158/0008-5472.CAN-10-0851.

[12] J. W. Hong, V. Studer, *et al.*, "A nanoliter-scale nucleic acid processor with parallel architecture," *Nature Biotechnology*, vol. 22, no. 4, pp. 435–439, 2004. DOI: 10.1038/nbt951.

[13] A. R. Wu, J. B. Hiatt, *et al.*, "Automated microfluidic chromatin immunoprecipitation from 2,000 cells," *Lab on a Chip*, vol. 9, no. 10, pp. 1365–1370, 2009. DOI: 10.1039/b819648f.

[14] B. Li, L. Li, *et al.*, "A smartphone controlled handheld microfluidic liquid handling system," *Lab on a Chip*, vol. 14, no. 20, pp. 4085–4092, 2014. DOI: 10.1039/C4LC00227J.

[15] T.-M. Tseng, M. Li, *et al.*, "Cloud columba: Accessible design automation platform for production and inspiration: Invited paper," in *2019 IEEE/ACM International Conference on Computer-Aided Design (ICCAD)*, 2019, pp. 1–6. DOI: 10.1109/ICCAD45719.2019. 8942104.

[16] H. Huang and D. Densmore, "Fluigi: Microfluidic device synthesis for synthetic biology," *ACM Journal on Emerging Technologies in Computing Systems*, vol. 11, no. 3, 26:1–26:19, 2015. DOI: 10.1145/2660773.

[17] B. Crites, C. Falzone, *et al.*, "Reducing microfluidic very-large-scale integration (mVLSI) chip area by seam carving," *IEEE Transactions on Computer-Aided Design of Integrated Circuits and Systems*, vol. 40, no. 10, pp. 2104–2116, 2021. DOI: 10.1109/TCAD.2020. 3033499.

Understanding Transistor Aging Impact on the Behavior of RRAM Cells

Seyed Hossein Hashemi Shadmehri[1], Supriya Chakraborty[1], Thiago Santos Copetti[1],
Fabian Luis Vargas[2] and Letícia Maria Bolzani Poehls[1]

[1] Chair of Integrated Digital Systems and Circuit Design, RWTH Aachen University, Germany
[2] IHP - Leibniz Institute for High Performance Microelectronics, Frankfurt (Oder), Germany
{hashemi, chakraborty, copetti, poehls}@ids.rwth-aachen.de, vargas@ihp-microelectronics.com

Abstract—**Resistive Random Access Memories (RRAMs) have been considered for implementing various emerging applications alongside storage. RRAMs offer the opportunity to address the Von Neumann bottleneck making the implementation of In-Memory Computing (IMC) possible. However, the exploration of RRAM potentials depends on being able to guarantee their reliability during lifetime. In this context, this paper investigates the impact of transistor aging on the behavior of 1T1R RRAM cells. A case study composed of an RRAM block, including peripherals, implemented using the 22 nm FDSOI GF technology is adopted. The obtained results indicate that the transistor aging, which was previously widely ignored, can lead to parametric degradation or even catastrophic faults. In addition, the obtained results show that the transistor type used for implementing the 1T1R RRAM cell, core or IO device, and operating temperature play important roles when analysing aging impact.**

Index Terms—**Resistive Random Access Memory, Aging, Reliability**

I. Introduction

Emerging applications, such as autonomous driving, image recognition and genomic data analysis, rely on large amounts of data. In a conventional Von Neumann computer, this data needs to be transferred between dedicated computation and memory units. Data movements are very costly in terms of delay and power consumption, leading to a severe loss of performance and energy efficiency [1]. In-Memory Computing (IMC) is a solution aimed at alleviating the Von Neumann bottleneck by integrating the computation inside the memory. There have been various attempts at realizing IMC, with some efforts augmenting the conventional memory types such as Flash with additional circuitry to embed processing capabilities [2]. Emerging Non-Volatile Memory (eNVM) technologies have also been heavily researched for their potential to enable IMC. Among them, Resistive Random Access Memories (RRAMs) have become one of the most promising candidates for this purpose. RRAMs can provide better latency and reduced energy consumption compared to Flash [3]. Furthermore, one can arrange RRAMs in a cross-point structure to facilitate $\mathcal{O}(1)$ execution of Vector Matrix Multiplication (VMM), which is at the core of neural network computations [4]. The most prevalent implementation of an RRAM cell is composed of one transistor and one memristive device (1T1R). In order to properly explore the potential of RRAMs, their reliability during lifetime needs to be assured. Reliability can be directly affected by temporal variations, a type of time-dependent deviations, which degrade circuits during lifetime depending on the circuit's workload and operating conditions. Consequently, understanding how the aging of the transistor present in the RRAM cell may degrade the memory lifespan becomes essential. It is worth noting that previous studies have only focused on analysing the novel device's endurance itself, not considering the possible degradation of the transistor and the peripheral circuitry [5]–[9]. In this paper, an analysis of the possible transistor aging impact on the behavior of 1T1R RRAM cell is presented. The main contributions of this work are listed below:

- Development of an RRAM block using the 22 nm FD-SOI GF technology and the memristive device model described in [10].
- Comparison of the transistor aging impact on the RRAM cell when using IO and core devices as access transistors, across different temperatures.
- Analysis of the combined effect of transistor and memristive device aging on 1T1R RRAM cell.
- Exploration of body biasing as an approach for minimizing aging impact on 1T1R RRAM cell.

The rest of the paper is organized as follows: Section II offers a brief overview of failure mechanisms that can affect CMOS and memristive devices during lifetime. In Section III, the implemented case study and the adopted experimental setup are described. Section IV presents and discusses the obtained results and finally, Section V concludes this paper.

II. Background

In this section, the failure mechanisms affecting CMOS and memristive devices during lifetime are summarized.

A. Transistor Degradation during Lifetime

The susceptibility of CMOS technology to temporal variations caused by phenomena, such as Hot Carrier Injection (HCI), Bias Temperature Instability (BTI) and Time-Dependent Dielectric Breakdown (TDDB), represents an important concern that becomes more prominent with technology scaling [11]. HCI occurs when charge carriers accelerated through channel area by the lateral field, gain enough energy to penetrate into the gate, creating traps at the gate-channel interface [12]. In general terms, BTI is caused by the stress induced by the vertical field in the presence of a zero to small

979-8-3315-3968-9/24 $31.00 © 2024 IEEE

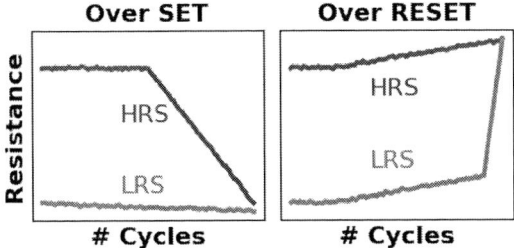

Fig. 1: Two types of RRAM endurance degradation [13].

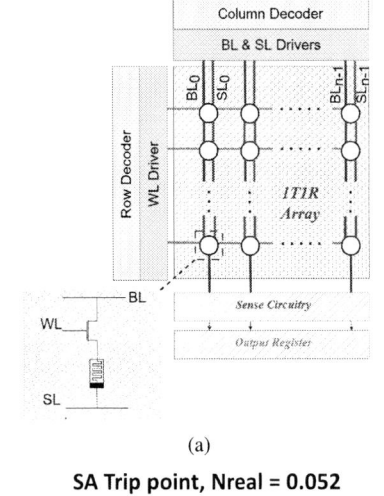

(a)

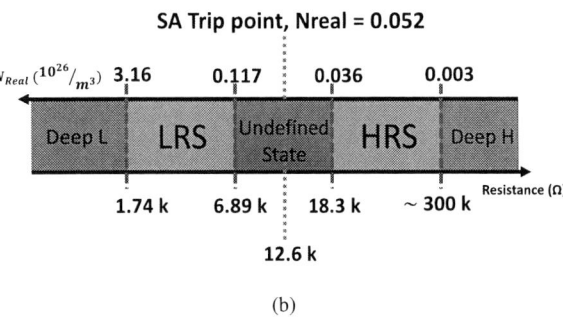

(b)

Fig. 2: (a) The RRAM block diagram and (b) N_{real} and corresponding resistive states: Deep L, LRS, Undefined State, HRS and Deep H.

lateral field. BTI can cause the Si-H bonds at the gate-channel interface to break and thus create charge traps. Depending on the considered BTI model, it is possible to assume that when the stress is removed, a recovery phase starts, decreasing the impact caused by this phenomenon. This recovery phase is not observed when considering the impact of HCI [12]. Finally, TDDB is caused by the formation of a conductive path inside the gate dielectric, destroying its role as an insulator. BTI and HCI degrade the CMOS devices in a gradual manner, while TDDB causes a sudden catastrophic failure [11]. It should be noted that in this paper BTI and HCI are the only degradation mechanisms being considered. BTI and HCI present similar behavior, being both caused by charge trapping in the gate dielectric, which causes the threshold voltage (V_{th}) to increase and carrier mobility to decrease, thereby reducing the current (I_{on}) and increasing the delay [11].

B. Memristive Device Degradation during Lifetime

In a Valence Change Mechanism (VCM)-based RRAM device, the formation and dissolution of a Conductive Filament (CF) governs the resistance. If oxygen ions are extracted from the intermediate layer and stored in the anode, the oxygen vacancies create a CF putting the device in Low Resistance State (LRS). If the stored ions are moved back and are recombined with the vacancies, the CF ruptures and the device will be in High Resistance State (HRS) [14]. The switching from HRS to LRS is called SET, and occurs when applying a V_{SET} voltage to the device. In contrast, the switching from LRS to HRS is called RESET, caused by applying a V_{RESET} voltage of opposite polarity to the device.

As observed in CMOS devices, RRAM devices are also susceptible to temporal variations that degrade their electrical parameters during lifetime. An analytical model for endurance degradation, based on measured data was proposed in [13]. This model includes two types of endurance degradation, as illustrated in Fig. 1. Over SET is the enlargement of the CF due to an excess in oxygen vacancy generation that leads to a drop in both HRS and LRS. Over RESET is the over-recombination of oxygen ions and vacancies, resulting in fewer oxygen vacancies, that cause an increase in both HRS and LRS. In both cases the conductance window shrinks and the ratio between HRS and LRS lowers significantly which is a reliability concern. The authors in [13], model endurance degradation by introducing a Δr and Δx into their RRAM

model, representing the filament radius enlargement and reduction in minimum number of oxygen vacancies, respectively. It should be noted that Δr and Δx are applied at the same time and may affect the device in opposite ways.

III. EXPERIMENTAL SETUP

In order to understand the impact of temporal variations on the behavior of 1T1R RRAM cells, a case study implemented using the 22 nm FDSOI technology from GF and the memristive device model available in [10] was adopted. Fig. 2a depicts the case study's block diagram. SET operation is performed by driving the Bit Line (BL) to V_{set} and grounding the Select Line (SL), while RESET is performed by driving SL to V_{reset} and grounding the BL. During read, BL is driven to a small V_{read} voltage, while SL is grounded and the sensing circuitry makes a decision by comparing the bit cell's current with a reference current. Fig. 2b illustrates the boundaries adopted for representing the possible RRAM resistive states. In more detail, N_{real}, the number of oxygen vacancies per cubic meter in the disc region [10], and the respective resistance value for each possible resistive state (Deep L, Low Resistance State - LRS, Undefined State, High Resistance State - HRS and Deep H) are depicted. Higher N_{real} indicates a higher conductance (lower resistance).

TABLE I: RRAM block specifications

	Case Study 1	Case Study 2
Block size	8×8	8×8
Cell Type	1T1R with IO transistor	1T1R with core transistor
$\mathbf{V_{WL}}$	2.0 V	1.2 V
$\mathbf{V_{read}}$	0.4 V	0.4 V
SET	$V_{set} = 1.2$ V for $1\mu s$	$V_{set} = 1.2$ V for $50\mu s$
Body Bias	$V_{BS} = 0$ V	$V_{BS} = 800$ mV
RESET	$V_{reset} = 2.5$ V for $20\mu s$	$V_{reset} = 1.3$ V for $170\mu s$

It is important to highlight that two different 1T1R RRAM cells were implemented: The first one using IO devices and the second one using core logic devices. Table I summarizes the case study's specifications for the RRAM blocks implemented based on both device types. There is an obvious trade-off between the applied voltages during SET and RESET, and the time required for performing both operations. In many cases, this leads to the access transistor, 1T present in the RRAM cell, being over-driven so that a suitable operation speed is achieved [15]. It should be noted that the experiments in section IV-A and IV-B are performed by solely considering transistor degradation. However, the results summarized in section IV-C, combine the impact of transistor aging with memristive device degradation, allowing a more comprehensive analysis of 1T1R RRAM cell. In this work, Cadence Relxpert tool was used to simulate the effects of BTI and HCI. Furthermore, in this work, we use the RRAM model introduced in [10] named JART VCM v1b and according to [16]; the endurance failure observed for this device is stuck-at-LRS, meaning a degradation of the HRS state happens. Hence, we will focus on the over SET case. The parameter in JART model representing filament radius is r_{det} and thus increasing r_{det} creates the over SET (HRS degradation) effect.

IV. RESULTS AND DISCUSSION

This section summarizes the results obtained when considering the RRAM block implemented using IO and core logic devices as access transistors of the 1T1R RRAM cell. In addition, this section also analyses the aging impact on 1T1R RRAM cells, assuming the degradation of not only transistors, but also the memristive device.

A. IO Device Aging Impact

Considering bit cells implemented using IO devices, Fig. 3a shows the result for 5 and 10 years of aging caused by HCI and BTI at three different temperatures: $-40°C$, $25°C$, and $125°C$. The analysis was conducted by executing the operation sequence 0w1r1w0r0. Fig. 3b illustrates the effect of CMOS degradation on SET operation under different temperatures. The figure shows that at room temperature ($25°C$), aging increases the resistance achieved after SET by 0.45% after 5 and 0.57% after 10 years (with respect to fresh case). A similar behavior, but with less significance, can be observed when considering $-40°C$. In this case, the resistance increase after 5 years is around 0.11% and 0.14% after 10. Finally, the most noticeable impact occurs when considering a temperature of $125°C$. In this case, the rise in resistance is 0.85% after 5 years and 1.08% after 10 years.

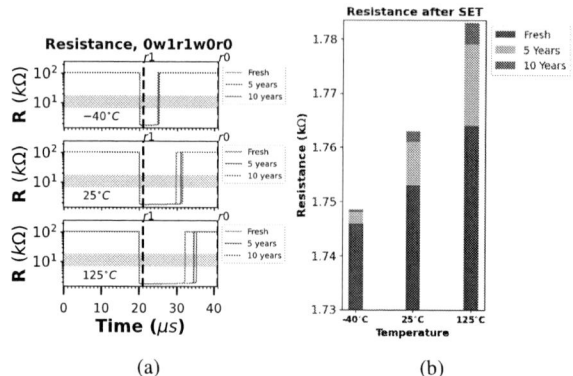

Fig. 3: Case study using IO devices (0w1r1w0r0): (a) Resistance vs Time (gray area shows region related to the Undefined State) (b) Resistance after w1 (SET operation).

TABLE II: Comparison of the RESET times (T_{reset}).

	-40°C		**25°C**		**125°C**	
	time	Δ	time	Δ	time	Δ
Fresh	$3.79\mu s$	-	$8.70\mu s$	-	$11.18\mu s$	-
5 years	$4.05\mu s$	6.9%	$10.02\mu s$	15.2%	$13.51\mu s$	20.8%
10 years	$4.12\mu s$	8.7%	$10.47\mu s$	20.3%	$14.27\mu s$	27.6%

In Fig. 3b it is observed that the resistance achieved after w1 operation was only slightly affected by the access device aging. However, the time required by the RESET operation (T_{reset}) was impacted more significantly. Table II summarizes the relation between access device aging and T_{reset}. In more detail, the access device aging increases T_{reset} and the effect is more pronounced at higher temperatures. For example, 10 years of aging would result in a 27.6% increase in T_{reset} at $125°C$, while the figure for $-40°C$ is only 8.7%.

B. Core Logic Device Aging Impact

The case study discussed in section IV-A was implemented using IO devices. A notable advantage of using IO devices is that they can tolerate much higher voltages before being affected by TDDB. However, they have the obvious disadvantage of being much bigger than core logic devices and hence, greatly sacrificing density. Also, the forming process has been a hurdle in the way of core device utilization in RRAMs, since the process requires high voltages to be applied. However, in recent years, methods have been proposed for reducing the forming voltage. For example, the authors in [17] propose to reduce the forming voltage by either increasing the temperature or the rise time of the forming pulse, while [18] suggests using bipolar incremental step pulse programming. Developing such techniques has become a part of RRAM research. Additionally, works like [19] and [20] represent efforts towards demonstrating forming-free RRAMs. Also, the forming step is done only once and it lasts a short time. Furthermore, [21] by Infineon states that the lifetime of the core logic devices used as access transistors is a reliability concern because the voltages applied during SET/RESET are much greater than the permitted values for logic implementation. Thus, in this work we assume that the forming voltage is

979-8-3315-3968-9/24 $31.00 © 2024 IEEE

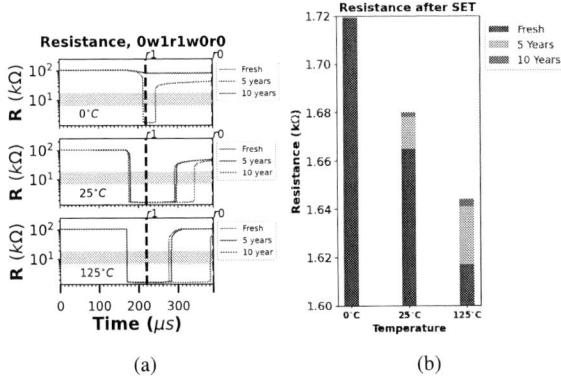

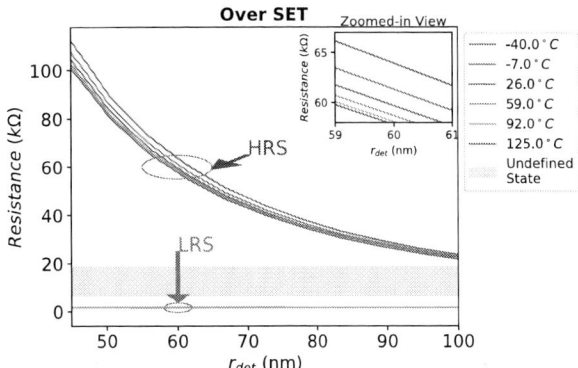

Fig. 4: RRAM block using core devices: (a) Resistance vs Time (gray area shows region related to the Undefined State), and (b) Resistance after w1, SET operation.

Fig. 5: Modeling over SET behavior by sweeping the r_{det} parameter from JART VCM v1b model [10].

not going to be an issue for the core devices and focus on the post-forming operations (SET and RESET) of the 1T1R RRAM cell.

Fig. 4a illustrates the resistance of the 1T1R RRAM cell while performing the operation sequence 0w1r1w0r0, across three different temperatures. Considering the operation at 0°C, it is possible to observe a fault that will be propagated to the output of the block, since the RRAM cell was not able to properly switch. It should be noted that when using IO devices in the 1T1R cell, no functional fault was observed and transistor aging was merely able to cause parametric faults. Another important aspect to be considered is that the minimum considered temperature in case of core access device is 0°C, in contrast to -40°C that was used in section IV-A. The reason is that the SET operation has a positive feedback with temperature. In other words, the heat generated in the RRAM device by applying the SET pulse makes the SET process happen faster. Thus, SET operation starts more easily in high temperatures and is more difficult in low temperatures [10]. Hence, when adopting core devices to implement 1T1R cells, the SET operation in low temperatures becomes even more difficult and in this case it would be time prohibitive in sub-zero temperatures. SET and RESET can be divided into two steps, a slow conductance change step followed by a fast one. In case of SET, during the slow conductance change step, N_{real} rises (resistance drops) slowly and then during the fast conductance change step an abrupt rise in N_{real} occurs. The time needed for the first step of the process is called delay time and the time required for the second step is called transition time [10].

Fig. 4b shows the resistance of the target RRAM cell after the SET operation (after the end of w1 in the operation sequence). Comparing Fig. 4b with Fig. 3b, opposite trends in terms of resistance after the SET operation can be observed, when considering transistors with no aging (fresh). In the case of IO access device (Fig. 3b), the resistance achieved after SET increases with temperature (albeit slightly). In contrast, in case of 1T1R RRAM cells using core devices (Fig. 4b), the resistance achieved after SET decreases with temperature. This

can be attributed to the portion of SET time (T_{set}) belonging to delay time or transition time. For the case of IO device, the higher voltage used causes the delay time to be very small and most of the T_{set} being transition time. In case of core device, however, the device shows a noticeable delay time at 0°C but an insignificant one at 125°C. In other words, the temperature effect is more visible when employing core devices (low voltages). In case of IO device, there is an 18Ω difference between the resistance after w1 at -40°C and 125°C when considering no aging. Meanwhile, the figure for core device is 102Ω between 0°C and 125°C.

A further analysis of the results depicted in Fig. 4a and Fig. 3a, shows that the trends for T_{reset} also differ depending on the type of adopted transistors (IO or core device). In more detail, when using IO devices for implementing the 1T1R cell, the time required for RESET operation increases with aging. However, the opposite behavior is observed when using core devices. This occurs due to the fact that the switching dynamics of the RRAM is not only voltage or temperature dependant. The switching dynamics is also state dependant [10], meaning that the state of the RRAM cell (stored resistance) before performing the SET/RESET operation affects the outcome and speed of the operation. In more detail, Fig. 4a shows that the higher resistance of the RRAM cell after w1 in case of aged transistors, leads to lower T_{reset} and consequently, the RRAM cell with an aged transistor is able to go to HRS sooner.

C. Combined RRAM and Transistor Aging Impact

As discussed in section II-B, RRAM aging is characterized by a shrinkage of the conductance window. Furthermore, considering the adopted case study, the aging effect on the HRS is much more significant than that on the LRS. Hence we consider the case of over SET, which is modeled by sweeping the r_{det} variable (filament radius) of the JART model. Fig. 5 shows the effect of r_{det} on both HRS and LRS.

Once again we consider the 1T1R RRAM cell with the IO devices and the operation sequence 0w1r1w0r0. Fig. 6a shows the resistance of the RRAM device after SET and the time required for RESET operation (T_{reset}), at different temperatures and for various filament sizes, in the case of

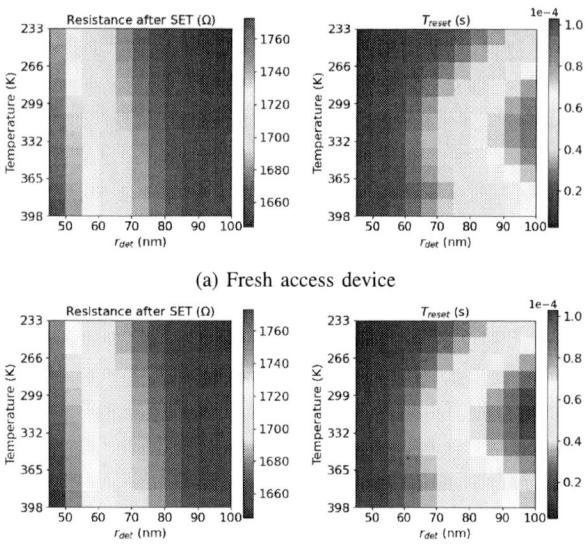

(a) Fresh access device

(b) 10 year aged access device

Fig. 6: RRAM block with IO devices: Resistance after SET and T_{reset} at different temperatures and various CF sizes when performing 0w1r1w0r0. (a) Fresh access device and (b)10 year aged access device.

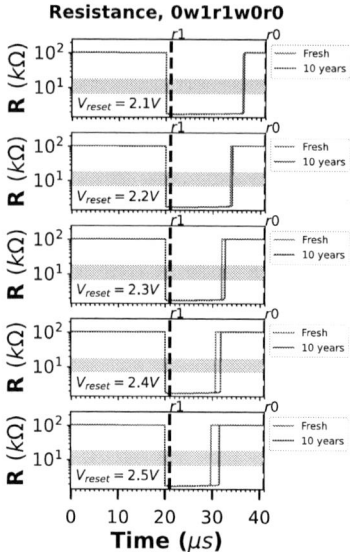

Fig. 7: RRAM block with IO devices: Sweeping the V_{reset}, while executing 0w1r1w0r0.

a fresh access device. In other words, in this case only the RRAM aging has been considered. It should be noted that filament radius was swept from the nominal value of 45 nm to the maximum value of 100 nm allowed by JART VCM v1b model [10]. It can be observed that having a bigger filament radius leads to achieving a lower resistance after the SET operation but at the same time it makes the following RESET operation more difficult. An increase in temperature leads to higher resistance after SET. From equal starting points, an increase in temperature also increases T_{reset}, however here we see the opposing effects of initial state (resistance after previous SET) and temperature. This is why in some cases we observe maximum (T_{reset}) at intermediate temperatures. Fig. 6a captures the interplay between filament radius (representing HRS degradation) and temperature change. A combination of high temperature and small filament radius is the worst case for the resistance after SET, while the combination of intermediate temperatures (42°C here) and big filament radius is the worst case for (T_{reset}).

Fig. 6b considers the case which the access device has experienced 10 years of aging. It can be seen that while the trends discussed above for r_{det} and temperature remain true, there is an overall increase in both the resistance after SET and the required time for the following RESET operation. This suggests that only considering RRAM aging in studying the reliability of the memory blocks, could lead to an underestimation of the aging effects.

D. Exploring Body Biasing

FDSOI is especially suitable for body biasing schemes thanks to the Buried Oxide (BOX) isolating the channel from the substrate, allowing the application of a wider range of body bias [22]. Considering the case study implemented using IO devices, Fig. 7 illustrates the effect of sweeping V_{reset}, while executing the operation sequence 0w1r1w0r0 at room temperature (25°C). It can be observed that decreasing V_{reset} reduces the difference between behavior of the fresh and aged RRAM cells. This behavior is expected, since the rates of mobility and V_{th} degradation due to BTI and HCI depend on the stress conditions. However, this robustness against aging has come at the price of reducing performance, such that there is an increase of 21.65% in the T_{reset}, when comparing the case study assuming a $V_{reset} = 2.5$ V with one assuming $V_{reset} = 2.1$ V.

Fig. 8 shows the comparison between the original RRAM block using IO devices (see specification characteristics in Table I) and the new RRAM block adopting $V_{reset} = 2.1$ V with forward body biasing of 250 mV during w0. Observing Fig. 8, it is visible that body biasing has elevated the performance with $V_{reset} = 2.1$ V, such that it's T_{reset} almost matches that of $V_{reset} = 2.5$ V with no body bias. It should also be noted that body bias was only applied during w0, because it's application during w1 would have lead to gathering more oxygen vacancies in the disc region (N_{real}), which in turn makes RESET more difficult. Table III summarizes the information regarding the application of body biasing.

V. CONCLUSION

In this paper, an analysis of the transistor aging impact on the behavior of 1T1R RRAM cells was presented. Two different case studies, one considering IO devices and another one considering core logic devices for implementing the access transistors of RRAM cells, were developed. The obtained results demonstrated that despite providing better density, core logic devices show slower operation and greater reliability

979-8-3315-3968-9/24 $31.00 © 2024 IEEE

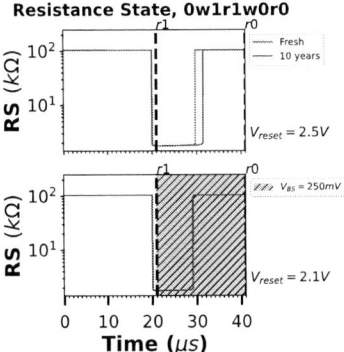

Fig. 8: RRAM block with IO devices, Top: Case Study 1 (Table I) vs. Bottom: Exploring body biasing and reduced voltage levels.

TABLE III: Comparison of the RESET times (T_{reset}).

	Fresh	10 years	Δ
$V_{reset} = 2.5V$ & $V_{BS} = 0V$	$8.7\mu s$	$10.47\mu s$	20.3%
$V_{reset} = 2.1V$ & $V_{BS} = 0V$	$15.12\mu s$	$15.37\mu s$	1.65%
$V_{reset} = 2.1V$ & $V_{BS} = 0 - 250mV$	$8.05\mu s$	$8.12\mu s$	0.87%

issues. In addition, the performed simulations show greater degradation in performance when considering combined effects of transistor and memristive device aging, compared to considering only RRAM endurance degradation. Preliminary results show that body biasing can be explored as an alternative for reducing the stress on the access transistor of the 1T1R RRAM cells, increasing cell's robustness against temporal variations.

ACKNOWLEDGMENTS

This work was supported in part by the Federal Ministry of Education and Research (BMBF, Germany) within the NEUROTEC project (project numbers 16ES1134 and 16ES1133K) and in part by the German Federal Ministry of Education and Research (Clusters4Future-NeuroSys) under Grant 03ZU1106CA.

Funded by the European Union

This project has received funding from the European Union's Horizon Europe research and innovation programme under grant agreement No. 101160182. Views and opinions expressed are however those of the author(s) only and do not necessarily reflect those of the European Union or the European Health and Digital Executive Agency (granting authority). Neither the European Union nor the granting authority can be held responsible for them.

REFERENCES

[1] F. Aguirre *et al.*, "Hardware implementation of memristor-based Artificial Neural Networks," *Nature Communications*, vol. 15, no. 1, Mar 2024.

[2] M. Hayashikoshi *et al.*, "Processing In-Memory Architecture with On-Chip Transfer Learning Function for Compensating Characteristic Variation," in *2020 IEEE International Memory Workshop (IMW)*, 2020, pp. 1–4.

[3] F. Zahoor *et al.*, "Resistive random access memory (RRAM): An overview of materials, switching mechanism, performance, multilevel cell (MLC) storage, modeling, and applications," *Nanoscale Research Letters*, vol. 15, no. 1, 2020.

[4] A. BanaGozar *et al.*, "ReMeCo: Reliable Memristor-Based In-Memory Neuromorphic Computation," in *2023 28th Asia and South Pacific Design Automation Conference (ASP-DAC)*, 2023, pp. 396–401.

[5] S. Zhang *et al.*, "Lifetime Enhancement for RRAM-based Computing-In-Memory Engine Considering Aging and Thermal Effects," in *2020 2nd IEEE International Conference on Artificial Intelligence Circuits and Systems (AICAS)*, 2020, pp. 11–15.

[6] p. pouyan *et al.*, "Memristive Crossbar Memory Lifetime Evaluation and Reconfiguration Strategies," *IEEE Transactions on Emerging Topics in Computing*, vol. 6, no. 2, pp. 207–218, 2018.

[7] Y. Cai *et al.*, "Long Live TIME: Improving Lifetime for Training-In-Memory Engines by Structured Gradient Sparsification," in *2018 55th ACM/ESDA/IEEE Design Automation Conference (DAC)*, 2018, pp. 1–6.

[8] A. Irmanova *et al.*, "Analog Self-Timed Programming Circuits for Aging Memristors," *IEEE Transactions on Circuits and Systems II: Express Briefs*, vol. 68, no. 4, pp. 1133–1137, 2021.

[9] W. Ye *et al.*, "Aging Aware Retraining for Memristor-based Neuromorphic Computing," in *2022 IEEE International Symposium on Circuits and Systems (ISCAS)*, 2022, pp. 3294–3298.

[10] F. Cüppers *et al.*, "Exploiting the switching dynamics of HfO2-based ReRAM devices for reliable analog memristive behavior," *APL Materials*, vol. 7, no. 9, p. 091105, 09 2019. [Online]. Available: https://doi.org/10.1063/1.5108654

[11] I. Hill *et al.*, "CMOS Reliability From Past to Future: A Survey of Requirements, Trends, and Prediction Methods," *IEEE Transactions on Device and Materials Reliability*, vol. 22, no. 1, pp. 1–18, 2022.

[12] J. Keane, X. Wang, D. Persaud, and C. H. Kim, "An All-In-One Silicon Odometer for Separately Monitoring HCI, BTI, and TDDB," *IEEE Journal of Solid-State Circuits*, vol. 45, no. 4, pp. 817–829, 2010.

[13] P. Huang *et al.*, "Analytic model of endurance degradation and its practical applications for operation scheme optimization in metal oxide based RRAM," in *2013 IEEE International Electron Devices Meeting*, 2013, pp. 22.5.1–22.5.4.

[14] B. Chen *et al.*, "Physical mechanisms of endurance degradation in TMO-RRAM," in *2011 International Electron Devices Meeting*, 2011, pp. 12.3.1–12.3.4.

[15] A. Levisse *et al.*, "Resistive Switching Memory Architecture Based on Polarity Controllable Selectors," *IEEE Transactions on Nanotechnology*, vol. 18, pp. 183–194, 2019.

[16] N. Kopperberg *et al.*, "Endurance of 2 Mbit Based BEOL Integrated ReRAM," *IEEE Access*, vol. 10, pp. 122 696–122 705, 2022.

[17] Y.-T. Su *et al.*, "A Method to Reduce Forming Voltage Without Degrading Device Performance in Hafnium Oxide-Based 1T1R Resistive Random Access Memory," *IEEE Journal of the Electron Devices Society*, vol. 6, pp. 341–345, 2018.

[18] H.-X. Zheng *et al.*, "Reducing Forming Voltage by Applying Bipolar Incremental Step Pulse Programming in a 1T1R Structure Resistance Random Access Memory," *IEEE Electron Device Letters*, vol. 39, no. 6, pp. 815–818, 2018.

[19] X. Ding *et al.*, "Forming-Free HfOx-Based Resistive Memory With Improved Uniformity Achieved by the Thermal Annealing-Induced Self-Doping of Ge," *IEEE Transactions on Electron Devices*, vol. 70, no. 4, pp. 1671–1675, 2023.

[20] W. Wang *et al.*, "Study on Multilevel Resistive Switching Behavior With Tunable ON/OFF Ratio Capability in Forming-Free ZnO QDs-Based RRAM," *IEEE Transactions on Electron Devices*, vol. 67, no. 11, pp. 4884–4890, 2020.

[21] C. Peters *et al.*, "Reliability of 28nm embedded RRAM for consumer and industrial products," in *2022 IEEE International Memory Workshop (IMW)*, 2022, pp. 1–3.

[22] P. Flatresse *et al.*, "Ultra-wide body-bias range LDPC decoder in 28nm UTBB FDSOI technology," in *2013 IEEE International Solid-State Circuits Conference Digest of Technical Papers*, 2013, pp. 424–425.

A New Control Law for N-Path Mixer Switches Enhancing Harmonic Rejection

Hasan MOUSSA, Estelle LAUGA-LARROZE, Laurent FESQUET
University of Grenoble Alpes, CNRS, Grenoble INP*, TIMA, 38000 Grenoble, France
* Institute of Engineering University of Grenoble Alpes

E-mail: {hasan.moussa | estelle.lauga-larroze | laurent.fesquet}@univ-grenoble-alpes.fr

Abstract— This paper presents a novel methodology to enhance harmonic rejection in N-Path Mixers thanks to a dedicated switch control law. The proposed method reduces the effective local oscillator distortion, in order to limit unwanted harmonics appearance at the mixer output. This is achieved by using a non-conventional control of the N-path mixer switches, based on level crossing sampling techniques. The methodology strength is its ability to be used with any N-path mixer, whatever the number of paths. The article details how to optimally design the switch control. To validate the effectiveness of the proposed approach, simulations have been conducted on 4-path and 8-path mixers and compared to the corresponding conventional N-path mixers. This analysis shows significant improvements in terms of harmonic rejection thanks to this new methodology.

Keywords— N-Path Mixer, Level Crossing Sampling Scheme, Control Law, Harmonic Rejection.

I. INTRODUCTION

In the dynamically evolving landscape of the wireless communications, there is a growing demand for multistandard radio receivers able to tune a wide frequency range, ensure avoidance of in-band interferences, and also withstand strong out-of-band (OOB) interferences. This requires a high selectivity and, thus high-Q filters. These latter are critical parts in wireless devices. Traditionally, achieving this level of filtering is accomplished using high-Q LC-filters or devices that exhibit similar characteristics. While these traditional devices are able to provide exceptional selectivity and linearity, they fall short in terms of the desired programmability essential for Software-Defined Radio (SDR) applications. As N-Path Mixers (NPM) exploit switches and capacitors, while offering high-Q filtering around a digitally programmable switching frequency, this explains the recent interest in NPM-first receivers for software defined radio applications [1]. NPM based receivers offer a wide tuning range and high-Q Radio Frequency (RF) filtering properties enhancing the robustness against OOB interferences.

The NPM in Fig. 1 performs a down-conversion of the RF input signal to the Intermediate Frequency (IF) by the means of a Local Oscillator (LO). Originally designed with 4 paths, these 4-phase mixers provide quadrature baseband outputs, help image rejection, and are of sufficiently low complexity to be scalable at high frequencies [1]. Moreover, their growing interest is due to advantages such as their ability to suppress the local oscillator harmonic mixing [2]. In conventional 4-path mixer-first receivers (see Fig. 2), 4-phase signals are generated from a LO. This LO signals consist of pulses with a 25% duty cycle and a 90-degree phase shift, which drive the NPM switches.

In this context, our work focuses on a novel approach to drive the switches with a non-conventional control law, unlike conventional approaches used in [2], [3], [4]. Moreover, this law is obtained thanks to a methodology defining the optimal configuration whatever the path number. The methodology exploits a non-uniform sampling technique, namely the Level Crossing Sampling Scheme (LCSS) [5], to obtain the best approximation of a sine wave with a limited processing. The purpose of this approximation is to obtain low distorted signals, since harmonics appear when the LO is not perfectly a sine wave. By appropriately using the sampling instants, a method is given for generating the required phases driving the NPM switches. This approach addresses the NPM challenges, especially the harmonic rejection.

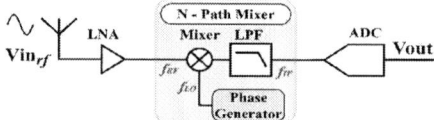

Fig. 1. RF Front-end Receiver

II. N – PATH MIXERS

A. Receiver Architecture

The principle of analog NPM circuits consists in sampling and integrating the RF input signal. An NPM is able to perform down conversion from the RF band to IF band. In addition, it also performs filtering of the IF signal through N paths (RC networks), successively selected through N switches driven by multiphase clocks. These clocks are digitally controlled and allow a fine tuning of the filter frequency. Fig. 2 shows a conventional 4-path mixer with clock-phases ($\varphi 1 \rightarrow \varphi 4$). Each phase φi is a delayed version of the LO, which is a square signal with a duty cycle equal to T/N. The signal is first down-converted via the switches then filtered by the RC low-pass filters formed by the resistances of the switches and the capacitors. After one switching cycle, the new sample coming from Vin_{rf} is hence added to the sample stored on the capacitor during the previous period. Thus, each capacitor contains the average of the part of the sine wave charging it at each cycle resulting in a low-pass filtered output signal at the IF. The In-Phase and Quadrature (IQ) demodulation is then performed by subtracting the opposite phases to obtain the I and Q components. The I and Q signals are summed to form the IQ demodulated signal.

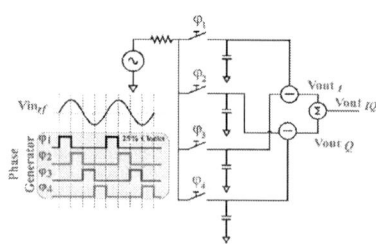

Fig. 2. Conventional N –Path Mixer

979-8-3315-3968-9/24 $31.00 © 2024 IEEE

B. Harmonics Suppression

As mentioned previously, square wave signals are used to drive the switches of the NPM, resulting in the generation of the LO frequency harmonics. The amplitude of these harmonics depends on the shape and duty cycle of the LO signal, and the filter characteristics of the NPM. These responses, mixed with the RF signal, result in out-of-band (OOB) blockers leading to the creation of unwanted frequencies, which can fold into the band of interest. Therefore, extensive researches have been conducted to mitigate these harmonics and generate a cleaner output signal spectrum [2], [4], [6]. Our novel approach, explored thereafter, generates dedicated LO signals (based on Level-Crossing Sampling) for driving the NPM switches and enhances the harmonic rejection.

III. NON-UNIFORM SAMPLING SCHEME AND METHODOLGY

A. Uniform Sampling

Conventional ADCs are mandatory blocks in data acquisition systems. They are based on a uniform sampling scheme, with a sampling frequency determined by at least twice the highest signal frequency following the Shannon theorem (for reconstruction): $f_s \geq 2.f_{max}$ (1)

B. Level Crossing Sampling Scheme

While traditional uniform sampling is well established, there are scenarios where non-uniform sampling techniques offer notable advantages. One of them is the LCSS [7], [8]. In that kind of non-uniform sampling, a predefined set of levels is used to cover the dynamic range of the input signal. Each time the continuous signal crosses one of these levels, a sample is captured along with the time elapsed since the previous captured sample, forming a pair $(x_n, \delta t x_n)$ where x_n is the signal amplitude at the nth sampling time, and $\delta t x_n$ is the time delay between the current and the previous samples, as shown in Fig. 3.

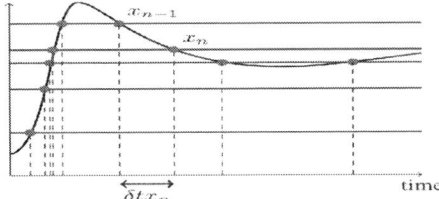

Fig. 3. Level Crossing Sampling Scheme [5]

This method contrasts with the classical uniform sampling, where time instants are known, and amplitudes are quantized; here, instants are quantized, and amplitudes are precisely known [5]. Thanks to these amplitudes and time delays, an interpolation is often used to reconstruct a continuous-time signal. Such a reconstruction is often sufficient for many applications, even at the zeroth or first order.

C. Proposed Methodology

The proposed methodology is based on the observation that harmonics appear when the LO is not perfectly a sine wave. This is typically the case with an NPM, because the switch system imperfectly emulates the LO. The resulting oscillator is called the effective LO (EFLO). This study focuses on a method, which optimally approximates a sine wave with square functions. For sake of clarity and concision, we use in the sequel a 4-path mixer.

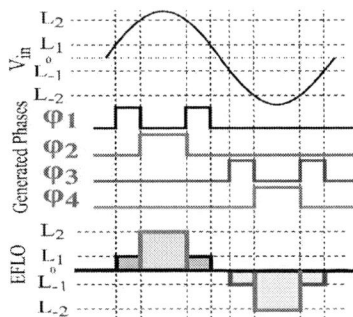

Fig. 4. Phase Generation Using LCSS and EFLO Obtained as an Approximated Sine Wave

In order to get a sinusoid approximation, a sine wave is sampled thanks to a LCSS (see Fig. 4). Let N, the number of paths, be also the threshold number used with the LCSS to determine the phases φ_1 to φ_N controlling the switches. The algorithm first explores all possible signal variations by defining a range of levels for each phase. Then, it generates every combination of these levels, ensuring no possibility is left unexamined. Then for each combination, the algorithm creates individual phase signals based on the chosen levels. On Fig. 4, notice that these generated phases are periodic signals but no longer with a duty cycle of T/N (25% for 4-path mixer). The generated phase signals are combined with their corresponding amplitudes for constructing the EFLO signal for each combination. The third step evaluates the harmonic rejection of each constructed EFLO signal. Then, the method identifies the optimal threshold placement for getting the best approximation by calculating the Mean Squared Error (MSE) between the approximated sine wave (EFLO signal) and an ideal sinusoid. This calculated MSE gives a distortion measurement of the reconstructed EFLO signal. The results reveal that the thresholds are not uniformly spaced. This can be seen on the bottom part of Fig. 4 where the threshold levels L_1, L_2, L_{-1} and L_{-2} are not evenly spaced. For the final step, the algorithm selects the levels of the best combination (lowest MSE). The winning combination is then used to generate the optimized EFLO signal. The resulting EFLO optimally mimics a sine wave built with pulse functions. It is important to notice that the threshold magnitude corresponds to an amplifier gain in front of each path.

D. Minimizing the Effective LO Distortion

The proposed method is based first on selecting the most appropriate thresholds of a LCSS in order to minimize the EFLO distortion. This is obtained by determining the lowest Mean Square Error (MSE) between the effective LO and an ideal sinusoid.

$$MSE = \frac{1}{n} \sum_{i=1}^{n} (Y_i - \hat{Y}_i)^2 \qquad (2)$$

where n is the number of samples in the dataset, Y_i is the actual (true) value of the ith sample, $\hat{Y}_i$ is the estimated value of the ith sample. For each set of threshold positions, the MSE is computed in order to determine the set of levels minimizing the MSE.

E. Effects of the Thresholds on the Effective LO Spectrum

Upon selecting the set of levels and after performing LCSS on the input sine wave, the next step consists in generating the phases aligned with these levels, where each level corresponds to a distinct phase. The generation of a phase occurs precisely when its associated level is crossed by the input signal. In order to check the effect of the selected thresholds on the EFLO, simulations have been made with different thresholds for 4 and 8 levels respectively. In addition to the MSE calculated for each combination of levels, the Total Harmonic Distortion (THD) is calculated to compare the resulting EFLOs for different combination of levels.

$$THD = \frac{\sqrt{V_2^2 + V_3^2 + V_4^2 + \cdots}}{V_1} \qquad (3)$$

where V_i is the amplitude of the ith harmonic. It is obvious that when the selected levels are considered having a better MSE, they can approximate better a sine wave. Thus, the obtained EFLO has a better shape and less distortion. The calculated THD is lower as the harmonics have less magnitude. A comparative analysis of the magnitude of harmonics in decibels (dB) for different combinations of levels used for this novel phase control has been done with 8 levels to better approximate the sine wave and compared with the conventional case. The conventional phases and EFLO used in [6] is based on specific gains that approximate the sine wave having least number of harmonics in its spectrum. These gains for 8 paths are respectively 1, $\sqrt{2}$, 1, 0, -1, -$\sqrt{2}$, -1, 0. The set of levels obtained thanks to the proposed methodology are: 0.2, 0.41, 0.63, 0.85, 0, -0.2, -0.41, -0.63, -0.85. Fig. 5 (a) shows the FFT of EFLO obtained in the conventional case, where Fig. 5 (b) shows the one obtained thanks to the proposed LCSS method.

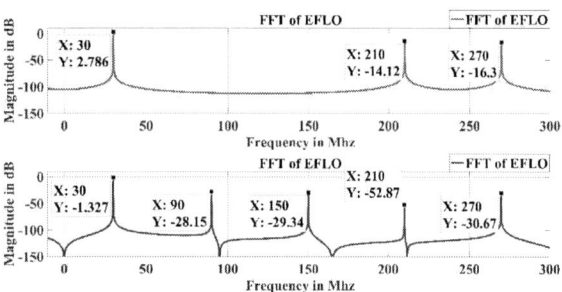

Fig. 5. FFT of EFLO (a) Obtained with 8 Conventional Phases (b) Obtained with LCSS Using Set of 8 Levels

While the proposed control law generates an EFLO with a greater number of harmonics compared to the FFT of the conventional method, these additional harmonics are significantly weaker. This decrease in harmonic magnitudes affects the THD making it lower for the new control law, where it is -14.85 dB in the conventional case and -23.16 dB for the new control law. The key advantage becomes clear in Fig. 6, which shows the overall magnitude spectrums of both EFLOs. The envelope of the conventional EFLO's magnitude is much higher than the one obtained using the proposed control law (LCSS), signifying less harmonic distortion.

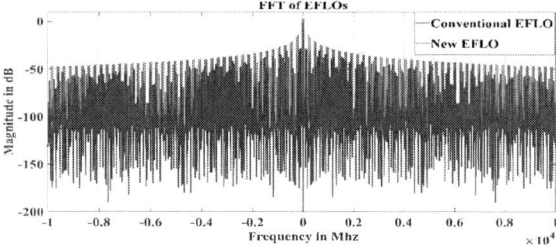

Fig. 6. EFLOs Comparison

IV. THE NEW SWITCH CONTROL APPLIED TO NPM

After presenting and comparing the EFLO using the conventional and the new LCSS switch control law, let us study the effects of both kind of phases on the mixer output and its harmonics. Therefore, in both cases, the input signal is an amplitude-modulated signal and the NPM architecture is the same. Their outcomes are subjected to a comparative analysis. The fundamental purpose of such a configuration is to perform a down conversion of an input signal transitioning from a higher band to a lower band, while rejecting as much as possible the harmonics. In addition, capacitors and switches acts as a low-pass filter, effectively filtering the output signal.

A. 4-Path Mixer Simulation Results

In the classical configuration, the four phases exhibit a 25% duty cycle and a phase shift of 90 degrees between two consecutive phases. The phases controlling the switches on each path are at the carrier frequency. Therefore, this result in a down-conversion of the RF modulated signal to the base band, followed by an IQ demodulation as illustrated in Fig. 2. The generated phases following the new switch control law are applied to the NPM switches illustrated in Fig. 2, substituting the utilization of the classical phases. In this innovative approach, each phase controls a specific path within the NPM, where each phase has different waveform from the conventional approach. The outcomes obtained from the conventional approach [6] (conventional phases driving the switches of the NPM) are given in Fig. 7 (a), while those resulting from the proposed approach (LCSS generated phases driving the switches of the NPM) are given in Fig. 7 (b).

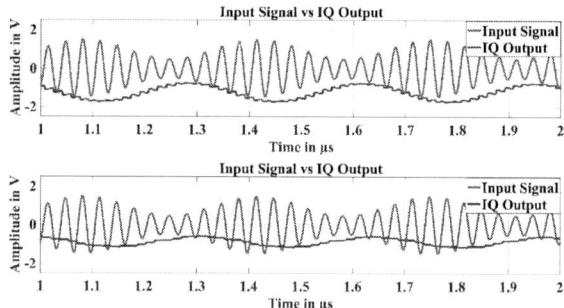

Fig. 7. Input vs IQ Output of (a) 4 Path Conventional Mixer (b) Proposed Control Phases for 4 Path Mixer

In both cases, the conventional [6] and the novel approach, the down conversion is correctly performed with a smoother output for the novel approach. This result is confirmed by the FFT where a gain of approximately 10 dB is observed at each harmonic. The new technique shows a clear improvement compared to the harmonics generated by the classical approach as frequency increased.

B. 8-Path Mixer Simulation Results

The setup for the 8-path mixer is similar to that of the 4-path mixer used, but each phase has a 12.5% duty cycle and a 45-degree phase shift with respect to the next phase. In the conventional case, each phase is amplified with its respective gain on each path. For the proposed new control law, the gains represented by the thresholds calculated are used. The input signal is also amplitude-modulated signal. Since the phases controlling the switches on each path share the carrier frequency, the signal is down-converted to the base band. Then, an IQ demodulation process follows. To compare the

979-8-3315-3968-9/24 $31.00 © 2024 IEEE

effectiveness of the LCSS generated phases against the classical approach, the time domain of the IQ demodulated signal is illustrated Fig. 8 while TABLE I. details their spectrum.

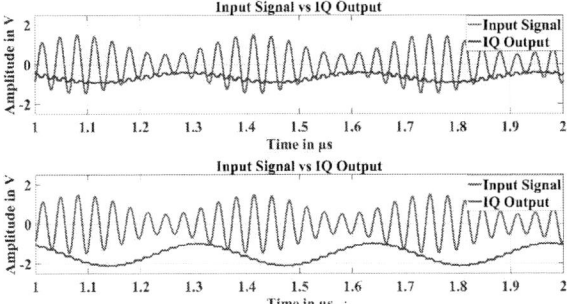

Fig. 8. Input vs IQ Output of (a) Conventional 8 Path Mixer (b) Proposed Control Law for 8 Path Mixer

TABLE I. IQ OUTPUT HARMONIC RESPONSE COMPARISON

| Frequency in MHz | Magnitude of Harmonics in dB | | | | |
| | 4 Path Mixer | | 8 Path Mixer | | |
	Classical	New Law	[6]	Classical	New Law
0	7.94	4.86	2.48	13.96	9.85
f_{fund}	**-6.3**	**-9.79**	**-11.52**	**-0.27**	**-4.81**
f_{LO}	-103.7	-115.3	-95.53	-95.46	-118
$f_{LO} - f_{fund}$	-70.63	-69.44	-61.96	-58.73	-62.6
$f_{LO} - f_{fund}$	-62.99	-75.3	-58.32	-55.78	67.59
$2f_{LO}$	**-99.2**	**-67.67**	**-93.05**	**-92.95**	**-56.66**
$2f_{LO} - f_{fund}$	-27.03	-36.91	-25.12	-23.73	-33
$2f_{LO} + f_{fund}$	-27.89	-35.73	-26.07	-24.72	-30.86
$3f_{LO}$	**-103.7**	**-135.1**	**-95.52**	**-95.59**	**-100.9**
$3f_{LO} - f_{fund}$	-71.16	-82.21	-66.53	-69.99	-74.5
$3f_{LO} - f_{fund}$	-81.71	-86.58	-71.78	-70.87	-79.27
$4f_{LO}$	**-107.7**	**-76.72**	**-95.5**	**-93.82**	**-80.01**
$4f_{LO} - f_{fund}$	-38.13	-46.06	-70.85	-34.97	-45.8
$4f_{LO} + f_{fund}$	-38.57	-42.51	-72.39	-35.4	-38.65
$5f_{LO}$	**-103.7**	**-128.3**	**-95.52**	**-95.59**	**-118.2**
$5f_{LO} - f_{fund}$	-84.97	-84.38	-74.42	-73.73	-81.13
$5f_{LO} - f_{fund}$	-76.57	-103.8	-72.3	-76.09	-87.74
$6f_{LO}$	**-99.2**	**-83.4**	**-93.05**	**-92.99**	**-70.22**
$6f_{LO} - f_{fund}$	-36.87	-50.75	-35.05	-33.71	-52
$6f_{LO} - f_{fund}$	-37.15	-47.9	-35.24	-33.86	-46.65
SNR	**10.94**	**16.24**	**5.55**	**14.31**	**17.69**

$f_{fund} = 3$ $f_{LO} = 30$

The results presented in Table 1 show the advantage of using the proposed switch control law over [6] and the classical one. As the frequency increases, this advantage becomes more salient, with the new control phases consistently showing lower magnitudes of harmonics. This advantage is extended to higher frequencies showing superior harmonic rejection capabilities of the new control phase configuration. Even though the rejection of some harmonics in [6] and in the classical NPM is better than the new proposed control law, the latter remains surpassing both. First, a gain of approximately 10 dB is observed at each harmonic. In addition, this enhancement is expressed when calculating the SNR of the IQ output in all cases. This SNR is computed over a specified band (0 to 200 MHz). The SNR of the NPM using the new control law is higher than that in [6] and the classical NPM whatever the number of path considered. In addition, the

significant reduction in harmonic distortion across the spectrum highlights again the effectiveness of the approach. It also demonstrates the applicability of the method to any number of paths. It is noticeable that the modulating signal is less attenuated with the proposed method, which is a clear advantage in term of SNR. These results affirm the efficiency of the novel proposed methodology for minimizing harmonic distortion and improving the performances of NPM in RF applications.

V. CONCLUSION

In this paper, a novel methodology for driving the switches of NPM has been presented. The proposed approach uses a unique switch control based on LCSS. This implies the generation of phase waveforms with varying duty cycles. These generated phases are different from the conventional phases that have the same duty cycle. By carefully selecting the number and amplitude of these levels, the generated phases produce an EFLO with improved harmonic rejection at the NPM output. The proposed methodology uses both the THD and the SNR of the EFLO as criteria to identify the optimal switch control law. This approach is not limited by the number of paths in the mixer and has been successfully applied to both 4-path and 8-path mixers. These demonstrations show the feasibility and benefits for the NPM's IQ demodulated output. Beyond this exploration, this study offers practical insights with direct implications RF systems, emphasizing the need for high linearity and online configurability of RF systems.

ACKNOWLEDGMENT

This work has been partially supported by the French Ministry of Higher Education, Research and Innovation.

REFERENCES

[1] E. A. M. Klumperink, H. J. Westerveld, and B. Nauta, 'N-path filters and mixer-first receivers: A review', in *2017 IEEE Custom Integrated Circuits Conference (CICC)*, Austin, TX: IEEE, Apr. 2017, pp. 1–8. doi: 10.1109/CICC.2017.7993643.

[2] S. Huang and A. Molnar, 'A 3.7-6.5GHz 8-Phase N-Path Mixer-First Receiver with LO Overlap Suppression Achieving 5dBm OOB B1dB', in *2021 IEEE Radio Frequency Integrated Circuits Symposium (RFIC)*, Atlanta, GA, USA: IEEE, Jun. 2021, pp. 87–90. doi: 10.1109/RFIC51843.2021.9490451.

[3] R. E. Struiksma, E. A. M. Klumperink, B. Nauta, and F. E. Van Vliet, 'A 500MHz– 2.7 GHz 8-path weaver downconverter with harmonic rejection and embedded filtering', in *ESSCIRC 2014 - 40th European Solid State Circuits Conference (ESSCIRC)*, Venice Lido, Italy: IEEE, Sep. 2014, pp. 223–226. doi: 10.1109/ESSCIRC.2014.6942062.

[4] A. A. Shakoush *et al.*, 'N-Path Mixer with Wide Rejection Including the 7 th Harmonic for Low Power Multi-standard Receivers', in *2022 20th IEEE Interregional NEWCAS Conference (NEWCAS)*, Quebec City, QC, Canada: IEEE, Jun. 2022, pp. 256–260. doi: 10.1109/NEWCAS52662.2022.9901392.

[5] B. Bidegaray-Fesquet and L. Fesquet, 'Levels, peaks, slopes... which sampling for which purpose?', in *2016 Second International Conference on Event-based Control, Communication, and Signal Processing (EBCCSP)*, Krakow, Poland: IEEE, Jun. 2016, pp. 1–6. doi: 10.1109/EBCCSP.2016.7605261.

[6] A. Al Shakoush, E. Lauga-Larroze, S. Subias, T. Taris, F. Podevin, and S. Bourdel, 'Low Complexity Architecture of N-Path Mixers for Low Power Application', in *2019 17th IEEE International New Circuits and Systems Conference (NEWCAS)*, Munich, Germany: IEEE, Jun. 2019, pp. 1–4. doi: 10.1109/NEWCAS44328.2019.8961234.

[7] N. Sayiner, H. V. Sorensen, and T. R. Viswanathan, 'A level-crossing sampling scheme for A/D conversion', *IEEE Trans. Circuits Syst. II*, vol. 43, no. 4, pp. 335–339, Apr. 1996, doi: 10.1109/82.488288.

[8] Y. Tsividis, 'Event-Driven Data Acquisition and Digital Signal Processing—A Tutorial', *IEEE Trans. Circuits Syst. II*, vol. 57, no. 8, pp. 577–581, Aug. 2010, doi: 10.1109/TCSII.2010.2056012.

A Scalable Hardware Architecture for Efficient Learning of Recurrent Neural Networks at the Edge

Yicheng Zhang*, Manil Dev Gomony, Henk Corporaal, and Federico Corradi*

Electrical Engineering Department, Eindhoven University of Technology, Eindhoven, The Netherlands

* *Correspondence to y.zhang1@student.tue.nl, f.corradi@tue.nl*

Abstract—Edge devices can execute pre-trained Artificial Intelligence (AI) models optimized on large Graphical Processing Units (GPU) but often need fine-tuning for real-world data. This process, known as edge learning, is crucial for personalized learning for tasks such as speech and gesture recognition and often requires recurrent neural networks (RNNs). However, training RNNs on edge devices faces challenges due to limited resources. We propose a system for RNN training through sequence partitioning using the Forward Propagation Through Time (FPTT) training method, facilitating edge learning. Our optimized HW/SW co-design for FPTT is the first of its kind. In our work, we have implemented the complete computational process for training Long Short-Term Memory (LSTM) networks using FPTT, and we have optimized and explored the hardware architecture leveraging the Chipyard framework. Our findings indicate considerable memory savings, with only a slight increase in latency, when training small-batch size sequential MNIST (S-MNIST) data.

Index Terms—Edge Learning, LSTM, HW/SW Co-design

I. INTRODUCTION

The standard approach for AI on edge devices involves training models in the cloud and then deploying optimized versions on embedded devices for tasks like speech recognition, video processing, surveillance, and biomedical signal analysis [1]. However, real-world data often differs from training data, necessitating further model training to handle specific user features, such as unique accents or patient-specific biomedical signals. Edge learning becomes crucial for personalized applications that process evolving data due to energy efficiency, latency, and privacy superiority over cloud re-training and re-deployment.

Learning RNNs for sequential applications at the edge presents challenges due to the traditional Back Propagation Through Time (BPTT) algorithm that requires substantial memory and computing resources. This is because BPTT stores all network states for the full sequence, which is impractical for embedded systems.

Kag et al. [2] proposed Forward Propagation Through Time (FPTT) that updates RNN parameters by optimizing an instantaneous risk function to solve long-range dependencies problems. FPTT, as a side effect, enables sequence partitioning to improve memory efficiency while maintaining accuracy, making it particularly promising for edge learning. However, its need for precision-sensitive RNNs and high computational demand means no existing edge hardware architecture efficiently accelerates FPTT.

In this paper, we address this problem by designing an optimized digital hardware architecture for FPTT-based edge learning without the need to transmit data to the cloud. Specifically, the contributions of this work are:

- We propose the FPTT-based training flow, and for efficient deployment optimize the edge hardware architecture composed of an accelerator for matrix multiplications with BF16 data types and multiple RISC-V cores.
- We demonstrate the effectiveness of the complete FPTT-based system for edge LSTM training. We present results on the S-MNIST benchmark for small batch sizes.

The paper is structured as follows: Section II presents the FPTT training algorithm and the framework for co-designing algorithms and architecture. Section III details the training process, the system architecture, and its optimizations and explorations. Section IV explains experimental set-up. Section V analyzes latency and memory footprint, highlighting the performance of the architecture. This work represents the first digital acceleration architectures for on-device training of RNNs, advancing efficient edge learning.

II. BACKGROUND

A. Forward Propagation Through Time

We exploit a forward-propagation algorithm called Forward-Propagation Through Time [2] in which the updates of the RNN parameters are obtained by optimizing an instantaneous risk function that incorporates a regularization penalty that evolves dynamically based on previously observed losses. This approach allows RNN parameter updates to converge to a stationary solution of the empirical RNN objective. Consequently, recurrent neural architectures like LSTMs can effectively address long-range dependency problems, and the optimization tasks are easily broken down into smaller steps, which enables memory savings. We refer the reader to this reference [2] for details on the algorithms.

In summary, the FPTT online formulation interleaves forward and backward passes with parameter updates for input sequences at each time step, dividing the sequence of T steps into T parts. The partition factor K enables support for coarser granularities, splitting the sequence into K parts. Additionally, incorporating the regularization term $R_{(t)}$ with the traditional loss $l_{(t)}$ ensures stabilization and convergence, with α as the hyperparameter.

$$L_{(t)} = l_{(t)} + R_{(t)} \tag{1}$$

979-8-3315-3968-9/24 $31.00 © 2024 IEEE

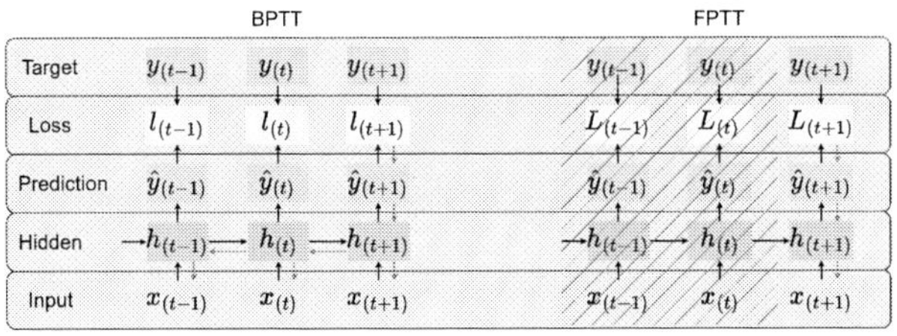

Fig. 1: Unrolling of computation graph: BPTT vs FPTT

$$R_{(t)} = \frac{\alpha}{2}||\theta - \overline{\theta}_{(t)} - \frac{1}{2\alpha}\lambda_{(t)}||^2 \qquad (2)$$

Figure 1 illustrates the distinction between BPTT and FPTT in updating the network parameter for time step $t + 1$. BPTT backtracks on all network states generated so far, while FPTT uses the states of the current sequence part. This means that the states of former parts can be released from the memory space.

B. Chipyard

To explore hardware and software co-design, we employed the Chipyard framework [3] for flexibly designing and evaluating full-system hardware. It comprises a collection of tools and libraries designed to integrate open-source and commercial tools for developing systems-on-chips. In the Chipyard frameworks, various open-source Hardware Intellectual Properties (IPs) in Chisel are available to compose customized systems. These IPs include RISC-V-based CPUs and domain-specific architectures like Gemmini.

III. METHODOLOGY

In this section, we first bridge the gap between FPTT theory and its application for edge learning by describing the training algorithm. Then, we present optimizations and explorations on the Chipyard-based hardware platform for efficient execution.

A. Computation Design

FPTT-based sequence training is presented in Algorithm 1.

Notation. Assuming that the training set B has N samples, i.e. $B = \{x^i, y^i\}_{i=1}^{N}$, expressed in a sequence of T time steps. To perform partitioned training, the sequence of T time steps can be divided into K parts, resulting in a stride of $st = \frac{T}{K}$ time steps. This leads to K iterations of processing, requiring K iterations to process all parts and update the network parameter K times. During iteration p, the network utilizes the p-th sequence part of st steps for the Forward Pass (FW). Following this, the Backward Pass (BW) occurs on the generated st network states to calculate the gradients for updating the network parameter. The value of K lies within the range of integers in $[1, T]$; divisions with a remainder are excluded. The extreme values for K, either 1 or T, imply no partition or the finest partition, respectively. In this subsection, we consider the latter scenario, where the subscript of the network parameter is synchronized with that of the network state.

Algorithm 1 Partitioned Sequence Training by FPTT

Input: learning rate η, hyper-parameter α
Input: training data $B = \{x^i, y^i\}_{i=1}^{N}$, sequence length T
Initialize: partition factor $K \in [1, T]$, stride $st = \frac{T}{K}$
Initialize: weight/bias θ_1 randomly in the domain θ
Initialize: running average $\overline{\theta}_1 = \theta_1$
Initialize: running estimate $\lambda_1 = 0$
Shorthand: $\theta, \lambda, \overline{\theta}$ as Network Parameters NP
Shorthand: Network States as NS
for $p = 1$ **to** K **do**
 $part = p$-th stride of $\{x^i\}_{i=1}^{N}$
 $NS_{[(p-1)\cdot st+1:p\cdot st]} = \mathbf{FW}(\theta_{(p)}, part, NS_{(p-1)\cdot st})$
 $\frac{\partial L_{(p\cdot st)}}{\partial\theta_{(p)}} = \mathbf{BW}(NP_{(p)}, \{y^i\}_{i=1}^{4}, NS_{[(p-1)\cdot st+1:p\cdot st]})$
 $NP_{(p+1)} = \mathbf{PU}(\frac{\partial L_{(t)}}{\partial\theta_{(p)}}, NP_{(p)})$
end for
Return: $\theta_{(K+1)}, Loss(\hat{y}_{(T)}, y_{(T)})$

Forward Pass (FW). FPTT particularly benefits from the gating structure inherent in recurrent models like LSTMs and GRUs [4]. Regarding terminal prediction problems, it is common to append a classifier layer for dimensional matching and inference.

Backward Pass (BW). Gradient calculation becomes twofold in FPTT-based training. As in Exp. 3, $L_{(t)}$ consists of the Loss $l_{(t)}$ and the regularization term $R_{(t)}$, their derivations are separated.

$$\frac{\partial L_{(t)}}{\partial\theta_{(t)}} = \frac{\partial(l_{(t)} + R_{(t)})}{\partial\theta_{(t)}} = \frac{\partial l_{(t)}}{\partial\theta_{(t)}} + \frac{\partial R_{(t)}}{\partial\theta_{(t)}} \qquad (3)$$

$l_{(t)}$ is network-dependent, and $\frac{\partial l_{(t)}}{\partial\theta_{(t)}}$ can be derived by Back Propagation Through Time (BPTT).

The regularization term $R_{(t)}$ is a function of network parameters independent of the topology, as shown in Exp.2, then $\frac{\partial R_{(t)}}{\partial\theta_{(t)}}$ can be determined by differentiation directly.

Running mean $\overline{\theta}_t$ and running estimate $\lambda_{(t)}$ are considered constants with regard to $\theta_{(t)}$ in the alternative optimization method [2], then in Exp.4 we have $\frac{\partial R_{(t)}}{\partial\theta_{(t)}}$.

$$\frac{\partial R_{(t)}}{\partial\theta_{(t)}} = \alpha(\theta_{(t)} - \overline{\theta}_{(t)}) - \frac{1}{2}\lambda_{(t)} \qquad (4)$$

Parameter Update (PU). With the gradients $\frac{\partial l_{(t)}}{\partial\theta_{(t)}}$ and $\frac{\partial R_{(t)}}{\partial\theta_{(t)}}$ that have been derived, we can update the network parameter θ, the running estimate λ, and the running average $\overline{\theta}$ in three sequential steps using optimizer, e.g. Gradient Descent, as shown in Exp.5, Exp.6 and Exp.7.

$$\theta_{(t+1)} = \theta_{(t)} - \eta(\frac{\partial l_{(t)}}{\partial\theta_{(t)}} + \frac{\partial R_{(t)}}{\partial\theta_{(t)}}) \qquad (5)$$

$$\lambda_{(t+1)} = \lambda_{(t)} - \alpha(\theta_{(t+1)} - \overline{\theta}_{(t)}) \qquad (6)$$

$$\overline{\theta}_{(t+1)} = \frac{1}{2}(\overline{\theta}_{(t)} + \theta_{(t+1)}) - \frac{1}{2\alpha}\lambda_{(t+1)} \qquad (7)$$

979-8-3315-3968-9/24 $31.00 © 2024 IEEE

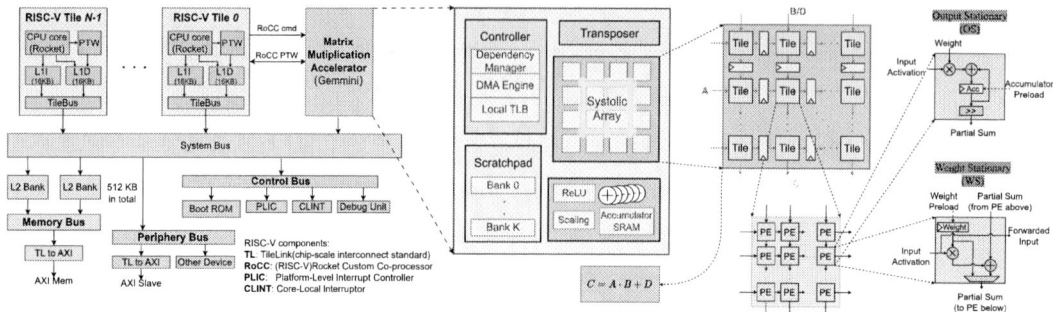

Fig. 2: Chipyard-based System Architecture

B. System Architecture

Leveraging the Chipyard framework, we propose a heterogeneous embedded system of multiple efficient RISC-V cores (Rocket) and one Matrix Multiplication Accelerator (Gemmini), as in Figure 2. Other functionalities in the system are organized by Memory Bus, Periphery Bus, and Control Bus.

The RISC-V tile, integrated into the system by Tile Link, contains one core and private L1 caches. The number of tiles in the system determines the speedup of the parallelizable application.

The Gemmini Accelerator, coupled to the RISC-V core by custom instructions, is essentially based on a systolic array of processing tiles with registers in between. Each tile is also an array of Processing Elements (PE), which can be configured to two dataflow modes to execute the Multiply-accumulate (MAC) between matrices.

C. Optimizations and Explorations

We propose optimizations on the Gemmini by compressing its area/resource utilization while maintaining the performance of the training process on it, as given in Table I.

TABLE I: Gemmini Architecture Compression

Division	Parameter	Default	Compressed
Arithmetic	inputType	FP32	BF16
	spatialArrayOutputType		
	accType		
Data Scaling	mvin_scale_args.mul_t		
	mvin_scale_acc_args.mul_t		
	acc_scale_args.mul_t		
Systolic Array	dataflow	WS, OS	WS
Scratchpad	sp_capacity	256KB	128KB
Accumulator	acc_capacity	64KB	32KB

BF16 is globally applied in calculations as [5] reported that deep learning training using BF16 tensors achieves the same state-of-the-art results across domains as FP32 tensors in the same number of iterations and with no changes to hyperparameters. Correspondingly, the memory space can be halved. In addition, only WS dataflow is instantiated instead of both to save area, as WS is reported to offer a speedup of 3× relative to the OS [6].

Besides, to investigate the trade-off between the system budget and application speedup, we explore the size of the systolic array and the number of RISC-V cores.

TABLE II: System Scale[1]

Division	Parameter	Value Set
Gemmini	tileRows, tileColumns	1
	meshRows, meshColumns	4,8
RISC-V core	number	1,2,4,8

[1] MS-C (MeshSize-Cores) distinguishes design points

IV. EXPERIMENTAL SET-UP

Benchmarks. We use the Sequential MNIST dataset, in which the sample has a sequence length T of 784 time steps. The model comprises a 2-layer network, which includes an LSTM with 128 cells and a fully connected classifier with 10 cells. To mimic a basic edge learning scenario, the model is initially pre-trained on the S-MNIST training set on the cluster and then fine-tuned using 4 samples from the test set on our proposed architecture. The partition factor K is set to $\{1,2,7,14,28,56\}$, in which large K values are avoided for trainability [2].

Baselines. To check the efficiency of the proposed edge system, we also run the benchmarks on server-level machines, i.e., NVIDIA GPU L4 and V100.

Simulation. We use `FireSim`, a fast FPGA-accelerated cycle-accurate simulation framework in Chipyard for performance and resource assessment of the proposed architecture. The FPGA used is the AMD UltraScale+ VCU118.

V. RESULTS

A. Effect of Compression

Table III compares the default and compressed Gemmini. It reveals that implementing BF16 globally in the architecture leads to nearly a 50% reduction in resource utilization across various areas. Additionally, it has a minor impact on the benchmarks.

B. Comparison to the cloud

For any benchmark of K in Figure 3, the latency on two high-end GPU platforms does not outperform the proposed

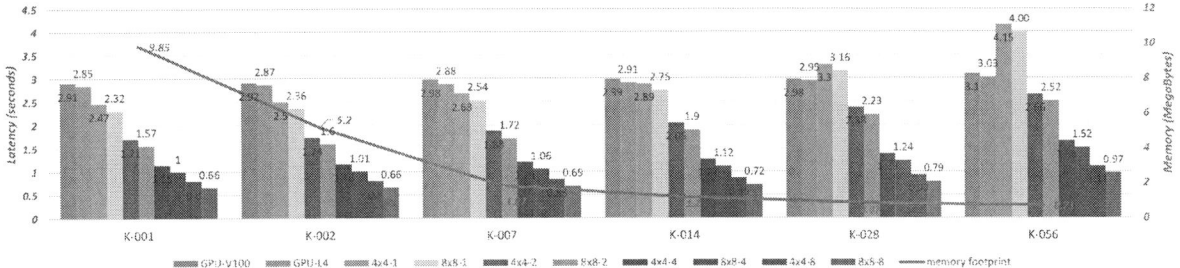

Fig. 3: Latency and Memory of FPTT benchmarks of six K values on ten architectures (All MS-C are normalized to 500MHz)

TABLE III: Effect of compression of Gemmini

	Resource Utilization				Loss Function					
	LUT	Register	BRAM	DSP	K-001	K-002	K-007	K-014	K-028	K-056
Default	94969	31067	80	256	0.059	0.058	0.058	0.056	0.054	0.052
Compressed	42022	23818	40	120	0.071	0.063	0.071	0.055	0.079	0.047

architecture with a 4x4 Gemmini and two CPUs (4x4-2), let alone the transmission time between the edge and the cloud. This shows the advantage of the dedicated embedded system in the efficiency of edge learning, i.e., small-scale fine-tuning, over the cluster. The reasons may involve higher overheads in communication and synchronization between the CPU and off-chip GPUs on the server for small matrices, which also can hardly make most of the parallelism of GPU.

C. Trade-off in design points

In Figure 3, each benchmark of K is also executed on 8 design points of the proposed architecture. We can observe an increase in speed for each set as the design becomes more complex, but the rate of increase is less than linear due to the limited number of parallelizable functions in the benchmark latency. Additionally, using an 8x8 mesh slightly reduces latency compared to a 4x4 mesh, as the larger size significantly speeds up matrix multiplications in the benchmark.

In addition, Table IV lists the resource utilization of 8 design points; scaling from 4x4-1 to 8x8-8 requires a 1.4-3.8x increase in various resources. The trade-off between benchmark speedup and resource increase can help balance performance requirements and system design budgets.

TABLE IV: Resource Utilization of Design Points

MS-C	LUT		Register		BRAM		DSP	
	Abs	Inc.	Abs	Inc.	Abs	Inc.	Abs	Inc.
4x4-1	86235	1.0	46766	1.0	168	1.0	135	1.0
8x8-1	123864	1.4	61105	1.3	168	1.0	135	1.0
4x4-2	115257	1.3	60224	1.3	180	1.1	150	1.1
8x8-2	152860	1.8	74561	1.6	180	1.1	150	1.1
4x4-4	172876	2.0	87117	1.9	196	1.2	180	1.3
8x8-4	210569	2.4	101454	2.2	196	1.2	180	1.3
4x4-8	288814	3.3	140893	3.0	228	1.4	240	1.8
8x8-8	326148	3.8	155229	3.3	228	1.4	240	1.8

D. Degree of partition

For any architecture in Figure 3, a benchmark of larger K, i.e., finer partition over a sequence, leads to a slight latency

increase but significant memory saving.

Taking 4x4-2 as an example, compared to K-001, i.e. in one go and no partition, K-014 can bring $\frac{9.85}{1.2} \approx 8.2$ times save in memory at the cost of $\frac{2.05-1.71}{1.71} \approx 20\%$ more latency.

VI. CONCLUSION

This work investigates the application of the FPTT algorithm for sequential learning at the edge by deploying it on an open-source Chipyard-based platform with dedicated optimizations and explorations. The results show that FPTT is a memory-efficient solution achieved by partitioning at a minor cost in latency. We refer to our public repository[1] for the source code and more details.

ACKNOWLEDGMENT

This work has been funded by the Dutch Organization for Scientific Research (NWO) with Grant KICH1.ST04.22.021.

REFERENCES

[1] Lalapura, Amudha, and Satheesh, "Recurrent neural networks for edge intelligence: a survey," *ACM Computing Surveys (CSUR)*, vol. 54, no. 4, pp. 1–38, 2021.

[2] Kag and Saligrama, "Training recurrent neural networks via forward propagation through time," in *Proceedings of the 38th International Conference on Machine Learning*, ser. Proceedings of Machine Learning Research, Meila and Zhang, Eds., vol. 139. PMLR, 7 2021, pp. 5189–5200.

[3] Amid, Biancolin, Gonzalez, Grubb, Karandikar, Liew, Magyar, Mao, Ou, Pemberton *et al.*, "Chipyard: Integrated design, simulation, and implementation framework for custom socs," *IEEE Micro*, vol. 40, no. 4, pp. 10–21, 2020.

[4] Yin, Corradi, and Bohté, "Accurate online training of dynamical spiking neural networks through forward propagation through time," *Nature Machine Intelligence*, pp. 1–10, 2023.

[5] Kalamkar, Mudigere, Mellempudi, Das, Banerjee, Avancha, Vooturi, Jammalamadaka, Huang, Yuen *et al.*, "A study of BFLOAT16 for deep learning training," *CoRR*, vol. abs/1905.12322, 2019. [Online]. Available: http://arxiv.org/abs/1905.12322

[6] Gookyi, Lee, Kim, Jang, and Lee, "Deep learning accelerators' configuration space exploration effect on performance and resource utilization: A gemmini case study," *Sensors*, vol. 23, no. 5, 2023. [Online]. Available: https://www.mdpi.com/1424-8220/23/5/2380

[1] https://github.com/federicohyo/HwRnnFPTT

Commercial Evaluation of Zero-Skipping MAC Design for Bit Sparsity Exploitation in DL Inference

Harideep Nair[*§], Prabhu Vellaisamy[*§], Tsung-Han Lin[†], Perry Wang[†], Shawn Blanton[*], and John Paul Shen[*]
* Carnegie Mellon University
† MediaTek USA Inc.

Abstract—**General Matrix Multiply (GEMM) units, consisting of multiply-accumulate (MAC) arrays, perform bulk of the computation in deep learning (DL). Recent work has proposed a novel MAC design, Bit-Pragmatic (PRA), capable of dynamically exploiting bit sparsity. This work presents *OzMAC* (*Omit-zero-MAC*), a modified re-implementation of PRA, but extends beyond earlier works by performing rigorous post-synthesis evaluation against binary MAC design across multiple bitwidths and clock frequencies using TSMC N5 process node to assess commercial implementation potential. We demonstrate the existence of high bit sparsity in eight pretrained INT8 DL workloads and show that 8-bit *OzMAC* improves all three metrics of area, power, and energy significantly by 21%, 70%, and 28%, respectively. Similar improvements are achieved when scaling data precisions (4, 8, 16 bits) and clock frequencies (0.5 GHz, 1 GHz, 1.5 GHz). For the 8-bit *OzMAC*, scaling its frequency to normalize the throughput, it still achieves 30% improvement on both power and energy.**

Index Terms—**Zero-skipping multiply-accumulate, commercial TSMC N5 evaluation, bit sparsity, deep learning inference**

I. INTRODUCTION AND BACKGROUND

General matrix multiply (GEMM) hardware, employing large arrays of multiply-accumulate (MAC) units, is the core compute fabric for modern deep learning accelerators (DLAs) [1]. A conventional bit-parallel MAC unit consists of a combinational array multiplier and an adder to accumulate the product, and a register to store the value. Any improvement on the MAC unit design is replicated many fold in the large MAC arrays, yielding potential for significant reduction in hardware complexity of DLAs [2]–[5]. Further, current industry standard for inference has moved from 32-bit floating-point (FP32) format to 16-bit floating-point (FP16) and 8-bit integer (INT8) formats. A recent study from IBM [6] summarizes the trend towards lower precision, highlighting imminent move towards 4-bit integer (INT4) and 2-bit integer (INT2) in the near future.

Recent work on Bit-Pragmatic (PRA) [3] has proposed a novel MAC design that leverages bit sparsity (i.e., the number of '0' bits within a binary value) to perform bit serial compute efficiently by skipping over zero bits using simple serial shift-and-add compute. In this work, we present a re-implementation of PRA with minor modifications for added hardware efficiency, called *OzMAC*, but the main contribution of this work is the rigorous state-of-the-art evaluation of *OzMAC* using commercial TSMC N5 (5nm) process node across multiple data precisions and clock frequencies, signif-

icantly extending beyond prior works utilizing TSMC 65nm technology and higher precision single-clock configurations. Key contributions of our work are:

- Present *OzMAC* based on Bit-Pragmatic (PRA), capable of exploiting dynamic bit sparsity by skipping over zero bits in binary values. This zero-skipping design in itself is not novel; the main focus of this work is its evaluation.
- Implement wide range of *OzMAC* designs using commercial design tools and TSMC N5 process design kit.
- Evaluate power-performance-area (PPA) for various data precisions (4-bits, 8-bits, 16-bits) and clock frequencies (500 MHz, 1 GHz, 1.5 GHz), against binary MAC.
- Demonstrate high bit sparsity in eight DL models leading to significant power reduction and how this can be used to increase *OzMAC*'s throughput via frequency scaling.

The paper is organized as follows. *OzMAC* microarchitecture is briefly summarized in Section II followed by hardware evaluation methodology in Section III. We present sparsity and corresponding PPA evaluation in Section IV, followed by bitwidth and frequency scaling analysis in Sections V and VI respectively. Finally, Section VII presents key conclusions.

II. OzMAC MICROARCHITECTURE AND DESIGN

OzMAC microarchitecture, derived from the Inner Product Unit within Bit-Pragmatic (PRA) [3], consists of three simple functional modules as shown in Fig. 1: 1) *Oz-encoder*, 2) *shifter*, and 3) *accumulator*. *Oz-encoder* is a Finite State Machine which keeps track of the current and next positions of '1' in the input bit pattern. Using this information, it outputs a one-hot encoded value capturing the bit positions of '1's every clock cycle for as many cycles as the number of '1's. For example, as illustrated in Fig. 1, the input '0101_2' is encoded as two one-hot values spanning two clock cycles: '0100_2' in the first cycle and '0001_2' in the next cycle. By doing this, it skipped over the two '0's and only incurs compute cycles for the '1's. The *Oz-encoded* input then goes to the shifter that determines the shift magnitude of the second input. The appropriately shifted second input is then added to the accumulator value. The minor modification to PRA employed in *OzMAC* is the *Oz*-encoder which feeds a 1-hot representation of the shift value to the shifter in contrast to PRA's oneffset generator which outputs a binary shift value. Employing 1-hot input simplifies the shifter hardware at the expense of more input lines (negligible overhead compared to reduction of shifter gate complexity).

[§]Equal contribution

979-8-3315-3968-9/24 $31.00 © 2024 IEEE

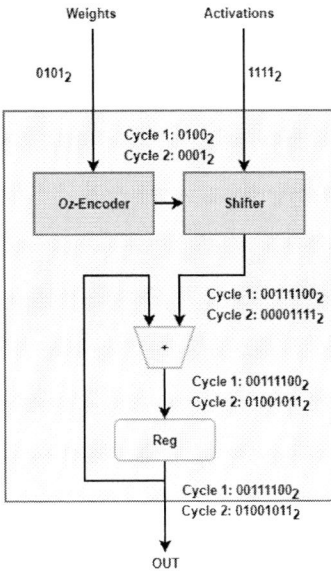

Fig. 1: *Oz*MAC (based on PRA [3]) with example compute.

TABLE I: Bit Sparsity and Cycle-Count Overhead for Pretrained Weights for Eight INT8 Quantized DL Benchmarks.

DL Benchmark	Average number of '1' bits (Actual cycle-count overhead)	Bit Sparsity Percentage
MobileNetV2	2.334	70.83%
MobileNetV3	1.711	78.61%
InceptionV3	2.430	69.62%
ShuffleNetV2	2.583	67.71%
GoogleNet	2.461	69.24%
ResNet18	2.398	70.02%
ResNet50	2.495	68.81%
ResNeXt101	2.289	71.39%

TABLE II: TSMC N5 PPA and Energy (averaged across eight DL benchmarks) for 8-bit OzMAC and bMAC at 500 MHz.

MAC Hardware	Area (μm^2)	Power (mW)	Latency (ns)	Energy (pJ)
bMAC	25.361	0.084	2	0.167
OzMAC	19.996	0.025	4.76	0.120
% Improvement	21.2	69.7	-	28.0

III. HARDWARE FRAMEWORK FOR EVALUATION

We perform rigorous, industry-standard evaluation of the *Oz*MAC design to get accurate PPA and energy results and compare against a conventional bit-parallel bMAC. The technology library used for evaluation is the commercial TSMC N5 (5nm) process node, with Synopsys design tools employed for simulation, synthesis, and power calculations.

First, *Oz*MAC RTL design is created in SystemVerilog, with functional verification performed using Synopsys VCS. A synthetic dataset with 1000 sample weights and activation values is developed, with the values reflecting the sparsity levels of the DL benchmarks under consideration. This allows for the appropriate switching activity to be captured, as well as resultant average *Oz*MAC compute cycles to be reported by the means of a testbench. Next, lint check is performed on the SystemVerilog source files using Synopsys SpyGlass and then synthesis is performed to convert the RTL-level design into a gate-level netlist using Synopsys Design Compiler, sourcing TSMC N5 library files. Gate-level netlist simulation is then performed for verification and collection of the switching activity of the design in the form of a SAIF dump. The SAIF dump is then sourced along with the netlist to perform accurate power calculations using Synopsys PrimeTime PX.

IV. SPARSITY SCALING ANALYSIS

Like PRA, *Oz*MAC performs highly efficient shift-and-add operations and trades off latency for lower area and power. The "omit-zero" capability of *Oz*MAC is key to mitigating this latency overhead by exploiting dynamic bit sparsity in input data. In other words, higher bit sparsity (i.e., more zero bits in the input data) will result in shorter compute latency and thereby lower energy consumption.

As illustrated in Table I, we use eight pretrained and quantized INT8 models, available as part of PyTorch's Torchvision library, that are widely used in state-of-the-art DL literature. Layer-by-layer analysis of the converged weights and activation values resulting from running ImageNet benchmark inputs illustrate the sparsities inherent in these benchmarks. For each model, we extract the average number of '0' bits in every 8-bit weight value across all the layers and calculate the bit sparsity as the percentage of '0' bits over the total number of bits. The DL models have close to 70% bit sparsity with MobileNetV3 having the highest sparsity of 78.61%. Table I also shows the number of '1' bits, which is equal to the compute latency in cycles. It can be seen that the effective compute latency for 8-bit *Oz*MAC owing to bit sparsity is between 1.7-2.5 cycles, much lower than the worst-case latency of 8 cycles. Comparing this to 1 cycle latency of bMAC, *Oz*MAC's power consumption must be 1.7-2.5x lower than that of bMAC to achieve similar energy efficiency.

Table II provides the die area, power, latency and energy consumption of *Oz*MAC and bMAC averaged across the eight DL benchmark models. Note that the power consumption values are obtained via PTPX using benchmark-specific test vectors that capture bit sparsity characteristics. The operating frequency for both designs is 500 MHz. An 8-bit conventional bMAC computes 1 MAC operation in 2 ns (1 cycle) while consuming about 25 μm^2 area, 84 μW power, and 167 fJ energy, whereas *Oz*MAC only consumes about 20 μm^2 area, 25 μW power, and 120 fJ energy while incurring 4.76 ns latency on average. Compared to conventional bMAC, this amounts to 21% less die area, 70% less power, and 28% less energy with 2.38x higher latency. This significant improvement in all three metrics can be attributed to three key factors: 1) simpler shift-and-add hardware with less area and leakage power footprint, 2) serial *Oz*-encoder that enables significant reduction in signal transitions at the input stage, thereby improving dynamic power, and 3) capability to exploit

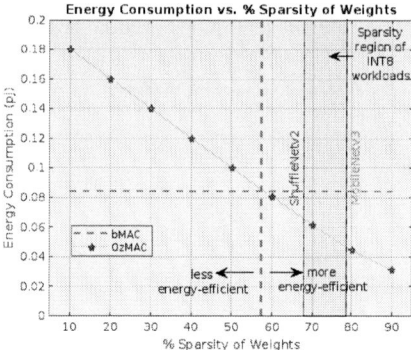

Fig. 2: Energy consumption vs. % bit-sparsity. Green-shaded region depicts the sparsity regions for Table I workloads.

TABLE III: TSMC N5 PPA at 500 MHz across varying bit precision of weights and activations: 4 bits, 8 bits and 16 bits.

MAC Hardware (wgt x act)	Area (μm^2)	Power (mW)	Latency (ns)	Energy (pJ)
bMAC (4x4)	5.451	0.015	2	0.031
OzMAC (4x4)	4.712	0.008	2.794	0.022
% Improvement	13.6	49.4	-	29.2
bMAC (4x8)	9.693	0.031	2	0.061
OzMAC (4x8)	8.3752	0.013	2.794	0.035
% Improvement	13.6	58.5	-	42.0
bMAC (8x8)	25.361	0.084	2	0.167
OzMAC (8x8)	19.996	0.025	4.76	0.120
% Improvement	21.2	69.7	-	28.0
bMAC (8x16)	45.282	0.177	2	0.355
OzMAC (8x16)	30.909	0.041	4.76	0.196
% Improvement	31.7	76.8	-	44.9
bMAC (16x16)	74.199	0.297	2	0.594
OzMAC (16x16)	60.608	0.065	9.28	0.601
% Improvement	18.3	78.2	-	-1.2

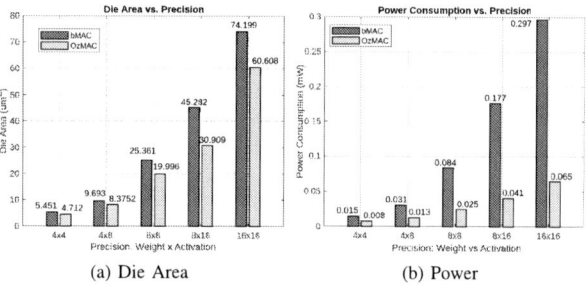

(a) Die Area   (b) Power

Fig. 3: Die area and power costs vs precision configurations.

the high bit sparsity present in the DL benchmarks (Table I).

Given the power consumption values from Table II, it can be seen that OzMAC reduces power by 3.36x on average. This implies that for an 8-bit OzMAC design, it can incur up to 3.36 clock cycles on latency overhead per MAC operation, before its energy consumption exceeds that of bMAC. We can calculate the minimum bit sparsity needed for OzMAC to maintain superior energy efficiency as $1 - \frac{3.36}{8} = 58\%$. Fig. 2 plots the energy consumption across varying bit sparsity, to demonstrate this cross-over point at 58% sparsity. Interestingly, all eight DL benchmarks exhibit bit sparsity higher than the threshold of 58% as can be seen from Fig. 2. Significant reduction in power consumption, coupled with sparsity-induced latency reduction, allows OzMAC to maintain superior energy efficiency over bMAC in spite of multi-cycle latency overhead. For throughput-sensitive applications, the higher latency of OzMAC can be addressed via frequency scaling, as will be demonstrated later in Section VI-B.

Key Takeaway: OzMAC achieves significant reduction in area, power and energy relative to bMAC for typical DL workloads, by exploiting inherent bit sparsity.

V. PRECISION SCALING ANALYSIS

Inference precision for DL workloads has been trending from 16-bits in the past to the current 8-bits with projection further down to 4-bits. Table III provides TSMC N5 PPA for five integer precision configurations: 1) 4-bit weights, 4-bit activations (4x4), 2) 4-bit weights, 8-bit activations (4x8), 3) 8-bit weights, 8-bit activations (8x8), 4) 8-bit weights, 16-bit activations (8x16), and 5) 16-bit weights, 16-bit activations (16x16). The mixed precisions, 4x8 and 8x16, are used to accommodate typical workloads that demand higher activation precision compared to weight precision. The corresponding area and power results are also plotted in Fig. 3.

Based on Table III, the smallest (4x4) OzMAC and bMAC designs consume 4.7 μm^2 area, 8 μW power, 22 fJ energy, and 5.4 μm^2 area, 15 μW power, 31 fJ energy, respectively. Compared to 4x4 designs, the largest (16x16) OzMAC incurs about 13x, 8x and 27x increase whereas 16x16 bMAC incurs

close to 14x, 20x and 20x increase in area, power and energy, respectively. Both 16x16 designs yield comparable energy while OzMAC still possesses area and power benefits. This indicates going beyond 16-bits for OzMAC is not beneficial.

From Fig. 3, area for OzMAC and bMAC scale up in a similar fashion almost linearly with respect to product of weight and activation bits. However, OzMAC's power consumption scales much better than that of bMAC, which incurs a much sharper increase with precision. Relative to bMAC, 8x16 OzMAC delivers the most area benefit (32% improvement), whereas the mixed precision 4x8 and 8x16 OzMAC designs offer the highest energy benefit up to 45%. Mixed precision designs deliver the highest energy improvements, as they can leverage the lower of the two precisions for Oz-encoding, incurring minimum latency (and thereby energy) overhead while taking advantage of the lower hardware complexity. Power benefits increase monotonically with precision due to the serial nature of Oz computation with signal transitions that get relatively sparser with higher precision.

Key Takeaway: OzMAC is more area and power-efficient than bMAC across all precision configurations, and more energy efficient across all but one (16x16) configuration. Energy consumption for both designs evens out at 16-bit weight precision, beyond which OzMAC becomes inefficient due to high latency overhead.

979-8-3315-3968-9/24 $31.00 © 2024 IEEE   154

VI. Frequency Scaling Analysis

In this section, we evaluate two types of frequency scaling to assess the effects on PPA trends between *Oz*MAC and bMAC.

TABLE IV: TSMC N5 PPA for INT8 (8-bits) OzMAC across varying frequencies: 500 MHz, 1 GHz and 1.5 GHz.

MAC Hardware	Power (mW)	Latency (ns)	Energy (pJ)
bMAC (0.5 GHz)	0.084	2	0.167
OzMAC (0.5 GHz)	0.025	4.76	0.120
% Improvement	69.7	-	28.0
bMAC (1 GHz)	0.166	1	0.166
OzMAC (1 GHz)	0.050	2.38	0.118
% Improvement	70.1	-	28.7
bMAC (1.5 GHz)	0.251	0.667	0.167
OzMAC (1.5 GHz)	0.075	1.587	0.119
% Improvement	70.2	-	29.0

A. Iso-Frequency Evaluation

As can be seen from Table IV, *Oz*MAC consumes 50 μW power and 118 fJ energy at 1 GHz, and only 75 μW and 119 fJ even at 1.5 GHz. At all three frequencies, *Oz*MAC improves power and energy by almost 70% and 29% respectively. As expected, power consumption scales linearly with frequency and energy stays almost constant since power increases and latency (due to clock period) reduces by similar amounts.

TABLE V: TSMC N5 PPA for OzMAC and bMAC across varying bit precisions at throughput-matching frequencies.

MAC Hardware (wgt x act)	Freq GHz	Power (mW)	Latency (ns)	Energy (pJ)
bMAC (4x4)	0.5	0.015	2	0.031
OzMAC (4x4)	0.7	0.011	2	0.022
% Improvement	-	29.2	Equal	29.3
bMAC (4x8)	0.5	0.031	2	0.061
OzMAC (4x8)	0.7	0.018	2	0.036
% Improvement	-	41.5	Equal	41.6
bMAC (8x8)	0.5	0.084	2	0.167
OzMAC (8x8)	1.2	0.059	2	0.118
% Improvement	-	29.5	Equal	29.6
bMAC (8x16)	0.5	0.177	2	0.355
OzMAC (8x16)	1.2	0.096	2	0.192
% Improvement	-	46.0	Equal	46.0

B. Iso-Latency Evaluation

*Oz*MAC's area-power-energy improvements are achieved at the cost of increased latency (1.4x for 4 bits and 2.4x for 8 bits). Such *Oz*MAC designs are ideal for edge inference applications that can tolerate the slight increase in latency (and reduction in throughput) but with stringent area/power/energy constraints. Here, we show that *Oz*MAC can even be used effectively for higher throughput with higher clock frequency.

To bridge the latency gap between *Oz*MAC and bMAC, we can scale *Oz*MAC's frequency by the corresponding ratio to match bMAC's compute latency and throughput. Table V provides TSMC N5 PPA for bMAC (0.5 GHz) and *Oz*MAC at throughput-matching frequencies. 16x16 *Oz*MAC incurs 4.6x higher latency and hence is not considered here.

For the same throughput, INT4 (4x4) and INT8 (8x8) designs deliver close to 30% improvement in power/energy, while mixed precision designs (4x8 and 8x16) achieve even higher improvements in power/energy by up to 46%. Note that *Oz*MAC can potentially deliver even higher throughput than bMAC by leveraging the remaining headroom in power reduction (29% to 46%) to further increase the frequency.

Key Takeaway: *Oz*MAC maintains superiority in area, power and energy efficiency at frequencies ranging from 500 MHz to 1.5 GHz, and can leverage relative frequency scaling to achieve equal or higher throughput compared to bMAC without adversely affecting its power or energy efficiency.

VII. Conclusions

This paper presents rigorous industry standard evaluation of *Oz*MAC, an updated re-implementation of previously proposed Bit-Pragmatic (PRA) MAC design that performs a series of simple shift-and-add operations. It accounts for only the '1' bits in input binary value (skipping the '0' bits) thus leveraging bit sparsity in DL workloads. The main goal of this work is to assess practical commercial implementation potential of dynamic bit sparsity exploitation through such zero skipping MAC designs. We demonstrate the presence of high bit sparsity in eight state-of-the-art DL benchmarks. We implement a wide range of *Oz*MAC designs using commercial design tools and latest TSMC N5 process node, and obtain PPA results across various data precisions and clock frequencies. *Oz*MAC shows substantial improvements in all three metrics: area (up to 30%), power (up to 80%) and energy (up to 46%) relative to conventional binary bMAC. Finally, we demonstrate the significant power reduction of *Oz*MAC and how this can be leveraged to increase throughput by increasing frequency without compromising area and energy efficiency benefits. Future work will evaluate a large array of *Oz*MAC units in an actual DLA at the system level. We believe all DLAs targeting low precision inference should adopt *Oz*MAC design.

References

[1] V. Sze, Y.-H. Chen, T.-J. Yang, and J. S. Emer, "Efficient processing of deep neural networks," *Synthesis Lectures on Computer Architecture*, vol. 15, no. 2, pp. 1–341, 2020.

[2] A. Delmas Lascorz, P. Judd, D. M. Stuart, Z. Poulos, M. Mahmoud, S Sharify, M. Nikolic, K. Siu, and A. Moshovos, "Bit-tactical: A software/hardware approach to exploiting value and bit sparsity in neural networks," in *Proceedings of the Twenty-Fourth International Conference on Architectural Support for Programming Languages and Operating Systems*, 2019, pp. 749–763.

[3] J. Albericio, A. Delmás, P. Judd, S. Sharify, G. O'Leary, R. Genov, and A. Moshovos, "Bit-pragmatic deep neural network computing," in *Proceedings of the 50th annual IEEE/ACM international symposium on microarchitecture*, 2017, pp. 382–394.

[4] P. Judd, J. Albericio, T. Hetherington, T. M. Aamodt, and A. Moshovos, "Stripes: Bit-serial deep neural network computing," in *2016 49th Annual IEEE/ACM International Symposium on Microarchitecture (MICRO)*. IEEE, 2016, pp. 1–12.

[5] H. Sharma, J. Park, N. Suda, L. Lai, B. Chau, V. Chandra, and H. Esmaeilzadeh, "Bit fusion: Bit-level dynamically composable architecture for accelerating deep neural network," in *2018 ACM/IEEE 45th Annual International Symposium on Computer Architecture (ISCA)*, 2018.

[6] N. Wang, J. Choi, and K. Gopalakrishnan, "8-bit precision for training deep learning systems," 2018.

979-8-3315-3968-9/24 $31.00 © 2024 IEEE

Compensating the Load Effect in Quadrature All-Pass Filters

U. Esteban Eraso, C. Sánchez-Azqueta, F. Aznar, C. Aldea, S. Celma

Group of Electronic Desing (GDE), Aragon Institute of Engineering Research (I3A), Zaragoza, Spain

{uesteban, csanaz, faznar, caldea, scelma}@unizar.es

Abstract— **This paper presents the design of a differential quadrature all-pass filter as a quadrature generator for being used in a phase shifter operating at 19.5 GHz, employing 65 nm CMOS technology. Various solutions have been investigated to address errors occurring in the phase and magnitude of I/Q signals at the output of the quadrature all-pass filter when connected to the next stage. The best result is obtained when combining asymmetry of the network with the inclusion of two additional resistive elements, achieving a quadrature error lower than 2.0° and a magnitude error lower than of 0.17 from 17 GHz to 22 GHz at the output of the quadrature all-pass filter. As an application, it has been proven that the RMS phase error at the output of a 5-bit active phase shifter has been decreased from 7.6° to 2.3°.**

Keywords—Active phase shifters, CMOS, I/Q signals, phased arrays, QAF, quadrature generators.

I. INTRODUCTION

Quadrature generators are very important components that enable the generation of signals with precise 90° phase shift, necessary for several applications in telecommunications and high-frequency systems, such as phased arrays [1].

Phased arrays are key elements in the new generation of wireless communication systems (5G/6G) as well as in satellite communications (SATCOM), since they allow the radiation pattern to be electronically directed in only a few microseconds [2], and to work with multiple beams simultaneously [3]. In certain sum-vector beamforming systems, the inclusion of a quadrature generator becomes necessary to generate two signals 90° phase shifted, to be subsequently weighed and added to obtain the desired phase shift [4]. Consequently, an efficient design and implementation of quadrature generators is critical to minimize errors in gain and phase at the output of the phase shifter, leading in an optimal performance of the complete receiver chain [5].

At RF frequencies, quadrature all-pass filters (QAF) based on RLC resonators can be used to obtain the desired phase shift minimizing power losses. In this paper, the topology of a quadrature generator for its use as the first stage of an active phase shifter is presented. The QAF is highly sensitive to the load effect imposed by the subsequent stage, which detrimentally impacts its performance.

In this work, the effect of loading and its compensation for a quadrature generator based on an RLC all-pass filter is investigated. In Section II the topology is described, and different solutions are proposed to diminish the effect of the load. These solutions will be compared among themselves and with another solution proposed in the literature [5]. A comparison of their application to a particular active phase shifter is addressed in Section III. Finally, in Section IV conclusions are drawn.

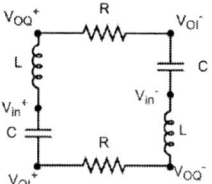

Fig 1. Differential quadrature all-pass filter (QAF) topology.

II. TOPOLOGY

Figure 1 shows a differential QAF based on RLC resonators. The 90° phase shift is achieved through the implementation of two 45° phase shifts within each I/Q path.

Once the QAF is connected to the next stage, the performance of this structure becomes very susceptible to the load effect. As in the final phase shifter topology the output of the QAF is connected to cascode structures, a capacitive load $C_L = 90$ fF provided by the following stage. This equivalent capacitive load can be considered invariant between the different phase configurations allowed by the phase shifter [6]. To assess its impact on the QAF phase and gain response, the errors in gain and phase are defined as follows.

$$\frac{\Delta G}{G} = \left| \frac{V_{OQ} - V_{OI}}{V_{OI}} \right| \qquad (6)$$

$$\Delta\phi = 90° - \left(\varphi_Q - \varphi_I \right) \qquad (7)$$

As the total gain varies across the frequency band due to resonance effects, the gain error is normalised with respect to in-phase signal for accurate assessment.

In this work, the solution proposed in [5] to compensate the negative effect of the load, has been tested for our topology of phase shifter. Also, three different solutions to this problem have been proposed trying to obtain similar quadrature results with lower losses. All the results presented from this point have been obtained from the average results of Montecarlo simulations based on 500 trials using the process design kit (PDK) of the TSMC 65 nm technology.

A. Symmetrical QAF with four additional resistors

Firstly, the solution proposed in [5] has been considered. In this case, the load problem is solved by introducing four additional series resistors R_S in the QAF. Each of these resistors is added in series with each inductor and capacitor of the QAF (see Fig. 2.a). This solution has been transposed to the topology and frequency band studied in this work. The values for the variables have been set considering a capacitive load C_L of 90 fF, obtaining the best performance for the following values of the components: $R = 100$ Ω, $C = 100$ fF, $L = 300$ pH, while the series resistor R_S is set to 50 Ω.

979-8-3315-3968-9/24 $31.00 © 2024 IEEE

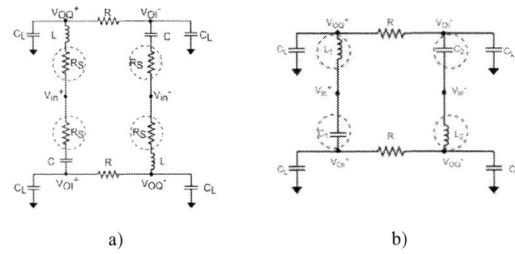 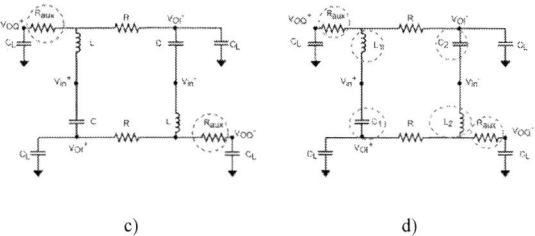

| a) | b) | c) | d) |

Fig. 2. Different load compensation techniques tested in this work.

Phase and gain errors using this solution are plotted as a blue line in Fig. 3 and Fig. 4, respectively. Note that the quadrature error diminishes to values lower than 1.4° in the whole frequency span, as well as the magnitude error which is lower than 0.13. However, the use of resistors entails significant insertion losses, obtaining that at the output of the QAF, the average gain is 0.56 at 19.5 GHz. These losses significantly compromise the signal-to-noise ratio, especially in the receiver chain.

B. Asymmetrical QAF without additional resistors

With the purpose of diminish the losses at the output of the QAF, an asymmetrical network configuration was explored. The objective was to address the disparity in gain magnitudes between the I/Q signals by employing an asymmetrical QAF without using any extra resistance, as depicted in Fig. 2.b. Through an iterative optimization process, the values of the components of the QAF were adjusted to achieve optimal phase shift between the I/Q signals, while trying to reduce the amplitude asymmetry. The values for the different variables have been set to: $R = 95\ \Omega$, $C_1 = 500\ \text{fF}$, $C_2 = 190\ \text{fF}$, $L_1 = 550\ \text{pH}$ and $L_2 = 225\ \text{pH}$.

The obtained results with this solution are represented with a red line in Fig. 3 and Fig. 4. The quadrature closely approaches 90° at the centre frequencies, with a maximum error of 5.4° in the frequency band. On the other hand, the gain magnitudes of the I/Q signals have a maximum error of 0.28. Regarding the gain at the output of the QAF, not having any extra resistive element in the network provides an average gain of 1.13 at 19.5 GHz. However, the errors are still high and can have a significant impact on the performance of the phase shifter, so it is necessary to further explore and refine the network configuration to attain the desired performance.

C. Symmetrical QAF with two additional resistors

The main problem of a symmetrical loaded QAF is that error in magnitude is high, being the quadrature signal higher than the in-phase one. A solution entails the introduction of two auxiliary resistors R_{aux} within the quadrature branch, to generate a slight attenuation and equalize the magnitudes of the signals. The proposed network configuration is depicted in Fig. 2.c. In this case, a symmetrical QAF network is used and through numerical computation, the optimal values were determined: $R = 100\ \Omega$, $C = 190\ \text{fF}$ and $L = 150\ \text{pH}$. The value of the auxiliary resistors is 50 Ω.

The results derived from implementing this solution are presented using a yellow line in Fig. 3 and Fig. 4. The values have been set to have a good quadrature in the target frequency with an error lower than 2.0°, and the error in gain improves to a value lower than 0.22. In terms of gain at the output of the QAF, an I/Q gain average of 0.91 is obtained at 19.5 GHz.

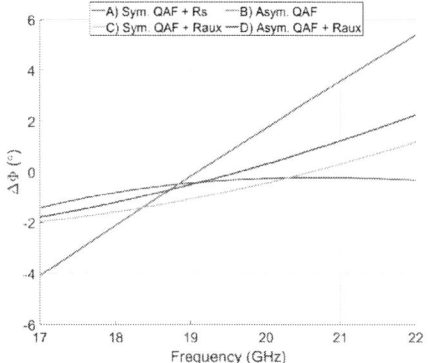

Fig. 3. Comparison of the phase error using each method.

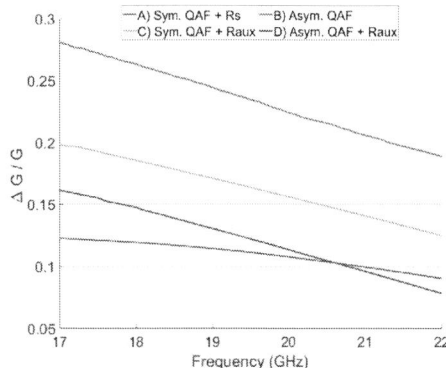

Fig. 4. Comparison of the gain error using each method

D. Asymmetrical QAF with two additional resistors

Finally, a hybrid solution combining elements from both (B) and (C) methodologies has been proposed to check if errors in gain and phase can be further reduced, without suffering as much loss as proposal (A). In this case, the QAF is composed of an asymmetrical network where there is also an auxiliary resistor in the Q-branch as shown in Fig. 2.d.

Through an iterative simulation process, the following values for the QAF components are obtained: $R = 95\ \Omega$, $C_1 = 190\ \text{fF}$, $C_2 = 240\ \text{fF}$, $L_1 = 120\ \text{pH}$ and $L_2 = 200\ \text{pH}$. Additionally, the auxiliary resistor R_{aux} in the Q branch was set to 50 Ω.

The results using this method are plotted using a purple line in Fig. 3 and Fig. 4. The obtained errors for the quadrature and the magnitude at the output of the QAF are lower than 2.2°

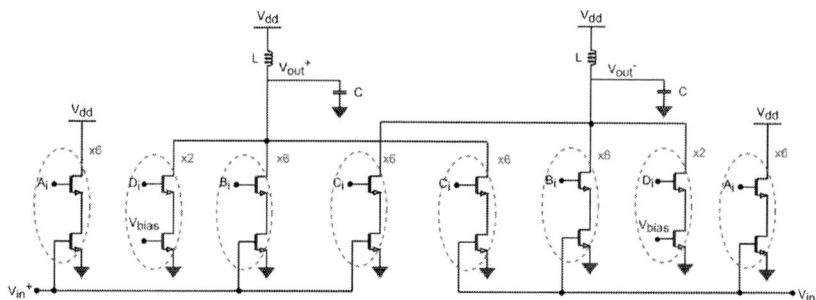

Fig 5. Topology of the employed VGAs.

and lower than 0.17, respectively, in the whole frequency span, with an average gain of 0.93 at 19.5 GHz.

III. RESULTS COMPARISON

To see the overall enhancement in phase shifter performance resulting from the different load compensation strategies, the resulting QAF topologies have been connected to two VGAs. The topology used for each VGA is represented in Fig. 5. Its role is to weigh each I/Q signal independently to obtain the desired phase shift at the output of the phase shifter. With this purpose, the VGAs include six blocks of six cascode structures (blocks A, B and C) and two blocks of two cascode structures (blocks D). Controlling which B and C structures are ON, the normalised gain of the I/Q signals can be independently set for −1 to 1 in steps of 1/6, while blocks A and D are, respectively, input and output dummy structures, used to keep input and output impedance constant between the different configurations [6]. So, calling A_i the gain of the VGA that weighs the I signal and A_j the gain of the VGA that weighs the Q signal, the ideal phase shift obtained will be given by:

$$\phi = tan^{-1}\frac{A_j}{A_i} \qquad (8)$$

The results for the RMS phase error at the output of the phase shifter for each solution are shown in Fig. 6. These graphs show the individual contributions of the asymmetric and symmetrical QAF networks accompanied by two additional resistors in the Q-branch, in reducing the RMS phase error at the output of the phase shifter. It can be seen that a higher improvement in the total RMS error is achieved with the series auxiliary resistor in the Q-branch. Nevertheless, when both methods are used simultaneously, this result can be improved, with the RMS phase error at the phase shifter output decreasing to 2.3° at 19.5 GHz. Furthermore, this reduced RMS error is consistently maintained below 3° across a broad frequency range from 17 to 22 GHz. The results obtained using both an asymmetrical QAF and an auxiliary resistor (D) are very close to the one obtained using a symmetrical QAF with a series resistor with each capacitor and inductor (A). It should be noted that all these results are significantly smaller than the tolerable maximum for 5-bit phase shifters (5.6°) [7].

Although the load effect is not totally compensated, the achieved reduction in RMS errors at the output of the phase shifter is remarkable. It is important to highlight that the VGA structure inherently introduces an RMS phase error of approximately 2.5° at 19.5 GHz when excited with two ideal I/Q signals (as indicated by the dotted line in Fig. 6). It can be

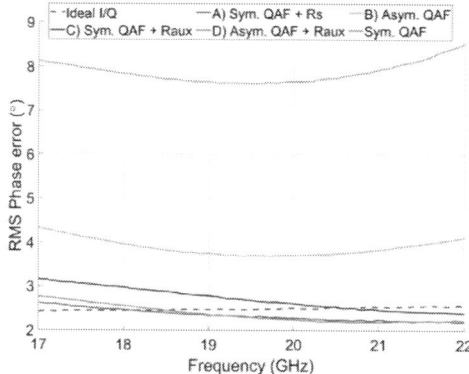

Fig 6. Comparison of the RMS results at the output of the phase shifter using each method.

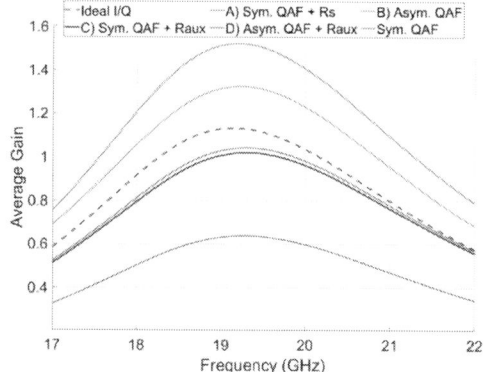

Fig 7. Comparison of the average gain at the output of the phase shifter using each method.

seen how the error associated with the QAF using solutions (A) and (D) partially reduces the error that is obtained when the VGAs of the phase shifter are excited using ideal I/Q signals.

On the other hand, Fig. 7 shows the average gain of the different phase configurations that can be given by the phase shifter for each of these solutions. The lowest RMS phase error is obtained when adding extra resistors to the network, these solutions entail important losses in the total gain of the system. Although both, the symmetrical QAF with four resistors (A) and the asymmetrical QAF with two auxiliary resistors in the Q-paths (D), have a good phase performance, in terms of gain the topology (D) is clearly preferable.

979-8-3315-3968-9/24 $31.00 © 2024 IEEE   158

TABLE I.   COMPARISION OF PROPOSED TOPOLOGIES

	QAF			Phase Shifter	
	$\Delta\phi_{max}$ (°)	$Gain$	$(\Delta G/G)_{max}$	$Gain$	RMS phase error (°)
A) Sym. QAF + R_s	1.5 @ 17 − 22 GHz	0.56 @ 19.5GHz	0.13 @ 17 − 22 GHz	0.63 @ 19.5GHz	2.3 @ 19.5GHz
B) Asym. QAF	5.4 @ 17 − 22 GHz	1.13 @ 19.5 GHz	0.3 @ 17 − 22 GHz	1.31 @ 19.5GHz	3.71 @ 19.5GHz
C) Sym. QAF + R_{aux}	2.0 @ 17 − 22 GHz	0.91 @ 19.5 GHz	0.2 @ 17 − 22 GHz	1.01 @ 19.5GHz	2.68 @ 19.5GHz
D) Asym. QAF + R_{aux}	2.2 @ 17 − 22 GHz	0.93 @ 19.5 GHz	0.17 @ 17 − 22 GHz	1.03 @ 19.5GHz	2.3 @ 19.5GHz

Moreover, it should be noted that the asymmetrical network solution without additional resistors has an RMS phase error associated of less than 4.5°, which is lower than the minimum allowed in 5-bit resolution phase shifters. So, when choosing which solution should be used in a particular case, it must be considered if the application of the phase shifter requires a higher accuracy in the phase shifts or a higher gain.

The numerical results obtained with these solutions implemented in the 5-bit phase shifter are presented in Table I. The smallest errors in phase and gain at the output of the QAF are obtained with solutions (A) and (D). In particular, the errors obtained with solution (A) are lower than the obtained with (D). However, in terms of gain, better results are obtained using solution (D). Regarding how these errors are reflected at the output of the phase shifter, the RMS phase error can be reduced to 2.3° at 19.5 GHz. Again, in terms of gain, solution (D) offers the best performance.

IV. CONCLUSIONS

This paper studies the performance of a differential quadrature all-pass filter (QAF) based on RLC resonators, when the load effect of the next stage (a VGA) is considered. The work evaluates its suitability for being used in a phase shifter operating at a frequency of 19.5 GHz, which means an approximated load of 90 fF in each output of the QAF.

Firstly, the solution proposed in [5] has been transferred to our design. The obtained results for the phase and gain error at the output of the QAF are low, and so are the phase RMS errors at the output of the phase shifter. High losses have been observed leading to a lower total gain at the output of the phase shifter, which degrades the signal-to-noise ratio in the signal reception path.

To address this issue, different techniques have been explored. The best performance is obtained when an asymmetrical QAF and two auxiliary resistors in the Q-branch (D) are used, obtaining at the output of the QAF quadrature error lower than 2.2°, and a magnitude error lower than 0.17 from 17 to 22 GHz. At the output of the phase shifter similar RMS phase errors are produced from both topologies (A) and (D), however, with topology (D), lower insertion losses are obtained.

It is noteworthy that, with solutions (A) and (D) the RMS phase error at the output of the phase shifter is smaller than the error obtained using an ideal quadrature source. This is because the small errors in magnitude and phase at the output of the QAF, partially compensate the phase error due to the VGA topology.

To sum up, it has been shown that the techniques that involve auxiliary resistors entail higher losses, that have a negative effect on the total gain. Moreover, an increase in the number of resistive elements would probably lead to a higher level of noise. On the other hand, when the asymmetrical network is used the total gain increases, but the RMS phase error gets worse. So, the best solution should be chosen depending on whether is needed in each application, phase accuracy or lower losses.

ACKNOWLEDGMENT

This work has been partially supported by the Spanish State Research Agency (PID2020-114110RA-I00 and PDC2023-145838-I00).

REFERENCES

[1] Z. Duan, Y. Wang, W. Lv, Y. Dai, F. Lin, "A 6-bit CMOS Active Phase Shifter for Ku-Band Phased Arrays," IEEE Microwave and Wireless Components Letters, vol. 28, no. 7, 2018.

[2] R. McMorrow, D. Corman, A. Creofts, "All Silicon mmW Planar Active Antennas: The Convergence of Technology, Applications, and Architecture," IEEE International Conference on Microwaves, Antennas, Communications and Electronic Systems (COMCAS), Tel Aviv, Israel, 2017.

[3] C. Changming, L. Wei, L. Ning, R. Junyan, "A 9–12 GHz 5-bit Active LO Phase Shifter with a New Vector Sum Method," Journal of Semiconductors, vol. 36, no 1, 2015.

[4] J. Kim, S. Oh, J. Oh, "A High Gain Vector-Sum Phase Shifter in 28-nm CMOS for 5G Communication", IEEE International Symposium on Radio-Frequency Integration Technology (RFIT), Busan, Korea, 2022.

[5] S. Y. Kim, D. -W. Kang, K. -J. Koh and G. M. Rebeiz, "An Improved Wideband All-Pass I/Q Network for Millimeter-Wave Phase Shifters," in IEEE Transactions on Microwave Theory and Techniques, vol. 60, no. 11, 2012.

[6] U. Esteban Eraso, C. Sánchez-Azqueta, C. Aldea, S. Celma, " A 19.5 GHz 5-Bit Digitally Programmable Phase Shifter for Active Antenna Arrays," Electronics, vol. 12, no 13, 2023.

[7] Y. Yu, P. G. Baltus, A. H. v. Roermund, Integrated 60 GHz RF Beamforming in CMOS, Springer.

Design Co-Processor based on Partially Homomorphic Encryption Execution Using Open-Source Tool

Mujahid Bilal[1]
mujahidbilal44@gmail.com

M. Kamran Bhatti[1]
mk.bhatti@yahoo.com

Muhammad Kahsif Minhas[2]
Kashifminhas934@gmail.com

Haroon Waris[2]
haroonwaris@gmail.com

Semiconductor Chip Design Fabrication Center, National institute of electronics Islamabad, Pakistan[1]

Center of Excellence in Science and Applied Technologies Islamabad, Pakistan[2]

Abstract— **Exploiting a weakness in OpenSSL's TLS protocol implementation, the Heartbleed vulnerability was found in April 2014 and exposed sensitive data. This study describes a co-processor architecture that reduces the danger of data leaking by applying partly homomorphic encryption. Using secure multiplexers and non-traditional arithmetic units as challenges, our fully functioning device is created in 130nm CMOS technology. Through UART, it exchanges data with the primary CPU. This paper describes an ASIC development process utilizing open-source methodologies, from RTL design to GDS-II.**

Index Terms—Data Privacy, Encrypted Execution, Partially Homomorphic Encryption, Hardware, ASIC, Skywater PDK.

I. INTRODUCTION

In an era where data has become the lifeblood of modern society, its protection and privacy have emerged as paramount concerns. The digital landscape is fraught with potential threats, ranging from cyberattacks on personal information to breaches of sensitive corporate and government data [2]. To combat these evolving security challenges, researchers and practitioners have continuously sought innovative solutions. Among the most promising advancements in this quest is the field of homomorphic encryption, which allows computations to be performed on encrypted data without ever decrypting it, thus preserving its confidentiality. In response to this challenge, our research embarks on a mission to bridge the gap between security and performance. We propose the development of a Co-Processor based on Partially Homomorphic Encryption Execution (COPHEE) – a solution that seeks to harness the power of homomorphic encryption while mitigating its computational overhead [3].

This research paper embarks on a journey to explore the landscape of homomorphic encryption and Open-Source Tool Design, delving into its principles, applications, and the security concerns it addresses. Partially Homomorphic Encryption (PHE) schemes provide a more practical and efficient alternative to Fully Homomorphic Encryption. Modern PHE cryptosystems like Paillier and RSA enable modular multiplication of ciphertexts, which is homomorphic to plaintext addition and multiplication, respectively. This feature, combined with efficient modular multipliers, suggests that dedicated hardware units could significantly benefit PHE-based algorithms. While PHE schemes do not offer the same computational completeness as FHE [4] schemes, as universal computation typically requires both addition and multiplication operations, this limitation primarily affects runtime decisions controlled by encrypted values. It can be

mitigated using a secure multiplexer as part of the root of trust. For instance, Cryptoleq, an architecture based on the One Instruction Set Computer (OISC) model, supports universal computation of PHE-protected algorithms with practical overhead [5]. Cryptoleq is internally founded on the Turing-complete Subleq abstract machine and implements the Paillier cryptosystem along with a secure multiplexer obfuscated in software using single-instruction self-modifying code.

Contribution:

In addressing the performance and universality limitations encountered when executing algorithms based on Partially Homomorphic Encryption (PHE), this research introduces a specialized processor known as the Co-processor for Partially Homomorphic Encrypted Execution (CoPHEE) which design using Open-Source Tool, tailored specifically for the Paillier encryption scheme. CoPHEE operates effectively with 2048-bit encrypted operands, making it well-suited to enhance a wide spectrum of secure applications. These applications encompass voting protocols, threshold cryptosystems, watermarking and secret sharing schemes, as well as server-aided polynomial evaluation protocols [6]. Since CoPHEE operates directly on encrypted data, any noticeable deviation in the execution flow could potentially leak sidechannel information about the values of corresponding control ciphertexts. Consequently, encrypted algorithms processed by CoPHEE must adhere to constant-time execution paths and predefined termination conditions to ensure security. Remarkably, the CoPHEE chip is design at Global Foundries with a 130nm technology node. Specifically, we used the Multi-Project Wafer (MPW) service from MOSIS, with a die area of 9mm2 and a target frequency of 58.8 MHZ.

II. PRELIMINARIES

A. Homomorphic Encryption and the Paillier Cryptosystem
Homomorphic encryption is a special form of encryption that allows meaningful manipulations directly on encrypted data. This property enables outsourcing to a third party the evaluation of a function that hides its inputs using encryption. Formally, if Enc denotes encryption, Dec denotes decryption, and f () is a regular function applied on plaintext values a, b, then homomorphic encryption supports the following property:

$$f(a,b) = Dec\ g(Enc(a), Enc(b)) \qquad (1)$$

Where g() is the homomorphic counterpart of f() that operates on encrypted values Enc(a), Enc(b). PHE cryptosystems support only one f (e.g., either addition or multiplication), while FHE cryptosystems support two orthogonal functions both addition and multiplication that form a functionally complete set of operations. Nevertheless, the corresponding g

979-8-3315-3968-9/24 $31.00 © 2024 IEEE

functions of PHE are significantly more efficient that the g1, g2 functions of FHE, so FHE remains prohibitive for practical applications.

B. Cryptoleq's Secure Multiplexer for Universal Computation

The homomorphic operation of Paillier enables the implementation of single-instruction abstract machines that can execute algorithms with runtime decisions that are not controlled by encrypted values [7]. In this case, to also support ciphertext-based runtime decisions, it is possible to extend the homomorphic properties of Paillier using a secure multiplexer [8]. The latter operates within a root-of trust and selects between two ciphertexts Y, Z using a cryptographic predicate P over a third ciphertext X. To support universal computation on Paillier ciphertexts, the aforementioned blueprint was introduced in Cryptoleq as function G, where the sign of Dec(X) is the predicate for selecting between Z = Enc(0) (i.e., the homomorphic identity element) and its second input Y :

$$G(X,Y) = P(X).Y + 1 - P(X).Enc(0) \qquad (2)$$

Where $P(X) = 0$ if $Dec(X) \leq 0$, otherwise $P(X) = 1$. In this case, P(X) can be computed efficiently using modular exponentiation of X to a private parameter FKF [9].

C. Threat Model with Hardware Root-of-Trust

The Paillier cryptosystem offers provable security guarantees against Chosen-Plaintext Attacks, based on the hardness of decisional composite residuosity problem (DCRP) [10]. According to DCRP, for a given integer x and composite n = pq, it is hard to decide whether there exists an integer y n such that x = y mod n2, without knowing the factors of n. Inheriting the security guarantees of DCRP, our threat model assumes that it is intractable to recover any value protected using Paillier when the bitsize of n is sufficiently large (i.e., at least 1024 bits).

III. COPHEE DESIGN FLOW OVERVIEW

The main objective of the CoPHEE co-processor is native support for modular operations over integers of size up to 2048 bits, which is essential for accelerating PHE cryptosystems such as Paillier. Our modular operations comprise multiplication, exponentiation and inversion. Moreover, our chip implements secure multiplexing, computes the GCD of 2048- bit integers, generates truly random numbers, and features a UART interface to communicate with the main processor.

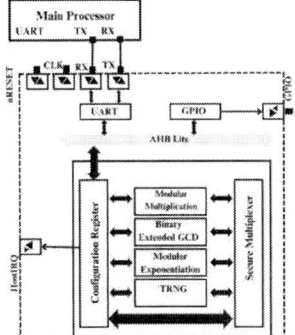

Fig. 1: Bus architecture diagram of CoPHEE.

A. External Interfaces

In given table we present the main IO pins of CoPHEE and how they interface with the host computer. Using the receiver line RX of the UART interface, the host computer can program the value of each operand and then trigger a desired operation (e.g., modular multiplication). As soon as the trigger bit in the CoPHEE configuration registers is set through UART, the requested operation starts executing on the co-processor. When the operation terminates, CoPHEE sets the interrupt line HostIRQ to signal the host that the requested output is ready. After receiving this interrupt, the host processor reads the computed result via UART using the transmitter line TX and clears the interrupt HostIRQ. Moreover, CoPHEE also features GPIO to assist debugging and post-silicon validation.

B. Internal Data Flow

In Fig. 1 we present the internal bus architecture diagram of CoPHEE that is based on a single-master two-slave system, Where the master communicates to the slaves using a 32-bit AHB-Lite bus protocol. Our goal is to make the AHB-Lite design parameterizable to facilitate the addition of masters or slaves to the bus. In CoPHEE, the UART is the only master on the bus, while the configuration registers unit and the GPIO are slaves. The configuration registers unit comprises special registers for the Paillier key, public modulus and encrypted operands, as well as registers for triggering operations, storing computed results and operation status.

C. Design Blocks

The design details of these blocks, as well as the design of our secure multiplexer and true random number generator (TRNG) blocks are presented in this section.

1) Modular Multiplication: Modular Multiplication refers to a mathematical operation performed by our chip, which involves multiplying two integers within a specific modular arithmetic framework. In our implementation, this process is carried out using an interleaved modular multiplier that supports integer values up to 2048 bits in length. This approach demonstrates greater efficiency when compared to more intricate algorithms like Montgomery multiplication, particularly in scenarios where the multiplication of the two inputs occurs only once.

2) Modular Exponentiation: Modular Exponentiation is a fundamental mathematical operation in Co-Processor Using Partial Homomorphic Encryption, executed through the implementation of a Montgomery multiplier. In contrast to interleaved multipliers which employ subtractions for modular reductions, Montgomery multiplication employs a more efficient bitwise right-shift operation. However, for Montgomery multiplication to be effective, its operands need to undergo a preliminary transformation into the Montgomery domain. This transformation entails multiplying each operand by 2N (achieved through bitwise left-shifting) and subsequently reducing the result modulo M, where N represents the bit width of an odd modulus, M. The final outcome of the Montgomery multiplication is transformed back into a standard integer using a

multiplication with the inverse of 2N modulo M. Modular exponentiation mathematical equation:

$$mod\_exp = (a * b) * r^{(-2N)} mod(m) \quad (3)$$

3) Modular Inversion: The modular inverse of an integer X over a modulus M the modular inverse of an integer X with respect to a modulus M exists when the greatest common divisor (GCD) of X and M equals 1. This mathematical property can be efficiently calculated using the binary extended greatest common divisor algorithm. Specifically, given input values X and M, the binary extended GCD algorithm calculates values A, B, and G that satisfy the equation A· X + B ·M = G, also known as Bézout's identity. If G is equal to 1, then A represents the modular inverse of X modulo M, denoted as X^{-1} mod M. Co-Processor Using Partial Homomorphic Encryption integrates a modular inversion block designed to compute $X^{\wedge}$ (-1) mod M using inputs X and M provided by the host processor. The modular inversion computation begins as soon as the "En" (Enable) input signal is toggled. Internally, this block establishes two instances of Equation 3 as follows:

$$-X_g = A_x.X + B_x.M \quad (4)$$
$$-Y_g = A_y.X + B_y.M \quad (5)$$

The variables {Ax, Bx, Ay, By} are initialized to {1, 0, 0, 1}, ensuring that the initial values of {Xg, Yg} are {X, M}.

4) True Random Number Generation:The True Random Number Generator (TRNG) design within Co-Processor Using Partial Homomorphic Encryption is grounded in a bi-stable circuit approach. This technique involves employing 16 discrete TRNG blocks distributed across the chip's architecture. This multiblock TRNG design significantly enhances the generated randomness. This improvement stems from 34 harnessing the inherent variations within the chip's designing process. Our resultant stream of random numbers is produced by applying an XOR operation to the outputs of all 16 individual TRNG blocks.

5) Secure Multiplexer: A Secure Multiplexer in Co-Processor Using Partial Homomorphic Encryption follows the framework established by Cryptoleq and encompasses the integration of various design 35 components. This secure multiplexer operates as a state machine, orchestrating the functionality of the design blocks mentioned earlier. Specifically, the secure multiplexer takes into consideration the following inputs: 1. Two encrypted inputs denoted as X and Y. 2. A function of the private (decryption) key denoted as FKF. 3.

6) Auxiliary Communication & Control Blocks:To support communication and control, CoPHEE incorporates the following auxiliary blocks: (1) UART master (used to interface with the external host computer), (2) configuration registers unit (used to store the operands, modulus and results), (3) GPIO (used to assist debugging during post-silicon validation), and (4) AHB bus interconnect (used to transfer data inside the chip). In our chip, the UART is the master, while the configuration registers unit and GPIOs are the slaves. Moreover, our configuration registers unit consists of 39 registers, with their size varying from 32-bit to 2048-bit.

D. Design Module Verification

Functional simulation is the process of running simulations to validate the correct behavior of a digital circuit or system. Simulation results help identify design bugs, ensure functional correctness, and gain confidence in the system's behavior before moving to subsequent design stages. We verified the functionality of our RTL design using both simulation and FPGA-based validation. The simulation was performed using on EDA Playground, Modelsim and Vivado at the top-level and block-level using random inputs (since the 2048-bit operand range cannot be exhaustively tested). Moreover, we test each and every module of our COPHEE design on Spartan-3.

IV. PHYSICAL DESIGN

The CoPHEE chip is design using the Multi Project Wafer (MPW) program of MOSIS using Skywater PDK. The PDK for the skywater 130pdk process provided us all the technology-related files, and we used the physical and timing libraries for standard cells and IO Pad libraries. Our chip was implemented flat, without any physical hierarchy.

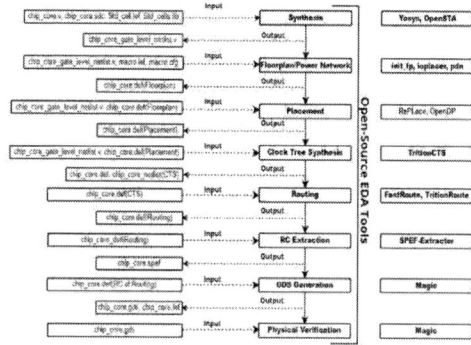

Fig. 2: Design Flow (Openlane) diagram

A. Synthesis

After successful functional verification, synthesis is carried out to convert the RTL code into a gate-level representation. Synthesis involves mapping the design to a library of standard cells and optimizing the circuit for area, power, and timing constraints.

B. Place and Route

1) Die size estimation: The physical parameters with respect to our layout after multiple iterations will be different. The minimum chip size supported by Global Foundries for 130pdk through MOSIS is (2.920um x 3.520um).

2) Floor Planning: Our layout outline along with IO pad placement adheres to the IO pad placement guidelines from Global Foundries. We have a total of 27 IO pads, where 8 of them are for VDD/VSS core power/ground supply, 8 DVDD/DVSS are for IO power/ground, while the remaining 11 are signal pads. One supply and one ground pad would be sufficient, but we utilize the empty spaces in the IO pad ring to improve the power structure robustness. The empty spaces between the pads are filled with filler pads to maintain continuity of the internal power/ground and other special signals.

TABLE I: Floorplan Statistics.

979-8-3315-3968-9/24 $31.00 © 2024 IEEE

Parameter	Value	Parameter	Value
Endcaps Inserted	278	VDD Voltage	1.800V
Tap cells Inserted	2115	PDN Nodes (VPWR)	10692
Tech. Layers created	13	PDN Nodes (VGND)	10692
Tech. Vias Created	25	All PDN Stripes Connected	Yes
Library Cells Created	441	Row Added to sites	139
Die Area(ums)	1600	VSRC location (um)	(129.870,129.920)
Output Delay Set	2	VSRC size (um)	10.000
Input Delay Set	2	VSRC relocated to (um)	(140.820,121.180)

3) Power Planning: We created a core power ring illustrated with the blue and green lines around the core in Fig. 3, which are located in metal layers (green horizontal) and (blue vertical) Fig. 3a shows the power straps distribution.

4) Placement and Optimization: Prior to the standard cell placement, we grouped and distributed our TRNG modules using bounds/regions in the chip to leverage on-chip variation. We note that the standard cells in the TRNG modules were not allowed to be optimized. In Fig. 3c we illustrate the TRNG module distribution, where the red highlighted standard cell groups compose our TRNG modules. After fixing these groups in position, the rest of the standard cells went through placement and optimization. Our design was then analyzed for timing, congestion, area and power, passing all requirements.

5) Routing: During GDS designing, the routing of the encrypted data co-processor entails establishing the physical connections between various functional blocks, interconnects, and metal layers of the IC layout.

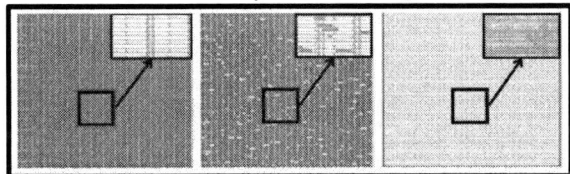

Fig. 3: Design Floorplaning (a), Placement (b) and Routing(c)

TABLE II: Power Statistics

Group	Internal power	Switching Power	Leakage Power	Total power	%
Sequential	5.07e-05	2.85e-05	1.61e-09	7.89e05	35.6%
Combinational	7.11e-05	7.14e-05	3.88e-09	1.43e04	64.4%
Macro	0.00e+00	0.00e+00	0.00e+00	0.00e+00	0.0%
Pad	0.00e+00	0.00e+00	0.00e+00	0.00e+00	0.0%
Total	1.22e-04	9.99e-05	5.49e-09	2.21e04	100.0%

6) Clock Tree Synthesis: Clock Tree Synthesis (CTS) plays a pivotal role in the design of an encrypted data coprocessor during the GDS (Graphic Database System) phase. It ensures and optimize that each part of the design get proper clock.

TABLE III: Post-CTS Statistics

Parameters	Value
Number of Standard Cells	242
Sequential power	1.07e-04
Combinational power	9.21e-05
Number of wire Segments	137808
Number of Signal nets	1275
Frequency	58.8MHZ
Design Area Utilization	11587um^2
Total Wire Length Library	10888um
delay max (Setup)	12.59 Slack

delay min (Hold)	3.67 Slack

7) Physical Verification: In this step, we verify the final layout against foundry manufacturing rules and insert dummy metal/poly fills using the Openlane software to meet the foundry's density requirements. The fill GDS obtained from Openlane is merged with the design GDS in our layout tool (magic), generating a final design ready for Design Rule Check (DRC) and Layout Versus Schematic (LVS) analysis. We used the Openlane magic tool to run DRC, LVS, and Antenna checks, and fixed any violations before re-running the checks to ensure a clean GDS for our tapeout.

Fig. 4: GDS-II of Design.

V. CONCLUSIONS AND FUTURE WORK

The design of an encrypted data co-processor in 130nm technology using open-source tools has culminated in a comprehensive and cost-effective solution for cryptographic applications. This endeavor has been a testament to the power of leveraging open-source resources in the field of chip design. The open nature of these tools allowed us to customize and fine-tune various aspects of the co-processor to meet our specific requirements. Given the silicon, future work will explore side-channel analysis and information extraction through power, timing, and electromagnetic emissions.

VI. REFERENCE

[1] M. Varia, S. Yakoubov, and Y. Yang, "HEtest: A Homomorphic Encryption Testing Framework," in Financial Cryptography and Data Security. Springer, 2015, pp. 213–230.

[2] C. Barron, H. Yu, and J. Zhan, "Cloud computing security case studies and research," in World Congress on Engineering, 2013, pp. 1287–1291.

[3] Nabeel, Mohammed, Mohammed Ashraf, Eduardo Chielle, Nektarios G. Tsoutsos, and Michail Maniatakos. "CoPHEE: Co-processor for partially homomorphic encrypted execution." In 2019 IEEE International Symposium on Hardware Oriented Security and Trust (HOST), pp. 131-10. IEEE, 2019.

[4] C. Gentry, "Fully homomorphic encryption using ideal lattices," in ACM Symposium on Theory of Computing, 2009, pp. 169–178. [5] R. L. Rivest, L. Adleman, and M. L. Dertouzos, "On data banks and privacy homomorphisms," Foundations of secure computation, vol. 4, no. 11, pp. 169–180, 1978.

[6] M. Naehrig, K. Lauter, and V. Vaikuntanathan, "Can homomorphic encryption be practical?" in Cloud Computing Security Workshop. ACM, 2011, pp. 113–124.

[7] O. Mazonka, N. G. Tsoutsos, and M. Maniatakos, "Cryptoleq: A heterogeneous abstract machine for encrypted and unencrypted computation," IEEE Transactions on Information Forensics and Security, vol. 11, no. 9, pp. 2123–2138, 2016.

[8] M. Varia, S. Yakoubov, and Y. Yang, "HEtest: A Homomorphic Encryption Testing Framework," in Financial Cryptography and Data Security. Springer, 2015, pp. 213–230.

[9] Y. Zhang, A. Juels, M. K. Reiter, and T. Ristenpart, "Cross-VM side channels and their use to extract private keys," in Computer and Communications Security (CCS), 2012, pp. 305–316.

[10] P. Paillier, "Public-key cryptosystems based on composite degree residuosity classes," in Advances in cryptology–EUROCRYPT'99. Springer, 1999, pp. 223–238.

Exploiting Functional Approximation on Decision-Tree based Multiple Classifier Systems

Mario Barbareschi*, Salvatore Barone*, Antonio Emmanuele* and Nicola Mazzocca*,
*Department of Electrical Engineering and Information Technologies, University of Naples Federico II, 80125, Naples, Italy
Email: {name.surname}@unina.it

Abstract—**Multiple Classifier Systems (MCSs) have been increasingly designed to take advantage of hardware features, such as high parallelism and computational power, to guarantee higher throughput and lower latency. Although the combination of multiple classifiers leads to high classification accuracy, the required area overhead makes the design of a hardware accelerator unfeasible, hindering the adoption of commercial configurable devices. For this reason, in this paper, we exploit the Approximate Computing (AxC) design paradigm to automatically generate approximated hardware implementations of MCSs by trading hardware area overhead off for classification accuracy. In particular, we propose an algorithm that identifies the resiliency source of the model and uses it to introduce approximation with minimum accuracy loss.**

Index Terms—**Ensemble Learning, Error Resiliency, Approximate Computing, Pruning, CAD**

I. INTRODUCTION

The need to process large volumes of information in a timely and reliable manner has led to the increasing adoption of machine learning models. In this context, hardware acceleration plays a significant role, since it enables energy and cost-efficient computation [1].

Among the machine learning models, Decision Tree based Mutiple Classifier Systems (DT MCSs) are one of the most adopted for classification and regression tasks. They allow for high accuracy by combining the outputs of multiple models to obtain the most likely result. Thanks to their simplicity and accuracy, these models are widely used in various applications, ranging from energy efficiency [2] to healthcare [3]. Due to their widespread adoption, many hardware implementations have been proposed for accelerating the inference phase. However, the need for high-accuracy models entails an increase in the number of trees and their depth, resulting in energy-inefficient hardware models, challenging to implement [4].

The contrasting goals of achieving accurate DT MCSs and feasible hardware implementations can be attained through the use of Approximate Computing (AxC) techniques, that leverages the inner error-resiliency of applications to enhance specific metrics, e.g., computational time and energy-efficiency. In other words, this design paradigm exploits the gap between the level of accuracy required by the application or the end-users, and that provided by the computing system – with the former being often far lower than the latter – for optimization. Thus, a wide range of applications benefit from

Authors contributed equally and are listed in alphabetical order.

AxC, including signal processing [5], artificial intelligence [6], [7], circuit design [8], [9], and so forth.

This paper proposes a new AxC-based design method for hardware accelerators processing DT MCSs. Our proposal exploits state-of-the-art AxC methods [8], [10], and is based on an iterative evaluation of the model resiliency, to maximize classification accuracy. We evaluate our proposal on models trained on three showing that, with minimum accuracy loss, significant savings can be obtained.

II. DECISION-TREE BASED CLASSIFIER SYSTEMS

A Decision Tree (DT) is a white-box classification model representing its decisions through a tree-like structure composed of an internal set of nodes containing *test conditions*, and leaf nodes which represent *class labels*. Nodes are joined by arcs symbolizing possible outcomes of each test condition, and classes can be either categorical or numerical. In the former case, we refer to classification trees, while in the latter case, we refer to regression trees. Significant improvements in classification accuracy have resulted from growing an ensemble of trees and letting them vote for the most popular class [11].

The need to accelerate the inference phase emerged inherently in some application fields: many tasks require high prediction speed, and software implementations may not meet the requirements, even if the multi-threading and General Purpose - Graphic Processing Unit (GP-GPU) technologies are adopted [4]. According to [12], architectures of accelerators for DT-based predictors can be categorized as comparator-centric and memory-centric. The former offers better performance for DT MCS while the latter offers better utilization of FPGA resources. In the following, we exploit the architecture being proposed in [13]. Particularly, this architecture evaluates which leaf is reached by computing in parallel all the nodes conditions, regardless of the position and depth at which nodes are located. A Boolean decision variable that indicates the fulfilling of a given condition is produced for each one of the evaluated predicates, and a Boolean function is defined to determine which class the input belongs to. Specifically, let

- $\mathcal{D}_\rho^t(f)$ be a node of the decision tree comparing the feature f against the constant threshold t using the operator $\rho \in \{<, \leq >, \geq, =, \neq\}$, i.e., a *decision box*;
- $\mathcal{BF}_c$ be the Boolean function that has to raise *logic-1* when the classification outcome results in "class c";

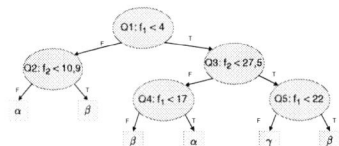

Fig. 1. Example of a decision tree.

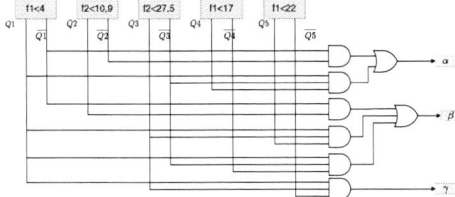

Fig. 2. Hardware implementation of a decision tree.

- $\mathcal{P}_c(i)$ be the i-th path leading to the i-th leaf node that is labeled as class c, with $i \in \{0, 1, \cdots, \text{leaves\_of}(c) - 1\}$;
- $\mathcal{A}_c(i)$ the logical intersection of all the decision boxes $\mathcal{D}_\rho^t(f)$ along $\mathcal{P}_c(i)$.

Each $\mathcal{D}_\rho^t(f)$ is a literal of $\mathcal{A}_c(i)$, and it can be taken either directly if the path continues towards the left child, or negated if the path continues towards the right child of the considered node, i.e., either as $\mathcal{D}_\rho^t(f)$ or $\overline{\mathcal{D}_\rho^t(f)}$. Thus, $\mathcal{A}_c(i)$ can be expressed as in Equation (1), and since the classification outcome is "class c" if any of the leaves labeled as c is reached during the tree visiting, $\mathcal{BF}_c$ can be expressed in the Sum-Of-Products (SOP) form, as in Equation (2).

$$\mathcal{A}_c(i) := \bigwedge_{\mathcal{D}_\rho^t(f) \in \mathcal{P}_c(i)} \begin{cases} \mathcal{D}_\rho^t(f) & \text{if the left child is taken} \\ \overline{\mathcal{D}_\rho^t(f)} & \text{if the right child is taken} \end{cases} \tag{1}$$

$$\mathcal{BF}_c := \bigvee_{i=0}^{\text{leaves\_of}(c)-1} \mathcal{A}_c(i) \tag{2}$$

For explanatory purpose, we report we report Boolean functions for classes of the model depicted in Figure 1.

$$\begin{aligned} \alpha &= (\overline{Q1} \wedge \overline{Q2}) \vee (Q1 \wedge \overline{Q3} \wedge Q4) \\ \beta &= (\overline{Q1} \wedge Q2) \vee (Q1 \wedge Q3 \wedge Q5) \vee (Q1 \wedge \overline{Q3} \wedge \overline{Q4}) \\ \gamma &= Q1 \wedge Q3 \wedge \overline{Q5} \end{aligned} \tag{3}$$

As for the classification outcome of multiple decision forest, we used the majority voter discussed in [13].

III. AUTOMATIC DESIGN OF APPROXIMATED DTMCSS

Focusing on the implementation introduced in Section II, we propose an automatic design flow for approximating the $\mathcal{BF}_c$. The choice to approximate $\mathcal{BF}_c$ is dictated by the observation that in a DT MCSs a single tree can misclassify a sample as long as it is correctly classified by the majority of trees in the ensemble.

A. Catalog-based Approximate Logic Synthesis

Since $\mathcal{BF}_c$ is essentially a boolean network, consolidated automatic approximation strategies can be used. The authors of [10] propose an approximation method of boolean circuits based on their And-Inverter Graph (AIG) representation.

AIG is a structural representation for boolean networks based on directed acyclic graphs. In an AIG, nodes can be either Primary Inputs (PIs) that have no *fan-in*, meaning that they have no incoming edges and hence, there is no node driving them, or logic-AND nodes, which have two incoming edges. Furthermore, nodes having no *fan-out*, i.e, nodes that do not drive any other node, are called Primary Outputs (POs). For what pertains to edges, they represent physical connections between nodes, and they can be indicated as complemented or not. Conventionally, the polarity of complemented edges is 0.

A k-feasible, or k-cut, of a node n, called *root*, is a set of at most K nodes of the AIG such that each path from a PI to n passes through at least one node of the set. The approach from [10], given the AIG of a combinational circuit, first enumerates k-feasible cuts, and then introduces approximation by superseding k-cuts with approximate ones, that are generated by exploiting the Exact Synthesis (ES).

Being f the Boolean function implemented by a k-cut, ES searches for those k-cuts implementing a function $f' \neq f$ that requires less resources w.r.t. f, and yet complies with error constraints. The latter constraints are expressed in terms of the Hamming distance between f and f'. The output of the ES is a list, referred as catalog, in which each element, corresponding to a k-cut, is a set of its approximated variants ordered by the Hamming distance that their implemented function f' has from the original function f.

The choice of the best superseding is then done in the Design-Space Exploration (DSE) phase that selects the best approximated variant of the circuit by minimizing its area size while keeping the introduced error to a minimum. Since these two are contrasting goals, Multi-objective Optimization Problem (MOP) algorithms are employed, using k-feasible as decision variables and catalog entries as their domain.

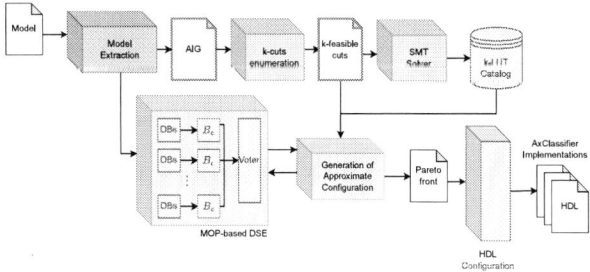

Fig. 3. Automatic approximation workflow targeting DT MCSs

B. Automatic Design Flow

The automatic workflow for generating approximate DT MCSs is represented in Figure 3. Starting from the model of the classifier, the assertion functions of the entire DT MCS,

979-8-3315-3968-9/24 $31.00 © 2024 IEEE

corresponding to each of the DTs, are extracted, in order to compute their AIG and to enumerate k-feasible cuts for each of them; then, the ES-based approximate cut generation is performed, as done in [10]. Once the catalog of approximate k-feasible cut has been computed, a MOP-based DSE is performed to select approximate configurations providing Pareto-optimal trade-offs between classification accuracy loss on a fraction of the test set T, referred as T_{dse}, and the number of Look-Up Tables (LUTs) of the $\mathcal{BF}_c$, both to be minimized. This phase is orchestrated by the Archived Multi-Objective Simulated Annealing (AMOSA) heuristic [14].

At the end of the DSE phase, the tool provides the set of Pareto-optimal trade-offs between classification accuracy loss and hardware overhead, which are exploited to configure the Hardware Description Language (HDL) source to implement the classifier. Since the DSE is carried out using only a fraction of the test set T_{dse}, the accuracy loss of the models in the Pareto-Front on unseen data is evaluated using a fraction of the test set, referred as T_{eval}, not overlapped with T_{dse}.

C. Approximation through resiliency evaluation.

The described workflow suffers from scalability issues as, when the number of classifiers in the ensemble increases, also the number of decision variables during the DSE increases, leading to computationally unfeasible problems. To overcome this issue, the workflow can be iteratively executed on a single DTs in the ensemble, limiting the number of decision variables. However, the accuracy loss being minimized during the DSE would be that of the considered DT, which may heavily impact the accuracy of the entire ensemble. In order to circumvent this problem, it is necessary to limit the approximation to the assertion functions contributing less to the accuracy of the ensemble. This is done by computing the correlation between the approximating assertion function and the ones of other DTs. Consider, for instance, two different leaves corresponding to the same class c, and let $\mathcal{A}_c(i)$ and $\mathcal{A}_c(j)$ respectively the $i-th$ and $j-th$ assertion function related to c. We define their correlation as in (4)

$$Q_{\mathcal{A}_c(i),\mathcal{A}_c(j)} = \frac{C_{\mathcal{A}_c(i),\mathcal{A}_c(j)} - D_{\mathcal{A}_c(i),\mathcal{A}_c(j)}}{C_{\mathcal{A}_c(i),\mathcal{A}_c(j)} + D_{\mathcal{A}_c(i),\mathcal{A}_c(j)}} \quad (4)$$

where $C_{\mathcal{A}_c(i),\mathcal{A}_c(j)}$ is the number of samples whose features make both assertions true, while $D_{\mathcal{A}_c(i),\mathcal{A}_c(j)}$ is the number of samples that activate only one of the two functions.

It is clear that, being the sum of these two values the set of samples activating at least one of the assertions, $Q_{\mathcal{A}_c(i),\mathcal{A}_c(j)}$ ranges from -1, indicating completely uncorrelated functions, to $+1$, denoting full correlation among two functions. Therefore, the resiliency of the ensemble when approximating $\mathcal{A}_c(i)$ is computed by summing its correlations with the assertion functions of the other DTs.

The workflow represented in Figure 3 can be iteratively executed for each tree in the ensemble by adding a resiliency evaluation step ensuring that only the assertions having the greatest correlation are enumerated for approximation. Correlation is evaluated on a portion of the test set T_{res}, not overlapped with

TABLE I
ACCURACY (%) ON T_{EVAL} OF THE TRAINED MODELS.

Dataset	5-DTs	10-DTs	15-DTs	20-DTs	25-DTs	30-DTs	35-DTs	40-DTs
Image Seg.	92.24	94.82	95.68	96.55	96.57	98.27	99.13	93.96
Spambase	94.34	92.17	96.95	94.78	97.39	93.47	94.34	95.21
Dry Bean	91.33	92.07	92.95	93.24	93.34	92.07	94.41	92.80

TABLE II
NUMBER OF LUTS USED FOR $\mathcal{BF}_c$ IN THE NON-APPROXIMATED TRAINED MODELS.

Dataset	5-DTs	10-DTs	15-DTs	20-DTs	25-DTs	30-DTs	35-DTs	40-DTs
Image Seg.	1770	2310	3380	8320	8070	11540	11510	13590
Spambase	16770	36610	55790	69240	90830	105520	229530	141910
Dry Bean	15670	31340	50500	70200	100590	125230	236730	275170

T_{dse} and T_{eval}. At each iteration, the DT in the Pareto-Front having the best accuracy on T_{eval} is substituted to its original counterpart to build the new approximated ensemble.

IV. IMPLEMENTATION AND EVALUATION

The evaluation was carried out using several models trained to address classification tasks involving benchmark datasets taken from the UC Irvine Machine Learning Repository[1]. In particular, we considered the Dry Bean [2], the Spambase [3], and the Image Segmentation [4] datasets. We resort to the Scikit-learn [5] for training DT MCSs while considering several datasets and generated and on Predictive Model Markup Language (PMML) file to export trained models. For each dataset, several random forest models were trained, varying the number of trees from 5 to 40 with a step of 5. The test set T is split into three equally sized non-overlapping sets: T_{res}, T_{dse} and T_{eval}; the accuracy of the original models on T_{eval} is reported in Table I.

Each model has been implemented in hardware, by using its PMML on the target AMD/Xilinx Zynq-7020 FPGA. These implementations have been used in order to evaluate the number of LUTs for the $\mathcal{BF}_c$ of the non-approximated models, reported in Table II. For each exact hardware model an approximated variant is generated using the proposed approach allowing, during each iteration, approximation only on the 40% of the assertions of a DT, using the resiliency evaluation procedure. We selected this percentage as it ensures the best trade-off between computational time of the DSE and quality of the results.

Results, reported in Figure 4, show how the accuracy loss varies w.r.t. the number of LUTs and the total LUTs saving. In that Figure, the ● denotes the FPGA LUTs of the synthesized circuits, while ⋆ represents LUTs savings, in percentage.

As the reader can observe, significant gains in terms of number of LUTs can be achieved through the proposed method, while the accuracy loss is negligible, or even equal to 0. This confirms that the metric successfully identifies the resiliency source within the model, allowing for a careful

---

[1] https://archive.ics.uci.edu/

[2] https://archive.ics.uci.edu/dataset/602/dry+bean+dataset

[3] https://archive.ics.uci.edu/dataset/94/spambase

[4] https://archive.ics.uci.edu/dataset/147/statlog+image+segmentation

[5] https://scikit-learn.org/stable/

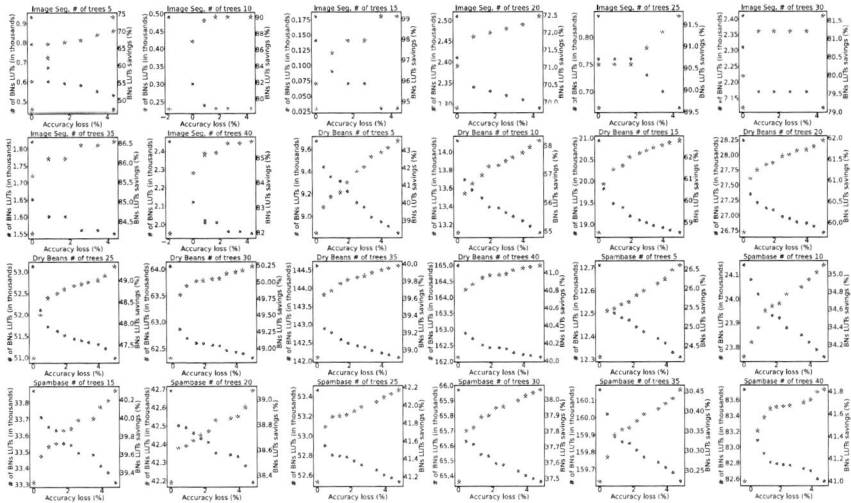

Fig. 4. Experimental results on the Image Segmentation, Spambase, Dry Beans. The ● denotes the hardware requirements of the synthesized circuits, in terms of FPGA LUTs, while ⋆ represents savings, in percentage.

approximation while simultaneously preserving the accuracy of the ensemble. Indeed, achieved savings peak more than 50% in many cases, and, additionally, the more DTs are involved in the classification process, the more LUTs can be saved. This, clearly, is due to a larger portion of error resiliency that can be exploited by our approach.

V. CONCLUSION

In this paper, we propose an automatic approach for the design and synthesis of approximate DT MCSs. Our proposal exploits the correlation between leaves corresponding to a same given class, but belonging to different DTs. Experimental results empirically prove that our approach successfully generates approximated variants of the original classifier achieving significant savings – up to 50% – with minimum accuracy loss.

REFERENCES

[1] W. Li and M. Liewig, "A Survey of AI Accelerators for Edge Environment," in *Trends and Innovations in Information Systems and Technologies*, ser. Advances in Intelligent Systems and Computing, . Rocha, H. Adeli, L. P. Reis, S. Costanzo, I. Orovic, and F. Moreira, Eds. Cham: Springer International Publishing, 2020, pp. 35–44.

[2] M. Zeki-Suac, S. Mitrovi, and A. Has, "Machine learning based system for managing energy efficiency of public sector as an approach towards smart cities," *International Journal of Information Management*, vol. 58, p. 102074, Jun. 2021. [Online]. Available: https://www.sciencedirect.com/science/article/pii/S0268401219302968

[3] C. Iwendi, A. K. Bashir, A. Peshkar, R. Sujatha, J. M. Chatterjee, S. Pasupuleti, R. Mishra, S. Pillai, and O. Jo, "COVID-19 Patient Health Prediction Using Boosted Random Forest Algorithm," *Frontiers in Public Health*, vol. 8, 2020. [Online]. Available: https://www.frontiersin.org/journals/public-health/articles/10.3389/fpubh.2020.00357

[4] B. Van Essen, C. Macaraeg, M. Gokhale, and R. Prenger, "Accelerating a Random Forest Classifier: Multi-Core, GP-GPU, or FPGA?" in *2012 IEEE 20th International Symposium on Field-Programmable Custom Computing Machines*, Apr. 2012, pp. 232–239.

[5] M. Barbareschi, S. Barone, A. Bosio, J. Han, and M. Traiola, "A Genetic-algorithm-based Approach to the Design of DCT Hardware Accelerators," *ACM Journal on Emerging Technologies in Computing Systems*, vol. 18, no. 3, pp. 1–25, Jul. 2022. [Online]. Available: https://dl.acm.org/doi/10.1145/3501772

[6] M. Barbareschi, S. Barone, and N. Mazzocca, "Advancing synthesis of decision tree-based multiple classifier systems: an approximate computing case study," *Knowledge and Information Systems*, pp. 1–20, Apr. 2021. [Online]. Available: https://link.springer.com/article/10.1007/s10115-021-01565-5

[7] S. Barone, M. Traiola, M. Barbareschi, and A. Bosio, "Multi-Objective Application-Driven Approximate Design Method," *IEEE Access*, vol. 9, pp. 86 975–86 993, 2021.

[8] M. Barbareschi, S. Barone, N. Mazzocca, and A. Moriconi, "FPGA Approximate Logic Synthesis through Catalog-Based AIG-Rewriting Technique," *Journal of Systems Architecture*, vol. (In Press), 2024.

[9] A. Piri, S. Pappalardo, S. Barone, M. Barbareschi, B. Deveautour, M. Traiola, I. OConnor, and A. Bosio, "Input-aware accuracy characterization for approximate circuits," in *2023 53rd Annual IEEE/IFIP International Conference on Dependable Systems and Networks Workshops (DSN-W)*. Porto, Portugal: IEEE, Jun. 2023, pp. 179–182. [Online]. Available: https://ieeexplore.ieee.org/document/10207125/

[10] M. Barbareschi, S. Barone, N. Mazzocca, and A. Moriconi, "A Catalog-based AIG-Rewriting Approach to the Design of Approximate Components," *IEEE Transactions on Emerging Topics in Computing*, 2022.

[11] L. Breiman, "Random forests," *Machine learning*, vol. 45, no. 1, pp. 5–32, 2001, publisher: Springer.

[12] X. Lin, R. S. Blanton, and D. E. Thomas, "Random Forest Architectures on FPGA for Multiple Applications," in *Proceedings of the on Great Lakes Symposium on VLSI 2017*, ser. GLSVLSI '17. New York, NY, USA: Association for Computing Machinery, May 2017, pp. 415–418. [Online]. Available: https://doi.org/10.1145/3060403.3060416

[13] M. Barbareschi, "Implementing Hardware Decision Tree Prediction: A Scalable Approach," in *2016 30th International Conference on Advanced Information Networking and Applications Workshops (WAINA)*. Crans-Montana, Switzerland: IEEE, Mar. 2016, pp. 87–92. [Online]. Available: http://ieeexplore.ieee.org/document/7471178/

[14] S. Bandyopadhyay, S. Saha, U. Maulik, and K. Deb, "A Simulated Annealing-Based Multiobjective Optimization Algorithm: AMOSA," *IEEE Transactions on Evolutionary Computation*, vol. 12, no. 3, pp. 269–283, Jun. 2008.

FortBoot: Fortifying Rooted-in-Device-Specific Security through Secure Booting

Sajeed Mohammad
Electrical and Computer Engineering
Univeristy of Florida
Gainesville, Florida 32608
Email: mlnu@ufl.edu

Farimah Farahmandi
Electrical and Computer Engineering
Univeristy of Florida
Gainesville, Florida 32608
Email: farimah@ece.ufl.edu

Abstract—With the complex System-on-Chip (SoC) architectures ever-increasingly used in security-critical applications, realizing a secure boot procedure is of at-most importance, in which the integrity and authenticity of the firmware (FW) on hardware (HW), as well as the trustworthiness of the software (SW), will be verified. Considering the complexity of the existing solutions and their vulnerability to emerging attacks, in this paper, we propose FortBoot, which is a comprehensive framework to strengthen secure boot using an integration of (1) *dynamic key generation* for unique randomness against brute-force and guessing attacks; (2) *Mutual FW-HW binding* for restricting unauthorized and malicious FW load/execution; (3) *anti-rollback measures* for preventing the use of (maliciously) flawed/downgraded/altered FW to be loaded/executed on unauthorized HW; and (4) *dynamic code attestation* for providing real-time code integrity confirmation. To show its effectiveness, FortBoot is implemented on OpenTitan SoC, whose comprehensive security assessment verifies its resistance against a wide range of advanced attacks.

Index Terms—Secure Boot, Attestation, Identity, Malware.

I. INTRODUCTION

Secure boot in SoCs is one of the countermeasures designed to cryptographically ensure the integrity and authenticity of the SW/FW on the HW [1], hence safeguarding it against FW attacks, malware (MW) infections, side-channel attacks [2], etc. By ensuring that only verified, trusted SW/FW is executed on HW, the secure boot process serves as the system's Root of Trust (RoT) and provides pivotal protection features, including device identity for asset management, secure manufacturer-led device update provisioning, data and code protection. Secure boot process must be a multi-staged affair, from pre-boot to bootloader validation before the operating system (OS) is loaded. Using this procedure, the secure boot process aids in establishing a robust RoT for the SoC, providing an impregnable defense against various attacks. However, considering the existing solutions in this breed, they almost suffer from their impractical complexity, susceptibility to advanced attacks, and their undesirable low performance. Additionally, none of the existing solutions focus on the comprehensiveness of the secure boot process, enabling all the security features provided by the secure boot, which is the main motivation behind our proposed holistic architecture. Additionally, with the recent proliferation of open-source (RISC-V-based) SoC architectures (e.g., lowRISC OpenTitan [3]), an advantageous environment for transparent and exhaustive security protocol evaluations becomes available. Using this, in this paper, we propose FortBoot, a novel secure boot protocol with several security features whose main aim is to build a scalable and high-performance secure boot capable of defending against a wider array of security threats. The main contribution of this paper is as follows. *(1) 1-to-1 FW-to-HW Mapping:* In FortBoot, we propose an exclusive 1-to-1 FW-to-HW mapping that allows an authorized FW to be executed only on an authorized HW (and vice versa). *(2) Elevated-only Upgrade Provisioning:* FortBoot is equipped with a mechanism preventing rollback and replay attacks. *(3) {FW+HW}-based Identity Derivation:* The device identity in FortBoot is a combination of HW&FW specification. *(4) Expirable Key Generation and Data Encryption:* We introduce a specific dynamicity into the key generation and data encryption in FortBoot capable of preventing unauthorized access and tampering-based attacks. *(5) Security Evaluation on lowRISC OpenTitan SoC:* We implement FortBoot on the OpenTitan SoC [3] and provide a detailed security and performance analysis.

II. BACKGROUND AND RELATED WORK

As security features of an SoC, like firewalls, HW RoT, trusted execution environment (TEE), antivirus, etc., are inactive at the boot process, this process must be secured to establish a trusted foundation for the whole system. Although BIOS FW is responsible for initiating HW and transitioning control to the OS [4], [5], its unauthorized (malicious) manipulations may result in emerging threats, e.g., permanent denial-of-service (DoS) or MW infections [6]. Even with the shift to UEFI-based boot, the threat of MW exploiting the BIOS is on the rise.

A. Key Features of a Secure Boot Process

For a secure boot process to be comprehensive enough and effective against today's attack vectors, three aspects are needed and require carefully balanced trade-offs between security, performance, and system constraints:
(Req1) Key Derivation: A trustworthy (cryptographically) key derivation ensures the secure handling and management of cryptographic keys used during the boot sequence. Here, methods like RoT and tamper-resistant HW are employed,

---

979-8-3315-3968-9/24 $31.00 © 2024 IEEE

TABLE I: Computational Adversarial Effort Range for Different IPs.

Attack	Description	Solution in the proposed FortBoot
BIOS/UEFI	BIOS/UEFI are vulnerable to malware attacks during manufacturing, supply chain, or remote access.	Code attestation, FW-HW mapping.
FW update	Attackers manipulate the FW update process to hijack the system through malicious code injection.	Code attestation, Dynamic key generation.
Physical access	Attackers gain physical access to the system and install malicious FW to undermine the system.	FW-HW mapping, Dynamic key generation.
Cold boot	Attackers access the system's memory post-reboot or shut down to extract sensitive information.	Dynamic key generation.
Man-in-the-middle	Attacks involve intercepting communication and altering/stealing data during FW updates.	Code attestation, Dynamic key generation.
FW rootkits	Malware alters FW to achieve system access, steal data, modify FW, and execute malicious code.	Rollback Prevention, Dynamic key generation.
Leakage-oriented	Attackers extract sensitive information or tamper with the FW at the boot stage.	FW-HW mapping, Dynamic key generation.
Malware-based	Malware exploits weak BIOS security controls or inherent vulnerabilities to modify system BIOS.	Code attestation, Dynamic key generation .

and non-volatile memory (NVM for storage) and physically unclonable functions (PUFs for generation) are used. *(Req2) FW Protection:* Since FW plays a crucial role in controlling system operations, there is a demand for safeguarding it from unauthorized access during the system startup phase. However, full encryption may not always be practical, and alternative methods, e.g., obfuscation [7]. Despite the numerous SW-based or HW-assisted FW protection methods that have been introduced so far [8], none of them is a one-size-fits-all solution for FW protection. *(Req3) Code Attestation:* To resist attacks like remote code execution, DoS, code attestation verifies the authenticity of the OS code. Various techniques could be used for attestation, e.g., digital signatures, TEE, remote attestation, and runtime analysis techniques.

B. Existing Attacks on Boot Process

To introduce a trustworthy secure boot process, several types of attacks must be considered, including BIOS/UEFI FW attacks [4], FW update attacks, MW-based attacks, physical access attacks, cold boot attacks, man-in-the-middle (MitM) attacks, FW rootkits, Bootkits [9], and leakage-oriented attacks. Table I provides a summarized definition of these attacks (+ the methods we have used to address these issues), showing why maintaining the integrity of system boot processes and FW is crucial. It is worth noting that many of the countermeasures try to resist a subset of these threats. However, none of them is a one-size-fits-all solution for FW protection.

C. Requirements of an Ideal Secure Boot Process

Ideally, a set of key components is required to guard against illicit modifications and MW targeted at boot (FW). It includes (i) *key management unit (KMU)* to generate/store/manage key and confirm FW's digital signatures; (ii) *code authentication unit* as an HW module that detects any tampering with the FW (bootloader) during the boot process; (iii) *protected ext memory* that safeguards the sourcing and storage of FW and SW (shielding them from any unauthorized tampering or modification); (iv) *RoT w/ TPM* as HW module in the SoC to instill trust in the FW, bootloader, and SW; and (iv) *cryptographic integrity check* to check the credentials - the authenticity and trustworthiness - of FW and SW.

To enhance robustness, the major components must be supported by a combination of other elements, such as TEE components, a secure boot chain, tamper-proof FW, a secure update protocol, and configurable security policies [10]. Building on this comprehensive description of requirements, the next section will illustrate how FortBoot effectively utilizes

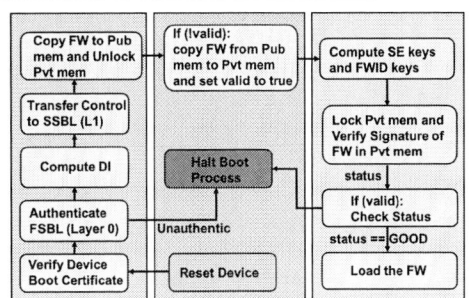

Fig. 1: Main Steps of Proposed Secure Boot Protocol, Highlighting the Sequence of Operations and Interactions Involved in System Boot-up.

a distinct subset of these components to achieve optimal robustness against the threats listed in Table I.

III. PROPOSED SCHEME: FORTBOOT

A. FortBoot

FortBoot leverages a security engine (SE) [10] and a host to establish a resilient chain of trust from the earliest stages of boot-up to the OS and SW (See Fig. 1). As FortBoot leverages SE, it begins with the 1st-stage bootloader for loading the FW from external memory into the locally trusted memory within the SE. The 1st bootloader verifies the digital signature of the 2nd-stage bootloader, confirming its authenticity and generates a *device identifier (DI)*, uniquely identifying the device SW/HW attributes. At 2nd-stage bootloader, the FW is transferred to secure private memory to mitigate violations associated with FW load protocols. Also, we keep the device's private memory locked during this phase, and an active flag ensures the FW code is copied only once. The device then generates the *SE Keys* and *FWID keys* needed for crypto operations. Once the SE FW has been successfully installed, the SE is responsible for authenticating the (untrusted) host FW from memory. This process is replicated at every stage of the boot process, with each stage verifying the digital signature of the next stage's code and generating crypto keys.

B. 1-to-1 FW-to-HW Mapping in FortBoot

In FortBoot, considering it on the manufacturing floor, every SOC possesses a unique key (K_D) securely stored in memory or the device's OTP. During provisioning, the server or asset management infrastructure (AMI) includes the hash of K_D (H_{K_D}) in the FW image. When the device boots, it generates its hash of K_D (H'_{K_D}), comparing it with the FW-provided H_{K_D}. If they match, the HW uses the public FW encryption key ($FW_{K_{pub}}$) to decrypt and run the FW. For HW-to-FW binding, the AMI in FortBoot encrypts the FW'

979-8-3315-3968-9/24 $31.00 © 2024 IEEE

using K_D, creates a hash of the encrypted FW ($H(E(FW))$), and combines it with the encrypted FW. This FW package is then encrypted using FWK_{pri} and stored in the device's NVM. During booting, the HW fetches the FW from the NVM, decrypts it with FWK_{pub}, and creates a hash of the encrypted FW ($H(E(FW'))$). This hash is compared with the stored hash, and if they match, the K_D key is used to decrypt the encrypted FW, allowing it to run on the device. By doing so, when the hash matching fails, the FW is declared unauthorized and is prevented from running. This bidirectional binding bolsters system security, ensuring the reliability and trustworthiness of the HW and FW together.

C. Elevated-only Upgrade Provisioning

The device HW in the FortBoot procedure is equipped with an anti-rollback counter, which allows only FW installations with a counter value higher than the existing one (elevated-only), thus preventing potential downgrades to insecure FWs. In this case, a real-time clock (RTC) facilitates FW validity verification during updates by keeping track of the current date and time. The bootloader cross-verifies the FW version and updates the timestamp to ensure validity and recent updates. Also, checks for authenticity and integrity are performed every time FW is accessed, ensuring no outdated versions are used, with the benchmark set by a minimum FW version determined at manufacturing. Adhering to this tripartite approach strengthens a device's defense against replay and rollback attacks, ensuring the system's security and integrity [9].

D. Device Identity using Dual-Binding

To mitigate the susceptibility of SoCs to cloning or spoofing attacks, FortBoot employs an innovative dual-binding device identifier that integrates HW and FW elements, fortified with encryption and HW-SW binding techniques. In FortBoot, each device is provisioned during manufacturing with a unique HW-based identifier *(HI)*, accessible only by the 1st-stage bootloader. This HI serves as the basis for generating the DI, a result of a HMAC applied to the HI and the digest of the 2nd-stage bootloader (L_1), i.e., *DI = HMAC(HI, H(L_1))*. There are two purposes for this newly formulated DI: (1) It is unique from HW and FW perspectives; (2) It acts as static code attestation evidence, verifying the 2nd-stage bootloader's authenticity. Even if the L_1 code is compromised, the intrinsic security of the original HI remains secure.

E. Dynamic Key Derivation and FW Protection

In FortBoot, we establish a dynamic key generation and FW protection during SoC's boot process. To do that, the ROM employs DI to calculate *SE keys*. In this case, even if an attacker replaced or modified L_1 (to extract SE keys), they only obtain keys associated with the malicious version, preventing them from decrypting data. The dynamic key generation process effectively counteracts potential attacks aimed at data extraction. Also, SE keys becomes related to the HI (uniqeness per HW), thus extracting SE keys from one device prevents the attacker

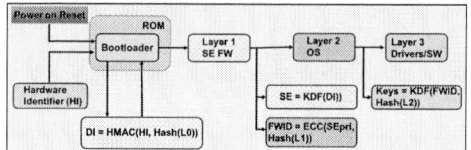

Fig. 2: Layered Boot Flow Diagram in FortBoot with Key Derivation.

to deduce the keys of another, eliminating all *"break-one-break-all"* threat scenarios. Also, another set of keys, FWID keys, is generated by L_1 from the private SE key and the subsequent stage code, L_2 ($FWID = ECC(SE_{pri}, H(L_2))$). Accordingly, each L_2 update receives FWID keys from L_1, allowing the decryption of previously encrypted data.

IV. SECURE BOOT FLOW IN FORTBOOT

FortBoot encompasses generating and utilizing keys throughout the boot flow, as depicted in figure 2. The structure of FortBoot allows the manufacturer to accurately identify the SW components of any 'nth' level, contingent on the key derived from the 'nth' stage. Notably, access to data protected by the 'nth' stage key is only permissible if the 'nth' stage code remains unaltered. Throughout the boot flow, DI, which is established during the bootloader stage, plays a fundamental role as a unique identifier derived from the HW and the crypto identity of the device's first mutable code (L_1). During the L_1 stage, corresponding to the operational FW of the SE, SE keys are generated and derived from the *DI* using a key derivation function (KDF). Furthermore, within the L_1 stage, FWID keys are also generated, reliant on the L_2 layer. This could be achieved through ECC. As the boot process progresses to the L_2 stage, corresponding to the host's application FW, FWID keys passed on from the previous stage are used for data protection within the L_2 layer. New keys are dynamically generated in case of a code update to ensure data protection and secure cryptographic operations. This intricate dance of keys and stages strengthens the secure boot protocol, enhances data security, and solidifies the trust chain, ensuring the device remains secure throughout its lifecycle.

V. IMPLEMENTATION AND EVALUATION

FortBoot counts on foundational elements such as a fully trusted processor and an initial set of trusted instructions such as OpenTitan [3], an open-source HW design initiative anchored in the RISC-V Ibex CPU architecture.

A. $L_{0/1/2}$ Implementation for FortBoot Setup

ROM code (L0) validates the ROM_EXT (L1) module through a series of security checks. The ROM_EXT authentication and *DI* derivation rely on two key inputs: the ROM_EXT manifest and a flash execution value (see Algorithm 1). Boundary checks are performed on the manifest fields, and the security version field is compared against a minimum security version stored in the memory. The manifest is then validated against the predetermined boot policy. Following this, the RSA key for the ROM_EXT is retrieved from the manifest. Subsequently, the HMAC-SHA256 digest computation is initiated. The HW device identifier is fetched from the OTP memory and, combined with the digest, is supplied to a KDF to generate the DI. A successful verification authenticates the ROM_EXT and generates SE and $FWID$ keys in 2nd stage boot loader using a KDF function.

Algorithm 1 DI Generation in FortBoot

1: **Input:** Manifest, Flash_exec;
2: **Output:** Signature Verification Error;
3: **if** $manifest.secversion < min\_secversion$ **then**
4:     Invalidate the signature;
5: boot_policy_manifest_check();
6: get rsa and spx keys for signature verification;
7: hmac_sha256_init();
8: hmac_sha256_update(manifest.start, manifest.length);
9: hmac_sha256_final(manifest.digest);
10: DI = HKDF(hardware device identifier, digest);
11: Store the boot measurements;

B. Security Evaluation and Analysis of FortBoot

Table II shows the build time, code size, and lines of code for the assembly and C code in FortBoot's two layers. For embedded systems with limited flash memory, code size is crucial. The ROM (L_0) assembly initializes exception handlers, and memory setup, while its C code handles counter initialization, DI computation, L_1 signature verification, and control transfer to L_1. At the L_1 stage, assembly code initializes interrupt vectors and global data structures, and C code manages counter initialization, SE and FWID key generation, digest generation, L_2 signature verification, and control transfer to L_2. This combination of assembly and C code ensures a secure boot process for FortBoot on OpenTitan.

TABLE II: Build Time (sec) and Line of Codes for L_0/L_1 in FortBoot.

Layer	Build Time (s)	Code Size (KB)		Line of Code	
		Assembly	C	Assembly	C
$ROM(L0)$	245.6	15.7	20.9	457	548
$ROM\_EXT(L1)$	164.4	5.0	8.6	179	251

1) Robustness against Device Identity Extraction Attacks: The *DI* is formulated by concatenating a HI with the hash of L_1 using an HMAC operation. So,

(a) For a brute force attack, an attacker must attempt all possible HI and L_1 hashes combinations. Assuming HI is 'n' bits and bits and the hash is 'm' bit, it results in 2^n and 2^m different HIs and L_1 hashes, leading to a total of $2^n \times 2^m$ computations for brute force the DI (directly exponential w.r.t. the HI's length and the hash function's output size - $ex_{|n \times m}$).

(b) For a code tampering attack, an attacker must generate a modified L_1' such that $HMAC(HI, H(L_1'))$ is equivalent to the original DI. Given an ideal hash function, this likelihood is negligible, at around $1/2^m$, where 'm' is the length of the hash output. Hence, the effort an attacker would have to expend to tamper with the code successfully can be 2^m ($ex_{|m}$).

(c) For a reverse engineering attack, the attacker requires to simultaneously know both the HI and the correct L_1 code, which is not possible as DI is a HMAC-oriented data. This

significantly escalates the difficulty of reverse engineering attempts and heightens the security of the boot process.

2) Robustness against Dynamic Keys Extraction Attacks: FortBoot utilizes dynamic key generation, creating keys from unique, interconnected data such as the device's HI, FW version, and boot stage. These keys are derived by hashing this concatenated data, ensuring each boot stage has a unique key reflecting the device's current state.

(a) For a key extraction attack, due to the one-way nature of KDF, reverse engineering the DI is impossible, even if SE keys are leaked. The dynamic nature of key generation means SE keys from one device cannot be used to infer keys for another device. The complexity of extracting DI given the SE is 2^b where 'b' is DI size. FWID keys are derived from the ECC of the private SE key (SE_{pri}) and L_2 digest ($H(L_2)$). The security of FWID is tied to the elliptic curve discrete logarithm problem (ECDLP), with computational complexity $O(2^{(b/2)})$ where b is the ECC key size.

(a) For a replay attack, SE and FWID keys are generated dynamically at each L0/1 update, making attacks impractical. An attacker replacing L1 with a malicious version would only obtain a new set of FWID keys, which cannot decrypt data encrypted with the older keys. The probability of a successful replay attack P_A is $1/(2^b)$, where 'b' is the key size ($ex_{|b}$).

VI. CONCLUSION

This paper introduced FortBoot, an integrated, robust strategy to enhance secure boot strength against cybersecurity threats. It incorporates dynamic key generation, mutual FW-HW binding, rollback prevention, and dynamic code attestation. In FortBoot, we use unique mutual authentication between HW and FW to build a strong barrier to unauthorized code execution. With its implementation on OpenTitan SoC, our experimental results show its performance efficiency, while the security analysis further proves its resilience against sophisticated attacks. Looking ahead, we aim to refine its features and develop a secure boot protocol for multi-processor systems, offering enhanced protection against cyber threats.

REFERENCES

[1] R. Wilkins and B. Richardson, "Uefi secure boot in modern computer security solutions," in *UEFI forum*, 2013, pp. 1–10.

[2] S. Guilley and R. Pacalet, "Soc security: a war against side-channels," *Annals of Telecommunications-annales des télécommunications*, 2004.

[3] lowRISC, "Opentitan: Open source silicon root of trust," https://opentitan.org/, 2021.

[4] D. Cooper, W. Polk, A. Regenscheid, M. Souppaya *et al.*, "Bios protection guidelines," *NIST Special Publication*, vol. 800, p. 147, 2011.

[5] A. Furtak *et al.*, "Bios and secure boot attacks uncovered," in *The 10th ekoparty Security Conference*, 2014.

[6] V. Costan and S. Devadas, "Intel sgx explained," *Cryptology ePrint Archive*, 2016.

[7] S. Mohammad and F. Farahmandi, "Dyfora: Dynamic firmware obfuscation and remote attestation using hardware signatures," in *Proceedings of the GVLSI 2024*, 2024.

[8] M. M. Hossain *et al.*, "Hexon: Protecting firmware using hardware-assisted execution-level obfuscation," in *IEEE ISVLSI*. IEEE, 2021.

[9] A. Matrosov *et al.*, *Rootkits and bootkits: reversing modern malware and next generation threats*. No Starch Press, 2019.

[10] S. Mohammad *et al.*, "Required policies and properties of the security engine of an soc," in *2021 IEEE International Symposium on Smart Electronic Systems (iSES)*, 2021, pp. 414–420.

FVDCLS: Functional Verification of RISCV based Dual-Core Lockstep Feature using Fault Injection Mechanism

Muhammad Kashif Minhas, Haroon Waris, and Sajid Baloch

Centers of Excellence in Science & Applied Technologies, Pakistan
kashifminhas934@gmail.com, haroonwaris@gmail.com, to_baloch@hotmail.com

Abstract— **This paper presents the framework to verify RISCV based Dual-Core Lockstep (DCLS), a feature that enhances the fault tolerance in processors. The DCLS feature, when two identical processors run in parallel, effectively detects the error and rolls back the core operations to a previously known state. The purpose-built fault injection mechanism is developed using universal verification methodology (UVM) and contains two modules: i) The fault injector (FI), ii) The virtual comparator (VC). The FI systematically injects the error in one of the core's internal states by flipping one or multiple bits. While the VC stores the internal states of both the RISCV cores on each clock cycle and reports the anomaly, if detected. The SweRV EH1 (an open-source RISCV core) is customized to add the DCLS functionality and later Google RISC-V DV test suite is used to validate the proposed FVDCLS. In total, nine different verification scenarios are developed consisting of transient and permanent errors using FI module. For all these test scenarios the VC compared hundred different internal state signals/registers. The result shows that the FVDCLS successfully detected and reported all the mismatches. Moreover, the FVDCLS incorporates an error debugging module that generates the summary of conducted test particularly, the type of fault injected, the number of errors detected, and the injected fault coverage. The proposed FVDCLS is portable and configurable; therefore, can be seamlessly integrated with any processor (with DCLS feature) under test.**

Index Terms—**Dual-core Lockstep, Fault injection mechanism, Universal verification methodology, RISC-V, Google RISC-V DV.**

I. INTRODUCTION

In recent years, the use of embedded processors for critical safety applications has increased. This is because the recently designed processors exhibit high-performance and are fault-tolerant. The ability of processors to maintain their functionality, dependability, and performance in the presence of defects and errors is referred to as fault tolerance. Fault tolerance can be applied at various abstraction levels, such as hardware, software, or system level. The Dual-core lock step (DCLS) is an error detection technique that ensures reliable computing and is employed in safety-critical systems such as aerospace and industrial control systems. In DCLS, two instances of the same or different processors execute identical instructions simultaneously. The result of each executed instruction on both the cores is compared on every clock cycle. A fault-handling mechanism is activated when a divergence is detected, indicating a potential error or failure in one of the cores. Moreover, delaying one core with respect to the other can result in additional resilience so that the same fault doesn't affect both

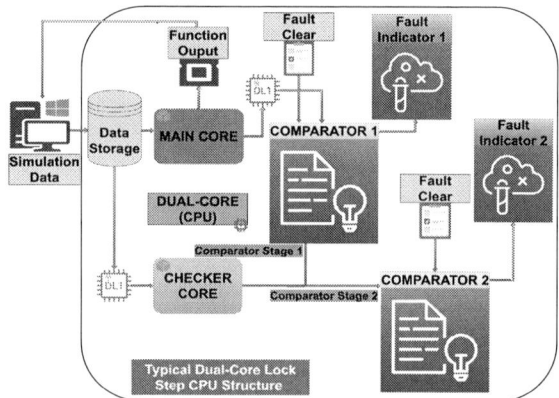

Fig. 1: Typical Dual-Core Lock Step Structural Block Diagram

processors. A general structural block diagram of DCLS is shown in Figure 1. In Figure 1, initially, the test vectors are put in data storage modules and are concurrently applied to both processors i.e., Main and Checker core. A delay of one cycle may be used before applying test vectors to the Checker core. In that case, Main core processed data is delayed by one cycle before sending it to the comparator unit for maintaining the synchronization. Finally, the faults are indicated by fault indicator unit. It's on user that how he handles the faults/errors detected.

Generally, the FI techniques fall into hardware and software-based techniques [ ]. The scope of the presented research work is on software-based fault injection mechanisms. The main advantages of adopting software methods are to reduce the hardware cost associated with the tools. Moreover, there are certain FI techniques that cannot be covered with hardware techniques [ ]. The software-based fault injection techniques [1] are further divided into: (i) Compile-time Injection and (ii) Run-time Injection categories. These techniques are described in detail in Section III of the paper. The proposed FI framework so-called FVDCLS is based on UVM methodology; therefore, FVDCLS is configurable and reusable with following novel features:

1) The proposed FVDCLS framework has the ability to induce both soft and hard errors in the design under verification (DUV) during run-time. The FVDCLS

979-8-3315-3968-9/24 $31.00 © 2024 IEEE

	Core	Platform/Tool	Language	Faults Injected	Errors Detected	Mode
Jingzhou [ ]	Falcon [RISC-V]	FPGA	SystemVerilog	NA	NA	Hardware
M.Peña [ ]	Cortex-M33,R4,R5	Zynq-Z7010	C/C++	871837	43769	Hybrid
Rafael [ ]	RI5CY [RISC-V]	Verilator	C++	2857	2228	Software
Krzysztof [ ]	CCRV32ST [RISC-V]	FPGA	C++	14068	7233	Hardware
Kasap [ ]	Cortex-A9	Zynq-7000	C	3712	3637	Hardware
Andrea [ ]	OpenRISC-1200	Synopsys-Z01X	C	1374	936	Software

TABLE I: Comparison of existing fault-injection techniques for DCLS verification

framework matches internal states of both the cores on every clock cycle.

2) The FVDCLS is successfully integrated with the open-source SweRV-EH1 (RISC-V core) testbench environment.

3) The Google RISC-V DV test-suites are used to validate the FVDCLS.

4) The proposed FVDCLS can be seamlessly integrated with any other open-source and customized processor verification environment.

II. RELATED WORK

A RISC-V based microprocessor, with embedded DCLS functionality, is presented in [ ]. The paper explains the implementation of a lockstep architecture, capable of detecting and correcting errors for safety-critical applications. Branch-target and Reorder buffer-based technique is used for DCLS verification. The already committed values of the microprocessor's internal registers are stored in reorder buffer and extracted for comparison. These microprocessors executed instructions are assessed against the reference model simulator and finally the outcomes are compared in the testbench's scoreboard class. Another DCLS-based architecture is presented in [ ] to mitigate the radiation-induced soft errors. The paper explains an emulator-based fault injection technique to verify the system's vulnerability to errors. For fault-injection, an Intellectual Protocol (IP) designed in Hardware Description Language (HDL) is used. The proposed technique randomizes the fault which is going to be injected and stores it in a register file. Then the processor's target-register is selected, whose data is going to be modified. The proposed DCLS system can mitigate up to 78% of the injected faults. Simulation-based DCLS design, for RISC-V processors is proposed in [ ]. For the evaluation of proposed design's software and hardware overhead, an injection routine is performed. The fault-injection is carried out in a comprehensive manner, causing all internal registers to undergo a slight flip throughout the entire campaign. A single fault is injected in one of the embedded cores, during each execution of the target application. The paper also proposed a module for comparing DCLS internal state registers upon calling fault-injection simulation routines. Finally, the presented algorithm generates the detected errors and injected-fault coverage report summary. All the fault-injection applications and routines are built using Object-Orient Programming (OOP) and C++ languages. The platform adopted for performing all the simulations is Verilator. Comparing all the fault-injection techniques discussed, every method has an

advantage over another, considering certain applications and features. The main advantage of our proposed fault-injection mechanism over all the existing techniques is re-usability and configurability, due to its state-of-the-art UVM based implementation. The proposed fault-injection mechanism also maximizes the DCLS verification coverage by defining all the possible scenarios of hard and soft errors. A brief comparison of different fault-injection techniques is described in Table 1.

III. PRELIMINARIES

The paper presents the dynamic or simulation-based fault-injection mechanism for DCLS verification, as mentioned before. Both hardware and software methods of injection have their own pros and cons depending upon the application where these techniques are going to be employed [ ]. The primary differences between the software and hardware approaches are the degree of perturbation, cost, and the fault-injection sites that they may access. Dynamic or software-based fault-injection techniques are the methods in which errors are introduced in systems Register-Transfer Level (RTL) or in application's assembly code. Software-based techniques are categorized based on the time of fault-injection, i.e, compile-time, and run-time. Also, the types of errors to be injected are also categorized as (i) Soft or Transient and (ii) Hard or Permanent faults. In the proposed mechanism, all the faults are introduced using run-time injection technique. These techniques are briefly discussed below.

1) Compile-Time Injection Methods

For injecting faults at compile-time, the instructions of the application program are altered before its execution. Errors are introduced into the processor's source code or assembly code of the application program to emulate the effect of transient faults. Compile-time injection methods offer a higher level of implementation simplicity but do not allow the flexibility of injecting faults as the workload program runs.

2) Run-Time Injection Methods

Triggering mechanisms are required for injecting faults during run-time. Commonly used mechanisms for run-time injections include (i) Time-out, (ii) Exception and (iii) Code insertion strategies. The Time-out injection technique generates unpredictable faults for handling transient and intermittent errors. In exception or trap technique, fault generation depends upon the occurrence of certain events, while timid instructions are added in the target program during implementing code-insertion techniques.

IV. METHODOLOGY

The proposed DCLS featured RISC-V core verification framework comprises SweRV-EH1 testbench environment, with integrated modules of Google RISC-V DV, Instruction-Set Simulator (ISS), tracer, RV-verification Interface (RVFI) and Fault-Injection (FI) routines. After initialization of FI framework, a RISC-V DV Instruction Stream Generator (ISG), generates the constraint assembly tests, which are cross-compiled, and run on the DUV and ISS modules simultaneously. Concurrently, the faults are also induced into the DUV through FI framework during the execution of core's simulation. The discrepancy between the main and faulty core is monitored throughout the simulation to detect any mismatch. The .csv (Comma Separated Values) files, generated after the ISS and DUV modules simulation, are then processed for comparison. Each value in both the .csv files is compared, and the framework eventually generates the DUV functional behavior report. The proposed DCLS core verification framework with integrated FI mechanism is shown in Figure 4.

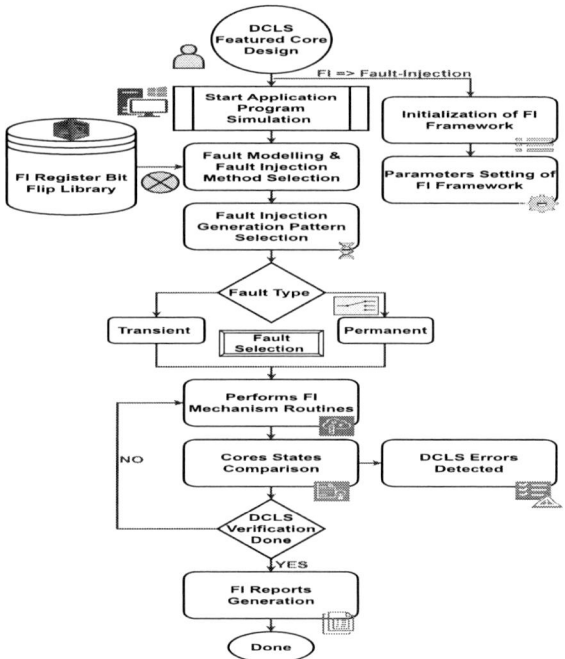

Fig. 2: FI Mechanism Based DCLS Core Verification Flow

A. Proposed Fault-Injection Framework

The proposed FI framework is designed to inject faults into the DCLS core architecture. The main objective of the proposed framework is to successfully flip the bits of the DUV, according to the fault type. The proposed FI framework accesses the faulty core internal registers, allowing it to customize the stored data at any point of execution. Any divergence observed by the virtual-comparator module between the main and faulty core behavior is indicated as an error.

Finally, the results of all the FI routines and detected errors are logged for report generation. The proposed FI mechanism flow hierarchy is shown in Figure 2. The details of the FI mechanism implementation steps are presented below.

1) Faults Modelling

The process of describing and characterizing the types, locations, and behaviors of faults that might arise in SoCs is known as fault modelling. Fault modelling can be done at several levels of abstraction, including physical, gate, RTL, and system. In the proposed FI framework, faults are modelled using *UVM-macros*. The *UVM-macros* allows backdoor access to DUV internal registers, via built-in Data Peripheral Interface (DPI) models. The framework categorized the faults into permanent, transient, and intermittent classes. No less than a hundred different DCLS internal state signals are accessed in the fault model, for customizing or flipping their existing stored data. The framework also gave the configurability of running one-time or multiple-time injection routines. The UVM macros used for faults modelling are (i) *uvm-hdl-deposit*, (ii) *uvm-hdl-force*, (iii) *uvm-hdl-force-time*, (iv) *uvm-hdl-release* and (v) *uvm-hdl-read*.

2) Faults Injection

In the proposed FI model, multiple injection campaigns are defined in *UVM-test-class*. Each campaign consists of a specific method, injection-candidate, pattern, time, and type of fault as described in Figure 2. The FI framework controls the system calls to injection campaigns. This exhaustive running of injection campaigns enhances the framework's fault-coverage and error detection capability.

3) Faults Simulation

In the proposed FI mechanism, simulation-oriented fault-injection technique is implemented. Application cases in the form of Google RISC-V DV test-suite are then applied to the DUV. Faults are induced during each application test case simulation. Simultaneously, the FI routines are also applied to the DUV. The resulting fault's latency, propagation and severity levels are then analysed using simulation waveforms data.

4) Faults Tolerance and Comparison

Fault tolerance capability of proposed DCLS featured SweRV core is finally evaluated using FI comparator module. The application case simulation data is fed to the *comparator-module*, which compares the state of each DUV's dump signal on-the-fly. On finding erratic behavior between the two cores, *comparator-module* automatically displays an error flag specifying that signal against whom the mismatch is found. After comparison, the module automatically generates the DCLS fault tolerance log report. Cumulative fault coverage is calculated using all the simulation data seeds.

V. RESULTS AND DISCUSSIONS

The following section discusses the results of the FVDCLS using FI framework. For evaluation of core's DCLS error detection capability each application case is used in an exhaustive manner, causing all the core's pipeline internal registers to undergo the state change upon any soft or hard fault.

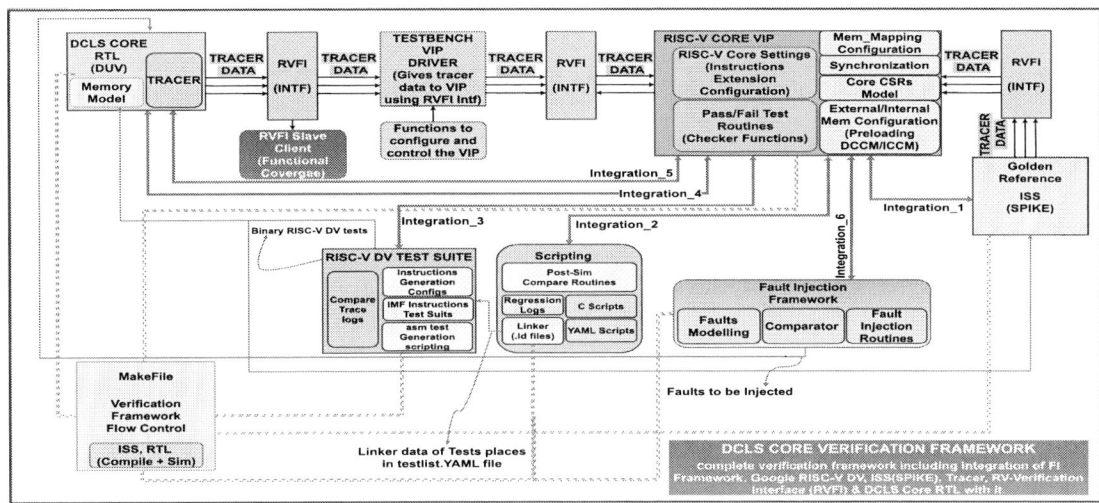

Fig. 3: DCLS Core Verification Framework with Integrated FI Mechanism

Approximately 20,000 hard and soft errors are injected by FI campaigns in all the application cases run. The FI framework successfully detects 98.7% of errors from the overall fault injection routines. Some of the application cases run and their number of detected versus injected faults are shown in Figure 5. The results clearly show that the proposed FI framework has high percentage of faults detection on comparing with existing solutions discussed in TABLE 1. The overall RISC-V core code and functional coverage achieved is 89.05% and 100% respectively. All the simulations are performed using Synopsys toolset while the FI framework is modelled using SystemVerilog and UVM as a medium.

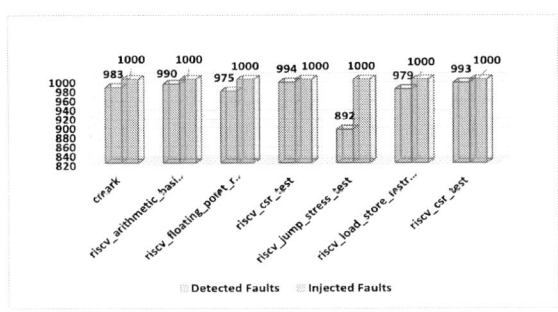

Fig. 4: Fault Injection Application Cases Results

VI. CONCLUSION AND FUTURE SCOPE

The paper proposed a software-oriented fault-injection (FI) methodology for specifically verifying the processor's embedded DCLS functionality. The applicability of the proposed solution is verified by using SweRV-EH1 core and Universal Verification Methodology (UVM) as a medium. In DCLS, the same set of instructions are run on two identical cores concurrently. Faults are injected into the faulty core while the other core is considered as golden reference. FI routines

implemented within a *UVM-test-class* induces the faults into the design under verification (DUV), during run phase of the application test case simulation. On monitoring any divergence between the two cores behavior, an error flag is raised, signalling DCLS functionality bug. The proposed verification environment integrates the FI framework with the already available open-source testbench environment of SweRV-EH1 processor. The proposed FI framework has the fault detection capability of 98.7% of overall errors injected. In future, we plan to integrate the proposed FI framework with other RISC-V cores verification environments.

REFERENCES

[1] M. C. Hsueh, T. K. Tsai and R. K. Iyer, "Fault injection techniques and tools," in Computer, vol. 30, no. 4, pp. 75-82, April 1997, doi: 10.1109/2.585157.

[2] J. Li, H. Chen, W. Liu, W. Zhang and H. He, "Falcon: A Dual-Core Lockstep Microprocessor Based on RISC-V ISA," *RISC-V Summit Europe, Barcelona*, June 2023.

[3] M. P. Fernandez et al., "Dual-Core Lockstep enhanced with redundant multithread support and control-flow error detection, *Microelectronics Reliability*, vol. 100–101, pp. 113447–113447, Sep. 2019, doi: https://doi.org/10.1016/j.microel.2019.113447.

[4] R. F. Viana, "Design and simulation of a RISC-V dual-core lockstep for fault tolerant systems," *http://www.repositorio.jesuita.org.br/handle/UNISINOS/9456, University of Vale do Rio dos Sinos*, July 2020.

[5] K. Marcinek and W. A. Pleskacz, "Variable Delayed Dual-Core Lockstep (VDCLS) Processor for Safety and Security Applications," *Electronics*, vol. 12, no. 2, pp. 464–464, Jan. 2023, doi: https://doi.org/10.3390/electronics12020464.

[6] S. Kasap, E. W. Wächter, X. Zhai, S. Ehsan and K. D. McDonald-Maier, "Novel Lockstep-based Approach with Roll-back and Roll-forward Recovery to Mitigate Radiation-Induced Soft Errors," *IEEE Nordic Circuits and Systems Conference*, Oslo, Norway, 2020, pp. 1-7, doi: 10.1109/NorCAS51424.2020.9265137.

[7] A. Floridia and E. Sanchez, "On-line self-test mechanism for Dual-Core Lockstep System-on-Chips," *Microelectronics Reliability*, vol. 112, p. 113770, Sep. 2020, doi: https://doi.org/10.1016/j.microel.2020.113770.

[8] Rodrigues, C. et al. (2019) 'Towards a heterogeneous fault-tolerance architecture based on ARM and RISC-V processors', IECON 2019 - 45th Annual Conference of the IEEE Industrial Electronics Society, doi:10.1109/iecon.2019.8926844.

Heterogeneous Approximation of DNN HW Accelerators based on Channels Vulnerability

Natalia Cherezova[1], Salvatore Pappalardo[2], Mahdi Taheri[1], Mohammad Hasan Ahmadilivani[1], Bastien Deveautour[2], Alberto Bosio[2], Jaan Raik[1], and Maksim Jenihhin[1]

[1]Tallinn University of Technology, Tallinn, Estonia
[2]Univ Lyon, ECL, INSA Lyon, CNRS, UCBL, CPE Lyon, INL, Ecully, France

Abstract—Since Deep Neural Networks (DNNs) gracefully withstands approximation due to its inherent redundancy, Approximate Computing (AxC) can be applied to reduce power consumption and execution time. In the literature, several works adopted the AxC paradigm to DNNs in the form of quantization, precision reduction, pruning, and functional approximation. Despite the promising results demonstrated so far, most of the existing works have applied homogeneous AxC techniques, meaning that the same degree of approximation has been applied to the entire DNN. However, different DNN components (i.e., channels, filters, layers, neurons) have different resiliency levels. This paper presents a framework for applying heterogeneous AxC to DNN hardware accelerators. The framework is based on the identification of channel resilience and applying a tailored degree of approximation per channel. Preliminary results carried out on the LeNet-5 model show that by using the proposed framework it is possible to decrease resource utilization by 65.2% and power consumption by 53.4% at the cost of a marginal drop of accuracy from 98.87% to 98.03%.

Index Terms—deep neural networks, data-flow architecture, hardware accelerator, approximate computing, circuit design

I. INTRODUCTION

DNNs are employed in many different fields, due to their outstanding computational capabilities. The current trend is pushing research and industry to deploy DNNs in embedded systems to be used at the edge in which power and computational resources are limited. To address the power and latency issue, the Approximate Computing (AxC) paradigm can be used to reduce the power and the area required by the circuit since it proved to be an effective technique [1].

Several works adopted the AxC paradigm to DNNs in the form of quantization, precision reduction, pruning and functional approximation [1]. Despite the encouraging results, most of the existing works applied homogeneous AxC techniques, meaning that the same degree of approximation has been applied to the DNN. However, DNN components (i.e., channels, filters, layers, neurons) show different resiliency levels. Therefore, instead of using a uniform degree of approximation to all components, it would be beneficial to exploit the heterogeneous AxC technique by adapting the approximation degree according to the resiliency level of each DNN component.

This paper presents a framework for applying heterogeneous AxC at the channel level to DNN hardware accelerators. The framework is based on identifying channels' resilience and

applies a specific degree of approximation per channel. In particular, we leverage precision reduction, which corresponds to neglecting the Least Significant Bits (LSBs) of the multiplier's input and effectively using a smaller bit-width.

The rest of this paper is structured as follows. In Section II, a brief overview of existing works is given. Section III describes the proposed methodology. Section IV presents the experimental setup and discusses the obtained results. Finally, Section V concludes the paper and proposes future directions.

II. RELATED WORKS

The advantages of implementing and deploying DNNs on FPGAs are advocated in several recent works [2], [3]. Dataflow is an important computation architecture for customized hardware accelerators, which enables the parallel temporal execution of multiple coarse-grained tasks [4]. The FINN framework [5] is released by Xilinx to explore quantized CNNs' inference on FPGAs that also provide customized data-flow architectures for each network.

To further enhance the performance of DNN accelerators, several techniques are introduced in the literature (e.g., quantization and approximate computing). In [6] a fully automated framework capable of applying various quantization-aware techniques, and hardware implementation is introduced to measure hardware parameters.

Multiplication is one of the primary arithmetic operations widely used in DNNs. Various approximate multipliers are proposed in the literature [7]. DeepAxe is a framework that enables the selective approximation for data-flow DNN accelerators to provide a design space exploration for the efficiency and accuracy of DNNs [8].

To the best of our knowledge, there is no work presenting customized heterogeneous channel-wise approximations for hardware accelerators leading to high efficiency in terms of resource utilization and power consumption. The approach proposed in this paper goes beyond the state-of-the-art by establishing a comprehensive methodology for enabling heterogeneous approximation of DNN HW accelerators based on channels' vulnerability analysis.

III. METHODOLOGY

The purpose of the proposed method (Fig. 1) is to systematically provide a hardware accelerator for CNNs in a way that

---

979-8-3315-3968-9/24 $31.00 © 2024 IEEE

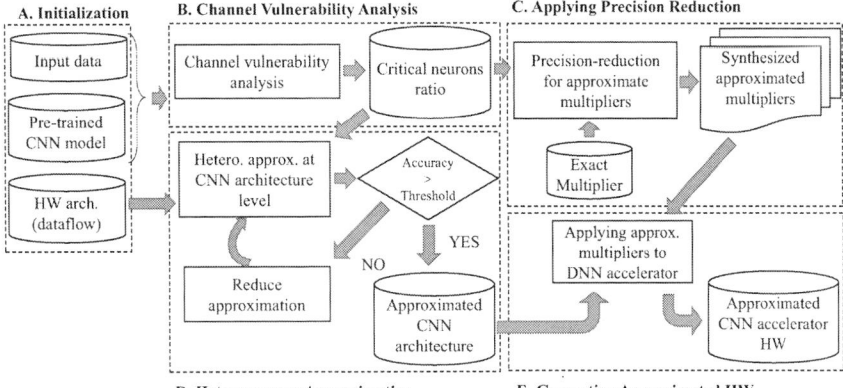

Fig. 1: Proposed flow to achieve heterogeneous approximation for data-flow CNN hardware accelerators

its multipliers are heterogeneously approximated through each layer, at the channel level, to achieve an efficient accelerator with highly accurate CNNs. In this regard, a vulnerability analysis for CNNs' channels is performed to enable error-aware approximation. Afterwards, an optimum point for the heterogeneous approximation throughout the channels and layers is obtained. Finally, the generated approximated multipliers are synthesized on an FPGA and hardware results are reported.

A. Initialization

The method receives a trained CNN model, test data, and the data-flow hardware accelerator as inputs. The method is presented for quantized CNNs, where the data-flow accelerator consists of a series of components connected according to the architecture of the selected neural network. Each component implements one layer of the network. Therefore, it can be optimized based on the layer parameters, e.g., the number of input and output channels, kernel size, etc. Outputs of the layers are stored in the intermediate memory buffers as the data is streaming through those components.

B. Channel Vulnerability Analysis

To obtain the maximum errors at the output of each channel/neuron leading to the minimum possible accuracy drop due to approximation, we adopt and extend the Deep-Vigor methodology presented in [9] and [10] that provides vulnerability analysis for DNNs and QNNs, respectively.

Let δ_k^l be a deviation that is an added positive or negative error value to an output Feature Map (FMap) by approximation for a k-th neuron at layer l with input data x. Let Δ_k^l be the minimum absolute deviation added to the corresponding neuron by approximation that misclassifies the input image x from its golden classification in the exact DNN. This deviation is defined as follows:

$$\Delta_k^l = min(|\delta_k^l|), \mathcal{E}_t < \mathcal{E}_i, i \neq t; \quad i, t \in C \quad (1)$$

where C is the set of all the classes of the classifier, i and t are elements belonging to C (more specifically t is the top class of the exact DNN) and $\mathcal{E}_t$ and $\mathcal{E}_i$ are the deviated output

logits corresponding to the respective output classes. Thus, Δ_k^l represents the maximum absolute deviation of a neuron's output from its error-free value due to the approximation that would not misclassify the DNN.

Approximation design exploits Δ_k^l to identify the bits that can be approximated in the calculations. Δ_k^l can be interpreted as the set of bits for each neuron to be approximated without misclassification. In the integer data type, approximating the LSB corresponds to $\Delta_k^l = 1$. As an example, if $\Delta_k^l = 16$, all the 4 least significant bits can be approximated, since any deviation less than this value would not affect the classification output. Therefore, the set of bits to be approximated is obtained by:

$$Bits\ to\ be\ approx. = [int(log_2(\Delta_k^l)) - 1 : 0] \quad (2)$$

Since approximation design is performed at the channel level of the DNN, the obtained Δ_k^l for each FMap in convolutional layers are aggregated channel-wise. To provide a preliminary analysis of the effect of approximating channels on the accuracy, we explore the level of approximation regarding each bit corresponding to the obtained Δ_k^l for the neurons inside that channel. This analysis expresses what portion of neurons in a channel has a higher Δ_k^l than the approximation error, which is called *critical neurons ratio*.

In this paper, we consider the *critical neurons ratio* to be lower than 5% as a safe approximation level. In this regard, the approximation in the multipliers of a channel is applied to the least significant bits up to the bit corresponding to the safe approximation level. It is noteworthy that each channel in a layer has its own safe approximation level, leading to a heterogeneous approximation within layers throughout the CNNs.

C. Applying Precision Reduction to Multipliers

As mentioned earlier, the multipliers approximation is based on precision scaling [11]. Specifically, the lower-order bits of the input operands of the multiplier are neglected. The number of neglected bits is denoted with n.

979-8-3315-3968-9/24 $31.00 © 2024 IEEE

However, any other customized library of approximate multipliers can be used. The approximation level is defined as the number n of lower-order bits that are disregarded and, consequently, the introduced Worst-Case Error (WCE) corresponds to $2^n - 1$ for the inputs and $2^{2n} - 1$ for the outputs.

D. Heterogeneous Approximation

This step attempts to identify an optimal configuration for a channel-wise heterogeneous approximated hardware accelerator. First, a separate approximation to each channel in a layer is applied to find a configuration in which the accuracy of the CNN does not drop more than 1% compared to the baseline exact CNN accuracy. To this end, the aggressive channel-wise approximation configuration is applied initially to a layer. Then, the approximation continues to increase by 1 bit for all channels until the accuracy drop is less than 1%.

After finding the best channel-wise approximation for each layer separately, all channels are approximated throughout the CNN based on the obtained configurations. Then, we reduce the level of channel-wise approximation from the first to the last layer, one by one, until the accuracy of the approximated CNN is not 1% less than the accuracy of the exact CNN. The output of this process is an optimal point for approximation where the accuracy of the approximated CNN is negligible.

E. Generating Approximated Hardware

The configuration obtained using step III-D defines what approximate multipliers should be used for each channel. The generated DNN accelerator is compared with the baseline accelerator, which utilizes exact multipliers, in terms of resource utilization, number of Look-Up Tables (LUT) and flip-flops (FF), and power consumption.

IV. EXPERIMENTAL RESULTS

A. Experimental Setup

A common CNN benchmark LeNet-5 is used as a case study. The structure of the network is presented in Table I. The network was trained on the MNIST dataset of handwritten digits [12] with PyTorch 16-bit integer data type with 98.87% accuracy.

TABLE I: Case-study DNN architecture

Layer	Input size	Kernel size	Output size
Conv1	$32 \times 32 \times 1$	$5 \times 5 \times 6$	$28 \times 28 \times 6$
Max pool	$28 \times 28 \times 6$	2×2	$14 \times 14 \times 6$
Conv2	$14 \times 14 \times 6$	$5 \times 5 \times 16$	$10 \times 10 \times 16$
Max pool	$10 \times 10 \times 16$	2×2	$5 \times 5 \times 16$
FC3	$5 \times 5 \times 16$	—	120
FC4	120	—	84
FC5	84	—	10

A data-flow accelerator for the LeNet-5 is developed considering the AMD-Xilinx FPGA platform as a target. Each convolutional layer consists of computational modules for each output channel. Thus, all output channels are processed in parallel. Each output channel, in turn, consists of computational modules for each input channel.

The input channel computational module includes a multiplier and an adder. Each fully-connected layer consists of computational modules for each output neuron. The output neuron computational module includes a multiplier and an adder as well (e.g. Conv1 layer utilizes $6 * 1 = 6$ multipliers, and FC5 layer 10 multipliers).

B. Results of Channel Vulnerability Analysis

An exploration is performed to obtain the portion of neurons in each channel of a layer such that their Δ_k^l is smaller than the error induced by approximation. The vulnerability of convolutional and fully-connected layers is analyzed since only those layers involve multiplication. The last fully-connected layer is excluded from the analysis because it is the output layer and any changes there will directly affect the output of the CNN.

Table II presents the results of channel vulnerability analysis for the first convolutional layer of LeNet-5. The table reports the percentage of approximated neurons in each channel that result in misclassification. The data is organized such that rows show analyzed channels and columns show the number of bits to approximate. The table is color-coded based on the set thresholds: green shows all the channel-approximation pairs that have less than 5% critical neurons, orange those that have less than 10%, and red all the rest.

TABLE II: Percentage of neurons in each channel of the first layer of the case-study DNN for error induced by approximation

Error induced by approximation	1	2	4	8	16	32	64	128
Channel 1	1.14	1.21	1.46	1.59	2.80	8.10	15.75	32.14
Channel 2	1.21	1.46	1.53	1.59	2.42	4.72	12.05	22.38
Channel 3	1.02	1.02	1.02	1.59	2.29	5.10	11.73	23.72
Channel 4	0.95	1.08	1.14	1.40	2.42	4.71	13.39	28.31
Channel 5	0.89	0.95	1.40	3.12	6.05	10.39	19.77	31.05
Channel 6	0.12	0.12	0.12	0.51	1.21	5.42	14.03	29.27

C. Approximate multipliers

A library of approximate multipliers is generated by applying the precision reduction technique to the exact 16-bit multiplier. The comparison of the resource utilization and power consumption is given in Table III.

D. Heterogeneous Approximation at the Architecture Level

Using the data obtained by channel vulnerability analysis and the critical neurons ratio set to be lower than 5%, a *safe approximation level* is derived for each channel in each layer. Safe approximation level defines the number of lower-order bits that can be approximated, in our case negated. First, the effect of individual layer approximation on the accuracy of LeNet-5 is studied using the network implementation in C language. Each layer is successively approximated while the rest of the network is calculated using precise multiplication.

For each layer, several tests are performed by adding an offset to the safe approximation level, which increases the number of negated bits. This way it is possible to observe how the degree of approximation affects the performance of the layer. The results of this study are given in Table IV. As

TABLE III: Resource utilization of generated multipliers

	16	15	14	13	12	11	10	9	8	7
LUT	416	377	301	264	223	171	146	117	81	64
Power (mW)	67.803	62.500	58.108	53.649	49.290	43.513	40.567	36.551	32.457	28.361

TABLE IV: The effect of individual layer approximation on the accuracy of the case-study DNN

Offset	No offset	Offset by 1	Offset by 2	Offset by 3	Offset by 4	Offset by 5	Offset by 6
Conv1	98.91	**98.23**	54.97	17.25	8.92	8.92	8.92
Conv2	98.88	98.88	98.80	98.79	**98.50**	95.26	84.57
FC3	98.88	98.83	98.75	**98.41**	97.27	92.52	69.18
FC4	98.85	98.88	98.89	98.82	98.79	**98.28**	96.82

observed, a safe approximation level results in a high accuracy for each single layer. Nonetheless, increasing the approximation level incurs a high accuracy drop. The selected configurations for each layer are **bold** in the table, as indicated in step III-D.

E. Generated Approximate DNN HW Accelerator

An optimal configuration is generated on step III-D. After applying the selected approximation in Table IV, the accuracy of the approximated CNN is 81.09%. By following step III-D, the final approximation for each layer is {Conv1: no offset, Conv2: offset by 3, FC3: offset by 2, FC4: offset by 4} resulting in 98.03% accuracy.

A comparison of the resource utilization and power consumption between the baseline configuration using exact multipliers and the selected configuration using heterogeneous approximate multipliers is given in Table V. It can be seen that by using the proposed methodology it was possible to decrease resource utilization by 65.2% (i.e. by the factor of 3x) and power consumption by 53.4% with the marginal drop of accuracy from 98.87% to 98.03%.

TABLE V: Resource utilization of generated accelerator

Parameters	Baseline	Selected config
Network accuracy, %	98.87	98.03
LUT	157,155	54,640
LUTRAM	148	148
FF	29,864	19,667
BRAM	77.5	77.5
Power, W	2.176	1.012
Resource savings, %	—	65.2
Power savings, %	—	53.4

V. CONCLUSIONS

This paper presents a framework for applying heterogeneous approximation to hardware DNN accelerators. The framework identifies the resilience of channels in convolutional and fully-connected layers and applies a specific degree of approximation per channel. A semi-analytical approach DeepVigor is adopted for vulnerability analysis of individual channels to identify a safe approximation level for each channel. As the approach for approximation, we leverage precision reduction that corresponds to neglecting the lower-order bits of the multiplier's

inputs and effectively using a smaller bit-width multiplier. Preliminary results demonstrate possibility to decrease resource utilization by 65.2% and power consumption by 53.4% at the cost of marginal accuracy drop from 98.87% to 98.03%.

VI. ACKNOWLEDGEMENT

This work was supported in part by the Estonian Research Council grant PUT PRG1467 "CRASHLESS" and by the Estonian-French PARROT project "EnTrustED".

REFERENCES

[1] A. Bosio, D. Ménard, and O. Sentieys, Eds., *Approximate Computing Techniques.* Springer International Publishing, 2022.

[2] M. H. Ahmadilivani, M. Taheri, J. Raik, M. Daneshtalab, and M. Jenihhin, "A systematic literature review on hardware reliability assessment methods for deep neural networks," *ACM Computing Surveys*, vol. 56, no. 6, pp. 1–39, 2024.

[3] M. H. Ahmadilivani, M. Barbareschi, S. Barone, A. Bosio, M. Daneshtalab, S. Della Torca, G. Gavarini, M. Jenihhin, J. Raik, A. Ruospo et al., "Special session: Approximation and fault resiliency of dnn accelerators," in *2023 IEEE 41st VLSI Test Symposium (VTS).* IEEE, 2023, pp. 1–10.

[4] H. Ye, H. Jun, and D. Chen, "HIDA: a hierarchical dataflow compiler for high-level synthesis," in *Proceedings of the 29th ACM International Conference on Architectural Support for Programming Languages and Operating Systems, Volume 1*, 2024, pp. 215–230.

[5] Y. Umuroglu, N. J. Fraser, G. Gambardella, M. Blott, P. Leong, M. Jahre, and K. Vissers, "Finn: A framework for fast, scalable binarized neural network inference," in *ACM/SIGDA international symposium on field-programmable gate arrays*, 2017, pp. 65–74.

[6] M. Taheri, N. Cherezova, M. S. Ansari, M. Jenihhin, A. Mahani, M. Daneshtalab, and J. Raik, "Exploration of activation fault reliability in quantized systolic array-based DNN accelerators," in *2024 25th International Symposium on Quality Electronic Design (ISQED)*, 2024.

[7] M. Taheri, N. Cherezova, S. Nazari, A. Rafiq, A. Azarpeyvand, T. Ghasempouri, M. Daneshtalab, J. Raik, and M. Jenihhin, "AdAM: adaptive fault-tolerant approximate multiplier for edge DNN accelerators," in *2024 IEEE European Test Symposium (ETS)*, 2024.

[8] M. Taheri, M. Riazati, M. H. Ahmadilivani, M. Jenihhin, M. Daneshtalab, J. Raik, M. Sjodin, and B. Lisper, "DeepAxe: a framework for exploration of approximation and reliability trade-offs in DNN accelerators," in *2023 24th International Symposium on Quality Electronic Design (ISQED)*, 2023, pp. 1–8.

[9] M. H. Ahmadilivani, M. Taheri, J. Raik, M. Daneshtalab, and M. Jenihhin, "DeepVigor: Vulnerability value ranges and factors for DNNs' reliability assessment," in *2023 IEEE European Test Symposium (ETS).* IEEE, 2023, pp. 1–6.

[10] ——, "Enhancing fault resilience of QNNs by selective neuron splitting," in *2023 IEEE 5th International Conference on Artificial Intelligence Circuits and Systems (AICAS).* IEEE, 2023, pp. 1–5.

[11] S. Mittal, "A survey of techniques for approximate computing," *ACM Computing Surveys (CSUR)*, vol. 48, no. 4, pp. 1–33, 2016.

[12] C. J. B. Yann, Y. LeCun, and C. Cortes, "The MNIST database of handwritten digits," http://yann.lecun.com/exdb/mnist/, [Online].

Industrial Contribution

Enhanced Diagnosis of Failing Bits in Memory Built-in Self-Test

Ali Shisha
Advanced Application Support
Engineer
Siemens EDA
Cairo,Egypt
Ali.shisha@siemens.com

Balajiraja Ravinarayanan
Application Support Engineering
Manager
Siemens EDA
Shannon, Ireland
balajiraja.ravinarayanan@siemens.com

Knut Mellenthin
Consultant Application Support
Engineer
Siemens EDA
Hamburg, Germany
knut.mellenthin@siemens.com

Abstract— **The diagnosis of failing bits in embedded memories pose significant challenges in the current SoC industry. Overcoming these challenges is crucial as it enables improved yield, enhanced product quality, and increased overall reliability of SoC devices. By developing robust diagnosis methods, the industry can address these challenges and ensures efficient production and reliable performance of memory components. In this paper, we present a case study that uses the enhanced stop-on-error (ESOE) approach of a Memory Built-In Self-Test (MBIST) controller IP. ESOE provides efficient access to key fail bit information in internal registers of the controller, enhancing diagnosis results accuracy. It also assists in streamlining the debugging procedures, ultimately promoting more efficient troubleshooting mechanisms.**

Keywords—MBIST, DFT, failure mapping, defects detection

I. INTRODUCTION

The rapid advancement of electronic systems has led to an exponential growth in the demand for high-performance and reliable memory components. However, the increasing complexity and density of memory architectures continues increasing new challenges in ensuring the quality and reliability of memory products.

MBIST is the state-of-the-art technique for detecting and diagnosing faults within (embedded) memories during the testing stages of the manufacturing process. The analysis of fail bit information in MBIST plays a pivotal role in identifying the root causes of memory failures, enabling targeted corrective actions. By accurately diagnosing failing bits within memory arrays, designers and engineers can undertake appropriate debugging, repair, and optimization measures.

In this case study we describe the usage of ESOE implemented by MBIST. We show the challenges encountered and required solutions during a manual diagnosis process, that may illustrate the internal complexity of available automated diagnosis tools. Section II provides a comprehensive illustration of the flow and implementation of the ESOE. Section III presents the state of the art which our case study relies on. Section IV presents a case study performed in a simple MBIST design and highlights specific difficulties encountered.

II. ENHANCED STOP-ON-ERROR TECHNIQUE

A. ESOE flow

The ESOE methodology within MBIST represents a sophisticated and proactive approach to collect crucial failure information during the testing phase of embedded memories. ESOE follows a series of steps to improve the diagnostic capabilities of observed failures in the memory [1].

Firstly, during the initialization phase, the necessary registers and configurations of the MBIST controller are set up. This includes initializing of an error counter register, which determines the desired error count that will trigger the stop-on-error condition.

Next, the MBIST controller executes the test sequence. It applies test patterns to the memory-under-test and compares the expected data with the actual readback data. As the test sequence progresses, the controller continuously monitors the readback data for miscompares. If a miscompare is detected, the error count is incremented.

After each miscompare detection, the controller checks the accumulated error count against the value specified in the error count register. Once the error count matches the specified value, the controller enters the stop-on-error condition. When the stop-on-error condition is triggered, it activates the FREEZE_STOP_ON_ERROR signal as shown in Figure 1. Further, the controller halts the ongoing test sequence and stops the ongoing test. This pause is done to allow for the scan-out of all pertinent fail bit information from the controller's internal registers. These internal registers store data such as the algorithm phase, address, and failing bit position, all important for a subsequent diagnosis.

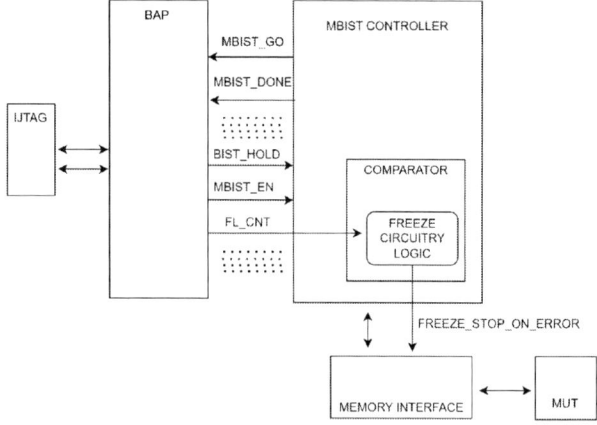

Fig. 1. MBIST ARCHITECTURE (ESOE IMPLEMENTED)

979-8-3315-3968-9/24 $31.00 © 2024 IEEE

Industrial Contribution

To ensure comprehensive diagnosis, the process is iteratively repeated. ESOE does not require the explicit increment value of the error counter to be shifted into the controller. Instead, a simple 'increment' command is sent. This simplifies the bits to shift in every single iteration to exactly the same bits.

B. Freeze circuitry overview

The architecture of the freeze circuitry in the controller for the ESOE technique encompasses several key components that work together to enable the timely detection and response to errors. While the specific implementation may vary, this section provides a general overview of the mechanism.

The freeze circuitry begins with error detection logic, which monitors various signals or conditions for the presence of an error. This logic employs comparators, error flags, or other mechanisms to evaluate the validity of data or check for specific error conditions. Upon detecting a miscompare, an error trigger signal is activated, serving as a control signal that initiates the freeze process.

The error trigger signal is then distributed to different parts of the system or test circuitry to propagate the freeze command. Through appropriate routing, the control signal reaches relevant components, such as instruction fetch units, memory access units, or test sequence controllers. A dedicated freeze control unit receives the error trigger signal and generates the necessary control signals to freeze the system. This unit comprises registers, flip-flops, or other control logic elements that respond to the trigger signal and initiate the freeze state.

The control signals generated by the freeze control unit are utilized to halt the system operation or suspend the ongoing test. By freezing the system in its current state, further instructions or tests are prevented from being executed.

The architecture of the freeze circuitry is designed to ensure a close to immediate response to errors, effectively halting the system or test operation upon error detection.

C. MBIST controller diagnostic capability

Each MBIST controller is equipped with internal register that extend from the Serial Input (SI) to the Serial Output (SO), passing through various registers. The presence of certain registers in this chain depends on the user's configuration choices during Design-for-Testability (DFT) insertion. These registers play a crucial role in enabling the diagnostic capabilities of the ESOE approach [2].

In Figure 2, the MBIST controller register chain is depicted, illustrating its internal registers [3]. Each register serves a specific purpose and functionality. Starting from the SI, we encounter the MEM_SEL register, responsible for selecting the memory to be tested in the current test step. Following that is the ALGO_MODE register, a two-bit register used for static retention testing of memories.

Next in line is the INST_POINTER register, which indicates the value of the current instruction where the controller stopped. The instruction pointer register follows the sequence of the algorithm applied by the controller. The value read from the instruction pointer reveals the specific algorithm phase and operations that were being executed when the controller halted. Subsequently, the

ADDRESS_COUNT register shows the memory address being accessed when the MBIST controller stops.

Continuing along the internal registers, we encounter the DATA_GEN register, which determines the write and expected data values. This is followed by the REPEAT_LOOP register, which represents the state of a two-loop counter. Moving further, we reach the GO_ID registers associated with each memory interface, and then return to the controller, where the ERROR_COUNT register is located. The ERROR_COUNT register stores the desired error count specified by the user and triggers the FREEZE_STOP_ERROR signal the condition is met. The GO_ID register then indicates the specific memory that is experiencing the miscompare. Additionally, the port_count and step_count are used to identify the test step in which the failing memory is located.

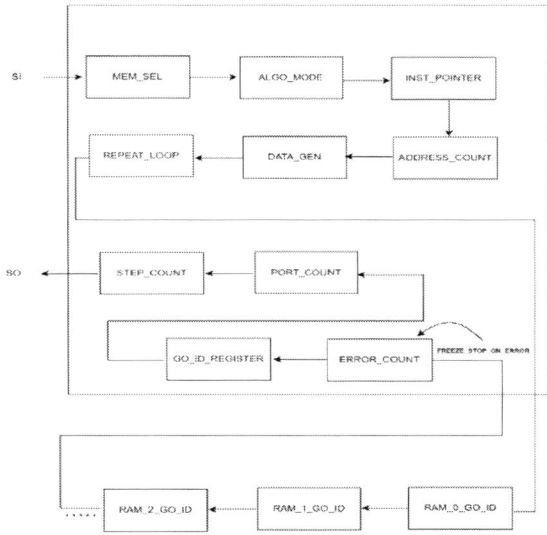

Fig. 2. MBIST CONTROLLER REGISTER CHAIN

III. STATE OF THE ART

Our case study is based on the implementation of the comprehensive and high-speed programmable MBIST architecture [3]. This architecture offers extensive support for a wide range of memory test algorithms and incorporates stop and restart diagnosis ESOE, and other options. Over time, it has undergone significant enhancements, resulting in the integration of all necessary registers within the controller chain to fulfill diagnostic requirements, as illustrated in Figure 2.

IV. CASE STUDY

In this case study, we undergo an in-depth technical approach to manually accessing and interpreting the collected ESOE data. This shall serve as an illustration on the details of the diagnosis process that otherwise is automated by DFT tools and hidden from the user.

A. Setup

The case study followed a systematic workflow to assess the effectiveness of the ESOE methodology in MBIST designs. Initially, we read in the design's RTL description, which represents the low-level hardware behavior of the design, and the associated memories to ensure they adhered

979-8-3315-3968-9/24 $31.00 © 2024 IEEE

Industrial Contribution

to design rules and specifications. This analysis aimed to validate the design's integrity and readiness for testing.

Subsequently, the design undergoes a comprehensive process of test instrumentation insertion. This involved integrating key MBIST components into the design, including the MBIST controller and the IJTAG instruments. The memory BIST controller includes the mentioned error count register that is initialized before each test run. The design depicted in Figure 3 reflects the implementation following MBIST insertion.

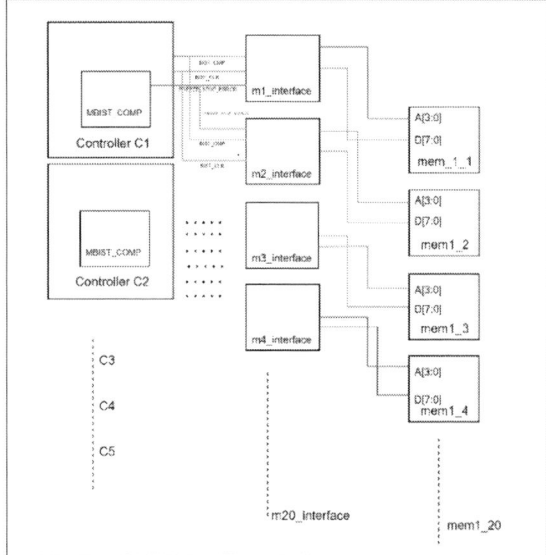

Fig. 3. Our MBIST design architecture

In our case study, the design comprises of 20 SRAM (SYNC_1RW_16x8) memories. We employed five MBIST controllers, each controller testing four SRAMs, using the SMarchCHKBci algorithm [4]. The memory mapping consists of 4 rows and 4 columns. As part of the test insertion process, a memory interface was created for each memory. These interfaces encompass the necessary diagnosis and repair logic tailored to each memory.

B. Fault injection

Most of the memories provided from various vendors include a fault injection module. This module serves the purpose of injecting faults during simulation into specific memory locations. In our case, the memories we utilized also possess a similar fault injection module. This allows us to simulate failures by injecting faults during the simulation phase before conducting ATE (Automatic Test Equipment) testing.

In our case study, we injected failures into simulation model of the first memory of the first controller. Subsequently, the ESOE methodology was employed to diagnose the failing memory, identify the failing address and data bit, as well as determine the specific memory that experienced the failure.

C. SMarchCHKBci algorithm

The default test algorithm employed by the MBIST controller to test Random Access Memories (RAMs) is SMarchCHKBci. This algorithm bears resemblance to the

SMarchCHKB test algorithm and is designed to support both synchronous and asynchronous SRAMs with single or multiple ReadWrite ports.

D. Implementation

Our pattern consists of three test steps designed to facilitate the diagnosis process, shown in Figure 4. The first test step is the initialization phase. During this step, the failure limit counter is configured with a specific integer value and the failure limit value is loaded into the MBIST counter. This counter serves as a parameter that determines the maximum number of failures that the memory BIST controller can record in the Stop-On-Nth-Error mode of diagnosis. By setting this counter, the diagnostic tool can effectively monitor and log the failures encountered during subsequent test steps.

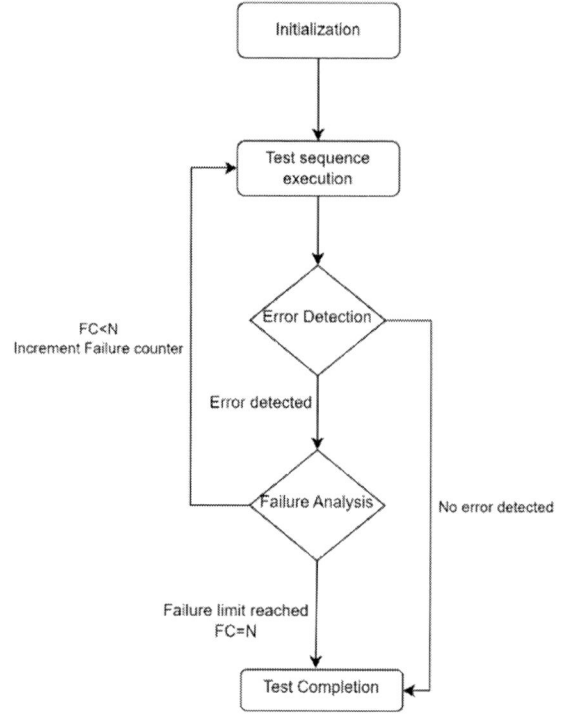

Fig. 4. ESOE flow

Transitioning to the subsequent test step, known as the execution phase, the test algorithm is executed, scanning for any mismatches or discrepancies. Whenever a failure is detected, the controller captures and scans out relevant failure data, including the failing memory id, the specific failing address, and the instruction where the algorithm identified the failure. This step ensures that comprehensive failure data is collected for further analysis and diagnosis.

The third and final test step is the Increment phase. During this step, the failure limit counter is incremented. This process continues until either the maximum number of ESOE failures is reached, or the execution pattern no longer encounters failures. We set the failure limit to one in our case which means that the controller will stop and shift out the failure data once the first failure occurs. Failures were injected in the first memory specifically to address 4 of the

979-8-3315-3968-9/24 $31.00 © 2024 IEEE

memory address. Where address 4 corresponds to Row1 Column1. So, when simulating our test patterns, we are expecting from the ESOE technique to scan out the fourth address of the memory.

E. Results

During the implementation of this approach, the controller, equipped with address registers for columns and rows, as well as Instruction Pointer (IP) registers, effectively responded when the initial miscompare was encountered. As expected, the controller promptly halted the MBIST process and retrieved all the necessary information for diagnosis. The address registers and IP registers played a crucial role in identifying the specific location and instruction where the failure occurred. This integrated functionality of the technique, along with the detailed register information, facilitates pinpointing and analyzing the problem for further troubleshooting.

These are the ICL registers that captured the fail bit information during the simulation: STEP_COUNTER_REG[0], Mem1_interface.GO_ID_REG[5], A_ADD_REG_X[0], A_ADD_REG_Y[0], INST_POINTER_REG[2], INSTPOINTER_REG[1], and INST POINTER_REG[0] .

The comparator identifies such miscompares and each internal register corresponds to a distinct functionality and operates as an individual entity within an array-like structure. For instance, REG[0] represents bit 0, while subsequent registers correspond to their respective bit positions. This arrangement enables accurate identification and isolation of failing bits [5].

At the outset, during the MBIST process, the controller initially shifts out the failing step where the failing memory exists. Even though the first step failed, this step contains all memories, necessitating the identification of the specific memory with failures. The GO signal and GO_ID registers play a crucial role in our system. When a failure is detected, the Go signal is triggered. In our inserted instruments, each Address bit in the memory has its own dedicated local comparator. These local comparators are assigned to the GO_ID registers. In the event of a memory failure, the GO_ID register helps identify the specific memory data bit associated with the failure. Based on the log file analysis, it was observed that the GO_ID corresponds to the failing data bit 5. Each local comparator is connected to a dedicated GO_ID register which allow us to identify the specific failing data bit accurately.

To determine the failing address bits of the memory, the address registers for both columns and rows are reported. In this case, row address 0 and column address 0 are failing, indicating that the fifth address bit (0101) is at fault, while our fault injection was applied to the fourth address bit (0100). The occurrence of this incorrect address is influenced by multiple factors that will be discussed in Section V.

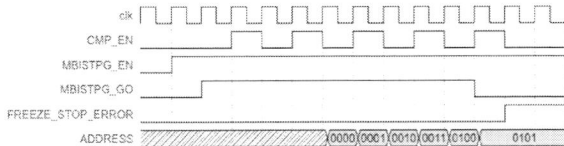

Fig. 5.   Waveforms

As seen in the waveforms, the FREEZE_STOP_ERROR (FSE) signal is triggered after the GO signal is going low

which means that the failure is detected. Once the FSE signal is triggered the controller stops the test immediately and the failing information is shifted out for the user and then the FL_CNT is incremented by one.

The failure limit counter has a limitation that must be taken into account when generating test patterns. The failure limit counter of the enhanced Stop-On-Nth-Error hardware does not check whether the maximum count specified with the failure limit reached. In some cases, within the hardware implementation of the MBIST controller, the error detection circuit is few clock cycles behind the address registers and instruction pointers. The number of clock cycles depends on a number of hardware factors such as pipeline stages, operation set and algorithm as in our case here, we can notice one clock cycle shift in the FSE signal.

Fig. 6.   One clock cycle shift of FREEZE_STOP_ERROR

This makes diagnosis more complex than expected. In our case study design this offset is caused by registers in the path of the FSE signal before shifting out the data from the TDO pin. To address this, a correction table is created. This manual correction table takes into account various factors, including the algorithm used, the operation set, pipeline stages, and freeze circuitry. By applying this correction table to the diagnosis output stream, the correct address of the failure could be obtained.

While the described manual methods offer valuable insights into the inner workings of memory diagnosis and may work for small designs where human interpretation is feasible, automated diagnosis tools are available and necessary, especially for large designs with complex IJTAG access networks and thousands of memories.

REFERENCES

[1] Xiaogang Du, N. Mukherjee, Wu-Tung Cheng and S. M. Reddy, "Full-speed field-programmable memory BIST architecture," IEEE International Conference on Test, 2005., Austin, TX, USA, 2005, pp. 9 pp.-1173, doi: 10.1109/TEST.2005.1584084.

[2] T.-P. Ho, E. Faehn, A. Virazel, A. Bosio and P. Girard, "An Effective Intra-Cell Diagnosis Flow for Industrial SRAMs," 2018 IEEE International Test Conference (ITC), 2018, doi: 10.1109/TEST.2018.8624799

[3] A. Singh, G. M. Kumar and A. Aasti, "Controller Architecture for Memory BIST Algorithms," 2020 IEEE International Students' Conference on Electrical,Electronics and Computer Science (SCEECS), Bhopal, India, 2020, pp. 1-5, doi: 10.1109/SCEECS48394.2020.43.

[4] N. Mukherjee, A. Pogiel, J. Rajski, and J. Tyszer, "High Volume Diagnosis in Memory BIST Based on Compressed Failure Data," IEEE Trans. Comput. Des. Integr. Circuits Syst., vol. 29, no. 3, pp. 441–453, Mar. 2010.

[5] G. Harutyunyan, S. Martirosyan, S. Shoukourian, and Y. Zorian, "Memory Physical Aware Multi-Level Fault Diagnosis Flow," IEEE Trans. Emerg. Top. Comput., vol. 6750, no. c, pp. 1–1, 2018.

[6] M. I. Masnita, W. H. W. Zuha, R. M. Sidek and I. A. Halin, "The data and read/write controller for March-based SRAM diagnostic algorithm MBIST," 2009 IEEE Student Conference on Research and Development (SCOReD), Serdang, Malaysia, 2009

Industrial Contribution

Exploring the Role of the Portable Stimulus Standard in Enhancing Security Property Verification

Jaimini Nagar
Infineon Technologies GmbH & Co. KG
Dresden, Germany
Jaimini.Nagar@infineon.com

Thorsten Dworzak
Infineon Technologies AG
Munich, Germany
Thorsten.Dworzak@infineon.com

Sebastian Simon
Infineon Technologies GmbH & Co. KG
Dresden, Germany
Sebastian.Simon@infineon.com

Ulrich Heinkel
Technical University of Chemnitz
Chemnitz, Germany
ulrich.heinkel@etit.tu-chemnitz.de

Djones Lettnin
Infineon Technologies AG
Munich, Germany
Djones.Lettnin@infineon.com

Abstract— The security of the hardware design has been a critical concern in the semiconductor industry in the recent years. Modern integrated circuit designs of the security-critical application contain numerous sensitive data that needs to be protected against adversary attacks. An Intellectual Property (IP) block such as security controllers is integrated into the System on Chip (SoC) to protect the sensitive asset and its integrity with certain measures that implement cryptographic algorithms. Verification of security properties of this type of IP block is especially important and crucial to ensure its trustworthiness. This paper presents an approach to enhance security property verification of the design using Portable Test and Stimulus Standard (PSS) along with existing testbench. PSS defines a specification to generate a single representation of stimulus and test scenarios, which is usable by a various user across different level of integration. We have modelled functionality and security property of the Security Controller IP at abstract level using PSS. Then, PSS model has facilitated to generate a set of concrete test cases in Specman/e for various scenarios. The generated test cases are executed in verification environment and analyzed the results that verify security properties.

Keywords— *Security property verification, PSS, security controller IP, simulation-based verification component*

I. INTRODUCTION

Modern computing systems and portable devices like smartphones, smart sensors, tablets, wearables, etc. are increasingly getting used in our personalized activities ranging from banking infrastructure to providing driving directions. Thus, modern SoC designs developed for automotive, computing devices or mobile contain plenty of sensitive data and personal information including contacts, location, banking credential that must be secured [1]. SoC designs encompass the security circuit that controls the access to sensitive assets by implementation of complex security properties. This emphasizes the importance of ensuring the trustworthiness of the hardware design used in security-critical applications; however, verifying the security properties of the integrated security circuits is particularly important and challenging. Security requirements can be formalized as taint-propagation properties and proved them using a formal verification approach [2]. Recently, the systematic use of the model checking based formal verification method has been published for the verification of properties for security-critical functionality of the design [3]. Simulation-based verification is also widely used in the

semiconductor industry to verify hardware designs. However, the limitation of employing this method is that corner case errors can be overseen. Writing a SystemVerilog assertion or directed tests to verify the security properties is extremely time consuming and difficult to compose. The Portable Test and Stimulus Standard (PSS) enhances the strength and quality of the existing verification approach by defining a specification for creating a single representation of stimulus and test scenarios that enables automation of stimulus generation [4]. PSS describes a domain-specific language (DSL) that is declarative in nature to specify the verification test intent. It is usable at various levels of integration and enables the generation of different implementations of a scenario that run on different execution platforms i.e., simulation, emulation, FPGA prototyping, and post-silicon [5]. The essence of the PSS lies in the action, which is the way in which the language expresses behavior. A PSS model contains a number of elements that define a set of feasible scenarios to be applied to the Design Under Verification (DUV) through the accompanying test environment. This paper presents a comprehensive study of the use of portable stimulus methodology for the verification of the security property and functionality of the hardware design.

The rest of the paper is organized as follows. In Section II, a summary of the existing approaches is described. An overview of the proposed methodology and PSS modelling for the security property verification is discussed in Section III. The experimental result of the implementation of the security property verification methodology using a simplified waveform is presented in the Section IV. In the Section V, the paper is concluded with a short summary.

II. RELATED WORK

During past few years, verification engineers and researchers have started to carry out an investigation on application of PSS to explore its features and benefits. The result of a case study shows that the capability of the industry standard UVM is strengthened multiple times and regression efficiency is increased by utilizing PSS modeling [6]. The PSS language offers traditional High-level Verification Language (HVL) features like assignments, expressions, control flow with loops, conditionals, function calls, sequential and concurrent execution, constraints and coverage constructs. For a complex multi agent testbench, the time required for coverage closure can be reduced by a factor of 10x using the PSS approach as reported [7]. These presented works use

979-8-3315-3968-9/24 $31.00 © 2024 IEEE

Industrial Contribution

portable stimulus based simulation for functional verification. Neither of these works explores the use of PSS approach for security property verification. In our proposed methodology, security properties are formulated as a PSS model and different implementations of scenarios are generated for an existing testbench.

The security verification of the hardware design has become a pivotal phase for the current SoC design because of various security threats and hardware attacks at different stages of the development cycle. Researchers have used the formal verification method to verify RISC-V architecture [3] and OpenRISC-1200 based SoC architecture [8]. Security analysis of the SoC design using an assertion based formal verification approach is presented in these works. However, scaling up the formal verification method for the complex SoC design is challenging and formalized properties could not be ported or reused for other verification platforms. The security properties have been formulated as a PSS model in our proposed methodology that can be reused across various verification platforms. Moreover, a PSS model facilitates the generation of meaningful and comprehensive tests for multiple test case scenarios.

III. PSS Supported Verification Methodology

This work proposes a methodology for verifying security properties based on the PSS that can be reused across different verification phase of the product development cycle. We selected Security Controller (SC) IP to apply the proposed methodology because of it implements security features and supports cryptographic operation using the Advanced Encryption Standard (AES). The crypto engine block is a functional sub-block of the SC IP which includes cryptographic engine used by the channels of the SC.

A. Design Under Verification

Fig. 1 depicts a block diagram of the design under verification called SC IP that provides access to the cryptographic engine. It implements security functions based on a cryptographic algorithm and master slave access control policy. SC IP has a key storage for storing keys that are required to perform the cryptographic operations. It contains separate input and output data storage as well as Control and Status Registers (CSRs) to configure the required operation and provide status information. The security properties of the

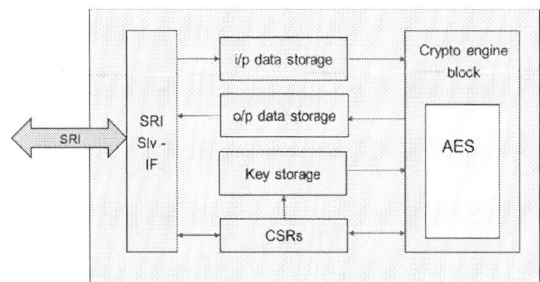

Fig. 1. Block diagram of Security Controller (SC) IP

design are developed by considering its functionality and analysis of potential security vulnerabilities considering integrity and confidentiality principles. SC IP has channels for providing the interface to software tasks to access the cryptographic engine and one of them is called "confidential channel". A bus master that has valid access to the channel is called channel master, while a bus master that has access rights granted to the confidential channel is called security master.

B. Overview of Methodology

We have deployed a portable stimulus-based verification methodology to enhance an efficiency of existing traditional simulation-based verification flow. As shown in Fig. 2, hardware detailed design derived from the hardware requirements is taken as an input for the RTL implementation and for the preparation of verification plan in the traditional verification process. Additionally, the generation of a PSS model starts to capture the security properties of the design in the declarative domain-specific language. The PSS specification encapsulates an abstract behavior of verification test intent of the design according to the requirements as well as target implementation binding for the test realization. A model is made up of two types of class definitions: i) *actions* that define elements of behavior, ii) *objects* as passive entities utilized by actions, such as resources, states, and data flow items. The behaviors associated with an action are specified as activities. *Components* encapsulate *actions* and *object* definitions in the form of reusable model pieces. All these PSS key elements can also be captured and extended in a package to allow for further reuse and customization. In parallel of the RTL design implementation by the design team

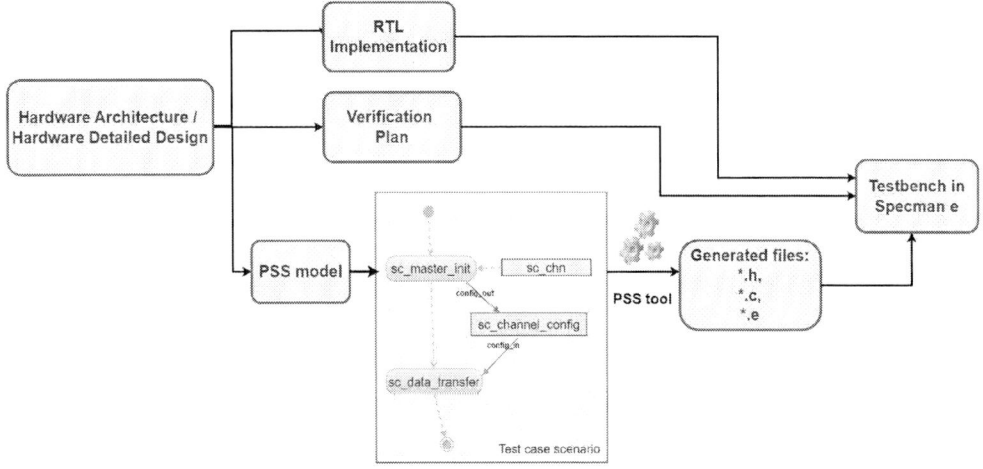

Fig. 2. PSS supported verification methodology

979-8-3315-3968-9/24 $31.00 © 2024 IEEE   185

Industrial Contribution

and ensuring the readiness of testbench environment, the test generation of different scenarios from the PSS model can be performed. The generated tests are directed to the existing testbench for verification and analyze the results.

C. PSS modelling of Security Property

Initially, we reviewed the design specification thoroughly to identify the test elements in context of security property. In natural language, a security property named SP_1 is stated as "*A channel master shall only be able to write keys if the security master explicitly enabled the channel master to write keys*". An abstract level model of SC IP in domain-specific language is created that is also called a PSS model. It must have one or more actions to define the system's behavior and set of attributes that is required to execute the actions. A snippet of an abstract model written in PSS is shown in Listing 1. There are multiple actions encapsulated into a component called 'sc_controller_c' and the actions contain data field as attributes. An interaction between actions is employed via *data flow objects*. SC IP performs cryptographic operation and it transfers the final data after computation. This behavior is modeled as an action called 'sc_data_transfer' and it needs cipher key and access to write a key in assigned CSR to complete the process. For property SP_1, a model should have an attribute for specifying the security master's CSR that enables the channel master to do the write key operation. It also needs to be added other responsible data fields in a model that are required to perform the correct operation of SC IP. A model should contain executables as *exec block* for mapping of atomic actions to their relevant implementation for a target platform. The positive check should be conducted to verify that the write enable (WRENA) bit field of the security master's CSR must be set to 'high' and this condition only allows channel master to write the key for the cryptographic operation. The negative check also must be done to ensure the trustworthiness of the design by setting the mentioned bit field to 'low' and this condition must not allow the channel master

```
resource sc_channel_r { };
// control and status register to write operation
struct sc_op_cmd_s {
    rand bit [1] CIPH ;
    rand bit [1] GEN ;
    rand bit [1] VER;
    .........
} ;
component sc_cotroller_c {
    pool [SC_NUM_CHANNELS] sc_channel_r sc_chn ;
    bind sc_chn *;

    action sc_master_init {
        lock css_channel_r  sc_chn ;
        rand int channel_number ;
        constraint channel_number==
                        sc_chn.instance_id;
        ...................
    };
    action sc_data_transfer {
        rand sc_op_cmd_s sc_op;
        rand bit [1] WRENA ;
        .............
        exec body {
            message (LOW, "calling sc_rand_op_seq");
            DVE.sc_rand_op( channel_number, WRENA,
                sc_op.CIPH, sc_op.GEN, sc_op.VER);
                };
            ...................
        };
};
```
Listing. 1. An abstract model in PSS

to write a key and must cause rising of the error and interrupt signals.

Another security property named SP_2 is stated in natural language as, "*The SC shall allow a key to be used only for operations allowed by the key attributes*". A particular register in the CSRs is dedicated to set/enable different operations performed by SC IP. A register is defined as 'sc_op_cmd_s' using 'struct' data type in the PSS model. SC IP implements multiple operations and one of them is encryption/decryption and the dedicated bit field for specifying this operation is CIPH. If the bit field for the cipher operation is set to 'high', then only the cryptographic operation will be executed. To verify this property, a PSS model must have an action including a data attribute to specify this behavior of the design. Therefore, an attribute called CIPH has been entered in the PSS model. To execute any cryptographic operation, CIPH must be set with 'high'. Otherwise, SC IP shall not allow to use the key for defined operation. Consequently, an error and interrupt signals will rise and terminate the operation. SC performs same behavior for each operation and their dedicated fields of a specific register in the CSRs.

The PSS model of SC IP is structured in such a way that it represents high level behavior of the hardware design and also captures verification stimulus. To drive the behavior signified by actions in a test scenario, an interaction of a PSS model with foreign language is done by calling the application programming interfaces (APIs) presented in the execution environment. The constructs used to specify the implementation of PSS entities are known as 'exec' blocks and they offer a mechanism to associate the specific functionality with a component, actions and flow objects. Additionally, the exec blocks are utilized to define targeted test case implementation and mapping a PSS model to the implementation code. Hence, we prepared a PSS specification that incorporates the definition of a verification target platform on which tests will be executed and key constructs for mapping it to the implementation code. The PSS model makes it effortless to create certain test scenarios for the functional verification as well as security property verification. To implement the test scenarios, the tests will be generated in the target language (*.e) and executed on the targeted platform (simulation).

IV. EXPERIMENTAL RESULT

The experimented verification methodology using portable stimulus modeling is explained in detailed in Section III-B. The PSS representation of SC IP should be compiled on a PSS processing EDA tool. It is important to mention that PSS does not create a test bench, so the infrastructure such as a Universal Verification Methodology (UVM) test bench must exist to run the generated tests. The generated tests are aimed to the test bench in Specman/e as shown in Fig.2. As the target platform is simulation for this work, the generated tests have been executed in the existing UVM test bench. We analyzed the result by means of waveform inspection in the simulator to verify if the security properties are met. A simplified view of the waveform is represented in Fig. 3 and Fig. 4. A result for the verification of security property SP_1 for the positive check is shown in Fig. 3 (a). We observed the standard functionality of SC IP without any error. To carry out a negative check, value of WRENA has been set to 'low' and consequently rising of error signal 'uninkey_err' and interrupt signal 'irq' could be observed in Fig. 3 (b).

979-8-3315-3968-9/24 $31.00 © 2024 IEEE

Industrial Contribution

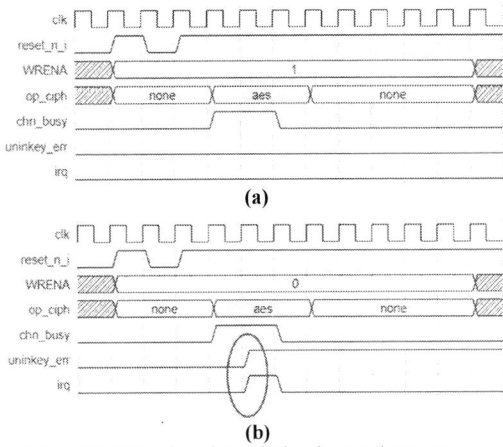

(a)

(b)

Fig. 3 Simplified Waveform for analyzing the security property verification for SP_1 positive check (a), SP_1 negative check (b)

In the same way, we have conducted positive check and negative checks to verify the security property SP_2. Only if the CIPH is set to 'high', a cryptographic operation of SC can be performed. The obtained result for SP_2 is shown as simplified waveform in Fig. 4 (a) and (b), respectively. Also, a triggering of error and interrupt signals is highlighted with a blue circle in Fig. 3 (b) and Fig. 4 (b). The proposed methodology strengthens the existing verification process by providing visuals of test scenario implementation in graphical user interface and automated stimulus generation for the functional and security property verification.

In addition to the tests for positive check and negative check, we were able to generate as many as possible solutions and tests based on constrained randomization for each test case scenario. A test case scenario for SP_2 is shown in Fig. 2. In PSS, randomizable type data field may be declared as random by preceding its declaration with 'rand' keyword and action can define 'constraint' expression on to set the range of data field. A scenario for SP_1 has facilitated generation of 42 tests. For the scenario of SP_2, total 168 tests were generated that cover all the possible values of randomizable data field of verification test intent. It also helps to resolve the problem regarding corner cases of simulation based verification.

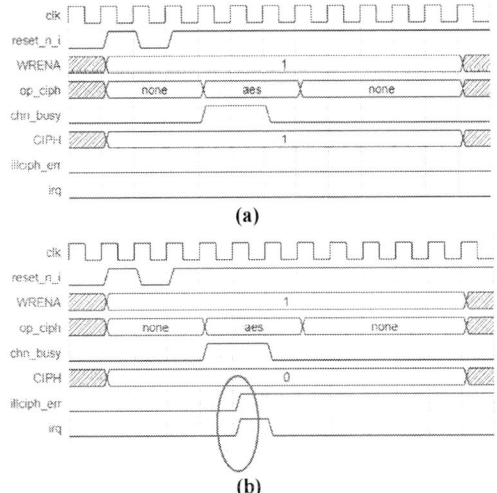

(a)

(b)

Fig. 4 Simplified Waveform for analyzing the security property verification for SP_2 positive check (a), SP_2 negative check (b)

V. CONCLUSION

The overall goal of this work is to explore the role of PSS for security property verification of the design to enhance traditional verification flow. We have formulated security properties of SC IP using PSS and this model encapsulates verification test intent as well as executables for mapping of atomic actions to the target implementation. The generated tests are able to verify that security properties are fulfilled. Moreover, the proposed methodology creates visuals of the test implementation since we could see the value of stimuli before the generated tests will run on simulator. Therefore, it makes easier to create a complex scenario implementation using PSS methodology, which is not possible with traditional simulation based verification. It is also advantageous to get the optimized number of tests based on constraint randomization to speed up coverage closure. Moreover, it is possible to port the stimulus across different verification platform by generating tests in respective language using this PSS model as a single reperestion.

We presented the methodology for security property verification using PSS at IP level for RTL simulation. In future work, a PSS model created for IP level would be used to generate C tests for SoC level verification to demonstrate vertical reusability. It can also be foreseen to generate the tests to be executed on other platforms such as virtual prototyping, emulation or post-silicon validation to exhibit its reusability in horizontal dimension.

ACKNOWLEDGMENT

This work has been developed in the project VE-VIDES (project label 16ME0243K) which is partly funded within the Research Programme ICT 2020 by the German Federal Ministry of Education and Research (BMBF).

REFERENCES

[1] S. Ray and Y. Jin, "Security policy enforcement in modern SoC designs," in *2015 IEEE/ACM International Conference on Computer-Aided Design (ICCAD)*, Austin, TX, USA: IEEE, Nov. 2015, pp. 345–350.

[2] P. Subramanyan and D. Arora, "Formal verification of taint-propagation security properties in a commercial SoC design," in *Design, Automation & Test in Europe Conference & Exhibition (DATE), 2014*, Dresden, Germany: IEEE Conference Publications, 2014, pp. 1–2.

[3] C. S. Chuah, C. Appold, and T. Leinmueller, "Formal Verification of Security Properties on RISC-V Processors," in *Proceedings of the 21st ACM-IEEE International Conference on Formal Methods and Models for System Design*, Hamburg Germany: ACM, Sep. 2023, pp. 159–168.

[4] Portable Test and Stimulus Standard Version 2.1. Accellera Systems Initiative. Accessed: Jan. 26, 2024. [Online]. Available: https://accellera.org/images/downloads/standards/pss/Portable_Test_S timulus_Standard_v2_1.pdf

[5] G. Moretti, "Accellera's Support for ESL Verification and Stimulus Reuse," in *IEEE Design & Test*, vol. 34, no. 4, pp. 69–75, Aug. 2017.

[6] N. K. Zubair and S. K. R. Sajja, "Increasing Regression Efficiency with Portable Stimulus". in *2019 proceeding of Design and Verification Conferemce and Exhibition (DVCon) Europe*.

[7] S. Henry and N. Regmi, "How to Close Coverage 10x Faster using Portable Stimulus Standard - A Case Study," in *2018 19th International Workshop on Microprocessor and SOC Test and Verification (MTV)*, Austin, TX, USA: IEEE, Dec. 2018, pp. 28–30.

[8] P. Bhamidipati, S. M. Achyutha, and R. Vemuri, "Security Analysis of a System-on-Chip Using Assertion-Based Verification," in *2021 IEEE International Midwest Symposium on Circuits and Systems (MWSCAS)*, Lansing, MI, USA: IEEE, Aug. 2021, pp. 826–831.

979-8-3315-3968-9/24 $31.00 © 2024 IEEE

High-Density Standard Cell Library
for Sequential 3D Integrated Circuits

Arturo Prieto and Joachim Rodrigues
Department of Electrical and Information Technology
Lund University, Lund, 22100 Sweden
Email: {arturo.prieto, joachim.rodrigues}@eit.lth.se

Abstract—Research efforts to push the integration density of circuits with technologies that transcend Moore's law have gained significant attention in recent years. This study investigates the silicon area gains of Sequential 3D technology, utilizing the third dimension of integrated circuits by accommodating nMOS and pMOS transistors in two stacked tiers with high-density and low-pitch 3D vias. The efficiency of the proposed integration strategy is exemplified through the design of a library with high-density 3D standard cells, including sequential and combinational logic. The integration of 3D vias within the standard cells mitigates the effort required for inter-tier connections during the routing of integrated circuits. Subsequent analysis indicated an average silicon area reduction of 36 % in comparison to commercially available libraries with purely planar cells. The proposed 3D cells have been incorporated into a commercial design flow for a 28 nm process technology and have been benchmarked using examples of large-scale integration designs, indicating an area and wirelength reduction of 44 % and 23 %, respectively.

Index Terms—More-than-Moore, Sequential 3D, Partitioning, High-Density, Library, 3D Cell.

I. INTRODUCTION

Technology scaling for integrated circuits (ICs) has slowed down considerably in recent years. However, transcending Moore's law, research on more-than-Moore technologies has gained increased attention, driven by the demand for high-performance computing, e.g., artificial intelligence, 6G networks and virtual reality. Consequently, higher computational demands require higher transistor density, and the corresponding complexity comes with significantly increased congestion and longer wire connections with inherent parasitics that impact the system performance. Moreover, core utilization will be reduced as signal routing may require a larger space between cells, and thus, the total silicon cost will increase.

A promising more-than-Moore technology that offers high integration density is Sequential 3D (S3D). This technology is realized by stacking a layer of transistors, which is fabricated in a sequential process, on top of another layer of transistors [1], [2]. In S3D, each active layer is referred to as a tier, realizing a 3D implementation by having tiers stacked above each other. Tiers are connected by 3D vias, known as Monolithic Inter-tier Vias (MIVs), which have the advantage of being the same size as conventional vias in a silicon process, and having up to 50 times smaller pitch than Through Silicon Vias (TSVs). These properties result in a significantly higher integration density, outperforming other 3D integration technologies [3].

The implementation of 3D circuits comes with challenges in tier integration and associated costs regarding stacking. On the bottom tier, standard nMOS and pMOS transistors are fabricated, including its Back-End-Of-Line (BEOL). Thereafter, 3D vias are integrated, followed by the manufacturing of the top tier. This procedure in the manufacturing process is extremely critical for achieving a sufficient yield. Therefore, the manufacturing temperature needs to be kept below 500 °C to avoid degradation of the bottom structures [4].

A performance limiter of today's circuits is the parasitics of lengthy interconnections, caused by the physical properties of ICs. S3D offers different implementation possibilities, based on various integration approaches, which reduce overall wirelength and silicon area. A cell library for S3D integrated circuits, designed with nMOS and pMOS transistors separation on top and bottom tier, respectively, is presented in [5]. The reduced pitch size offered by MIV enables the division of tiers for a specific transistor type, i.e., nMOS and pMOS, creating 3D cells with all transistors of the same type concentrated in the same tier. Alternatively, a division of logic cells in tiers, rather than transistor separation, is evaluated in [6]. This analysis exposes the required research on placement techniques, taking into account routing congestion in 3D integrated circuits for efficient connections between tiers. In [7], arithmetic logic blocks for multiplications are evaluated, making a circuit integration with several layers stacked above each other, and employing vertical pillars for the connection of multiplier elements. These works show different possibilities with 3D technology to obtain high integration density. However, the aforementioned 3D ICs are realized by accommodating individual logic cells, or transistor types, in each tier. An alternative is the evaluation of 3D cell designs by partitioning the cell circuit across tiers.

This work investigates the integration density and area-wirelength reduction efficiency of S3D technology by evaluating the design and integration of 3D cells, where nMOS and pMOS transistors are freely accommodated on both tiers. A library of 3D standard cells is created for large-scale designs.

The remainder of this manuscript is organized as follows: Section II presents the proposed integration for the library of high-density 3D cells, and Section III discusses silicon area gains and the use of the presented cells as part of benchmarked examples. Finally, Section IV concludes this study.

979-8-3315-3968-9/24 $31.00 © 2024 IEEE

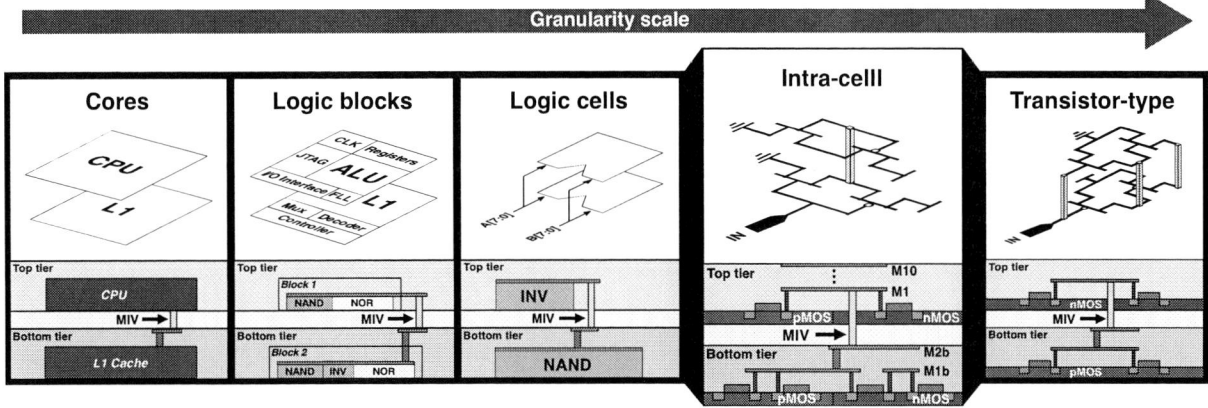

Fig. 1: Granularity scale for different partitioning possibilities with Sequential 3D integration technology, including high-density 3D cells integrated as intra-cell partition with nMOS and pMOS transistors accommodated freely on two tiers.

II. HIGH-DENSITY 3D STANDARD CELL LIBRARY

Sequential 3D technology offers various levels of partitioning. This section will provide background information on the integration levels and details on the defined intra-cell partition for the generation of our 3D standard cell library.

A. Sequential 3D Integration

3D circuits can be realized by considering different implementation styles, as shown in Fig. 1. At the lowest granularity level, an entire core is placed on one tier, and data communication is realized with vertical connections between tiers. Conventional 2D cells are used, increasing the engineering overhead for finding an optimal partition of the design that reduces routing congestion on both tiers and achieves an efficient area implementation. On the other end of the scale, reaching the finest granularity, the nMOS and pMOS transistors are strictly separated, i.e., the pull-down and pull-up networks of a logic cell are in the top and bottom tier, respectively, and MIVs realize transistor connections. By making a transistor-type partition, higher integration density can be achieved using more MIVs, at the cost of increased parasitics arising from more inter-tier connections. The technique proposed in this study closes the design gap between separating the tiers by logic cells or transistor type, seeking high integration density with efficient inter-tier connections. This is accomplished by freely integrating nMOS and pMOS on either tier, with MIVs for intra-cell connections.

The stack of transistors and metal layers for intra-cell partition with Sequential 3D integration technology is shown in Fig. 1. Both top and bottom tiers include nMOS and pMOS transistors, with two metal layers (M1b and M2b) dedicated to the bottom for internal cell connections including power rails, and ten (M1-M10) to the top for routing. The connection between tiers is realized by MIVs, which have the same physical dimensions as other vias between metal layers.

With intra-cell partition approach, transistors are efficiently integrated in 3D cells with the same distance between power rails as conventional 2D cells, providing a large design space for routing tracks with reduced congestion of inter-tier connections. Consequently, the bottlenecks derived from 1) routing congestion of MIVs produced by using 2D cells on 3D stacking, and 2) reduced cell height with more MIVs that limit routing tracks on transistor-type partition, are addressed with a solution that offers high-density integration with MIVs included in cells.

In Sequential 3D integration, the stacking of transistors generates a coupling effect that makes the top tier more sensitive to voltage variations in the bottom tier. The introduction of an isolation plane between both tiers for avoiding undesirable interaction has been evaluated in [8], where a polysilicon Ground Plane (GP) is introduced sequentially in the manufacturing of 3D integrated circuits. When inserting an isolation plane, extra design space rules are required for including MIVs, which reduces the integration density. However, results in [9] demonstrate that there is minor coupling between top and bottom tiers for digital ICs, therefore the introduction of an inter-tier GP is not needed for purely digital Sequential 3D circuits. For the design of high-density 3D cells, the GP is not considered in order to achieve high integration density. The trade-off between inter-tier coupling effect and integration density for 3D cells will require further technology analysis.

An engineering overhead for S3D technology is produced by IC design tools and their ability to efficiently synthesize and route in the 3rd dimension. However, the design technique proposed in this study, integrating MIVs in the cell design, has the advantage of a routing effort that is comparable to 2D technology. The routing step is fulfilled at the top tier, while at the bottom tier only power rails are connected. The use of a conventional IC design flow reduces the design overhead and tool dependency for S3D circuits.

B. 3D Cells Library Design

An extensive number of transistors are employed in the process of designing large-scale integrated circuits. In order to create digital designs, pre-designed and pre-characterized building blocks are commonly used to implement different logic functions. A selection of combinational and sequential standard cells, as fundamental building blocks, are compiled in

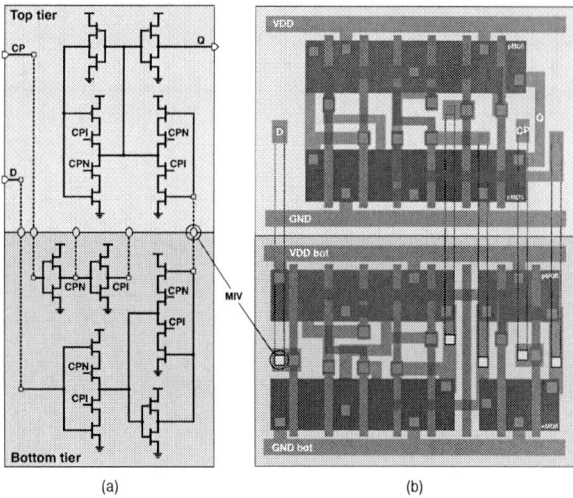

Fig. 2: Circuit partition of a DFF in two tiers showing MIVs for the 3D cell and circuit connections. (a) Schematic, and (b) layout.

a library. In the generation of large-scale designs, the building blocks defined in the library are selected and combined for producing the logic of the design.

The realization of sequential and combinational logic cells is evaluated to study the efficiency of the proposed 3D integration strategy. The partition is based on dividing the structure of standard cells in top and bottom tiers. When using MIVs, intermediate metal layers between tiers are introduced adding extra metal connections in the cell. For a standard cell with intra-cell partitioning, the cell structure is divided targeting the efficient use of MIVs by limiting inter-tier connections and achieving high-density integration.

As an illustrative example, the 3D integration of a flip-flop (DFF) cell, including the partition of the transistors in the schematic, is shown in Fig. 2. The schematic and layout views show MIVs employed for inter-tier connections. The input-output interface, Data-in (D), Clock Pulse (CP) and Data-out (Q), is on the top tier, while the bottom tier accommodates transistors for obtaining high-density cell integration. The routing step for digital designs is performed by connecting 3D cells interface ports at the top tier. The presented cells are fully characterized for inclusion in a large-scale design flow.

III. EXPERIMENTAL SETUP AND RESULTS

This section details the advantages of S3D technology for silicon area and wirelength gains by evaluating design optimizations with high-density 3D standard cells.

A. Methodology

Different considerations for Sequential 3D integration technology are taken into account for this evaluation. Advances in S3D implementation enable the use of the same technology, i.e., 28 nm, in both stacked tiers [10]. Design rules for MIVs are considered to be the same as for other vias employed to connect metal layers, with the same area and space distance. As mentioned in section II.A, a ground plane is not required

between tiers, considering negligible the coupling effect due to tier-to-tier interference for digital circuits.

The 3D cells compiled to a library are designed on schematic level with their correspondent physical layout. The standard cells partition for top and bottom tiers are created in independent views using Cadence Virtuoso, and verified with commercial 28 nm design rules. MIVs are modeled as connections between cell sub-circuits accommodated in each tier, introducing a 20 Ω resistance between M2b and M1, with irrelevant lateral coupling.

Different large-scale digital circuits, with the number of cells employed in the integration of each design, are presented in the benchmark evaluations of [5], [6]. Silicon area gains are estimated by evaluating the proposed implementations with logic cells of our designed 3D standard cell library, and comparing to a commercial library. These benchmarks will provide representative results regarding the integration density achievable with our 3D standard cell library. Impact on power and timing exceeds the scope of this evaluation, and will be considered for future analysis.

The 3D cells are characterized to be included in a commercial design flow. The top and bottom tier cell partitions are separated for an individual realization, adapting the use of 3D cells with conventional IC design tools. For a digital circuit implementation, the cells position is defined by the placement of the top tier partitions. Then, routing of the signals in the top will finalize this design step. The power rails included in the bottom tier of standard cells define the mesh for the power of the bottom transistors.

B. Results

In [11], benchmark implementations with different partition granularities and S3D optimizations are evaluated. For this work, the focus is on large-scale designs that will benefit from increased area and wirelength efficiency, performing the evaluation in simulation.

The silicon area gain, presented as area reduction percentage, of the 3D cells library compared to a commercial library, is shown in Fig. 3. The average area reduction considering all logic cells in the library is estimated as 36 %. Logic cells with a larger transistor count achieve higher area gains, since a larger cell layout benefits more from the higher integration density achieved with the proposed technology.

Comparison with other works using an academic benchmark circuit, s38584, and a RISC-V processor, is shown in Table I. The area reduction is estimated as the area difference between using our 3D standard cell library and a commercial 2D library. The wirelength reduction is estimated as the difference in the total length of wires in a layout generated for each case. The density of inter-tier connections is measured as MIV/cell, which determines the number of MIV employed in the circuit integration as a function of the number of cells of the design. A smaller factor indicates lower congestion of 3D vias, with reduced inherent parasitics of inter-tier connections, benefiting routing.

979-8-3315-3968-9/24 $31.00 © 2024 IEEE

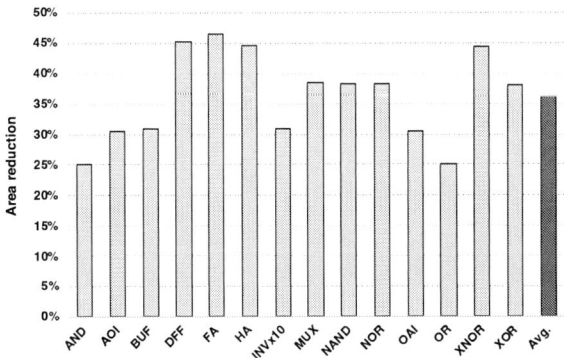

Fig. 3: Area reduction of designed 3D standard cell library compared to a commercial library.

TABLE I: Area, wirelength and MIV density comparison using benchmark circuit designs.

Source	[5]	[6]	**This work**	
Granularity	Transistor	Logic cells	Intra-cell	
Circuit	s38584	RISC-V	s38584	RISC-V
Frequency [MHz]	500	555	500	
Area red. [%]	29,5 - 38,3	23,6	39,0	44,0
Wirelength red. [%]	12,4 - 17,2	6,5	20,5	23,1
MIV/cell	4	2 - 9	3	

In [5], different large-scale integration designs are presented, and each of them is evaluated with three types of 3D cells based on transistor level partitioning with transistors accommodated on type-specific tiers (nMOS on top and pMOS on bottom). The three proposed implementations of 3D cells, each considering a different number of routing tracks, represent different percentages of area and wirelength reduction gathered in a range for the case of the benchmark circuit s38584 included in the comparison. In [6], two cases of Monolithic 3D (M3D) implementation for a RISC-V circuit are presented. The partitioning method accommodates SRAM memories in one tier and a RISC-V processor in the other, using MIVs for inter-tier connections. The implementation of the logic that is part of the RISC-V processor is considered for our evaluation.

Considering the circuit s38584, the use of the proposed 3D standard cell library achieves an area gain similar as the most efficient area implementation of [5]. However, the number of MIV per cell is reduced by 25 %, reducing the congestion of inter-tier connections. Considering the RISC-V design, the area gain is 20 % higher with up to 67 % less MIV/cell compared to [6]. Employing the same IC design flow as a planar implementation, our 3D standard cell library reduces the wirelength 23 % compared to a commercial 2D library. The proposed 3D integration method achieves higher wirelength reduction than the aforementioned works, showing an efficient Sequential 3D integration solution for reducing routing congestion on large-scale circuits.

Wire delay grows quadratically with the routing wirelength [12], which motivates the use of high-density cells to reduce distances between cells. They have the advantage of a reduced area and routing cost, which results in higher efficiency in terms of density, performance and energy. The presented evaluation is considered as a baseline scenario, where similar gain possibilities are expected for larger designs.

IV. CONCLUSION

This study presents the design of a 3D standard cell library using MIVs for intra-cell connections, which facilitates a high-density Sequential 3D integration in two tiers. The designed library is compatible with commercial IC design tools with a routing effort comparable to conventional technologies. The proposed design of 3D cells has an average of 36 % silicon area gain compared to commercial libraries. The implementation method is benchmarked using examples of large-scale integration circuits, showing up to 44 % silicon area gain, and 23 % wirelength improvement, with reduced congestion of 3D vias compared to other works.

V. ACKNOWLEDGMENT

This work has been supported by the EU's Horizon 2020 funding scheme in the project 3D-MUSE.

REFERENCES

[1] S. Bobba et al., "CELONCEL: Effective design technique for 3-D monolithic integration targeting high performance integrated circuits," *16th Asia and South Pacific Design Automation Conference (ASP-DAC)*, pp. 336-343, 2011.

[2] P. Vivet et al., "Monolithic 3D: an alternative to advanced CMOS scaling, technology perspectives and associated design methodology challenges," *25th IEEE International Conference on Electronics, Circuits and Systems (ICECS)*, pp. 157-160, 2018.

[3] H. Sarhan, S. Thuries, O. Billoint and F. Clermidy, "An Unbalanced Area Ratio Study for High Performance Monolithic 3D Integrated Circuits," *IEEE Computer Society Annual Symposium on VLSI*, pp. 350-355, 2015.

[4] I. Radu et al., "Ultimate Layer Stacking Technology for High Density Sequential 3D Integration," *International Electron Devices Meeting (IEDM)*, pp. 1-4, 2023.

[5] C. Yan and E. Salman, "Mono3D: Open Source Cell Library for Monolithic 3-D Integrated Circuits," *IEEE Transactions on Circuits and Systems I: Regular Papers*, vol. 65, no. 3, pp. 1075-1085, 2018.

[6] S. Thuries et al., "M3D-ADTCO: Monolithic 3D Architecture, Design and Technology Co-Optimization for High Energy Efficient 3D IC," *Design, Automation & Test in Europe Conference & Exhibition (DATE)*, pp. 1740-1745, 2020.

[7] E. Giacomin, F. Catthoor and P. -E. Gaillardon, "Area-Efficient Multiplier Designs Using a 3D Nanofabric Process Flow," *IEEE International Symposium on Circuits and Systems (ISCAS)*, pp. 1-5, 2021.

[8] P. Sideris, A. Peizerat, P. Batude, C. Theodorou and G. Sicard, "Inter-tier Coupling Analysis in Back-illuminated Monolithic 3DSI Image Sensor Pixels," *International Conference on Modern Circuits and Systems Technologies (MOCAST)*, Thessaloniki, pp. 1-4, 2021.

[9] P. Sideris et al., "Inter-tier Dynamic Coupling and RF Crosstalk in 3D Sequential Integration," *IEEE International Electron Devices Meeting (IEDM)*, pp. 3.4.1-3.4.4, 2019.

[10] T. Mota-Frutuoso et al., "3D sequential integration with Si CMOS stacked on 28nm industrial FDSOI with Cu-ULK iBEOL featuring RO and HDR pixel," *International Electron Devices Meeting (IEDM)*, pp. 1-4, 2023.

[11] S. Bobba, A. Chakraborty, O. Thomas, P. Batude, V. F. Pavlidis and G. De Micheli, "Performance analysis of 3-D monolithic integrated circuits," *IEEE International 3D Systems Integration Conference (3DIC)*, pp. 1-4, 2010.

[12] R. Ho, K. W. Mai and M. A. Horowitz, "The future of wires," *Proceedings of the IEEE*, vol. 89, no. 4, pp. 490-504, 2001.

Holistic Framework for Evaluating the Trustworthiness of Integrated Circuits

Mouadh Ayache[†‡*], Enkele Rama[§*], Saleh Mulhem[‡], Mladen Berekovic[‡], and Matthias Korb[§]

[†]Synopsys GmbH, Germany,
[§]Institute for Integrated Systems, Universität der Bundeswehr München, Germany
[‡]Institute of Computer Engineering, Universität zu Lübeck, Germany
{*mouadh.ayache*}@*synopsys.com*, {*enkele.rama, matthias.korb*}@*unibw.de,*
{*saleh.mulhem, mladen.berekovic*}@*uni-luebeck.de*

Abstract—New applications such as autonomous driving, cyber-physical systems, or remote surgeries demand integrated circuits (ICs) with an ever-lower tolerance for failure. Typical IC design focuses on the targets of functionality and power, performance, and area. An emerging topic in IC design is trustworthiness. It attempts to unify the various interdependent functional and non-functional aspects, such as correct functionality, reliability, security, and functional safety. Existing methodologies and standards focus on evaluating trustworthiness issues (TIs), i.e., causes of faults, and their effects on only one particular attribute. Instead, TIs should be evaluated on their effect on trustworthiness as a whole, which demands a holistic approach. In this paper, we make two main contributions. The first contribution is a framework with a set of unified evaluation criteria that can be applied across all trustworthiness attributes, and a metric, called the Residual Risk Value (RRV). The latter can be used to assess the residual risk of a TI, where a low RRV indicates low risk remaining, and vice versa. RRV considers the impact and likelihood, and the potential for risk reduction enabled by implementing countermeasures. The second contribution is a questionnaire-based measure that ranks TIs according to the priority of addressing them. The results highlight that TIs that emerge during the early stages of IC development should be treated with greater priority. Further, there is a tendency to prioritize security-related TIs as a greater risk to trustworthy ICs. Meanwhile, TIs affecting well-established aspects of IC design and verification are given a lower priority.

Index Terms—evaluation framework, integrated circuits, trustworthiness, functional safety, security, reliability

I. INTRODUCTION

To perform the increasingly complex functions in applications with very low tolerance for failures (e.g. autonomous driving, cyber-physical systems, and remote surgeries), it becomes important to have a holistic approach to the development of integrated circuits (ICs) that considers not only the correct functionality and optimization of power, performance, and area (PPA), but also additional non-functional aspects, such as security and reliability, required to operate the IC in the target application. *Trustworthiness* attempts to bring together

This work was supported by the German Ministry of Education and Research under the VE-VIDES project (16ME0251), project partners who participated in the survey and edacentrum who hosted the survey.
* These authors contributed equally to this work (Co-First Author).

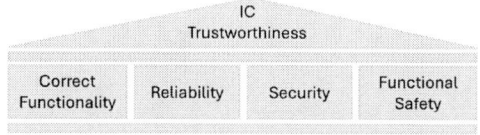

Fig. 1. Main attributes of IC trustworthiness per [9].

these various aspects. As an emerging concept, trustworthiness is investigated in several research projects [1], [2].

Initially, dependability [3] emerged as an attempt to reconcile the functional and non-functional aspects of computing systems. Trustworthiness is either considered equivalent to dependability [4], as its *"twin"* [5], or as a concept more focused on security aspects [6]. Standardization-wise, [7] defines trustworthiness for system level, while [8] confines it to security and reliability.

In the context of ICs, [9] provides a practical definition of trustworthiness with four attributes: *correct functionality, reliability, security,* and *functional safety,* as shown in Fig. 1. In this paper, we adopt this definition. Furthermore, we use the term trustworthiness issue (TI) to describe causes that can lead to or indirectly allow fault occurrence, thus undermining trustworthiness. In Table II, we present a list of significant TIs.

To ensure IC trustworthiness, it is important to identify the most critical TIs that need to be tackled. As indicated in [9], there is no evaluation framework that considers the effect of a TI on all IC trustworthiness attributes. To close this gap, we:

- Propose a set of unified evaluation criteria that can be applied across all trustworthiness attributes
- Propose the Residual Risk Value (RRV), a metric to assess the residual risk after considering countermeasures
- Provide results from a questionnaire with experts who rate identified TIs based on the proposed evaluation criteria

These amount to a holistic evaluation framework that provides engineers from different domains a unified way to assess TIs.

II. BACKGROUND AND RELATED WORK

There are various established methodologies for assessing specific non-functional attributes of IC trustworthiness.

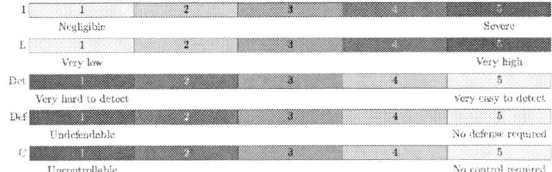

Fig. 2. Rating scale between 1 and 5 for the proposed evaluation criteria.

TABLE I
MAPPING OF RATINGS FOR LIKELIHOOD FROM OTHER METHODOLOGIES.

Methodology (Criteria)	Scale				
CWSS (Likelihood of Exploit)	Low		Medium	High	
HARA (Exposure)	E1	E2	E3	E4	
FMEA (Occurance in RPN)	1/2	3/4	5/6	7/8	9/10
Proposed (Likelihood)	1	2	3	4	5

In the **reliability** domain, Failure Modes and Effects Analysis (FMEA) and its variants are used to identify failure modes, their cause and potential effects [10]. The metric Risk Priority Number (RPN) [11], computed in FMEA, is used to prioritize failure modes:

$$\text{RPN} = S \cdot O \cdot D \quad (1)$$

with severity (S), occurrence (O), and detection (D). The risk is considered proportional to RPN, where $S, O, D \in [1, 10]$.

In the **security** domain, various methodologies focus on different aspects. The Common Weakness Scoring System (CWSS) [12] scores weaknesses using $S_{\text{CWSS}} \in [0, 100]$:

$$S_{\text{CWSS}} = S_{\text{Base}} \cdot S_{\text{AttackSurface}} \cdot S_{\text{Environment}} \quad (2)$$

with the subscores *base finding* $S_{\text{Base}} \in [0, 100]$, *attack surface* $S_{\text{AttackSurface}} \in [0, 1]$, and *environmental* $S_{\text{Environment}} \in [0, 1]$. Similarly, Common Vulnerability Scoring System (CVSS) [13] assesses vulnerabilities, while the Common Criteria Framework evaluates attack potential [14]. For threat modeling, STRIDE [15] and DREAD [16] can be applied.

In the **functional safety** domain, industry standards based on IEC 61508:2010 [17] define the requirements that must be met to avoid harm due to malfunctions. During Hazard Analysis and Risk Assessment (HARA), the appropriate Safety Integrity Level (SIL) is assigned using allocation tables. Automotive SIL (ASIL) [18] considers *exposure* ($E \in \{1, 2, 3, 4\}$), *controllability* ($C \in \{1, 2, 3\}$), and *severity* ($S \in \{1, 2, 3\}$).

III. HOLISTIC FRAMEWORK FOR EVALUATING THE TRUSTWORTHINESS OF INTEGRATED CIRCUITS

Existing evaluation methodologies assess the effect of a TI on specific attributes. However, attributes are interdependent and can be simultaneously undermined. For example, the *insufficient verification of all scenario* undermines reliability and correct functionality. Thus, it is critical to evaluate the effects of a TI on all the attributes of trustworthiness. Therefore, in this paper, we propose a set of unified evaluation criteria for all attributes and a metric to measure the residual risk of a TI.

A. Proposed Unified Evaluation Criteria

A prerequisite for analyzing TIs holistically is a set of unified evaluation criteria that consider all attributes. We propose five criteria, which can be rated per the scale in Fig. 2.

- **Impact** (I) indicates the scale of the negative effect of a TI. In CWSS, I can be the *technical impact* of exploiting a weakness. In HARA, it can be *severity*, which asses the

extent to which a hazard can cause harm. In FMEA, it can be *severity*, which evaluates the negative outcomes of a failure.
- **Likelihood** (L) indicates the probability that a TI can occur. This criterion is considered by other methodologies, such as *likelihood of exploit* in CWSS, *exposure* in HARA, and *occurance* in FMEA.
- **Detectability** (Det) indicates the difficulty of detecting the occurrence of a particular TI. DREAD considers *discoverability*, evaluating how difficult it is to discover a threat. FMEA considers *detection*, i.e. the chance that the effect of failure is detected before affecting the user.
- **Defendability** (Def) indicates the difficulty of defending against the effect of a TI, i.e. the capability of internal IC prevention mechanisms to maintain IC trustworthiness. It is typically a security-relevant factor. CWSS considers *internal control effectiveness*, concerned with rendering the weakness unexploitable. As security vulnerabilities can be exploited to cause hazards, defending against attacks could prevent functional safety-related TIs.
- **Controllability** (C) indicates the difficulty of controlling the negative implications of a TI during operation. It is related to external measures, such as actions by the user. The *external control effectiveness* evaluates the ability to prevent the exploitation of vulnerabilities in CWSS. In functional safety, *controllability* reflects how challenging it is for users to prevent harm arising from the manifestation of hazards.

Since each proposed criterion maps to certain criteria in the evaluation methodologies in Section II, rating based on those methodologies could be transferred to our framework. Table I shows how this could be done for *likelihood*.

B. Proposed Metric: Residual Risk Value

Based on the proposed unified criteria, the following metrics can be computed to holistically evaluate the negative effect of TIs. The product of L and I is used to quantify risk [19]. Using $L \in [1, 5]$ and $I \in [1, 5]$, we define Risk Value (RV) $\in [1, 25]$ for a TI, $i \in \mathcal{I}$, where $\mathcal{I}$ is the set of considered TIs:

$$\text{RV}_i = I_i \cdot L_i. \quad (3)$$

Risk can be reduced by implementing countermeasures or by the user during operation. To quantify the potential for risk reduction and mitigation, the following combination of Det $\in [1, 5]$, Def $\in [1, 5]$, and $C \in [1, 5]$ results in Risk Reduction Potential (RRP) $\in [1, 25]$ of $i \in \mathcal{I}$:

$$\text{RRP}_i = min\left(\text{Det}_i, \text{Def}_i\right) \cdot C_i \quad (4)$$

where the minimum of Det and Def is introduced to avoid overestimating their effect since they are non-orthogonal, i.e.

TABLE II
SELECT INTEGRATED CIRCUITS TRUSTWORTHINESS ISSUES [9].

Trustworthiness issue (TI)	
1 Architectural flaws	12 Unsecure on-chip bus communication
2 Insufficient PDK quality	13 Hardware Trojans within specification
3 Insufficient specification of PPA parameters	14 Hardware Trojans in third-party IP blocks
4 Integration of counterfeited IP blocks	15 Hardware Trojans introduced by rogue in-house designers
5 Integration of blackbox IP blocks	16 Insufficient specification of safety mechanisms
6 Integration of malfunctioning IP blocks	17 Improper implementation or missing safety mechanisms
7 Insufficient specification of the test concept	18 Integration of unsafe third-party IP blocks
8 Improper implementation or missing DFT logic	19 Insufficient implementation and verification of LP & multi-voltage designs
9 Deficient power and signal integrity	20 Insufficient verification of all scenarios
10 Insufficient specification of security measures	21 Insufficient verification of corner cases
11 Integration of unsecure third-party IP blocks	22 Insufficient verification of configurable IP blocks

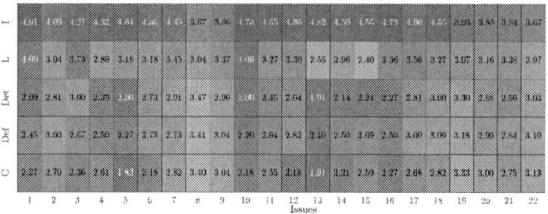

Fig. 3. Questionnaire results or the rating of each criterion across all TIs.

Fig. 3 provides an overview of how respondents evaluated the TIs presented in Table II, based on the proposed criteria. The darker the shade of blue in the heatmap, the worse the rating, which corresponds to the colors of the scale in Fig. 2.

A. Comparison between Related Trustworthiness Issues

To gain insight into the questionnaire results, we compared the ratings of three related groups. Fig 4(a) compares TIs related to hardware Trojans (HTs). *HTs within specification* was rated worse in all aspects, except *L*. This is expected as a HT already embedded in specification will not be flagged by subsequent design and verification steps. Meanwhile, *HTs in third-party IP blocks* received the highest *L* rating, potentially because malicious attacks are usually perceived as external.

For safety and security-related TIs, shown in Fig. 4(b), those that occur during specification were also rated worse. This highlights the fact that the earlier a TI occurs, the higher the difficulty of addressing it. Furthermore, the *insufficient specification of security mechanisms* was rated slightly worse than that of safety, especially for *L*. This is likely because IC safety is well covered by industry standards, unlike IC security. The same applies for TIs related to third-party Intellectual Property (IP) blocks, shown in Fig. 4(c). There, the *integration of blackbox IP blocks*, a security TI, was rated worse than the *integration of unsafe IP blocks*, a functional safety TI.

B. RRV Metric for Prioritizing Trustworthiness Issues

Fig. 5 shows the ranked TIs based on normalized RRV, and their respective RV and RRP. Most likely, all TIs have $RRV_i < 0.5$ because we provided potential countermeasures in the questionnaire, leading to increased RRP_i.

Four of the top five ranked TIs occur during the specification and architecture definition stages, indicating the priority of avoiding potential TIs early in IC development. The bottom five ranked TIs are related to well-established topics in IC design and verification, such as Design for Testability (DFT) and signal and power integrity, which have higher RRP due to established countermeasures and best practices. Furthermore, considering both RV and RRV is important as indicated by:

- HTs (TIs #13, #14, #15) rank in the bottom half for their RV, likely a result of their difficulty to implement. Despite their low RV, the RRV for TIs #13 and #14 is still high (4th and 6th) and they should be treated with priority.
- Although *the integration of unsafe third-party IP blocks* (TI #18) has the 7th highest RV, its RRV is not high, ranked 16th, due to established standards and IC design methodologies.

detection may be sufficient for certain TIs, while detection and defense are required for others.

Finally, we propose the Residual Risk Value (RRV) as a metric to assess the residual risk of $i \in \mathcal{I}$:

$$\text{RRV}_i = max\left(log\left(\frac{\text{RV}_i}{\text{RRP}_i}\right), 0\right). \quad (5)$$

In the case where RV $<$ RRP, the max function ensures that RRV $\geq$ 0. If RRV is normalized between 0 and 1, then:

- RRV $=$ 0 indicates no residual risk since RV $\leq$ RRP.
- 0 $<$ RRV $<$ 1 indicates the risk that is not adequately covered by existing countermeasures since RV $>$ RRP $>$ 1.
- RRV $=$ 1 indicates maximum risk with a complete lack of risk reduction measures since RV $=$ RV$_{max}$ and RRP $=$ 1.

RRV could be utlizied in the following two use-cases:

1) Pre-development Prioritization: To prioritize TIs that represent the highest residual risk, and should be treated with higher priority during IC development.
2) In-development Evaluation: To evaluate the residual risk after the implementation of countermeasures during design. Here the objective would be to ensure that $\text{RRV}_j = 0$, $\forall j \in \mathcal{J}$, where $\mathcal{J}$ is the set of TIs against which countermeasures were implemented.

IV. EVALUATING TRUSTWORTHINESS ISSUES USING THE PROPOSED FRAMEWORK

To validate the *pre-development prioritization* use case, we conducted a questionnaire to prioritize the TIs from Table II, described in detail in [9]. The respondents selectively rated the using the scale in Fig. 2. To ensure their practical significance, the TIs were identified through discussions with IC architects, designers, and verification engineers from the VE-VIDES project for trustworthy electronics [20]. In total, 43 respondents participated: 25 from industry, 15 from academia, and 3 undisclosed. To ensure a common understanding among respondents, the TIs were briefly described, including their main negative effects and selected countermeasures. On average, a TI received 21 ratings, with a range of 11 to 32.

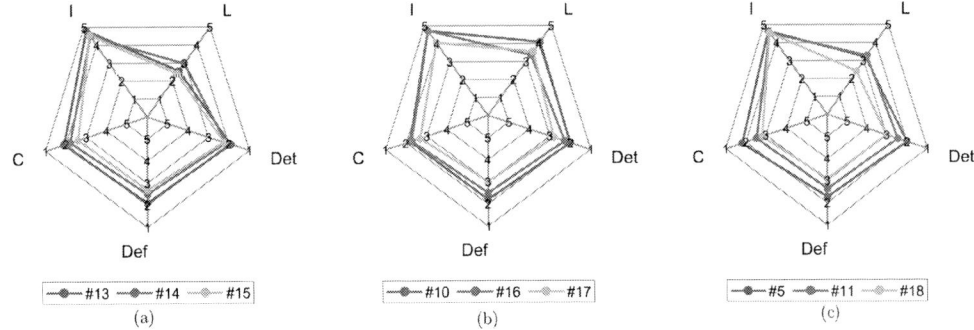

Fig. 4. Comparison of the criteria rating of related TIs based on the conducted questionnaire.

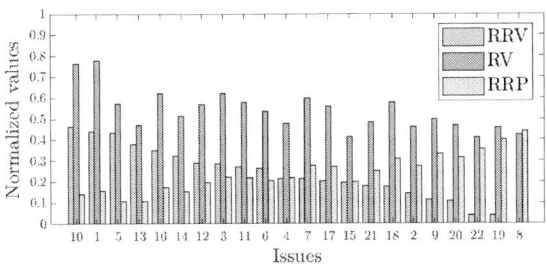

Fig. 5. Questionnaire results for the normalized RRV, RV and RRP values for the TIs presented in Table II. Ranking is based on their RRV.

- Developing effective countermeasures is pivotal for the trustworthy development of ICs. This is evident by the results shown in Fig. 5, where TIs with high RRV have consistently very low RRP (TIs #10, #1, #5, #13, #14).
- The relatively high residual risk for *integration of blackbox IP blocks* (TI #5), ranked 3rd, indicates that the community should further address it. Specifically, effective countermeasures are necessary, as it is rated the lowest overall for RRP.

V. CONCLUSION

We present a holistic evaluation framework that assesses the effect of TIs on IC trustworthiness. The proposed framework uses a set of unified evaluation criteria applicable across all trustworthiness attributes. Moreover, we introduce a metric to measure the inherent risk that a TI presents and the residual risk after countermeasure implementation. To demonstrate its applicability for TIs prioritization, we conducted a questionnaire to rank among a list of significant TIs. The results emphasize the importance of mitigating TIs that arise in early development. The respondents prioritize TIs related to security, over well-established topics like DFT and PPA. This should guide IC development teams to prioritize efforts effectively. Future work should apply this framework to an IC design and explore methodologies for trustworthy IC development.

REFERENCES

[1] DARPA. (2018, 11) DARPA announces next phase of electronics resurgence initiative. Accessed: 2024-03-01. [Online]. Available: https://www.darpa.mil/news-events/2018-11-01a

[2] Bundesanzeiger. (2020, 3) Vertrauenswürdige Elektronik (ZEUS). BMBF. Accessed: 2024-03-01. [Online]. Available: https://www.elektronikforschung.de/foerderung/bekanntmachungen/zeus

[3] J.-C. Laprie, *Dependability: Basic Concepts and Terminology*. Vienna, Austria: Springer Vienna, 1992.

[4] A. Avižienis, J.-C. Laprie, and B. Randell, "Dependability and its threats: a taxonomy," in *Building the Information Society: IFIP 18th World Computer Congress Topical Sessions 22–27 August 2004 Toulouse, France*. Springer, 2004, pp. 91–120.

[5] B. Bauer *et al.*, "On the dependability lifecycle of electrical/electronic product development: The dual-cone v-model," *Computer*, vol. 55, no. 9, pp. 99–106, 9 2022.

[6] K. Atsushi, V. Bellinghausen, and J. Fujita, "IIoT value chain security – the role of trustworthiness," 4 2020, Accessed: 2024-03-01. [Online]. Available: https://www.plattform-i40.de/IP/Redaktion/EN/Downloads/Publikation/IIoT_Value_Chain_Security.pdf

[7] ISO and IEC, *ISO/IEC TS 5723:2022 Trustworthiness - Vocabulary*, ISO and IEC Std., 2022.

[8] EC Directorate-General for Communications Networks, Content and Technology, *Study on trusted electronics*. European Union, 2024.

[9] E. Rama *et al.*, "Trustworthy integrated circuits: From safety to security and beyond," *IEEE Access*, vol. 12, pp. 69 603–69 632, 2024.

[10] G. Popov and B. Lyon, *Failure Mode and Effects Analysis*. John Wiley & Sons, Ltd, 2021, ch. 8, pp. 153–169.

[11] B. Lyon, *Defining Risk Assessment Criteria*. John Wiley & Sons, Ltd, 2021, ch. 4, pp. 65–91.

[12] MITRE. (2014, 9) Common Weakness Scoring System (CWSS). Accessed: 2024-03-01. [Online]. Available: https://cwe.mitre.org/cwss/cwss_v1.0.1.html

[13] NIST. National vulnerabilities database - vulnerability metrics. Accessed: 2024-03-01. [Online]. Available: https://nvd.nist.gov/vuln-metrics/cvss

[14] Common Criteria, *CEM:2022 Revision 1 Common Methodology for Information Technology Security Evaluation*, CCRA Std., 11 2022, Accessed: 2024-03-01. [Online]. Available: https://www.commoncriteriaportal.org/files/ccfiles/CEM2022R1.pdf

[15] J. Geib, *et al.* (2022, 5) Microsoft threat modeling tool threats: STRIDE model. Accessed: 2024-03-01. [Online]. Available: https://learn.microsoft.com/en-us/azure/security/develop/threat-modeling-tool-threats

[16] EC-Council. DREAD threat modeling: An introduction to qualitative risk analysis. Accessed: 2024-03-01. [Online]. Available: https://www.eccouncil.org/cybersecurity-exchange/threat-intelligence/dread-threat-modeling-intro/

[17] IEC, *IEC 61508:2010 - Functional safety of electrical/electronic/programmable safety-related systems*, IEC Std., 2010.

[18] ISO, *ISO 26262:2018 Road Vehicles – Functional Safety*, Std., 2018.

[19] T. M. Chen, "Information security and risk management," in *Encyclopedia of Multimedia Technology and Networking, Second Edition*. IGI Global, 2009, pp. 668–674.

[20] edacentrum. (2021) Design methods and HW/SW co-verification for the unique identifiability of electronic components (VE-VIDES). BMBF. Accessed: 2024-03-01. [Online]. Available: https://www.edacentrum.de/vevides/

Lightweight Active Fences for FPGAs

Anis Fellah-Touta
Laboratoire Hubert Curien
Saint-Étienne, France
anis.fellah.touta@univ-st-etienne.fr

Lilian Bossuet
Laboratoire Hubert Curien
Saint-Étienne, France
lilian.bossuet@univ-st-etienne.fr

Vincent Grosso
Laboratoire Hubert Curien
Saint-Étienne, France
vincent.grosso@univ-st-etienne.fr

Carlos Andres Lara-Nino
Universitat Rovira i Virgili,
Tarragona, Spain
carlos.lara@fundacio.urv.cat

Abstract—**The use of active fences has been proposed as a protection against remote power analysis attacks. This countermeasure relies on reserving a reconfigurable space within the FPGA which will separate it into sub-regions. These "fences" will then generate some electrical interference to hinder the performance of an attack. As FPGAs can be configured in multiple ways, there are different approaches for connecting the hardware inside the fence. In this work, we describe a LUT-based configuration which can achieve the same instantaneous power drop as a ring oscillator bank with less LUTs. This contributes to reducing the hardware costs of active fences.**

I. INTRODUCTION

Remote Power Analysis (RPA) is an emerging threat against FPGAs [1], [2]. In this paradigm, an adversary employs internal sensors to remotely study the power of the system. Several countermeasures have been proposed against this attack. They include cryptographic methods like masking and shuffling, software methods like bitstream checking, and physical methods like generating electric interference. In this work, we are interested in the latter methods given their simplicity and generality. One of the most promising of such strategies involves the use of active fences [3], [4]. These circuits are composed of LUT-based ring oscillators placed between the victim circuit and the rest of the FPGA. These oscillators add noise to the power consumption patterns, thereby reducing the signal-to-noise ratio (SNR) and reducing the ability of an adversary for retrieving useful information about the victim.

In this paper, we present a new LUT-based power waster circuit useful to build active fences. The proposed design can generate an equivalent instantaneous voltage drop as a conventional ring oscillator array, but with fewer LUTs. Given that neural networks (NN) are one of the usual victims of RPA, we have applied the proposed active fence to protect one of such implementations and evaluate the effectiveness of the proposed configuration. Our proposed lightweight fence is effective in reducing the SNR, offering a promising countermeasure against side-channel attacks while providing improved resource utilization.

The rest of the paper is structured as follows. In Section II we review some relevant works from the literature. Section III describes the proposed active fence design and analyzes its characteristics. Section IV describes our evaluation methodology. In Section V we provide and discuss experimental results. Lastly, Section VI concludes the paper.

II. BACKGROUND AND RELATED WORK

Preemptive security against RPA attacks can be provided through bitstream scanning. This strategy seeks to prevent the deployment of bitstreams containing malicious circuits that could enable RPA [5]. One means of doing this is through identifying LUT-based oscillator designs in the configuration bitstream. FPGA vendor tools already give critical warnings in case combinatorial loops are detected during the bitstream generation. Other tools, like FPGADefender [6], go further and warn against the use of other potentially malicious circuits like non-combinatorial loops and time-to-digital converters (TDC). This is a virus-scanner-like tool specifically designed for FPGAs. It scans the bitstreams for the presence of self-oscillator circuits used in RPA, such as ROs. Once such elements are found, the tool blocks the loading of the bitstream onto the FPGA, hence effectively stopping possible attacks in the early stages of deployment. The main drawback of these tools is that, like antivirus software, they must be periodically updated to catch up with emerging internal sensor designs [7].

Other RPA countermeasures focus on enhancing the resilience of the victim against side channel analysis. This is primarily achieved through masking [8] and shuffling [9]. These techniques break the relationship between the sensitive data and the side channels, effectively thwarting the attack even if the adversary manages to retrieve some measurements. However, applying these protections to NN implementations is complicated and resource-consuming [10], [11]. Neural networks usually require a great number of computations and data transfers, and the addition of masking or shuffling can reduce the performance of the NN. Additionally, these solutions must be tailored to every particular NN architecture.

Another approach for protecting neural networks relies on disrupting the side-channels of the device. These strategies aim at reducing the SNR, making it difficult for the attacker to extract useful information from the leakages. Some of these approaches include de-synchronization [12], addition of spurious computations [13], hardware reconfiguration [14], and noise generation through hardware circuits [15]. A rather common technique is to place a grid of ring-oscillators as an "active fence" between the victim and the attacker. The ring-oscillators are randomly activated to induce noise and increase the number of measurements required to perform an attack. The main advantage of this approach is that it does not require changes in the victim implementation itself, making it

979-8-3315-3968-9/24 $31.00 © 2024 IEEE

a generic form of protection.

III. PROPOSED LIGHTWEIGHT ACTIVE FENCE

Power wasters are a type of hardware trojan sometimes used in fault injection attacks. However, they can be used in the construction of active fences. By activating power wasters randomly it is possible to introduce electrical noise into the side channels. This can obfuscate the data-associated leakages of the device. The activation pattern can be generated with internal circuits, for example a pseudo-random number generator (PRNG). Optimizing the cost of the active fence involves designing power wasters that achieve the desired level of voltage drop with minimal resource cost. The magnitude of the instantaneous drop should be restricted by the operational threshold of the FPGA (0.9-1.1V). Therefore, what we seek are smaller power wasters. In Fig. 1 we illustrate the power waster proposed in this work. The circuit is a series of inverters driven by the output of a ring oscillator. In an FPGA, the ring oscillator and the inverters can be implemented using LUTs. We refer to both components as logic elements (LE).

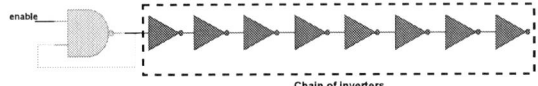

Fig. 1. Proposed LUT-based power waster design.

The voltage drop induced by the proposed power waster depends on the number of inverters connected to the ring oscillator. We tested multiple configurations to determine the optimal chain length which maximized the instantaneous voltage drop. We fixed an upper bound of 1,100 LUTs to build an active fence and implemented multiple "copies" of the power waster to fit this area. This upper bound was selected based on the results from [3]. The NewAE CW305 test board, featuring an AMD-Xilinx Artix-7 FPGA (XC7A100-2TFTG256), was used to implement the active fences. A digital oscilloscope with a sampling rate of 10GSpS was used to observe the power supply of the FPGA. The AMD-Xilinx 2020.2 toolchain was used to generate all the bitstreams and to configure the test board. The results for this experiment are shown in Fig. 2.

In our active fences, a single power waster consists of a ring oscillator and N inverters. Therefore, with one LE the chain length is zero and with 16 LE we have one ring oscillator and 15 chained inverters. The reader should note that the 01 LE configuration is equivalent to the active fences from [3]. From Fig. 2 we observe that adding inverters results in a higher voltage drop compared to the configuration consisting solely on ring oscillators. However, the relationship between the number of inverters and the magnitude of instantaneous voltage drop appears to be non-linear. For example, the configuration with eight LE has in a larger voltage drop than the configuration with 16 LE. Fig. 2 shows that the maximum voltage drop (0.91 V) was achieved with a configuration of 9 LE (one ring oscillator and eight chained inverters), suggesting this as the optimal setup for active fence applications.

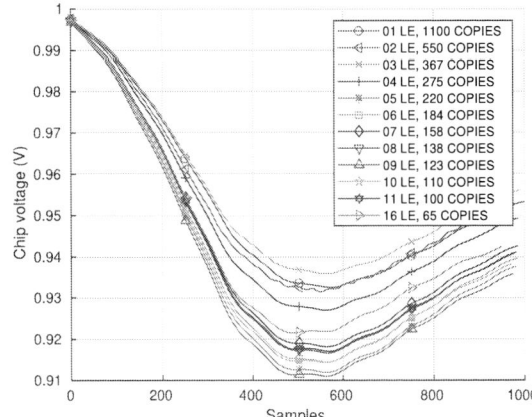

Fig. 2. Instantaneous voltage drop generated with active fences of at most 1,100 LUTs. Each fence consists of several copies of the proposed power waster.

With the same experimental setup we studied the placement impact of the inverter chains of the power wasters. We deployed an active fence with 123 optimal power wasters (9LE) and compared the produced instantaneous power drop when the circuits are manually placed in the FPGA versus the case where the toolchain handles the placement. Fig. 3 shows the results for this experiment. This analysis demonstrates that the voltage drop is greater when the inverters are manually placed compared to when they are placed by the toolchain. Manual placement allows for optimized spatial distribution and connectivity of the inverters, which may lower parasitic capacitance. A decrease in parasitic capacitance results in an increase in switching activity of the inverters, thus increasing the current draw and consequently the instantaneous voltage drop.

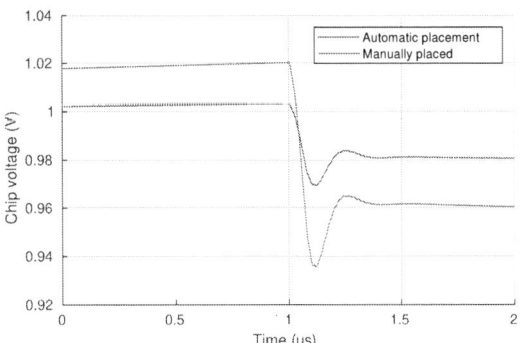

Fig. 3. Impact of manual vs. automatic placement on the voltage drop.

IV. EXPERIMENTAL EVALUATION

We devised a second experimental setup to evaluate the effectiveness of the proposed active fence. We used a Trenz

979-8-3315-3968-9/24 $31.00 © 2024 IEEE

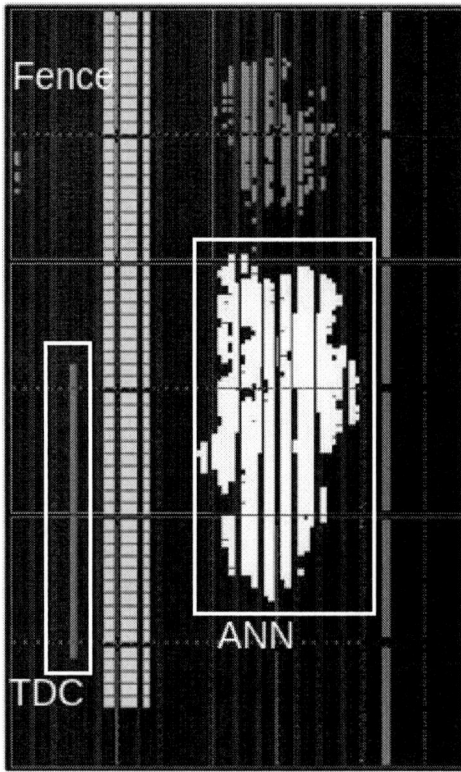

Fig. 4. Floor-planning used in the experimental setup

TE0802 development board, equipped with an AMD-Xilinx Zynq UltraScale+ SoC FPGA (XCZU2CG-1SBVA484E). The AMD-Xilinx Vivado 2020.2 toolchain was used to generate the FPGA bitstreams and launch applications through Vitis.

The aim of this experiment was to estimate the SNR of a NN design when protected with the proposed countermeasure. The SNR can serve as a metric for estimating the difficulty of an attack by evaluating the strength of the exploitable signal in comparison to the background noise. This estimation was conducted both in the presence and absence of the counter-measure, in order to assess the utility of the proposed design. A handwritten-number recognition application was used as the victim architecture. We employed an internal sensor based on a TDC array with a sampling rate of 200 MSpS to emulate an optimal configuration for RPA [2]. The active fence was positioned between the victim and the sensor and controlled by a PRNG implemented using a 6-bit linear feedback shift register (LFSR). Fig. 4 illustrates the physical layout of the experimental setup.

We conducted a series of three experiments, each involving the calculation of the SNR. The first experiment served as a baseline, where the countermeasure was disabled. In the second experiment we enabled a RO-based fence as a point of comparison. Finally, in the third experiment we evaluated the performance of the active fence proposed in this work.

A. NN implementation

We implemented the handwritten-number recognition NN architecture following the approach described in [16]. The implemented network has 400 inputs (receiving a 20x20 pixel image), one hidden layer with 25 nodes, and 10 outputs, following the [400, 25, 10] configuration. It was tested on a subset of the MNIST database and achieved an accuracy of 99%. The NN IP was created using Vivado HLS 2020.2. On the Trenz TE0802 development board it occupies 12% of the total available area with an operating frequency of 50 MHz.

B. Active fence implementation

Following the approaches described in [3], [4], the active fence was designed to match the resource utilization of the circuit under protection. In our case this amounts to 5,760 LUTs. The fence was divided into 64 banks, with each bank containing 10 instances of the proposed optimal power waster configuration (9 LE). Within each bank, all wasters were controlled by a single dedicated enable signal. However, each bank could be independently enabled. This was achieved by using a PRNG to generate the enable signals that drove the banks. For the conventional active fence, we follow to the same organization of 64 banks, with each bank containing 90 ROs implemented using NAND gates. This configuration ensures that both fences consume a total of 5,760 LUTs, facilitating a fair comparison of their effectiveness in mitigating power side-channel attacks. The reader should note that this setup allows for a fair comparison, however the proposed active fence could be implemented with less resources.

V. RESULTS AND DISCUSSION

We evaluated the impact of different active fences on the SNR during the operation of the NN. For each configuration we collected one thousand power traces corresponding to the NN operation. Mathematically, the SNR was derived using the following formula:

$$\text{SNR}_{dB} = 10 \log_{10} \frac{E[S^2]}{E[N^2]}$$

Where S represents the data-dependent leakages and N is the noise. The former was computed as the average of unprotected power traces using the same input. The latter was computed by subtracting N from all the observations. The comparative analysis of the SNR values can be observed in Fig. 5.

Without any fence applied the NN implementation shows an average SNR of $-2dB$, indicating a huge amount of leakage of exploitable information from the system. The average SNR was reduced to $-30dB$ with the RO-based fence. The proposed fence configuration managed a further reduction of the SNR, bringing its average down to $-31dB$ when using the same hardware amount as the RO-based fence. The reader should note that a reduction of $1dB$ implies a ratio of $10^{1/10}$ between the state of the art and the proposed fence. A lower value of the SNR indicates an increased complexity for an attack. Therefore, the SNR results show a clear improvement

979-8-3315-3968-9/24 $31.00 © 2024 IEEE

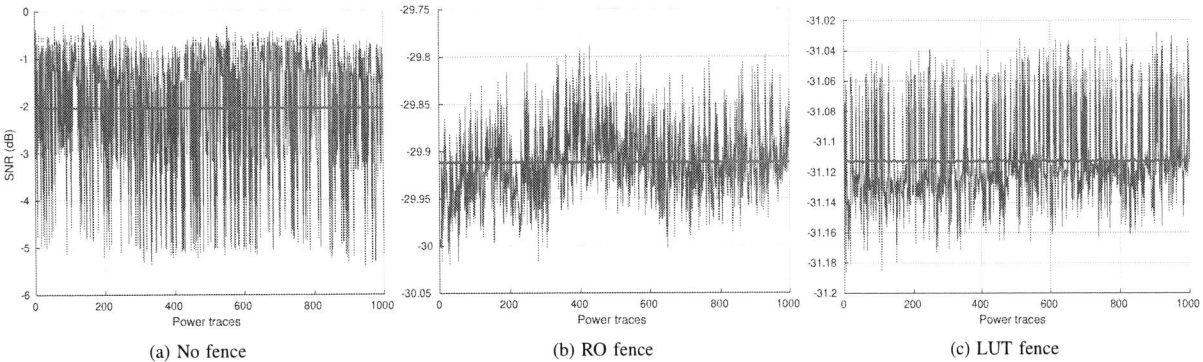

(a) No fence   (b) RO fence   (c) LUT fence

Fig. 5. SNR plots for the three scenarios

in terms of the performance of the proposed active fence countermeasure with respect to the conventional active fences in terms of the leakage of exploitable information from the victim architecture.

Analyzing the SNR provides valuable insights into the effectiveness of a countermeasure. However, it does not guarantee a perfect mitigation. A significant reduction in SNR, such as dividing the original SNR without the fence by a significant factor, would point to a substantial decrease in exploitable leakage. It hence makes the amount of information an attacker can gain from it considerably reduced, thereby enhancing the security. Nonetheless, it is important to recognize that a lower SNR does not completely eliminate the risk of side-channel attacks. It only makes the attacks harder. A reduced SNR should hence not be considered the sole indicator of improved security, but it should be accompanied by other security measures and constant monitoring to ensure full protection.

VI. CONCLUSIONS

In this paper, we have proposed a novel power waster design which allows to build lightweight active fences. These circuits can be used as a countermeasure against remote power side-channel attacks on FPGAs. The proposed active fences offer the same instantaneous drop in voltage as conventional fences while reduced the amount of hardware resources in the circuit. The key advantages of this design lie in its optimized resource efficiency and effective noise generation to obfuscate power consumption patterns, thereby mitigating side-channel leakages and reducing the exploitable SNR for potential attacks. Future work could focus on investigating the scalability of this approach for larger FPGA designs and exploring its integration with other security measures for increased protection.

ACKNOWLEDGMENT

This work has been supported by the French government through the *Agence Nationale de la Recherche* in the framework of the *France 2030* initiative under project ARSENE (ANR 22 PECY 0004).

REFERENCES

[1] M. Zhao and G. E. Suh, "FPGA-based remote power side-channel attacks," in *S&P*, pp. 229–244, 2018.

[2] F. Schellenberg, D. R. E. Gnad, A. Moradi, and M. B. Tahoori, "An inside job: Remote power analysis attacks on FPGAs," *IEEE Des. Test*, vol. 38, no. 3, pp. 58–66, 2021.

[3] J. Krautter, D. R. Gnad, F. Schellenberg, A. Moradi, and M. B. Tahoori, "Active fences against voltage-based side channels in multi-tenant FPGAs," in *ICCAD*, pp. 1–8, 2019.

[4] O. Glamočanin, A. Kostić, S. Kostić, and M. Stojilović, "Active wire fences for multitenant FPGAs," in *DDECS*, pp. 13–20, 2023.

[5] D. R. E. Gnad, S. Rapp, J. Krautter, and M. B. Tahoori, "Checking for electrical level security threats in bitstreams for multi-tenant FPGAs," in *FPT*, pp. 286–289, 2018.

[6] T. M. La, K. Matas, N. Grunchevski, K. D. Pham, and D. Koch, "FP-GADefender: Malicious self-oscillator scanning for Xilinx UltraScale + FPGAs," *ACM Trans. Reconfigurable Technol. Syst.*, vol. 13, no. 3, 2020.

[7] A. Fellah-Touta, L. Bossuet, and C. A. Lara-Nino, "A lightweight non-oscillatory delay-sensor for remote power analysis," in *HOST*, pp. 343–348, 2024.

[8] Y. Ishai, A. Sahai, and D. Wagner, "Private Circuits: Securing Hardware against Probing Attacks," in *CRYPTO*, pp. 463–481, 2003.

[9] N. Veyrat-Charvillon, M. Medwed, S. Kerckhof, and F.-X. Standaert, "Shuffling against side-channel attacks: A comprehensive study with cautionary note," in *ASIACRYPT*, pp. 740–757, 2012.

[10] M. Brosch, M. Probst, M. Glaser, and G. Sigl, "A masked hardware accelerator for feed-forward neural networks with fixed-point arithmetic," *IEEE Trans. Very Large Scale Integr. Syst.*, vol. 32, no. 2, pp. 231–244, 2024.

[11] A. Dubey, A. Ahmad, M. A. Pasha, R. Cammarota, and A. Aysu, "ModuloNET: Neural networks meet modular arithmetic for efficient hardware masking," *IACR Trans. Cryptogr. Hardware Embedded Syst.*, vol. 2022, no. 1, p. 506–556, 2021.

[12] J. Breier, D. Jap, X. Hou, and S. Bhasin, "A desynchronization-based countermeasure against side-channel analysis of neural networks," in *CSCML*, pp. 296–306, 2023.

[13] H. Chabanne, J.-L. Danger, L. Guiga, and U. Kühne, "Parasite: Mitigating physical side-channel attacks against neural networks," in *SPACE*, pp. 148–167, 2022.

[14] M. M. Ahmadi, L. Alrahis, O. Sinanoglu, and M. Shafique, "FPGA-Patch: Mitigating remote side-channel attacks on FPGAs using dynamic patch generation," in *ISLPED*, pp. 1–6, 2023.

[15] X. Yan, C. H. Chang, and T. Zhang, "Defense against ML-based power side-channel attacks on DNN accelerators with adversarial attacks," Preprint 2312.04035, arXiv, 2023.

[16] H. Mittal, A. Sharma, and T. Perumal, "FPGA implementation of handwritten number recognition using artificial neural network," in *GCCE*, pp. 1010–1011, 2019.

MCS-NTT: Multi-Chip System Design for NTT Acceleration

Mohammed Nabeel, Homer Gamil, Johann Knechtel, Michail Maniatakos

New York University Abu Dhabi (NYUAD), Abu Dhabi, UAE

Abstract—**Hardware implementations of Number Theoretic Transform (NTT), especially ASIC designs, have provided significant speed improvements for lattice-based cryptography schemes used by Post-Quantum Cryptography (PQC) and Fully Homomorphic Encryption (FHE). While most of the existing solutions are tailored for fixed polynomial degrees and modulus sizes, both parameters can vary considerably depending on the application and scheme. Toward this end, our paper introduces MCS-NTT, the first hardware architecture for NTT acceleration that is based on a multi-chip-system (MCS) design approach. Our proposed solution provides scalability to existing NTT accelerators by seamlessly integrating multiple accelerator units around an FPGA-based centralized unit. This configuration effectively establishes a customized star network tailored to meet specific use cases. The experimental results indicate that MCS-NTT offers considerable flexibility with better performance metrics.**

Index Terms—**NTT, PQC, FHE, Hardware accelerator, Multi Chip System, FPGA**

I. INTRODUCTION

In recent years, hardware acceleration of Number Theoretic Transform (NTT) has gained considerable attention from the scientific community. This is mainly driven by the fact that most of the Post-Quantum Cryptography (PQC) and Fully Homomorphic Encryption (FHE) schemes [1], [2], [3] are based on RLWE, where the plaintexts and ciphertexts are represented by polynomials and among all polynomial arithmetic, the costliest operation is the polynomial multiplication, which has a time complexity of $O(n^2)$, when performed naively. NTT brings down the runtime from $O(n^2)$ to $O(n\ logn)$ [4]. In light of this, numerous studies and works have been proposed that focus on implementing NTT directly in hardware platforms [5], such as GPUs, FPGAs, and ASICs.

Depending on the scheme, security level, and application, the polynomial size can vary for both PQC and FHE. There is no one size that fits all. For FHE schemes, ciphertexts are inherently large, represented as high-degree polynomials with large integer coefficients. Also, it is desirable to have a large modulus size for the polynomial coefficients to delay the costly bootstrapping operation. Bootstrapping is done to remove the accumulated error so far in FHE computation, as further accumulation of error can result in wrong decryption of the ciphertexts. The error accumulation is mainly contributed by the ciphertext multiplication operation and when to perform bootstrapping is decided by the number of multiplications performed in sequence, termed as multiplication level (L).

A popular way to increase L is to perform *Modulus-switching* [6]. It slows the accumulation of errors by dividing the intermediate ciphertext and its modulus by a constant after each multiplication. It is hence desirable to have a high modulus size to start with so that one can perform multiple *Modulus-switching*, thereby increasing L.

Computation-wise, a polynomial with a bigger coefficient modulus is manageable using the Residual Number System (RNS) [7] . RNS helps to divide such polynomials into multiple polynomials of the same degree, but each with reduced modulus size. However, to maintain the security level, increasing the size of the modulus should be accompanied by increasing the degree of polynomial. Depending on the parameters chosen, the polynomial degree can vary from 1024 to 65,536 or more [8]. For PQC schemes, the polynomial degree is decided by the algorithm chosen, ranging from 256 to 1,024. Thus, an NTT hardware accelerator designed for a particular scheme or application can be underfit or overfit for others. Though FPGA-based designs give flexibility as they can be reprogrammed as per the new application requirement, the Power-Performance-Area cost of FPGA designs is way higher than its corresponding ASIC design. ASIC-based accelerators have the highest performance per unit area.

The research question we pose is: can we use already fabricated ASIC hardware accelerators to support higher polynomial degrees and coefficient modulus sizes?.

Contributions: This paper addresses the challenges previously outlined. We introduce MCS-NTT, a novel design aimed at NTT acceleration through a multi-chip-system approach. A centralized FPGA unit facilitates and manages the communication both between the host and the NTT accelerators and among the accelerators themselves.

In the scope of this research, we present the following contributions:

- We propose a generic architecture to integrate heterogeneous NTT accelerator chips to improve scalability.
- To offload the host from communication overhead, we design a specialized FPGA-based centralized unit (CU) to seamlessly manage communication flows between the accelerators, streamlining data exchanges and enhancing overall system efficiency. RTL design of the CU will be open-sourced.
- We evaluate the performance of the proposed method using RTL simulation.

II. PRELIMINARIES

A. Lattice-based Cryptosystems

Ring Learning With Errors (RLWE) is a mathematical problem that serves as the foundation for various cryptographic schemes, including BFV [1], and CKKS [2] for FHE, as well as Crystals-dilithium [9], Crystals-Kyber [10] and Falcon [11] for PQC. RLWE-based cryptosystems work over two polynomial rings, one for the plaintext space and another for the ciphertext space. The plaintext space is usually defined over the polynomial ring $\mathcal{P} = \mathbb{Z}_t[x]/(x^N + 1)$, while the

979-8-3315-3968-9/24 $31.00 © 2024 IEEE

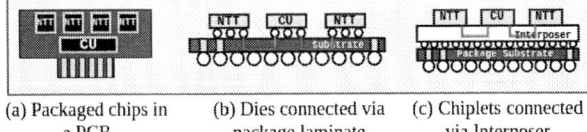

(a) Packaged chips in a PCB (b) Dies connected via package laminate (c) Chiplets connected via Interposer

Fig. 1: MCS-NTT integration options.

ciphertext space is defined over $\mathcal{C} = \mathbb{Z}_q[x]/(x^N+1)$, where N is the polynomial degree, and t, q, and x^n+1 are the plaintext, ciphertext, and polynomial moduli, respectively.

B. NTT and Polynomial Multiplication

The Number Theoretic Transform (NTT) [4] is a generalized form of Discrete Fourier Transform (DFT). Polynomial multiplication using traditional methods requires $O(N^2)$ operations. NTT helps in transforming the coefficient representation of a polynomial to its point representation, and when implemented using the Fast Fourier Transform (FFT) algorithm [12], this transformation can be done in $O(N\log N)$ runtime. Polynomial multiplication in their point representation is just a Hadamard multiplication, i.e., runtime complexity of $O(N)$.

Using NTT is absolutely critical as the polynomial degree can be as high as 65,536, depending on the cryptographic scheme and the target application. Furthermore, implementing NTT in modern CPUs is not fast enough because of the high polynomial degree, and the underlying operation is modular arithmetic, which is not natively supported by modern CPUs. For this reason, hardware accelerators are preferred, particularly ASIC-based accelerators, for NTT computations.

C. Multi-Chip System Design and Packaging Techniques

Chiplets are smaller modular chips that can be assembled to form larger integrated systems or platforms often referred to as multi-chip modules (MCMs). Such systems are already in widespread adoption in the industry [13]. Instead of crafting a single, monolithic chip that integrates all functions, chiplets break down the functionalities into distinct modules that can be developed and manufactured separately and then integrated onto an interposer [13]. The chiplet approach offers flexibility in mixing and matching modules based on customer requirements, upgrading/swapping one module with another.

III. ARCHITECTURE & METHODOLOGY

A. Overview

Hardware accelerators are typically connected to the host via the PCI Express (PCIe) interface. To support a higher polynomial degree (N') than the accelerator's native capacity (N), there are two primary approaches:

1) *Sequential Division:* The host interfaces with a single accelerator and divides the polynomial coefficients into N'/N parts. It then performs NTT operations on each part sequentially and handles the remaining $log_2(N) - N'/N$ NTT stages independently. However, this approach introduces significant communication and computation costs for the host.

2) *Parallel Accelerators:* Alternatively, the host can interface with multiple accelerators (e.g., N'/N accelerators) and parallelize the NTT operations on the coefficient parts. After this initial parallel processing, the host

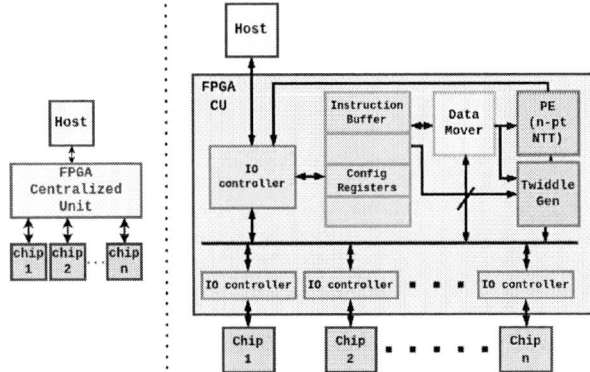

Fig. 2: MCS-NTT system-level connection and CU architecture.

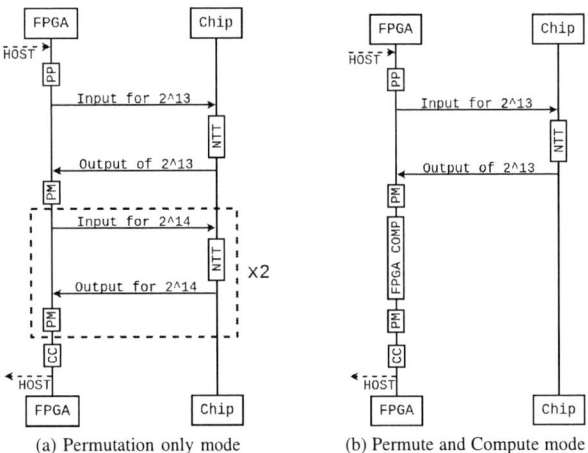

(a) Permutation only mode (b) Permute and Compute mode

Fig. 3: Operational modes.

manages the remaining $log_2(N) - N'/N$ NTT stages itself. While this option reduces sequential workload, it may require more PCIe links, adding to the overall cost.

Both of these approaches come with trade-offs in terms of communication and computation costs, depending on the specific requirements and constraints of the system. In our proposed architecture, we introduce an FPGA-based centralized unit (CU) to connect the host and multiple such accelerators and also perform the remaining $log_2(N) - N'/N$ stage in a pipelined manner before sending the result to the host. In this section, we present modes of operation the CU supports, followed by a generic architecture for such CUs, and also present different ways of integrating the CU and the accelerators and their implementation practicality.

B. Operational Modes and Data Flow

To give flexibility and for performance-cost trade-offs, there are two modes of operation. The permutation-only mode ($Mode1$) and the permute and compute mode ($Mode2$). In $Mode1$, the CU only performs data permutation between the accelerators and hence requires fewer FPGA resources compared to $Mode2$ where CU can also perform n-stage NTT. These modes also help us to examine the effects of

979-8-3315-3968-9/24 $31.00 © 2024 IEEE

delays between computation and data transfers on the overall workload. Both the modes are depicted in Figure 3, for the parameter values $(N', N, n) = (2^{14}, 2^{13}, 2)$.

In both $Mode1$ and $Mode2$, CU routes the first half of polynomial coefficients to one accelerator and the second half to the rest. then triggers the accelerators to perform 2^{13}-point NTT. Then, in $Mode1$, to perform the last stage NTT, the first half of the result from the second accelerator and the second half of the result from the first accelerator are swapped, then trigger the accelerator to perform only the first stage of the N-point NTT after which the results are ready and can be read back to the host. In $Mode2$, instead of swapping, results are read back and passed through the processing element (PE) inside the CU that performs radix-2 butterfly, and the output of the PE is passed to the host.

C. Centralized Unit

The architecture diagram for the Centralized Unit (CU) is displayed in Figure 2. The host programs the configuration registers inside the CU with parameters including the number of connected accelerators (n), the polynomial degree supported by each accelerator, the modulus of the polynomial coefficients, and the initial values of the twiddle factors.

Following this configuration, a limited number of instructions are loaded into the instruction buffer to guide data movement by the data mover. These instructions specify details such as source and destination addresses, increment values, and whether the instruction is a broadcast to all accelerators. The broadcast instruction facilitates parallel loading of twiddle factors and issuing read commands when the destination is the host. They also indicate if addresses should be bit-reversed, which supports various NTT algorithms. Depending on the mode ($Mode1$ or $Mode2$), the data mover block either permutes data between accelerators or transfers it to the PE for pipelined n-radix butterfly operations, with the output being sent as the final NTT result to the host. The data mover also synchronizes data reads from all accelerators to address PLL clock mismatch among chiplets.

The Twiddle-Gen block generates the twiddle factors for the accelerators if they rely on the twiddle factors loaded in its on-chip memory. This also offloads the host from computing them. The Twiddle-Gen block also generates twiddles for n-radix butterfly unit within CU, depending on the chronology of the coefficient sets that are loaded to PE from the accelerator. Modular multipliers in Twiddle-Gen and PE blocks make use of the Barrett reduction algorithm [14] for modular reduction.

D. Integration Options

As shown in Figure 1, the CU and the NTT accelerator chips can be integrated either on a printed circuit board (PCB), or as a system in package (SiP). SiP gives the possibility of either 2D or 2.5D integration. In 2.5D, inter-chip connections are made through an interposer, that is an organic substrate. Connected chips are called chiplets. The benefits of such an approach have been discussed in Section II. For this work at hand, we only envision the benefits of such advanced package integration. In future work, we will study the different integration options in more depth, e.g., via system simulation considering practical bandwidth and throughput ranges reported for such advanced packed systems. For practical implementation and proof-of-concept implementation, for now, we resort to

TABLE I: MCS-NTT CU Implementation Results

Platform	n	LUT/ REG	Clock (Mhz)	II Latency CC	II Latency ns
Artix-7	2	1633/1897	198	18	90
	4	2721/2818	198	27	136
	8	5116/4688	182	36	197
	16	11375/9108	174	45	258
	32	22756/17693	178	54	303
Kintex-7	2	1664/1897	347	18	51
	4	2759/2818	347	27	77
	8	5189/4688	300	36	120
	16	11515/9118	325	45	138
	32	23114/17699	296	54	182
Kintex-UltraScale+	2	1657/1891	604	18	29
	4	2739/2806	604	27	44
	8	5063/4664	581	36	61
	16	11612/9070	604	45	74
	32	23145/17600	516	54	71

TABLE II: MCS-NTT System Level Performance Result

Work	Freq Mhz	$log_2 q$	N'	n	Computation Latency (μs) CPU	Computation Latency (μs) MCS-NTT
[17]	72	14	1024	1	81.00	81.00
			2048	2	93.63	95.47
			4096	4	128.50	95.59
			8192	8	225.46	95.72
			16384	16	447.57	95.84
			32768	32	1011.00	95.97
[18]	300	32	512	1	1.60	1.60
			1024	2	7.20	3.36
			2048	4	26.87	3.39
			4096	8	72.85	3.42
			8192	16	194.21	3.45
			16384	32	459.81	3.48
[19]	25	14	1024	1	409.00	409.00
			2048	2	421.64	450.68
			4096	4	456.50	451.04
			8192	8	553.46	451.40
			16384	16	775.57	451.76
			32768	32	1339.00	452.12
[20]	250	128[†]	8192	1	214.10	214.10
			16384	2	305.64	246.84
			32768	4	586.00	246.87
			65536	8	1363.00	246.91

[†]For performance analysis, modulus size of 32 bits is assumed.

commonly available FPGA technologies as discussed next.

Interfacing the complete system with the host processor can be done using PCIe. PCIe is not as commonly used for chip-to-chip communication because it is a heavyweight protocol in terms of both area and complexity. A popular high-speed interface for chip-to-chip communication, supported by modern FPGAs, is Interlaken [15]. There are FPGAs available with both PCIe and up to 12 lanes Interlaken IO interfaces [16], and some of these FPGAs are also ready to be integrated as chiplets . Thus, our proposed solution can be realized with existing FPGA solutions.

IV. EVALUATION

In this section, we report the implementation result for different CU configurations in terms of the number of accelerators connected to it. We also evaluate the efficacy of MCS-NTT through RTL stimulation and hardware prototyping.

A. Experimental Setup

The CU design is written in Verilog RTL and is highly parameterized to select the required maximum number of

979-8-3315-3968-9/24 $31.00 © 2024 IEEE

accelerators supported, pipeline depth of the modular multiplier, maximum modulus size, etc. The design is functionally verified by RTL simulation using Synopsys VCS. Following functional verification, the design is synthesized using Xilinx Vivado tools targeting Artix-7 FPGA xc7a100tcsg324-1 (mid-range), Kintex-7 FPGA xc7k160tfbg676-2 (mid-to-high) and Kintex Ultrascale CLB xcku5p-ffvb676-2-e (high-range). The Modulus size is fixed to 32-bit, and the modular multiplier pipeline stage is set to three for all the synthesis runs.

B. Implementation Results

Table I shows the implementation results for the selected parameters. As the number of accelerators (n) increases, the frequency decreases, though the critical path is within the modular multiplier, which is the same in all the variants. This performance reduction is primarily because of the net delay, as there will be more design logic to place with a higher value of n, making their inter-distance larger. There are no BRAMs used as the computation is performed on the fly upon the data in transit from the accelerators to the host. In other words, if the IO bandwidth between the accelerator-to-CU and CU-to-host is the same, then there is no throughput loss but only initial latency because of the initiation interval (II) of the butterfly stages decided by the number of accelerators connected and the pipeline depth of the modular multiplier. Table I lists the II values of various CU configurations. As expected, the number of LUTs and Flip flops used increases with the increase in n primarily because of the increase in the number of butterfly stages.

C. Performance Modeling

In order to model the NTT accelerator, a bus functional model (BFM) is coded in Verilog, which can be constrained to behave as an NTT accelerator of a given polynomial degree and coefficient modulus, and it also has interface ports through which the CU can read or write with a defined throughput rate. Multiple BFMs can be integrated around the CU to form the MCS-NTT. To evaluate the performance, we constrain the BFM with the parameters corresponding to the ASIC-based NTT accelerators proposed in the literature, mainly [17], [18], [19] and [20]. For the evaluation purpose, we assume that the CU works at the same clock frequency as the accelerators and that the CU-to-accelerator interface throughput is such that the CU gets one operand every clock cycle. The performance results to perform NTT of higher polynomial degree (N') than the accelerator size obtained from the RTL simulation are shown in Table II. We compare our performance with a CPU connected to n accelerators in parallel. To obtain the CPU numbers for NTT, we conducted the experiment using a single core of an AMD Ryzen 7 5800H processor with Radeon Graphics (16 cores) with 16.0 GiB of RAM and clocked at at least 2.8 Ghz. It can be seen that MCS-NTT outperforms CPU except for the cases where the target $N' \leq 2048$ and the operating frequency of ASIC NTT accelerator is low. When N' increases, CPU computation time also increases; this can be attributed to the complex data access pattern during NTT and the increase in the number of modular multiplication operations with respect to the polynomial degree.

V. Conclusion

In this study, we introduced MCS-NTT, the first NTT acceleration hardware design using a Multi-Chip-System (MCS) approach. By integrating multiple units with a centralized FPGA unit, our design boosts the scalability of NTT accelerators without compromising performance metrics.

References

[1] J. Fan and F. Vercauteren, "Somewhat practical fully homomorphic encryption," Cryptology ePrint Archive, Report 2012/144, 2012. [Online]. Available: https://eprint.iacr.org/2012/144

[2] J. H. Cheon, A. Kim, M. Kim, and Y. Song, "Homomorphic encryption for arithmetic of approximate numbers," in *Advances in Cryptology – ASIACRYPT 2017*, T. Takagi and T. Peyrin, Eds. Cham: Springer International Publishing, 2017, pp. 409–437.

[3] T. N. P. Team, "Pqc standardization process: Announcing four candidates to be standardized, plus fourth round candidates," 2022. [Online]. Available: https://csrc.nist.gov/News/2022/pqc-candidates-to-be-standardized-and-round-4

[4] M. Scott, "A note on the implementation of the number theoretic transform," in *IMA International Conference on Cryptography and Coding*. Springer, 2017, pp. 247–258.

[5] A. C. Mert, E. Karabulut, E. Öztürk, E. Savaş, and A. Aysu, "An extensive study of flexible design methods for the number theoretic transform," *IEEE Transactions on Computers*, vol. 71, no. 11, 2020.

[6] Z. Brakerski, C. Gentry, and V. Vaikuntanathan, "(leveled) fully homomorphic encryption without bootstrapping," *ACM Transactions on Computation Theory (TOCT)*, vol. 6, no. 3, pp. 1–36, 2014.

[7] JamesPommersheim, *Number Theory: A Lively Introduction with Proofs, Applications, and Stories*. John Wiley & Sons, 2010.

[8] M. Albrecht, M. Chase, H. Chen, J. Ding, S. Goldwasser, S. Gorbunov, S. Halevi, J. Hoffstein, K. Laine, K. Lauter, S. Lokam, D. Micciancio, D. Moody, T. Morrison, A. Sahai, and V. Vaikuntanathan, "Homomorphic encryption standard," 2018. [Online]. Available: https://eprint.iacr.org/2019/939

[9] V. Lyubashevsky, L. Ducas, E. Kiltz, T. Lepoint, P. Schwabe, G. Seiler, D. Stehlé, and S. Bai, "Crystals-dilithium," *Algorithm Specifications and Supporting Documentation*, 2020.

[10] J. Bos, L. Ducas, E. Kiltz, T. Lepoint, V. Lyubashevsky, J. M. Schanck, P. Schwabe, G. Seiler, and D. Stehle, "Crystals - kyber: A cca-secure module-lattice-based kem," in *2018 IEEE European Symposium on Security and Privacy*, 2018, pp. 353–367.

[11] T. Prest, P.-A. Fouque, J. Hoffstein, P. Kirchner, V. Lyubashevsky, T. Pornin, T. Ricosset, G. Seiler, W. Whyte, and Z. Zhang, "Falcon," *Post-Quantum Cryptography Project of NIST*, 2020.

[12] J. W. Cooley and J. W. Tukey, "An algorithm for the machine calculation of complex fourier series," *Mathematics of computation*, vol. 19, no. 90, pp. 297–301, 1965.

[13] J. Kim, G. Murali, H. Park, E. Qin, H. Kwon, V. Chaitanya, K. Chekuri, N. Dasari, A. Singh, M. Lee *et al.*, "Architecture, chip, and package co-design flow for 2.5 d ic design enabling heterogeneous ip reuse," in *Proceedings of the 56th Annual Design Automation Conference 2019*, 2019.

[14] D. Soni, M. Nabeel, R. Karri, and M. Maniatakos, "Optimizing constrained-modulus barrett multiplier for power and flexibility," in *2023 IFIP/IEEE 31st International Conference on Very Large Scale Integration (VLSI-SoC)*. IEEE, 2023, pp. 1–6.

[15] I. Alliance, "Interlaken protocol definition," 2008. [Online]. Available: https://4b1b46.a2cdn1.secureserver.net/wp-content/uploads/2019/12/Interlaken_Protocol_Definition_v1.2.pdf

[16] Intel, "F-tile interlaken intel® fpga ip user guide," 2023. [Online]. Available: https://www.intel.com/content/www/us/en/docs/programmable/683622/22-4-6-0-0/about-the-f-tile-interlaken-fpga-ip-core.html

[17] U. Banerjee, T. S. Ukyab, and A. P. Chandrakasan, "Sapphire: A configurable crypto-processor for post-quantum lattice-based protocols," *IACR Transactions on Cryptographic Hardware and Embedded Systems*, vol. 2019, no. 4, p. 17–61, Aug. 2019. [Online]. Available: https://tches.iacr.org/index.php/TCHES/article/view/8344

[18] S. Song, W. Tang, T. Chen, and Z. Zhang, "Leia: A 2.05 mm 2 140mw lattice encryption instruction accelerator in 40nm cmos," in *2018 IEEE Custom Integrated Circuits Conference (CICC)*. IEEE, 2018, pp. 1–4.

[19] T. Fritzmann and J. Sepúlveda, "Efficient and flexible low-power ntt for lattice-based cryptography," in *2019 IEEE International Symposium on Hardware Oriented Security and Trust (HOST)*. IEEE, 2019.

[20] M. Nabeel, D. Soni, M. Ashraf, M. A. Gebremichael, H. Gamil, E. Chielle, R. Karri, M. Sanduleanu, and M. Maniatakos, "Cofhee: A co-processor for fully homomorphic encryption execution," in *2023 Design, Automation & Test in Europe Conference & Exhibition (DATE)*, 2023.

MEAN: Mixture-of-Experts Based Neural Receiver

Bram van Bolderik, Vlado Menkovski, Sonia Heemstra, and Manil Dev Gomony

Eindhoven University of Technology, The Netherlands

Abstract—Dynamic Neural Networks (DyNN) adapt its network architecture at run-time compared to static neural networks. DyNNs benefit in wireless receivers based on neural networks (or *neural receivers*), that need to adapt their performance under varying channel conditions. Mixture-of-Experts (MoE) is one efficient way to realize a DyNN in neural receivers in which several smaller *expert* networks are dynamically combined in the run-time to create a dedicated network according to the channel requirements. This paper presents a novel hard-gated (also known as sparsely-gated) MoE-based neural receiver architecture called MEAN for Single Input Multiple Output (SIMO)-based wireless communication systems. Our proposed MEAN architecture for a SIMO wireless system shows a reduction of up to 50% in the number of active layers during runtime.

I. INTRODUCTION

Recently, neural network (NN)-based wireless receivers, in short *neural receivers*, have shown improved system-level performance compared to classical receivers at different channel conditions [1]. Despite the improved system-level performance, the computational complexity of neural receivers is significantly higher compared to classical receivers [2]. This is mainly because they use excessive computational resources (and power consumption) by deploying one large NN to deal with all noise conditions in the communication channel. One promising way to reduce the computational complexity of neural receivers is by adapting the NN architecture dynamically in the run-time (*Dynamic Neural Networks (DyNN)* [3], [4]) according to the varying channel conditions. There are several aspects of an NN, such as width, depth, parameters, routing, etc, that can be adapted dynamically in a DyNN [3]. Dynamism in DyNNs can be implemented either on a fine-grained (e.g. adding/removing neurons) or coarse-grained (e.g. adding/removing layers) level. Fine-grained implementation of dynamism increases hardware implementation overhead.

Mixture-of-Experts (MoE) is a coarse-grained DyNN in which multiple smaller NNs are combined dynamically to realize a dedicated NN according to input conditions [3]. This is especially interesting for neural receivers to reduce the run-time computational complexity as it allows combining different expert NNs according to input channel conditions. In this paper, we propose a MoE-based neural receiver architecture (MEAN), shown in Figure 1, that has multiple smaller experts, called *SNR experts*, each trained to perform in a different channel condition. A gating network is used to dynamically select the correct SNR expert for each input sample. Since only fewer SNR experts needs to be active at a time depending on channel conditions, the run-time computational complexity can be reduced. We compare the performance of our proposed MEAN architecture and classical receivers by evaluating the

performance in a SIMO wireless communication system model under different channel conditions.

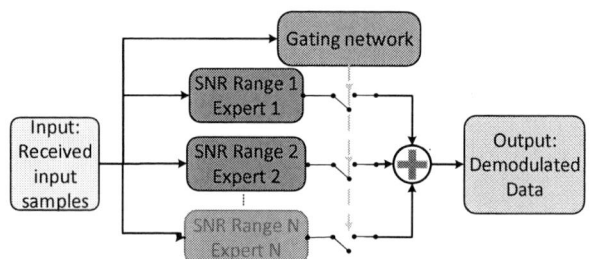

Fig. 1: High level overview of MEAN architecture for a neural receiver. Multiple SNR experts are combined dynamically by the gating network according to the input channel conditions.

The rest of the paper is structured as follows. Section II presents the related work, and Section III provides background information on state-of-the-art neural receivers. The proposed MEAN architecture is presented in Section IV, and Section V the system-level performance evaluation results. Finally, the paper is concluded in Section VI.

II. RELATED WORK

There are different types of DyNNs [3], that implement techniques such as early exiting, layer skipping, dynamic depth etc. Layer-skipping has been proposed for image classification applications, that skips convolutional blocks with the help of a gating network [5], and that decides the layer to be skipped for each input image on the fly [6]. Layer skipping is also proposed for Neural receivers by skipping layers based on the Signal-to-Noise Ratio (SNR) of a given input sample [2]. A fine-grained approach is proposed in [7] for dynamic precision adaptation in a NN-based demodulator. Hardware overhead for implementing fine-grained techniques increases for larger NNs such as neural receivers. MoE-based architectures that selectively executes a branch in a CNN based on the input [8] and sparsely-gating MoE network in a long short-term memory (LSTM) network for multilingual machine translation applications [9] have been proposed in the past. However, existing DyNN architectures cannot be applied directly to neural receivers as they need to adapt according to varying channel conditions.

III. BACKGROUND

The baseline wireless system in this paper is a Single Input Multiple Output (SIMO) system based on an implementation proposed in *Sionna* [1], as shown in Figure 2. The Transmitter - Base station(BS) consists of a Forward Error Correction (FEC) encoder, a (de)modulator and an OFDM (Orthogonal

979-8-3315-3968-9/24 $31.00 © 2024 IEEE

#Layers	Name	#Input channels	#Output channels	Parameter count	Kernel size
1	Input Conv2D	5	128	5706	3x3
8	Conv2D	128	128	147456	3x3
8	Layer Normal - ization	128	128	458752	None
1	Output Conv2D	128	2	2304	3x3

TABLE I: Layer parameters traditional neural receiver.

Frequency Division Multiplexing) mapper. The FEC encoder uses a Low-Density Parity-Check (LDPC) code for data encoding, enhancing data reception quality and reliability in noisy environments. We consider two types of receivers: a classical receiver and a neural receiver. The classical receiver includes a Linear Least Mean Squared (LLMSE) equalizer, which leverages Channel State Information (CSI) to mitigate noise effects caused by the channel. CSI provides details on the channel's impact on signal quality, factoring in scattering, fading, and power loss over distance. It employs two CSI estimation methods: one using exact (ideal) CSI and the other using the Least Squares (LS) channel estimation, which is commonly used in real systems [10].

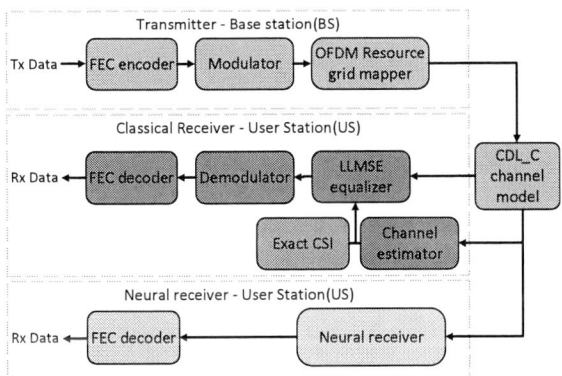

Fig. 2: SIMO system overview with classical and neural receiver [1].

A. Architecture of static neural receiver

The architecture of the static neural receiver, based on state-of-the-art [1], [10], [11], is shown in Figure 3. It consists of eighteen layers with a combined total of approximately 4.8 million parameters for all weights and biases. The details of these parameters are shown in Table I. The NN is trained for the complete noise range with randomized input data for each noise point. The dataset for training is generated by feeding randomized data packets through the transmitter. This generates data for each of the SNR points for the network to train on. The downside of such an NN is that to cover the whole noise range and still have good performance the network needs to be large resulting in a large number of parameters. Different modulation schemes or code rates(CR) can be used with these networks by changing the modulation scheme or CR and retraining the network. This generates a different set of weights, by switching the weights in the neural network it can be used for different modulation schemes or CR.

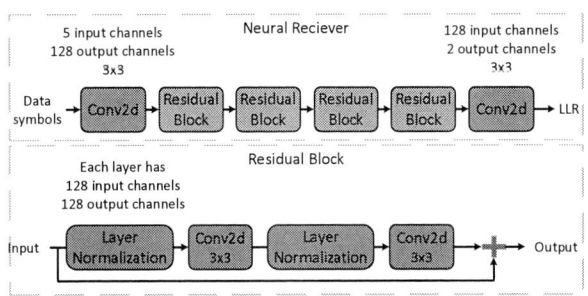

Fig. 3: Architecture of the traditional neural receiver.

IV. PROPOSED MEAN ARCHITECTURE

There are three ways by which we can apply dynamism in a NN: *sample-wise, spatial-wise, and temporal-wise* [3]. Sample-wise dynamism allows the network to modify its architecture based on individual input samples. Spatial-wise dynamism adjusts the network according to the specific location of data points within a sample, such as a pixel's position in an image. Temporal-wise dynamism, on the other hand, changes the network based on the temporal aspects of data that unfold over time such as in text or video sequences.

Sample-wise dynamism is relevant for neural receivers since the input data fed to the NN consists of received data samples that lack temporal and spatial characteristics. The changing SNR can serve as a basis to modify the network's architecture. There are two types of sample-wise dynamism: *dynamic depth* and *dynamic width*. Adapting dynamic depth includes methods such as skipping layers or early exiting, while methods implementing dynamic width skip a number of neurons or channels in the run-time. MoE is a dynamic width adaptation method. MoE consists of two types of sub-networks: *experts* and *gating networks*. Experts are several smaller networks that are trained to specialize in a specific feature. In a neural receiver, this could be different SNR ranges. A gating network decides which experts contribute to the output of the MoE.

There are two types of MoEs: soft-gated and hard-gated (also known as sparsely-gated). A hard-gated MoE only has one expert that contributes to the output at a time while the others are disabled. In a soft-gated network, one or more experts can be active at a time and not all experts have to contribute equally to the output. For example, 30% of expert 1 could be used and 70% of expert 2. This work mainly focuses on hard-gated MoE because this will be easier to implement in hardware with less overhead (for e.g., only 3 gating mechanisms are needed for 3 experts). On the other hand, soft gating would require either to have at least two experts active reducing the power savings. Another option would be to have partial gating of the experts, which is more difficult to implement and results in more overhead depending on the amount of partial gating possible in each of the experts. So for example if one expert can be partially gated in 4 sections so 25, 50, 75, 100% of the expert active. Four gating mechanisms would need to be used per expert which would result in twelve gating mechanisms for a three expert MOE introducing significant overhead. While this would also make the choice for the gating network a significantly more complex

This paper focuses on a MoE implementation with three SNR experts called MEAN. SNR experts are small NNs trained on exploiting a specific feature such as SNR. Figure 4 shows the layer configuration of an SNR expert, which consists of four residual blocks with each residual block containing one layer-normalization layer and one conv2D layer. Note that this is only half of the number of layers in the static neural receiver. The architecture of MEAN is shown in figure 4. MEAN uses the input data of the system so the gating network consisting of dense layers can determine the best SNR expert according to channel conditions. Only one of the experts is active at a time so the number of active layers during runtime is significantly reduced. Each of the SNR experts is trained in a certain SNR range so it can specialize in this region.

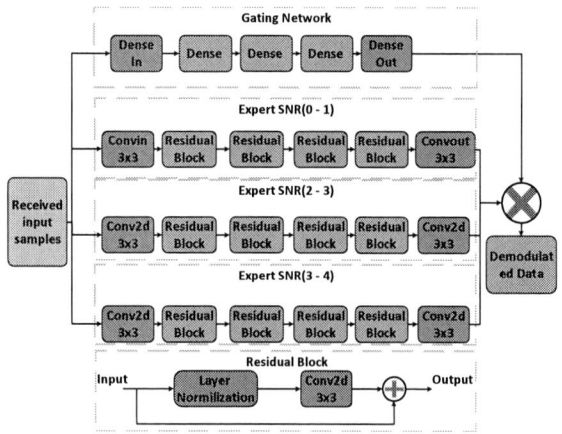

Fig. 4: Overview of the MEAN architecture, showing the layer configurations of experts and the layers used in the gating network.

A. Training of SNR Experts

Each SNR expert is trained separately on its own noise range. The experts are trained individually to not let the training of the SNR experts influence each other so they can specialize in their own noise region. Once the experts are trained, their performance is evaluated in the system model for the specific noise range and only the best-performing ones are selected. Experts overlapping in performance to the chosen experts or worse-performing experts get discarded. Keeping them in the network would result in extra silicon area overhead in hardware, without benefiting in the performance gain.

B. Training of Gating network

This network is trained along with pre-trained SNR experts. The weights of the SNR experts are loaded in and locked (So that the gating network does not influence the SNR experts). Then the gating network is trained over the entire noise range of the system. The gating network parameters are then trained to select the best-performing SNR expert for certain input data after training. Essentially the training process tries to determine the SNR expert that needs to be selected for a given input sample. The output of a gating network is a coefficient with values between 0 and 1, for each of the SNR experts. The higher the coefficient value for an expert determined by the gating network for an expert, it is most likely suitable for

the SNR range at the current input. An *argmax* function is used in the gating network to detect the largest coefficient and set them to 1, while the remaining coefficients are set to 0. This is done so only one SNR expert can be active at a time. One issue with using argmax is that it is a non-differentiable function, which means that back-propagation with a gradient descent training method is not possible. To fix this a Straight-through-estimator (STE) [12] was used to approximate the differential during back-propagation. When a backwards pass is executed the gradient is passed through and the *argmax* acts as an identity function which replicates the gradient of the layer above and passes it to the next layer. Therefore the gradient is passed through the *argmax* operation as if it were a continuous function. During the forwarded pass *argmax* is used. With this method, the gating network can be used to determine which SNR expert should be active based only on the input data of the system.

V. PERFORMANCE EVALUATION OF MEAN

For performance analysis, we use the wireless system model of a SIMO system [1], as introduced in Section III, with the following parameters: FFT size of 128, sub-carrier spacing of 30 kHz, 14 OFDM symbols, and carrier frequency of 3.5 GHz. We used QAM 16 modulation scheme and two different Code Rates (CR) of 0.66 and 0.25 for the FEC decoder. For performance comparison, we used two baselines: (1) Baseline-Perfect CSI, which is the theoretical best performance that can be achieved, (2) Baseline-LS Estimation, a classical technique used to estimate CSI used in commercial systems, and (3) Static neural receiver, the state-of-the-art neural receiver (introduced in Section III).

	SNR expert 1	SNR expert 2	SNR expert 3
QAM16 CR 0.66	3-3.5 dB	4-4.5 dB	6-6.5 dB
QAM16 CR 0.25	0-0.5 dB	1-1.5 dB	3-3.5 dB

TABLE II: Table with SNR ranges for experts.

#Layers	Name	#Input channels	#Output channels	Para-meter count	Kernel size
1	Input Conv2D	5	128	5706	3x3
4	Conv2D	128	128	147456	3x3
4	Layer Normal-ization	128	128	458752	N/A
1	Output Conv2D	128	2	2304	3x3
1	Dense In	5	16	96	N/A
3	Dense	16	16	272	N/A
1	Dense Out	16	3	1088	N/A

TABLE III: Layer parameters of MEAN architecture.

For the MEAN architecture, we have trained up to to 20 SNR experts each trained on its own SNR range. The SNR experts were trained with increments of 0.5 energy-per-bit to noise power (Eb/No), which is the normalized SNR. From the trained SNR experts, the best-performing experts were selected by performing a design-space exploration (by performing system-level simulation explained in next section) of experts with different SNR ranges. Finally, we selected only

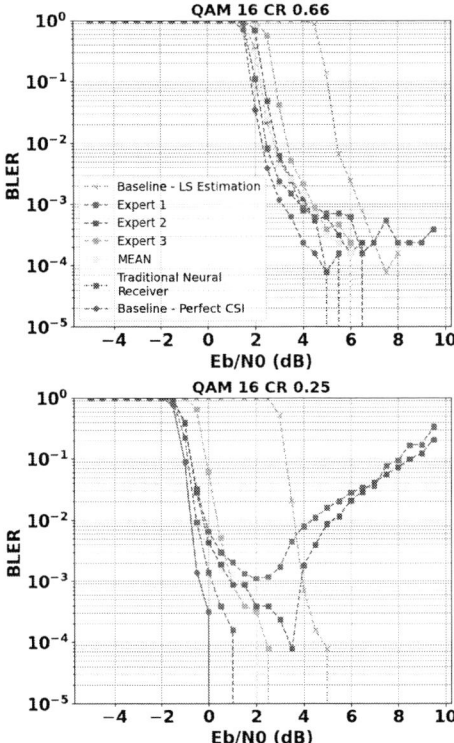

Fig. 5: System level performance at different CR for SNR experts, MEAN, traditional receiver and baseline.

the minimal number of experts needed to meet the overall system performance. This means that there are some experts that perform in a different SNR range other than at which they are trained for. The noise ranges of the selected SNR experts for the MEAN network are shown in table II. After this, the gating network is trained with the selected SNR experts (weights locked) for the entire SNR range, i.e. -5 to 10 dB. The layer parameters of MEAN are shown in table III. The dataset for training gating networks and SNR experts is generated by feeding randomized data packets through the transmitter similar to the method for training the static neural receiver as was described in section III-A.

A. Performance results

Figure 5 shows the system-level performance comparison of MEAN with respect to state-of-the-art implementations for QAM16 modulation with a CR of 0.25 and 0.66. The performances of individual SNR experts used in the MEAN architecture are also shown. The y-axis represents the system-level performance in terms of Block Error Rate (BLER), which indicates the number of erroneous blocks (contains multiple data samples) versus total transmitted blocks. Overall, we can see that MEAN architecture outperforms the Baseline - LS Estimation and have similar performance as traditional neural receiver despite using fewer layers. Using MEAN results in a system that reduces the total active layers during runtime by 50%. Note that there will be addition overhead for the gating network which consists of five dense layers. However, since

the parameter count of gating network is not significant (as shown in Table III) compared to the experts, we expect the impact to be minimal.

When comparing to individual experts, at CR of 0.66, the overall performance of individual SNR experts is inferior to the MEAN architecture as expected. For SNR range of 0-4 dB, the MEAN architecture has the similar performance of Expert 0 as the gating network has selected this expert which is the best-performing expert in this channel condition. Similarly, Expert 2 is selected for SNR range of 4.5-6 dB, and Expert 3 from 6 dB or larger (after 6 dB the BLER becomes very small). This shows that the gating network is able to select the best-performing SNR expert for a certain noise region. Similar results is also observed with CR of 0.25 as shown in Figure 5, which shows MEAN architecture selects Expert 1 for SNR until 0.5 dB, Expert 2 for SNR range of 1-1.5 dB and Expert 3 from 1.5 dB and larger. As is clear from the plot this modulation scheme and CR operate in a different Eb/No region. This is also the reason why the experts are trained for different noise regions as was shown in table II.

VI. CONCLUSIONS AND FUTURE WORK

DyNN architectures benefit neural receivers by reducing computational complexity in the run-time by adapting their structure according to varying channel conditions. This work presented a novel MoE-based DyNN architecture called MEAN that dynamically selects a number of experts during runtime. The experts in MEAN are trained to specialize in a dedicated SNR region and a gating network is trained to select the best expert according to the channel conditions. Compared to static neural receivers, our proposed MEAN architecture uses only 50% of the layers in a static network to achieve similar performance and with minimal overhead. As a future work, hardware implementation of MEAN will be made and evaluated against state-of-the-art in terms of silicon area usage and power consumption.

REFERENCES

[1] Hoydis and *et.al*, "Sionna: An Open-Source Library for Next-Generation Physical Layer Research," *arXiv preprint*, Mar. 2022.

[2] B. van Bolderik and *et.al*, "Agile Design-Space Exploration of Dynamic Layer-skipping in Neural Receivers," *DSD*, 2024.

[3] Y. Han and *et.al*, "Dynamic Neural Networks: A Survey," *IEEE Transactions on Pattern Analysis and Machine Intelligence*, 2 2021.

[4] M. D. Gomony, , and *et.al*, "PetaOps/W edge-AI μ Processors: Myth or reality?," in *DATE*, 2023.

[5] X. Wang and et.al, "Skipnet: Learning dynamic routing in convolutional networks," 2018.

[6] A. Veit and et.al, "Convolutional networks with adaptive computation graphs," *CoRR*, vol. abs/1711.11503, 2017.

[7] P. S. Allwin and *et.al*, "Run-time Non-uniform Quantization for Dynamic Neural Networks in Wireless Communication," in *ASPDAC*, 2024.

[8] N. Shazeer and *et.al*, "HydraNets: Specialized Dynamic Architectures for Efficient Inference," in *2018 IEEE/CVF Conference on Computer Vision and Pattern Recognition*, 2018.

[9] N. Shazeer and *et.al*, "Outrageously Large Neural Networks: The Sparsely-Gated Mixture-of-Experts Layer," *CoRR*, vol. abs/1701.06538, 2017.

[10] M. Honkala and *et.al*, "DeepRx: Fully Convolutional DL Receiver," *IEEE Trans. Wireless Commun.*, 5 2021.

[11] A. Aoudia and et.al, "End-to-end learning for ofdm: From neural receivers to pilotless communication," *IEEE Transactions on Wireless Communications*, 2022.

[12] P. Yin and et.al, "Understanding Straight-Through Estimator in Training Activation Quantized Neural Nets," 2019.

Industrial Contribution

Resource Management of Automotive Engine Control Units

Istvan Andras Gergely[‡], Sebastian Rausch[‡], Nahla Elaraby[*†], Axel Jantsch[*]
*TU Wien, Vienna, Austria
†Canadian International College-CIC, Cairo, Egypt
‡ Robert Bosch AG, Powertrain Solutions, Vienna, Austria

Abstract—Embedded systems play a crucial role in the contemporary automotive industry, where advanced vehicles integrate numerous electronic control units responsible for diverse functionalities. This paper specifically focuses on the engine control unit within the power-train. With increasing ambitions on reducing emission values and a growing functional environment, Robert Bosch AG anticipates that the resources of the current generation of Engine Control Units (ECUs) will reach their limits within the next 3-5 years. The current development workflow relies on resource optimizations with an iterative expert-based approach, success is achieved through significant manual effort. In this work, we propose a methodology that leverages an existing measure based on task splitting from the early stages of ECU development. Our goal is to streamline the design process for engineers and accelerate ECUs development through a cutting-edge software architecture workflow, aiming to minimize manual effort and expedite project timelines. The proposed approach was integrated and validated on a real-world control unit of the power-train, demonstrating highly significant results, with savings of around $50\% \pm 10\%$. This considerable reduction in time can be attributed to the increased investment required during the architectural design stage, which subsequently reduces the labor-intensive manual integration effort. Additionally, there has been an achieved reduction of approximately 10% in core load.

Index Terms—Embedded systems, automotive industry, engine control unit, multi-core, software architecture, load balancing

I. INTRODUCTION

In recent years, the automotive industry has experienced a significant increase in vehicle functional requirements, leading to a rise in embedded systems and electronic control units. This shift highlights the importance of software-intensive solutions alongside mechanical components. Software engineering is now crucial for innovation and maintaining competitiveness in the industry [1]. In today's automotive market, "software is the number-one decisive competitive factor" [2]. Advanced technologies in vehicles, like driving-assistance systems and safety features, require complex software. With the industry becoming saturated, rapid release cycles and continuous cost and efficiency optimization are essential for success. The software within an automotive vehicle is executed by multiple control units, each responsible for specific domain control functions. Over time, the Electric and Electronic (E/E) architecture of these control units has developed to a highly sophisticated level. Internal research conducted at Robert Bosch AG indicates that the existing hardware may not meet long-term requirements. Consequently, the microcontroller devices currently in use will either undergo updates or necessitate optimal software distribution and resource utilization.

The literature [3] highlights increased system complexity, which is also observed in practice. Runtime measurements on functional software in a test vehicle validate this, assessing the effectiveness of Operating System (OS) scheduling practices and enabling optimization of runnables and code sequences. Runnables, in this context, represent individual functionalities encapsulated within a subsystem.

Following a series of runtime measurements conducted on multiple productive software releases, load disparity among cores is evident, with core 1 bearing a significantly higher load than other cores. This pattern is consistent across multiple projects in the safety-critical automotive embedded system domain.

Figure 1 shows that some resource categories are nearing their predefined limits. Key definitions in this paper include CPU load, which varies with software execution paths, and core load, calculated as the function's runtime divided by its repetition rate. High core load issues in later development stages constrain viable solutions and increase rectification costs. Given the finite hardware resources of a microcontroller in an ECU, precise resource allocation is crucial to prevent exceeding maximum thresholds throughout the product lifecycle. Robert Bosch AG's current approach involves core load splitting and task scheduling on different cores, but verifying these partitions and resolving interfaces is manual-intensive and requires expertise. Due to the limited number of experts in each project, the associated efforts could be more cost-effective for the company. Delegating such design decisions to other colleagues is not feasible as experts are already engaged in other critical tasks. In this work, we propose a methodology that leverages software architecture to aid early design and established projects with minimal effort. Using graph theory to map software components, we support decisions on allocating these components to different cores. Post-implementation, we verify our assumptions by integrating a software image onto an ECU family control unit, evaluating core load balance and reduced manual effort.

The remainder of this paper is organized as follows: the next section discusses related work and introduces the novelty of our proposal. We then present our approach, system model, and tools. Section 4 details our integration efforts, and finally, we conclude and discuss future research directions.

979-8-3315-3968-9/24 $31.00 © 2024 IEEE

Industrial Contribution

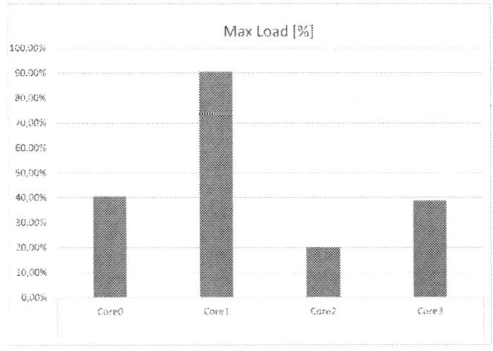

Fig. 1. Result of runtime measurement concerning core utilization

II. RELATED WORK

Both graph theory and load balancing are subjects with extensive literature; however, their intersection remains relatively unexplored. Load balancing, including the parallelization of heavy single-core applications, has been extensively researched in the automotive domain, with both static and dynamic techniques explored in various papers, such as [6], [7], and [10]. This paper does not examine the feasibility of task splitting on the Engine Management System (EMS) of Robert Bosch AG, as this possibility has been exhaustively studied in [8] and [9], and verified on the same EMS. Therefore, we assume that by creating a concept for task splitting and validating it using the integrational toolchain methodology from [8] and [9], we can execute the splitting process.

In [4] and [5], similar approaches to ours were researched, utilizing graph theory as the basis for system model analysis and exploring strategies to maximize load balancing based on predefined criteria. However, our approach differs in that we propose the utilization of software architecture, aligned with the reference architecture approach within Robert Bosch AG. We assume that the software components, typically inherited from the reference architecture and then modified to suit specific project needs, adhere to a fixed task sequence. Consequently, inter-task dependencies need not be considered when splitting at the border of a software component. To our knowledge, there is currently no literature incorporating software architecture into the decision-making process of task splitting. In a related work [11], a similar approach was identified as a further field of research, fitting into our system model, wherein tasks are not split due to the potential loss of task sequence, leading to data corruption.

III. PROPOSED APPROACH

This section proposes a methodology for optimizing resources based on software architecture. The solution focuses on a subset of ECU resources, such as ports and computational entities, without requiring a complete set. It manages these limited resources by concentrating on cores and their workloads, addressing the most critical aspects of the ECU resource problem. The architecture is analyzed at a high abstraction level, termed the software module architecture. Our approach uses the software architecture as a foundation for task allocation, employing a graph representation for programmatic analysis. In this representation, the software architecture's software modules are considered the graph's nodes, and the graph's edges represent the interdependency between the modules. As automotive software commonly allows for bidirectional communication between components, the resulting graph is considered to be an undirected graph. Thus, formally describing a software architecture in a graph representation entails: $G(a) = (n, v)$ with A being a set of architecture, hence a ϵ A, n being a Software Component (SWC) from the available set of SWCs, n ϵ N and v is the set of vertices that present the connection(interface) between two SWCs v ϵ V. The proposed solution aims to investigate potential trade-offs. Minimal dependencies while considering the implementation constraints specified in III-A. This is accomplished by identifying an optimal cut in a graph that minimizes the number of connections between the start and end nodes. Subgraphs are compared to identify the most suitable one for core balancing. The methodology uses an expert-in-the-loop approach to tackle resource scarcity in software architecture. Due to the complexity of software architecture, human analysis alone cannot easily find the optimal edge for splitting tasks. Graph optimization algorithms are employed to determine ideal cuts, with their limitations defined by their big O notation.

A. Restrictions of our software model

We focused on specific aspects of production architectures to maintain a manageable approach, applying assumptions carefully to preserve system functionality, as noted in [15]. Complexity reductions were discussed with and reviewed by a lead system architect to ensure no detrimental trade-offs.

In hardware design, we concentrated on the microcontroller's cores, assuming three fixed cores with uniform computational capabilities and peripheral access speeds. Initial utilization was set at 100% for all cores, with unrestricted on-chip communication among them.

In developing the software architecture, we focused on the relationships between SWCs, as these connections define component dependencies. An interface, linked to a port, specifies the data or operations a software component offers or requires [17]. We model the software architecture as an undirected, unweighted graph, where nodes represent SWCs and edges denote dependencies. Only the module-level abstraction of the architecture is considered.

B. Methodology and runtime analysis

The proposed methodology consists of three main parts. First, it checks for all minimum cuts in the software architecture to identify leaf elements, which are defined as elements with only one connection to another element. The second part involves designing a split based on load distribution and minimum dependencies. In the third part, a split is proposed based solely on minimum dependencies. For the initial stages, we use

979-8-3315-3968-9/24 $31.00 © 2024 IEEE

functions from the Networkx library [13] to perform global balanced cuts. Initially, we used a Brute-Force approach, but empirical runtime analysis showed it was inadequate. As the architecture's nodes and connections increased, the Brute-Force method became impractical, requiring excessive time even for moderately sized architectures.

Our developed methodology is outlined in the following

- Finding all minimum cuts: The initial stage of the implementation involves identifying all minimum cuts for leaf elements. The underlying concept of the implemented script is to systematically remove edges from a duplicated graph to avoid changes to the original one.
- Partitioning concerning load and dependencies: In the second implementation stage, a script is developed to partition the software architecture based on node load and dependencies. Initially, we used a Brute-Force approach, but our analysis (Table I) showed it was unsuitable for production-scale architectures. We then switched to the Metis library [18], a widely regarded partitioning tool often used as a benchmark for validating new methods, as seen in [19], [20], and [21].
- User-defined partitioning: The first two parts of the methodology aim to provide a globally optimal solution for resource scarcity by splitting tasks from the software architecture without human intervention. In the third step, the software system designer determines if a locally optimal solution is more suitable or technically easier to implement on the ECU's task set. The user specifies the number of edges to cut, generating and outputting all possible bipartitions. The system designer then evaluates these options based on the required criteria.

As previously mentioned, the efficiency of graph-splitting algorithms is largely determined by their Big O notation. To assess their performance empirically, we conducted a runtime analysis. The present section involves executing the methodology's scripts on graphs of varying sizes. Table I displays the runtime of each approach for different input sizes. Initially, a small input size was utilized, and we increased it gradually. To conduct the empirical runtime analysis, the measured runtime values for each method were normalized by plotting the number of nodes against the elapsed time on a logarithmic scale using a base of two. This approach resulted in a linear relationship, where the line's slope represents the runtime's polynomial degree. From this, the computational time complexity can be determined.

Runtime measurements revealed that the Brute-Force approach for finding a globally optimal solution in graph partitioning of automotive software architectures is impractical. Calculations showed that for architectures with 32 software modules, finding an optimal solution would take about 32 hours. A lead software architect confirmed that this runtime significantly exceeds reasonable expectations.

Although Metis employs heuristic algorithms, which may not always yield the optimal solution, it strives to achieve load balancing across partitions. However, perfect balance may only sometimes be attainable, particularly for complex or irregular graphs or meshes.

IV. INTEGRATION AND VERIFICATION

To evaluate the criteria for reducing the development cycle through early design decisions, this subsection applies the methodology to a real product example from Robert Bosch AG. We selected the Dosing Control Unit (DCU) from the EMS as a proof of concept. Although less complex than the ECU, the DCU allows us to implement the methodology with a manageable number of SWCs and interfaces, minimizing integration and configuration effort. This integration and verification aim to demonstrate:

- It is feasible to formulate design determinations concerning the system resources of the EMS predicated upon the software architecture.
- The design of the methodology can be integrated into the development pipeline of Robert Bosch AG with a manageable level of effort.
- The application of a methodology can result in a tangible reduction in manual development efforts.

In developing the proof-of-concept, we carefully partitioned the software using the architectural framework. For validation, we integrated this partitioned architecture into the DCU integration pipeline, which automates the examination of task scheduling and data consistency. The result was a deployable flashable software image for the DCU.

V. RESULTS

At the beginning of this conference paper, two primary objectives were established. The first was to reduce the development cycle duration and manual workload in ECU development. The second was to optimize core utilization. The results for the first objective were evaluated by domain experts using the proposed methodology and current practices. We compared core utilization before and after implementing our methodology versus the current software solution to evaluate the percentage difference. Expert interviews on software deliveries and regression loops indicated that our architectural approach yields time savings of approximately $50\% \pm 10\%$ by reducing manual integration effort through upfront design investment.

Regarding core utilization, as shown in Figure 2, we initially identified the issue in the broader project context and focused on the DCU. Runtime measurements on our proof-of-concept demonstrated a roughly 10% reduction in load on the more heavily used core.

This outcome can be regarded as a significant result, given that the current iterative expert-based approach typically requires substantial time and effort for even a 1-2% improvement. Additionally, it is noteworthy that the load originally present on the core has been successfully reallocated to another core, resulting in a more balanced distribution of core utilization. Demonstrating the feasibility of making design decisions about resource allocation through meticulous front-end planning and task segmentation at the SWC level.

TABLE I
EMPIRICAL RUNTIME ANALYSIS

Input size (No. of nodes)	All leaf cuts	Load distribution	User defined cuts	PyMetis
4	0.0009925 sec	0.0000931 sec	0.0004625 sec	0.0000689 sec
8	0.0020006 sec	0.0020108 sec	0.0039902 sec	0.0000879 sec
16	0.0049994 sec	0.8968558 sec	0.0168388 sec	0.0001738 sec
32	0.0160007 sec	-	0.6357337 sec	0.0002651 sec
64	0.0915489 sec	-	25.829432 sec	0.0004410 sec
128	0.1960001 sec	-	111.10125 sec	0.0011162 sec
256	1.5181146 sec	-	969.09581 sec	0.0017340 sec
512	8.8497076 sec	-	9590.9342 sec	0.0048061 sec

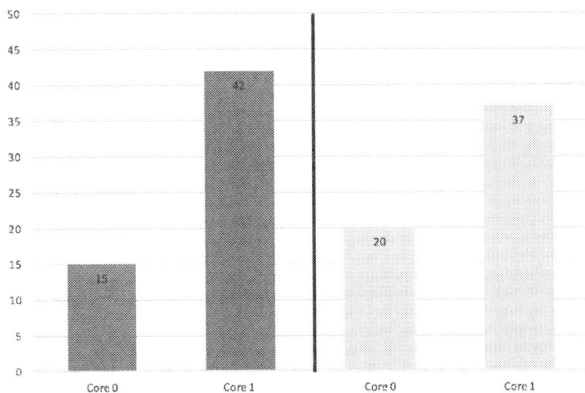

Fig. 2. The final runtime measurement results comparing two software images are presented here. The left side shows the original software, while the right side displays the software with our methodology applied.

VI. CONCLUSION

As functional and non-functional requirements for embedded vehicle systems grow, early-stage resource planning becomes increasingly challenging, particularly in the cost-constrained, long-cycle automotive industry. Our research evaluates the feasibility of informed decision-making early in software development. Using graph partitioning theory and established algorithms, we developed mathematically optimized software architecture partitions, addressing constraints such as complex, time-critical interfaces relevant to the automotive sector. To assess our approach's effectiveness in reducing development cycle time and achieving load balancing, we implemented it on a product within the Robert Bosch ECU family. The proposed approach has demonstrated highly significant results, with savings of around $50\% \pm 10\%$. This considerable reduction in time can be attributed to the increased investment required during the architectural design stage, which subsequently reduces the labor-intensive manual integration effort. Additionally, there has been an achieved reduction of approximately 10% in core load.

REFERENCES

[1] A. Haghighatkhah, M. Oivo, A. Banijamali, and P. Kuvaja, "Improving the state of automotive software engineering," IEEE Software, vol. 34,
no. 5, pp. 82–86, 2017.
[2] C. Ebert and J. Favaro, "Automotive software," IEEE Software, vol. 34, no. 03, pp. 33–39, may 2017.
[3] V. Antinyan, "Revealing the complexity of automotive software," in Proceedings of the 28th ACM Joint Meeting on European Software Engineering Conference and Symposium on the Foundations of Software Engineering, ser. ESEC/FSE 2020. New York, NY, USA: Association for Computing Machinery, 2020, p. 1525–1528.
[4] Y. Shoukry, A. Kumar, M. W. El-Kharashi, G. Bahig, and S. Hammad, "Graph-based approach for software allocation in automotive networked embedded systems: A partition-and-map algorithm," in Proceedings of the 2013 Forum on specification and Design Languages (FDL). IEEE, 2013, pp. 1–6
[5] R. Hegde and K. Gurumurthy, "Load balancing issues in automotives," in 2009 International Multimedia, Signal Processing and Communication Technologies. IEEE, 2009, pp. 138–141.
[6] J. Kienberger, S. Schmidhuber, C. Saad, S. Kuntz, and B. Bauer, "Parallelizing highly complex engine management systems," Concurrency and Computation: Practice and Experience, vol. 29, no. 15, p. e4115, 2017
[7] H. Hussain, M. Shoaib, M. B. Qureshi, and S. Shah, "Load balancing through task shifting and task splitting strategies in multi-core environment," in Eighth International Conference on Digital Information Management (ICDIM 2013), 2013, pp. 385–390.
[8] M. Lowinski, D. Ziegenbein, and S. Glesner, "Partitioning embedded real-time control software based on communication dependencies." in MASE@ MoDELS, 2015, pp. 3–12.
[9] M. Lowinski, Parallelization of legacy automotive control software for multi-core platforms. Technische Universitaet Berlin (Germany), 2019.
[10] G. Mishra and K. Gurumurthy, "Dynamic task scheduling on multicore automotive ecus," International Journal of VLSI Design & Communication Systems, vol. 5, no. 6, p. 1, 2014.
[11] G. Mishra and R. Hegde, "Task scheduling with load balancing on automotive multicore ecus," in 2018 International Conference on Recent Innovations in Electrical, Electronics Communication Engineering (ICRIEECE), 2018, pp. 1993–1995.
[12] https://uk.mathworks.com/
[13] https://networkx.org/
[14] https://www.python.org
[15] C. Yang, P. Liang, and P. Avgeriou, "A survey on software architectural assumptions," Journal of Systems and Software, vol. 113, pp. 362–380, 2016.
[16] https://elearning.vector.com/mod/page/view.php?id=291
[17] https://elearning.vector.com/mod/page/view.php?id=291
[18] G. Karypis and V. Kumar, "A fast and high quality multilevel scheme for partitioning irregular graphs." SIAM Journal on scientific Computing, vol. 20, no. 1, pp. 359–392, 1998.
[19] A. Pacaci and M. T. Özsu, "Experimental analysis of streaming algorithms for graph partitioning," in Proceedings of the 2019 International Conference on Management of Data, 2019, pp. 1375–1392.
[20] R. Banos, C. Gil, J. Ortega, and F. G. Montoya, "Multilevel heuristic algorithm for graph partitioning," in Applications of Evolutionary Computing: EvoWorkshops 2003: EvoBIO, EvoCOP, EvoIASP, Evo-MUSART, EvoROB, and EvoSTIM Essex, UK, April 14–16, 2003 Proceedings. Springer, 2003, pp. 143–153.
[21] H. Meyerhenke, B. Monien, and S. Schamberger, "Graph partitioning and disturbed diffusion," Parallel Computing, vol. 35, no. 10-11, pp. 544–569, 2009.

SystemVerilog-SystemC TestBench Architecture for VLSI Chip Design Verification

Mohammad Ismael[§], Ayman Hroub[*], Nasib Naser[§]

Department of Electrical and Computer Engineering, Birzeit University, Birzeit, Ramallah, Palestine[*]
Orion VLSI Technologies, Rawabi, Palestine[§]

Abstract—Simulation-based verification of Very-Large-Scale Integration (VLSI) chip design is inevitable. However, the simulation speed can be a bottleneck in the verification process productivity. People tried to accelerate simulation-based verification through parallelization, emulation, and reducing the number of test scenarios, etc. In this paper, we exploited the power of abstraction of SystemC to design fast and accurate golden functional reference models. These reference models can be integrated in the Universal Verification Methodology (UVM) based TestBench through Transaction Level Modeling (TLM). Thus, the UVM based TestBench interacts with the SystemC reference model to get the golden values, and use them for checking. We used the proposed methodology to verify the execution unit (EXU) of the open-source RISC-V based VeeR EL2 processor core. The experimental results showed that the proposed TestBench architecture with SystemC reference model is up to 15x faster than the TestBench with SystemVerilog reference model.

Index Terms—Design Verification, UVM TestBench, Transaction Level Modeling, SystemC, Co-Simulation.

I. Introduction and Background

VLSI chip designs are getting complex and harder to verify. The VLSI design process comprises many steps. Each design step is a transformation of the design from one representation to another. For example, the chip's business requirements are turned into architectural specifications. Then, the architectural specifications are transformed into micro-architectural design. After that, the micro-architecture is translated into a synthesizable RTL (Register Transfer Level) design code, and so on. Each design step has a corresponding verification step, to ensure that the the original design intent has been preserved during these transformations.

Design verification is a crucial process to ensure that the chip's RTL design behaves as expected, before it is fabricated on silicon. Thus, the verification process aims to hunt the design bugs as early as possible to avoid the high cost and time-to-market delays resulting from bugs escape and chip respin. The verification process comprises two steps, namely, driving and checking:

1) Driving the DUT (Design under Test) with an interesting, legal, and valid stimulus.
2) Capturing the DUT's actual response for this stimulus, and comparing it with the expected correct response obtained from the golden reference model.

The reference model is a high-level executable version of the design specifications. It predicts the expected correct response of the DUT. It can be written in any language, such as, C, C++, MATLAB, SystemVerilog, etc.

Simulation-based verification is a dynamic verification technique, where the DUT is instantiated within a TestBench [1]. In this technique, the TestBench drives the DUT with constrained pseudo-randomly generated stimuli. The simulator, such as, Synopsys VCS or Cadence Xcelium simulates the behavior of the DUT for each stimulus. Then, the TestBench samples the DUT's response. Finally, the DUT's response is compared with the expected correct response obtained from the reference model for the same stimulus. If the two values do not match, then a bug has been detected. Figure 1 depicts the role of the reference model in the simulation-based verification environment.

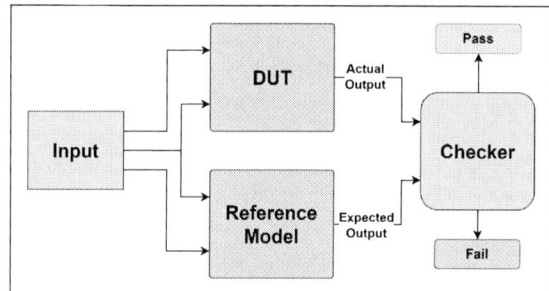

Fig. 1: High Level TestBench

SystemC [2]–[4] is a powerful C++ based system-level hardware modeling language. It has native support for TLM which provides standardized communication models and seamless integration with hardware and software components.

On the other hand, UVMC (UVM Connect) is an extension of UVM that provides a standardized way to connect simulation environments using TLM through a look-up string. In other words, it is a bridge between SystemVerilog UVM and SystemC TLM [5]. It enables seamless communication between SystemVerilog and SystemC models. UVM [6] is a SystemVerilog class library, and a methodology to build a TestBench using this library. The UVM TestBench comprises several verification components, such as, agents, drivers, monitors, sequencers, coverage, and scoreboards. Moreover, UVM uses TLM [6] as a communication abstraction to exchange data between its different components.

979-8-3315-3968-9/24 $31.00 © 2024 IEEE

In this paper, we introduce a new UVM-based verification environment that uses a SystemC based golden functional reference model.

The reference model can be either, (1) run offline to generate an execution trace that can be used as a golden reference model, (2) or connected with the TestBench, such that the TestBench can interact with the reference model and obtain the reference values at simulation time. We went for the latter in this paper. There are different ways to connect the SystemC reference model with a verification environment, such as, Direct Programming Interfaces (DPIs) or Verilog wrappers using **syscan**. However, these ways have some limitations, such as, limited reusability, and limited coverage. This makes them unsuitable for complex designs.

The DUT used to prove the efficiency of the proposed TestBench architecture is the execution (EXU) unit of the open-source RISC-V-based VeeR EL2 processor core [7]. This unit is responsible for supporting a wide range of arithmetic and logic operations. Moreover, it implements RISC-V bit manipulation extension instructions. Also, it handles branches, pipeline flushes, pipeline control, and optimizations for small-number arithmetic.

The simulation results showed that the proposed methodology resulted in a simulation speedup of up to 15x compared with the verification environment that uses a SystemVerilog-based reference model. This is thanks to the abstract and fast SystemC-based reference model, and the TLM-based communication between the UVM TestBench and the reference model.

The rest of this paper is organized as follows. Section II surveys the related work. Section III shows the proposed TestBench Architecture. Section IV discusses and analyzes the experimental results. Finally, we conclude in section V.

II. RELATED WORK

A. Moursi et al. [8] discussed the effect of using high-level languages over SystemVerilog for building reference models. They used implementation complexity, memory footprint, runtime, and UVM interface parameters to compare the high-level language reference model's performance.

Murti et al. [9] presented SystemC as an example of Executable Specification Languages (ESL) to show the concrete illustration that can be applied in real-world scenarios, which will enhance the understanding of the theoretical concepts and architectural exploration. Moreover, Abdulhameed et al. [10] talked about the benefits of using SystemC, such as, the Analog and Mixed-Signal good accuracy in modeling and simulating complex systems.

Baskar et al. [11] used the basic building blocks of TLM connection examples provided with the UVMC library to build and connect SystemC with SystemVerilog using UVMC. The paper shows a co-simulation of two send-receive packets examples. A packet is sent from the SystemVerilog to the SystemC side, which works as a monitor that unpacks the received data and compares the bits using a comparator to check that the received data are the same as the sent data.

F. Poppen et al. [12], [13] talked about how UVM can work with SystemC and their own Integrated Functional verification Script environment (IFS) in the verification process without relying on a specific software. In [12], they used Mentor and Cadence simulators to show that their approach worked with UVMC without depending on a certain proprietary technology. In [13], they extended their work to cover the VCS-MX simulator from Synopsys to use UVM inside a SystemC TestBench for the verification of a Bus Arbiter with codes provided. However, to continue their productivity, they needed to modify the verification strategy carefully. E. Erichsen [14] Used TLM 2.0 to interface a SystemC reference model of a complex memory controller DUT with a UVM verification environment to verify its functionality. SystemC and UVM have their own TLM structure that is used as a communication abstraction, where transactions are sent as communication objects between components. However, SystemC TLM and UVM TLM are not compatible with each other without a communication protocol.

Baskar et al. [15] proposed an event sampling solution to export functional coverage from the SystemC model to SystemVerilog each time the event is triggered. The coverage data are sent as packets from the SystemC side to the SystemVerilog side via TLM. On the SystemVerilog side, the data can be de-serialized for functional coverage analysis. Table I summarizes the related work.

III. THE PROPOSED TESTBENCH ARCHITECTURE

Figure 2 shows a high-level block diagram of the proposed TestBench architecture. The DUT here is a RISC-V-based processor execution unit connected to the verification environment via a virtual interface. The SystemC EXU unit reference model has similar functionality to the DUT, but with a high level of abstraction. It is used to calculate the expected output and send it back to the UVM TestBench for checking. The primary objective is to convert our native UVM TestBench which uses a built-in SystemVerilog reference model to a UVM TestBench that can be interfaced with a SystemC reference model using TLM. On the UVM TestBench side, the workflow is as follows:

- The top TestBench module is renamed to **sv_main**, serving as the entry point of the SystemVerilog side.
- During the **connect phase** of the **uvm_env**, UVMC TLM ports are connected and registered using lookup strings.
- These ports handle generic payload transactions via a TLM-blocking transport interface defined in the scoreboard's **run phase**.
- The **uvm_scoreboard** has three TLM connections:
  - A native **uvm_analysis_imp** import to receive the data from the monitor using the **write()** method.
  - A **uvm_tlm_b_initiator_socket,** which is a socket that sends generic payload transactions to the SystemC **simple_target_socket** via the **b_transport** method.
  - Another **uvm_analysis_imp** import to receive the expected output from the reference model.

979-8-3315-3968-9/24 $31.00 © 2024 IEEE   213

TABLE I: Related Work Summary

Paper	Key Contributions	Limitations	Our Proposed Improvement
[8]	Compared high-level languages with SystemVerilog for building reference models	Generic payload data array size length limitation in SystemC.	A Python script that takes the **uvm_sequence_item** class file as an input, and generates the required code for sending and receiving data.
[11]	Presented methods for data transfer between SystemC and SystemVerilog, including TLM 1.0 and TLM 2.0 approaches.	Generic payload data array size length limitation in SystemC. Solved by designing a custom-made conversion algorithm for each connection using TLM-2.0 libraries	A Python script that does not require any custom-made conversions as it uses the UVM native generic payload class.
[12], [13]	Investigated UVM integration with SystemC and their IFS environment, showcasing compatibility with various simulators without reliance on proprietary technologies.	Requires verification environment changes.	Does not require such changes.
[14]	Interfacing SystemC reference models with UVM verification environments using UVMC.	TLM compatibility issues and integration with SystemVerilog.	Easier way to connect with SystemC reference models.

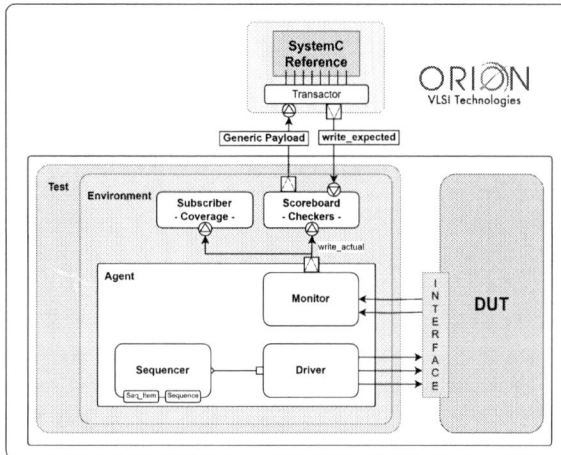

Fig. 2: TestBench architecture and connections

On the SystemC reference model side, the workflow is as follows:

- The entry point for the SystemC side is the **sc_main** function where the UVMC TLM ports with lookup strings are defined.
- A **simple_target_socket** receives transactions from the scoreboard.
- Upon receiving a transaction, an event signals that the reference model's inputs are ready.
- After calculating the expected output, another event triggers a thread that uses a **tlm_analysis_port** to send the output back to the scoreboard.

During the **connect phase**, the registration of the TLM lookup strings takes place when calling the UVMC TLM instances. This lookup string works as a reference to the instance and is registered globally across both SystemC reference model and UVM TestBench.

Before UVM's **end_of_elaboration** phase, the UVMC will establish the actual cross-language connection and match the

similar lookup strings of the registered ports. The matched ports will interact with each other to send and receive transactions. If there is a port with no match, an error is reported, and the simulation is dropped. Because we are working with cross-language verification, synchronization inside the UVM TestBench is very important, as we need to make sure what is being compared is what was initially driven to the DUT. To achieve this synchronization, two points were taken into consideration:

- The initial synchronization occurs during the driving and sampling of transactions. The UVM driver employs a clocking block that waits for the clock's positive edge to drive the signals. As the driver sends signals to the DUT,

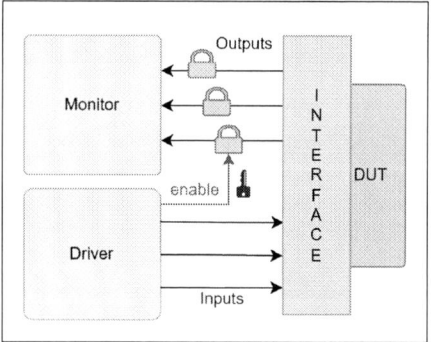

Fig. 3: Driver-Monitor Synchronization Workflow

it simultaneously changes the value of **enable** signal to high at the monitor to signal a driving operation as shown in Figure 3. The monitor will only sample if the **enable** signal is set to high and after a certain number of clock cycles. This prevents any sampling without driving, leading to improved reporting.

- The second synchronization takes place in the **uvm_scoreboard**. Figure 4 shows two queues that are defined in the scoreboard, and are used to make sure we are comparing the right data. The first queue stores the received actual response (DUT response) gathered

979-8-3315-3968-9/24 $31.00 © 2024 IEEE   214

by the monitor. On the other hand, the second queue stores the received expected response obtained from the SystemC reference model. When the **write ( )** method of the native TLM receives the DUT's response, it pushes it on the first queue.

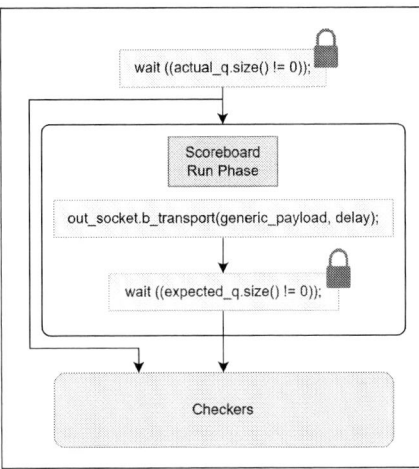

Fig. 4: Queue Synchronization Workflow

This queue triggers the **run phase** as it waits for the queue's size to be greater than zero. The data inside this queue will be pulled out and sent to the reference model via the b_transport method after some tweaks. At the same clock cycle of sending the data to the reference model, the reference model will react and send back an expected output via a customized **write_expected**() method that pushes the data on the second queue. When the second queue's size is greater than zero, this means that the actual response is inside the first queue and the expected response is inside the second queue. Now, the actual and expected responses can be sent for checking. SystemC elaborates faster than UVM. This means that the SystemC reference model will start sending data to the UVM TestBench before it is ready due to synchronization issues. This was solved by the queue **wait**() synchronization method shown in Figure 4. The generic payload includes a data attribute represented as a byte array. This means that each index of this array takes only eight bits. The DUT in the verification environment has signals with varying data widths (32 and 64 bits). To address this UVMC limitation, a Python script was written to partition the signals that are wider than eight bits into equivalent bytes within the data array.

IV. EXPERIMENTAL RESULTS

To evaluate the efficiency of using SystemC reference models instead of SystemVerilog reference models in the design verification environment, we compared our SystemC reference model with its SystemVerilog counterpart using the execution unit (EXU) of the open-source RISC-V-based VeeR EL2 processor core as our DUT.

We used Synopsys VCS for simulation. The simulation results revealed that the verification environment with the SystemC reference model is up to 15x faster than the verification environment with the SystemVerilog reference model.

V. CONCLUSION

VLSI chip design verification is a process of making sure that what has been designed is what has been specified. This process is complex and resources-consuming. In this paper, we tried to accelerate the simulation-based design verification process through using fast SystemC-based reference models. We tested the proposed methodology on the execution unit (EXU) of the open-source RISC-V-based VeeR EL2 processor core. We replaced the EXU SystemVerilog reference model with a new SystemC reference model. We integrated the SystemC reference model with the UVM TestBench through UVMC. The simulation results revealed that the UVM TestBench with a SystemC reference model is up to 15x faster than the corresponding TestBench with a SystemVerilog reference model.

REFERENCES

[1] G. Fey and R. Drechsler, "Improving simulation-based verification by means of formal methods," in *ASP-DAC 2004: Asia and South Pacific Design Automation Conference 2004 (IEEE Cat. No. 04EX753)*, pp. 640–643, IEEE, 2004.

[2] D. C. Black, "The definitive guide to systemc: The systemc language," 2015. Available at: https://newport.eecs.uci.edu/~doemer/eee_uci_edu/17f/16905/DAC15_SystemC_Training.pdf.

[3] Wikipedia, "Systemc - wikipedia," 2023. Available at: https://en.wikipedia.org/wiki/SystemC.

[4] A. standards and TLM2.0, "www.accellera.org/downloads/standards/systemc (visited: 17.02.2024).."

[5] Verification Academy, "Uvm-connect and tlm-2.0 primer," 2023.

[6] "Uvm 1.2 user guide and reference manual," 2015. [online] Available in https://www.accellera.org/downloads/standards/uvm.

[7] Chips Alliance, "Cores-VeeR-EL2: A collection of risc-v cores and other ips targeting different levels of performance, size, and features.." https://github.com/chipsalliance/Cores-VeeR-EL2, Accessed: 2024-03-04.

[8] A. Moursi, R. Samhoud, Y. Kamal, M. Magdy, S. El-Ashry, and A. Shalaby, "Different reference models for uvm environment to speed up the verification time," in *2018 19th International Workshop on Microprocessor and SOC Test and Verification (MTV)*, pp. 67–72, IEEE, 2018.

[9] K. Murti and K. Murti, "Specification languages: Systemc," *Design Principles for Embedded Systems*, pp. 85–117, 2022.

[10] A. Abdulhameed, B. AlKindy, and B. Al-Mahdawi, "Smart system modeling and simulation design of hemodialysis machine by sysml with systemc-ams," 2022.

[11] V. Baskar, "Why not "connect" using uvm connect: Mixed language communication got easier with uvmc," *DVCon United States*, 2022.

[12] F. Poppen, M. Trunzer, and J.-H. Oetjens, "Connecting a company's verification methodology to standard concepts of uvm," *DVCon Europe*, 2014.

[13] F. Poppen, M. Trunzer, and J.-H. Oetjens, "Using synopsys vcs to connect a company's systemc verifiacation methodology to standard concepts of uvm," *SNUG Germany*, 2015.

[14] E. S. Erichsen, "Interfacing systemc inside uvm using tlm 2.0," Master's thesis, NTNU, 2020.

[15] V. Baskar, "Do not forget to 'cover' your systemc code with uvmc," *DVCon United States*, 2023.

Time-to-Digital Converter based Self-Timed Ring Oscillator: an FPGA Implementation

Assia El-Hadbi[1], Oussama Elissati[1], and Laurent Fesquet[2]

[1]Institut National des Postes et Télécommunications, STRS Lab., Rabat, Morocco.
E-mail: {elhadbi, elissati}@inpt.ac.ma
[2]Université Grenoble Alpes, CNRS, Grenoble INP, TIMA Lab., Grenoble, France.
E-mail: laurent.fesquet@univ-grenoble-alpes.fr

Abstract—This paper proposes an FPGA implementation of Self-Rimed Ring Oscillator (STRO) based Time-to-Digital Converter (TDC). Thanks to the STRO features, this TDC is able to achieve a time resolution as fine as needed by simply increasing its number of stages. Even if the achievable time resolution is close to the tenths of picoseconds, the approach does not require a high frequency oscillator. This resolution is only limited by the STRO-stage noise in the FPGA. As a hardware proof-of-concept, this TDC has been implemented in a Cyclone IV FPGA to further confirm and evaluate the advantages of our proposed TDC architecture. Most of the measurements are perfectly in accordance with our theoretical claims.

Index Terms—Time-to-digital converter, self-timed ring oscillator, time resolution, time measurement, FPGA.

I. INTRODUCTION

Time-to-digital converters (TDCs) are widely used for precise time measurements. TDC quantifies the measured time interval T and provides its digital value as a function of the time resolution, which often represents the time step of the digital clock or a gate delay. In the simplest TDC architectures, the time resolution is bounded by the gate delay. In order to overcome this technological limitation, subgate delay resolution solutions have been proposed. Nevertheless, while many of the TDC architectures proposed in the literature can achieve high measurement precision, they often require repetitive measurements. Therefore, on-the-fly measurements for fast non-periodic signals (a few tens of picoseconds) are extremely challenging in TDC designs. TDCs are implemented as Application-Apecific Integrated Circuits (ASIC) or Field Programmable Gate Arrays (FPGA). Most of TDCs are implemented as ASIC in order to benefit from the high advances of CMOS technologies and to reach fine time resolutions [1]–[3]. Moreover, the total control of the place-and-route process helps to limit the measured differential nonlinearity (DNL) and the integral nonlinearity (INL). However, ASICs have a long development cycle and are expensive, especially for small productions and dedicated products. Conversely, the FPGA-based TDCs benefit from lower cost due to their short development cycle and offer good flexibility for further improvements, reconfigurations, and verifications. Additionally, many other digital blocks can be included on the same chip to perform more complex applications. Most FPGA-based TDCs are based on the dedicated carry chains, which are powerful resources for implementing delay lines [4]–[8]. The tapped delay line (TDL) is the most proposed FPGA-TDC architecture in the literature. This structure has been developed to achieve a fine resolution using interpolation techniques such as the Vernier delay line (VDL). TDLs are good candidates for multichannel TDCs and are frequently used because they offer a good trade-off between ASIC and processor-based systems [6]–[8]. In this paper, we present an FPGA implementation of a new and compact TDC architecture based on a Self-Timed Ring Oscillator (STRO), as previously described in [9]–[11]. This architecture has been adapted for implementation in an Intel Cyclone IV FPGA. As a hardware proof-of-concept, we implemented 23-stage and 141-stage STRO-based TDCs, achieving time resolutions of 108 ps and 20 ps, respectively. Measurement results are promising, even though the place-and-route operations are not fully optimized. The paper is organized as follows. Section II summarizes the principle of the STRO. The proposed STRO-based TDC architecture and its FPGA implementation are presented in Section III. Measurement results are presented in Section IV. Finally, Section V concludes the paper.

II. SELF-TIMED RINGS AND OPERATING PRINCIPLES

Self-Timed Ring Oscillator (STRO) architecture is a looped line of stages, composed by a C-element (Muller gate) and an inverter. Fig. 1 presents the global L-stage STRO architecture. Due to its unique timing features, it is considered as a promising solution to exploit the phase difference between events propagating in the same structure to generate a sub-gate delay time resolution. The STRO behavior and operation have already been presented in our previous works [9]–[12]. The oscillator is modeled with a token game. The stage contains a token if its output is different from the output of the next stage and contains a bubble in the other case. The token propagates to the next stage, if the latter contains a bubble. Oppositely, the bubble replaces this token. The events in the STRO correspond to the tokens or the bubbles according to the initialization of the ring [12].

The oscillation frequency increases with the number of events, and then starts dropping when the number of free stages is lower than the number of events with regards to the

979-8-3315-3968-9/24 $31.00 © 2024 IEEE

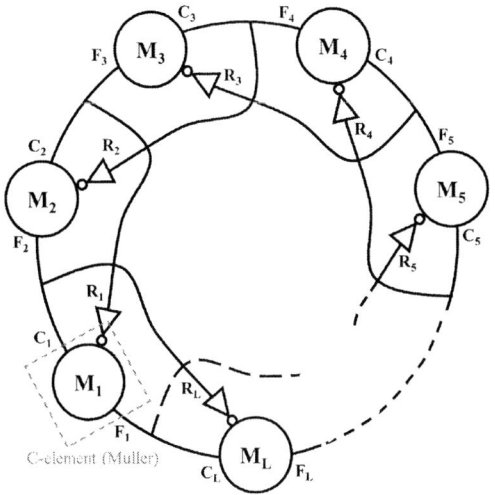

Figure 1. Global architecture of a L-stage STRO.

propagation delays of the stages. The evenly-spaced oscillation mode in the STRO provides a uniformly distributed events with a sub-gate time resolution. A special case occurs when the number of stages and the number of events are co-prime. In this case, the STRO outputs produces L different equidistant phases distributed over the half period of the oscillation T_{STR} with a time spacing $\Delta\varphi = \frac{T_{\mathrm{STR}}}{2L}$ lower than the stage delays, which makes it sub-gate time resolution [12]. Hence, increasing L will accordingly improve the ring resolution since T_{STR} does not depend directly on L. Obviously, the time resolution enhancement will be limited by the level of jitter which affects the phase order. These features have been exploited to propose a new TDC architecture based on a STRO. This STRO-based TDC benefits from the uniform distribution of phases.

III. THE PROPOSED TDC ARCHITECTURE AND IMPLEMENTATION

There are many different types of FPGAs [2]. They differ in the number of the configurable logic blocks (CLBs), also called logic array block (LAB), which are the main specification of the FPGA. In our case, an Intel Cyclone IV is used, which is considered as a low cost hardware solution. In this board, each LAB contains just 16 logical cells containing one look-up table (LUT) and one flip-flop. As Intel devices do not offer a routing access, it is required to place carefully components and blocks to avoid random delays and enhance routing regularity. In addition, different implementations can be observed after compilation, which are susceptible to distort slightly the study.

A. STRO-based TDC architecture

According to the temporal parameters of the stages implemented in the FPGA, a fixed frequency is obtained. Therefore, the STRO length is determined in accordance with the targeted time resolution. Fig. 2 illustrates our proposed STRO-based TDC architecture in which an STRO of L stages is used to

generate uniform distributed phases. The events in the ring are initialized using *Init_STR* signal generated by the finite state machine (FSM).

To initialize the STRO, two types of stages are used. Stages with a *SET* signal allowing to initialize to 1 the stage output and the others with a *RESET* signal allows to set to 0 the stage output. The signal *Init_STR* allows to apply this initialization at the beginning, according to the number of events to be inserted, by setting or resetting each stage before leaving them in a free running mode, which lets oscillating the STRO. The FSM is also used to generate the time interval T by providing the *Start* and *Stop*, which are generated using the internal clock of the FPGA. Two states are used to generate these signals spaced with a time equivalent to a multiple of the clock period. Each measurement requires to reset the FSM in order to ensure a one-shot measurement.

One arbitrary STRO phase is connected to a n-bit counter while the $L - 1$ others are connected to 2-bit counters. The TDC output ($\mathrm{TDC_{out}}$) is given by a data processing algorithm based on the collected data from the n-bit counter, Hamming, and Parity blocks blocks (cf. Fig. 2). The hamming and parity codes are based on the information collected from the two least significant bits of the counters. They are given by the Hamming and Parity blocks.

B. STRO implementation

Conventionally, the output C of a C-element gate, with A and B as inputs, is modeled by: $C = A \cdot B + A \cdot C^{-1} + B \cdot C^{-1}$, where C^{-1} represents the previous state of the output C. A simple implementation using combinational logic is possible. This requires implementing an STRO stage in a single LUT.

As already mentioned, the routing is determined by the FPGA routing matrix, which is not programmable by the users. To minimize delays that can affect the STRO frequency, the programmed LABs have to be chosen the closest from each other. Since a LAB can only accept two clock inputs, each STRO stage is implemented with its first counter bit to ensure nearly identical delays. In addition, a looped structure of STRO is chosen. The connection between the STRO and TDC is also critical for timing. It is important to make the routing delays between the STRO stages and the corresponding counters the closest as possible to limit the measurement errors. The TDC architecture has been adapted to meet these specifications compared to to ASIC implementation [11].

C. Counter implementation

An L-stage STRO, with a number of events co-prime with its number of stages, provides L signals evenly distributed over its half oscillation period $T_{\mathrm{STR}}/2$. Quantifying each STRO output with a step of $T_{\mathrm{STR}}/2$ gives a resultant value M which verifies: $M \cdot \frac{T_{\mathrm{STR}}}{2} \leq T < (M+1) \cdot \frac{T_{\mathrm{STR}}}{2}$

The quantification error of this coarse measurement is less than $T_{\mathrm{STR}}/2$. Since the coarse measurement result of each STRO phase will be either M or $M+1$, the quantification can be improved by aggregating the results in order to achieve a measurement of T with $\Delta\varphi$ precision. Therefore, connecting

979-8-3315-3968-9/24 $31.00 © 2024 IEEE   217

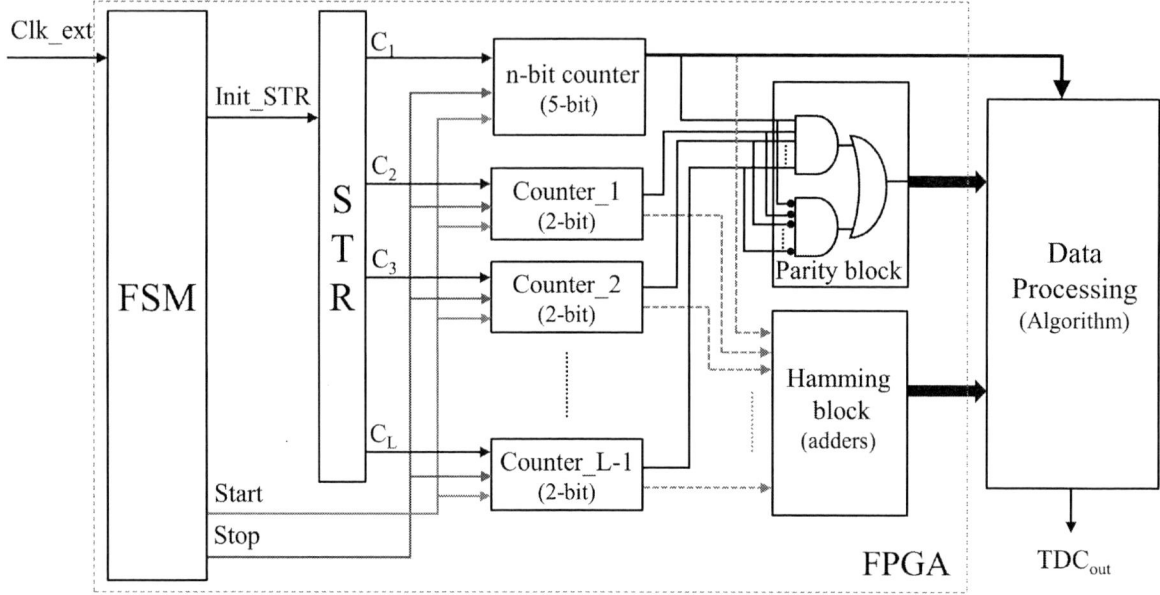

Figure 2. Composition of the proposed TDC architecture using STRO of a L stages.

all STRO outputs to n-bit counters shows that k counters count the value M while $L-k$ counters count the value $M+1$. Then, the measured time value T_m is $T_m = M \cdot \frac{T_{\mathrm{STR}}}{2} + k \cdot \Delta\varphi = (M \cdot L + k) \cdot \Delta\varphi$, where M corresponds to the coarse measurement of the number of $T_{\mathrm{STR}}/2$ steps, while k corresponds to the finer measurement (the number of $\Delta\varphi$).

A simple study of the counter bits reveals that M and $M+1$ is given by the two least significant bits of the counters. Consequently, the coarse conversion is carried out by a large n-bit counter which is connected to a single STRO output while the other $L-1$ outputs are connected to a 2-bit counters in order to determine both M and k. Fig. 2 shows the proposed architecture based on a L-stage STRO oscillator with a 5-bit counter. In order to consider the falling and the rising edges of the signals, the counters have been duplicated so that one is operating with the falling edge and the other one for the rising edge. This may increase the area of the proposed TDC but, at the same time, ensures less problems related to the counter blocks [9].

D. Parity and Hamming blocks

Let N_v be the number of transitions counted by the n-bit counter during the time interval T. To determine if the value N_v corresponds to M or $M+1$, we evaluate the parity of M using the 2-bit counter outputs. In fact, the MSBs of M and $M+1$ differ only when the LSB of M is '1' (M is odd). Therefore, if all the MSBs of the 2-bit counters differ, then M is odd, and if they are equal, then M is even. This information is computed using the parity block of Fig. 2. The parity block output is set to '1' when its inputs are identical and to '0' otherwise. As a result, M is even when the result of the parity block is '1', and odd otherwise. If the LSB of M

is '0' (M is even), the number k of counters having the value $M+1$ is the count of ones among the LSBs of the counters. The hamming block uses the LSBs to compute the hamming weight H which is equal to k when M is even and $L-k$ when M is odd [9].

The Hamming and Parity blocks can be hardware or soft logic blocks. Since hardware adders are much faster than soft adders, hardware adders have been used for the hamming block. Likely, the Parity block is a simple circuit because it just requires a few combinatorial gates as shown in Fig. 2.

IV. MEASUREMENT RESULTS

A 23-stage and 141-stage STRO-based TDCs have been implemented in the EP4CE115F29C7 Intel Cyclone IV. The 50 MHz on-board crystal oscillator has been used as the clock of the FSM. For an accurate frequency measurement, Low Voltage Differential Signaling (LVDS) outputs have been used with differential oscilloscope probes. This technique aims to reduce the impact of slow input/output circuitry and parasitic effects of the output. The TDC measurement is obtained without any calibration, except the STRO frequency measurement. The results are directly displayed on 7-segment displays.

A. 23-stage STRO-based TDC

The 23-stage STRO-based TDC has been initialized with 14 events. A theoretical time resolution of 108.7 ps is obtained with a frequency of 200.0 MHz ($T_{\mathrm{STR}} = 5.0\,\mathrm{ns}$). As 14 and 23 are co-prime, 23 different evenly-spaced phases are obtained. Repetitive measurements for time intervals of 9.5 ns and 13.0 ns have been done in order to characterize the TDC output. 300 measurements have been collected for each time interval.

979-8-3315-3968-9/24 $31.00 © 2024 IEEE   218

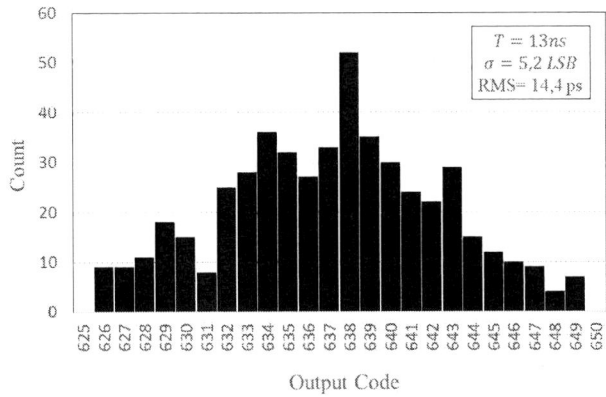

Figure 3. Measurement of 13.0 ns time interval using 141-STRO based TDC.

The measured values are more concentrated around 88 for 9.5 ns (with a standard deviation of 1.4 LSB) and around 119 for 13.0 ns (with a standard deviation of 1.6 LSB). According to this, we can express differently the results by identifying the percentage of measurements which resulted in an error lower than $\Delta\varphi$, $2\Delta\varphi$ or greater than this. A classification according to the ratio error over the time resolution $\Delta\varphi$ reveals that more than 75% (resp. 70%) of the measurements are correct ($\|Err\| \leq 2\Delta\varphi$) and the mean value is around 9.59 ns (resp. 12.94 ns) in the case of $T = 9.5$ ns (resp. $T = 13.0$ ns).

The errors observed in the TDC output can be attributed to a fundamental source of error inherent to any converter, known as quantification error. In fact, for an ideal TDC, the input and output are related by : $T = T_m + \varepsilon \ (-\Delta\varphi/2 \leq \varepsilon \leq \Delta\varphi/2)$. Otherwise, the time interval T produced by the FSM can introduce some timing noise due to the routing effects.

B. 141-stage STRO-based TDC

As explained in the Section II, the time resolution can be reduced as fine as needed by increasing the number of stages. Thus, a 141-stage STRO-based TDC initialized with 88 events presents a theoretical time resolution of 20.3 ps with a frequency of 174.06 MHz ($T_{STR} = 5.745$ ns). As a result, the total number of LUTs has increased by 31% while the number of registers has quadrupled. The STRO exhibits a measured jitter standard deviation of 20.0 ps. The level of time resolution indicates that the TDC is operating in a noisy environment, which can disrupt the evenly-spaced mode. Therefore, the proposed measurement algorithms may no longer be suitable.

Time intervals of 9.5 ns and 13.0 ns are measured 500 times. As expected, due to the presence of jitter with the obtained time resolution, only 50% (resp. 40%) of samples presents an error lower than $2\Delta\varphi$ for the measurement of 9.5 ns (resp. 13.0 ns). Therefore, the corresponding timing uncertainty of the time interval 9.5 ns (resp. 13.0 ns) measurements is 4.0 [LSB] (resp. 5.2 [LSB]). Fig. 3 plotted the histogram of the digital TDC outputs of the time interval 9.5 ns. However, the mean value of the 500 samples shows a high precision of measurement: it is equal to 9.51 ns for the measurement of 9.5 ns and equal to 12.98 ns for the

measurement of 13.0 ns. Thus, using a calibration technique, controlling counter outputs, with high number of samples may improve measurements of this TDC.

V. CONCLUSIONS

A STRO-based TDC with sub-gate delay time-resolution has been proposed and implemented in a low-cost FPGA. A hardware proof-of-concept implementation has been presented using a 23-stage and 141-stage STRO-based TDCs. The implementation is straightforward and the results are promising for future research to achieve optimal nonlinearity without needing calibration. A 20.0 ps time resolution has been achieved using 141-stage. Moreover, thanks to the STRO unique features, this TDC architecture can achieve as fine as desired time resolution by increasing the number of stages. This TDC is only limited by the STRO stage noise in the FPGA. Furthermore, higher performances can be enhanced with recent and faster FPGA boards, making it very suitable for Time-of-Flight measurement applications.

REFERENCES

[1] M. Z. Straayer and M. H. Perrott, "A multi-path gated ring oscillator TDC with first-order noise shaping," *IEEE Journal of Solid-State Circuits*, vol. 44, no. 4, pp. 1089–1098, April 2009.

[2] Z. Cheng, X. Zheng, M. J. Deen, and H. Peng, "Recent developments and design challenges of high-performance ring oscillator CMOS time-to-digital converters," *IEEE Transactions on Electron Devices*, vol. 63, no. 1, pp. 235–251, Jan 2016.

[3] A. El-Hadbi, O. Elissati, and L. Fesquet, "Time-to-digital converters: A literature review and new perspectives," in *2019 5th International Conference on Event-Based Control, Communication, and Signal Processing (EBCCSP)*, May 2019, pp. 1–8.

[4] J. Y. Won and J. S. Lee, "Time-to-digital converter using a tuned-delay line evaluated in 28-, 40-, and 45-nm FPGAs," *IEEE Transactions on Instrumentation and Measurement*, vol. 65, no. 7, pp. 1678–1689, July 2016.

[5] P. Kwiatkowski and R. Szplet, "Efficient implementation of multiple time coding lines-based TDC in an FPGA device," *IEEE Transactions on Instrumentation and Measurement*, vol. 69, no. 10, pp. 7353–7364, 2020.

[6] J. Kalisz, R. Szplet, J. Pasierbinski, and A. Poniecki, "Field-programmable-gate-array-based time-to-digital converter with 200-ps resolution," *IEEE Transactions on Instrumentation and Measurement*, vol. 46, no. 1, pp. 51–55, Feb 1997.

[7] Y. Wang, Q. Cao, and C. Liu, "A multi-chain merged tapped delay line for high precision time to digital converters in FPGAs," *IEEE Transactions on Circuits and Systems II: Express Briefs*, vol. 65, no. 1, pp. 96–100, Jan 2018.

[8] P. Chen, Y. Y. Hsiao, Y. S. Chung, W. X. Tsai, and J. M. Lin, "A 2.5-ps bin size and 6.7-ps resolution FPGA time-to-digital converter based on delay wrapping and averaging," *IEEE Transactions on Very Large Scale Integration (VLSI) Systems*, vol. 25, no. 1, pp. 114–124, Jan 2017.

[9] A. El-Hadbi, A. Cherkaoui, O. Elissati, J. Simatic, and L. Fesquet, "On-the-fly and sub-gate-delay resolution TDC based on self-timed ring: A proof of concept," in *2017 15th IEEE International New Circuits and Systems Conference (NEWCAS)*, June 2017, pp. 305–308.

[10] A. El-Hadbi, A. Cherkaoui, O. Elissati, J. Simatic, and L. Fesquet, "An accurate time-to-digital converter based on a self-timed ring oscillator for on-the-fly time measurement," *Analog Integrated Circuits and Signal Processing*, vol. 97, 06 2018.

[11] A. El-Hadbi, O. Elissati, and L. Fesquet, "Self-timed ring oscillator based time-to-digital converter: A 0.35 μm CMOS proof-of-concept prototype," in *2021 IEEE International Instrumentation and Measurement Technology Conference (I2MTC)*, 2021, pp. 1–6.

[12] O. Elissati, A. Cherkaoui, A. El-Hadbi, S. Rieubon, and L. Fesquet, "Multi-phase low-noise digital ring oscillators with sub-gate-delay resolution," *{AEU} - International Journal of Electronics and Communications*, vol. 84, pp. 74 – 83, 2018.

979-8-3315-3968-9/24 $31.00 © 2024 IEEE

Special Session

Reliability Assessment of Large DNN Models: Trading Off Performance and Accuracy

Junchao Chen[1], Giuseppe Esposito[2], Fernando Fernandes dos Santos[3], Juan-David Guerrero-Balaguera[2],
Angeliki Kritikakou[3], Milos Krstic[1,4], Robert Limas Sierra[2], Josie E. Rodriguez Condia[2],
Matteo Sonza Reorda[2], Marcello Traiola[3], and Alessandro Veronesi[1]

[1]IHP – Leibniz Institute for High-Performance Microelectronics, Germany
[2]Politecnico di Torino, Department of Control and Computer Engineering (DAUIN), Turin, Italy
[3]Univ Rennes, CNRS, Inria, IRISA - UMR 6074, F-35000 Rennes, France
[4]University of Potsdam, Germany

Abstract—The adoption of Deep Neural Networks (DNNs) in several domains allows for increased effectiveness in applications that deal with massive data-intensive and complex data inputs. When employed in safety-critical scenarios, such as automotive, aerospace, healthcare, and autonomous robotics, assessing the DNNs' reliability and functional safety is crucial to ensure their correct in-field operation, even in the presence of hardware faults. However, the system complexity and the massive amounts of data to be processed by DNNs prevent the effective adoption of traditional strategies for reliability characterization and for identifying the most fault-sensitive structures. Accurate fault assessment strategies usually require unacceptable computational power and large evaluation times. On the other hand, faster strategies commonly lack accuracy in correctly representing system faults. Consequently, it is necessary to develop effective strategies that trade-off between performance and accuracy.

This work analyses three reliability assessment strategies for deep neural networks and their underlying hardware, highlighting the main solutions and challenges in terms of evaluation performance and fault characterization accuracy. We overview different solutions to evaluate the hardware accelerators implementing DNNs at three abstraction levels: *i)* by physically injecting faults on a GPU running DNNs, *ii)* by performing microarchitectural characterization of GPUs to develop application-accurate error models, and *iii)* by using structure-aware cross-layer error modeling on DNN hardware accelerators. Our experimental results indicate that accurate error representation requires structural features from the targeted hardware.

I. INTRODUCTION

Deep Neural Networks (DNNs) are widely used in various fields to enhance the performance of complex applications, such as autonomous driving, natural language processing, and industrial robots. These applications typically involve processing complex operations and large volumes of data (i.e.,

This work was partially supported by the Italian National Resilience and Recovery Plan (PNRR) through the National Center for HPC, Big Data and Quantum Computing, and by the Federal Ministry of Education and Research of Germany under the programme of "Souverän. Digital. Vernetzt." Joint project 6G-RIC, project identification number: 16KISK026.
This activity received funding from the European Union's 2020 research and innovation programme under grant agreement No 101008126 (RADNEXT project), and by the ANR FASY (ANR-21-CE25-0008-01). ChipIR provided and supported neutron beam time experiments (DOI https://doi.org/10.5286/ISIS.E.RB2300036).

$1x10^6$ operations per second) [1], [2]. When employed in safety-critical applications, such as aerospace, automotive, and healthcare, DNN-based solutions must match strict reliability requirements mandated by industrial standards (i.e., ISO 26262 and ISO/IEC 22989 [3], [4]). In safety-critical applications, reliability characterization and evaluation are mandatory to ensure that any fault/defect arising in the system can be controlled and properly handled. These reliability assessments may guide designers during early design stages and contribute to achieving the target fault tolerance.

DNN accelerators, including *Graphics Processing Units* (GPUs), are built with advanced technology node processes that enable the design of smaller, faster, and energy-efficient devices [5]. Unfortunately, miniaturization (7nm or below) seriously increases the reliability concerns in a system connected to temporal (aging and wear-out) and environmental variations (over-stress, environmental harshness due to temperature, voltage, or radiation) [6]–[8]. These variations increase the fault rate, impacting the hardware and propagating the faults to the application as *Silent Data Corruptions* (SDCs), jeopardizing the in-field operation of the complete system [9]–[11].

To assess the impact of faults in a DNN-based system, several reliability evaluation strategies have been developed, such as formal techniques, simulation-based architectural evaluations, and physical/beam experiments. Formal evaluation and software simulation strategies focus their analyses on the DNN model with feasible evaluation times but neglect the underlying hardware. In contrast, other strategies inject faults at the hardware level and can provide fine-grain accuracy, i.e., at the gate or micro-architectural level. However, hardware simulations for large DNNs often lead to years/decades of expected simulation time. Hence, DNN's complexity, operational density, and underlying hardware demand the development of effective strategies to provide efficient evaluation times under feasible levels of fault characterization accuracy [12]. Inspired by the performance bottlenecks in some evaluation strategies and the challenges in correctly characterizing fault impacts on DNN-based systems, *we analyze and discuss the main advantages and challenges of three strategies to*

979-8-3315-3968-9/24 $31.00 © 2024 IEEE

Special Session

assess the reliability of DNNs. These strategies consider the structural hardware characteristics while keeping a good trade-off between evaluation time and fault simulation details.

The first strategy (**Section III**) combines physical characterization using a beam of neutrons and software-based fault simulation on real devices to evaluate the impact of transient faults on large DNN models for image classification. Additionally, we show that an evaluation that considers the hardware characteristics is mandatory to deploy efficient software fault tolerance for large DNNs. A second strategy (**Section IV**) combines the structural features of a GPU and its units (i.e., schedulers and *'TCUs'*) to characterize fault effects at the system level. This strategy determines representative hardware-aware fine-grain corruptions to identify accurate error models for evaluating large DNNs at the software level. Finally, the third strategy (**Section V**) develops a cycle-accurate micro-architecture model of the NVDLA architecture, incorporating fine-grained structural details to ensure accurate fault characterization for large DNN workloads, and shows that variations in hardware parameters significantly impact both the types and occurrence of observed errors, depending on the specific DNN model.

We explore the benefits of combining two or more layers of fault injection abstraction to effectively assess and harden large and complex DNNs within acceptable evaluation times. The strategies we discuss are intended to support quick system improvements and efficient hardening.

II. CHALLENGES IN DNN RELIABILITY ASSESSMENT

When GPUs are used as DNN accelerators in autonomous systems, they are susceptible to multiple sources of failures, such as ionizing radiation, permanent faults, performance degrading faults, timing errors, and intermittent faults [13], [14]. These sources of faults may not make the GPU completely inoperative, but they can significantly impact the execution of DNN models, potentially changing the final inference result. When not **masked**, faults can become errors that can propagate to the software level and lead to failures such as the **Detected Unrecoverable Errors (DUEs)**, which hang the program or crash the entire system, and **Silent Data Corruptions (SDCs)**, that allow the application to complete its execution but with an incorrect output. *Without a proper reliability assessment guiding the introduction of proper fault tolerance solutions, the resulting failure remains undetected.* When considering DNN models, SDCs can be further categorized into **Tolerable SDCs**, which modify the model output but not the inference outcome (i.e., the classification, detection, or segmentation), or **Critical SDCs**, which cause the model to change the inference probabilities, resulting in mis-classification, mis-detection, or mis-segmentation.

In order to characterize the hardware faults that reach the highest levels of a system and become failures, different approaches can be used, each one having its own advantages and disadvantages. We can separate the reliability assessment methods into three categories: *physical fault injection*, *fault simulation techniques*, and *software-based fault simulation.*

Physical fault injection require special facilities to expose the system to a beam of particles and induce faults in hardware. This strategy is an efficient and realistic evaluation approach for safety/mission-critical systems that operate in harsh environments, such as space and avionics applications. As the faults are physically injected into the circuit, physical fault injection effectively characterizes radiation-induced transient faults and permanent faults from accumulated radiation dose [15]–[17]. However, the strategy can hardly provide information regarding specific sub-structures in a system or a DNN model. For instance, radiation experiments do not allow fault propagation analysis since failures are only observed at the output, making it hard to identify the error sources in hardware or software parts in a system, as well as hindering the error modeling [18]. These constraints impede the straightforward identification of the most vulnerable parts of the system.

Hardware fault simulation techniques can be performed at several abstraction levels, from low-level micro-architecture (i.e., Register-Transfer or gate levels) [19] to architecture/functional evaluations [20], [21]. These strategies measure the fault propagation probability on the system, i.e., the *Architectural Vulnerability Factor* [22]. Hardware fault simulations have a fine-grain accuracy and can support the identification of the most fault-vulnerable hardware structures. However, for the hardware fault simulations, the computational power and evaluation times are proportional to the fine-grain abstraction, the system's size, and complexity. Thus, the complexity and time required for a detailed evaluation can require an unacceptable time, e.g., $> 10,000$ days for a small DNN evaluation [14].

Software-based fault simulation consists of code instrumentation or/and modification strategies that add instructions or routines to a targeted application to allow faults injection in the underlying hardware architecture [23], [24]. Faults injected in software fault simulations can be used to estimate the Program Vulnerability Factor (PVF), i.e., the probability for a fault to propagate [25]. Software-based strategies usually use real execution time and can be performed on actual hardware platforms, leading to fast evaluations. The main challenge of software fault simulation strategies is the accuracy of identifying and describing faults from the underlying hardware. The user defines the fault model, which risks not being realistic, leading to inaccurate results. Additionally, the evaluated faults can only be injected into a subset of available and accessible resources at the software level (e.g., registers or variables).

The following sections demonstrate how to effectively and efficiently evaluate large DNN models by combining two or more reliability assessment approaches without compromising accuracy or incurring prohibitively long evaluation times.

III. DNN'S RELIABILITY EVALUATION THROUGH PHYSICAL AND SOFTWARE FAULT INJECTION

In this section, we show how combining two experimental evaluation methods – physical fault injection using a neutron beam and software fault simulation – enables the efficient evaluation and hardening of large DNN models without compromising accuracy. We start by summarizing the experimental

979-8-3315-3968-9/24 $31.00 © 2024 IEEE

methodologies, and then we discuss the general trends observed in the experiments and their importance.

A. Experimental Methodologies

We performed fault injection using a software fault simulator and a neutron beamline. For all the experiments, we evaluated 5 ViT models without any protection, as well as the same models protected by a range restriction strategy. In the employed range restriction, the ViT parameters are checked to see if they fall within a range of accepted values when propagating through the ViT's Identity layers. If they do, they are propagated as usual. If not, they are replaced by 0. We evaluate both versions of the ViT models to demonstrate why it is mandatory to consider both hardware and software analysis when proposing a fault tolerance method for DNNs.

1) Software fault simulation: We used the NVIDIA Bit Fault Injector (NVBitFI) [24] for instruction-level fault simulation. NVBitFI enables the simulation of faults at the Shader Assembly level (SASS), which means at the assembly level of GPU kernels. With NVBitFI, we could select different fault models and sites to evaluate the ViT models.

We injected faults in general-purpose registers, memory load instructions, and arithmetic floating-point operations. We simulated 1,750 faults per ViT model. The failures (SDCs and DUEs) were counted similarly to beam experiments. Through fault simulations, we calculated the Program Vulnerability Factor (PVF) for each ViT model. The PVF represents the probability of an injected fault propagating from the assembly instruction to the application output [25].

We performed fault simulations with single-bit flip, random value, and warp random value fault models. The first two fault models change an instruction's output register by flipping a bit or replacing it with a random value. While injecting single-bit flips at software is a widely used fault model, it has been demonstrated to incorrectly model fault impact for complex applications [26]. Injecting an experimentally tuned fault model in software (i.e., the observed manifestation of the hardware fault in a software visible state) has been proven to be accurate for GPUs [12], [27]. Consequently, faults are also injected at the warp level using NVBitFI. The warp level fault model is based on recently proposed fault models for GPUs [19], [27], where the outputs of the same instruction in all the threads within a warp are corrupted.

2) Beam experiments: Experimental tests were conducted at ChipIr in Rutherford Appleton Laboratory (RAL, UK). The facility provides a neutron beam to replicate atmospheric neutron effects on electronic devices [28]. This allows for the realistic measurement of device failure rates while running a code. Figure 1 shows the installed setup, which consists of GPUs aligned with the neutron beam and connected to the motherboard. Python scripts run on a server outside the beam monitor and launch the ViT models on the devices inside the beam room. The software is designed to recover from device hangs and restart the program if it fails to respond within a specified timeframe. The same ViT model is run on the GPU for several iterations, and any differences between the output

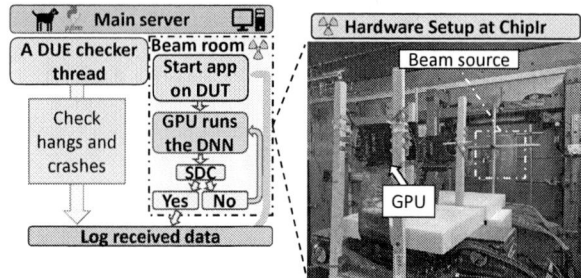

Fig. 1: Software and hardware setups for the neutron beam experiments at ChipIr. The server located outside the beam room controls the devices exposed to the neutron beam. Scripts monitor any disturbances (SDCs and DUEs) while the ViT models perform inferences on the GPUs.

TABLE I: ViT MODELS SIZE, ACCURACY ON IMAGENET DATASET, AND EXECUTION TIMES FOR PASCAL GPU.

	Config.	Size (MB)	Accuracy (%)	Time (ms)
ViT-H [29]	H14-224	2479	88.20	1644
EVA2 [30]	L14-448	1176	89.95	2686
SwinV2 [31]	L-256	787	86.94	404
MaxViT [32]	L-384	845	87.98	938

and a previously saved output (fault-free golden) are recorded as Tolerable SDC or Critical SDC. We make all the codes used in the beam experiments available [1].

The beam experiments measured the probability of a neutron causing a failure in the GPU. The failure rate determined in the experiments can be used to estimate the terrestrial failure rate caused by neutrons on a GPU. The beam experiments provide the Failure In Time (FIT) - the number of faults expected in $10^9 h$ of operation. FIT is calculated by dividing the number of errors by the neutron fluence, then multiplying by the terrestrial neutron flux ($13n/(cm^2 \times h)$) and by 10^9.

3) Device and DNNs Under Test: For the beam experiments, the NVIDIA GPU Pascal architecture (Quadro P2000) was used. The Quadro P2000 is built with TSMC $16nm$ FinFET, featuring an L1 cache of 48KB per Streaming Multiprocessor (SM), an L2 cache of 1280 KB, and 1024 CUDA cores. The GPU has 256 KB registers per SM and a power consumption of up to 75W. Our beam experiments only focus on GPU core errors (beam spot set to 2cm diameter to avoid affecting onboard DRAM).

We evaluated 5 ViT models from the HuggingFace library [33]. The models belong to 4 families: Original ViT [29], EVA2 [30], SwinV2 [31], and MaxViT [32]. The models differ in size and input patches. For the experiments, we used a Python program with PyTorch to load the ViT and perform inferences on a batch of random images from the ImageNet dataset [34].

B. Software Fault Simulation vs Beam Experiments

Fault simulation and lower levels of fault injection (such as physical or microarchitectural) can yield significantly different

[1]https://github.com/diehardnet/maximals

Special Session

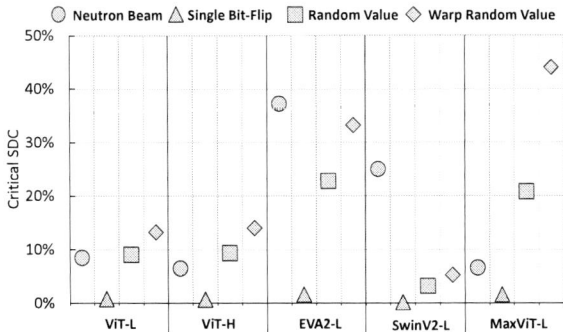

Fig. 2: Comparison of Critical SDC percentages between software fault simulations and neutron beam experiments. We used various fault models for broad analysis.

TABLE II: Comparison of the results obtained from software fault simulation and beam experiments for both the unhardened models (Baseline) and the models protected by range restriction (Hardened).

		NVBitFI [PVF]	ChipIr [FIT]	Reduction NVBitFI	Reduction ChipIr
ViT-L	Baseline	6.29%	1.77±0.45	3.67×	1.69×
	Hardened	1.71%	1.05±0.26		
ViT-H	Baseline	6.11%	1.61±0.33	2.74×	3.20×
	Hardened	2.23%	0.50±0.13		
EVA2-L	Baseline	15.20%	3.95±1.18	10.64×	–
	Hardened	1.43%	0.00±1.99		
SwinV2-L	Baseline	2.17%	0.87±0.60	1.23×	1.74×
	Hardened	1.77%	0.50±0.21		
MaxViT-L	Baseline	15.89%	1.75±0.36	1.94×	4.61×
	Hardened	8.17%	0.38±0.07		

results [26]. Recent studies have shown that this difference can be orders of magnitude [26], [35]. To measure the differences between the beam experiments and the software fault injection for ViTs, we start our analysis by evaluating the impact of the faults on the Critical SDCs.

Figure 2 shows the percentages of Critical SDCs for both NVBitFI and ChipIr neutron beam experiments of all the ViTs listed in Table I. In all cases, the single-bit flip underestimated the percentages of Critical SDCs compared to the beam experiments. On average, 0.80% of the injections with a single bit flip generated a Critical SDC, while 13.95% of the observed SDCs in the beam experiments were critical. Similar to other types of DNNs, ViTs are resistant to single-bit flips caused by transient faults [36]. The effect of a single bit in a single parameter on models containing millions of parameters is expected to be minimal. More complex fault models resulted in a significantly higher percentage of Critical SDCs, with 10.87% for random value and 18.27% for warp random value. Our data indicates that more complex fault models are mandatory to evaluate large ViTs' reliability accurately.

Single-bit flips injected at the instruction level do not provide a realistic ViT evaluation since most are masked and produce a Critical SDC rate close to zero, well off the rate obtained with beam experiments. We observed on the fault simulation that only 3% of single-bit flip fault injections resulted in values higher than 10^6 after the fault mask was applied to the target register. In contrast, when random values were used, 45% of the fault mask immediately produced values higher than 10^6, NaN, or infinity values. If these faults spread to ViT structures, they may lead to Critical SDCs. To protect the ViTs against those faults, we employed float value restriction on the Identity layers of the selected models. Value restriction limits the maximum and minimum values the DNN parameters can propagate, filtering the values corrupted by faults. Value restriction is a standard method to improve the fault tolerance of DNNs with a low overhead [37].

Table II presents the Critical SDC comparison between the Baseline and Hardened ViT models, assessed using NVBitFI (Critical SDC PVF) and the ChipIr (Critical SDC FIT rate)

neutron beam. The Critical SDC reduction (Baseline/Hardened) is presented for both NVBitFI and ChipIr experiments. During the beam experiment campaign, no Critical SDC was observed for EVA2-L.

By employing a complex fault model in the software simulation, we are able to design a fault tolerance that is capable of preventing Critical SDCs on the beam experiments. Table II shows that the hardened versions of the ViT models produced, on average, 2.81× less Critical SDCs compared to the unhardened version on the ChipIr beam experiments. Similarly, the NVBitFI, using a set of complex fault models (including single-bit flip, random values, and warp random values), shows a 2.40× reduction on the Critical SDC.

Physical fault injection with a beam of particles is a method widespread in industries such as space exploration and avionics, and they are also used as a validation method for academic research. However, the high cost (thousands of dollars per hour), complexity, and limited availability (facilities have scheduling windows of months) hinder a more detailed evaluation of complex systems such as DNN-based systems. As a result, fault simulation is employed to model faults and errors at various hardware and software abstraction levels. Using fault simulation, researchers and engineers are not limited to transient or permanent faults caused by ionizing particles but also transient and permanent faults from other sources, such as aging, voltage variations, and performance degrading faults. The following sections will explore different methods for simulating faults on DNNs.

IV. Modeling Errors from Faulty TCUs in GPUs

A. TCUs in GPUs

Modern GPUs are equipped with specialized in-chip accelerators (e.g., Tensor Core Units, or TCUs) to boost the computation of *General Matrix Multiplication* (GEMM) algorithms, representing the fundamental building blocks for the efficient implementation of DNNs, including cutting-edge models, such as transformers or Large Language Models [38], [39]. Moreover, optimized libraries (such as cuBLAS for NVIDIA GPUs) use the GEMM algorithm to reshape the kernel weights and feature maps as matrices [40] for their later deployment in GPUs. In particular, these libraries split

979-8-3315-3968-9/24 $31.00 © 2024 IEEE

Special Session

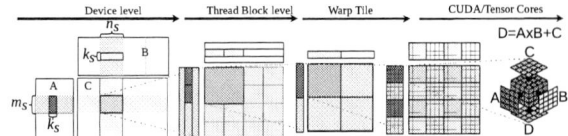

Fig. 3: A scheme of the tiling-based GEMM execution in TCUs.

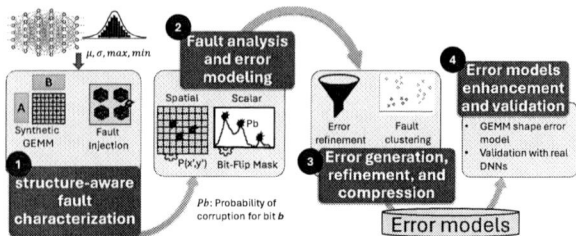

Fig. 4: A general scheme of the proposed strategy.

matrices into *tiles* and exploit the TCUs inside a GPU, splitting tiles at the device level into sub-tiles at the thread block, warp, and TCU levels, as depicted in Fig. 3 for the matrix operation A×B+C [41]. In detail, each *tile* is assigned to a unique cooperative thread group (CTA), which accelerates the computations by distributing all CTAs among the available parallel cores (SMs) in a GPU.

The widespread use of TCU-based GPUs for GEMM acceleration raises reliability concerns due to the large size of DNNs, and the increasing complexity of the underlying hardware [14], [42]. This section shows an efficient and accurate strategy to characterize the corruption effects from faults in the TCUs in terms of software-level error models, thus supporting application-level evaluations.

B. A method to model errors produced by faults in TCUs

Our strategy for software-level error modeling consists of determining *bit-flip-masks* that represent the corruption effects of the faults affecting the TCUs on the outputs of GEMM operations. Our strategy includes four steps: *i)* structure-aware fault characterization, *ii)* fault analysis and error modeling, *iii)* error generation, refinement and compression, and *iv)* model enhancement and validation (Fig. 4).

1) Structure-aware fault characterization: a seed GEMM operation describes a normal distribution of synthetic data with statistical information on the weights and the intermediate features from a target DNN model. The GEMM operation is deployed and executed on an instruction-accurate architectural simulator of TCUs in GPUs, which includes fault injection (FI) infrastructure for reliability evaluation [38], [39], [43]. Then, FI campaigns are conducted on the TCU's data-path structures.

2) Error generation: The fault effects are analyzed to determine **spatial** and **scalar** effects on the GEMM's outputs. The spatial evaluation finds the likelihood of every GEMM's output element (*location*) when corrupted by a faulty TCU. Furthermore, the scalar evaluation measures the bit-flip rate of the corrupted outputs on the GEMM computation. In particular, we consider two factors that corrupt a GEMM's output: *i)* the matrix tile's size, and *ii)* the ability of the data to activate and propagate faults. Thus, propagation effects are observed at the GEMM output as data corruptions in some locations of the executed tiles (i.e., from 1 to 8 corrupted elements per GEMM's tile at the warp level [38], [39]). Other factors, such as the number of SMs and the scheduling policy, directly depend on the targeted GPU (i.e., a faulty TCU inside an SM might induce data corruption only into those CTAs/tiles executed in the affected SM) [43].

The spatial effects are computed as *corruption probabilities* through a bi-dimensional representation in which every tile

location reports the probability of a fault to induce corruption. In detail, we employ a common coordinate system (x', y'), an indicator function $\mathbb{1}_{(x',y')}(x, y)$ [44], and the experimental results to indicate spatially the corrupted GEMM's locations. Then, the probability per fault $P(x', y')$ is calculated as the ratio between the number of observed effects in the (x', y') element and the total number of executed tiles in a faulty SM.

At the scalar level, we evaluate the impact of permanent faults in TCUs as the changes on one or multiple bits of the final value of the computed GEMM operation by resorting to *bit-flip probabilities*. These correspond to the probability of corruption per bit (P_b), i.e., to the number of times the bit b changes w.r.t. the golden values divided by the total number of corrupted bits. It must be noted that the *bit-flip probabilities* can be adapted to specific number formats (e.g., floating-point or integer). Both probabilities (*corruption* and *bit-flip*) are combined to generate corruption masks (*bit-flip-masks*) to describe errors from hardware faults directly on the outputs of the GEMM operation.

In this work, we focused on floating-point formats, so bits are grouped as *Sign*, *Exponent* and *Mantissa*, which allow the identification of bit probabilities per group (e.g., *Probability of Exponent Corruption* or PEC, which is the ratio between faulty values with at least a one-bit flip in the exponent over the total number of elements affected by the fault) and simplify the definition of injection masks correctly describing identical effects from permanent faults in TCUs. From experimental results, we calculate metrics to identify corruption masks in the exponent and mantissa, such as PEC. Then, we combine error generation and refinement steps by first processing any propagated fault as a GEMM error. Afterward, we evaluate the error quality in a refinement routine. A complementary error compression step optimizes the number of errors for application-level evaluations. In our strategy, each fault causing a spatial GEMM corruption in the generic *element* x, y is evaluated through an iterative process. The strategy classifies the corruption probability per affected location ($P(x', y') \neq 0$) as a feasible target for error generation. Then, a *bit-flip-mask* is generated for the given error and associated with the allocated CTAs/tiles (*GEMM coordinate*) on a faulty TCU.

It must be noted that one or more *bit-flip-masks* can be associated with each fault. Moreover, some error models might produce equivalent spatial corruptions (i.e., sharing an identical spatial distribution but corrupting a value differently). In both cases, refinement and compression processes compact

Special Session

the errors while preserving the quality and accuracy for the represented scalar error corruptions as *bit-flip-masks*.

3) Error refinement: in this step we use two complementary and focused evaluations using new randomly generated GEMM seeds. The first evaluation employs FI campaigns targeting all faults that caused error effects, while the second one uses the initial *bit-flip-masks* to place corruptions on GEMM operations. Then, we compute the *Mean Absolute Error* or (MAE) for corruptions from the FI campaigns (MAE_{faulty}) and from the error models (MAE_{masks}), evaluating the error replication accuracy by comparing (MAE_{faulty} and MAE_{masks}). Errors over- or under-estimating the scalar corruption effects (i.e., larger or lower MAE's magnitude than standard) are discarded, and their faults are considered *"unmodelable"*. Then, a refinement *Threshold* (Th, maximum discrepancy between corruptions from the proposed strategy w.r.t. the standard evaluation) is applied as ($1/Th < MAEs_{mask}/MAEs_{fault} < Th$) and experimentally tuned to trade-off corruption accuracy and number of represented errors from hardware faults.

4) Error Compression: this step compacts similar error models into clusters by considering their GEMM location and corruption magnitude. First, we group error models with the same corrupted GEMM region. Then, we compare the MAE per *bit-flip-masks* with the (Th) to organize them as clusters. A low value of *Th* provides higher error corruption accuracy (more clusters) but a lower compression, while a large value of *Th* allows higher compression (fewer clusters) at the expense of some loss in the error modeling accuracy. This compression also optimizes the overall error evaluation time since its main idea is to employ only one error per cluster instead of several errors with equivalent effects during application-level evaluations.

5) Model enhancement: this step focuses on improving and validating the effectiveness of the error models for different GEMM scenarios (i.e., sizes) against the conventional FI campaigns. Since the GEMM size, the number of tiles (T), and the reuse of faulty TCUs impact the distribution and accuracy of erroneous outputs, we resort to shape-wise error models to enhance the accuracy of errors according to the GEMM's shape. For this purpose, we compress error models for several GEMM shapes (e.g., $200 \times 200 \times 200$) using GEMM seeds. Then, we validate the accuracy of the shape-wise error models using convolutional layers from typical DNNs.

The validation involves two evaluations per layer. *1)* conventional FI campaigns in the TCUs and *2)* error evaluations using the shape-wise error models, injecting only one *bit-flip-mask* per cluster. The MAEs are collected for both evaluations to compute the vector cross-correlation coefficient and quantify the equivalence between the strategies (FI campaigns and the proposed one). When the error model is accurate enough to represent hardware faults, it can evaluate complete CNNs.

C. Results and analyses

In the experiments, we use a tool named *PyOpenTCU*, which combines the behavioral operation of the schedulers

TABLE III: DNN LAYERS FOR ERROR MODEL VALIDATION.

DNN layer	GEMM shape (A×B×C)
ResNet18 RB1	$64 \times 147 \times 12,544$
ResNet18 RB2	$64 \times 576 \times 12,544$
MobileNetV2	$3 \times 3 \times 12,544$
LeNet5	$6 \times 75 \times 576$
YoloV5	$32 \times 27 \times 102,400$

TABLE IV: NUMBER OF FAULTS MODELED AS ERRORS (*bit-flip-masks*) AND COMPRESSED CLUSTERS PER GEMM SEED.

Golden GEMM seed	faults as *bit-flip-masks*	Clusters
100X	7,998 (94.1%)	1,350
200X	7,052 (83%)	1,159
300X	7,740 (91.1%)	1,052
400X	7,602 (89.4%)	939

and general GPU's hierarchy with the instruction-accurate TCU architecture [38], [39]. For the purpose of this work, *PyOpenTCU* is configured with 7 SMs and 28 TCUs.

For the shape-wise error model generation, eight statistical FI campaigns injected $6.8x10^4$ permanent (*stuck-at*) faults on the TCUs of one SM, with the 95% of confidence level and 1% error margin, and observed the corruptions only at the GEMM's outputs. The first 4 FI campaigns resort to GEMM seeds with shapes: **100X** ($100 \times 100 \times 100$), **200X** ($200 \times 200 \times 200$), **300X** ($300 \times 800 \times 300$), and **400X** ($400 \times 1,200 \times 400$) to identify corruptions and define the initial error models. Then, a second set of 4 campaigns performs the error refinement. We use 5 representative convolutional layers with different GEMM shapes for validation purposes (*LeNet5, ResNet18-RB1, ResNet18-RB2, Yolov5,* and *MobileNetv2*), as reported in Table III. Five additional FI campaigns (one per layer) injected around $5x10^3$ faults to generate the reference corruptions for validation against the proposed shape-wise error models. All experiments used a workstation HP Z2G5 with a 20-core Intel i9-10800 CPU and 32 GB of RAM.

A preliminary analysis indicates that a considerable percentage of the evaluated faults in TCUs (around 70% to 90%) produced corruption effects (SDCs).

We refined the initial scalar error models for each fault that causes corruptions (*bit-flip-masks*) during the FI campaigns. Then, we compare the MAEs from the additional FI campaigns (randomly generating new GEMM seeds for each fault) and those obtained through the error models.

Our exploration of different thresholds (Th) indicates that a narrow Th (i.e., 1) reduces the number of *bit-flip-masks* representing the corruption effects from hardware faults and only groups those errors with high corruption accuracy. However, when Th moves from 3 to 10 the total amount of *bit-flip-masks* describing corruptions under acceptable levels increases from about 66% to around 93% on all analyzed GEMM seed shapes under the same order of magnitude. This proves that the refinement can provide acceptable accuracy to represent errors, as reported in Table IV with compression results on each shape-wise error model for *Th*=10.

In some cases, e.g., 100x GEMM seed, the shape-wise error model (*bit-flip-masks*) covers a considerable percentage

979-8-3315-3968-9/24 $31.00 © 2024 IEEE   225

Special Session

of effectively represented errors (up to 94.1%) from hardware faults under an acceptable number of clusters (1,350). Thus, instead of using a conventional FI campaign (7,998 faults), an application-level error evaluation provides the equivalent corruption effects by evaluating only 1,350 errors with our approach. Moreover, the experimental results support the idea that compression can effectively reduce the number of corruption errors to evaluate in large GEMM workloads with a similar impact on their evaluation times (5X to 8X).

Finally, we validated the effectiveness of the shape-wise error models by determining the cross-correlation factor (the higher, the better) between conventional FI campaigns and our error modeling strategy for the DNN layers. The results, illustrated in Figure 5 *(Left)*, show that the GEMM seed's shape is vital for the effective representation of errors at the software level. The results suggest that small-shape GEMMs (100X with correlations from 55% to 92%) are more accurate than large ones (e.g., 400X with correlations from 15% to 52%) in representing errors on all evaluated layers. A comparison of MAEs from the FI campaign and the shape-wise error model for *ResNet18 RB1* layer in Figure 5 *(Right)* shows that up to 92% of the *bit-flip-masks* accurately represented errors. In contrast, large-shape GEMMs are affected by the amount and size of operations in the layers and the GEMM seeds (i.e., a similar amount and size of operations in 100X shape-wise error models and the evaluated layers, while completely different for the 300X and 400X models).

The results indicate that our strategy can represent with a good accuracy the effect of faults and optimize application-level reliability evaluations with performance improvements of up to 225X (from 8.82 h in conventional FIs to 2.26 min using our error modeling strategy with **100X** *bit-flip-masks*).

V. CROSS-LAYER SIMULATION METHODOLOGIES IN AI ACCELERATORS

Deep Learning Accelerators (DLAs) are typically optimized during design-time to meet stringent area, power, and performance requirements. However, these architectural parameters can also impact the propagation of faults through the system, affecting overall dependability. While existing studies have explored the correlation between architectural parameters and SDC errors, they remain in the preliminary stages and lack comprehensive methodologies to evaluate the impact of design choices on system reliability [45] [46]. Simulation-based and emulation-based techniques offer high fault-injection control but often suffer from long simulation times, limiting their applicability to early design phases [47] [48].

In this section, we present a cross-layer simulation methodology designed to evaluate the reliability of DLAs by injecting faults at various levels of abstraction. Our approach leverages a cycle-accurate microarchitectural simulation tool that allows for fault injection in software-visible registers while maintaining the time-awareness of RTL simulations [49]. This enables us to model how architectural parameters influence SDCs and assess their impact on DNN applications in terms of performance and dependability.

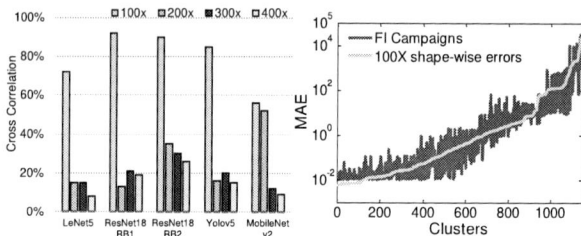

Fig. 5: Cross-correlation on the shape-wise error models for the evaluated layers.

A. Case Study: The NVIDIA Deep-Learning Accelerator

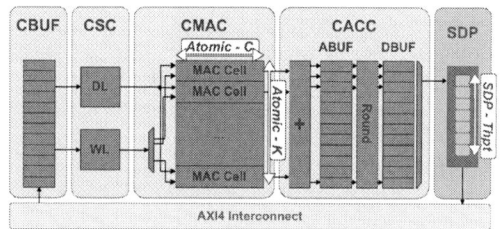

Fig. 6: The NVDLA pipeline as block diagram

The *NVIDIA Deep-Learning Accelerator (NVDLA)* [50] is a domain-specific accelerator designed for 2D convolution operations. It features a five-stage convolution pipeline, comprising key components such as a dedicated memory interface *(CDMA)*, tightly coupled input buffers *(CBUF)*, a control unit for data and weight loading *(CSC)*, a multiply-and-accumulate unit *(CMAC)*, and an accumulation unit *(CACC)*. The pipeline is followed by a reconfigurable post-processing unit *(SDP)*, which supports element-wise operations like activations.

The system's flexibility stems from its configurability. Key parameters include the buffer sizes, the number of input/output channels handled by the CMAC *(Atomic-C, Atomic-K)*, and the precision of integer pipelines (8, 16, or 32 bits). By analyzing different configurations, we assess how hardware choices impact fault propagation and system reliability during deep learning operations.For further details on the NVDLA architecture, refer to our previous work [49].

B. Cross-Layer workflow

To overcome RTL simulation speed, without loosing the time information, we adopt in our workflow a cycle-accurate microarchitectural model extended with hardware information.

The example in Pseudocode 1 shows a possible implementation of the NVDLA CMAC unit, performing vector-to-matrix multiplications of sizes *Atomic-C* $\times$ *Atomic-K*. While Pseudocode 2 presents a possible extension to model it as an array of multipliers, followed by an adder tree, which critical path is reduced by inserting a register barrier in the middle.

```
cmac(weights, inputs, psums):
    for outCh in range(0, Atomic-K):
        for inCh in range(0, Atomic-C):
```

Special Session

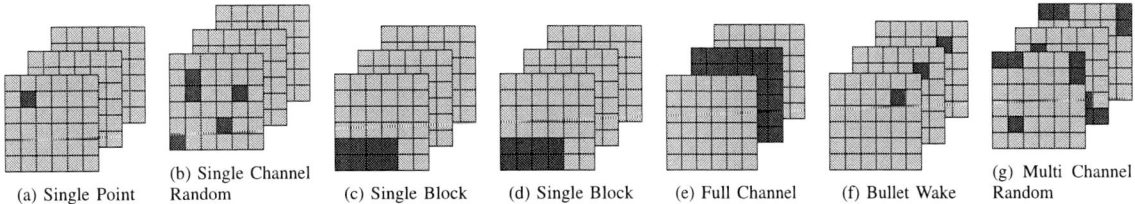

Fig. 7: Observed error patterns in the output tensor of single DNN layers.

```
4    // Multiply-and-Accumulate
5    psums[outCh] += inputs[inCh] * weights[outCh][
     inCh];
```

Pseudocode 1: "Plain" implementation of NVDLA MAC array.

```
1  cmac(weights, inputs, psums):
2    for outCh in range(0, Atomic-K):
3      // Multiplier Array
4      for inCh in range(0, Atomic-C):
5        mulout[inCh] = inputs[inCh] * weights[inCh];
6      // Adder Tree - Level 1
7      for inCh in range(0, Atomic-C):
8        reg[inCh / regnum] += mulout[inCh];
9      // Adder Tree - Leve 2
10     for rit in range(0, regnum):
11       psums[outCh] += reg[rit];
```

Pseudocode 2: "Extended" implementation of NVDLA MAC array. `regnum` is the number of registers in the middle of the adder tree.

By modeling hardware registers as software-visible variables, we can observe the register content at simulation time and, thus, inject it with faults in the form of bitflips or corruption masks thanks to the support of saboteurs. The latter can be easily built after an accurate RTL description inspection, thus closely matching the hardware behaviour.

Last, to characterize the error models observed in the accelerator, we feed the simulation results to CLASSES [12]. This state-of-the-art tool is composed of an analyser, extracting error models statistics, and by a network error simulator, which replicates the observed error models in order to asses the full-system fault tolerance [45].

C. Error Modeling Results

TABLE V: ANALYSED NVDLA CONFIGURATIONS. BUFFER SIZES ARE NORMALIZED ON THE DATAPATH BITWIDTH.

Configuration	CBUF size	Atomic-C	Atomic-K	CACC size	SDP thpt
nv_small64	131072	8	8	8192	1
nv_small256	262144	32	8	8192	1
nv_medium512	262144	32	16	32768	4
nv_medium1024	262144	32	32	65536	8
nv_large2048	262144	64	32	65536	16

Table V reports the analyzed NVDLA configurations (available at [50]). They majorly differ in MAC array sizes and post-processing unit parallelism. However, buffers don't linearly scale up with the configuration complexity. In particular, the CBUF is constant through most of the tested designs. We performed a SEU injection campaign in the form of bitflips in registers for every combination of designs under test and tested the network layer. The layers were selected from AlexNet

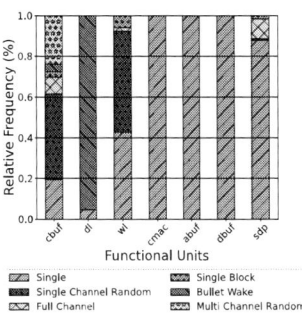

Fig. 8: Relative frequency of error patterns across hardware units in the *nv_small256* configuration, quantized to 16-bit precision.

and ResNet-18 and trained on CIFAR-10 and CIFAR-100, respectively. Each fault injection campaign consists of 10.000 SEU injections, for a total of 10K simulations for 15 designs and 24 convolutional layers.

The observed error models (Figure 7) present in the output tensors belong to six different geometries previously characterized for GPUs [12]. They are referred to as:

- **Single Point:** A single output tensor value is corrupted;
- **Single Channel Random:** Multiple corrupted values in the same channel;
- **Single Block:** Corrupted elements sharing the same channel having contiguous X-Y locations;
- **Full Channel:** An entire channel is corrupted;
- **Bullet Wake:** Multiple corrupted values across different channels sharing the same X-Y coordinates;
- **Multi Channel Random:** Multiple values across different X-Y coordinates and channels.

Figure 8 presents the relative distribution of the observed error models into the nv_small256 configuration running with integer 16-bit precision. *Single Point* errors are by far the most common in the pipeline and are associated with corrupted data that are not reused during the computation. On the other hand, data from the CBUF, DL, or WL easily turns into error models associated with multiple erroneous values in the output tensor.

Full Channel, *Single Block*, and *Single Channel Random* error patterns are always associated to a weight data corruption, where the difference between the three is determined by the error activation probability. An erroneous weight can impact even all the output partial sums associated with a given neuron, but they always share the same output channel. In fact,

Special Session

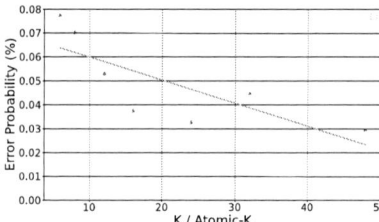

Fig. 9: Probability of output tensor corruption over the number of tiled convolutions (K / *Atomic-K*).

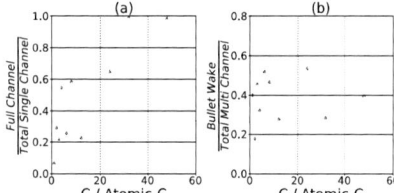

Fig. 10: Relative frequencies of error patterns for faults in the weights (a) and faults in the features(b), against the number of CBUF entries (C / *Atomic-C*).

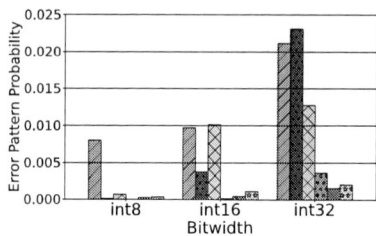

Fig. 11: Error pattern probability in relation to datapath bitwidth. The colors correspond to the error patterns shown in Figure 8.

due to the weight-stationary data reuse policy of NVDLA, we know how a corrupted weight won't be prefetched until the accelerator processes all the corresponding input features, thus impacting multiple X-Y coordinates. Differently, *Multi Channel Random* and *Bullet Wake* error patterns are associated with feature data corruption. In fact, corrupted feature data are broadcasted to multiple neurons, impacting multiple output channels. Thanks to the hardware parallelism of NVDLA, this also suggests how a multi-channel pattern cannot impact more than *Atomic-K* channels at a time. Last, we can also observe a minor presence of *Full-Channel*, *Single Block*, and *Single Channel Random* error patterns in the post-processing unit. They are majorly associated with biased data corruption. Since biases are loaded at the beginning of the execution and are added to all the elements of the same output channel.

The probability that, given an SEU, this translates into any of the observed error patterns is plotted in Figure 9. As can be observed, this exposes a negative trend against the ratio $\frac{K}{Atomic\text{-}K}$, where K is the number of output channels of a layer and *Atomic-K* is the number of hardware neurons. It is easy to observe how this ratio represents the number of hardware convolutions the layer is tiled into. Therefore, since the more the tiles, the smaller they are, with the growth of the $\frac{K}{Atomic\text{-}K}$ ratio, we are computing fewer data together, so the probability that corrupted data is computed with sensitive data decreases.

Second, Figure 10 plots the relative frequency of error patterns associated with weights and feature corruption in the CBUF. As illustrated, there is a growing trend in association with the growth of the $\frac{C}{Atomic\text{-}C}$ ratio, where C is the number of layer's input channels, and *Atomic-C* is the number of hardware input channels of the CMAC. This can be explained if we observe how the hardware execution time is proportional

to the ratio $\frac{C}{Atomic\text{-}C}$. In fact, by increasing the total execution time, it also increases the time window for the CBUF to be exposed. Therefore, it increases the incidence of error patterns associated with CBUF data corruption.

Last, Figure 11 illustrates the probability of different error patterns across the analyzed hardware datapath precisions. It can be observed that the probability of error patterns does not scale uniformly across all datapath precisions. While *Single Point* patterns are by far the most common in 8-bit pipelines, error patterns related to weight data corruption rapidly scale up with larger bit widths. This effect is directly caused by the quantization scheme. Our analysis adopted a symmetric post-training quantization scheme, which quantizes the weights at 8 bits to reduce memory usage. This means that the remaining bits in the pipeline are used for sign bit extension. Faults among the sign bits are way more impactful than bitflips among the information bits. Furthermore, quantization schemes are not designed to account for robustness against faults but to preserve the highest network accuracy with the minimum memory usage. Therefore, pipelines designed to support large data (i.e., 32-bit) computation expose a higher amount of sign bits, and thus a higher probability to generate error patterns different from *Single Point*.

It is possible to observe how different error patterns expose different impacts on the final application [12], [45]. In particular, while *Single Point* errors are the most common, other patterns may be way more dangerous (in particular, *Bullet Wake* patterns). As Figure 9 and Figure 10 illustrate, the relationship between DNN hyper-parameters and DLA hardware parameters has a significant impact on the type of observed error patterns and the probability with which they appear, opening the door for future reliability-driven hardware configuration space explorations.

VI. CONCLUSIONS

In this paper we first assessed the effects of faults induced by radiation on large ViTs for image classification. While ViT models provide high accuracy, we observed in our experiments a high percentage of critical SDCs (misclassifications). We then designed a fault tolerance approach based on value range restriction to reduce/remove the critical SDCs. Our approach combines analysis from beam experiments and data from software fault simulation. Furthermore, we analyzed the effectiveness of multi-abstraction assessments to characterize

979-8-3315-3968-9/24 $31.00 © 2024 IEEE

Special Session

the impact of hardware faults in TCUs and propose software error models able to preserve accuracy while significantly speeding up reliability evaluation. Finally, we highlighted how different hardware parameters in an NVDLA core can significantly change the observed error models and their occurrence, depending on the target DNN model. Our work opens future perspectives in investigating performance-dependability-cost tradeoffs.

REFERENCES

[1] A. Khan *et al.*, "A survey of the recent architectures of deep convolutional neural networks," *Artif. Intell. Rev.*, vol. 53, pp. 5455–5516, 2020.

[2] L. Alzubaidi *et al.*, "Review of deep learning: concepts, cnn architectures, challenges, applications, future directions," *J. Big Data*, vol. 8, pp. 1–74, 2021.

[3] ISO 26262-5, "Road vehicles — functional safety — part 5: Product development at the hardware level," pp. 1–90, 2018.

[4] ISO/IEC 22989, "Information technology — artificial intelligence — artificial intelligence concepts and terminology," pp. 1–60, 2022.

[5] I. Hill *et al.*, "Cmos reliability from past to future: A survey of requirements, trends, and prediction methods," *IEEE Trans. Device Mater. Reliab.*, vol. 22, no. 1, pp. 1–18, 2022.

[6] J. D. Guerrero Balaguera *et al.*, "Understanding the effects of permanent faults in gpu's parallelism management and control units," in *Int. Conf. for High Performance Computing, Networking, Storage and Analysis (SC'23)*, 2023.

[7] J. Rajski *et al.*, "The future of design for test and silicon lifecycle management," *IEEE Design & Test*, pp. 1–1, 2023.

[8] IEEE, "The international roadmap for devices and systems: 2022," in *Institute of Electrical and Electronics Engineers (IEEE)*, 2022.

[9] H. D. Dixit *et al.*, "Silent data corruptions at scale," 2021. [Online]. Available: https://arxiv.org/abs/2102.11245

[10] P. H. Hochschild *et al.*, "Cores that don't count," in *Workshop on Hot Topics in Operating Systems*, 2021, pp. 9–16.

[11] G. Papadimitriou and D. Gizopoulos, "Silent data corruptions: Microarchitectural perspectives," *IEEE on Trans. Comput.*, vol. 72, no. 11, pp. 3072–3085, 2023.

[12] C. Bolchini *et al.*, "Fast and accurate error simulation for cnns against soft errors," *IEEE Trans. Comput.*, vol. 72, no. 4, pp. 984–997, apr 2023.

[13] N. Mahatme *et al.*, "Comparison of Combinational and Sequential Error Rates for a Deep Submicron Process," *IEEE Trans. Nucl. Sci.*, vol. 58, no. 6, pp. 2719–2725, 2011.

[14] J. E. R. Condia *et al.*, "A multi-level approach to evaluate the impact of gpu permanent faults on cnn's reliability," in *EEE Int. Test Conf. (TC)*, 2022, pp. 278–287.

[15] K. Ito *et al.*, "Analyzing due errors on gpus with neutron irradiation test and fault injection to control flow," *IEEE Trans. Nucl. Sci.*, vol. 68, no. 8, pp. 1668–1674, 2021.

[16] J. M. Badia *et al.*, "Comparison of parallel implementation strategies in gpu-accelerated system-on-chip under proton irradiation," *IEEE Trans. Nucl. Sci.*, pp. 1–1, 2021.

[17] M. B. Sullivan *et al.*, "Characterizing and mitigating soft errors in gpu dram," in *54th Ann. IEEE/ACM Int. Symp. on Microarchitecture*, 2021, pp. 1–10.

[18] P. R. Bodmann *et al.*, "Soft error effects on arm microprocessors: Early estimations versus chip measurements," *IEEE on Trans. Comput.*, vol. 71, no. 10, pp. 2358–2369, 2022.

[19] J. E. R. Condia *et al.*, "Flexgripplus: An improved gpgpu model to support reliability analysis," *Microelectronics Reliability*, vol. 109, p. 113660, 2020.

[20] A. Chatzidimitriou *et al.*, "RT level vs. microarchitecture-level reliability assessment: Case study on ARM(r) cortex(r)-a9 CPU," in *DSN Workshop*, 2017.

[21] C. Constantinescu *et al.*, "Error injection-based study of soft error propagation in amd bulldozer microprocessor module," in *DSN*, 2012.

[22] S. S. Mukherjee *et al.*, "A Systematic Methodology to Compute the Architectural Vulnerability Factors for a High-Performance Microprocessor," in *MICRO*, 2003.

[23] A. R. Anwer *et al.*, "Gpu-trident: Efficient modeling of error propagation in gpu programs," in *SC*, 2020.

[24] T. Tsai *et al.*, "NVBitFI: Dynamic Fault Injection for GPUs," in *Ann. IEEE/IFIP Int. Conf. on Dependable Systems and Networks (DSN)*, 2021, pp. 284–291.

[25] V. Sridharan and D. R. Kaeli, "Eliminating microarchitectural dependency from Architectural Vulnerability." IEEE HPCA, 2009.

[26] G. Papadimitriou and D. Gizopoulos, "Demystifying the system vulnerability stack: Transient fault effects across the layers," in *IEEE ISCA*. IEEE ISCA, 2021, p. 902–915.

[27] F. F. d. Santos *et al.*, "Characterizing a neutron-induced fault model for deep neural networks," *IEEE Trans. Nucl. Sci.*, vol. 70, no. 4, pp. 370–380, 2023.

[28] C. Cazzaniga and C. D. Frost, "Progress of the scientific commissioning of a fast neutron beamline for chip irradiation," *Journal of Physics: Conference Series*, vol. 1021, no. 1, p. 012037, may 2018.

[29] A. Dosovitskiy *et al.*, "An Image is Worth 16x16 Words: Transformers for Image Recognition at Scale." 9th Int. Conf. on Learning Representations (ICLR), 2021.

[30] Y. Fang *et al.*, "Eva-02: A visual representation for neon genesis," *arxiv*, 2023.

[31] Z. Liu *et al.*, "Swin transformer v2: Scaling up capacity and resolution." IEEE IEEE/CVF Computer Vision and Pattern Recognition Conf. (CVPR), 2022, pp. 12 009–12 019.

[32] Z. Tu *et al.*, "MaxViT: Multi-axis vision transformer." ECCV, 2022, pp. 459–479.

[33] R. Wightman, "Huggingface," huggingface.co/timm.

[34] J. Deng *et al.*, "ImageNet: A large-scale hierarchical image database," in *IEEE CVPR*. IEEE CVPR, 2009, pp. 248–255.

[35] F. F. d. Santos *et al.*, "Demystifying gpu reliability: Comparing and combining beam experiments, fault simulation, and profiling," in *IPDPS*, 2021.

[36] G. Gavarini *et al.*, "Evaluation and mitigation of faults affecting swin transformers." IEEE 29th Int. Symp. on On-Line Testing and Robust System Design (IOLTS), 2023.

[37] Z. Chen *et al.*, "A Low-cost Fault Corrector for Deep Neural Networks through Range Restriction." IEEE/IFIP DSN, 6 2021.

[38] R. L. Sierra *et al.*, "Analyzing the impact of different real number formats on the structural reliability of tcus in gpus," in *IFIP/IEEE 31st Int. Conf. on Very Large Scale Integration (VLSI-SoC)*, 2023, pp. 1–6.

[39] R. Limas Sierra *et al.*, "Exploring hardware fault impacts on different real number representations of the structural resilience of tcus in gpus," *Electronics*, vol. 13, no. 3, 2024.

[40] V. Sze *et al.*, "Efficient processing of deep neural networks: A tutorial and survey," 2017.

[41] V. Thakkar *et al.*, "CUTLASS," 2023.

[42] A. Ruospo *et al.*, "A pipelined multi-level fault injector for deep neural networks," in *IEEE Int. Symp. on Defect and Fault Tolerance in VLSI and Nanotechnology Systems (DFT)*, 2020, pp. 1–6.

[43] R. Limas Sierra *et al.*, "Analyzing the impact of scheduling policies on the reliability of gpus running cnn operations," in *42nd IEEE VLSI Test Symp. (VTS 2024)*, 2024, pp. 1–7.

[44] K. Q. Ye, "Indicator function and its application in two-level factorial designs," *The Annals of Statistics*, vol. 31, no. 3, pp. 984 – 994, 2003.

[45] A. Veronesi *et al.*, "Cross-layer reliability analysis of nvdla accelerators: Exploring the configuration space," in *IEEE European Test Symp. (ETS)*, 2024, pp. 1–6.

[46] B. Reagen *et al.*, "Ares: A framework for quantifying the resilience of deep neural networks," in *55th ACM/ESDA/IEEE Design Automation Conf. (DAC)*, 2018, pp. 1–6.

[47] S. Pappalardo *et al.*, "A fault injection framework for ai hardware accelerators," in *24th Latin American Test Symp. (LATS)*, 2023, pp. 1–6.

[48] X. Feng *et al.*, "Runtime fault injection detection for fpga-based dnn execution using siamese path verification," in *2021 Design, Automation & Test in Europe Conf. & Exhibit. (DATE)*, 2021, pp. 786–789.

[49] A. Veronesi *et al.*, "Exploring software models for the resilience analysis of deep learning accelerators: the nvdla case study," in *Int. Symp. on Design and Diagnostics of Electronic Circuits and Systems*, 2022, pp. 142–147.

[50] "The nvidia deep-learning accelerator." [Online]. Available: www.nvdla.org

Special Session

qCrop: An IoT based Framework to Enhance Crop Productivity in Smart Agriculture

Mahdi Shamsa
Dept. of Computer Science and Engineering,
University of North Texas,
Denton, USA
mahdishamsa@my.unt.edu

Laavanya Rachakonda
Dept. of Computer Science,
University of North Carolina Wilmington,
Wilmington, USA
rachakondal@uncw.edu

Saraju P. Mohanty
Dept. of Computer Science and Engineering,
University of North Texas,
Denton, USA
saraju.mohanty@unt.edu

Elias Kougianos
Dept. of Electrical Engineering,
University of North Texas,
Denton, USA
elias.kougianos@unt.edu

Abstract—**Food is crucial in our life. Despite food manufacturers' efforts to meet consumers' needs with manufactured food, it cannot attain the identical level of quality and flavor as natural food. Some fruits and vegetables are classified as food that cannot be manufactured. Also, they play an essential role worldwide. This paper focuses on how farmers can protect their farms from an object that might threaten the crops and cause damage to them. It proposes a framework that farmers may utilize to safeguard the crops against any species of birds. Considering the adverse impact of bird attacks on crop production, this system effectively addresses this problem and consistently enhances the quality and quantity of them. It utilizes computer vision technology to establish a secure environment for the crop. This system is called Quality of Crop Device (qCrop), and it works with a You Only Look Once (Yolov8m) model to detect birds with high accuracy to protect farms.**

Index Terms—**Quality of Crop (qCrop), Protect Crops, Avert Bird, Issue a Sound, Bird Species.**

I. INTRODUCTION

Although the recent emphasis on technology, promoting smart agriculture is an effective method to offer consumers a portion of nutritious and healthy food [22]. fruits and vegetables represent the same aspect of the other food. They have an essential role in the human body.

Conversely, food manufacturers endeavor to obtain customer satisfaction with the food they provide [1], [24]. Currently, ensuring the quality of food continues to be a multifaceted challenge for food manufacturers [25]. Although individuals may have developed a sophisticated palate for food, the original flavors of natural resources persist [2].

Studies have projected that the global population is predicted to reach roughly 9.7 billion people by the year 2050 [22]. furthermore, In 2022 and 2023, people consumed about 791 million metric [20]. So, Proper care for crops is necessary due to the expected many million metric tons rise in fruits and vegetables consumption in 2023 and 2024. This implies that consumers will contribute to an increase in the proportion of natural resource demand. Preserving natural resources is the sole effective method to maintain the superior quality and flavor of food. Currently, there is a growing emphasis among researchers on the field of smart agriculture due to its significant significance.

Farmers frequently have the challenge of bird species targeting their crops [19]. Furthermore, farmers around the world encounter these challenges due to the specific geographical location of crops, which serve as stops for migratory birds. Periodically, birds immigrant emigrate, and farmers begin utilizing conventional methods, such as ropes or their vocalizations, to safeguard crops from potential harm caused by birds. Protecting crops from natural disasters, stray animals, and potential thefts incurs significant financial expenses for farm owners [8], [11]. During the period of bird emigration, farmers initiate preparations to safeguard their crops through various means, including seeking individuals to employ them to evict birds manually [18].

Usually, people who obtain compensation utilize traditional approaches, such as hurling stones and raising loud voices, to remove birds from agricultural land or erect scarecrows along the perimeter of the farm, which is a demanding job. Furthermore, the insufficient educational attainment of farmers has resulted in their deaths [12]. Due to the severity of the situation, an additional problem arises during the birds' migration, namely the current scarcity of human personnel.

From 2018 to 2019, 44.3% of the human population resides in rural areas [4], which means 55% will be located in smart cities. This will diminish the available agricultural land, particularly in rural areas. Economic inflation within the agricultural sector poses a significant challenge, leading to a reduction in crop yield. Fig. 1 depicts the adverse impact of diminishing yield on farmers. Insufficient crop's production in farms hampers fruits and vegetables availability from suppliers, markets, and consumers. qCrop is a capable system that has the ability to protect crops by emitting sounds

979-8-3315-3968-9/24 $31.00 © 2024 IEEE

Special Session

to evict attacked birds.

The proposed system aims to create a conducive environment for farmers, enhancing their comfort with their crops. Additionally, it seeks to reduce costs by eliminating the necessity for human labor in bird eviction, improving crop quality, and facilitating the provision of flour to suppliers and bread to consumers in bakeries.

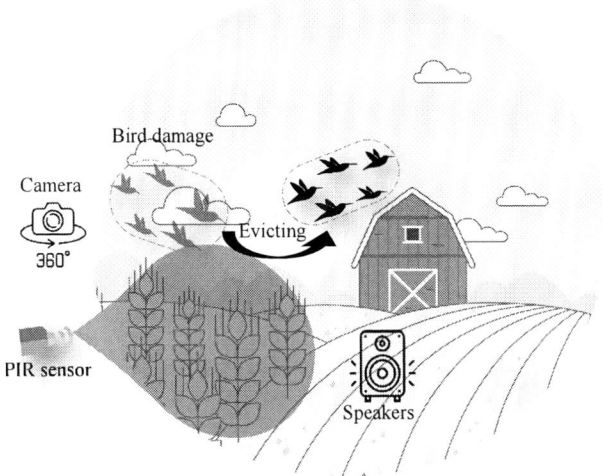

Fig. 1. Motivation for q-Crop to Keep the Crop Safe From Birds

Due to the natural preference of birds for fruits and vegetables as their primary diet, significant crop waste and damage are caused. The suggested system aims to improve crop protection and contribute to several aspects of the country, including farmers' income and the overall economy, through the monitoring and management of farms.

The rest of this paper has been organized as follows: Sec. II presents the contribution of the current paper. It includes the problem being addressed, the proposed solution, and its significance of this solution. Sec. III includes the related work to provide an explanation and comparison of the proposed work's existence with qCrop. Sec. IV depicts the proposed novel methodology for elucidating the system architecture, qCrop's algorithms, and the levels of qCrop. Sec. V provides the experiment and validation process, including the model details and used dataset. Sec. VI presents the conclusion of the work, and discusses potential future research suggestions.

II. PROPOSED CONTRIBUTION OF THE QCROP SYSTEM

This study introduces a state-of-the-art approach to safeguard farms in various aspects:

- The paper proposes an Internet of Things (IoT) [10] system that serves multiple functions and includes a full automation system.
- The system is designed specifically to address the difficulties encountered by farmers in protecting their crops. It operates totally automatically, conserves power by operating during daylight, and minimizes labor costs.

- Crop monitoring is the primary method used for protection.
- By integrating Light Dependent Resistor (LDR), Camera, Passive Infrared sensor (PIR), and speaker, the proposed approach ensures an accurate model.

This technique is a completely automated system that does not necessitate any intervention from the farmers to prevent the birds from harming the farms. The reason why farmers do not need to interact with the qCrop system is that it utilizes a variety of sounds to effectively evict birds from the farms. The system operates during the diurnal period, commencing at sunrise when bird flight commences, and transitions into a dormant state upon sunset.

A. Problem Addressed

This research aims to examine a significant challenge consistently encountered by farmers. Given that the primary objective of crops is to supply consumers with food of high-quality, it is essential for farmers to safeguard their crops against any potential assault by animals that could compromise the crop's resources.

The objective of this study is to safeguard the health of crop production over time from any diseases that might impact the fruits and vegetables to be healthy for the consumers. Consequently, the global scarcity of wheat (for example) poses a significant danger to markets, bakeries, and food suppliers [6], [7]. In order to enhance the quality and quantity of crops' production, farmers must safeguard their crops from bird attacking, as birds pose the greatest hazard to fruits and vegetables. Currently, in certain regions, farmers utilize manual methods to safeguard their crops.

B. Proposed Solution Through qCrop

There are a multitude of techniques that can be implemented to establish a secure environment for agriculture in different approaches like weather forecasting to mitigate soil and biological damage. Upon closer examination, it becomes evident that it is necessary to prioritize crop production by addressing potential threats such as soil quality, wild animals, and birds. The qCrop system employs advanced technology to monitor and mitigate the detrimental impact of birds on agricultural production. It achieves this through a fully automated system that operates independently without requiring any human intervention.

C. Significance of the Proposed Solution

This study presents a way for farmers to safeguard their farms from potential harm caused by birds or the deterioration of fruits and vegetables quality by evicting birds. Furthermore, this study aims to enhance the accuracy, low cost, and reduce power consumption of crops cultivation. It is equipped with diverse sensors for crop monitoring. In order to ensure a secure environment for the crop, it is recommended that those sensors be strategically located around the farm. This will assist the system when it begins to sense and observe until it detects any bird species. Subsequently, the system autonomously initiates operations to remove birds from the confines of the farm.

979-8-3315-3968-9/24 $31.00 © 2024 IEEE

Special Session

TABLE I
EXISTING SYSTEMS METHODOLOGIES

Authors	Years	Animals Detected	Methodology	Drawbacks
Ranparia, et al. [14]	2020	wild boar, nilgai, deer	ML Model	Those animals don't eat fruits farm's produces
Reddy, et al. [15]	2022	Not Listed	ML Model	Detects wildlife using ML needs time and power
Manikandan, et al. [9]	2021	Not Listed	Motion Sensor	Animal might be a bird and used sensors can't detect it
Deotale, et al. [3]	2021	Not Listed	Arduino+sensors	The wild animals don't eat fruits and vegetables
Shaik, et al. [21]	2022	Not Listed	Arduino+sensors	No dataset. The system evicts all objects
Upadhyay, et al. [23]	2020	cattle	Arduino+camera	The system needs an action done by the farmer

III. STATE-OF-THE-ART LITERATURE

Given the rapid advancements in technology, researchers are interested in assisting people in resolving world issues. One of the primary challenges in the field of agriculture pertains to safeguarding farms in order to ensure the provision of nutritious and abundant food to consumers. Table I displays some systems that have been developed with the purpose of safeguarding farms. However, these systems mostly focus on addressing non-crop-threatening animals.

An approach to protect crops against Wild Boar, Nilgai, and Deer through the use of machine learning technology (CNN) YOLOv3 in [14].

Another way to present a concept that focuses on guarding the crop from animal damage. It concentrates on particular species of wildlife. The efficacy of this technique is optimal when utilized for the purpose of preventing wild animals from a specified region, such as food warehouses [15].

To continuously monitor all animals, a motion sensor operates around the clock. This motion sensor alone is insufficient for protecting the crops, and a 24/7 system is unnecessary as the activity of animals is low [9].

Another system of crop safeguarding has been proposed. However, the inclusion of additional elements that are unrelated to the primary objective of this system may have an impact on its primary purpose and efficiency, such as the incorporation of temperature sensors and soil moisture monitoring [3].

A proposed system that operates continuously, 24 hours a day, seven days a week. The author specifically identified certain species as targets of the system, although most of the mentioned animals are not considered harmful as they mostly consume meat and do not have a negative impact on farms. [21].

Utilizing a manual eviction method and having the farmer perform the act is deemed inconvenient. In the present era, it is imperative for the system to be completely automated [23].

The proposed qCrop system addresses significant challenges and how their systems contribute to the advancement of Artificial Intelligence in the Smart Agricultural domain on a global scale. Table II depicts the quality and capabilities of qCrop as it presents a proposed significant method that aims to protect crops against bird species. The objective of this study is to explore strategies for safeguarding crops against birds. Safeguarding crops from bird attacks is crucial as they are the sole perilous species capable of causing harm to fruits and vegetables. The bole and height of some fruits and vegetables trees render them impervious to threats or harm from other wildlife, with the exception of birds.

IV. PROPOSED NOVEL METHODOLOGY

A. Methodology

The proposed system, as depicted in Fig. 2, has been designed to operate with optimal efficiency and guarantee the crop's safety and quality of the produce. Once a bird assaults the crop, qCrop emits a sound to evict the bird and thereafter sends a status report to the owner at the end of the day. The system operates on three levels. The interconnection of these three levels serves to enhance proficiency inside the system. These levels are commonly referred to as the software/hardware level, the cloud level, and the end user level and are further elaborated upon as follows:

B. Hardware/Software level

There are two algorithms, Algo [1, 2], that are used to split the hardware and software levels. These two algorithms demonstrate the approach of the qCrop system and its effectiveness in protecting crop production through using machine learning technology in combination with adaptive sensors.

C. The Cloud Level

The second level is the cloud. The cloud performs three primary functions, namely farmer notification, data storage, and report delivery. Upon receiving the data, the cloud promptly informs the farm owner, proceeds to count the captured objects, records the date and time of capture, and then stores the information. Ultimately, the cloud sends a report to the intended recipient.

D. The End User

The final level of this system is the end-user level. There are two nodes in this level. The first node represents the owner of the farm. The Ministry of Agriculture is the second node, provided the farm is registered. At the end of each working day, two soft copies of reports are dispatched to the aforementioned nodes. This report furnishes the nodes with data regarding the identified transactions during the day. The Ministry of Agriculture requires the storage of this information in its database for convenient backup when necessary. The interconnection of these three levels is facilitated through internet communications.

979-8-3315-3968-9/24 $31.00 © 2024 IEEE

Special Session

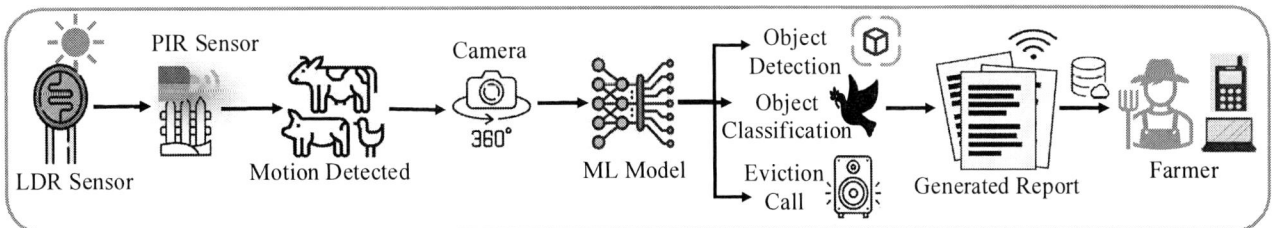

Fig. 2. Architecture of the Proposed qCrop system

Algorithm 1 qCrop Object Detection Methodology

Require: Light Dependent Resistor LDR, Passive Infrared Sensor PIR, Object Caption OC.

Ensure: Running or Sleepy Mode and Object Caption

1: LDR is running for light detection
2: **if** LDR detects light **then**
3:    qCrop turns on running mode
4:    qCrop in Running Mode and sensors begin to sense
5:    PIR starts sensing the heat energy
6:    **if** PIR sensed a heat energy **then**
7:       PIR notifies the system for an object is passing
8:       The system turns on the camera
9:       The system takes a OC
10:      Internal system takes OC and introduces it to DL Model
11:   **else**
12:      PIR is sensing
13:   **else**
14:      qCrop maintains on Sleepy Mode
15:   **end if**
16: **end if**

V. EXPERIMENT AND VALIDATION OF THE PROPOSED qCROP SYSTEM

A. Model (software)

The YOLOv8m model is used in the qCrop device to attain a notable level of efficiency [16]. The T4 GPU was utilized to train this model. The T4 GPU is a processor used in Google Colab [5] for training models. The T4 GPU operates using 40 GB of RAM. The model was evaluated on a processor with a Core-i5 CPU, 16 GB of RAM, and a 512 GB solid-state drive (SSD) for storage. The assessment of the model is depicted in Fig. 5. It displays the training model's loss with precision and recall metrics, while the lower part displays the validation/testing model loss. Fig 6 depicts the iteration of Mean Average Precision mAP50, precision, and recall as a graph of metrics [13].

The final crucial component of the trained model's outcome is the confusion matrix. Fig. 3 displays the accuracy of the projected bird results in the dataset and their level of truthfulness. The conversion from PyTorch to TensorFlow lite in YOLOv8m was achieved by the utilization of Open Neural Network Exchange model (ONNX). Fig. [4, 6] display

---

Algorithm 2 qCrop Object Classification Methodology

Require: Object Caption OC.

Ensure: Activation Status of the Alert System

1: OC contains the detected object
2: Internal system receives OC and forward it to our Deep Learning Model
3: Deep Learning Model uses OC as an input
4: The model classifies the object in OC
5: The model outputs the bird species
6: **if** The bird species is harmful to the farm **then**
7:    qCrop changes the activation status to risk mode
8:    Internal system chooses randomly bird distress calls
9:    qCrop produces bird distress calls
10:   **if** The chosen sound has been issued recently **then**
11:      Choose another sound from a list of [tones] = [ton1, tone n]:
         if n == tone1:
         n= tone2
         tone1++
12:      Alert System keeps altering for a specified period of time
13:      qCrop turns activation status to safe mode
14:   **else**
15:      qCrop plays the chosen bird distress calls
16:   **else**
17:      qCrop maintains the activation status as a safe mode
18:   **end if**
19: **end if**

---

the conclusive result of the YOLOv8m model's exceptional accuracy and its annotation at the end of this section. In order to mitigate the risk of overfitting, mAP50 and precision values were stated as 93% after 100 epochs.

B. Dataset

The system has been proposed utilizing an online dataset of bird species. The dataset is referred to as the 30birds-dataset [17]. In order to attain a result of high accuracy, it is essential to utilize a dataset including a large number of images depicting various species of birds. The dataset contains 905 original images displaying 30 distinct bird species, devoid of any null images. "Null photos" refers to an image that has been uploaded to the dataset but does not belong to any specific category within the dataset classes. The second

Special Session

TABLE II
qCROP VS. EXISTING SYSTEMS

Authors	Accuracy	ML model type	connectivity (WiFi?)	Automation	Cost-effcient System?
Ranparia, et al. [14]	NA	Detection	No	Fully automation	No
Reddy, et al. [15]	Up to 77%	Classification	Isn't mentioned	Fully automation	No
Manikandan, et al. [9]	NA	None	No	Isn't automatic	No
Deotale, et al. [3]	NA	None	No	Fully automation	No
Shaik, et al. [21]	NA	None	WiFi OR LAN	Isn't automatic	No
Upadhyay, et al. [23]	NA	None	No	Isn't automatic	No
qCrop	87%	Classification	WiFi	Fully automation	Yes

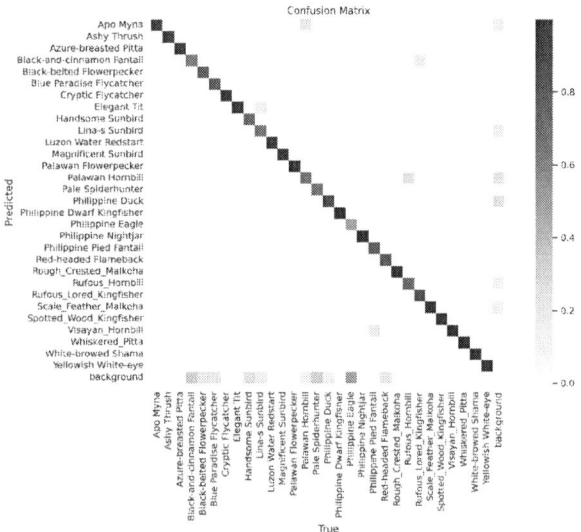

Fig. 3. Confusion Matrix of the Proposed qCrop Model

Fig. 4. Obtained Confidence Interval from the Proposed qCrop Object Detection Model

significant component of this dataset relates to the health of the dataset. Health, as a whole, exhibits a high degree of sensitivity to various factors. From a technical standpoint, the condition of images significantly impacts any dataset. The presence of a variety of bird species contributes to enhanced farm protection and the preservation of crops against bird threats. The images have been high-quality annotated. The dataset has been expanded from 905 initial photographs to 2171 pictures after performing the augmentation process. In datasets, augmentation is utilized to increase the size of the dataset which helps to avoid any overfitting. To provide further explication on the dataset that was used in this study, it has been divided into three primary components.

- Dataset split. The training set consists of 1899 photos, and an accuracy of 87% has been attained.
- Processing. An Auto-Oriented approach was implemented. During the processing stage, the dimensions of the images have been adjusted to 416*416.
- Augmentations. The saturation of the whole photo ranges from -25% to +25%.

The utilized dataset contains high-quality images. The great quality of the image is achieved through the use of several image positions, facilitating easy bird detection and ensuring the protection of farms from birds regardless of their location. The testing set images have been annotated to a total of 929 photographs, with an average size of 0.32 megapixels. The picture size ranges from 0.02 megapixels for the smallest size to 12.44 megapixels for the greatest size. Integrating Hardware and Software in qCrop certifies its ability to effectively safeguard crops' lives beyond the existing systems' capabilities as depicts in table II.

VI. CONCLUSION AND FUTURE RESEARCH

The present research introduces an IoT system named qCrop, which aims to enhance the quantity of produce from farms. The qCrop system offers continuous monitoring and automated control functionalities for farming on a global scale.

Primarily, it operates during the daytime, transitions into sleep mode throughout the nighttime, and awaits a command from the processor. Continuous Issuing of different sounds randomly by the system causes birds to be evacuated from the farms each time they attack crops. Getting used to the same sound by birds will not evacuate them again.

In future studies, the qCrop system might be enhanced with a more recent iteration of the YOLO models and an expanded dataset, hence facilitating the advancement and general utility of the qCrop device.

REFERENCES

[1] A. Alkinani, A. Mitra, S. Mohanty, and E. Kougianos, "Fruitpal 2.0: A smart healthcare framework for automatic monitoring of fruit consumption," 12 2023, pp. 422–425.

Special Session

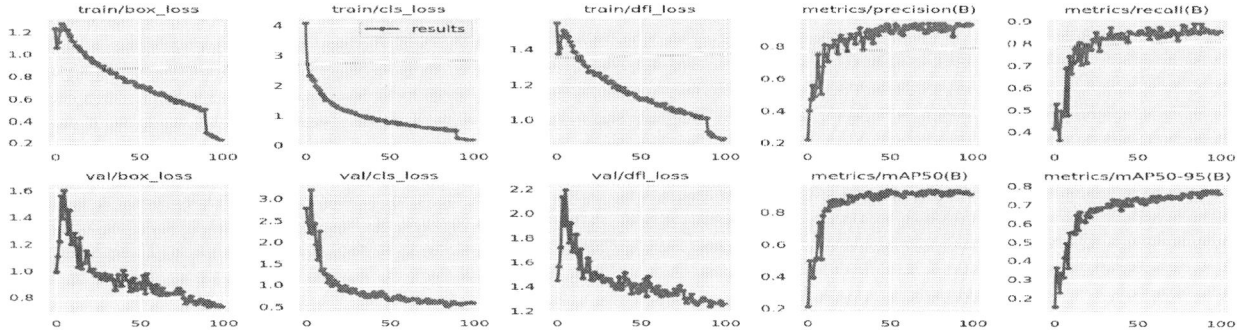

Fig. 5. Precision, Accuracy, Loss Generated from qCrop Classification Model for 100 Iterations.

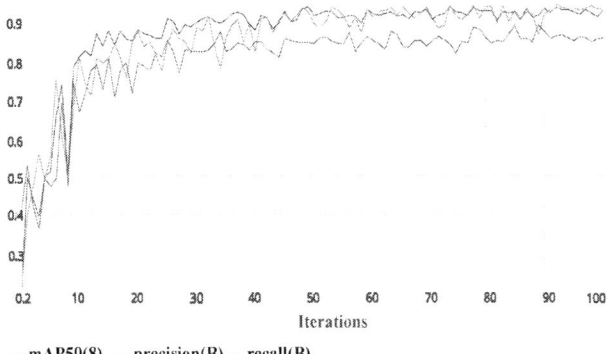

-- mAP50(8) -- precision(B) -- recall(B)

Fig. 6. Performance Metrics of the Proposed qCrop Classification Model.

[2] A. Chalmers, D. Zholzhanova, T. Arun, and A. Asadipour, "Virtual flavor: High-fidelity simulation of real flavor experiences," *IEEE Computer Graphics and Applications*, vol. 43, no. 02, pp. 23–31, mar 2023.

[3] P. Deotale and P. Lokulwar, "Smart crop protection system from wild animals using iot," in *International Conference on Computational Intelligence and Computing Applications (ICCICA)*, 2021, pp. 1–4.

[4] Geoadaptive, "Smart villages for azerbaijan," storymaps.arcgis.com/stories/c1cdeeb00ffe4200a98ccd30cbe521c4, Jan 2021.

[5] Google, "Explore the gemini api," colab.research.google.com/, 2017

[6] A. Konidena, M. Shanbhog, S. Singh, V. Sharma, A. K. Jain, and N. Sharma, "Efficient disease detection in wheat crops: A hybrid deep learning solution," in *3rd International Conference on Technological Advancements in Computational Sciences (ICTACS)*, 2023, pp. 648–654.

[7] D. Kumar, Y. Kumar, V. Kukreja, A. Bansal, and A. Bhattacherjee, "High performance eda and lda analysis: An application for wheat yield estimation," in *3rd International Conference on Advances in Computing, Communication, Embedded and Secure Systems (ACCESS)*, 2023, pp. 163–167.

[8] V. A. Kumar, A. Renaldo maximus, S. Vishnupriyan, K. Sheikdavood, and P. Gomathi, "Iot and artificial intelligence-based low-cost smart modules for smart irrigation systems," in *International Conference on Automation, Computing and Renewable Systems (ICACRS)*, 2022, pp. 254–260.

[9] P. Manikandan, A. Thenmozhi, G. Ramesh, T. R. K. Naidu, K. V. Reddy, and K. B. K. Reddy, "Crops protection system from animals using arduino," in *3rd International Conference on Advances in Computing, Communication Control and Networking (ICAC3N)*, 2021, pp. 682–685.

[10] S. P. Mohanty, U. Choppali, and E. Kougianos, "Everything you wanted to know about smart cities: The internet of things is the backbone," *IEEE Consumer Electronics Magazine*, vol. 5, no. 3, pp. 60–70, 2016.

[11] G. S. S. Preethi, K. Kavya, T. Monish, P. Poul, and B. Jayanag, "Internet of things based smart farm security system," in *2nd International Conference on Smart Electronics and Communication (ICOSEC)*, 2021, pp. 77–83.

[12] L. Rachakonda, "Agri-aid: An automated and continuous farmer health monitoring system using iomt," in *Internet of Things. IoT through a Multi-disciplinary Perspective*, L. M. Camarinha-Matos, L. Ribeiro, and L. Strous, Eds. Cham: Springer International Publishing, 2022, pp. 52–67.

[13] L. Rachakonda, S. P. Mohanty, E. Kougianos, and P. Sundaravadivel, "Stress-lysis: A dnn-integrated edge device for stress level detection in the iomt," *IEEE Transactions on Consumer Electronics*, vol. 65, no. 4, pp. 474–483, 2019.

[14] D. Ranparia, G. Singh, A. Rattan, H. Singh, and N. Auluck, "Machine learning-based acoustic repellent system for protecting crops against wild animal attacks," in *IEEE 15th International Conference on Industrial and Information Systems (ICIIS)*, 2020, pp. 534–539.

[15] D. R. Reddy, M. Kavya, S. Dharani, S. S. Tumpudi, P. Kodali, and N. Sandhya, "Design and development of a low-cost crop protection system using the internet of things and machine learning," in *IEEE International Symposium on Smart Electronic Systems (iSES)*, 2022, pp. 610–614.

[16] J. Redmon, "Home," docs.ultralytics.com/, IEEE Conference, Jan 2023.

[17] Roboflow, "dataset," universe.roboflow.com/bird/30birds_dataset, Nov 2021.

[18] A. S, A. Jose, C. Bhuvanendran, D. Thomas, and D. George, "Farmcopter: Computer vision based precision agriculture," 09 2020, pp. 1–6.

[19] C. Sausse, A. Baux, M. Bertrand, E. Bonnaud, S. Canavelli, A. Destrez, P. E. Klug, L. Olivera, E. Rodriguez, G. Tellechea, and S. Zuil, "Contemporary challenges and opportunities for the management of bird damage at field crop establishment," *Crop Protection*, vol. 148, p. 105736, 2021. [Online]. Available: https://www.sciencedirect.com/science/article/pii/S0261219421002064

[20] M. Shahbandeh. (2024, Jan) Total wheat consumption worldwide 2023/2024. accessed: 2023-02-15. [Online]. Available: www.statista.com/statistics/1094056/total-global-rice-consumption/

[21] M. F. Shaik, R. Mounika, A. D. Prasad, I. R. Raja, B. Sekhar, and D. Sampath, "Intelligent secure smart crop protection from wild animals," in *8th International Conference on Advanced Computing and Communication Systems (ICACCS)*, vol. 1, 2022, pp. 321–325.

[22] P. Sundaravadivel, E. Kougianos, S. P. Mohanty, and M. K. Ganapathiraju, "Everything you wanted to know about smart health care: Evaluating the different technologies and components of the internet of things for better health," *IEEE Consumer Electronics Magazine*, vol. 7, no. 1, pp. 18–28, 2018.

[23] A. Upadhyay and S. K. Maurya, "Protecting the agriculture field by iot application," in *International Conference on Power Electronics IoT Applications in Renewable Energy and its Control (PARC)*, 2020, pp. 411–414.

[24] V. Vageesan, M. K. Chakravarthi, V. B. Kumar, and G. Charan, "Anoxic microbial methanogenic detection for food safety sustentation," in *International Conference on Recent Trends on Electronics, Information, Communication Technology (RTEICT)*, Aug 2021, pp. 495–498.

[25] W. Xi and X. Tian, "Risk state analysis of food manufacturing quality risk and safety management model," in *2nd IEEE International Conference on Emergency Management and Management Sciences*, 2011, pp. 410–413.

Special Session

SanaSolo 2.0: Edge-Based Monitoring and Management of Soil Fertility Using IoT

Laavanya Rachakonda
Department of Computer Science
University of North Carolina Wilmington
Wilmington, NC, USA
rachakondal@uncw.edu

Samuel Stasiewicz
Department of Computer Science
University of North Carolina Wilmington
Wilmington, NC, USA
srs1996@uncw.edu

Abstract—**In order to achieve optimal and sustainable produce, soil fertility monitoring and management are needed. A fully automated IoT-enabled edge-based system, SanaSolo 2.0, is proposed through this paper, which continuously monitors soil moisture, soil temperature, electrical conductivity (EC), pH, and the fertility levels of nitrogen (N), phosphorus (P), and potassium (K). The system is applied to two different types of soils - infertile and fertile. The vermicompost is added to the soils to observe the improvements in the nutrition content. With calibration and analyses, a significant improvement was observed when infertile soil was added with vermicompost. These findings not only demonstrate the importance and need of this system but also prove that the system is reliable, thereby supporting smart agriculture practices.**

Index Terms—**soil fertility, worm castings, management, smart agriculture, IoT.**

I. INTRODUCTION

The ability to provide and supply essential nutrients to plants for their well-being and growth through soil can be termed soil fertility. The plant nutrients include soil's physical, chemical, and biological properties that determine the growth of the plant yield [1]. Thus, monitoring and managing soil health becomes a major component in the ecological paradigm of smart agriculture, which in turn promotes environmental sustainability [2].

Soil health and fertility have a direct relationship with factors like crop yield, quality, and resilience to environmental stresses [3]. Healthy and fertile soil supports healthy plant growth, which leads to healthy and higher agricultural yields, contributing to the food supply chain [4]. With fertile soil, there will be an abundance of diverse soil organisms that play a vital role in organic matter decomposition and natural fertilizer compositions [5]. Thus, the health of ecosystems also depends on soil fertility while mitigating the effects of climate change [6].

In order to maintain a healthy ecosystem, soil fertility management is essential [7]. Sustainable soil fertility management practices such as crop rotation, organic amendments, and precision agriculture help in maintaining soil fertility while minimizing the need for chemical fertilizers which often lead to pollution and degradation of natural resources

Supported by Center for the Support of Undergraduate Research & Fellowships Grant Award, UNCW

[8], [9]. Incorporating such practices also helps in water conservation, greenhouse gas emission reductions, and natural habitat preservation [10].

With the growth in technological advancements, there has been an improvement in deploying smart agricultural applications for daily practices of farming [11]. Smart agriculture refers to a combination of advanced technologies, data-driven techniques, IoT devices, sensors, and artificial intelligence to monitor, manage, and maintain several aspects of farming, including soil fertility and soil health [12], [13]. Soil health monitoring and management plays a vital role as such systems provide the data needed to make informed decisions [14], [15].

This paper introduces SanaSolo 2.0, a fully automated IoT-enabled edge-based system, an advanced version of the previously published SanaSolo system, designed to further enhance soil fertility monitoring and management within the framework of smart agriculture. SanaSolo 2.0 is an extension of the original system, which monitors the seven parameters and signifies the impact of vermicompost or a natural fertilizer in converting infertile soil to fertile soil. The monitored data are sent to the cloud for storage, which the user can access. The broad perspective of the system is represented in Figure 1.

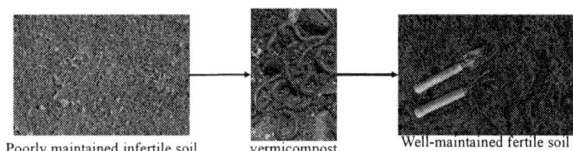

Fig. 1: Broad Representation of the Proposed SanaSolo 2.0 System

The organization of this paper is as follows: Section II summarizes the existing state-of-the-art literature for soil fertility monitoring and management. Section 2 lists the novel contributions along with the motivation behind this research. Section IV represents the broad perspective of the SanaSolo 2.0 System. Section V represents the methodology of the proposed system and discusses its implementation and validation, followed by conclusions and future research in Section VI.

979-8-3315-3968-9/24 $31.00 © 2024 IEEE

Special Session

II. Literature Review

The soil fertility monitoring and management field has experienced significant advancements with the integration of smart technologies aimed at optimizing agricultural practices. This section reviews the key literature that aligns with the approach adopted in SanaSolo 2.0, highlighting state-of-the-art systems, key methodologies deployed, and drawbacks.

Several studies emphasize the importance of real-time soil nutrient monitoring using advanced sensor networks. Precision agriculture techniques, such as in-field soil sensors, monitor crucial parameters like nitrogen (N), phosphorus (P), and potassium (K), and wireless sensor networks (WSN) in real-time soil moisture and nutrient monitoring proposed through [16], [17].

An IoT-based smart soil systems that track temperature, moisture, pH, and electrical conductivity to provide actionable insights for farmers is discussed in [18]. A study [12] further discussed how Industry 4.0 technologies, including IoT, are transforming agriculture by improving efficiency and sustainability.

AI and ML methods have been deployed to monitor the nutrient content in soil and soil health, in general, [19]–[22] Studies showed that vermicompost enhances soil structure, nutrient availability, and microbial activity, supporting the approach used in SanaSolo 2.0 [23], [24]. However, the studies mentioned above do not have a system that performs continuous monitoring and management of soil health. A detailed explanation of a few of the state-of-the-art literature is mentioned in Table I.

A. Need for Soil Fertility Monitoring and Management

The need for advanced soil fertility monitoring systems is driven by the increasing demand for sustainable agriculture and the need to optimize resource use. As modern farming practices evolve, there is a growing recognition that precise, real-time data on soil health is crucial for making informed decisions about soil management. Proposing a smart, fully automated Edge-enabled IoT smart system through SanaSolo 2.0 offers a promising solution to this challenge.

III. Motivation for SanaSolo 2.0

The major factors that motivate a fully automated IoT-enabled edge-based system SanaSolo 2.0 are mentioned in this section and are represented in Figure 2.

A. Soil Nutrition

Nutrition depletion in the produce that we are currently consuming is highly concerning. One way to address this issue is by maintaining a proper soil fertility monitoring and management plan. Soil needs adequate amounts of nutrients to sustain the rapidly changing environments. The need to continuously monitor and manage soil health remains the motivating factor for SanaSolo 2.0.

B. Impact on Yield/Plant Growth

Poorly maintained soil can result in stunted growth, reduced yields, and lower-quality produce. By providing real-time, continuous, and precise data to the farmers, Sanasolo 2.0 allows them to make timely interventions, thereby optimizing nutrient levels and improving plant growth outcomes. The system helps farmers to produce high-nutrient crops.

C. Carbon Footprint

SanaSolo 2.0 strives to optimize soil health and reduce the dependency on synthetic fertilizers, thereby promoting the use of natural fertilizers. The use of synthetic fertilizers has a huge impact on the carbon footprint of agricultural practices. The proposed system promotes natural fertilizers that contribute to lower greenhouse gas emissions and maintain and monitor soil health.

D. Usage of Natural Fertilizers

Natural fertilizers like compost and vermicompost improve the quality of soil over time. This addition will reduce the need to depend on synthetic fertilizers. SanaSolo 2.0 supports this practice by monitoring the direct impact of natural fertilizers on soil health and also provides information to farmers so they can make the needed adjustments, which promotes environmentally friendly farming practices.

E. Increasing Barren/Infertile Land and Conversion to Fertile Land

A major challenge for global agriculture is the expansion of barren and infertile land due to poor maintenance of soil health and fertility [2]. SanaSolo 2.0 is motivated by the potential to reverse this trend by providing the tools needed to rehabilitate degraded soils. The system helps to transform the infertile land into fertile land by continuously monitoring soil health.

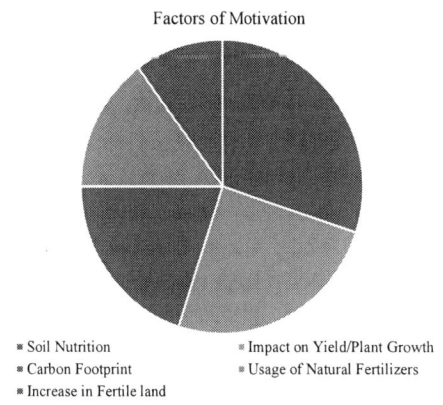

Fig. 2: Motivation for the Proposed SanaSolo 2.0 System

The novel contributions that are proposed through the fully automated IoT-enabled edge-based system, SanaSolo 2.0 system are listed as follows:

979-8-3315-3968-9/24 $31.00 © 2024 IEEE

Special Session

TABLE I: Literature Review.

Study	Methodology	Drawbacks/Weaknesses
Adamchuk et al. (2010) [16]	On-the-go soil sensors for precision agriculture and emphasizes the use of in-field sensors for real-time monitoring of soil nutrients.	Limited to specific crops and conditions; may require significant calibration for different soil types.
Jawad et al. (2019) [17]	Demonstrates the effectiveness of WSNs in real-time soil moisture and nutrient monitoring.	High cost of deployment and maintenance; potential connectivity issues in remote areas.
Zhang et al. (2020) [18]	IoT-based smart soil monitoring systems	Data security and privacy concerns; reliance on stable internet connectivity.
Zambon et al. (2019) [12]	Discuss the role of Industry 4.0 in agriculture	Requires significant investment in infrastructure adoption barriers in developing regions.
Arancon et al. (2006) [23]	Reviews the impact of vermicompost on soil health	Inconsistent results depending on the source of organic material.
Pathma & Sakthivel (2012) [24]	Explores the agricultural benefits of vermicompost.	Limited scalability; potential for contamination if not properly managed.
Gebbers & Adamchuk (2010) [19]	Stresses the importance of calibration and validation in precision agriculture	No system or prototype designed to test the theory.
Sun et al. (2021) [20]	Explores the integration of AI for predictive analysis in soil health monitoring.	Requires high computational resources; potential bias in AI models if not properly trained.
Iqbal et al. (2022) [21]	Discusses machine learning models for predicting soil nutrient levels.	No system or prototype designed to test the theory.
Gao et al. (2023) [22]	Integrate IoT with blockchain for secure and transparent soil data management.	High complexity in system design; potential challenges in scalability.
SanaSolo (2023) [15]	Smart soil fertility monitoring system	System does not perform soil fertility management.

- A Fully Independent Automated System operates automatically without human intervention, ensuring healthy soil.
- IoT-Enabled Edge Computing: Integrates IoT technology with edge computing, allowing on-site data processing for real-time decision-making and more responsive soil management.
- Real-Time Soil Monitoring and Management: Provides continuous, real-time data on key soil parameters (moisture, temperature, pH, nutrient levels), enabling informed decisions to optimize soil fertility and improve crop yields.
- Sustainable Agriculture and Carbon Footprint Reduction: Promotes sustainable farming practices by minimizing reliance on synthetic fertilizers and optimizing resource use, contributing to a reduced carbon footprint in agriculture.
- Rehabilitation of Infertile Land: Aids in the rehabilitation of infertile land by providing the necessary tools for monitoring and improving soil health, supporting the conversion of degraded land into productive farmland.
- A fully Connected System: A fully connected system with an easy user interface for farmers to analyze the soil parameters.

IV. DESIGN PERSPECTIVE OF THE PROPOSED SANASOLO 2.0 SYSTEM

The proposed fully automated IoT-enabled edge-based system's detailed representation is in Figure 3. The system starts with performing an initial fertility check on the infertile soil, followed by the addition of vermicompost or natural fertilizer. The transformation of the soil nutrient properties is displayed through the user interface to the farmers, enabling them to adjust soil parameters as needed to optimize plant growth and yield, thus allowing soil fertility monitoring and management.

The steps involved in the logic of the proposed system are mentioned in Algorithm I.

Algorithm 1 Proposed Methodology for IoT enabled Edge SanaSolo 2.0 Soil Fertility Monitoring and Management System

---
1: Develop the system with integrated sensors and IoT components.
2: Calibrate the sensors to ensure accurate measurements of soil parameters.
3: Collect two soil samples: garden soil (infertile) and vermiculture soil (fertile).
4: Measure the soil nutrient parameters for both samples using the calibrated system.
5: Add vermiculture compost to the garden soil sample.
6: Measure the soil nutrient parameters after adding compost.
7: **if** nutrient readings are not optimal **then**
8:     Add more vermiculture compost and repeat the process.
9: **end if**
10: Transfer the collected data to the user interface (UI) for convenient analysis.

A. Parameters Considered

A total of seven parameters are monitored through the proposed system.

1) Moisture Content: Soil moisture is crucial for plant growth as it directly affects the availability of water to plant roots, supporting the uptake of nutrients and overall plant health.
2) Soil Temperature: Soil temperature impacts seed germination, root development, and microbial activity, making it a key factor for planting and fertilization.

Special Session

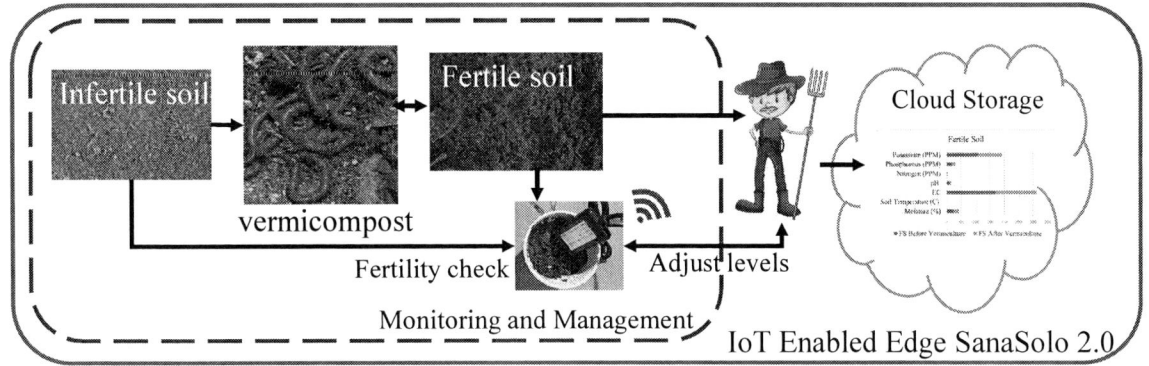

Fig. 3: Detailed Representation of the Proposed SanaSolo 2.0 System

3) Electrical Conductivity (EC): EC is an indicator of soil salinity, which affects the soil's ability to retain nutrients and water. High EC can lead to nutrient imbalances and reduced plant growth. Low EC levels indicate that there are insufficient nutrients available in the soil.

4) pH: Soil pH determines the availability of essential nutrients and affects microbial activity in the soil. A balanced pH is essential for plants to uptake optimal nutrients.

5) Nitrogen (N) Fertility Level: Nitrogen is a vital nutrient for plant growth, for leaf development, and overall biomass production. Adequate nitrogen levels are necessary for healthy plant growth.

6) Phosphorus (P) Fertility Level: Phosphorus is essential for plant energy transfer, root development, and flowering. It plays a critical role in forming DNA and RNA within the plant.

7) Potassium (K) Fertility Level: Potassium is important for plant water regulation, enzyme activation, and disease resistance. It also helps strengthen plant cell walls and improve overall plant health.

V. METHODOLOGY OF THE PROPOSED FULLY AUTOMATED IOT-ENABLED EDGE-BASED SANASOLO 2.0 SYSTEM

A. SanaSolo 2.0 System Calibration

The SanaSolo system has been calibrated and tested on two soil samples - garden soil (infertile) and vermiculture soil (fertile) using a 7-in-1 soil sensor and a microprocessor as shown in Figure 4.

The system uses capacitance to measure the soil's dielectric permittivity. The sensor creates a voltage proportional to permittivity, which can be affected by different soil parameters. Higher moisture content in the soil increases the dielectric permittivity, which enhances soil composition parameters. After calibrating, the system has observed significant measurements as represented in Figure 5.

Fig. 4: Device Calibration for the Proposed SanaSolo 2.0 System

Before Calibration	After Calibration
Moisture: 7.30 %	Moisture: 25.00 %
Temperature: 21.50 C	Temperature: 21.20 C
EC: 0 us/cm	EC: 1820 us/cm
ph: 9.00 ph	ph: 11.00 ph
Nitrogen: 0 mg/kg	Nitrogen: 43 mg/kg
Phosphorous: 0 mg/kg	Phosphorus: 60 mg/kg
Potassium: 0 mg/kg	Potassium: 121 mg/kg

Fig. 5: Soil Parameters Before and After Calibration for Infertile Soil

B. SanaSolo 2.0 Validation

In order to verify the accuracy of the data obtained from the SanaSolo 2.0 system, soil samples were sent to the University of Massachusetts (UMass) for independent testing. The results from UMass matched the readings generated by the system, confirming the reliability of our measurements. The UMass samples for infertile (6a) and fertile (6b) soils are represented in Figure 6.

C. SanaSolo 2.0 Implementation

After calibrations were made and used to collect different data sets for each, the fertile soil calibration set had a range of 60 – 90 % accuracy, and the infertile calibration set ranged from 75 - 90 % accuracy. The percentage of true positives

979-8-3315-3968-9/24 $31.00 © 2024 IEEE   239

Results

Analysis	Value Found	Optimum Range	Analysis	Value Found	Optimum Range
Soil pH (1:1, H2O)	5.3		Cation Exch. Capacity, meq/100g	7.8	
Modified Morgan extractable, ppm			Exch. Acidity, meq/100g	7.0	
Macronutrients			Base Saturation, %		
Phosphorus (P)	0.5	4-14	Calcium Base Saturation	9	50-80
Potassium (K)	14	100-160	Magnesium Base Saturation	1	10-30
Calcium (Ca)	134	1000-1500	Potassium Base Saturation	0	2.0-7.0
Magnesium (Mg)	12	50-120	Scoop Density, g/cc	1.27	
Sulfur (S)	2.9	>10	Optional tests		
Micronutrients *			Soluble Salts (1:2), dS/m	0.03	<0.6
Boron (B)	0.0	0.1-0.5	Nitrate-N (NO3-N), ppm	0	
Manganese (Mn)	1.9	1.1-6.3			
Zinc (Zn)	6.9	1.0-7.6			
Copper (Cu)	1.4	0.3-0.6			
Iron (Fe)	16.6	2.7-9.4			
Aluminum (Al)	55	<75			
Lead (Pb)	3.2	<22			

* Micronutrient deficiencies rarely occur in New England soils; therefore, an Optimum Range has never been defined. Values provided represent the normal range found in soils and are for reference only

Soil Test Interpretation

Nutrient	Very Low	Low	Optimum	Above Optimum
Phosphorus (P):				
Potassium (K):				
Calcium (Ca):				
Magnesium (Mg):				

(a) UMass Test Results for Infertile Soil.

Results

Analysis	Value Found	Optimum Range	Analysis	Value Found	Optimum Range
Soil pH (1:1, H2O)	3.6		Cation Exch. Capacity, meq/100g	51.7	
Modified Morgan extractable, ppm			Exch. Acidity, meq/100g	33.3	
Macronutrients			Base Saturation, %		
Phosphorus (P)	54.1	4-14	Calcium Base Saturation	29	50-80
Potassium (K)	375	100-160	Magnesium Base Saturation	5	10-30
Calcium (Ca)	2978	1000-1500	Potassium Base Saturation	2	2.0-7.0
Magnesium (Mg)	304	50-120	Scoop Density, g/cc	0.27	
Sulfur (S)	59.5	>10	Optional tests		
Micronutrients *			Soluble Salts (1:2), dS/m	0.21	<0.6
Boron (B)	0.3	0.1-0.5	Nitrate-N (NO3-N), ppm	0	
Manganese (Mn)	5.4	1.1-6.3			
Zinc (Zn)	4.9	1.0-7.6			
Copper (Cu)	0.2	0.3-0.6			
Iron (Fe)	2.4	2.7-9.4			
Aluminum (Al)	5	<75			
Lead (Pb)	2.5	<22			

* Micronutrient deficiencies rarely occur in New England soils; therefore, an Optimum Range has never been defined. Values provided represent the normal range found in soils and are for reference only

Soil Test Interpretation

Nutrient	Very Low	Low	Optimum	Above Optimum
Phosphorus (P):				
Potassium (K):				
Calcium (Ca):				
Magnesium (Mg):				

(b) UMass Test Results for Fertile Soil.

Fig. 6: Test Results from UMass Soil Sampling Lab.

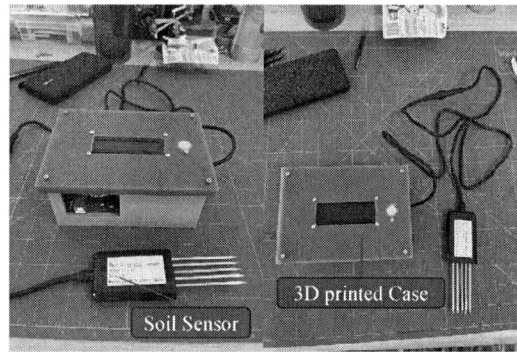

Fig. 7: Designed Prototype of the Proposed SanaSolo 2.0 System

SanaSolo 2.0 a valuable tool for enhancing productivity and supporting the future of smart agriculture by helping convert infertile lands to fertile lands.

For future research, the SanaSolo system will be further improved and will be tested for more soil types. The system will create a relationship between the nutrient contents of the produce that is grown in different soil types. The system will remain fully automatic and will be deployed on Edge for scalability and reliability.

VII. ACKNOWLEDGEMENT

We thank the University of Massachusetts Sample Testing Lab for their prompt collaboration with us.

from the different data sets was calculated accurately. From the results generated by the UMass Soil Sampling lab, the interpretation of the parameters has been defined by referring to manuals [25], [26]. The fertility rates for the soil parameters are represented in Table II.

A system prototype was designed using CAD tools and subsequently 3D printed for further testing and validation on edge, as represented in Figure 7.

After implementing the interpreted values as shown in Table II, the results are represented in Figure 8 on the edge processor. The data are also transferred to a cloud space with user access.

VI. CONCLUSION AND FUTURE RESEARCH

SanaSolo 2.0 offers an advanced IoT-enabled fully automated solution for soil fertility monitoring, integrating real-time data on key soil parameters such as moisture, temperature, EC, pH, and nutrient levels. Enabling precise soil management helps farmers optimize crop yields and promote sustainable practices. The system's reliability has been confirmed through successful calibration and external validation, making

REFERENCES

[1] R. Weil and N. Brady, *The Nature and Properties of Soils. 15th edition*, 04 2017.

[2] R. Lal, "Restoring Soil Quality to Mitigate Soil Degradation," *Sustainability*, vol. 7, no. 5, pp. 5875–5895, 2015.

[3] M. G. Kibblewhite, K. Ritz, and M. J. Swift, "Soil health in agricultural systems," *Philos Trans R Soc Lond B Biol Sci*, vol. 363, no. 1492, pp. 685–701, Feb. 2008.

[4] D. Tilman, K. G. Cassman, P. A. Matson, R. Naylor, and S. Polasky, "Agricultural sustainability and intensive production practices," *Nature*, vol. 418, no. 6898, pp. 671–677, 2002.

[5] R. D. Bardgett and W. H. van der Putten, "Belowground biodiversity and ecosystem functioning," *Nature*, vol. 515, no. 7528, pp. 505–511, 2014.

[6] V. Girija Veni, C. Srinivasarao, K. Sammi Reddy, K. L. Sharma, and A. Rai, "Chapter 26 - Soil health and climate change," in *Climate Change and Soil Interactions*, M. N. V. Prasad and M. Pietrzykowski, Eds. Elsevier, 2020, pp. 751–767.

[7] P. Smith, M. F. Cotrufo, C. Rumpel, K. Paustian, P. J. Kuikman, J. A. Elliott, R. McDowell, R. I. Griffiths, S. Asakawa, M. Bustamante, J. I. House, J. Sobocká, R. Harper, G. Pan, P. C. West, J. S. Gerber, J. M. Clark, T. Adhya, R. J. Scholes, and M. C. Scholes, "Biogeochemical cycles and biodiversity as key drivers of ecosystem services provided by soils," *SOIL*, vol. 1, no. 2, pp. 665–685, 2015.

[8] K. W. T. Goulding, "Soil acidification and the importance of liming agricultural soils with particular reference to the United Kingdom," *Soil Use and Management*, vol. 32, no. 3, pp. 390–399, 2016.

[9] J. C. Dawson, D. R. Huggins, and S. S. Jones, "Characterizing nitrogen use efficiency in natural and agricultural ecosystems to improve the performance of cereal crops in low-input and organic agricultural systems," *Field Crops Research*, vol. 107, no. 2, pp. 89–101, 2008.

Special Session

TABLE II: Fertility Ranges Considered for SanaSolo 2.0 System.

Parameter	Deficient	Below Target	Optimal	Above Target	Excessive
pH	0-3	4-5	6-8	9-10	11+
Nitrogen	0-9	10-29	30-50	51-70	71+
Phosphorus	0-3	3-5	6-14	15-45	46+
Potassium	0-49	50-99	100-160	161-300	301+
Electrical Conductivity	-1	0.0-0.7	0.8-1.8	1.8-2.4	2.5+

```
-Garden TopSoil-

Moisture: 25.00 %
Temperature: 21.20 C
EC: 1820 us/cm : SUFFICIENT EC
ph: 11.00 ph : INSUFFICIENT - Soil pH is Basic
Nitrogen: 43 mg/kg : SUFFICIENT Nitrogen Content
Phosphorous: 60 mg/kg : SUFFICIENT Phosphorous Content
Potassium: 121 mg/kg : SUFFICIENT Potassium Content
```

(a) Soil Parameters with Garden (infertile) soil.

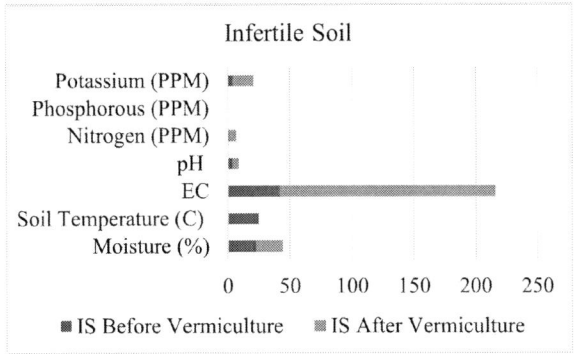

(b) Soil Parameters with Garden (infertile) soil.

```
-Vermiculture-

Moisture: 25.20 %
Temperature: 21.10 C
EC: 2500 us/cm : SUFFICIENT EC
ph: 7.00 ph : SUFFICIENT - Soil pH is Neutral
Nitrogen: 84 mg/kg : OVERLY SUFFICIENT Nitrogen Content
Phosphorous: 118 mg/kg : OVERLY SUFFICIENT Phosphorous Content
Potassium: 237 mg/kg : OVERLY SUFFICIENT Potassium Content
```

(c) Soil Parameters with Vermiculture (fertile) soil.

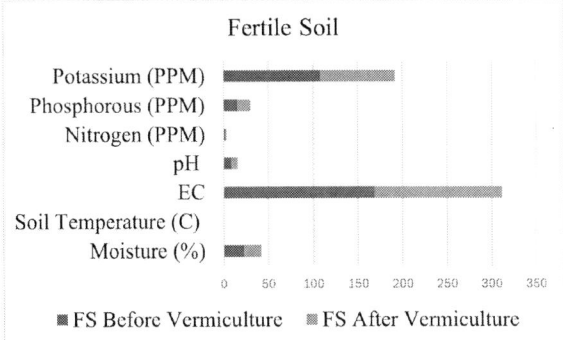

(d) Soil Parameters with Vermiculture (fertile) soil.

Fig. 8: SanaSolo 2.0 Results for Two Types of Soil.

[10] D. S. Powlson, A. P. Whitmore, and K. W. T. Goulding, "Soil carbon sequestration to mitigate climate change: a critical re-examination to identify the true and the false," *European Journal of Soil Science*, vol. 62, no. 1, pp. 42–55, 2011.

[11] N. Zhang, M. Wang, and N. Wang, "Precision agriculture - A worldwide overview," *Computers and Electronics in Agriculture*, vol. 36, pp. 113–132, 11 2002.

[12] I. Zambon, M. Cecchini, G. Egidi, M. G. Saporito, and A. Colantoni, "Revolution 4.0: Industry vs. Agriculture in a Future Development for SMEs," *Processes*, vol. 7, no. 1, 2019.

[13] L. Rachakonda, "Agri-Aid: An Automated and Continuous Farmer Health Monitoring System Using IoMT," in *Internet of Things. IoT through a Multi-disciplinary Perspective*, L. M. Camarinha-Matos, L. Ribeiro, and L. Strous, Eds. Cham: Springer International Publishing, 2022, pp. 52–67.

[14] J. Anderson and G. Feder, "Agricultural Extension," R. Evenson and P. Pingali, Eds. Elsevier, 2007, vol. 3, ch. 44, pp. 2343–2378.

[15] L. Rachakonda and S. Stasiewicz, "Sana Solo: An Intelligent Approach to Measure Soil Fertility," in *Internet of Things. Advances in Information and Communication Technology*, D. Puthal, S. Mohanty, and B.-Y. Choi, Eds. Cham: Springer Nature Switzerland, 2024, pp. 395–404.

[16] V. I. Adamchuk, J. W. Hummel, M. T. Morgan, and S. K. Upadhyaya, "On-the-go soil sensors for precision agriculture," *Computers and Electronics in Agriculture*, vol. 44, no. 1, pp. 71–91, 2004. [Online]. Available: https://www.sciencedirect.com/science/article/pii/S0168169904000444

[17] H. M. Jawad, R. Nordin, S. K. Gharghan, A. M. Jawad, and M. Ismail, "Energy-Efficient Wireless Sensor Networks for Precision Agriculture: A Review," *Sensors*, vol. 17, no. 8, 2017.

[18] N. Zhang, M. Wang, and N. Wang, "Precision agriculture—a worldwide overview," *Computers and Electronics in Agriculture*, vol. 36, no. 2, pp. 113–132, 2002.

[19] R. Gebbers and V. Adamchuk, "Precision Agriculture and Food Security. Science327(5967), 828-831," *Science (New York, N.Y.)*, vol. 327, pp. 828–31, 02 2010.

[20] S. Meghwanshi, "ARTIFICIAL INTELLIGENCE IN AGRICULTURE: A REVIEW," *International Research Journal of Modernization in Engineering Technology and Science*, vol. 6, pp. 4358–4363, 03 2024.

[21] K. Gurubaran, p. S, S. S, D. M B, V. Nagavel, S. Parveen, P. Ramesh, and P. T. V. Bhuvaneswari, "Machine Learning Approach for Soil Nutrient Prediction," 11 2023.

[22] J. Lin, Z. Shen, A. Zhang, and Y. Chai, "Blockchain and IoT based Food Traceability for Smart Agriculture," in *International Conference on Crowd Science and Engineering*, ser. ICCSE'18. New York, NY, USA: Association for Computing Machinery, 2018.

[23] N. Arancon, C. A. Edwards, and P. Bierman, "Influences of vermicomposts on field strawberries: Part 2. Effects on soil microbiological and chemical properties," *Bioresource technology*, vol. 97, pp. 831–40, 04 2006.

[24] J. Pathma and N. Sakthivel, "Microbial diversity of vermicompost bacteria that exhibit useful agricultural traits and waste management potential," *SpringerPlus*, vol. 1, no. 1, p. 26, 2012.

[25] UnivOfMass, "Soil and Plant Nutrient Testing Laboratory," Online, Jul. 2013. [Online]. Available: https://ag.umass.edu/soil-plant-nutrient-testing-laboratory/fact-sheets/interpreting-your-soil-test-results

[26] UMass, "UMass Extension Crops, Dairy, Livestock and Equine Program," Online, Jul. 2014. [Online]. Available: https://ag.umass.edu/crops-dairy-livestock-equine/fact-sheets/nutrient-recommendations-for-field-crops-in-massachusetts

Special Session

Transforming Agriculture: A Mini-Review of IoT Innovations and Their Impact

Laavanya Rachakonda
Department of Computer Science
University of North Carolina Wilmington
Wilmington, NC, USA
rachakondal@uncw.edu

Abstract—The Internet of Things (IoT) has revolutionized traditional agricultural practices, making its impact in the era of intelligent farming. This mini-review explores the role of IoT-enabled electronics and smart systems and the need to deploy them to transform agriculture through real-time data collection, automation, and resource optimization. The paper also discusses the critical technologies grouped into sensors, drones, and data processing techniques for the contributions of IoT-enabled electronics to precision and vertical farming, livestock monitoring, autonomous irrigation systems, soil, and farmers. The paper also highlights the benefits of IoT devices by addressing global challenges such as data security and privacy, climate change, labor shortages, and sustainable energies while discussing the technical and socioeconomic challenges to widespread adoption across the agricultural community. Finally, the review further emphasizes the potential of collaborative robots, swarm robots, AI and ML applications for regenerative agriculture, biometric sensors, and personalized farming to enhance the effectiveness and scalability of IoT-enabled smart agriculture.

Index Terms—Smart and Intelligent Systems, Smart Agriculture, IoT, IoT Technologies.

I. INTRODUCTION

The agricultural industry has experienced significant growth in applying advanced technologies to improve efficiency, accuracy, productivity, and sustainability. Among these technological innovations, incorporating the Internet of Things (IoT) has transformed traditional farming practices into more innovative and more efficient methodologies by utilizing intelligent and connected systems. The shift in this growth was motivated by the need to address global challenges such as environmental concerns, food scarcity, land scarcity, resource limitations, and increasing food demand [1]. This paper provides an overview of the role of IoT-enabled electronics, highlighting its key technologies and trends through a review of current practices and emerging technologies in smart agriculture as represented in Figure 1.

A transformative technological paradigm that integrates physical devices to collect, exchange, analyze, and process data autonomously while connected to digital networks, all by maintaining individual IP addresses, can be termed the Internet of Things (IoT) [2]. In simple terms, IoT can be called the interconnection of everyday devices through the internet, allowing them to communicate and perform tasks without direct human intervention. These devices are equipped with sensors, actuators, and communication protocols to gather

Fig. 1: Illustration of Smart Systems at Various Aspects in Smart Agriculture.

and share the data over the network, enabling real-time monitoring and control [3]. This connectivity allows for seamless integration among the physical and digital worlds, enhancing the capabilities and efficiency of various sectors, including agriculture [4].

Smart agriculture is the sector that uses advanced IoT technologies and data analytics to optimize and improve various agricultural productions. Smart agriculture integrates automation and smartness into agricultural operations by incorporating sensors and data analytics to enable intelligent systems for irrigation, livestock management, farmer health management, precision, and vertical agriculture [5]. This approach enhances crop yield, reduces waste, and optimizes the limited available resources [6].

Smart systems, or systems of systems, are complex networks in which each subsystem interacts and collaborates to achieve a common goal. These systems are characterized based on data-driven insights like the ability to adapt, learn, and optimize their performance. For innovative agricultural operations, these systems embed diverse components such as sensors, actuators, and data platforms to create a coherent framework enabling more productive management of resources and processes [7], [8].

IoT-enabled electronics encompass a range of devices and

979-8-3315-3968-9/24 $31.00 © 2024 IEEE

systems designed to interact within the IoT framework. These devices include a variety of sensors, actuators, and communication protocols to collect and transfer information to centralized systems for data processing and analysis. Smart irrigation systems, livestock management systems, soil and farmer health monitoring systems, drones, and climate systems are a few of the many IoT-enabled electronics for agriculture. These devices drive decision-making processes and operational efficiency in farming [5], [9]

This paper introduces an overview of such systems, technologies, and practices in smart agriculture. The organization of the paper is as follows: Sec II describes the need for such improved systems, Sec III discusses the technologies that are currently deployed in such systems, Sec IV discusses various state-of-the-art applications and smart products, Sec V discusses the possible difficulties that may arise while deploying the systems, and Sec VI states the conclusion while bringing the possible future directives to improve and enhance the agricultural practices.

II. THE NEED FOR IoT IN SMART AGRICULTURE

The growth in complex modern agricultural practices brings challenges requiring innovative solutions that perform beyond traditional methods. IoT enables addressing such challenges by providing real-time, data-driven insights. Therefore, the need for IoT in intelligent agriculture can be understood by examining the critical global pressures and the advantages IoT delivers.

A. Global Challenges

The agricultural sector faces significant challenges due to global trends such as population growth, climate change, and water scarcity [10]. With the world population projected to reach 9.7 billion by 2050, the demand for food will increase by nearly 70%, according to the United Nations [11]. By incorporating IoT, optimal resource usage, mitigation of environmental aspects, and an increase in resilience against climate-related risks can be achieved.

B. Precision Agriculture

A farming management technique using IoT-based data-driven technologies like drones and connected systems to analyze and utilize real-time data to improve crop yields and farm management efficiency can be called precision agriculture [1]. Studies have shown that precision farming techniques can reduce water use by 20–40% and improve crop yields by 10–20% [12].

C. Livestock Management

IoT in livestock management market size was estimated at 3.21 USD Billion in 2023. This industry is expected to grow from 3.89 USD Billion in 2024 to 18.31 USD Billion by 2032. A 21.35% growth is expected in the period 2024-2032 [13].

D. Farmer Health

IoT-enabled electronics can improve farmer health by tracking vital physiological parameters while considering environmental parameters like UV exposure and air quality [14], [15].

E. Labor Shortages

Agriculture is increasingly affected by labor shortages, particularly in developed countries, because of fewer immigrants, an aging workforce, rising wages, and fewer younger individuals entering the industry. IoT-enabled automation offers solutions to this labor gap [16], [17]. IoT enables automated 24/7 monitoring and management by reducing the dependency on manual labor while increasing the consistency and speed of operations, thereby maintaining crop health and productivity.

F. Sustainability Goals

As there is growing awareness of the environmental impact of conventional farming methods, sustainability has become an essential aspect of agricultural operations. IoT enables sustainability by optimizing the limited resources and their usage, reducing chemical inputs, and minimizing waste by following diverse compost techniques. For example, IoT-enabled smart irrigation systems significantly lower water consumption, and automated monitoring systems can reduce overall pesticide usage by targeting affected areas across the field [6], [18]. These IoT-driven practices contribute to achieving global sustainability standards such as the United Nations' Sustainable Development Goals (SDGs) [19].

G. Government Initiatives and Policies

Governments worldwide recognize the importance of IoT in addressing food security and sustainability issues and have launched initiatives to promote the adoption of IoT-enabled smart agriculture technologies. The European Union's Common Agricultural Policy (CAP) and the U.S. Department of Agriculture's (USDA) Climate-Smart Agriculture program are encouraging farmers to adopt IoT-enabled systems [20], [21]. These initiatives aim to enhance food production efficiency, reduce carbon emissions, improve soil health, reduce greenhouse emissions, and conserve water. Government incentives include research grants for developing new IoT technologies and subsidies for precision farming equipment.

H. Economic Impact

The economic benefits of IoT in agriculture stem from increased efficiency and cost reduction. According to statistics, the global agricultural IoT market is projected to value 13 USD Billion by 2026. IoT enabled with robotics, automated crop harvesting, on-field navigation, remote sensing, and drones can improve operations, allowing farmers market access and enhancing income opportunities through resource management and market visibility [22].

I. Technological Integration

Rural connectivity is a significant barrier as many farms have no reliable internet access [23]. Cost-effective, reliable solutions that enable connectivity at any location are very much needed.

The impact of IoT on various sectors discussed above is represented in Figure 2.

979-8-3315-3968-9/24 $31.00 © 2024 IEEE

Special Session

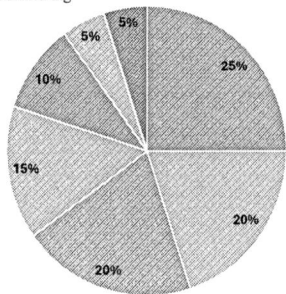

- Precision Agriculture
- Crop Monitoring and Management
- Livestock Management
- Farmer Health Monitoring
- Soil Health Monitoring
- Water Management
- Supply Chain Optimization

Fig. 2: Impact of IoT-Enabled Electronics on Various Sectors.

III. IoT Technologies in Agriculture

IoT technologies are transforming traditional agriculture by providing innovative and smart solutions that enhance productivity, sustainability, and efficiency. This section explores vital IoT technologies employed in agriculture, grouped into sensor technologies, data processing, and connectivity solutions.

A. Sensor Technologies

Sensor networks consist of interconnected sensors deployed across various devices on agricultural fields, enhancing data collection and monitoring capabilities. These networks provide real-time data, a comprehensive and continuous view of field conditions, and allow farmers to make informed decisions. For example, a soil moisture and weather sensor network will enable farmers to make data-driven irrigation, fertilization, and pest control decisions [2], [24].

B. Drones

Drones embedded with advanced imaging technologies offer aerial monitoring and management of crops and livestock. By bringing that information to farmers, they can assess plant and animal health, monitor crop growth, and identify areas needing attention. They improve precision agriculture by enabling farmers to conduct aerial surveys quickly and efficiently, leading to better resource allocation and management [25].

C. Data Processing Technologies

Data processing technologies confine the most commonly used paradigms- edge and cloud computing. These computing techniques are essential in analyzing the vast data collected from sensor units and drones. Edge computing processes the data closer to the source or user, allowing for real-time decision-making without relying solely on cloud infrastructure, thereby reducing the latency. This capability benefits embedded systems such as automated irrigation systems, where timely responses significantly impact crop health [26], [27].

Cloud computing platforms store and process data from IoT devices across the fields. They provide farmers with access to analytics tools, visualization tools, and dashboards for remote monitoring and management of agricultural operations.

These platforms support collaboration, cost reduction, and supply chain optimization, enabling farmers to make informed decisions [27].

D. Connectivity Solutions

Robust and reliable connectivity is essential for IoT applications in agriculture. IoT leverages wireless technologies such as LoRaWAN, Zigbee, and cellular networks to facilitate communication between devices, ensuring seamless data transfer to monitor and manage agricultural operations more effectively. They also enable real-time monitoring, enhancing responsiveness and efficiency [28].

IV. IoT-Enabled Electronics and their Applications for Smart Agriculture

This section explores the real-world systems and market-available products that enhance productivity, sustainability, and resource management for Smart Agriculture.

A. Smart Irrigation Systems

IoT-enabled smart systems revolutionize water management techniques in smart agriculture. These systems automatically adjust the irrigation schedules, optimizing water consumption and reducing waste. For example, the CropX agronomic farm management system integrates soil data with weather forecasts to deliver precise irrigation recommendations, improving crop yield while conserving water resources [29]. Automated irrigation systems like Rachio [30] and Netafim [31] offer similar benefits, with Netafim's drip irrigation solutions achieving up to 30% water savings.

B. Livestock Monitoring

IoT-enabled smart devices like wearables and smart collars help farmers track the health and behavior of their farm animals. These devices collect data based on physiological signals like vital signs, activity levels, and location to monitor health and well-being continuously. IoT solutions may also reduce veteran costs as they allow early disease diagnosis [32]. For example, Allflex Smart Tags [33] and Cowlar [34] offer smart collars to monitor the health and reproductive status of cows. These devices have been shown to reduce veterinary costs by up to 30% while improving overall herd health. IoT solutions such as automated feeding systems can be deployed to ensure rich nutrition and timely meals are provided to the animals, minimizing waste.

C. Precision Farming Solutions

IoT-enabled precision farming techniques enable real-time data to optimize agricultural operations. For example, Trimble [35] and John Deere's Precision Ag Technology [36] offer systems that gather data from the GPS-guided machinery and soil sensor units, allowing the farmers to make informed decisions on planting, fertilization, and pest control, thereby increasing the overall productivity. Farmers using these technologies have reported water savings of 20-40% and yield improvements of 10-20%. Such solutions also help prepare for any natural or environmental disasters.

979-8-3315-3968-9/24 $31.00 © 2024 IEEE   244

Special Session

D. Soil Health Management

IoT devices provide farmers with detailed real-time data on soil fertility like soil moisture, pH, and nutrient levels, enabling them to decide on timely resource or pesticide additions [37]. For example, FarmLogs [38] is a software platform that integrates soil data to help farmers manage field variability efficiently. IoT-enabled solutions can also support crop and soil rotation schedules and strategies [39].

E. Farmer Health Monitoring

For sustainable agricultural practices, the health and well-being of the farmers is essential. Smart wearables such as Fitbit 40 and Garmin [41] monitor the vital signs of the farmers to help manage their health. For agricultural settings, wearables from Zebra's Technologies [42] are explicitly designed to monitor and provide alerts for user fatigue or stress levels.

F. Vertical Farming

With the rapid growth in population and the limited availability of resources, IoT-enabled systems can help improve indoor farming techniques, thereby increasing the potential for urban agriculture. For example, Plenty [43] uses IoT sensors to track temperature, humidity, light levels, and nutrient concentrations in their vertical farms. Aerofarms [44] employs IoT technology and data analytics to optimize yield. Vertical farms utilizing these IoT solutions have reported a 20% increase in crop yield and improved food security, sustainability, and safety compared to traditional farming methods [45].

A brief review of the state-of-the-art products and technologies, along with their cost and scope of customers, is mentioned in Table I.

V. CHALLENGES AND FUTURE DIRECTIONS

A. Technical Challenges

1) Data Security and Privacy: IoT devices gather vast amounts of data, including sensitive information about farmer health, farming practices, crop yields, and livestock health. Unauthorized access to this information can lead to financial losses, mental health issues, operational disruptions, and compromises in food safety.

Implementing robust and reliable cybersecurity measures is critical, but many farms need more resources and expertise. Clear regulations regarding data ownership and privacy to protect farmers are essential. Therefore, establishing data governance frameworks with guidelines and rules on data usage, sharing, and consent is necessary to mitigate risks and enhance trust in IoT solutions.

2) Interoperability: As different manufacturers use varied communication protocols and data formats, integration across multiple devices over the network may be a challenge for IoT devices or systems for agriculture. This can limit the scope of building a cohesive system, which can maximize efficiency and data utility, leading to increased operational complexity.

3) Energy Constraints: Many IoT systems across the agriculture field may require continuous and reliable electricity access, which may be a challenge given the location and nature of the farm. This acts as a driving force for developing energy-efficient devices or alternate power solutions like solar energy to ensure smooth agricultural operations.

B. Socioeconomic Challenges

1) Cost of Adoption: IoT devices often include initial investments for software, hardware, installation, and maintenance, which may be a challenge for small-scale farmers. Financial barriers may create a digital divide within the agricultural community where large-scale farms benefit while smaller operations need help to compete.

2) Digital Literacy: The unfamiliarity or the lack of necessary digital skills can slow adoption rates and limit the potential benefits of IoT systems. To eliminate this challenge, educational programs and training initiatives are essential.

C. Emerging Technologies and Their Challenges

1) 5G Integration: To achieve real-time communication across numerous connected devices, incorporating high-speed, low-latency 5G networks plays a crucial role. This integration can significantly enhance decision-making and operational efficiency, allowing farmers to use drones and autonomous vehicles for agricultural operations. However, the high infrastructure cost and limited coverage in rural areas can challenge providing equitable benefits across the farming communities [53].

2) AI and Predictive Analytics: Integrating AI into IoT systems can significantly enhance crop and livestock management decision-making processes by utilizing predictive analytics to forecast yield outcomes, early disease diagnosis, and optimize resource allocation. Extensive technical expertise and understanding of AI, the need for high-quality, comprehensive datasets to train AI models may be challenges [54].

3) Sustainable Energy Solutions: Due to the limited nonrenewable energy sources, the development of energy-efficient IoT devices is required. As these devices solely depend on renewable energy sources, the initial investment cost and the variable in the resource type may cause challenges. As these solutions are expected to maintain their performance in diverse environmental conditions, collaborative efforts between technology providers and farmers will be essential [55].

4) Blockchain for Supply Chain: Integrating IoT and blockchain technology into IoT-enabled innovative systems can significantly enhance traceability and transparency in the agricultural supply chain, creating secure, fixed records of products from farm to table. The complexity of implementing blockchain solutions and the need for industry-wide collaboration may cause challenges. Cooperation among stakeholders, farmers, manufacturing companies, and regulatory bodies will be required to ensure the integrity of the deployed blockchain system [56].

979-8-3315-3968-9/24 $31.00 © 2024 IEEE   245

Special Session

TABLE I: Overview of IoT-Enabled Products and Companies in Agriculture.

Company Name / Product Name	Type	Cost	Customers	Rating
Blue River Technology [46]	Software	Varies by solution	Farmers, Agricultural Enterprises	4.6/5
Aerobotics [47]	Software	Varies	Farmers, Crop Managers	4.5/5
Ranch Systems [48]	Hardware	Varies	Ranchers, Livestock Farmers	4.3/5
FarmLogs [38]	Software	Subscription-based	Farmers, Agronomists	4.4/5
Harvest Croo Robotics [49]	Hardware	Varies by system	Farmers, Agricultural Enterprises	4.5/5
DroneDeploy [50]	Software	Subscription-based	Farmers, Agricultural Analysts	4.3/5
John Deere Precision Ag Technology [36]	Hardware	Varies by system	Farmers, Agricultural Enterprises	4.5/5
Trimble Ag Software [35]	Software	Subscription-based	Farmers, Crop Consultants	4.2/5
SoilMoisture Sensing [51]	Hardware	$150-$300 per sensor	Agricultural Producers	4.0/5
Rachio Smart Sprinkler Controller [52]	Hardware	$249-$299	Homeowners, Farmers	4.6/5
Netafim Drip Irrigation Solutions [31]	Hardware	Varies by project	Farmers, Irrigation Specialists	4.3/5
Allflex Smart Tags [33]	Wearable	$30-$50 per tag	Dairy and Livestock Farmers	4.7/5
Cowlar [34]	Wearable	$200 per collar	Dairy Farmers	4.5/5
Plenty [43]	Hardware	Varies	Urban Farmers, Retailers	4.8/5
AeroFarms [44]	Hardware	Varies by system	Urban Farmers, Researchers	4.7/5
Fitbit [40] / Garmin [41]	Wearable	$100-$200	Farmers, Agricultural Workers	4.4/5

VI. CONCLUSION AND FUTURE DIRECTIONS

Integrating IoT-enabled intelligent systems or electronics in agricultural operations represents a significant shift towards more efficient, data-driven farming practices. Critical challenges, such as food security and environmental sustainability, can be addressed by deploying connected devices, enhancing productivity, and optimizing resource use. As discussed in this mini-review, IoT-enabled technologies promote precision farming, improve soil health management, enable automated irrigation, and enhance livestock monitoring, contributing to a more resilient agricultural ecosystem.

Despite its promising potential and growth, several challenges, primarily data security and interoperability, cost, and digital literacy, pose significant barriers to the widespread adoption of IoT technologies. With the development of technology in the agricultural community, stakeholders—including farmers, technology providers, and policymakers- must collaborate to bridge the gaps.

Key future directions for IoT in agriculture include:

- Biometric Sensors for Livestock: advanced biometric sensors could be used for livestock health monitoring and management to detect early signs of disease stress and improve overall welfare.
- Collaborative Robotics (Cobot Integration): The future of agricultural operations may have cobots working alongside humans, assisting in harvesting, seeding, and crop care. This helps reduce labor shortages while maintaining precision and efficiency in farm operations.
- Swarm Robotics for Large-Scale Farming: A group of small autonomous machines can work together for planting, weeding, or pest control while communicating with each other to optimize coverage and productivity, thereby

transforming large-scale farming into a more automated, seamless operation.
- Regenerative Agriculture with Smart Systems: IoT systems can support regenerative agriculture, which focuses on restoring soil health and biodiversity. They can measure soil regeneration rates, track carbon sequestration, and maintain nutrient levels while aligning the practices with conservation goals.
- Personalized Farming Insights with Predictive Analytics: Tailored insights based on historical data and real-time conditions will be provided to the farmers by incorporating predictive analytics.

These innovative future directions may revolutionize agriculture, making it more efficient, sustainable, and resilient.

REFERENCES

[1] USGA, "Precision Agriculture: Benefits and Challenges for Technology Adoption and Use," Jan. 2024. [Online]. Available: https://www.gao.gov/products/gao-24-105962#:∼:text= Challengeslimitingthebroaderadoption,datasharingandownershipissues.

[2] L. Rachakonda, "ETS: A Smart and Enhanced Topsoil Health Monitoring and Control System at Edge using IoT," 12 2022, pp. 689–693.

[3] J. Gubbi, R. Buyya, S. Marusic, and P. Muthukaruppan, "Internet of Things (IoT): A vision, architectural elements, and future directions," *Future Generation Computer Systems*, vol. 29, no. 7, pp. 1645–1660, 2013.

[4] L. Atzori, A. Iera, and G. Morabito, "The Internet of Things: A survey," *Computer Networks*, vol. 54, no. 15, pp. 2787–2805, 2010.

[5] V. Kumar, K. V. Sharma, N. Kedam, A. Patel, T. R. Kate, and U. Rathnayake, "A comprehensive review on smart and sustainable agriculture using IoT technologies," *Smart Agricultural Technology*, vol. 8, p. 100487, 2024.

[6] S. Wolfert, L. Ge, C. Verdouw, and M.-J. Bogaardt, "Big Data in Smart Farming – A review," *Agricultural Systems*, vol. 153, pp. 69–80, 2017.

[7] M. Jamshidi, *System of Systems Engineering: Principles and Applications*. CRC Press, 2008. [Online]. Available: https://doi. org/10.1201/9781420065893

[8] M. Mennenga, F. Cerdas, S. Thiede, and C. Herrmann, "Exploring the Opportunities of System of Systems Engineering to Complement Sustainable Manufacturing and Life Cycle Engineering," *Procedia CIRP*, vol. 80, pp. 637–642, 2019, 26th CIRP Conference on Life Cycle Engineering (LCE) Purdue University, West Lafayette, IN, USA May 7-9, 2019.

[9] V. K. Quy, N. V. Hau, D. V. Anh, N. M. Quy, N. T. Ban, S. Lanza, G. Randazzo, and A. Muzirafuti, "IoT-Enabled Smart Agriculture: Architecture, Applications, and Challenges," *Applied Sciences*, vol. 12, no. 7, 2022.

[10] Food and Agriculture Organization of the United Nations, *Water for Sustainable Food and Agriculture*, 1 ed. Food and Agriculture Organization, Nov. 2016. [Online]. Available: https://openknowledge. fao.org/handle/20.500.14283/i7959en

[11] U. Nations, "World Population Prospects 2019: Highlights," Online, 2019. [Online]. Available: https://www.un.org/ development/desa/pd/news/world-population-prospects-2019-0#:~: text=PDF-,Highlights,-PDF

[12] R. Akhter and S. A. Sofi, "Precision agriculture using IoT data analytics and machine learning," *Journal of King Saud University - Computer and Information Sciences*, vol. 34, no. 8, Part B, pp. 5602–5618, 2022.

[13] W. Reports, "IoT in Livestock Management Market," Online, Jun. 2024. [Online]. Available: https://www.wiseguyreports.com/reports/ iot-in-livestock-management-market

[14] L. Rachakonda, "Agri-Aid: An Automated and Continuous Farmer Health Monitoring System Using IoMT," in *Internet of Things. IoT through a Multi-disciplinary Perspective*, L. M. Camarinha-Matos, L. Ribeiro, and L. Strous, Eds. Cham: Springer International Publishing, 2022, pp. 52–67.

[15] L. Rachakonda, P. Rajkumar, S. P. Mohanty, and E. Kougianos, "imirror: A smart mirror for stress detection in the iomt framework for advancements in smart cities," in *2020 IEEE International Smart Cities Conference (ISC2)*, 2020, pp. 1–7.

[16] USDA, "Farm Labor," Online, Aug. 2023. [Online]. Available: https://www.ers.usda.gov/topics/farm-economy/farm-labor/

[17] M. Shamsa, L. Rachakonda, S. Mohanty, and E. Kougianos, "qCrop: An IoT based Framework to Enhance Crop Productivity in Smart Agriculture," in *IFIP/IEEE 32nd International Conference on Very Large Scale Integration (VLSI-SoC)*, 2024.

[18] L. Rachakonda and S. Stasiewicz, "Sana Solo: An Intelligent Approach to Measure Soil Fertility," in *Internet of Things. Advances in Information and Communication Technology*, D. Puthal, S. Mohanty, and B.-Y. Choi, Eds. Cham: Springer Nature Switzerland, 2023, pp. 395–404.

[19] U. Nations, "Sustainable Development: The 17 Goals," Online, Jan. 2024. [Online]. Available: https://sdgs.un.org/goals

[20] E. Commission, "The Common Agricultural Policy (CAP): A Strategic Plan for Agriculture and Rural Development," Online, 2020. [Online]. Available: https://ec.europa.eu/info/food-farming-fisheries/key-policies/ common-agricultural-policy_en

[21] USDA, "Climate-Smart Agriculture and Forestry," Online, Jan. 2018. [Online]. Available: https://www.farmers.gov/conservation/ climate-smart

[22] S. Srivastava, "Impact of IoT in the Agriculture Industry: Everything You Need to Know," appinventiv, Aug. 2024. [Online]. Available: https://appinventiv.com/blog/iot-in-agriculture-industry/

[23] F. Taheri, M. D'Haese, D. Fiems, and H. Azadi, "Facts and fears that limit digital transformation in farming: Exploring barriers to the outreach of wireless sensor networks in Southwest Iran." *PLoS One.*, vol. 17, no. 12, Dec. 2022.

[24] A. Soussi, E. Zero, R. Sacile, D. Trinchero, and M. Fossa, "Smart Sensors and Smart Data for Precision Agriculture: A Review," *Sensors (Basel)*, vol. 24, no. 8, Apr. 2024.

[25] A. Rejeb, A. Abdollahi, K. Rejeb, and H. Treiblmaier, "Drones in agriculture: A review and bibliometric analysis," *Computers and Electronics in Agriculture*, vol. 198, p. 107017, 2022.

[26] L. Rachakonda, S. P. Mohanty, and E. Kougianos, "Good-Eye: A Device for Automatic Prediction and Detection of Elderly Falls in Smart Homes," in *2020 IEEE International Symposium on Smart Electronic Systems (iSES) (Formerly iNiS)*, 2020, pp. 202–203.

[27] C. R. Kalyani Y, "A Systematic Survey on the Role of Cloud, Fog, and Edge Computing Combination in Smart Agriculture," *Sensors (Basel).*, vol. 21, no. 17, Sep. 2021.

[28] J. Xu, B. Gu, and G. Tian, "Review of agricultural IoT technology," *Artificial Intelligence in Agriculture*, vol. 6, pp. 10–22, 2022.

[29] CropX, "CropX — Precision Irrigation for Every Farmer." [Online]. Available: https://cropx.com/

[30] R. Technologies, "Rachio Smart Sprinkler Controllers." [Online]. Available: https://rachio.com/

[31] Netafim, "Netafim — global leader in smart irrigation solutions." [Online]. Available: https://www.netafim.com/en/

[32] G. S. Karthick, M. Sridhar, and P. B. Pankajavalli, "Internet of Things in Animal Healthcare (IoTAH): Review of Recent Advancements in Architecture, Sensing Technologies and Real-Time Monitoring," *SN Computer Science*, vol. 1, no. 5, p. 301, 2020.

[33] Allflex, "Allflex — Livestock Identification and Animal Management Solutions." [Online]. Available: https://www.allflex.global/na/

[34] Cowlar, "Cowlar — Precision Livestock Management." [Online]. Available: https://cowlar.com/

[35] T. Inc., "Trimble Agriculture Software," 2024, accessed: 2024-09-21. [Online]. Available: https://ptxtrimble.com/en/products/software/ trimble-agriculture-software

[36] Deere and Company, "Deere and Company," 2024, accessed: 2024-09-21. [Online]. Available: https://www.deere.com/en/

[37] L. Rachakonda and S. Stasiewicz, "SanaSolo 2.0: Edge-Based Monitoring andManagement of Soil Fertility Using IoT," in *IFIP/IEEE 32nd International Conference on Very Large Scale Integration (VLSI-SoC)*, 2024.

[38] I. FarmLogs, "FarmLogs," 2024, accessed: 2024-09-21. [Online]. Available: https://m.farms.com/agriculture-apps/crops/farmlogs

[39] R. S. Upendra, M. R. Ahmed, A. V. Omkar, J. Goyal, V. Chaitra, H. Muskan, P. Kamath, and K. Thirumala Akash, "Smart Approaches to Measure Soil Fertility for Sustainable Agriculture," in *2021 Second International Conference on Smart Technologies in Computing, Electrical and Electronics (ICSTCEE)*, 2021, pp. 1–6.

[40] Fitbit, "Fitbit Official Website," 2024, accessed: 2024-09-22. [Online]. Available: https://www.fitbit.com/global/us/home

[41] Garmin, "Garmin Official Website," 2024, accessed: 2024-09-22. [Online]. Available: https://www.garmin.com/en-US/

[42] Z. Technologies, "Zebra Technologies — The Future of Enterprise Visibility." [Online]. Available: https://www.zebra.com/us/en.html

[43] Plenty, "Plenty Official Website," 2024, accessed: 2024-09-22. [Online]. Available: https://www.plenty.ag/

[44] AeroFarms, "AeroFarms Official Website," 2024, accessed: 2024-09-22. [Online]. Available: https://www.aerofarms.com/

[45] X. G. Avgoustaki DD, "How energy innovation in indoor vertical farming can improve food security, sustainability, and food safety?" *Advances in Food Security and Sustainability.*, vol. 5, pp. 1–51, Sep. 2020.

[46] I. Blue River Technology, "Blue River Technology," 2024, accessed: 2024-09-21. [Online]. Available: https://bluerivertechnology.com/

[47] I. Aerobotics, "Aerobotics," 2024, accessed: 2024-09-21. [Online]. Available: https://www.aerobotics.com/

[48] I. Ranch Systems, "Ranch Systems," 2024, accessed: 2024-09-21. [Online]. Available: https://www.ranchsystems.com/home/

[49] L. L. C. Harvest Croo Robotics, "Harvest Croo Robotics," 2024, accessed: 2024-09-21. [Online]. Available: https://www. harvestcroorobotics.com/

[50] I. DroneDeploy, "DroneDeploy," 2024, accessed: 2024-09-21. [Online]. Available: https://www.dronedeploy.com/

[51] SoilMoisture, "SoilMoisture Official Website," 2024, accessed: 2024-09-22. [Online]. Available: https://soilmoisture.com/

[52] Rachio, "Rachio Official Website," 2024, accessed: 2024-09-22. [Online]. Available: https://rachio.com/

[53] P. Majumdar, D. Bhattacharya, and S. Mitra, *Utilities of 5G Communication Technologies for Promoting Advancement in Agriculture 4.0: Recent Trends, Research Issues and Review of Literature*. Singapore: Springer Nature Singapore, 2023, pp. 111–125.

[54] S. A. Bhat and N.-F. Huang, "Big Data and AI Revolution in Precision Agriculture: Survey and Challenges," *IEEE Access*, vol. 9, pp. 110 209–110 222, 2021.

[55] Y. Majeed, M. U. Khan, M. Waseem, U. Zahid, F. Mahmood, F. Majeed, M. Sultan, and A. Raza, "Renewable energy as an alternative source for energy management in agriculture," *Energy Reports*, vol. 10, pp. 344–359, 2023.

[56] G. Mirabelli and V. Solina, "Blockchain and agricultural supply chains traceability: research trends and future challenges," *Procedia Manufacturing*, vol. 42, pp. 414–421, 2020, international Conference on Industry 4.0 and Smart Manufacturing (ISM 2019).

979-8-3315-3968-9/24 $31.00 © 2024 IEEE

Special Session

A Novel Current Comparator Enabling Large RRAM Crossbars for BNNs and PUFs

Gokulnath Rajendran[*§], Debajit Basak[†§], Suman Deb[‡], Siyi Wang[*], Anupam Chattopadhyay[*]

[*]NTU, Singapore, [†]Atom Semiconductor, Hong Kong,[‡]CREATE, Singapore

debajit.basak@atom-semiconductor.com, {gokulnat002,siyi002}@e.ntu.edu.sg,
suman.deb@imperial.ac.uk, anupam@ntu.edu.sg

Abstract—Emerging non-volatile memory (NVM) device technologies are advancing in-memory computing (IMC) applications by providing faster computation speeds and reducing resource overhead. Crossbar structures are commonly used in IMC with NVM devices such as memristors to perform matrix-vector multiplication for deep learning and security primitive applications. Large crossbars are required to implement today's deep learning models, especially for implementing specialized Binarized Neural Networks (BNNs) architectures and to construct security primitives such as physical unclonable functions (PUFs). To digitize crossbar currents, current-sense comparators based on current mirrors are widely used in BNN crossbars. However, conventional comparators have a limited current range due to the decrease of the bit-line voltage as the number of active crossbar elements connected to the bit-line increases. This paper presents a current comparator that employs a regulated cascode sensing stage which boosts the input current range by stabilizing the bit-line voltages. By increasing the current range, a larger crossbar size can be supported. Extensive simulations were carried out to verify the proposed technique, demonstrating that a significantly larger crossbar size can be achieved with the proposed comparator compared to a traditional current-sense comparator for the same area and resolution. Designed in a 180 nm technology, the proposed comparator achieves a resolution of 50nA and 100μA for PUF and BNN applications, respectively, and dissipates 198 μW.

Index Terms—RRAM, BNN, PUF, Sense amplifier

I. INTRODUCTION

Advancements in emerging non-volatile memory (NVM) device technologies, such as Resistive Random Access Memory (RRAM), Phase Change Memory (PCM), Spin-Transfer Toque (STT), and Spin-Orbit Torque (SOT) Magnetic Random Access Memories (MRAM) are mainly driven by their in-memory computing (IMC) applications [1], [2].The motivation behind IMC is to integrate memory and processing units into a single entity, thereby reducing data transfer costs. Therefore, the resource-intensive operations in Von Neumann architectures, where memory and processing units are separate, can be executed with fewer instructions using IMC accelerators [3]. The standard technique in IMC involves constructing crossbars with NVM devices placed at their intersections, as depicted in Fig.1. IMC crossbars have several applications, including matrix-vector multiplication (MVM) [4], memristor-aided logic (MAGIC) design [5], and strong Physical Unclonable Function (PUF) [6], [7].

MVM is a fundamental operation in all types of neural networks (like, BNNs) and PUFs. It is directly implemented

§ Equally contributed

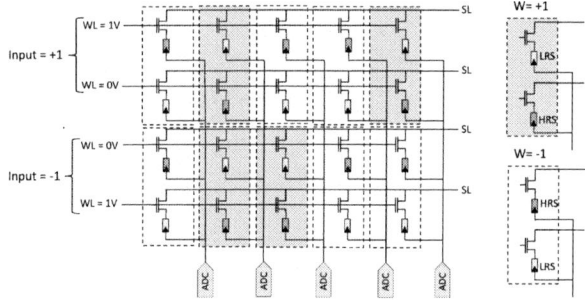

Fig. 1. An RRAM crossbar for Matrix Vector Multiplication (MVM)

as a multiply-and-accumulate (MAC) block using a crossbar. In the case of neural networks, its weights are mapped to the resistance states of the RRAM device. The input signals are applied to the crossbar to perform analog computing, where the conductance state of the RRAM device $G_{j,k}$ is multiplied by the input voltage V_j, as described by Ohm's law, $I_k = V_j G_{j,k}$, where j and k represent the row and column index of the crossbar. The generated I_k is digitized with a comparator or analog-to-digital converter (ADC) [1], [8] as shown in Fig.1.

A. Implementing BNN Using RRAM

BNNs are a class of neural networks in which weights and activations are constrained to take +1 and -1, unlike full-precision neural networks [9]. BNNs employ full precision only during training and are ideally suited for near-sensor computing applications. The forward propagation of BNN is detailed in Algorithm 1. The sign function is used as the activation function in BNN, and it can be directly implemented by a comparator. Consequently, the energy- and area-intensive ADC used in standard crossbars is replaced by a more efficient comparator or sense amplifier in BNN crossbars. Implementing BNNs in RRAM crossbars has been shown to outperform conventional CMOS implementations in terms of faster computation speed, lower area, and reduced energy consumption [8], [10], [11]. In the BNN topology, the batch normalization (BN) layer is often placed before the activation function (line 3 of Algorithm 1). For optimized RRAM crossbar-based BNN implementation using a comparator, the BN layer can be approximated to a threshold value B_{th}, as presented in

979-8-3315-3968-9/24 $31.00 © 2024 IEEE   248

Special Session

equation (1). Here β, σ, ϵ, γ, and μ are the parameters of the BN layer. There are two methods to implement B_{th} in the crossbar for achieving the output of the activation function as defined in equation (2), effectively combining steps 3 and 5 of Algorithm 1 into one [12]. The first method uses separate columns of the crossbar for weights and B_{th}, while the second method integrates them within a single column and compares them to a reference value of -1 as illustrated in Fig.2. The drawback of this optimized crossbar integrations of combining BN and MVM layer weights is that the weight matrix of one layer has to be implemented entirely in a single crossbar. The method of obtaining multiple MVM partial outputs from smaller crossbars and adding them together to finally get y_k is not supported. This design requirement arises from using a comparator as the activation function in BNNs [12]. Hence, the crossbars used for BNNs are expected to be larger. This necessitates a current comparator that can work with larger crossbars without loss of resolution. For example, to implement a weight matrix of size 512×512, we require crossbars of size 1024×1024 and 1536×512 with double-column and single-column BNN approaches, respectively. Additionally, there may be instances where the values of B_{th} and X_k are identical, resulting in identical inputs to the comparator. In such scenarios, it is critical for the comparator to reliably manage this equality by consistently producing an output of +1.

$$B_{th} = -\beta \frac{\sqrt{\sigma^2 + \epsilon}}{\gamma} + \mu \tag{1}$$

$$X_k^b = \begin{cases} +1, & \text{if } X_k \geqslant B_{th,k} \\ -1, & \text{otherwise} \end{cases} \tag{2}$$

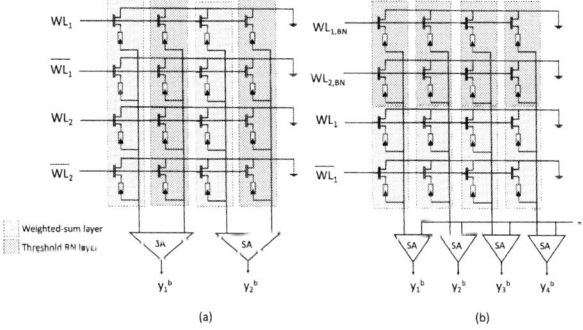

Fig. 2. BNN implementation in RRAM crossbar using (a) double column and (b) single column methodologies

B. Implementing PUF Using RRAM

The fabricated RRAM devices undeniably have device-to-device (D2D) variations in resistance states and switching voltages. The D2D variations of RRAM are mainly used to construct PUFs. A simple technique is to compare the resistance variations of two selected RRAM devices with a comparator to generate the output bit. Notably, all the RRAM PUFs,

---

Algorithm 1 Forward propagation in Binarized Neural Network

Require: Minibatch inputs X, Trained weights W, Batch normalization paramters γ
Ensure: $y_k^b = ForwardPropogation(X_{k-1}^b)$
1: **for** $k \leftarrow 1$ to N **do**
2:      $y_k \leftarrow X_{k-1}^b W_k^b$
3:      $X_k \leftarrow BatchNorm(y_k, \gamma_k)$
4:      **if** k < N **then**
5:          $y_k^b \leftarrow Binarize(X_k)$
6:      **end if**
7: **end for**

---

regardless of the scheme, require a comparator, and the output of the comparator is often treated directly as the PUF response. A high-resolution current comparator is especially necessary to construct strong PUFs based on MVM implemented in a crossbar. In MVM operation-based PUFs, the word lines of the crossbars are selected based on the input challenge bits, and the resistive mismatches of the selected groups of NVM devices generate different currents in the crossbar columns [6], [13]. The current comparator compares the two column currents to generate the response bit. The RRAM devices generate distinct resistive mismatches influenced by their inherent material properties and the specifics of their fabrication process. Thus, the resolution of the comparator becomes a primary concern, and it decides how well the minor ohmic differences of selected RRAMs can be reliably digitized to response bits [14]. The use of a poor-resolution comparator leads to biased responses, rendering it unsuitable for PUF construction. To construct larger PUFs, we need a comparator that maintains high resolution for an increase in crossbar size. In addition, conventional comparators produce unreliable responses when the input currents for comparison are below the resolution of the comparator.

II. BACKGROUND AND MOTIVATION

Current comparators based on current mirrors are widely used for direct current sensing [15]. Fig.3(a) shows a conventional current-mirror based current comparator. Input currents flow into a matched current mirror, and the differential current is then mirrored into a clocked latch, producing the comparison result. The bit-line voltages across the RRAM cells, V1 and V2, can be set by an external reference voltage (V_{REF}) and the cascode transistors M1 and M2 as shown in Fig.3(a). In order to support a large RRAM crossbar, the current comparator is required to handle large input currents. This is difficult to achieve with the conventional topology due to the variation of the bit-line voltages with the input current magnitude. As the input current increases, the bit-line voltages decrease sharply due to the quadratic voltage-to-current relationship of the cascode transistors. This issue is particularly problematic for resistive RRAM cells because firstly it reduces the dynamic range of the comparator. Secondly, at high current inputs, the increase in the input current for every additional

979-8-3315-3968-9/24 $31.00 © 2024 IEEE   249

Special Session

RRAM cell drops below the resolution of the comparator. This greatly limits the input current range. Thirdly, as the transistor threshold voltage drifts with temperature, the gate reference voltage (V_{REF}) would have to be trimmed with temperature if a stable bit-line voltage is desired.

The main contributions of this work are as follows:

- We propose a current comparator with a regulated cascode input sensing stage which greatly reduces the bit-line voltage sensitivity to the input current and thereby widens the input current range.
- We present simulation analyses to demonstrate that the proposed comparator accommodates a larger crossbar compared to the traditional comparator without compromising resolution. We tuned the comparator and analyzed its performance for BNN and PUF applications separately.
- Additionally, we introduce a methodology for managing scenarios where RRAM crossbars encounter input currents that are nearly equal and fall below the comparator's resolution.

III. PROPOSED CURRENT COMPARATOR

The proposed comparator with a regulated-cascode sensing first stage as shown in Fig.3(b). The negative feedback loop ensures a fixed bias voltage for the RRAM cells regardless of the input current. This not only supports a wide current input range but also improves current sensing accuracy. The read voltage is set at 200 mV in this work. To save area

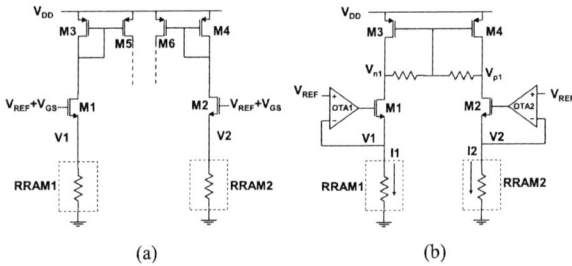

(a)   (b)

Fig. 3. Schematic of the first stage of a (a) traditional current sense comparator; (b) proposed high-precision current sense comparator.

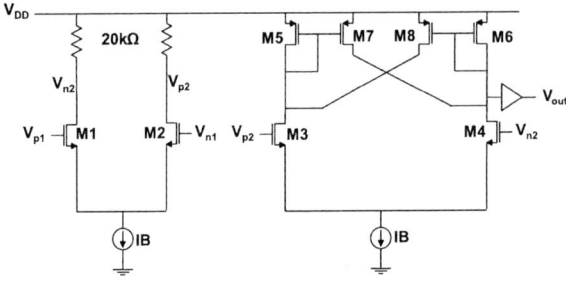

Fig. 4. Schematic of the second and third stages of the proposed comparator.

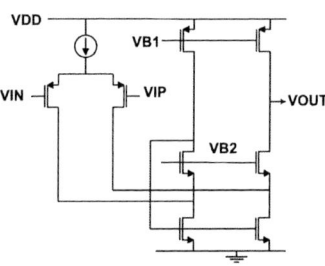

Fig. 5. Schematic of Operational Transconductance Amplifier (OTA) in the proposed comparator.

and power, a voltage output is generated instead of mirroring the input current like the conventional architecture as shown in Fig.3(a). However, at such high currents, the intrinsic output resistance (r_o) of the transistors M3 and M4 severely degrades, diminishing the gain of the first stage of the proposed comparator. Therefore, a preamplifier is added as a second stage as shown in Fig.4. The preamplifier gain is around 9, which is sufficient to suppress the threshold voltage variations of the subsequent latch stage as confirmed by a 3-sigma monte-carlo simulation. The final stage is a traditional voltage comparator with internal hysteresis as shown in Fig.4. Transistors M7 and M8 are used to create hysteresis through positive feedback [16]. The hysteresis is used to handle the equality condition explained in section II ($B_{th} = X_k$). By adjusting the ratio of the transistor aspect ratios of M5/M7 and M6/M8, the upper and lower tripping points can be fixed. A single-stage folded-cascode topology is chosen for the operational transconductance amplifiers (OTA1 and OTA2) which are designed for low power consumption as shown in

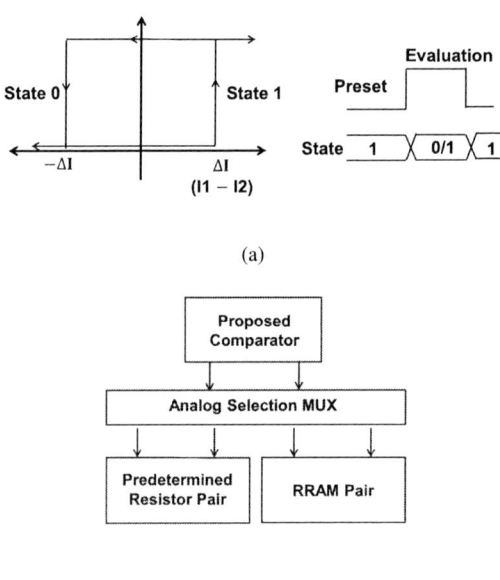

(a)

(b)

Fig. 6. (a) Two-phase timing of the proposed comparator; (b) comparator system diagram.

979-8-3315-3968-9/24 $31.00 © 2024 IEEE   250

Special Session

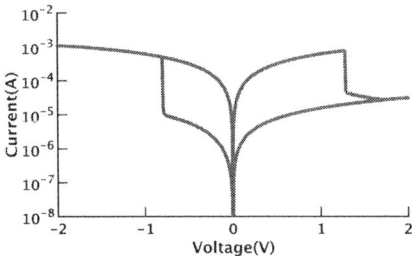

Fig. 7. Simulated voltage-current characteristic of RRAM

Fig.5. Each OTA consumes about $5\mu A$ current from a 1.8 V supply.

While hysteresis eliminates sensitivity to noise, for input currents smaller than the thresholds, the comparator's decision relies on its previous state, as illustrated in Fig.6(a). However, for BNN applications of RRAM crossbar, a fixed output is required for currents below the threshold, addressing the equality condition discussed in Section.II. To achieve that, a two-phase operation is proposed in this design. The two phases are preset and evaluation as shown in Fig. 6(b). In the preset phase, the comparator is connected to a predetermined resistor pair, setting the comparator to state 1. In the next phase, which is the evaluation phase, the comparator is connected to the crossbar. This arrangement is achieved using an analog multiplexing scheme, as shown in Fig.6(b).

IV. SIMULATION RESULTS

RRAM is a two-terminal device and works on a resistive switching mechanism. For our simulations, we utilized the RRAM model described in [17], with its voltage-current characteristics shown in Fig. 7. The model parameters are set to the default values as given in [17]. In this study, we established the set and reset voltages at -0.8 V and 1.5 V, respectively, based on calculations from Fig.7, to ensure reliable switching. Both write and read operations are performed on the same terminals of RRAM. Therefore, the read operation is performed at a much lower voltage than the write operation to avoid disturbing the state of the RRAMs. A higher read voltage will create a switching probability. To minimize the probability of unintentional switching, it is strongly recommended to keep the read voltage below 0.5V [17]. We set the read voltage at 0.2V, a level that does not switch the RRAM under any conditions, including variabilities—an assertion experimentally verified in [17]. This requirement dictates the use of a current comparator to generate and maintain a low read voltage.

The proposed comparator is designed in a 180nm CMOS process technology. It is simulated for a large number of numbers of RRAM cells and compared with a conventional comparator designed for the same target resolution.

A. Comparator tuned for BNN crossbar

For BNN applications, the input current range can be on the order of several hundreds of milliampers. Thanks to the feedback structure, the RRAM read voltage is held at a fixed

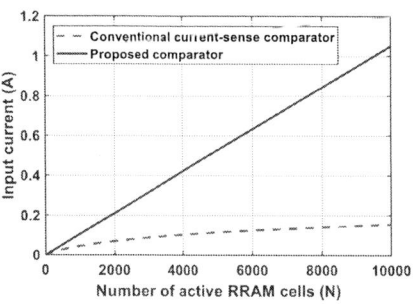

(a)

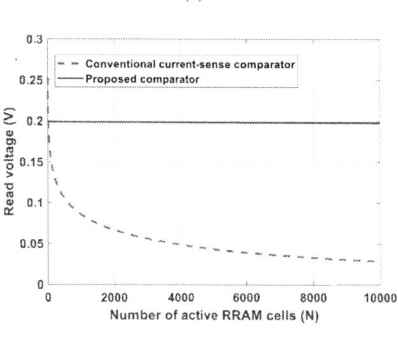

(b)

Fig. 8. Comparison of the (a) input current through the sense amplifier; (b) bit-line read voltage of the proposed comparator with that of the conventional comparator for RRAM BNN implementations.

reference voltage independent of the input current. As seen in Fig.8(a), the input current magnitudes increase with the number of active RRAM cells (N) for both the comparators. However, the input current is significantly more linear for the proposed comparator compared to that of the conventional one for a large number of crossbar cells. The input current for the conventional comparator plateaus for large N due to the decrease in the read voltage as shown in Fig.8(b). On the otherhand, due to the regulated cascode structure, the voltage across the RRAM cells remains nearly fixed for the proposed comparator as seen in Fig.8(b). Although a dip can be seen for high RRAM cells, it is much smaller than that of the conventional comparator as seen in Fig.8(b). This enables the proposed comparator to handle a larger crossbar size as high current gain can be maintained for a wider range. The hysteresis threshold (ΔI) is set as follow:

$$\Delta I < \frac{0.2}{R_{RRAM}} \qquad (3)$$

where R_{RRAM} is the effective DC impedance of each RRAM cell. Fig.9(a) and Fig.9(b) show the hysteresis profile of the proposed current comparator at 3 mA and 300 mA input currents, respectively. At nominal corner, the hysteresis threshold of the comparator is 50 µA. With process variation and mismatch, the thresholds are below 100 µA. The current gain or the increase in the input current for every additional

979-8-3315-3968-9/24 $31.00 © 2024 IEEE   251

Special Session

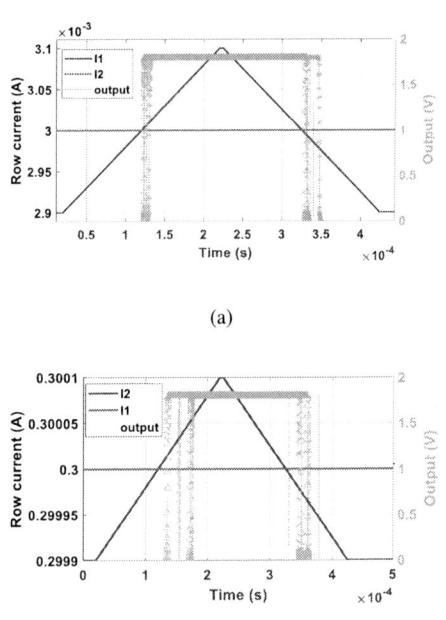

(a)

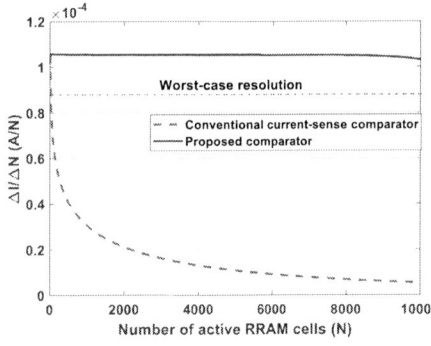

(b)

Fig. 9. Monte-carlo simulation of the proposed comparator showing hysteresis at (a) 3 mA; (b) 300 mA input current.

RRAM cell (N) should be above the hysteresis threshold for the entire input range. Thanks to the regulated-cascode input stage, the current gain of the proposed comparator is above the target resolution for a crossbar size of 10000, which is significantly higher than that of the conventional comparator as shown in Fig.10. Fig.11 shows the 2-phase operation of the proposed comparator. The preset and evaluation phases are clearly shown. The settling speed is determined by the finite OTA bandwidth. Since the OTA's power is negligible compared to the whole comparator's power with the input current, the bandwidth can be increased if faster settling is desired without much overhead.

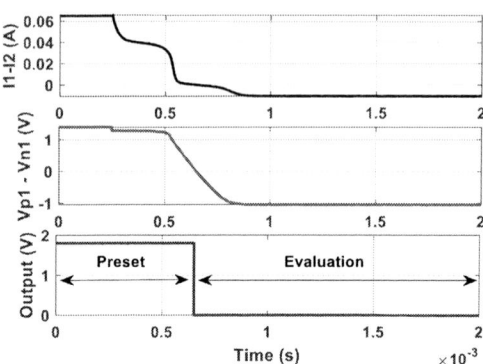

Fig. 11. Transient simulation comparing the proposed comparator with the conventional comparator during preset and evaluation phases.

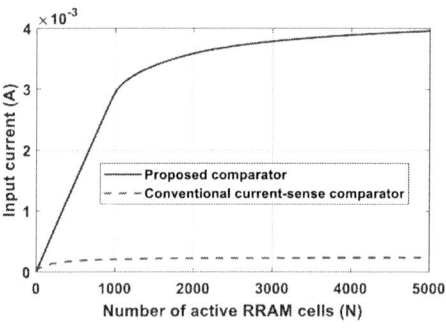

(a)

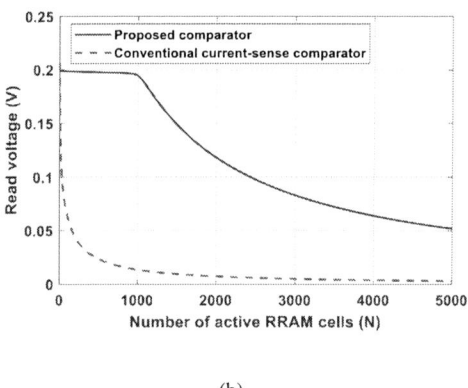

(b)

Fig. 12. Comparison of the (a) input current through the sense amplifier; (b) bit-line read voltage of the proposed comparator with that of the conventional comparator for RRAM PUF.

We analyzed the proposed comparator within the BNN topology using the MNIST dataset. We found that the comparator can enable the same accuracy of 96.74% as that of the one we obtained for baseline. This indicates that our comparator can handle larger crossbars without any performance degradation.

Fig. 10. Current gain of the conventional and the proposed comparators for RRAM BNN implementations.

Special Session

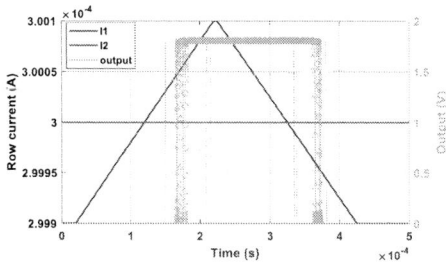

Fig. 13. Monte-carlo simulation of the proposed comparator showing hysteresis at 300uA.

B. Comparator tuned for PUF crossbar

In case of RRAM crossbar for PUF, the input current range is much lower compared to that of the BNN case. Thus, the W/L ratio of the transistors M1-M4 is scaled down by 500 times compared to those of the comparator for the BNN case. Fig.12 (a) and (b) show the input current and read voltage variation of the proposed comparator compared to a traditional comparator. The resolution of the comparator for the PUF case should be set as small as possible to detect small mismatches in the differential input current. However, this is limited by device mismatch and noise. Fig.13 shows the hysteresis profile of the proposed current comparator at 300 μA input current. At nominal corner, the hysteresis threshold of the comparator is 50 nA. With process variation and mismatch, the thresholds are below 100 nA. Similar to the BNN case, the current gain of the proposed comparator is above the worst-case resolution for a crossbar size of 4000, which is significantly higher than that of the conventional comparator as shown in Fig.14.

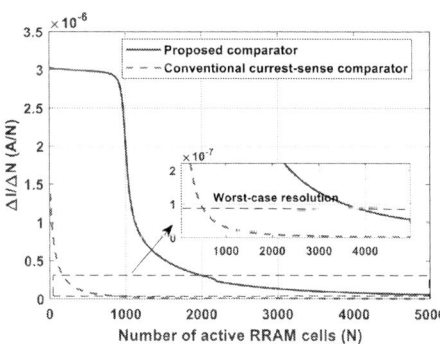

Fig. 14. Current gain of the conventional and the proposed comparators for RRAM PUF.

V. Conclusion

We proposed a current comparator to accommodate a wider input current range. The technique employs a regulated cascode stage, which fixes the read voltage across the RRAM cells. This, in turn, helps extend the input current range, facilitating the construction and use of larger crossbars. Design ex-

amples for BNN and PUF applications are presented to verify the proposed technique. In comparison with traditional current-sense comparators, our design accommodates a significantly larger crossbar while achieving the same target resolution. Operating on a 1.8V supply, our comparator designs achieve 50nA and 100uA resolutions for PUF and BNN applications, respectively, with a power dissipation of 198μW.

Acknowledgment

This work is supported by NTU-DESAY SV Research Program under Grant 2018-0980. This work is partially supported by the National Research Foundation, Prime Minister's office, Singapore under its CREATE programme.

References

[1] W. Wan et al., "A compute-in-memory chip based on resistive random-access memory," *Nature*, vol. 608, no. 7923, pp. 504–512, 2022.

[2] S. Yinet al., "High-throughput in-memory computing for binary deep neural networks with monolithically integrated rram and 90-nm cmos," *IEEE Transactions on Electron Devices*, vol. 67, no. 10, pp. 4185–4192, 2020.

[3] Y. Chen et al., "A survey of accelerator architectures for deep neural networks," *Engineering*, vol. 6, no. 3, pp. 264–274, 2020.

[4] L. Xia et al., "Technological exploration of rram crossbar array for matrix-vector multiplication," *Journal of Computer Science and Technology*, vol. 31, no. 1, pp. 3–19, 2016.

[5] S. Kvatinsky et al., "Magic—memristor-aided logic," *IEEE Transactions on Circuits and Systems II: Express Briefs*, vol. 61, no. 11, pp. 895–899, 2014.

[6] M. R. Mahmoodi et al., "Rx-puf: Low power, dense, reliable, and resilient physically unclonable functions based on analog passive rram crossbar arrays," in *2018 IEEE Symposium on VLSI Technology*. IEEE, 2018, pp. 99–100.

[7] ——, "A strong physically unclonable function with $> 2^{80}$ CRPs and $<$ 1.4% ber using passive ReRAM technology," *IEEE Solid-State Circuits Letters*, vol. 3, pp. 182–185, 2020.

[8] X. Sun et al., "Xnor-rram: A scalable and parallel resistive synaptic architecture for binary neural networks," in *2018 Design, Automation & Test in Europe Conference & Exhibition (DATE)*. IEEE, 2018, pp. 1423–1428.

[9] I. Hubara et al., "Binarized neural networks," *Advances in neural information processing systems*, vol. 29, 2016.

[10] T. Hirtzlin et al., "Digital biologically plausible implementation of binarized neural networks with differential hafnium oxide resistive memory arrays," *Frontiers in neuroscience*, vol. 13, p. 1383, 2020.

[11] L. Ni et al., "An energy-efficient digital reram-crossbar-based cnn with bitwise parallelism," *IEEE Journal on Exploratory solid-state computational devices and circuits*, vol. 3, pp. 37–46, 2017.

[12] H. Kim et al., "In memory batch-normalization for resistive memory based binary neural network hardware," in *Proceedings of the 24th Asia and South Pacific Design Automation Conference*, 2019, pp. 645–650.

[13] R. A. John et al., "Halide perovskite memristors as flexible and reconfigurable physical unclonable functions," *Nature Communications*, vol. 12, no. 1, p. 3681, 2021.

[14] P.-Y. Chen et al., "Exploiting resistive cross-point array for compact design of physical unclonable function," in *2015 IEEE International Symposium on Hardware Oriented Security and Trust (HOST)*. IEEE, 2015, pp. 26–31.

[15] M.-F. Chang et al., "An offset-tolerant fast-random-read current-sampling-based sense amplifier for small-cell-current nonvolatile memory," *IEEE Journal of Solid-State Circuits*, vol. 48, no. 3, pp. 864–877, 2013.

[16] P. E. Allen et al., *CMOS analog circuit design*. Elsevier, 2011.

[17] C. Bengel et al., "Variability-aware modeling of filamentary oxide-based bipolar resistive switching cells using spice level compact models," *IEEE Transactions on Circuits and Systems I: Regular Papers*, vol. 67, no. 12, pp. 4618–4630, 2020.

Special Session

Secure Software/Hardware Hybrid In-Field Testing for System-on-Chip

Saleh Mulhem*, Christian Ewert*, Andrija Nešković, Amrit Sharma Poudel,
Christoph Hübner, Mladen Berekovic, and Rainer Buchty *Institute of Computer Engineering, Universität zu Lübeck, Lübeck, Germany*
{name.surname}@uni-luebeck.de

Abstract—**Modern Systems-on-Chip (SoCs) incorporate built-in self-test (BIST) modules deeply integrated into the device's intellectual property (IP) blocks. Such modules handle hardware faults and defects during device operation. As such, BIST results potentially reveal the internal structure and state of the device under test (DUT) and hence open attack vectors. So-called result compaction can overcome this vulnerability by hiding the BIST chain structure but introduces the issues of aliasing and invalid signatures. Software-BIST provides a flexible solution, that can tackle these issues, but suffers from limited observability and fault coverage. In this paper, we hence introduce a low-overhead software/hardware hybrid approach that overcomes the mentioned limitations. It relies on (a) keyed-hash message authentication code (KMAC) available on the SoC providing device-specific secure and valid signatures with zero aliasing and (b) the SoC processor for test scheduling hence increasing DUT availability. The proposed approach offers both on-chip- and remote-testing capabilities. We showcase a RISC-V-based SoC to demonstrate our approach, discussing system overhead and resulting compaction rates.**

Index Terms—**Secure Built-In Self-Test, System-on-Chip, KMAC, In-Field Testing**

I. INTRODUCTION

The increasing complexity and functionality of System-on-Chip (SoC) devices require extensive and detailed testing to ensure reliable behavior during their operation [1]. For this, built-in self-test (BIST) provides a mechanism for testing without physical access to the device under test (DUT). BIST is usually a hardware mechanism, but can be software-based (SBIST) [2]. It interrogates the internal device status via a scan chain, thus giving a detailed insight into the system state to be analyzed by automated test equipment (ATE).

From a security perspective, the adversary can exploit the BIST's full observability of the DUT internal status to compromise the SoC [3]. Therefore, it is considered a side-channel information leakage [4]. This is particularly harmful for cryptographic SoCs and safety-critical systems. Relying on output-response analyzers (ORA) to compact the scan-chain output responses into a signature might sometimes help with hiding internal DUT data. However, this is not always the case

---
* These authors contributed equally to this work.
This work has been supported by DAIS (https://dais-project.eu/), which has received funding from the ECSEL Joint Undertaking (JU) under grant agreement No 101007273. The JU receives support from the European Union's Horizon 2020 research and innovation program and Sweden, Spain, Portugal, Belgium, Germany, Slovenia, Czech Republic, Netherlands, Denmark, Norway, and Turkey.

---

for classical signature generation as it reveals sensitive data or information [5]. Another challenge of traditional methods is the generation of invalid signatures when $L < d$, where L and d are the bit length of the DUT output response and its corresponding signature, which in this case directly reveal internal DUT information. Countermeasures to overcome such vulnerabilities are not only specific to the DUT, but potentially also introduce a vast overhead, making them not suitable for resource-constrained SoC [5].

Further, hardware testing interferes with the SoC's availability; it hence cannot be tested during normal system operation. SBIST [2] provides a suitable solution for testing under operational conditions by only scheduling the test when the DUT is not on operational demand. By nature, the SBIST approach is self-contained. As such, the DUT observability is restricted to the self-testing device. A detailed DUT analysis also requires an extensive database, making SBIST-only unsuitable for resource-constrained embedded devices.

The mentioned test methods exhibit the following shortcomings:

S1: Traditional BIST methods are vulnerable to *data mapping* and *signature analysis* attacks [5]. Thus, once a single device is compromised, all other in-field devices are vulnerable to the same attacks.

S2: Although the test methods achieve almost zero *signature aliasing rate* (collision rate) when $L \geq d$, they generate invalid and insecure signatures when $L < d$.

S3: To harden the test methods, extra hardware resources must be adapted to the specific device, introducing additional overhead. In the case of SBIST, also an extensive database is needed for detailed DUT analysis.

In this paper, we tackle these issues and introduce a secure hybrid software/hardware approach for in-field SoC testing. The proposed approach takes advantage of available hash function primitives in modern SoCs for secure signature generation and employs the SoC processor (CPU) to coordinate and schedule the test in a similar manner to SBIST. The main contribution of this work can be listed as follows:

- We propose a new software/hardware hybrid approach as ORA for SoC in-field testing based on keyed-hash message authentication code (KMAC) for signature generation. This mechanism generates valid signatures even

979-8-3315-3968-9/24 $31.00 © 2024 IEEE
254

Special Session

when $L < d$. Furthermore, a sufficiently large device-specific key ensures that potential attacks are limited to a dedicated device, and compromising one makes it difficult to compromise other devices.

- Inspired by SBIST, we deploy the SoC's processor (CPU) to coordinate the test in conjunction with KMAC-based ORA.
- The proposed SoC test is flexible: It allows course-grained functional diagnosis via on-chip testing and a detailed DUT fault analysis as a new remote testing mechanism with a fault dictionary for detailed fault diagnosis.
- We demonstrate the proposed hybrid approach featuring a RISC-V ecosystem.

To the best of our knowledge, this is the first work introducing this combination of KMAC and SBIST.

II. BACKGROUND & RELATED WORK

This section partially reviews state-of-the-art SoC in-field test approaches, particularly emphasizing solutions based on cryptographic primitives.

A. SoC In-field Test Approaches

BIST is a technique for performing in-field tests detecting SoC defects during device operation. Additional scan chains are employed, applying test vectors on the DUT and collecting responses for subsequent analysis by the ATE. During testing, the SoC is unavailable. This test-induced downtime needs hence to be minimized. Resulting, new test methods were introduced: SBIST [2] provides a suitable solution to this challenge. A self-test library (STL) is stored beside the normal execution programs in memory. The processor executes the test program for STL-based tests and collects the responses compared to a golden reference. This mechanism reduces SoC downtime as the SBIST is scheduled only when the corresponding component or system operation is available for testing.

To perform remote SoC in-field testing and diagnosing, the JTAG debug interface is often used to enable device access where local testing is not feasible [6]. The JTAG interface is connected to the local BIST scan chain. The ORA then compacts this test result into a signature which is sent to the remote tester for evaluation using a signature database.

B. SoC Test Vulnerabilities and Countermeasures

The provided deep SoC access for remote system diagnosis and testing via JTAG could reveal internal and sensitive information of SoC components via the BIST scan-chain [5]. This is not just restricted to, e.g., structural information, but can potentially include sensitive data of, e.g., a cryptographic chip: An attacker can, for instance, access dedicated registers via internal scan-chain states to leak sensitive data such as cipher secret keys in a data mapping analysis attack [7] [8]. In signature analysis attacks, the attacker targets the compacted test response to reveal the secrets of AES [9], [10], DES [11] and RSA [12]. In both attack scenarios, the attacker knows the internal scan-chain structure and can control the test at a fine granularity.

SoA countermeasures provide limited protection against the mentioned attacks or are rather costly [5], [13]. Obfuscation techniques have been deployed to hide the internal scan-chain state and protect the DUT against data mapping analysis attacks. Obfuscation techniques incorporate an XOR-scan and added randomization technique via physically unclonable functions (PUF) [14], state-dependent randomization [15], or dynamic obfuscation provided by random number generators [16]. To protect the DUT against signature analysis attacks further, encrypting [17] or masking the interface via a PUF [18] were introduced. It must be noted that all reviewed countermeasures require an individual solution for each possible attack.

C. Cryptographic Primitives-based Solutions for SoC Test Vulnerabilities

Hash-based scan-chain obfuscation was first proposed in [4]. A salted SHA3-512 is utilized to hash the scan-chain output. The salt is mainly deployed to protect the inference of small output responses from a possible rainbow attack. Two steps are required to generate the salt: First, a random selection of bits from the scan-chain output is applied to create a *seed*, and then XORing the *seed* bits results in the salt. Although the salt is usually a public value [19], the salt function of the salted SHA3-512 is assumed to be confident. A secret salt, or so-called pepper, typically requires a salt size of 112 bits [20]; however, it significantly slows down the hash verification. In [21], the JTAG interface and its transmitted data were protected by AES for encrypting the communication channel. Since the encryption of JTAG data is not required for all device life-cycles (e.g., testing, assembling, and shipping), this solution relies on privilege management to perform the encryption only when needed. The mentioned countermeasures for protecting the scan-chain data and communication channels introduce a non-negligible overhead to the SoC, which is unsuitable for resource-constrained devices.

In this paper, we introduce hybrid software/hardware SoC in-field testing to overcome the shortcomings of state-of-the-art solutions, particularly S1, S2, and S3 as stated in the previous section.

III. THE PROPOSED METHODOLOGY

The proposed test method relies on the following pillars: (a) the available KMAC on the SoC to provide device-specific secure signatures, (b) a virtual prototype (VP) of SoC to generate the golden reference signatures, and (c) the CPU to schedule the test, increase DUT availability, and support the remote testing approach. This section covers and explains these pillars in detail.

A. Golden Reference Signature Generation

The golden reference signature is generated in the early SoC design phase, based on reference values extracted from an SoC VP built for testing purposes at a high abstraction level [22].

Special Session

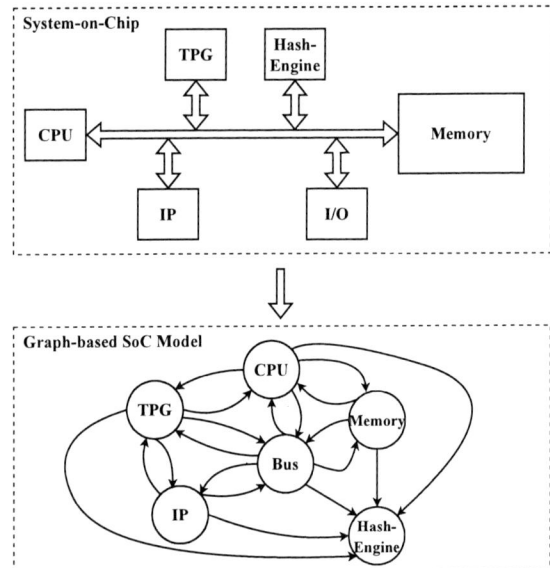

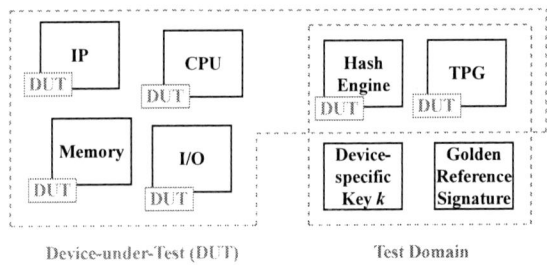

Fig. 1. Signature Generation based on SoC Graph Model

Fig. 2. DUT & Test Domain Separation

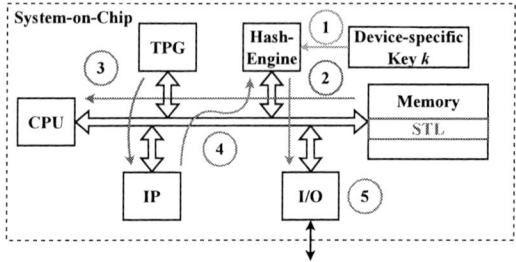

Fig. 3. The proposed SoC Hash-based Test Method

Using the hash function, signatures of (a) each component, (b) subsystems, and (c) the complete SoC VP are generated. Fig. 1 shows a graph-based system modeling approach representing the complete SoC and its components. This includes the processor, memory, IPs, peripherals, and interconnection buses [23]–[25]. The graph-based system modeling could provide a formal and scalable approach. Here, the interaction between the SoC and its components is a directed acyclic graph $G = (V, E)$, in which a vertex V could determine an overall system state, a system component, a dedicated register and its state, or a logic gate. An edge E could represent a system's state transition, a component, a register, a bus connecting components, or a wire connecting registers and logic gates. The granularity of the graph-based model determines the level of detail representing the SoC on the logic level, the component functionality, or the interaction of the SoC components.

B. KMAC-based Test Method: Setup

Fig. 2 outlines the separation into test domain and DUT. It must be noted that every system component or a composition thereof, like the entire SoC, is a DUT. This includes parts of the test domain as both KMAC hash engine and test pattern generator (TPG) also can be tested individually. These two

units together with the device specific keys and the golden reference signature comprise the entity of the test domain.

C. KMAC-based Test Method: Concept

The test mechanism is performed during normal system operation, but without disrupting it. For the proposed test methodology illustrated in Fig. 3, the following steps should be carried out:

1) KMAC is initialized with a device-specific key k. A key manager can perform the key derivation, which is hidden from the system bus.
2) The main CPU coordinates the hardware-assisted SBIST approach using an STL. The CPU schedules the test when a component or the system is not on operational demand.
3) The CPU requests test patterns from a linear feedback shift register (LFSR) deployed as a pseudo-random TPG and feeds them to the DUT.
4) The output responses of the DUT are sent to the KMAC.
5) Finally, KMAC generates the corresponding digest as a device-specific DUT signature, compared to a stored golden reference.

Due to the hash function's high collision resistance, numerous output responses can be fed to the hash, providing high test coverage while preserving zero-aliasing of the signature. In contrast to state-of-the-art in-field test approaches, our proposal provides secure testing of resource-constrained devices, where adding hardware-intensive scan chains and their countermeasures against the presented attacks is unfeasible. If necessary, the test coverage can be further increased by extending the hybrid SBIST with low-overhead LBIST hardware [26] including the hash engine for response compaction without being exploited to an external tester.

D. Two Configurations of the Proposed Test Method

Let $\mathcal{S} = \{s_1, \ldots, s_n\}$ be the seeds which are deployed to generate the test vector patterns $\mathcal{V} = \{v_1, \ldots, v_n\}$ by the TPG and $\mathcal{R} = \{r_1, \ldots, r_n\}$ the responses by the targeted DUT, where $|r_i| = L_i$ bits for every $i = 1, \ldots, n$. The set $\mathcal{H} = \{h_1, \ldots, h_n\}$ represents the corresponding signatures obtained:

$$KMAC^A(r_i) = H(k_A || r_i) = h_i. \qquad (1)$$

SoC^A indicates an SoC with a device-specific key k_A, and H is a standard hash function with $|h_i| = d$ bits for

979-8-3315-3968-9/24 $31.00 © 2024 IEEE   256

Special Session

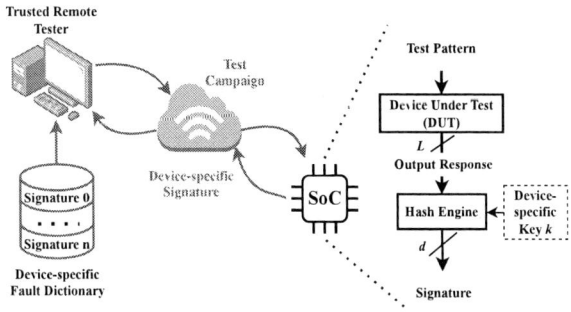

Fig. 4. Proposed KMAC-based Test Method

TABLE I
REMOTE TESTING PROTOCOL

	Device-Specific SoC Remote Testing Protocol		
	Trusted Remote Tester		SoC^A: DUT
(1)	Chooses s_j from $\mathcal{S}$.	$\xrightarrow{s_j}$	$TPG(s_j) = v_j$ $DUT(v_j) = r'_j$
(2)	**If** $h'_j = h_j$, the diagnose can be found based on the corresponding r_j in $\mathcal{R}$ **else** h'_j is an invalid signature.	$\xleftarrow{h'_j}$	$KMAC^A(r'_j) =$ $H(k_A \| r'_j) = h'_j$

every $i = 1, \ldots, n$. Every response signature h_i determines specific diagnoses of all DUTs, such as fault-free, faulty, etc. Therefore, we call the database including $\mathcal{S}$ and $\mathcal{H}$ the device-specific fault dictionary D_A for the DUTs in SoC^A. The in-field test can be proposed for two different system configurations based on the targeted applications of SoC and its design decisions.

1) On-Chip Testing: On-chip testing is sufficient if it ensures functional correctness of the SoC or component, reducing the database to a locally stored small device-specific fault dictionary D_A. For the on-chip test, the CPU selects a seed s_j from $\mathcal{S}$ stored in the memory and sends it to the TPG, generating the test vectors v_j. The test is performed based on the method shown in Fig. 3. The generated signature h'_j then is compared to the locally stored golden reference hash h_j to ensure the functional correctness of the DUT in SoC^A.

The fixed hash size enables the device to determine its complex status without requiring an excessive amount of memory. This provides predictable test requirements from early design steps and helps integrate such in the overall design phases.

2) Remote Testing: The signature can be compared based on an exhaustive fault dictionary by involving an external trusted authority, enabling a detailed fault analysis. This configuration of the proposed test method can be performed as shown in Fig. 4 and illustrated step-by-step in Table I :

(1) The trusted remote tester sends seed s_j from $\mathcal{S}$ to SoC^A. The TPG of SoC^A generates test vectors v_j from $\mathcal{V}$ based on s_j.

(2) The test vectors v_j are applied to the DUT. The corresponding response signature h'_j is generated via KMAC hardware engine and SoC^A sends back h'_j to the remote tester. Then, the DUT diagnosis can be performed by comparing the computed signature h'_j and stored golden reference h_j in the device-specific fault dictionary.

The remote testing leverages the test from a fast but coarse-grained on-chip approach testing the functional correctness of the DUT to a more sophisticated mechanism, which can determine specific faults. For every detectable fault, a dedicated response signature h_j is stored in the remote tester's fault dictionary enabling a detailed fault analysis of all DUTs in SoC^A.

IV. HASH SELECTION & TEST SECURITY ANALYSIS

This section introduces the selected hash function and provides a detailed security evaluation of the proposed test methodology.

A. Cryptographic Hash Selection

According to NIST [27], hash functions ideally provide a collision resistance of $\frac{d}{2}$ with d being the digest's bit length. We chose KECCAK [28] for our work. KECCAK consists of a state of b bits divided into a rate of r bits and capacity of c bits, so that $b = r + c$. Deploying KECCAK for ORA in BIST delivers a constant size digest as a signature for arbitrary output responses with the length L. KMAC is KECCAK with a device-specific key utilized to prevent signature analysis attacks in case of the attacker's knowledge about the internal test structure or rainbow table attacks when the size of the output response is small. The signature generation, in this case, is performed by the KMAC [29] mode of KECCAK. KMAC builds upon the extendable output function (XOF) SHAKE, which delivers two modes with different collision resistance properties: SHAKE128 with a collision resistance of $\min(d/2, 128)$ and SHAKE256 of $\min(d/2, 256)$, where d can be chosen arbitrary. Thus, the digests should have a minimum length of $d_{\text{SHAKE128}} = 256$ and $d_{\text{SHAKE256}} = 512$ to provide the maximum collision resistance, respectively. We feed the KMAC-KECCAK by the DUT output response. The hash digest of size d represents a resulting signature. This enables using KMAC for complex BIST of SoC devices while providing a compact and device-specific signature with high collision resistance.

B. Test Security Analysis

The state-of-the-art ORAs are distinguished between space and time compactors. One way of space compaction is XOR reduction [30], providing either a high compaction rate with low area overhead at the price of a high aliasing rate, or high area consumption when requiring lower signature aliasing. By utilizing a serial-input signature register (SISR) or multiple-input signature register (MISR), time compaction provides a suitable solution for more significant tests while reducing hardware consumption. We analyze our approach in comparison with SISR and MISR compaction.

979-8-3315-3968-9/24 $31.00 © 2024 IEEE   257

Special Session

TABLE II
Hardware Consumption @ ZCU102 FPGA Evaluation Board

	LUT	Register	Slices
PULPissimo SoC	48639	42321	11289
KMAC128	3618	1646	534
TPG	59	71	26

The aliasing probability PA for SISR and MISR is calculated as follows [31]:

$$PA_{SR}(n, L) = \frac{2^{L-n} - 1}{2^L - 1} \tag{2}$$

With this methodology, the SISR or MISR generates a signature for the DUT output response with the length L. The SISR or MISR size n determines the signature length d. Thus, the compaction rate CR of SISR and MISR is calculated as follows:

$$CR_{SR}(n, L) = 1 - \frac{n}{L} \tag{3}$$

(2) & (3) only hold when $L > n$. A minimum DUT output response length is required to generate a valid signature. When providing signatures with $L \leq n$, the uncompacted output response from the DUT is revealed enabling data mapping analysis attacks. For this, ORAs are specifically adapted to fit the output size of a DUT to provide circuit-specific signatures. Deploying the KMAC mode of Keccak as an ORA, the aliasing probability is equal to the collision resistance based on the digest size d:

$$PA_{\text{KMAC}}(d) = \frac{1}{2^{d/2}} \tag{4}$$

The output response of the DUT is compacted to a fixed-size signature. For this, the compaction rate of KMAC generated signature is calculated as follows:

$$CR_{\text{KMAC}}(d, L) = 1 - \frac{d}{L} \tag{5}$$

Utilizing a SISR or MISR as an ORA with equal aliasing probability to KMAC would require a signature register of size $n = d/2$. For example, deploying KMAC128 as a hash engine with a digest size $d = 256$ to match the maximum collision rate, an equal SISR/MISR-based compactor would require a register of size $n = 128$, thus for all output responses $L \leq 128$ an invalid signature would be generated.

(3) & (5) show the requirement for having a larger signature when deploying KMAC as an ORA in comparison to SISR/MISR-based signature for equal aliasing probability. For large test responses $L \gg d$, the compaction rate of KMAC never the less is still sufficient, especially when performing complex SoC tests.

V. Implementation & Results

We chose a RISC-V SoC as a case study to demonstrate the proposed method. First, we extended the RISC-V VP

TABLE III
Compaction & Aliasing Rate Analysis utilizing ISCAS-85 Benchmark

	Primary Outputs	# Test Pattern	Test Response Length	Compaction Rate	Aliasing Rate
c17	2	7	14	-1728.57%	0%
c432	7	63	441	41.95%	0%
c499	32	55	1760	85.45%	0%
c880	26	148	3848	93.35%	0%
c1355	32	100	3200	92.00%	0%
c1908	25	128	3200	92.00%	0%
c2670	140	444	62160	99.59%	0%
c3540	22	264	5808	95.59%	0%
c5315	123	599	73677	99.65%	0%
c6288	32	33	1056	75.76%	0%
c7552	108	455	49140	99.48%	0%

[32] with C++ KMAC128 and TPG modules to generate the golden references. STL was also designed and verified in the RISC-V VP. Then, we integrated a KMAC128 engine into the PULPissimo RISC-V SoC [33]. We selected KMAC128 and a digest size of $d = 256$ bits to match the maximum collision resistance while the hash output (signature) is kept as minimal as possible. The signature aliasing probability in this case is $PA_{\text{KMAC128}} = \frac{1}{2^{128}}$, which is comparable to state of the art solutions [5]; in contrast, however, this low aliasing probability persists even for the case of $L < d$. Furthermore, we selected a key k with a length of $|k| = 64$ bits. Thus, at least 2^{64} possibilities are provided to generate any signature with an arbitrary output response length of the DUT, making it a hard-to-achieve task for an attacker to perform signature analysis attacks. To provide an integrated TPG without requiring additional memory for storing the test pattern, an LFSR with a polynomial of 32nd degree is deployed. The size of the TPG is adapted to match the 32-bit bus size, generating sufficient test pattern for all SoC components. The PULPissimo RISC-V SoC with the integrated KMAC128 and TPG as memory-mapped IPs were implemented on the Xilinx ZCU102 Evaluation Board utilizing a Zynq UltraScale+ FPGA using Vivado 2022.1. Table II shows the SoC hardware small overhead for both the KMAC128 and the TPG, indicating 7.56% of look-up-tables (LUT), 4.06% registers, and 4.98% of slices.

To showcase the scalability for several sizes of circuits (DUTs) and test responses, we utilized circuits from the ISCAS-85 benchmark and performed automated test pattern generation (ATPG) by the *ATALANTA* tool. The test responses were compacted into the KMAC128-generated signatures with a digest size of $d = 256$ bits. To analyze signature aliasing, we performed stuck-at-fault simulations utilizing Synopsys *Z01X* for all fault locations of the benchmark circuits. The results presented in Table III underline the scalability of our approach. Our results show compaction rates up to 99.65% with no aliasing occurring after signature generation. Even for negative compaction rates, i.e., $L < d$, valid signatures are generated without revealing circuit response data. Thus utilizing KMAC for response compaction in combination with SBIST provides a powerful mechanism to locally and remotely test SoC and achieve a high fault coverage without exposing the internal BIST to a malicious tester.

VI. Conclusion

In this paper, we tackled the problem of insecure built-in self-test (BIST) relying on a scan chain mechanism. Thus, we proposed a (KMAC)-based response compaction featuring unique device keys. The proposed test method is reliant on (a) a virtual prototype (VP) to generate test golden references, (b) KMAC to generate the signatures of the device under test (DUT), and (c) CPU to schedule the test, increase the DUT availability, and provide a remote testing capability. The proposed test method waives the need for a dedicated compaction unit since an existing cryptographic hash engine can be used. Therefore, sufficiently detailed system-state analysis can be achieved without revealing system internals. The compaction process based on KMAC is sufficiently secured against inference due to the use of a unique device key. We demonstrate the feasibility of our approach using a RISC-V-based SoC. The results demonstrate that the presented approach could meet all set goals, providing a solid solution for a secure test method.

References

[1] B. Bauer, M. Ayache, S. Mulhem, M. Nitzan, J. Athavale, R. Buchty, and M. Berekovic, "On the dependability lifecycle of electrical/electronic product development: The dual-cone v-model," *Computer*, vol. 55, no. 9, pp. 99–106, 2022.

[2] Wang, *System-on-Chip Test Architectures: Nanometer Design for Testability*. San Francisco, CA, USA: Morgan Kaufmann Publishers Inc., 2007.

[3] E. Rama, M. Ayache, R. Buchty, B. Bauer, M. Korb, M. Berekovic, and S. Mulhem, "Trustworthy integrated circuits: From safety to security and beyond," *IEEE Access*, vol. 12, pp. 69 603–69 632, 2024.

[4] A. Cui, M. Li, G. Qu, and H. Li, "A guaranteed secure scan design based on test data obfuscation by cryptographic hash," *IEEE Transactions on Computer-Aided Design of Integrated Circuits and Systems*, vol. 39, no. 12, pp. 4524–4536, 2020.

[5] X. Li, W. Li, J. Ye, H. Li, and Y. Hu, "Scan chain based attacks and countermeasures: A survey," *IEEE Access*, vol. 7, pp. 85 055–85 065, 2019.

[6] M. Portolan, "Packet-based jtag for remote testing," in *2012 IEEE International Test Conference*, 2012, pp. 1–6.

[7] B. Yang, K. Wu, and R. Karri, "Scan based side channel attack on dedicated hardware implementations of data encryption standard," in *2004 International Confernce on Test*, 2004, pp. 339–344.

[8] S. Ali, O. Sinanoglu, S. Saeed, and R. Karri, "New scan attacks against state-of-the-art countermeasures and dft," in *2014 IEEE International Symposium on Hardware-Oriented Security and Trust (HOST)*. Los Alamitos, CA, USA: IEEE Computer Society, may 2014, pp. 142–147. [Online]. Available: https://doi.ieeecomputersociety.org/10.1109/HST.2014.6855585

[9] B. Ege, A. Das, S. Gosh, and I. Verbauwhede, "Differential scan attack on aes with x-tolerant and x-masked test response compactor," in *2012 15th Euromicro Conference on Digital System Design*, 2012, pp. 545–552.

[10] J. DaRolt, G. Di Natale, M.-L. Flottes, and B. Rouzeyre, "Scan attacks and countermeasures in presence of scan response compactors," in *2011 Sixteenth IEEE European Test Symposium*, 2011, pp. 19–24.

[11] H. Kodera, M. Yanagisawa, and N. Togawa, "Scan-based attack against des cryptosystems using scan signatures," in *2012 IEEE Asia Pacific Conference on Circuits and Systems*, 2012, pp. 599–602.

[12] J. Da Rolt, A. Das, G. Di Natale, M.-L. Flottes, B. Rouzeyre, and I. Verbauwhede, "A new scan attack on rsa in presence of industrial countermeasures," in *Constructive Side-Channel Analysis and Secure Design*, W. Schindler and S. A. Huss, Eds. Berlin, Heidelberg: Springer Berlin Heidelberg, 2012, pp. 89–104.

[13] G. Vishwakarma and W. Lee, "Exploiting jtag and its mitigation in iot: A survey," *Future Internet*, vol. 10, no. 12, 2018. [Online]. Available: https://www.mdpi.com/1999-5903/10/12/121

[14] S. Banik, A. Chattopadhyay, and A. Chowdhury, "Cryptanalysis of the double-feedback xor-chain scheme proposed in indocrypt 2013," in *Progress in Cryptology – INDOCRYPT 2014*, W. Meier and D. Mukhopadhyay, Eds. Cham: Springer International Publishing, 2014, pp. 179–196.

[15] Y. Atobe, Y. Shi, M. Yanagisawa, and N. Togawa, "State dependent scan flip-flop with key-based configuration against scan-based side channel attack on rsa circuit," in *2012 IEEE Asia Pacific Conference on Circuits and Systems*, 2012, pp. 607–610.

[16] X. Wang, D. Zhang, M. He, D. Su, and M. Tehranipoor, "Secure scan and test using obfuscation throughout supply chain," *IEEE Transactions on Computer-Aided Design of Integrated Circuits and Systems*, vol. 37, no. 9, pp. 1867–1880, 2018.

[17] M. Da Silva, M.-L. Flottes, G. Di Natale, and B. Rouzeyre, "Preventing scan attacks on secure circuits through scan chain encryption," *IEEE Transactions on Computer-Aided Design of Integrated Circuits and Systems*, vol. 38, no. 3, pp. 538–550, 2019.

[18] W. Li, J. Ye, X. Li, H. Li, and Y. Hu, "Bias puf based secure scan chain design," in *2018 Asian Hardware Oriented Security and Trust Symposium (AsianHOST)*, 2018, pp. 31–36.

[19] M. S. Turan, E. B. Barker, W. E. Burr, and L. Chen, "Sp 800-132. recommendation for password-based key derivation: Part 1: Storage applications," Gaithersburg, MD, USA, Tech. Rep., 2010.

[20] P. Grassi, J. Fenton, E. Newton, R. Perlner, A. Regenscheid, W. Burr, J. Richer, N. Lefkovitz, J. Danker, Y.-Y. Choong, K. Greene, and M. Theofanos, "Digital identity guidelines: Authentication and lifecycle management," 2020-03-02 2020.

[21] P. Kumar, "Jtag architecture with multi level security," *IOSR Journal of Computer Engineering*, vol. 1, pp. 54–59, 01 2012.

[22] A. Jayasena and P. Mishra, "Directed test generation for hardware validation: A survey," *ACM Comput. Surv.*, vol. 56, no. 5, jan 2024. [Online]. Available: https://doi.org/10.1145/3638046

[23] P. Mishra and N. Dutt, "Specification-driven directed test generation for validation of pipelined processors," *ACM Trans. Des. Autom. Electron. Syst.*, vol. 13, no. 3, jul 2008. [Online]. Available: https://doi.org/10.1145/1367045.1367051

[24] M. Grosso, W. Perez-Holguin, E. Sanchez, M. Sonza Reorda, A. Tonda, and J. velasco medina, "Software-based testing for system peripherals," *Journal of Electronic Testing*, vol. 28, 04 2012.

[25] E. Larsson and Z. Peng, *An Integrated System-on-Chip Test Framework*. Dordrecht: Springer Netherlands, 2008, pp. 439–454.

[26] A. Floridia, G. Mongano, D. Piumatti, and E. Sanchez, "Hybrid on-line self-test architecture for computational units on embedded processor cores," in *2019 IEEE 22nd International Symposium on Design and Diagnostics of Electronic Circuits & Systems (DDECS)*, 2019, pp. 1–6.

[27] Q. Dang, "Recommendation for applications using approved hash algorithms," 2012. [Online]. Available: https://tsapps.nist.gov/publication/get_pdf.cfm?pub_id=911479

[28] M. Dworkin, "Sha-3 standard: Permutation-based hash and extendable-output functions," 2015-08-04 2015.

[29] J. Kelsey, S. jen Chang, and R. Perlner, "Sha-3 derived functions: cshake, kmac, tuplehash and parallelhash," 2016. [Online]. Available: https://tsapps.nist.gov/publication/get_pdf.cfm?pub_id=922422

[30] S. Mitra and K. S. Kim, "X-compact: an efficient response compaction technique," *IEEE Transactions on Computer-Aided Design of Integrated Circuits and Systems*, vol. 23, no. 3, pp. 421–432, 2004.

[31] L.-T. Wang, C.-W. Wu, and X. Wen, *VLSI Test Principles and Architectures: Design for Testability*. San Francisco, CA, USA: Morgan Kaufmann Publishers Inc., 2006.

[32] V. Herdt, D. Große, H. M. Le, and R. Drechsler, "Extensible and configurable risc-v based virtual prototype," in *2018 Forum on Specification Design Languages (FDL)*, 2018, pp. 5–16.

[33] P. D. Schiavone, D. Rossi, A. Pullini, A. Di Mauro, F. Conti, and L. Benini, "Quentin: an ultra-low-power pulpissimo soc in 22nm fdx," in *2018 IEEE SOI-3D-Subthreshold Microelectronics Technology Unified Conference (S3S)*, 2018, pp. 1–3.

Special Session

The Impact of Logic Synthesis and Technology Mapping on Logic Locking Security

Lilas Alrahis[†], Mohammed Thari Nabeel[‡], Johann Knechtel[‡], and Ozgur Sinanoglu[‡]

[†]*Khalifa University, Abu Dhabi, UAE*
[‡]*New York University Abu Dhabi (NYUAD), Abu Dhabi, UAE*
Email: lilas.malrahis@ku.ac.ae, {mtn2, johann, os22}@nyu.edu

Abstract—Logic locking is a design-for-trust solution, safeguarding the intellectual property of integrated circuits within the global semiconductor supply chain. Traditionally, logic synthesis has been relied upon to enhance the security of logic locking. However, recent research has unveiled vulnerabilities inherent in this approach, as logic synthesis is not security-aware by design. On the other hand, state-of-the-art logic-locking techniques leveraging specific locking structures, such as routing networks, were initially presumed secure by design. However, the optimization capabilities of logic synthesis have been shown to compromise these structures, diminishing their security assurances and rendering logic locking vulnerable to attacks. This ongoing interplay between logic locking and logic synthesis necessitates thorough reevaluation. This paper discusses the vulnerabilities and challenges that have emerged at the intersection of logic locking and logic synthesis, offering insights into future research directions aimed at mitigating these issues.

Index Terms—Logic locking, Logic synthesis, IP piracy

I. INTRODUCTION

Logic locking has been proposed as a comprehensive countermeasure against security risks related to integrated circuits (ICs), such as intellectual property (IP) piracy [1] or hardware Trojans [2], [3]. With the semiconductor industry adopting a globalized supply chain model where IC design companies outsource fabrication and other major steps to offshore facilities, logic locking has gained significant interest. Logic locking secures IP designs by introducing modifications to the circuit, typically by adding key-controlled logic elements (e.g., logic gates, referred to as key-gates). The functionality of the design is disrupted unless the correct key is provided via the newly added key-inputs.

There has been significant research on designing methods for key-gate placement (i.e., locking strategies) to optimize both security and performance (i.e., aiming to reduce the overhead of logic locking). In addition to the various proposed locking schemes, substantial research has focused on testing the security of logic locking by developing attacks, primarily aimed at extracting the secret locking key or removing the protection logic altogether. This ongoing cat-and-mouse game between logic-locking schemes and attacks has significantly shaped the history and development of logic locking [4].

Until recently, logic locking was considered an independent step in the electronic design automation (EDA) flow, typically applied either before or after the logic synthesis step. However,

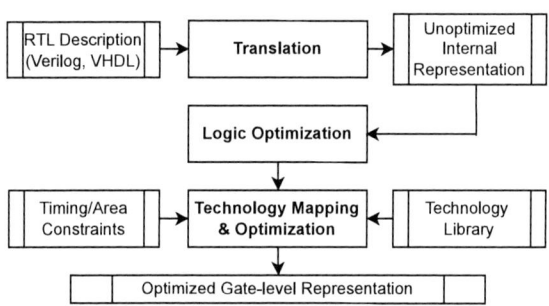

Fig. 1. Logic synthesis flow.

it has become clear that other EDA steps, including logic synthesis, have a significant effect on the security of logic locking (and other design-for-trust solutions [5]).

The aim of this paper is to discuss and summarize selected instances in logic locking research that demonstrate the impact of logic synthesis on the security of logic locking and to explore future research directions necessary for ensuring a secure integration of logic locking into the EDA workflow. Toward that end, we also discuss empirical case studies.

II. BACKGROUND

In this section, we provide the necessary background on logic synthesis and the threat models of logic locking.

A. Logic Synthesis

Logic synthesis is the process of converting a specification of desired circuit behavior, modeled typically using register-transfer level (RTL), into a design implementation using logic gates. This process involves translating a design representation into a circuit made up of transistors. It is a crucial step in hardware design and implementation. Typically, synthesis is carried out using an EDA tool like Synopsys' Design Compiler (DC). These tools translate designs specified in hardware description languages like `Verilog` or `VHDL` into an optimized gate-level representation.

As shown in Fig. 1, the logic synthesis process mainly consists of three steps: translation, optimization, and technology mapping. In the translation phase, the RTL code is translated to a technology-independent representation. For example, in Synopsys DC, this step is called the elaboration phase, where

979-8-3315-3968-9/24 $31.00 © 2024 IEEE

260

Special Session

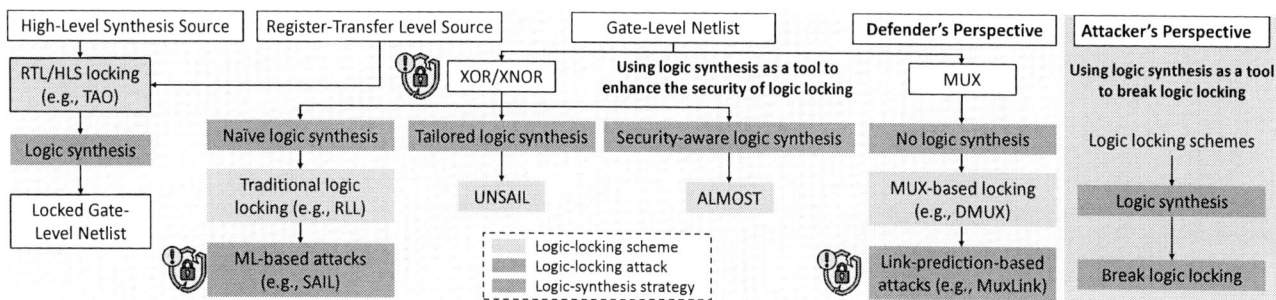

Fig. 2. Proposed taxonomy, organizing selected works at the intersection of logic locking and logic synthesis into defender's and attacker's perspectives. From a defender's perspective, logic synthesis has been relied upon to enhance the security of logic locking. In contrast, from an attacker's perspective, logic synthesis can be exploited as a tool to break logic locking.

the RTL is transformed into a set of instances from a generic technology (GTECH)-independent library. These components are unmapped representations of Boolean functions and act as placeholders for the technology-dependent library. In the optimization step, these Boolean equations are optimized and then mapped to technology-dependent library logic gates based on design constraints and the available logic gates in the target technology library. In Synopsys DC, optimization and technology mapping is done during the compilation step.

Two important constraints that are given for synthesis are the area and timing constraints. Timing constraint sets the target clock frequency the circuit is supposed to operate at with some extra margin for the clock jitter, circuit aging, and the fabrication process uncertainty. The target area constraint is usually set to zero, allowing the tool to minimize the area as much as possible without compromising the timing constraint. Furthermore, the target technology library used is typically the so-called worst corner library, which is characterized for the worst temperature, voltage, and fabrication process; this will ensure the synthesized circuit can work at the target frequency in all operating scenarios without any setup violation.

B. Logic Locking Threat Models

The two seminal threat models for logic locking are the oracle-guided [6] and oracle-less [7]–[13] attacks. Both assume the attacker has access to a gate-level, reverse-engineered but still locked netlist and can trace the key-inputs and locate the key-gates or protection logic. However, the secret key required for proper activation of functionality remains unknown. The attacker could be either within the foundry or at the end-user side.

The key difference is that the oracle-guided model assumes the attacker has access to a functional, unlocked chip with the key inside, which can be purchased from the market. This functional chip can serve as an oracle, allowing the attacker to input patterns and observe correct responses, thus aiding in deciphering the secret key. In contrast, the oracle-less model assumes no such access and aims to extract the key solely from the structure of the locked gate-level netlist. Oracle-less attacks are considered more potent due to their ability to break logic locking with limited resources and earlier in the supply chain (e.g., during fabrication).

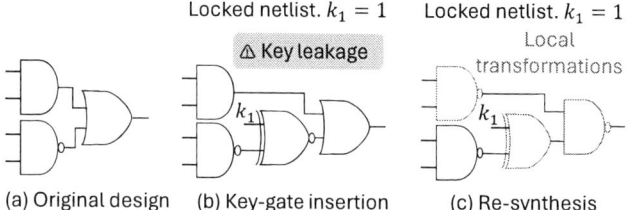

Fig. 3. Traditional XOR/XNOR logic locking. (a) Original netlist. (b) XNOR key-gate with a correct key-bit value of 1. (c) Re-synthesis performed, transforming the key-gate type to XOR while the correct key remains 1, along with other local transformations of gates.

III. CAUSAL NEXUS OF LOGIC LOCKING AND LOGIC SYNTHESIS : A TAXONOMY

In this paper, we propose a taxonomy to categorize the related work at the intersection of logic locking and logic synthesis. Specifically, we divide the work into two perspectives: the defender's perspective and the attacker's perspective, as shown in Fig. 2.

From the defender's perspective, the list of works includes: (i) logic-locking schemes that rely on logic synthesis to secure locking, whether using naïve synthesis, tailored synthesis techniques including bubble pushing, or high level synthesis [14]–[18]; (ii) logic locking attacks inspired by these schemes [7]–[9]; (iii) logic locking solutions that have evolved based on observations from points (i) and (ii) and been designed to eliminate the use of logic synthesis for obfuscation [19]. Importantly, note that the defender's perspective in this paper encompasses both defenses and attacks that have guided the development of state-of-the-art logic locking solutions referred to as "learning resilient." From the attacker's perspective, the list includes attacks that use logic synthesis as an optimization step to break logic locking [10], [20]–[22]. In other words, logic synthesis (or similar optimization tools or concepts) is employed as an attack tool.

A. Defender's Perspective

Traditional Logic Locking and Naïve Logic Synthesis: The first logic-locking scheme, referred to as random logic locking (RLL) [14], randomly inserts XOR and XNOR key-gates into the circuit, as illustrated in Fig. 3. To maintain

Special Session

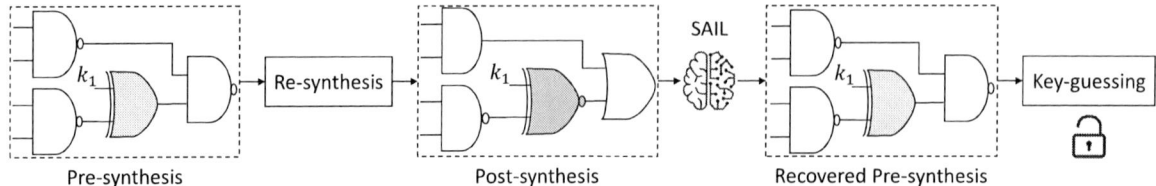

Fig. 4. SAIL attack illustration [9]. ML models are trained on a set of pre-synthesized and post-synthesized locked netlists to undo synthesis optimization, enabling the attacker to predict key-bit values based on the original inserted key-gates prior to logic synthesis. Figure sourced from [16].

correct functionality, the key-bit values should be set as follows: 1'b0 (referred to as 0 for short) for an XOR key-gate, and 1'b1 (referred to as 1 for short) for an XNOR key-gate. This way, both types of key-gates act as buffers. Because of this direct mapping, however, this scheme is inherently vulnerable. Consequently, researchers have relied on logic synthesis to optimize the locked circuits, with the aim of transforming the key-gate types and performing other local gate transformations, as demonstrated in Fig. 3(c). This represents the first reliance on logic synthesis to enhance the security of logic locking. However, the effect of the logic synthesis script/recipe was not studied.

Following RLL, researchers explored different key-gate insertion algorithms and considered various key-gate types, such as AND and OR gates. These initial logic locking schemes are referred to as traditional logic locking. Researchers have also developed attacks to extract the key, focusing on the oracle-guided model and relying on the correlation between input-output patterns and the correct key, with less emphasis on the structure of the locked netlist itself [4].

Machine Learning-based Attacks: The SAIL attack [9] was introduced as one of the first oracle-less attacks on logic locking. It demonstrated that logic synthesis optimization is deterministic and can be learned and undone by machine learning (ML) models. This is because logic synthesis uses fixed algorithms and consistently optimizes for area, power, and delay, resulting in a predictable outcome. Once this optimization is reversed, the attacker can directly predict the key-bit values based on the key-gate types, as depicted in Fig. 4. SAIL showed that naïve reliance on logic synthesis for security purposes is not sufficient.

Later, the OMLA [7] and SnapShot [8] attacks further revealed that attackers do not even need to reverse synthesis optimization. ML models can directly predict the key-bit values by analyzing the locality of gates around the key-gates. This is because logic synthesis is not only deterministic but also often localized. Hence, not only a key-gate itself but also its surrounding are influenced by the correct key-bit value.

Tailored Logic Synthesis and Bubble Pushing: UNSAIL logic locking [16] thwarts the SAIL attack. It creates similar key-gate localities that correspond to both keys 0 and 1, thereby confusing ML models and other attack analysis techniques. UNSAIL incorporates a tailored logic synthesis step in its design procedure to help transform some of the key-gates and generate similar key-gate localities. Although no successful attacks on UNSAIL have been reported to date,

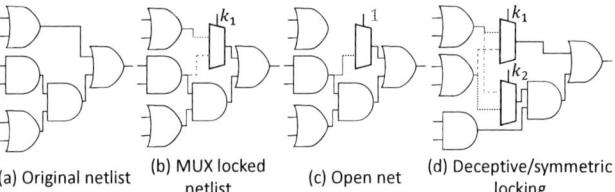

Fig. 5. Different MUX locking strategies. (a) Original unlocked netlist. (b) Naïve MUX insertion, with the true wire denoted in green and the false wire in red; the correct key-bit in this case is 0. (c) Structural leakage when the incorrect key is applied. (d) Symmetric [20] and deceptive MUX locking [19], ensuring no structural leakage for any key combination.

its reliance on logic synthesis may introduce vulnerabilities that could be exploited in the future.

Truly random logic locking (TRLL) [18] is an XOR/XNOR logic-locking solution proposed to thwart SAIL and other ML-based attacks without relying on a logic synthesis step for security. UNSAIL and TRLL were the first schemes proposed as "learning resilient." TRLL leverages the circuit's structure and characteristics, in addition to manual bubble-pushing techniques, to guide the insertion of key-gates. It generates similar key-gate localities with different corresponding key-bit values. Although TRLL is secure in principle, its security relies on certain characteristics of the circuit, including the number of inverters (since it is based on bubble pushing and inversion absorbing). This dependency could limit the number of key-gates and, more concerning, correct keys might still be obtained from the overall structure of the locked netlist.

Multiplexer-based Logic Locking: Deceptive logic locking (DMUX) [19] moves away from XOR/XNOR locking and instead utilizes multiplexer (MUX) gates for locking. MUX locking was initially considered more secure because a MUX key-gate does not have a corresponding key-bit value (i.e., no logic synthesis is required after locking), as illustrated in Fig. 5. More specifically, a MUX key-gate takes two inputs from the circuit, a true wire and a false wire, and the key-input acts as a select line, see Fig. 5(b). Upon applying the correct key, the true wire will pass through the output of the MUX, maintaining the correct functionality. Depending on whether the true wire is connected to the first or second input of the MUX, the correct key will be 0 or 1, respectively. However, naïve selection of the false wires can lead to vulnerabilities in this approach. For example, for the wrong key-bit value selected, the true wire may be left dangling, causing an entire logic cone to become disconnected, as shown in Fig. 5(c). Therefore, DMUX introduces specific rules for selecting true

979-8-3315-3968-9/24 $31.00 © 2024 IEEE   262

Special Session

and false wires to ensure that no disconnected cones occur for any key-bit value, as illustrated in Fig. 5(d).

Concurrent with DMUX, symmetry-focused logic locking schemes were proposed [20], [23], achieving the same desired security outcome and following a similar locking strategy.

Full-Lock [24] is a MUX-based logic-locking scheme designed to specifically thwart Boolean Satisfiability (SAT)-based attacks [6]. It achieves this by using configurable routing and look-up tables to obfuscate a set of gates and their corresponding input connections. Configurable routing is implemented using Banyan networks, a class of logarithmic networks that permutes connections based on a key. The design of the Banyan network makes the circuit SAT-hard, meaning that it is difficult to solve for the correct key, thereby providing security by construction. However, if an attacker performs a round of logic synthesis on the locked gate-level netlist, the synthesis process optimizes and simplifies the network, thereby breaking its SAT hardness. Consequently, circuits that were previously unbreakable within 15 days of runtime can be solved in seconds.

In short, XNOR/XOR locking relies on logic synthesis for security, which leads to vulnerabilities due to its lack of security-aware implementation. MUX locking showed promise in overcoming this dependency; however, these schemes later proved vulnerable to ML-based attacks like MuxLink and UNTANGLE [25], [26], which are focused on link prediction.

Security-Aware Logic Synthesis: To address these challenges, ALMOST [15] was introduced. It is the first logic-locking platform that optimizes logic synthesis to maximize security. It is also the first to explicitly demonstrate the direct impact of logic synthesis on the security of logic locking. In a motivational experiment, the authors of ALMOST used an RLL locked circuit, which was initially 100% vulnerable, and synthesized it using different logic synthesis recipes. They then attacked each generated netlist using the OMLA attack and demonstrated how the accuracy of the attack varied across the samples, even though all samples were derived from the same initial locked circuit, as shown in Fig. 6. This experiment effectively isolated and illustrated the impact of logic synthesis on security. Furthermore, the authors employed simulated annealing to identify an optimal logic synthesis recipe to achieve the highest level of security. This represents a significant step towards security-aware EDA. Still, other attack approaches can still be promising against this scheme, as also shown in our case study in the next subsection.

High-Level Synthesis: Alternative high-level locking techniques aim to obscure semantic information before it is incorporated into the gate-level netlist through logic synthesis [17]. For instance, the TAO approach introduces obfuscations during high-level synthesis (HLS) [17]. Thus, TAO requires access to the HLS source code for integrating these obfuscations and is not capable of obfuscating pre-existing IP. An alternative approach is to apply obfuscations at the RTL [27], [28]. While most semantic details, such as constants, operations, and control flows, remain evident at the RTL stage, this method allows for obfuscation to be applied to existing RTL IPs.

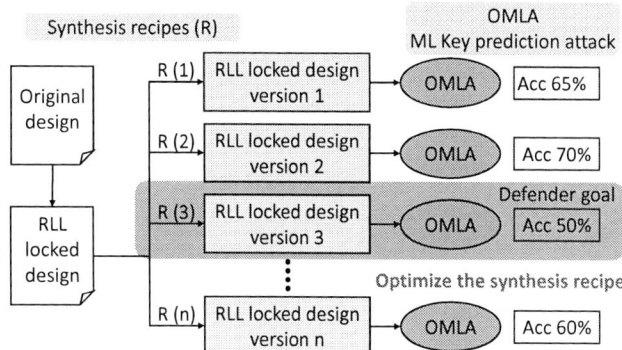

Fig. 6. In ALMOST [15], the authors initially employed the same vulnerable RLL locked design and explored various synthesis recipes, then launched the OMLA attack on the generated netlists. Different attack results were reported, isolating and demonstrating the effect of logic synthesis on security.

B. Attacker's Perspective

Removal-based Attacks: The functional reverse engineering (RE) attack [29], a removal-based method, targeted TT-Lock [30]. The attack employed a functional RE tool to identify and remove a Hamming-distance circuit module used by the locking scheme. The authors reported that when the logic-locked circuit was implemented with a full technology library it was challenging to identify that specific module, due to the complex technology-dependent optimizations performed by synthesis. However, once the attackers re-synthesized the design using a simpler, constrained library, the performance in detecting the specific module increased significantly.[1] Thus, both the logic synthesis recipe and technology mapping can considerably impact the security of logic locking.

Synthesis-based Attacks: SCOPE [20] is a synthesis-based constant propagation attack. SCOPE is a variation of the SWEEP attack [21] but operates in an unsupervised manner, unlike SWEEP. SCOPE utilizes a synthesis tool to analyze each key-input port for critical features that can reveal the correct key-bit value. Specifically, each key-input is examined individually by assigning it a 0 in one instance and a 1 in another instance. The two modified locked netlists are then synthesized, and the resulting reports are compared. The structural and physical properties reported by the synthesis tool for both circuits are investigated to find any correlation with the correct key-bit value. SCOPE is applicable to various locking schemes; however, it yields the best results against MUX-based locking schemes, where the circuit's characteristics, such as area, vary significantly for different key options.

The Desynthesis attack [10] exploits the fact that most logic-locking schemes apply locking after the initial logic synthesis phase (i.e., locking the gate-level netlist rather than the RTL). This attack demonstrates that information about

---

[1] Specifically, for TTLock [30] with 32 bits key-size, the detection accuracy for that specific module was all 0% across the considered ISCAS-85 circuits for the full library, namely a 65nm low-power library with ARM standard cells. For the constrained library (containing only XOR, INV, BUF, NAND, and NOR gates), however, the detection accuracy increased to 100%, 100%, and 93.75% for the c2670, c5315, and c7552 circuits, respectively.

979-8-3315-3968-9/24 $31.00 © 2024 IEEE   263

Special Session

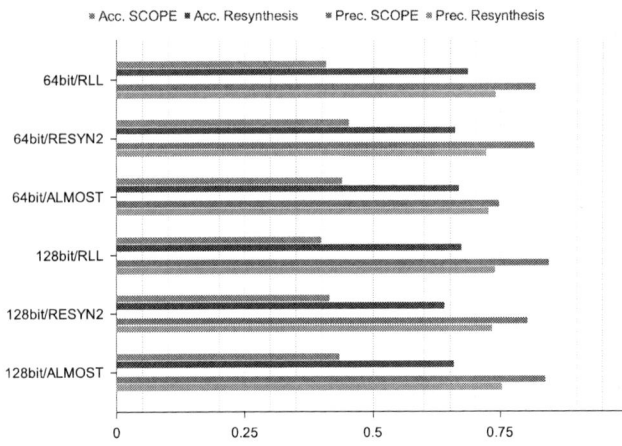

Fig. 7. Accuracy (Acc.) and precision (Prec.) metrics for SCOPE versus re-synthesis attack runs, averaged across the `ISCAS-85` benchmarks.

the intended Boolean functionality is already embedded in the synthesized locked netlist and can be extracted. The attack begins by assigning a randomly selected key to the circuit and performing a synthesis-based analysis. The applied key is then slightly modified, followed by another round of synthesis-based analysis. The key is optimized based on the observed structural properties and guided by a correlation process between the different key-bit values and the physical properties of the circuit.

The re-synthesis attack also leverages logic synthesis and optimization to defeat logic locking [22]. During a pre-attack stage, the logic-locked netlist under attack is re-synthesized using various synthesis parameters, resulting in a large number of structurally different netlists that maintain the same functionality. The SCOPE attack is then applied to each generated netlist to predict the key-bit values. Finally, a majority voting process is used to determine the most likely key for the circuit.

Case Study: Consider Fig. 7 for a case study on the re-synthesis attack. Next, we describe the setup. We contrast SCOPE versus re-synthesis attack runs on `ISCAS-85` benchmarks, the latter being locked through different means: RLL [14], RESYN2, and ALMOST [15]. Note that RESYN2 is a baseline/naïve technique that was proposed along with ALMOST. Also note that, for metrics, we consider the well-established notions of accuracy and precision, as follows: accuracy = # correctly inferred key-bits over # total key-bits, precision = (# correctly inferred key-bits + # undecided key-bits) over # total key-bits. We obtain all the locked benchmark instances from the authors of ALMOST. We utilize our extended re-synthesis attack setup [31]. We employ default settings for both the extended re-synthesis setup and for the SCOPE setup. For SCOPE, we only report numbers for the superior key variant, i.e., Variant 2 here for all runs.

For analysis and interpretation of Fig. 7, consider the following. First, the accuracy for the re-synthesis attack is notably higher than for the baseline SCOPE attack. In fact, while the accuracy for SCOPE always remains below 50%

(i.e., is worse than a random guess), the re-synthesis attack pushes this to around 66% across all benchmarks, locking techniques, and key-sizes. This means that the number of correctly inferred key-bits is notably larger for the re-synthesis attack. Second, the precision for the re-synthesis attack is, in most cases, reducing for the re-synthesis attack over the baseline SCOPE attack. Given that the number of correctly inferred key-bits is always larger, this reduction implies that the number of undecided key-bits is reducing as well, and even to a larger degree than the increase of correctly inferred key-bits. In other words, the re-synthesis attack does not only succeed to infer more key-bits correctly but also to reduce the attacker's uncertainty by (correctly) deciding on more key-bits.

Along with the strength of the re-synthesis attack, these two findings clearly demonstrate the key point of this paper: synthesis holds a major impact on the security (or rather lack thereof) of logic locking. Even techniques that were specifically devised with this in mind can mitigate this adversarial impact only to some degree.[2]

Redundancy-based Attacks: In the redundancy attack [11], logic synthesis is not directly used for the attack; however, it exposes certain vulnerabilities of logic locking implementations. Specifically, if logic locking is applied after logic synthesis (i.e., no optimization is performed on the circuit post key-gate placement), a vulnerability arises. By default, the circuit tends to be more optimized for the correct key than for any incorrect key. The redundancy attack examines each key-gate, applies both possible key options, and counts the number of untestable faults resulting from each key option. The key with the fewest untestable faults is likely to be the correct key. Thus, the choice of whether to apply logic locking before or after logic synthesis has a significant impact on security.

IV. FURTHER CONSIDERATIONS

1) Physical Synthesis: In most logic-locking works, attacks are implemented on locked gate-level netlists after logic synthesis. However, this overlooks the fact that attackers work with circuits after physical synthesis, an additional optimization stage often neglected by the research community. This stage involves further circuit modifications. Therefore, both attackers and defenders must consider the impact of layout generation on the security of logic locking [2], [3].

2) Program Synthesis: The authors of [32] present a locking mechanism called higher-order logic locking (HOLL), combining ideas from logic locking, program synthesis, and programmable devices. Unlike traditional logic locking, which relies on a sequence of independent key-bits, HOLL introduces a functional relationship among key-bits. By leveraging program synthesis (which generates programs from input-output examples or specifications, focusing on behavioral patterns), HOLL converts the original design into a locked version. This differs from logic synthesis (which transforms high-level

---

[2]Here, the gains obtained by the re-synthesis attack over SCOPE for ALMOST and RESYN2 were less than for RLL, but still considerable: accuracy increased by around 22 percentage points (PPs) for ALMOST and RESYN2 versus around 27 PPs for RLL.

979-8-3315-3968-9/24 $31.00 © 2024 IEEE

Special Session

hardware descriptions into lower-level gate representations), allowing for more complex key dependencies, likely making it harder for attackers to bypass the locking.

V. CONCLUSION

In this paper, we presented selected works on logic locking that emphasize the impact of logic synthesis on the security of locking schemes from both defender's and attacker's perspectives. Additionally, we implemented a case study demonstrating how re-synthesis from an attacker's perspective can significantly enhance the success rate of attacks on locked netlists. Our discussion and case study highlight that logic locking should not be treated as an isolated step in the EDA flow. Instead, it must be integrated into a comprehensive security-aware approach to effectively mitigate vulnerabilities. This call for action aligns well with prior community efforts toward secure IC design [5], which are still to be realized.

REFERENCES

[1] K. Shamsi, M. Li, K. Plaks, S. Fazzari, D. Z. Pan, and Y. Jin, "IP protection and supply chain security through logic obfuscation: A systematic overview," *ACM Trans. Des. Autom. Electron. Syst.*, vol. 24, no. 6, sep 2019. [Online]. Available: https://doi.org/10.1145/3342099

[2] F. Wang, Q. Wang, B. Fu, S. Jiang, X. Zhang, L. Alrahis, O. Sinanoglu, J. Knechtel, T.-Y. Ho, and E. F. Young, "Security closure of IC layouts against hardware Trojans," in *Proc. Int. Symp. Phys. Des.*, 2023, pp. 229–237.

[3] F. Wang, Q. Wang, L. Alrahis, B. Fu, S. Jiang, X. Zhang, O. Sinanoglu, T.-Y. Ho, E. F. Y. Young, and J. Knechtel, "Trolloc: Logic locking and layout hardening for ic security closure against hardware trojans," 2024. [Online]. Available: https://arxiv.org/abs/2405.05590

[4] M. Yasin and O. Sinanoglu, "Evolution of logic locking," in *IFIP/IEEE International Conference on Very Large Scale Integration (VLSI-SoC)*, 2017, pp. 1–6.

[5] J. Knechtel, E. B. Kavun, F. Regazzoni, A. Heuser, A. Chattopadhyay, D. Mukhopadhyay, S. Dey, Y. Fei, Y. Belenky, I. Levi, T. Güneysu, P. Schaumont, and I. Polian, "Towards secure composition of integrated circuits and electronic systems: On the role of EDA," in *Proc. Des. Autom. Test Europe*, 2020.

[6] P. Subramanyan, S. Ray, and S. Malik, "Evaluating the security of logic encryption algorithms," in *IEEE International Symposium on Hardware Oriented Security and Trust (HOST)*, 2015, pp. 137–143.

[7] L. Alrahis, S. Patnaik, M. Shafique, and O. Sinanoglu, "OMLA: An oracle-less machine learning-based attack on logic locking," *IEEE Transactions on Circuits and Systems II: Express Briefs*, vol. 69, no. 3, pp. 1602–1606, 2022.

[8] D. Sisejkovic, F. Merchant, L. M. Reimann, H. Srivastava, A. Hallawa, and R. Leupers, "Challenging the security of logic locking schemes in the era of deep learning: A neuroevolutionary approach," *J. Emerg. Technol. Comput. Syst.*, vol. 17, no. 3, may 2021. [Online]. Available: https://doi.org/10.1145/3431389

[9] P. Chakraborty, J. Cruz, and S. Bhunia, "SAIL: Machine learning guided structural analysis attack on hardware obfuscation," in *Asian Hardware Oriented Security and Trust Symposium (AsianHOST)*, 2018, pp. 56–61.

[10] M. E. Massad, J. Zhang, S. Garg, and M. V. Tripunitara, "Logic locking for secure outsourced chip fabrication: A new attack and provably secure defense mechanism," *arXiv preprint arXiv:1703.10187*, 2017.

[11] L. Li and A. Orailoglu, "Redundancy attack: Breaking logic locking through oracleless rationality analysis," *IEEE Transactions on Computer-Aided Design of Integrated Circuits and Systems*, vol. 42, no. 4, pp. 1044–1057, 2023.

[12] L. Alrahis, S. Patnaik, F. Khalid, M. A. Hanif, H. Saleh, M. Shafique, and O. Sinanoglu, "GNNUnlock: Graph neural networks-based oracle-less unlocking scheme for provably secure logic locking," in *Design, Automation & Test in Europe Conference & Exhibition (DATE)*, 2021, pp. 780–785.

[13] L. Alrahis, S. Patnaik, M. A. Hanif, H. Saleh, M. Shafique, and O. Sinanoglu, "GNNUnlock+: A systematic methodology for designing graph neural networks-based oracle-less unlocking schemes for provably secure logic locking," *IEEE Transactions on Emerging Topics in Computing*, vol. 10, no. 3, pp. 1575–1592, 2022.

[14] J. A. Roy, F. Koushanfar, and I. L. Markov, "EPIC: Ending piracy of integrated circuits," in *Design, Automation and Test in Europe*, 2008, pp. 1069–1074.

[15] A. B. Chowdhury, L. Alrahis, L. Collini, J. Knechtel, R. Karri, S. Garg, O. Sinanoglu, and B. Tan, "ALMOST: Adversarial learning to mitigate oracle-less ML attacks via synthesis tuning," in *ACM/IEEE Design Automation Conference (DAC)*, 2023, pp. 1–6.

[16] L. Alrahis, S. Patnaik, J. Knechtel, H. Saleh, B. Mohammad, M. Al-Qutayri, and O. Sinanoglu, "UNSAIL: Thwarting oracle-less machine learning attacks on logic locking," *IEEE Transactions on Information Forensics and Security*, vol. 16, pp. 2508–2523, 2021.

[17] C. Pilato, F. Regazzoni, R. Karri, and S. Garg, "TAO: Techniques for algorithm-level obfuscation during high-level synthesis," in *Design Automation Conference*, 2018, pp. 1–6.

[18] N. Limaye, E. Kalligeros, N. Karousos, I. G. Karybali, and O. Sinanoglu, "Thwarting all logic locking attacks: Dishonest oracle with truly random logic locking," *IEEE Transactions on Computer-Aided Design of Integrated Circuits and Systems*, vol. 40, no. 9, pp. 1740–1753, 2021.

[19] D. Sisejkovic, F. Merchant, L. M. Reimann, and R. Leupers, "Deceptive logic locking for hardware integrity protection against machine learning attacks," *IEEE Transactions on Computer-Aided Design of Integrated Circuits and Systems*, vol. 41, no. 6, pp. 1716–1729, 2022.

[20] A. Alaql, M. M. Rahman, and S. Bhunia, "SCOPE: Synthesis-based constant propagation attack on logic locking," *IEEE Transactions on Very Large Scale Integration (VLSI) Systems*, vol. 29, no. 8, pp. 1529–1542, 2021.

[21] A. Alaql, D. Forte, and S. Bhunia, "Sweep to the secret: A constant propagation attack on logic locking," in *Asian Hardware Oriented Security and Trust Symposium (AsianHOST)*, 2019, pp. 1–6.

[22] F. Almeida, L. Aksoy, Q.-L. Nguyen, S. Dupuis, M.-L. Flottes, and S. Pagliarini, "Resynthesis-based attacks against logic locking," in *International Symposium on Quality Electronic Design (ISQED)*. IEEE, 2023, pp. 1–8.

[23] S. Elsharief, L. Alrahis, J. Knechtel, and O. Sinanoglu, "IsoLock: Thwarting link-prediction attacks on routing obfuscation by graph isomorphism," *Cryptology ePrint Archive*, 2022.

[24] H. M. Kamali, K. Z. Azar, H. Homayoun, and A. Sasan, "Full-lock: Hard distributions of SAT instances for obfuscating circuits using fully configurable logic and routing blocks," in *ACM/IEEE Design Automation Conference (DAC)*, 2019, pp. 1–6.

[25] L. Alrahis, S. Patnaik, M. Shafique, and O. Sinanoglu, "MuxLink: Circumventing learning-resilient MUX-locking using graph neural network-based link prediction," in *Design, Automation & Test in Europe Conference & Exhibition (DATE)*, 2022, pp. 694–699.

[26] L. Alrahis, S. Patnaik, M. A. Hanif, M. Shafique, and O. Sinanoglu, "UNTANGLE: Unlocking routing and logic obfuscation using graph neural networks-based link prediction," in *IEEE/ACM International Conference On Computer Aided Design (ICCAD)*, 2021, pp. 1–9.

[27] Y. Lao and K. K. Parhi, "Obfuscating DSP circuits via high-level transformations," *IEEE transactions on very large scale integration (VLSI) systems*, vol. 23, no. 5, pp. 819–830, 2014.

[28] R. S. Chakraborty and S. Bhunia, "RTL hardware IP protection using key-based control and data flow obfuscation," in *International Conference on VLSI Design*. IEEE, 2010, pp. 405–410.

[29] L. Alrahis, M. Yasin, H. Saleh, B. Mohammad, and M. Al-Qutayri, "Functional reverse engineering on SAT-attack resilient logic locking," in *IEEE International Symposium on Circuits and Systems (ISCAS)*, 2019, pp. 1–5.

[30] M. Yasin, B. Mazumdar, J. J. V. Rajendran, and O. Sinanoglu, "TTLock: Tenacious and traceless logic locking," in *IEEE International Symposium on Hardware Oriented Security and Trust (HOST)*, 2017, pp. 166–166.

[31] J. Knechtel, "Resynthesis + SCOPE Setup," 2023, https://github.com/DfX-NYUAD/resynthesis_attack.

[32] G. Takhar, R. Karri, C. Pilato, and S. Roy, "HOLL: Program synthesis for higher order logic locking," in *International Conference on Tools and Algorithms for the Construction and Analysis of Systems*. Springer, 2022, pp. 3–24.

Special Session

BlockShield: A TPM-Integrated Blockchain-based Framework for Shielding Against Deepfakes

Venkata K. V. V. Bathalapalli
Dept. of Computer Science and Engineering
University of North Texas.
Email: vb0194@unt.edu

Aakarshan Kumar
Texas Academy of Mathematics and Science
University of North Texas.
Email: aakarshankumar@my.unt.edu

Saraju P. Mohanty
Dept. of Computer Science and Engineering
University of North Texas.
Email: saraju.mohanty@unt.edu

Elias Kougianos
Dept. of Electrical Engineering
University of North Texas
Email: elias.kougianos@unt.edu

Venkata P. Yanambaka
School of the Sciences
Texas Woman's University
vyanambaka@twu.edu

Abstract—**The increasing threat to individual privacy and personalized digital content on social media posed by Deepfakes has highlighted the importance for a secure and reliable multimedia content integrity mechanism. In this paper a novel Blockchain driven hardware secure video attestation scheme, BlockShield is proposed for Deepfake mitigation. The proposed system includes a novel approach that ensures digital content traceability and privacy using Blockchain smart contracts and TPM's digital signature mechanism. The proposed work explores the scope of hardware-assisted security for Deepfake mitigation through a hardware TPM working together with Blockchain for enhanced digital media protection and sharing. The proposed work is experimentally validated and presented which validates the scope of hardware-assisted and blockchain integrated Deepfake mitigation framework.**

Index Terms—**Deepfake, Blockchain, Trusted-Platform-Module (TPM)**

I. INTRODUCTION

The growing era of Artificial Intelligence (AI) enhances the capability for fake content creation and manipulation. Deepfake, an emerging AI application threat uses neural network architectures to perform audiovisual content modification. Initially these techniques were deployed in entertainment industry and referred to as Synthetic media. As the usage and influence of social media and its content is increasing along with readily available applications that use deep learning techniques for content modification, the AI based digital content modification creation and sharing on social media is increasing at an alarming rate. To counter this, various approaches were considered as potential solutions ranging from regulatory frameworks to digital signature based content attestation. Blockchain emerges as a feasible solution for digital content attestation and verification due to its immutability, decentralized working flow and potential in emerging global technological advancements in AI, Machine learning, Internet-of-Things (IoT) [1], [2].

Any person uploading digital content on social media cannot be sure of its authenticity as an adversary can access it and use various techniques to perform gesture swapping, facial attribute manipulation, or face swapping, and lip syncing. Facial attribute manipulation could be performed by manipulating the original face in the video/image with the target's facial attributes. Deepfake detection and mitigation emerged as alluring research areas for countering the impact of fake content creation and sharing. Deepfake detection and mitigation have become more challenging with advancing Deep learning enhanced techniques and mobile applications enabling a user to perform modification of digital media more realistically. This research work focuses on tracebility of digital content and a presents a sustainable approach for digital media sharing and copyright protection using Blockchain and TPM hardware security module. The conceptual overview of proposed research work is presented in Fig. 1

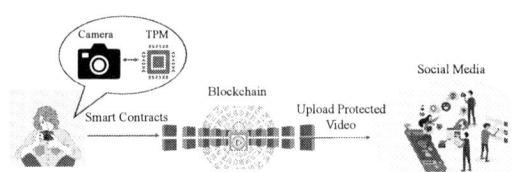

Fig. 1: Overview of BlockShield

Deepfake detection involves using AI-assisted techniques to identify the authenticity of uploaded digital content on social media. Various techniques include monitoring breathing pattern of source in video, analyzing eye blinking rate, and audio inconsistencies. These techniques help in verifying the legitimacy of digital content. However, mitigation can also be a sustainable solution to counter Deepfakes for ensuring trustability. Due to the emergence of various applications and easy access to AI, super realistic synthesis of digital content becomes easy which make mitigation more suitable option than detection to address the digital content integrity issues.

979-8-3315-3968-9/24 $31.00 © 2024 IEEE

Special Session

Video-based modification techniques at the frame level, pixel, and block levels could really question the integrity of videos. Video broadcasting can also impact the authenticity due to the unreliable transmission channel. Furthermore, the video tampering also could be done within a specified region of video frame called a block through various techniques like resizing and pixel intensity variation [3]. Due to this, the temporal consistency in a video and connection between consecutive frames is disrupted, which can lead to inconsistency.

Deepfake mitigation involves combating unauthorized access and modification to personalized digital content. The approaches include regulation, digital rights protection, security through watermarking and forensic techniques, and trusted ownership attestation of multimedia content. Digital rights management (DRM) and protection techniques have been the most sought research topics in Deepfake mitigation. Trusted attestation techniques include watermarking, encryption of multimedia, and authenticated communication. However, the computational and performance overhead make it challenging to sustain these resource-intensive cryptographic and watermarking techniques. Furthermore, compression of digital media make it challenging to decode watermarks and could pose a question on watermarking solutions. To address these issues, trusted attestation of video/image through a secure and reliable hardware-integrated distributed ledger framework can guarantee data protection and sharing ensuring authenticity.

This research work presents a broad vision of ensuring a hardware-assisted secure and decentralized framework to attest the digital content with a Blockchain-enhanced trusted content-sharing framework. The TPM hardware security module can be used to perform various tasks, ranging from remote attestation, cryptographic key generation, random number generation, and secure booting [4] in computing applications. It is a secure microprocessor which offers default security and privacy to emerging computing applications and can be an effective approach for ensuring root-of-trust in digital media applications [5]. The proposed research idea to the best of authors knowledge is the first work on Deepfake mitigation that proposes a TPM and Blockchain-integrated framework. The proposed TPM-integrated Blockchain framework securely attests the digital content source by attesting the device capturing video/image using TPM and then securely storing the captured video metadata and the device's TPM-generated digital signature for video inside the blockchain. The proposed work can ensure the integrity of digital content at the hardware source capturing the video/image. The device can be a smartphone or camera with an embedded TPM chip for security.

The rest of this paper is organized as follows. Section II discusses related works on Blockchain based Deepfake mitigation. Novel contributions of the proposed work are presented in Section III. A detailed explanation of the proposed Deepfake mitigation approach is discussed in section IV. Experimental validation results are presented in section V and finally, the conclusion and future research directions are discussed in Section VI.

II. RELATED RESEARCH

This section discussed state-of-art research on blockchain-based Deepfake mitigation approaches. Table. I presents various research works for Deepfake detection and mitigation.

A blockchain-based framework as a proof-of-concept for Deepfake mitigation is presented in [2] where a blockchain-based approach is used for securely uploading videos on social media and to verify video integrity using hash value and video metadata. A Deepfake mitigation approach through Blockchain smart contract is proposed in [6], where a video is uploaded onto a decentralized file storage system and can be edited by secondary artists using smart contract which would be approved by the primary artist. The above work claims to counter Deepfake with a robust smart contract-based video editing process.

Video source tracking and integrity verification using blockchain which allows video source tracking using timestamp is presented in [7] which works based on the Inter-Planetary file system-based approach for video source verification. Machine learning and Blockchain integrated framework to perform Deepfake detection, similarity, and fake news using IPFS storage is proposed in [8]. A smart contract-based video sharing and content copyright protection framework is presented in [9] which works on storing video in a decentralized file storage system that has video metadata and can be used to trace back the original artist. A secondary artist can connect to the smart contract and request video editing permission. This work claims to counter unauthorized access and copyright protection using Blockchain and Hyper ledger fabric. The smart contract based approach in this work is inspired from the above cited works and proposes a TPM-based hardware attestation of video for sustainable content sharing and traceability.

In [10], a Deepfake news mitigation framework based on watermarking and blockchain is proposed which presents a watermarking framework to attest the video. Blockchain supports the forensic analysis of the videos to perform verification of video's legitimacy by retrieving stored video data from decentralized IPFS file storage system. For authenticity of digital media, a Hyperledger Fabric 2.0 and CNN LSTM model as a Proof of Concept is presented in [11] which works on obtaining a hash of digital media captioning which is encoded with CNN LSTM model and securely stores data inside Blockchain.

Blockchain and Non-Fungible Token-based digital content traceability system in [12] works on smart contract-based digital content creation and minted NFT tokens are created for media in Non-Fungible Token Marketplace (NFTM), thereby guaranteeing authenticity to digital media. Furthermore, movie streaming can also be facilitated, video metadata and artifacts can also be stored on Blockchain through smart contracts as discussed in [13].

III. NOVEL CONTRIBUTIONS

The objective of the paper aims to address the video and its metadata integrity ensuring Deepfake resistant personalized

979-8-3315-3968-9/24 $31.00 © 2024 IEEE

Special Session

TABLE I: Related Research On Deepfake Detection and Mitigation

Work	Technique	Methodology	Tools
[8]	Deepfake Detection	ML and Blockchain integrated Fake news detection	Efficient Net, Smart Contracts
[14]	Mitigation(Image)	PUF and ML framework for facial feature attestation	Dlib 68 (Facial detection and keypoint prediction), PUF
[10]	Fake news mitigation	Watermarking and Blockchain for Deepfake Video protection	IPFS, MTCNN algorithm, and Face Alignment Network (FAN) algorithm
[15]	Audio Deepfake Mitigation	Fragile speech watermarking with Blockchain	MTCNN, Wav2Lip
BlockShield (Current Work)	Deepfake Video Mitigation	Blockchain and TPM-based video attestation	Hardware TPM, Smart Contracts

digital content sharing through hardware-based multimedia attestation.

A. Research Problems

- Authenticity of digital content
- Reliable digital content traceability mechanism.
- Lack of trusted multimedia content sharing mechanism
- Conventional video protection mechanism using Blockchain incurring high storage and transaction fees.
- Lack of trusted camera authentication and hardware-assisted content attestation mechanism

B. Contributions of the Paper

- A sustainable Deepfake mitigation approach using state-of-art TPM and Blockchain technologies.
- A secure visual Deepfake mitigation approach for individual content privacy and security on social media.
- An energy efficient solution that integrates TPM and Blockchain using smart contracts.
- A secure digital content sharing framework using Blockchain to provide integrity and authenticity.
- An approach based on TPMs digital signature mechanism facilitating hardware root-of-trust for the video/image.

IV. METHODOLOGY

This section includes details of the proposed Blockchain and TPM based video attestation technique.

Peer-to-Peer decentralized Inter-Planetary File storage system (IPFS) can be used to the store video and its metadata which is more efficient than Blockchain-based storage which is expensive and computationally infeasible. IPFS generates a unique identifier or hash for a bundle of files and the unique identifier can be used to access and retrieve files from IPFS. Furthermore, Trusted Platform Module (TPM) is a hardware security primitive that performs video attestation by digitally signing the video file before uploading to IPFS. It generates primary root keys like the endorsement key (EK) and attestation identity keys (AIK) which can further generate RSA key pairs for digitally fingerprinting TPM. Furthermore, the primary RSA key pair is generated and used to perform TPM attestation operations using public portion of AIK. The RSA key pair is generated for digitally attesting the video file and stored inside TPMs NVRAM where the sig.rssa contains the generated digital signature of video. The video and its

corresponding digital signature file from TPM are stored in IPFS and are included in smart contract as attributes. Proposed attestation framework is illustrated in Fig. 2.

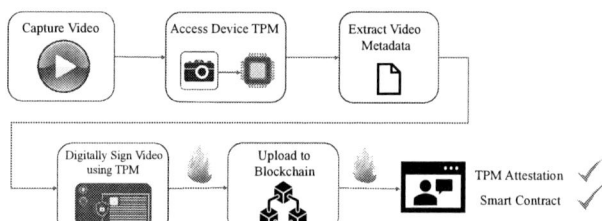

Fig. 2: Secure Video Attestation in BlockShield

Additionally, the video metadata of IPFS is also digitally signed and stored in IPFS for verification. The uploaded video on social media if prone to Deepfake and modified can be easily traced back and verified using TPM since TPM guarantees tamper-proof fingerprinting of digital content. The working flow of TPM-Digital content attestation is illustrated in Algorithm 1, and 2.

Algorithm 1 Performing TPM attestation of digital content

Input: Video File Fi
Output: Digital Signature of Video D_{Fi}
1: Access TPM hardware security module at the camera
   Video file Fi .mp4 $\rightarrow$ TPM 2.0 $\rightarrow$ Response D_{Fi}
2: *tpm2_createprimary -C e -c primary.ctx*
   Create Primary Key
3: tpm2_evictcontrol -C o -c primary.ctx 0x81010001
   Assign a unique identifier in TPM NV RAM to make it persistent
4: tpm2_create -G rsa -u rsa.pub -r rsa.priv -C 0x81010001
5: tpm2_load -C 0x81010001-u rsa.pub -r rsa.priv -c rsa.ctx.
6: tpm2_evictcontrol -C o -c rsa.ctx 0x81010002
   Create RSA keys using primary key and make it persistent
7: Load the video file and Hash it $Fi \rightarrow SHA256(Fi) \rightarrow Fi.hash$
   Hash the video file
8: tpm2_sign -c 0x81010002 -g sha256 -o sig.rssa $Fi.hash$
   Digitally sign the video hash file using TPM
9: tpm2_hash -C e -g sha 256 -o sig.rssa.hash -t ticket.sig.rssa sig.rssa
   Genrate SHA 256 hash of Digital signature for video file

The video, its extracted metadata, and the TPM signature file are then uploaded to IPFS for secure and public access. This action also triggers the smart contract to store the video

Special Session

and generate the IPFS hash along with the video editing permissions file. From here, an editor can obtain the video from IPFS and request editing permissions to the video by accessing the contract address of original source, triggering a 'requestPermission' event from the smart contract. When permission is requested, the source user is notified and can either accept or deny permissions to edit. If permission is given, the 'grantPermission' event of the smart contract is triggered which adds the editor address to the list of editors for the video. Then, the editor can send the edited video to the original artist for authorization. The original source will generate a digital signature of the edited video uploaded to IPFS.

Finally, the uploaded IPFS unique identifier is shared with the editor or secondary user. The edited video's metadata will contain a hash that points back to the original artist's address. This is a crucial step in verifying the chain of edited videos and their integrity. Videos that are not approved by the owner will not contain this point-back hash and are not considered as trusted videos. The working flow of BlockShield is presented in Fig. 3.

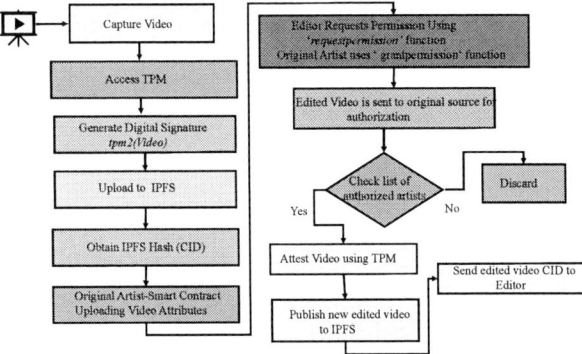

Fig. 3: Working flow of BlockShield

V. EXPERIMENTAL EVALUATION

For hardware evaluation, GeeekPi TPM 2.0 module based on Infineon Optiga™ SLB 9670 and compatible with Raspberry pi through is considered. Once a primary key is created, a unique handle in TPM non-volatile memory storage is allocated to the primary key to make it persistent. The primary key in primary.ctx file is assigned 0x81010001 as identifier for accessing in future. Once the primary key is created and persisted, a signing key pair is generated from the primary key and persisted by assigning 0x81010002 as identifier at NVRAM. Once the TPM signing keys are made persistent, they can be used to verify signature since the signing key and primary key have unique identifiers in TPM non-volatile memory. The TPM attestation and digital signature for video are shown in Fig. 4

The tools required for the implementation include metamask, Ganache, Remix IDE, and the Volta Test Network. The implementation used Solidity 0.8.0 inside Remix IDE

Algorithm 2 Blockchain video access control framework

Input: Digital Signature of Video file D_{Fi} and Video File Fi
Output: Digital content is securely stored in Blockchain and secure accessed using smart contract
1: **for** Each Primary artist **do**
2:    Individual contract is called by Primary artist to manage access
3:    Primary Artist attested Video file information on to IPFS system
4:    **for** Each Video File **do**
5:       Upload video file Fi and Digital signature D_{Fi} on IPFS.
6:       IPFSfile $Ii \leftarrow$ IPFS.upload(Fi, D_{Fi})
7:    **end for**
8:    ArtistContract.addIPFSHash(Ii)
      Return hash is added as an attribute in the newly created Primary artist contract
9: **end for**
10: Share the contract address and IPFS hash of the video provide access to video file
11: Editor initiates a 'requestpermission' function to primary artist contract address which is accessible online.
12: Update list of artists at the primary artist side using 'grantpermission' function
13: Editor submits a new edited version of video file Fi to primary artist
14: Edited video file VFi is attested using TPM $VFi \rightarrow$ TPM $\rightarrow F = D_{FVi}$
15: Upload Edited and attested video file VFi and D_{FVi} to IPFS and share it with secondary artist or editor.

---

to write the smart contract, the metamask wallet was used to import the two test accounts. These test accounts were imported from Ganache, a local environment that allows for simulated execution of the blockchain environment locally and smart contract testing. and transactions were made through the Volta Test Network. The two addresses used were '0x250805540e2978852a011   237a2bb77e33a0729a4' for the editor and '0x6c27c94191c630438ace1 2e123164a1b628882a6' for the original artist or source.

The front-end deployment of proposed work which involves artists connecting to meta mask wallet uploading video attributes and deploying smart contract is shown in Fig. 5. The performance analysis of BlockShield is evaluated for 5 opensource videos which initially have TPM attestation and then the digital signature file along with video uploaded to IPFS. As soon as it is uploaded, a CID is generated and finally, a unique transaction hash is obtained for each video. Table II presents the obtained evaluation results for BlockShield.

The source user initiates the smart contract with 4 functions: 'uploadMedia', 'requestPermission','grantPermission', and 'editMedia'. The uploadMedia function takes in the parameters video title, video IPFS hash, and video metadata IPFS hash and stores this data in a hash map with the key as the video IPFS hash. This triggers the MediaUploaded event. Fig. 6 shows smart contract function calls and the transaction validation outputs from viola test network. The 'requestPermission' function takes in the IPFS hash of the video and triggers the 'PermissionRequested' Event The function notifies the source user of the permission request. The 'grantPermission' function takes in the IPFS hash of the video

---

979-8-3315-3968-9/24 $31.00 © 2024 IEEE   269

Special Session

TABLE II: BlockShield Performance Analysis

Video	Duration (s)	Frame Rate	Bitrate	TPM-Signature	TX Hash	IPFS CID
48255-453831896.mp4	16	24	1633.922	e206fc334fa 48ae5917cac93dff 260d0fc0f0535f4e2 25c932466c 2291833df9	0x4c99d2c8f26 5498b09c53b372 e94f3cefc89d17be 12b401de25bf2b db892609b	QmQjxbmrjFbS7Xz4WXSigtkP msAKRi5FEXdDDpPK1Zn5g1
61299-498228517.mp4	26.559867	29.97	3526.986	fa60e6faf0f64 c50846ac74ca185ffc d83d89fbd68fb 9d2985a6bb5a454eab1a	0x631719a0744dd 4d880924ac7ad 57b98d5d 385a73af7e8e5 4e55039d 4612e723b	QmS5PjDzYqGtfTCYYCm9QrW Fam6ZHZVKU2uw5vJgN6abqf
61706-500316063.mp4	15.65	29.97	937.89	e206fc334fa48a e5917cac93dff260d0fc0f 0535f4e225c93 2466c2291833df9	0x539dfe6fad 4015a6b4ed84 85717614f b9091f7ec0ff1 aa5fc7c93 3a76821b0fe	QmTtS4J4GHG4n2tMEPsfNbVT MuTEmw3uKni5x4djUPEqSN
73711-549547411.mp4	25.2	29.27	1098.274	9c1f7e38f1528cc1876 5b79e28fa76f3fab662d01 63cd309260939 b208d7dcee	0xbc61d12a 52c0815799da10 e0ea8806 28c287a538eb aa65fe857e8 aa0b1435cc	QmNw2PUDAtZ Mt8twnVQ4Kjw7Py 1P4NsZ7dWnak71eVHAK
44645-439940290.mp4	10.88	25	5596.436	9bee3bef81c8ac 3f596a 6f4c44b b218cb713d2 ba2541e89c487a 59362729f60f	0x067ee279182e24 372e49ecd7e5 22551169d23 1a30da152f7 651c1c4ac2 945a6d	QmWVAzf4EJJS 7NHtVweTDyk7gq724y RyqGFCQqTkrVeABY

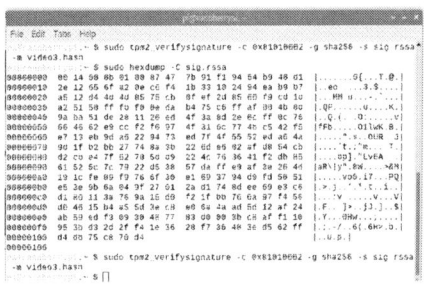

Fig. 4: TPM-Attestation

Metadata Hash:

QmRvppbvgME1u7rEqmJ2v6i5SrWKB7j1FooHiwBJTnfyto

Video Hash:

QmeDLUDoD5Cb1X4Sr3jBV1ZZGqHMZT6jVBkcoWUppBRfFP

Fig. 5: Frontend Deployment of BlockShield

and the editor address, sets the editor as a valid editor for the video, and triggers the 'PermissionGranted' event. The editMedia function takes in the IPFS hash of the video, the IPFS hash of the edited video, and the editor's deployed blockchain contract address and appends the edited video's hash to the chain of edits for the original video. It also triggers the MediaEdited event. If the permissions are given, then the editor can upload a video to IPFS, and once again the video metadata will automatically be extracted from the video and uploaded to IPFS, triggering the grantPermission event of the

smart contract and adding the editor address to the list of editors for the video. The metadata will include a hash that points back to the original owner of the video. This is a crucial step in verifying the chain of edited videos and their integrity.

Threat analysis of proposed BlockShield shows that the editing and modifying the video contents can be done only after the authorization of the original source. The editor or secondary user can only get the modified video attested at the original user or source through TPM attestation. The original source generates the digital signature and securely uploads video contents to IPFS, and video metadata is extracted and updated with the secondary user smart contract deployment address. This shows the robustness of proposed approach in enhancing the trustworthiness of digital content source. Furthermore, when uploaded on social media, the media links can also be included in IPFS and accessed using smart contract to ensure source integrity.

VI. CONCLUSION AND FUTURE RESEARCH

This research work presented and experimentally validated a Blockchain and TPM integrated approach for Deepfake mitigation through TPM-based hardware digital content attestation. The proposed work with state-of-art TPM-digital signature approach ensures hardware based digital content source attestation facilitated through Blockchain smart contract-based access control approach ensuring digital content authenticity. This is a novel work with TPM attestation and blockchain smart contract for access control and digital content sharing with the substantial performance indicators showcasing the robustness of the proposed Deepfake mitigation approach.

Furthermore, proposed research work could be further applied to Deepfake mitigation for images with effective mechanism for facial feature and biometric-based user authentication. Additionally, this work could be extended to smart cities surveillance applications which work in untrusted environ-

Special Session

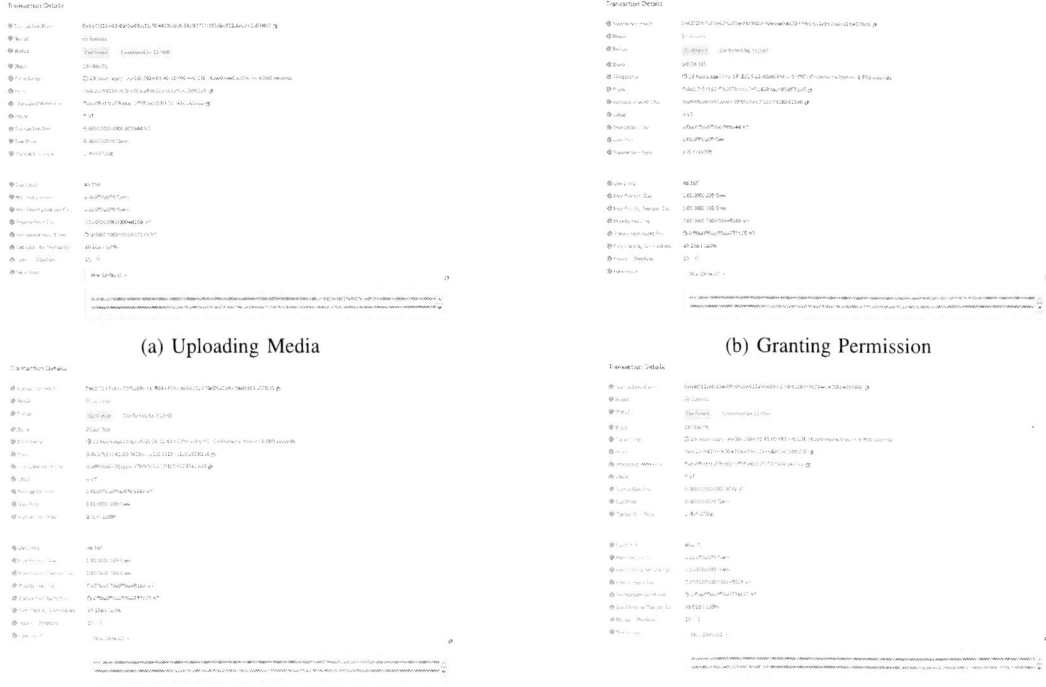

(a) Uploading Media

(b) Granting Permission

(c) Request Permission

(d) Editing Media

Fig. 6: Smart Contract Deployment

ments to guarantee privacy, security and traceability to digital content.

REFERENCES

[1] D. Garg and R. Gill, "Deepfake Generation and Detection - An Exploratory Study," in *Proc. 10th IEEE Uttar Pradesh Section International Conference on Electrical, Electronics and Computer Engineering (UPCON)*, vol. 10, 2023, pp. 888–893.

[2] U. Patil, P. Chouragade, and P. Ambhore, "An Effective Blockchain Technique to Resist Against Deepfake Videos," in *Proc. Third International Conference on Inventive Research in Computing Applications (ICIRCA)*, 2021, pp. 1646–1652.

[3] U. Patil and P. Chouragade, "Deepfake Video Authentication Based on Blockchain," in *2021 Second International Conference on Electronics and Sustainable Communication Systems (ICESC)*. IEEE, 2021.

[4] V. K. V. V. Bathalapalli, S. P. Mohanty, E. Kougianos, V. Iyer, and B. Rout, "iTPM: Exploring PUF-based Keyless TPM for Security-by-Design of Smart Electronics," in *Proc.IEEE Computer Society Annual Symposium on VLSI (ISVLSI)*, 2023, pp. 1–6.

[5] S. P. Mohanty, V. P. Yanambaka, E. Kougianos, and D. Puthal, "PUFchain: Hardware-Assisted Blockchain for Sustainable Simultaneous Device and Data Security in the Internet of Everything (IoE)," 2019.

[6] H. R. Hasan and K. Salah, "Combating Deepfake Videos Using Blockchain and Smart Contracts," *IEEE Access*, vol. 7, pp. 41 596–41 606, 2019.

[7] J. A. Costales, S. Shiromani, and M. Devaraj, "The Impact of Blockchain Technology to Protect Image and Video Integrity from Identity Theft using Deepfake Analyzer," in *Proc. International Conference on Inno-*

vative Data Communication Technologies and Application (ICIDCA), 2023, pp. 730–733.

[8] M. Taeb, H. Chi, and S. Bernadin, "Targeted Data Extraction and Deepfake Detection with Blockchain Technology," in *Proc. 6th International Conference on Universal Village (UV)*, 2022, pp. 1–7.

[9] M. M. Rashid, S.-H. Lee, and K.-R. Kwon, "Blockchain Technology for Combating Deepfake and Protect Video/Image Integrity," *Journal of Korea Multimedia Society*, vol. 24, pp. 1044–1058, 08 2021.

[10] A. Alattar, R. Sharma, and J. Scriven, "A System for Mitigating the Problem of Deepfake News Videos Using Watermarking," *Electronic Imaging*, vol. 32, no. 4, pp. 117–1–117–10, 2020.

[11] C. C. Ki Chan, V. Kumar, S. Delaney, and M. Gochoo, "Combating deepfakes: Multi-lstm and blockchain as proof of authenticity for digital media," in *2020 IEEE / ITU International Conference on Artificial Intelligence for Good (AI4G)*. IEEE, Sep. 2020.

[12] H. R. Hasan, K. Salah, R. Jayaraman, I. Yaqoob, and M. Omar, "NFTs for combating deepfakes and take metaverse digital contents," *Internet of Things*, vol. 25, p. 101133, 2024.

[13] A. Yazdinejad, R. M. Parizi, G. Srivastava, and A. Dehghantanha, "Making Sense of Blockchain for AI Deepfakes Technology," in *2020 IEEE Globecom Workshops (GC Wkshps*. IEEE, 2020.

[14] V. K. V. V. Bathalapalli, V. P. Yanambaka, S. Mohanty, and E. Kougianos, "PUFshield: A Hardware-Assisted Approach for Deepfake Mitigation Through PUF-Based Facial Feature Attestation," in *Proceedings of the Great Lakes Symposium on VLSI 2024*, ser. GLSVLSI '24. ACM, 2024.

[15] A. Qureshi, D. Megías, and M. Kuribayashi, "Detecting Deepfake Videos using Digital Watermarking," in *Asia-Pacific Signal and Information Processing Association Annual Summit and Conference (APSIPA ASC)*, 2021, pp. 1786–1793.

979-8-3315-3968-9/24 $31.00 © 2024 IEEE

Special Session

Fortified-Edge 5.0: Federated Learning for Secure and Reliable PUF in Authentication Systems

Seema G. Aarella
University of North Texas
Denton, Texas, USA
seema.aarella@unt.edu

Venkata P. Yanambaka
Texas Woman's University
Denton, Texas, USA
vyanambaka@twu.edu

Dr. Saraju P. Mohanty
University of North Texas
Denton, Texas, USA
saraju.mohanty@unt.edu

Dr. Elias Kougianos
University of North Texas
Denton, Texas, USA
elias.kougianos@unt.edu

Abstract—Physical Unclonable Functions (PUFs) are widely studied for the security of devices in the largely heterogenous Internet-of-Things ecosystem. The need for low-power and low-cost yet robust and reliable security systems is of prime importance in resource-constrained environments like smart villages. Using PUFs as a security primitive has the limitation of environmental effects that lead to bit flipping in the PUF response, the challenge in using PUFs is to overcome the bit errors without adding to the area overhead or computational overhead. This research proposes a novel bit error detection and correction algorithm implemented using Federated Learning (FL). The error detection and correction model uses the N-gram concept of Natural Language Processing (NLP). The FL model is implemented on Flower AI, the global model gets the locally trained model's parameters, updates itself, and shares the updated models with all the local models. At the edge, the use of FL for model training and updating enhances the efficiency of the authentication system that uses PUF Challenge-Response Pairs (CRPs), reduces the area overhead and power consumption, and improves the security of the PUF-based authentication system.

Index Terms—Physical Unclonable Functions(PUFs), Smart Village, Collaborative Edge Computing, Federated Learning, Secure Authentication, Natural Language Processing

I. INTRODUCTION

According to the statistics on IoT devices, it is expected that the number of IoT devices is expected to triple from 2020 to 2030, estimating the numbers to be in billions. Based on the emerging applications in areas such as smart healthcare, manufacturing, automotive industries, autonomous vehicles, gaming, and so on, there will be huge data generation that will exhaust the existing infrastructure. With the rise of time-sensitive applications, the processing is moving closer to the user edge. In the future, it is forecasted that the Edge Data Centers (EDC) could move closer to the data sources [1].

EDCs in a collaborative environment are a resourceful solution for faster computing with reduced latency and reduced load on the processors, through task offloading. Security of the EDC during load balancing requires a robust authentication and authorization system. The EDCs closer to user devices demand a security application that is low power consuming and computationally accommodatable. A suitable authorization and authentication system using PUFs was proposed in the research [2].

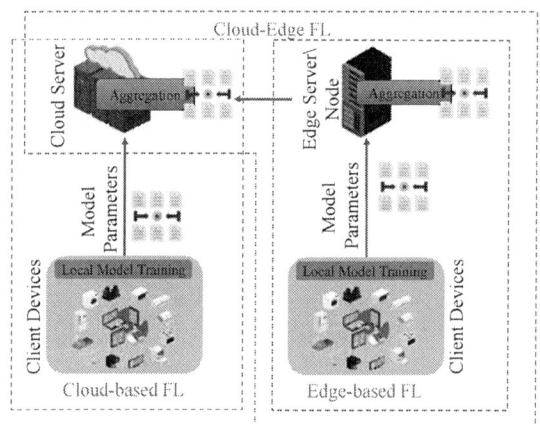

Fig. 1: Overview of Federated Learning Structure.

Although PUFs have been long studied as a robust solution for device identification and authentication in security applications, they are not without challenges. One of the prime challenges in implementing PUF is reliability, only the stable PUFs are considered secure, the main cause for instability in PUFs is environmental effects, the variations in the environment cause the bits in the PUF response to flip, thus reducing the stability and reliability of the PUF. To address this issue a novel ML-based PUF bit error detection and correction was proposed in the research [3]. This model is suitable for implementation at the edge of an IoT ecosystem that uses PUFs for authentication and authorization as it consumes low power, is computationally less intensive, and does not increase the area overhead.

Machine Learning has been a valuable enabling technology to use at the edge for developing security applications, it helps make intelligent and informed decisions concerning attack detection and prevention and mitigates security and privacy threats. ML can be leveraged to develop models to prevent or detect attacks like Distributed Denial-of-Service (DDoS), Intrusion, Eavesdropping, Malware, Man-in-the-Middle, Phishing, Spoofing, and so on [4]. Taking forward the capacity of ML is the emerging method FL that helps build models

979-8-3315-3968-9/24 $31.00 © 2024 IEEE   272

Special Session

without exposing private data. FL with the capability of deploying and training on user devices where data is generated reduces the communication overhead and preserves privacy. FL allows the participation of a large number of devices enabling collaborative learning that will help in providing cloud-like computing capabilities at the edge [5].

FL structure is shown in Figure 1, there are 3 distinct frameworks for FL, it can be Cloud-based FL, where the model aggregation takes place at the cloud servers, Edge-based FL where the model aggregation is done at the edge servers without cloud dependency, and a hierarchical framework that involves both cloud and edge servers based on the distribution of aggregation. ML models that are low on processing and power requirements are most suitable for aggregation at the edge servers. The challenges in using FL at the edge lie in tackling the heterogeneity of the ecosystem, handling the volume of data, frequency of model updates, and so on. One must design the model in a way that avoids the communication overhead and uses sufficient data for model updates. This research uses the FL model for collaborative deployment and updation of the ML model for bit error correction, which overcomes the challenges.

II. RELATED PRIOR RESEARCH

The security and privacy of PUF CRPs need to be considered when designing PUF-based authentication system. The CRP dataset needs to be stored and the data is looked up when authentication requests come in. The dataset is a must for a verifier that identifies the corresponding response for a given challenge and rightly verifies the requesting device. The dataset needs to be secure and prevent any illegal access, that will compromise the security of the whole application. The CEC ecosystem that uses a PUF-based authentication system requires the CRP dataset to be stored in the participating servers, in scenarios like load balancing when a mutual authentication between servers or EDCs is needed on the go. However, the local availability of CRP datasets in multiple servers or EDCs could become a security threat. To address this issue research proposes the use of a Certificate Authority-based PUF authentication system that removes the need to store dataset in every EDC in CEC [6].

Error correction codes and Fuzzy extractors are largely used for bit error correction to increase the reliability of PUFs. These methods need publicly available helper data which is vulnerable to data manipulation attacks. A repetition code-like error correction protocol using machine learning is proposed in the research [7] that uses simulation challenges, the mechanism can predict corrected responses with 100% probability.

To improve the reliability of SRAM PUF, a novel lightweight one-layer convolution scheme is proposed in the research [8]. The scheme uses verification matrices and compares the verified and unverified matrix to generate PUF response. The reliability of the PUF responses reached upto 100%.

Studies have shown that ML techniques are effective in PUF-based authentication systems where they can be employed for error correction, simulated challenge, response generation, and so on. It can be said that ML can greatly contribute to enhancing security and privacy, reducing area overhead, and computation overhead. FL framework allows

a network of devices to train a model without the need for centralized data, analysis of the performance of FL at the edge with various types of system heterogeneity, statistical heterogeneity, system statistical heterogeneity, and communication bandwidth, can be used to study the impact on model convergence, the knowledge of which can be used for optimizing the FL systems [9]. Combining the aspects of ML and using the FL framework the bit error correction model can be further enhanced to suit the needs of edge computing in a network while preserving data privacy, which is crucial for the security of the CRP dataset.

The following Table I shows the comparative study of various applications that use the FL framework. All these various research uses different datasets related to healthcare, medical, Image classification, and Intrusion detection, and apply FL for better results in detection and classification with enhanced data security and privacy.

III. NOVEL CONTRIBUTIONS OF CURRENT RESEARCH

Enhancing the reliability of the PUF-based secure authentication and authorization system is the prime focus of this research. This research uses a 64-bit Arbiter PUF for CRP generation [15]. The use of Machine Learning (ML) and Artificial Intelligence (AI) to support the security applications not only improves the security it also makes the system powerful against external attacks by providing attack detection and prevention options through various protocols across several levels of the IoT infrastructure. In the process of leveraging the benefits of ML and AI and reinforcing the PUF-based authentication system, the following are the novel contributions of the current research:

- Exploring FL for edge computing in a collaborative environment
- FL model with model training and deployment which is efficient in computation and power consumption
- Proposing a FL based framework for PUF bit error detection that uses ML algorithm
- ML model training using NLP approaches
- Global Model aggregation through parameters received from local models
- Global and Local Model training and testing on edge devices for computational efficiency
- Power analysis and computation analysis of the FL model

IV. FEDERATED LEARNING AT THE EDGE FOR PUF BASED AUTHENTICATION

FL is a distributed approach used to train AI and ML models that allows collaboration while preserving data privacy and security. The key aspects of FL are decentralized training, privacy preservation, and collaborative learning. FL allows a baseline model to be shared with participating devices or clients and allows the clients to train on their local model with enhanced learning quality. Another advantage of FL is that it enables access to diverse datasets without data sharing with reduced data communication and storage requirements. Edge computing as we know is growing to get processing closer to the user device, and the attributes of FL are suitable for such an environment. Based on the data distribution FL is classified into Horizontal FL (HFL), Vertical FL (VFL) and Federated Transfer Learning (FTL) [16].

979-8-3315-3968-9/24 $31.00 © 2024 IEEE

Special Session

TABLE I: Comparative Table for State-of-the-Art Literature.

Research	Year	ML Algorithm	Dataset	Metrics
Karim et. al. [10]	2023	RainForest	WEKA-Hypothyroid	Accuracy, Precision, Recall, F1 Score
Jain et. al. [11]	2023	SGD	Adobe Stock	Accuracy
Korkmaz et. al. [12]	2022	Inception-v3	Medical Image Dataset	Accuracy
Chen et. al. [13]	2020	GRU (gated Recurrent Unit) and SVM	KDD CUP99	Accuracy, F1 Score
Mahadik et. al. [14]	2024	CNN	CICDDoS2019	Accuracy
Current Research Fortified-Edge 5.0	2024	K-mer Sequence	100K PUF Response Dataset	Accuracy

- *Horizontal FL:* The database has same feature space but different sample spaces, the clients can use the same AI or ML model to train data locally.
- *Vertical FL:* The database has different feature space but the sample space is same, enable shared AI or ML model training.
- *Federated Transfer Learning:* The database has different feature space and different sample space, enables to train data aggregated from multiple clients.

FL applications in an edge computing environment include computation offloading and content caching, Malware and Anomaly Detection, Task scheduling, and resource allocation. Some of the challenges that FL at Edge poses are communication and computation efficiency, heterogeneity management, privacy and security preservation, client selection, and resource allocation [17].

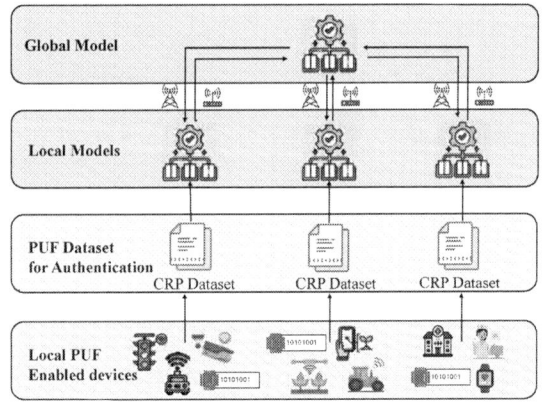

Fig. 2: Federated Learning Framework for Bit Error Correction in PUF enabled Authentication System.

An FL framework for edge that can be used for PUF bit error correction is shown in Figure 2. The data-generating devices are at the edge which are using PUF modules for authentication and authorization, in a CEC the EDCs can be replaced as the local devices. the local devices generate local data and save it locally as their dataset, each participant can train their local model on their data and send parameters to the global model, the global model aggregates the clients, updates the global model, and sends the updated model to local devices for updation.

V. PROPOSED FEDERATED LEARNING FRAMEWORK

The proposed FL framework is shown in Figure 3. The client devices are the participants that use PUF for device authentication and authorization. In the use case for this research, we consider EDCs as the participants in a collaborative environment. Each PUF module generates its own PUF responses for a given set of challenges, each module will have its own CRP dataset represented in the figure as Dataset A, Dataset B, and so on. Each dataset is trained using a local model using K-mer sequencing and Count Vectorization, for classification we are using MultinomialNB. The local ML model is responsible for generating the vectors for the extracted features from the PUF response and classifying them. The local model classifies the responses into unique classes based on which it is trained to predict the class of any new response. Flower Framework is used to implement the federated client-server model [18].

The local model is further modified to detect responses with errors, predict the correct class, and correct the response as shown in Figure 4.

Whenever a new challenge is used in any one of the clients, and a new response is generated the local parameters of the client are sent to the global model and the trained global model will send updates to all the local models. In this way, the other clients will be updated to correctly predict the class of the new response if the same challenge is given to them. The workflow of the proposed FL framework is as follows:

- *Initialization:* The Server initializes the global model hosted in CPU
- *Client Selection:* The Server can select to train on all available clients of a subset of the clients, maximum number of clients available in this research is 10.

Special Session

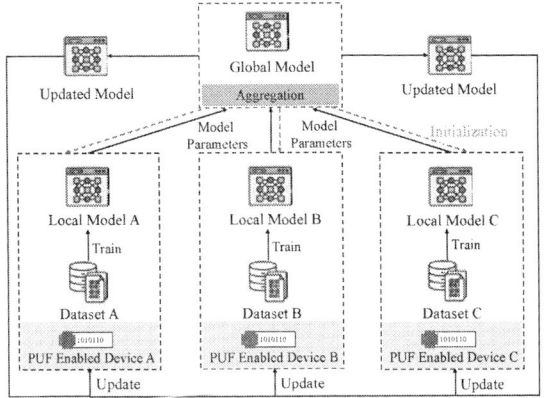

Fig. 3: Proposed Federated Learning Framework for Machine Learning based PUF bit Error Correction System.

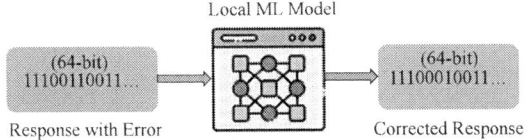

Fig. 4: Local Machine Learning Model for PUF Bit Error Correction.

- *Model Distribution:* The participating clients receive a global model update and the same is made available for other clients.
- *Local Training:* The local ML model is trained using the CRP dataset locally available for each participating client. The model is trained over certain epochs to improve performance and KFold cross-validation is done.
- *Model Aggregation:* The updated models from clients are sent to the server after local training.
- *Model Averaging:* The central server aggregates the model by averaging the parameters received from the clients.

The operation of the Client-Server model of the FL system is split into two parts, local training using the ML algorithm and Federated averaging using the FL framework for model updation. The local training steps are shown in Algorithm 1. Each participating client is trained on a local 10K response dataset.

Algorithm 1: Local Model Training

Input: 64-bit Binary Response Dataset stored in CSV file

Result: Trained model and predictions

1. Read CSV File;
2. Convert Binary data to string;
3. Label the data;
4. Apply K-mers of size 6;
5. Use CountVectorizer() for feature extraction;
6. Split data into train and test set;
7. Classify using MultinomialNB();
8. Predict;

The process of FL is discussed as the client-side and server-

side processes. The Algorithm 2 shows the steps involved in the server-side model.

Algorithm 2: Server Side Evaulation

Input: Number of Clients, Model Parameters

Result: Aggregation and Averaging

1. Set the Number of clients;
2. Start flower server;
3. Request initial parameters from random client;
4. **if** *received parameters* **then**
   > Evaluate initial global parameters;
   > Evaluate loss and accuracy;
   > Start fit;

end

5. update the global model;
6. Send updated global model to all clients;

The process involved in the client-side model training and evaluation is shown in Algorithm 3.

Algorithm 3: Client Side Evaulation

Input: Response dataset CSV file

Result: Updated Model

1. Load data;
2. Preprocess data for client_n;
3. Train Local model;
4. **if** *Trained* **then**
   > Start flower client;
   > Send model parameters to server;
   > Wait;

end

5. **if** *received updated model from server* **then**
   > Start fitting;
   > Evaluate model;
   > End model update;

end

The process of local training, sending parameters to the server, aggregation, and new model updates are repeated over several counts until the model converges and performs well across all the clients.

VI. EXPERIMENTAL SETUP

The FL framework proposed in this research is Horizontal Federated Learning (HFL), where all clients train a global FL model using their local dataset. The feature space of each dataset is the same, but the sample space is different. Flower provides the infrastructure to perform FL in an easy, scalable, and secure way, allowing federation, analytics, and evaluation of any ML framework.

A 64-bit Arbiter PUF architecture using PUFs. PYNQ™ Z2 FPGA which is based on Xilinx Zynq C7Z020 SoC, Xilinx BASYS3 FPGA was used to build the PUF. A 100K dataset of PUF responses is generated from this Arbiter PUF.

Each participating client is individually trained on 10K unique dataset of responses. The graph in Figure 5, shows the size of data consumed by each of the 10 clients. To suit the HFL model, the feature space for local models is the same, that is 506 features, but the dataset for each client is different.

979-8-3315-3968-9/24 $31.00 © 2024 IEEE

Special Session

That is, each local model is trained with a different set of responses.

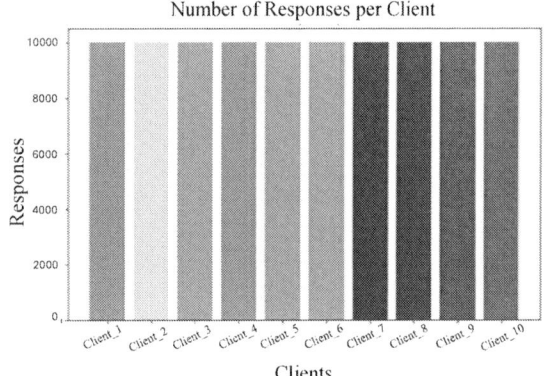

Fig. 5: Data Partitioning of 100K dataset.

To emulate the edge computing environment, Raspberry Pi 4 is used to act as the 10 clients and the server is the CPU, with a 64-bit Operating System, Intel i7 processor, 16GB RAM, and 2.80 GHz.

10K dataset is pre-processed, the binary data is converted to a string, and K-mers of size 6 are applied to convert the data to sequences, the classifier groups the sequences into unique classes. The classification of the sequences for the 10K dataset for Client_1 is shown in Figure 6. The sequences are vectorized using the CountVectorizer class from the scikit-learn library in Python, the vector-matrix columns represent the unique n-gram.

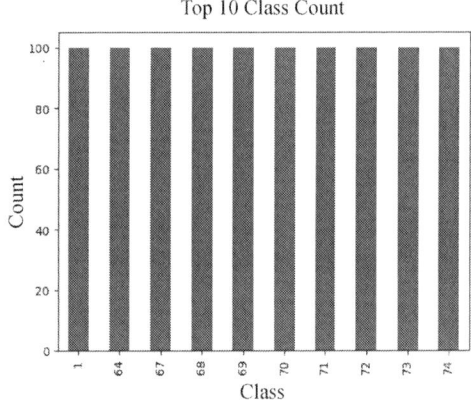

Fig. 6: Classification of sequences into classes.

MultinomialNB classifier is used for classification and the local model is fitted. The confusion matrix showing the actual and predicted class is shown in Figure 7.

VII. RESULTS AND ANALYSIS

The local model is trained on a 10K dataset, 80% of the data is used for training and 20% is used for testing. The model can efficiently predict classes of new responses and be ready for client-side evaluation. To test for overfitting of the local model KFold cross-validation is done to study the performance over multiple folds. The accuracies obtained over 5-fold cross-validation are 98.75%, 99.3%, 99.65%, 99.8%, and 99.35%, with a mean accuracy of 99.37%.

Predicted Actual	31	8	1	75	12	65	83	88	18	22
31	36	0	0	0	0	0	0	0	0	0
8	0	30	0	0	0	0	0	0	0	0
1	0	0	28	0	0	0	0	0	0	0
75	0	0	0	28	0	0	0	0	0	0
12	0	0	0	0	27	0	0	0	0	0
65	0	0	0	0	0	27	0	0	0	0
83	0	0	0	0	0	0	27	0	0	0
88	0	0	0	0	0	0	0	27	0	0
18	0	0	0	0	0	0	0	0	26	0
22	0	0	0	0	0	0	0	0	0	26

Fig. 7: Confusion matrix.

TABLE II: Comparative Table of Results for State-of-the-Art Literature.

Research	Year	ML Algorithm	Accuracy(%)
Karim et. al. [10]	2023	RainForest	0.99
Jain et. al. [11]	2023	SGD	0.94
Korkmaz et. al. [12]	2022	Inception-v3	0.8-0.99
Chen et. al. [13]	2020	GRU (gated Recurrent Unit) and SVM	0.99
Mahadik et. al. [14]	2024	CNN	0.99
Current Research Fortified-Edge 5.0	2024	K-mer Sequence	0.99

After training the local model the client and server are set up for FL using the Flower FL system.

The server-side evaluation results show a total of 3 server rounds are repeated in fitting the model parameters from 10 clients with 0 failures. The total time taken by the server for fitting the global model is 154.62. The server evaluation is increased for 10 rounds, time taken to complete is 202.32s.

The client-side evaluation shows an average of 99.45% accuracy with 0.0 loss for all 10 clients. The total time taken for local model training with initial parameter update is 6s, total time taken for model update over 10 rounds is 130.42s.

The idle power of the Raspberry Pi was an average of 3.7W, and the average power consumed for local model training was 4.5W.

A comparison of the results from various research listed in the comparative table for state-of-the-art Literature is shown in Table II. The accuracy of the baseline models used in an FL framework is high and is used by a variety of IoT applications handling different datasets.

VIII. CONCLUSIONS

FL framework is easy, scalable, and secure and enables the use of any ML algorithms for local model training. The use of FL for PUF bit error correction has shown enhanced performance and prediction accuracy while providing data privacy and security. In a collaborative environment authentication

979-8-3315-3968-9/24 $31.00 © 2024 IEEE

Special Session

system using PUFs can benefit from this technique where the CRP dataset need not be stored locally. The accuracy and power consumption evaluations also prove that the model is suitable for edge deployment. The model can be further improved with secure ML model development strategies and the research can be taken forward to explore applications like Deepfake detection, Secure Communication, and Secure Authentication protocols with minimum data exposure.

REFERENCES

[1] P. Arroba, R. Buyya, R. Cárdenas, J. L. Risco-Martín, and J. M. Moya, "Sustainable edge computing: Challenges and future directions," *Software: Practice and Experience*, vol. n/a, no. n/a, 2024. [Online]. Available: https://onlinelibrary.wiley.com/doi/abs/10.1002/spe.3340

[2] S. G. Aarella, S. P. Mohanty, E. Kougianos, and D. Puthal, "PUF-based Authentication Scheme for Edge Data Centers in Collaborative Edge Computing," in *IEEE International Symposium on Smart Electronic Systems (iSES)*, 2022, pp. 433–438.

[3] S. G. Aarella, V. P. Yanambaka, S. P. Mohanty, and E. Kougianos, "Fortified-Edge 4.0: A ML-Based Error Correction Framework for Secure Authentication in Collaborative Edge Computing," in *Proceedings of the Great Lakes Symposium on VLSI*, ser. GLSVLSI '24. New York, NY, USA: Association for Computing Machinery, 2024, p. 639–644. [Online]. Available: https://doi.org/10.1145/3649476.3660384

[4] S. Singh, R. Sulthana, T. Shewale, V. Chamola, A. Benslimane, and B. Sikdar, "Machine-Learning-Assisted Security and Privacy Provisioning for Edge Computing: A Survey," *IEEE Internet of Things Journal*, vol. 9, no. 1, pp. 236–260, 2022.

[5] Q. Duan, J. Huang, S. Hu, R. Deng, Z. Lu, and S. Yu, "Combining Federated Learning and Edge Computing Toward Ubiquitous Intelligence in 6G Network: Challenges, Recent Advances, and Future Directions," *IEEE Communications Surveys and Tutorials*, vol. 25, no. 4, pp. 2892–2950, 2023.

[6] S. G. Aarella, S. P. Mohanty, E. Kougianos, and D. Puthal, "Fortified-Edge: Secure PUF Certificate Authentication Mechanism for Edge Data Centers in Collaborative Edge Computing," in *Proceedings of the Great Lakes Symposium on VLSI*, ser. GLSVLSI '23. New York, NY, USA: Association for Computing Machinery, 2023, p. 249–254. [Online]. Available: https://doi.org/10.1145/3583781.3590249

[7] A. Ali-pour, F. Afghah, D. Hely, V. Beroulle, and G. Di Natale, "Secure PUF-based Authentication and Key Exchange Protocol using Machine Learning," in *IEEE Computer Society Annual Symposium on VLSI (ISVLSI)*, 2022, pp. 386–389.

[8] R. Cao, N. Mei, and Q. Lian, "Method for Improving the Reliability of SRAM-Based PUF Using Convolution Operation," *Electronics*, vol. 11, no. 21, 2022. [Online]. Available: https://www.mdpi.com/2079-9292/11/21/3493

[9] P. K. Quan, M. Kundroo, and T. Kim, "Experimental Evaluation and Analysis of Federated Learning in Edge Computing Environments," *IEEE Access*, vol. 11, pp. 33 628–33 639, 2023.

[10] M. Karim, N, K Kundu, D. Saha, S. Kabir, S, A Mim, and D. Md. Farid, "Implementing Federated Learning based on RainForest Model," in *IEEE 8th International Conference for Convergence in Technology (I2CT)*, 2023, pp. 1–6.

[11] S. Jain and K. R. Jerripothula, "Federated Learning for Commercial Image Sources," in *IEEE/CVF Winter Conference on Applications of Computer Vision (WACV)*, 2023, pp. 6523–6532.

[12] A. Korkmaz, A. Alhonainy, and P. Rao, "An Evaluation of Federated Learning Techniques for Secure and Privacy-Preserving Machine Learning on Medical Datasets," in *IEEE Applied Imagery Pattern Recognition Workshop (AIPR)*, 2022, pp. 1–7.

[13] Z. Chen, N. Lv, P. Liu, Y. Fang, K. Chen, and W. Pan, "Intrusion Detection for Wireless Edge Networks Based on Federated Learning," *IEEE Access*, vol. 8, pp. 217 463–217 472, 2020.

[14] S. S. Mahadik, P. M. Pawar, and R. Muthalagu, "Edge-Federated Learning-Based Intelligent Intrusion Detection System for Heterogeneous Internet of Things," *IEEE Access*, vol. 12, pp. 81 736–81 757, 2024.

[15] P. K. Sadhu, V. P. Yanambaka, S. P. Mohanty, and E. Kougianos, "Easy-sec: Puf-based rapid and robust authentication framework for the internet of vehicles," 2022.

[16] D. C. Nguyen, M. Ding, P. N. Pathirana, A. Seneviratne, J. Li, and H. Vincent Poor, "Federated Learning for Internet of Things: A Comprehensive Survey," *IEEE Communications Surveys and Tutorials*, vol. 23, no. 3, pp. 1622–1658, 2021.

[17] H. G. Abreha, M. Hayajneh, and M. A. Serhani, "Federated learning in edge computing: A systematic survey," *Sensors*, vol. 22, no. 2, p. 450, Jan 2022.

[18] "Flower: A Friendly Federated Learning Framework," https://flower.ai/, accessed: 2024-07-31.

Special Session

Performance Analysis of Greedy and Auction-Based Resource Allocation Algorithms in Ubiquitous Computing Environments

Akshay Nagpal
IEEE Member
New Jersey, USA

Vivekananda Jayaram
IEEE Senior Member
Texas, USA

Manjunatha Sughaturu Krishnappa
Oracle America Inc
California, USA

Nikhil Jagdish Bangad
IEEE Senior Member
Texas, USA

Darshan Mohan Bidkar
IEEE Member
Texas, USA

Manoj Jayantilal Kathiriya
Gartner Inc
Connecticut, USA

Seema G Aarella
University of North Texas
Texas, USA

Abstract—In the era of ubiquitous computing, efficient resource allocation is critical to managing the diverse and dynamic environments created by interconnected devices. This paper presents a comprehensive comparative analysis of greedy and auction-based resource allocation algorithms within ubiquitous computing systems. The study examines the performance of these algorithms under three implementation approaches: centralized, decentralized, and hierarchical. We use a process-based discrete-event simulation model to evaluate the algorithms based on key performance metrics, including throughput, average response time, and energy utilization. Our results demonstrate that auction-based algorithms, particularly in a hierarchical implementation, consistently outperform greedy algorithms in both throughput and energy efficiency while maintaining competitive response times. The findings highlight the advantages of auction-based strategies in dynamically adapting to changing conditions and optimizing resource use in complex, decentralized environments. This research contributes valuable insights into optimal resource management strategies for ubiquitous computing, guiding system designers and researchers toward more efficient and scalable solutions. Future work should address the limitations of simplified network and energy models to enhance the applicability of these findings in real-world scenarios.

Index Terms—Ubiquitous Computing, Resource Allocation, Greedy Algorithms, Auction-Based Algorithms, Distributed Computing, Performance Analysis

I. INTRODUCTION

Ubiquitous computing, a concept pioneered by Mark Weiser [1] in the early 1990s, envisions a world where computing is seamlessly integrated into our everyday environment. This paradigm shift from traditional desktop computing to embedded, interconnected devices has given rise to numerous challenges, particularly in resource management. As computing becomes more pervasive, the need for efficient allocation of limited resources such as processing power, memory, and network bandwidth becomes increasingly critical [2]. In ubiquitous computing environments, resources are often heterogeneous, distributed, and subject to dynamic changes. Devices ranging from smartphones and wearables to smart home appliances and industrial sensors compete for shared resources, creating a complex ecosystem that demands sophisticated allocation strategies [3]. The effectiveness of resource allocation directly impacts system performance, energy efficiency, and user experience in these environments. Resource allocation algorithms in ubiquitous computing encompass a diverse range of approaches, each with its own strengths and limitations. Greedy algorithms stand out as the most practical due to their simplicity and speed, making them highly effective in dynamic environments where quick decisions are crucial [4]. While they may not always yield optimal results, they often produce sufficiently good solutions in practice [5]. Auction-based algorithms [6] are both practical and effective, especially in distributed systems, adapting well to changing conditions and achieving a good balance between efficiency and fairness [7]. These are particularly useful in scenarios with diverse resources and tasks. Machine Learning-based methods have shown promise, especially in stable environments with predictable patterns, becoming more practical as systems accumulate data over time [8]. However, their effectiveness depends heavily on the quality and quantity of historical data [9]. Game Theory-based approaches and Genetic Algorithms, while powerful in certain scenarios, are often less practical for real-time, dynamic resource allocation due to their complexity and computational requirements [10], [11]. Among these varied approaches, greedy and auction-based algorithms have gained significant attention due to their balance of simplicity, adaptability, and effectiveness in real-world ubiquitous computing scenarios. These algorithms are particularly important as they can handle dynamic environments and decentralized decision-making, making them well-suited to the challenges posed by ubiquitous computing systems. This paper aims to provide a comprehensive comparison of greedy and auction-based resource allocation algorithms in the context of ubiquitous computing. We will examine their underlying principles and performance characteristics through

---

979-8-3315-3968-9/24 $31.00 © 2024 IEEE   278

Special Session

extensive simulations. Our experimental approach involves multiple runs of both greedy and auction-based algorithms under various scenarios, focusing on key performance metrics including throughput, response time, and energy utilization. By analyzing these algorithms across these critical dimensions, we seek to offer insights into their relative strengths and weaknesses in different ubiquitous computing contexts. This comparative study will contribute to the ongoing discussion on optimal resource management strategies for the ever-expanding world of ubiquitous computing, providing valuable guidance for system designers and researchers in the field.

II. BACKGROUND AND RELATED WORK

A. Resource Allocation Algorithms and Implementation Approaches

1) Algorithm types: Various algorithm types are discussed here.

Greedy algorithms

In ubiquitous computing environments perform resource allocation by assigning tasks to the best available resource at each decision point [12]. When a new task enters the system, the algorithm evaluates the currently available resources and selects the one that best fits the task's requirements based on predefined criteria such as processing power, energy efficiency, or network connectivity [13]. This process is repeated for each task as it arrives, without reconsidering previous allocations. Greedy algorithms are valued in ubiquitous computing for their simplicity and speed, allowing for rapid decision-making in dynamic environments where conditions may change quickly [14]. However, their myopic nature - focusing only on the current best option - means they may not always yield globally optimal results [15]. While this approach can lead to quick and efficient allocations in many scenarios, it may sometimes result in suboptimal overall resource utilization across the system [16].

Auction-based algorithms

In ubiquitous computing environments approach resource allocation by allowing devices to "bid" for tasks based on their current resource availability [17]. When a task needs to be allocated, the system initiates an auction where available devices (such as processors, sensors, or storage units) submit bids. These bids typically reflect the device's current capacity, energy levels, processing capabilities, or other relevant factors. The task is then allocated to the device with the most suitable bid, which could be the highest or lowest depending on the specific auction mechanism used. This approach enables a more dynamic and adaptive allocation process, as devices' bids naturally reflect their changing conditions and capabilities [18]. Auction-based algorithms can adapt well to the fluctuating nature of ubiquitous computing environments, where device availability and capacity may vary over time due to factors like mobility, energy constraints, or concurrent task execution. While this method can lead to more balanced and efficient resource utilization, it may involve more computational and communication overhead compared to simpler greedy approaches [19].

2) Implementation Approaches: The implementation of resource allocation algorithms in ubiquitous computing environments can be categorized into three main approaches: centralized, decentralized, and hierarchical. Each approach has its own characteristics, advantages, and challenges, making them suitable for different scenarios and scales of ubiquitous computing systems.

Centralized Approach

In a centralized implementation, a single controller is responsible for making all allocation decisions. This central entity has a global view of the system, including information about all available resources and pending tasks. It processes this information to make optimal or near-optimal allocation decisions. The centralized approach can potentially achieve globally optimal solutions but may face scalability issues in large-scale ubiquitous computing environments.

Decentralized Approach

The decentralized approach distributes the decision-making process among individual devices or nodes in the network. Each device makes local allocation decisions based on limited information available to it, such as its own resource availability and the requirements of nearby tasks. This approach can be highly scalable and robust, as it doesn't rely on a single point of control. However, due to limited global information, it may lead to suboptimal overall resource utilization.

Hierarchical Approach

The hierarchical approach combines elements of both centralized and decentralized approaches to balance their respective advantages and disadvantages. In this model, the system is organized into multiple levels. Lower levels make local decisions (similar to the decentralized approach), while higher levels coordinate and optimize allocations across broader sections of the network. This approach aims to achieve better scalability than purely centralized systems while maintaining a degree of global optimization not possible in fully decentralized systems.

Each of these implementation approaches can be applied to both greedy and auction-based algorithms, with the specific choice depending on the scale of the ubiquitous computing environment, the nature of the tasks and resources involved, and the particular requirements of the system in terms of efficiency, scalability, and fault tolerance.

III. MODELING AND IMPLEMENTATION

A. Simulation Model and Setup

The simulation model is designed to analyze the performance of greedy and auction-based resource allocation algorithms in ubiquitous computing. The model evaluates three implementation approaches for each algorithm: centralized, decentralized, and hierarchical. The simulation is set up using the SimPy library, a process-based discrete-event simulation framework, which allows the modeling of complex systems with dynamic behavior. The entire simulation framework is implemented in Python 3.9.6. Implementation code available at [20].

The simulation is configured using parameters defined in a JSON file, which specifies the simulation duration, task

Special Session

arrival rates, and device characteristics. Table I summarizes the specifications of each device, including its processing power, memory capacity, and energy consumption. The configuration is loaded to initialize the simulation time, task arrival rates, and a set of devices, each represented by processing power, memory capacity, and energy consumption attributes.

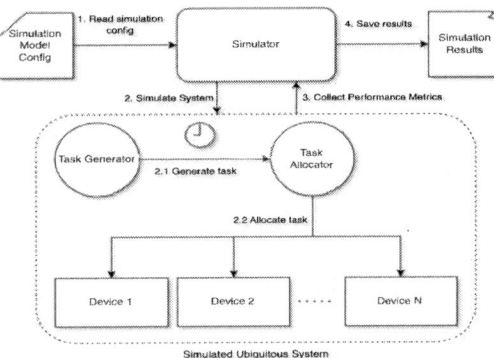

Fig. 1: Simulation setup and architecture

The simulation environment is constructed using the UbiquitousSystem class, which models a system of devices that receive and execute tasks according to specific allocation algorithms. Tasks are generated dynamically based on an exponential distribution of interarrival times, simulating a realistic arrival process in ubiquitous computing environments. The task_generator process continuously generates tasks, while the allocator process allocates these tasks to devices based on the selected algorithm. Figure 1 shows the high-level architecture.

For each combination of algorithm and implementation approach, the simulation is run for a predefined duration. Performance metrics, including throughput, average response time, and energy utilization, are computed and stored for analysis. The results are saved to a CSV file for further examination and visualization.

TABLE I: Device specification for the simulation environment

Device ID	Processing (units)	Memory (MB)	Energy (units)
1	1000	512	10
2	2000	1024	20
3	1500	768	15
4	800	256	5
5	3000	2048	30
6	1000	512	10
7	2000	1024	20
8	1500	768	15
9	800	256	5
10	3000	2048	30

B. Performance Metrics

Throughput (T) represents the number of tasks processed by the system per unit of time. In the simulation, it's calculated as the total number of completed tasks (N) divided by the simulation time.

$$T = \frac{N}{T_{\text{sim}}} \tag{1}$$

Average Response Time (R) is the average time taken to complete a task from the moment it arrives in the system until it is fully processed. In the simulation, it's computed as the mean of the response times for all completed tasks.

$$R = \frac{1}{N} \sum_{i=1}^{N} \left(t_{\text{end},i} - t_{\text{start},i} \right) \tag{2}$$

Energy Utilization (E) measures the proportion of energy consumed by the system relative to the total available energy capacity of all devices. It is calculated as the ratio of total energy consumed by all devices to the total energy capacity of the devices.

$$E = \frac{\sum_{j=1}^{D} E_{\text{consumed},j}}{\sum_{j=1}^{D} E_{\text{capacity},j}} \tag{3}$$

C. Implementation

The implementation is divided into several Python modules that encapsulate different aspects of the simulation:

simulate.py: This is the main driver script that orchestrates the simulation. It initializes the simulation environment, loads configuration settings, and iterates over various task arrival rates. For each rate, it runs the simulation across all algorithm implementations—Greedy (Centralized, Decentralized, Hierarchical) and Auction (Centralized, Decentralized, Hierarchical). Each algorithm is represented as a function in the algorithms.py module. The script calculates key performance metrics such as throughput (tasks processed per unit time), average response time (average time from task arrival to completion), and energy utilization (proportion of energy consumed by the system). These results are printed to the console and saved to a CSV file for subsequent analysis.

algorithms.py: This module defines the various resource allocation algorithms tested in the simulation. The greedy algorithms (centralized, decentralized, hierarchical) focus on optimizing task allocation based on available device resources, aiming to minimize processing and memory utilization. The auction-based algorithms (centralized, decentralized, hierarchical) simulate a bidding process where devices bid for tasks based on their capacity to handle processing and memory demands, with the goal of maximizing system efficiency. Each algorithm function receives a list of devices and tasks and returns a list of task-device allocations.

ubiquitous_system.py: This module contains the UbiquitousSystem class, which represents the simulated environment. The class manages the processes for task generation and allocation, tracks completed tasks and computes energy consumption. The task generator creates new tasks at intervals determined by an exponential distribution, mimicking real-world task arrivals. The allocator process assigns tasks to

Special Session

devices based on the selected allocation function. The system monitors each task's execution and records completion times and energy usage, providing data necessary for performance evaluation.

IV. RESULTS

Table II shows the results of simulating task arrival rate of 0.2 up to 1 across the algorithm and implementation approaches for each of the 3 performance metrics.

A. Throughput

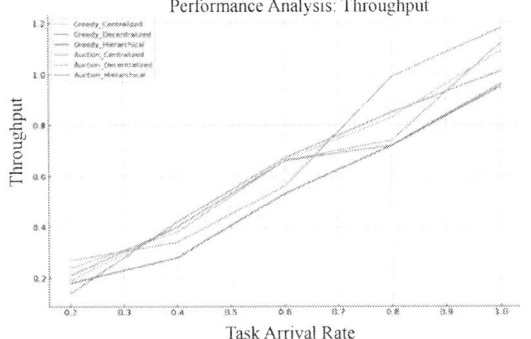

Fig. 2: Performance Analysis-Throughput

Based on Figure 2, Auction Hierarchical shows the highest throughput, especially as the task arrival rate increases. Auction Centralized and Auction Decentralized also demonstrate strong throughput performance but are slightly lower than the Hierarchical approach. Greedy approaches generally have lower throughput across all configurations, with Greedy Hierarchical performing the best among them.

B. Average Response Time

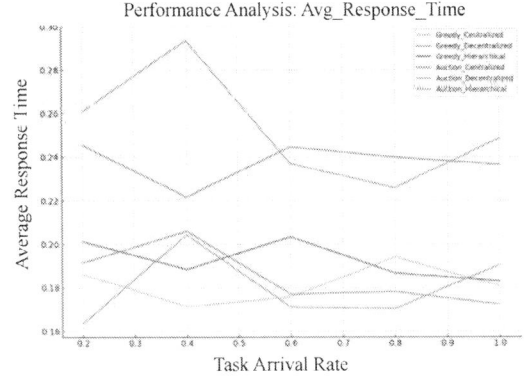

Fig. 3: Performance Analysis - Average Response Time

Based on Figure 3, Auction Hierarchical and Auction Centralized maintain the lowest response times, indicating faster task processing capabilities. Greedy Decentralized and Auction Decentralized tend to have higher response times, especially as the task arrival rate increases. Hierarchical approaches (both Greedy and Auction-Based) provide more

stable response times compared to their centralized and decentralized counterparts.

C. Energy Utilization

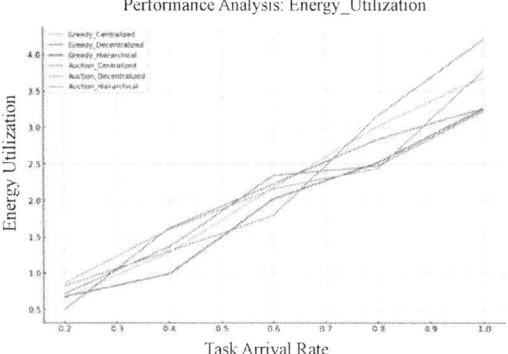

Fig. 4: Performance Analysis - Energy Utilization

Based on Figure 4, Greedy Hierarchical is the most efficient in energy utilization as task arrival rates increase, indicating better optimization. Auction Hierarchical is efficient for low to mid task arrival rates, but becomes less efficient as task arrival rate increases. Decentralized approaches (both Greedy and Auction-Based) tend to have higher energy utilization, suggesting less efficiency. Centralized approaches show moderate energy utilization, with the Auction-Based strategy being slightly more efficient than the Greedy strategy.

D. Overall Trends

Hierarchical approaches (both Auction-Based and Greedy) are generally more efficient in terms of balancing throughput, response time, and energy utilization. Auction-based strategies outperform Greedy strategies across all metrics, particularly in throughput and energy efficiency. Decentralized Approaches tend to be less efficient in energy usage and response times but can still provide reasonable throughput. Centralized Approaches offer a balance but do not excel in any single metric as much as the Hierarchical or Auction-Based approaches do. Auction-Based Hierarchical approach comes out as the best choice, effectively balancing all three performance metrics. Table III summarizes these trends across scenarios.

V. CONCLUSIONS

We compared the performance of greedy and auction-based resource allocation algorithms in the context of ubiquitous computing across centralized, decentralized, and hierarchical implementation approaches using a simulation model. The results of our simulation model show that auction-based algorithms outperform greedy algorithms across all evaluation metrics, particularly in throughput and energy efficiency. Overall, the auction-based hierarchical approach came out as the best choice, effectively balancing all three performance metrics when simulated in our model. In addition to providing these insights for various distributed computing scenarios in ubiquitous systems, our simulation model also provides a

Special Session

TABLE II: Simulation results (T=Throughput, R = Average Response Time, E = Energy utilization)

Task Arrival Rate	Greedy Centralized			Greedy Decentralized			Greedy Hierarchical			Auction Centralized			Auction Decentralized			Auction Hierarchical		
	T	R	E	T	R	E	T	R	E	T	R	E	T	R	E	T	R	E
0.20	0.19	0.19	0.66	0.21	0.25	0.72	0.18	0.20	0.68	0.14	0.19	0.50	0.24	0.26	0.86	0.27	0.16	0.83
0.40	0.40	0.17	1.29	0.40	0.22	1.36	0.28	0.19	0.99	0.42	0.21	1.62	0.38	0.29	1.61	0.34	0.20	1.30
0.60	0.66	0.18	2.18	0.66	0.24	2.34	0.53	0.20	2.02	0.67	0.18	2.22	0.66	0.24	2.16	0.56	0.17	1.80
0.80	0.83	0.19	3.02	0.72	0.24	2.47	0.72	0.19	2.52	0.85	0.18	2.84	0.74	0.23	2.44	0.99	0.17	3.16
1.00	1.09	0.18	3.69	0.96	0.24	3.23	0.95	0.18	3.26	1.01	0.17	3.26	1.12	0.25	3.78	1.18	0.19	4.21

TABLE III: Overall Trends

Scenario	Algorithm and Implementation Approach
High Throughput and Low Average Response Time	Auction-Based Hierarchical
High Throughput and Low Energy Utilization	Auction-Based Hierarchical
Low Average Response Time and Low Energy Utilization	Greedy Hierarchical
Balanced across Throughput, Energy Utilization & Average Response Time	Auction-Based Hierarchical

framework that can be used and extended by researchers to test and evaluate the right algorithm and implementation approach for their specific performance requirements.

VI. LIMITATIONS AND DISCUSSION

A. Inter Device Network Latency

The simulation assumes uniform network latency between all devices, which is a significant simplification of real-world network conditions. In actual ubiquitous computing environments, network latencies can vary greatly depending on factors such as physical distance, network congestion, and the type of connection (e.g., WiFi, cellular, or wired). This assumption may lead to unrealistic task allocation decisions, particularly in scenarios where communication costs play a crucial role. Future work should consider implementing a more realistic network model that accounts for variable latencies, bandwidth limitations, and potential network failures. This could involve creating a dynamic network topology that evolves over time, simulating network congestion, or incorporating real-world network traces.

B. Energy Model

The current simulation employs a simplified energy consumption model where energy usage is directly proportional to the execution time and the device's energy capacity. This approach, while computationally efficient, does not capture the complexities of real-world energy consumption in ubiquitous computing environments. It doesn't account for factors such as idle power consumption, dynamic voltage and frequency scaling (DVFS), or the energy costs of data transfer and storage operations. Future work could explore more sophisticated energy models that incorporate these factors, as well as device-specific power profiles.

REFERENCES

[1] M. Weiser, "The Computer for the 21st Century," *Scientific American*, vol. 265, no. 3, pp. 94–104, 1991.

[2] M. Satyanarayanan, "Pervasive Computing: Vision and Challenges," *IEEE Personal Communications*, vol. 8, no. 4, pp. 10–17, 2001.

[3] J. Hahner, C. Becker, and K. Rothermel, "Approaches to Resource Management in Ubiquitous Computing Environments," *IEEE Transactions on Mobile Computing*, vol. 4, no. 4, pp. 366–381, 2005.

[4] E. M. Feller, L. Rilling, and C. Morin, "Energy-aware Ant Colony Based Workload Placement in Clouds," in *Proceedings of the 12th IEEE/ACM International Conference on Grid Computing*, 2011, pp. 26–33.

[5] S. Singh and I. Chana, "Cloud Resource Provisioning: Survey, Status and Future Research Directions," *Knowledge-Based Systems*, vol. 93, pp. 12–26, 2016.

[6] N. Nisan, T. Roughgarden, E. Tardos, and V. V. Vazirani, Eds., *Algorithmic Game Theory*. Cambridge University Press, 2007.

[7] Y. Wang, L. Toka, and M. Caballero, "Auction-Based Resource Allocation in Cloud Computing with Variable Demand," *IEEE Transactions on Cloud Computing*, vol. 8, no. 2, pp. 489–501, 2019.

[8] H. Mao, M. Alizadeh, I. Menache, and S. Kandula, "Resource Management with Deep Reinforcement Learning," in *Proceedings of the 15th ACM Workshop on Hot Topics in Networks*, 2018, pp. 50–56.

[9] J. Xu and Y. Zhao, "Reinforcement Learning and Its Applications in Modern Networking: A Survey," *Journal of Communications and Networks*, vol. 21, no. 6, pp. 539–551, 2019.

[10] D. P. Palomar and M. Chiang, "A Tutorial on Decomposition Methods for Network Utility Maximization," *IEEE Journal on Selected Areas in Communications*, vol. 24, no. 8, pp. 1439–1451, 2006.

[11] B. D. Martino, G. Cretella, and A. Brancaleoni, "Genetic Algorithms for Dynamic Resource Management in Cloud Computing," *Journal of Cloud Computing*, vol. 5, no. 1, p. 16, 2016.

[12] S. Albers and S. Schmidt, "On the Performance of Greedy Algorithms in Dynamic Networks," *Theoretical Computer Science*, vol. 527, pp. 1–15, 2014.

[13] C. Tang and C. Z. Xu, "QoS-aware Replica Placement for Content Distribution," *IEEE Transactions on Parallel and Distributed Systems*, vol. 17, no. 10, pp. 1296–1308, 2006.

[14] M. AlFares, A. Loukissas, and A. Vahdat, "A Scalable, Commodity Data Center Network Architecture," *ACM SIGCOMM Computer Communication Review*, vol. 38, no. 4, pp. 63–74, 2008.

[15] D. S. Hochbaum, *Approximation Algorithms for NP-Hard Problems*. PWS Publishing Co., 1997.

[16] A. L. Barabási and E. Bonabeau, "Scale-Free Networks," *Scientific American*, vol. 288, no. 5, pp. 60–69, 2003.

Special Session

[17] D. C. Parkes, "Iterative Combinatorial Auctions: Achieving Economic and Computational Efficiency," Ph.D. dissertation, University of Pennsylvania, 2001.

[18] H. Ma, Y. Wang, and L. Cheng, "A Bidding-Based Resource Allocation Framework for Mobile P2P Computing," *IEEE Transactions on Parallel and Distributed Systems*, vol. 19, no. 2, pp. 169–182, 2008.

[19] D. P. Bertsekas and R. G. Gallager, *Data Networks*. Prentice-Hall, 1987.

[20] A. Nagpal, "Ucrasimulation," https://github.com/UCRASimulation, 2024.

Industrial Contribution

A Unified Functional Safety EDA Framework for Accurate Diagnostic Coverage Estimation

Abhiroop Bhowmik*[†], Subin Babukutty[†], Mottaqiallah Taouil*, Moritz Fieback*

*Delft University of Technology, Delft, The Netherlands, {a.bhowmik, m.taouil, m.c.r.fieback}@tudelft.nl
[†]NXP Semiconductors, Eindhoven, The Netherlands, {subin.babukutty_1}@nxp.com

Abstract—As electronics and software become more integrated into automobiles, Functional Safety (FuSa) per ISO 26262 becomes important. It assesses the risk level of automotive chips, reflected by the Automotive Safety Integrity Level (ASIL). Fault injection simulation verifies the FuSa of a design by injecting faults and classifying them based on whether safety mechanisms detect them. Discrepancies in classification results from FuSa EDA tools can lead to varying ASIL assignments and misrepresent associated risk. Thus, we evaluate two FuSa EDA tools, Cadence® XFS and Synopsys® VC Z01X, for RTL designs. We find that the fault space covered by the tools is not complete. Hence, we propose a novel verification methodology combining both tools to achieve maximum fault space coverage. We apply this approach to the AutoSoC benchmark suite and achieve a more accurate Diagnostic Coverage (DC) of 97.79%, over the baseline verification methodology of 98.36%, at the cost of injecting 1.31 times more faults. Our work ensures that the correct ASIL level is assigned through accurate DC estimation.

Keywords—Functional Safety, ISO 26262, Fault Injection Simulation, EDA, Verification methodology

I. Introduction

The automotive industry is experiencing a significant shift towards advanced electronic and software integration due to the increasing demand for self-driving and autonomous vehicles [1]. As automotive systems become more complex, the risk of malfunctions increases, necessitating robust safety measures. Functional Safety (FuSa), as defined by standards such as ISO 26262 [2], addresses these concerns by incorporating Safety Mechanisms (SM) to mitigate failures and hazards that could be life-threatening. ISO 26262 requires exercising faults on various design locations and analysis of safety mechanisms' detection capabilities. The fault space in modern designs is typically quite large and is associated with millions of design components, making FuSa verification a complex and time-consuming process [3]. Therefore, it becomes important to develop verification methodologies capable of covering the entire fault space while also achieving run-time efficiency.

Fault Injection (FI) simulation is the suggested ISO 26262 verification methodology for FuSa Verification and is widely seen in different solutions [4–6]. Formal methods and ATPG solutions are also seen in combination with FI simulation [5, 6] to identify safe faults that are undetectable. Testbench-based fault simulation [7, 8] involves utilizing existing functional verification testbenches with additional modifications for manual fault injection. However, this approach is not scalable to larger, complex designs and requires higher manual effort. Emulation-based techniques [9–11] can also be utilized for

fault analysis using dedicated emulators. However, they also require additional hardware in the form of FPGAs to produce accelerated results in comparison to simulation platforms. Therefore, our work primarily focuses on simulation-based platforms for fault injection.

Leading vendors such as Cadence, Synopsys, and Siemens offer dedicated EDA tools for FI simulation. However, there is a lack of research on why one tool should be favored over another. It has been shown that there are inconsistencies when identifying safe faults while using different technologies like ATPG, formal and FI simulation [5, 6]. Hence, it is not clear whether different FI simulation tools will generate the same results. Further, as noted in [12], an important concern regarding the accuracy of results arises in FuSa verification at higher abstraction levels like RTL, contrasting with the predominant focus of most verification solutions on gate-level designs. Discovering bugs at the gate level would require subsequent changes to the RTL, prompting a repetitive and time-consuming process to achieve the desired coverage and stability of the design.

Considering the lack of research, we compare two FuSa EDA tools on RTL designs. We analyze the disparities in results, their limitations, and propose a verification solution to address the issues. The tools considered are: Xcelium™ Fault Simulator (XFS) by Cadence® and VC Z01X by Synopsys®. The main contributions of our work are:

- We compare the tools based on metrics such as correctness of results, fault space coverage, and simulation run-time to evaluate their performance, capabilities, and limitations.
- We propose a unified EDA framework combining the strengths of the tools to develop a verification methodology that maximizes the possible fault space while minimizing simulation time.
- We validate the proposed methodology on reference designs and an industrial-grade automotive SoC.

The rest of the paper is organized as follows: Section II discusses FI simulation along with FuSa concepts. Section III presents an analysis of the strengths and weaknesses of the FuSa EDA tools under consideration, highlighting the results produced by the tools. Section IV describes the proposed FuSa verification methodology. Section V discusses the results of the proposed methodology on reference designs. Sections VI and VII present discussion and conclusions.

979-8-3315-3968-9/24 $31.00 © 2024 IEEE

Industrial Contribution

TABLE I
TARGET VALUES FOR DC BASED ON ASIL LEVEL [2]

ASIL	DC
B	>90%
C	>97%
D	>99%

TABLE II
FAULT CLASSIFICATIONS

	Detected Functional	Undetected Functional
Detected Checker	Observed Diagnosed (OD)	Not observed Diagnosed (ND)
Undetected Checker	Observed Not Diagnosed (ON)	Not Observed Not Diagnosed (NN)

II. BACKGROUND

This section introduces key concepts related to FuSa verification as per ISO 26262 and outlines general verification flows used with FI simulation tools.

A. ISO 26262 FuSa concepts

One of the first steps in the FuSa lifecycle involves performing a *Hazard Analysis and Risk Assessment (HARA)* for different automotive components. HARA is used to identify and categorize hazards associated with components. Possible hazards are categorized based on three parameters: **severity** of a potential injury, **exposure** or how frequently an operational situation arises, and **controllability** of whether a situation can be managed to avoid injury. Based on these three parameters, the Automotive Safety Integrity Level (ASIL) is defined for each component. ASIL quantifies the risk associated with a component and determines the level of risk reduction necessary. ASIL is categorized into four levels: A, B, C, and D, with D representing the highest ASIL, requiring the most significant risk reduction measures.

Once ASILs are defined, safety goals are formulated followed by the implementation of SMs. An SM is an additional piece of logic designed to identify faulty behavior within the circuit and subsequently detect or correct these faults. The efficiency of these mechanisms is quantified by Diagnostic Coverage (DC). DC denotes the percentage of faults detected by SMs and is calculated based on fault simulation results. Table I presents the target values for DC for different ASILs.

B. Fault Injection (FI) Simulation

As mentioned earlier, FI simulation is the suggested ISO 26262 methodology for FuSa verification. A typical FI simulation flow consists of the following steps:

1) Fault targets (locations for fault injection) are defined along with available fault model types (Stuck-At (SA), transient faults)
2) A fault-free simulation is run to generate a reference database. Users must also define *functional* and *checker* strobes in the design. Functional strobes capture information related to functional outputs that directly impact the design output. Checker strobes are integral to SMs and can be regarded as signals for fault detection or alarms.
3) Fault simulations are run by injecting faults from the fault target list. Classification results are generated based on differences in strobe values of good and faulty simulations. Classifications are made based on whether the fault propagates to functional and/or checker strobes, as illustrated in Table II.

The DC can then be estimated by Equation 1. However, EDA tools offer additional classifications beyond these fundamental ones, allowing for modifications to the equation accordingly. For instance, there is also a class of **Safe** faults, which do not affect functional and checker strobes, for example, because they are unused or blocked by other signals, Not Controllable (**NC**) faults, which are signals that do not toggle during simulation) and Impossible x-state (**IX**) faults, which are transient faults injected on signals at an unknown state.

$$\text{Diagnostic Coverage (DC)} = \frac{OD+ND}{OD+ND+ON} \times 100\% \quad (1)$$

Our work determines differences in the classification results of FI simulation EDA tools when applied to a design. To the best of our knowledge, no previous studies have conducted a comparative analysis on this aspect. If discrepancies in classifications exist, they would result in varying DCs, consequently leading to varying ASILs. This will, in turn, misrepresent the associated risk of the component.

III. CAPABILITIES AND LIMITATIONS OF EXISTING FuSa EDA TOOLS

This section discusses the results of applying the tool flows from the two vendors on several reference designs. We describe the experimental setup and comparison metrics. We also analyze discrepancies in simulation results and their causes.

A. Tool setup and reference designs

The tool flows are tested first on two simple RTL designs: a full adder and 4-bit up counter. These examples serve as simple cases to analyze FI simulation results and enable manual inspection of classifications due to their small fault space. However, since they lack SMs, we also examine a FIFO design, provided by Synopsys® and illustrated in Figure 1. This design incorporates SMs like ECC for memories and module duplication. The FIFO is implemented using a dual port RAM, and encoded with ECC. A flags module is utilized to determine the status of the FIFO. The Read/Write Pointer modules are used to calculate the addresses for FIFO read/write. These three modules are duplicated to provide redundancy and generate errors in the event of faults.

We develop scripts to automate the tool flows, including placeholders for fault injection targets, fault models, and strobing options related to the different designs. Fault injection targets refer to design points where we want to inject faults. All locations are enabled for fault injection across the three designs. Further, the fault models employed include SA and

Industrial Contribution

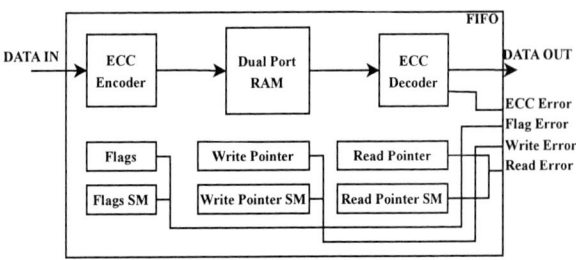

Fig. 1. FIFO with ECC and module duplication

TABLE III
CLASSIFICATION RESULTS FOR SA FAULTS ON ADDER, COUNTER AND FIFO

	Adder		Counter		FIFO	
	XFS	VC Z01X	XFS	VC Z01X	XFS	VC Z01X
ND	6	6	4	4	77	94
NN	0	0	0	0	98	1
OD	8	8	8	8	119	74
ON	2	2	8	8	84	112
Safe	-	-	-	-	44	88
NC	-	-	-	-	-	53
Total	**16**		**20**		**422**	

transient faults. Regarding strobing points, in the adder design, the sum and carry out signals are respectively designated as functional and checker strobes. Similarly, for the counter, the upper and lower 2 bits of the output work as functional and checker strobes respectively. The selection of these strobing points is arbitrary to verify fault classification results. However, in the case of the FIFO design, a more informed decision is made by assigning the data out signal as a functional strobe, and the error signals as checker strobes.

VC Z01X offers various options for instrumenting faults across different location types, including PORT, PRIMITIVE, FLOP, ARRAY, WIRE, and VARIABLE. XFS, on the other hand, does not support all of these distinct options for fault instrumentation. Consequently, the fault space of VC Z01X is larger than that of XFS due to the possibility of a signal instrumented as different location types. Therefore, for initial comparisons, we make the same fault lists for both tools to facilitate a fair comparison. Later, we extend the fault space to include all possible options supported by the tools.

B. Comparison metrics

We consider the following metrics to compare the tools:

1) **Correctness of results**: The tool results must be consistent and in line with expected fault classifications. There are rules defined by tools regarding fault propagation on different locations such as PORTS, FLOPS, WIRES etc. Based on how faults propagate in a design, the tool classification should match the expected result.

2) **Fault space coverage**: The tools should be able to cover the entire SA and transient fault space to provide an accurate estimation of DC.

3) **Run-time**: For large, complex designs, the fault space increases drastically, requiring more FI simulation time. We measure the fault simulation run-time for equal fault lists.

The first priority when comparing results is to thoroughly check their correctness. If the results are not correct, the tool cannot be trusted. Further, the tools should be capable of covering the entire fault space without any coverage limitations. This is important for obtaining an accurate DC metric. Finally, run-time is important when there are thousands of faults to be tested, as verification tests can take long simulation times. In the next section, we compare the tools based on these metrics and present the results.

C. Analysis of classification results

1) *Correctness of results:* Table III illustrates the classification results of the two tools applied on the adder and counter designs for SA faults. As seen from the table, the fault classifications obtained from the tools are the same and in line with the manual analysis done on the injected faults. However, for SA faults instrumented on the ports of the FIFO design, we observe multiple differences in fault classifications.

The first major difference lies in NN fault classification. NN faults often require manual analysis to determine whether they are actually safe or if they remain undetected due to test limitations. The difference of 97 faults in this regard is due to the classification of an additional 44 Safe and 53 NC faults by VC Z01X. While both tools identify 44 common Safe faults, VC further categorizes 97 additional faults, saving users' debugging time for NN fault classification.

The disparities in other fault classification categories (ON, OD, and ND) arise primarily due to variations in how faults in ports are modeled by the two tools. To better understand this, let us consider the example of the *FLAGS* module, which is duplicated as an SM to detect faults, as illustrated in Figure 2. Any difference in the status signals arising from the two modules is triggered as a *FlagErr*. According to fault propagation rules of the two tools, faults injected at input ports should propagate inwards towards lower hierarchies, whereas output port faults should propagate outwards towards higher hierarchies. However, when we inject a fault at an input port, for example, *FLAGS_SM.Write*, XFS propagates this fault on the wire connecting the port. Thus, it affects the value of *FLAGS.Write* as well. Technically, such behavior is not incorrect, but the tool should be able to consider the effect of both fault types. Otherwise, we would lose out on a particular section of the fault space. The disparities in fault classifications are essentially due to this difference in fault modelling.

The SA fault space is then further extended to include all types of signals - intermediate nets, registers, variables, etc. The total number of faults instrumented for XFS and VC Z01X increased to 706 and 1408 respectively. This difference is seen as a result of VC Z01X instrumenting signals as multiple location types. For example, faults are injected at *FLAGS_SM.Write* as two types: PORT and WIRE. For the latter, the classification is the same as the one obtained from XFS, and is correct for the considered location type. XFS,

979-8-3315-3968-9/24 $31.00 © 2024 IEEE

Industrial Contribution

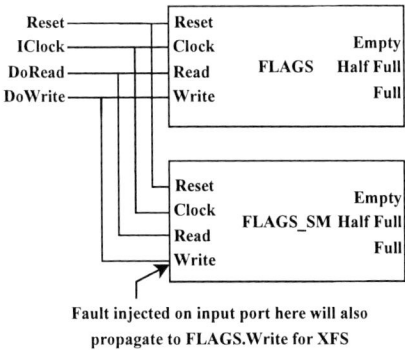

Fault injected on input port here will also
propagate to FLAGS.Write for XFS

Fig. 2. Difference in fault modeling of input/output ports of tools

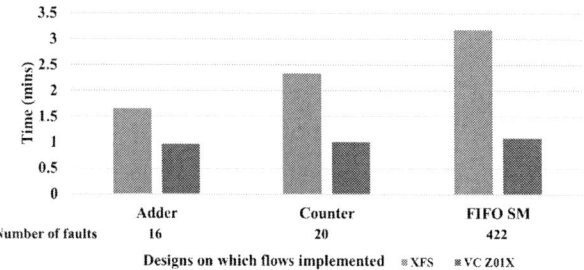

Fig. 3. Run time comparisons of tools on Adder, Counter and FIFO

on the other hand, instruments a signal as a single type only. The effect of a fault and its corresponding classification can vary depending upon whether it is injected at the port itself, or the wire connecting the port. This is not taken into consideration by XFS. There are no further discrepancies in fault classifications for the extended list.

The analysis is also extended to the transient space by injecting faults on all locations within a range of timestamps. There are no additional discrepancies seen in terms of fault classifications. However, there are differences in how transient faults are modeled by the two tools. Modifications are made to fault injection commands to make similar fault lists. The strobing mechanism also makes a difference in the final result depending on whether the signals are strobed at every timestamp or at clock edges. Such differences can also result in different classifications for a given transient fault. Further, the IX classification is observed for certain signals that remain in an unknown state because the testbench fails to assign any value after a certain timestamp. Upon further analysis, it is seen that they can be reclassified as Detected. However, neither tool offers a feature to manually update fault classifications, presenting a limitation common to both.

2) Fault space coverage: The SA fault space for VC Z01X is quite extensive and takes into account the effect of different fault types. XFS does not instrument a signal as different types, resulting in the omission of certain fault effects and thereby not contributing to the final diagnostic coverage. However, VC Z01X cannot inject transient faults on inputs and intermediate wires, leaving a portion of the fault space uncovered. On the other hand, XFS does not restrict transient fault placement on inputs or nets.

3) Run-time: As shown in Figure 3, VC Z01X fares better than XFS in terms of fault simulation run-time owing to its concurrent engine support. Although XFS does provide a concurrent flow, it exhibits limitations, including limited compatibility with RTL designs, unsupported constructs, and the incapability to handle VHDL and SystemVerilog designs. Also, VC Z01X exhibits better scalability in terms of run-time, particularly for larger and more complex designs, as it can instrument thousands of faults in one shot.

D. Comparison conclusions

Both tools do not cover the entirety of the fault space required for FuSa verification. So, the first step in a verification flow should be to address all faults required for FI simulation. Second, if there is an overlap of fault space coverage between the tools, the runtime of the tools needs to be taken into consideration. Therefore, in the next section, we introduce a novel methodology aimed at providing an extensive and robust FuSa verification framework to address these issues.

IV. UNIFIED FuSa VERIFICATION FRAMEWORK

This section discusses the concept of the proposed verification methodology along with its design and implementation.

A. Concept

One of the key objectives of our work is to offer a precise evaluation of DC by covering all faults within the design space. VC Z01X is extensive for SA faults, but this is not the case for transient faults. On the other hand, XFS does not restrict fault placement for transient space, but lacks options for location types as compared to VC Z01X. Thus, the main aim of the proposed verification methodology is to combine the strengths of the tools to cover the entire fault space. If there is any fault space overlap, VC Z01X is chosen owing to its faster simulation capabilities. Further, both tools share a limitation in their inability to update fault classifications of signals when necessary. We take care of this aspect by introducing a feature in the flow to update classifications with an external file.

B. Design and Implementation

Figure 4 provides an overview of the proposed verification methodology. The different steps of the flow are outlined below:

1) We develop automated VC Z01X scripts to concurrently run stuck-at and transient fault campaigns in the first step. All possible location types and fault targets are enabled for SA faults. Further, all supported configurations for transient faults are also enabled. Start and end cycles for fault injection are specified based on design toggle activity and strobe detection possibilities.

2) Signals not subjected to transient fault injection with VC Z01X are identified from reports generated in the first stage.

979-8-3315-3968-9/24 $31.00 © 2024 IEEE   287

Industrial Contribution

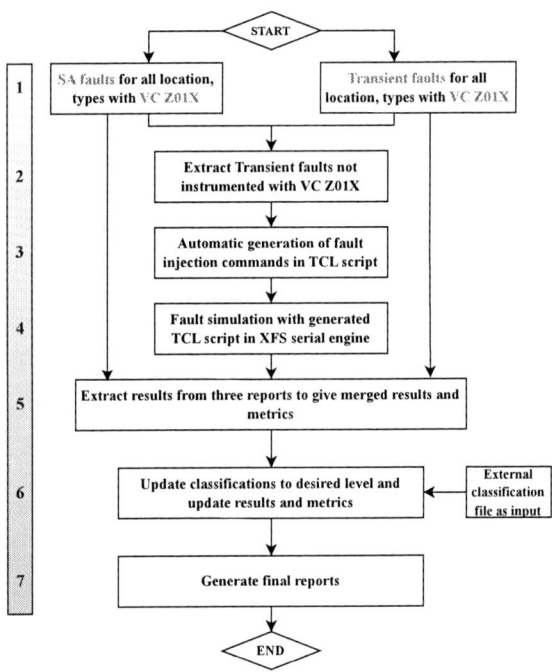

Fig. 4. Proposed verification methodology

3) Fault injection commands are automatically generated using developed scripts in Tool Command Language (TCL) format, considering parameters like clock period, reset time, start cycle, end cycle, and hold time range.
4) Fault simulation is run with generated TCL script using XFS serial engine.
5) The three reports generated are consolidated together to provide individual results along with the merged results, containing the updated DC in a final report.
6) A feature is introduced to manually update fault statuses. An external classification file is used to specify signals for status conversion along with source and destination classifications. Fault classes are updated for specified signals along with classification numbers for individual reports and the final report.
7) Both original and modified merged reports are kept as separate files to track differences before and after using the update classification feature.

The proposed flow covers the maximum possible fault space supported by the tools, improving the accuracy of the calculated DC. Additionally, using VC Z01X's concurrent engine ensures efficient simulations, minimizing runtime. There is also the option of triggering multiple runs with the XFS serial engine, which can help speed up the remainder of the fault simulation, albeit at the cost of using multiple licenses. This methodology is further validated in the next section.

V. VALIDATION

This section discusses the validation of the proposed methodology on the discussed designs and an automotive SoC.

A. Experimental setup

The proposed methodology is validated on the three designs previously discussed. Further, to evaluate the flow's effectiveness at a larger scale, we select the Automotive SoC (AutoSoC) [13, 14], an open-source benchmark suite designed for automotive SoC applications. AutoSoC consists of configurable hardware IPs integrated into an SoC, offering diverse SMs (Dual core lock step processor, bus parity, checkpoint control, ECC on memories, Software Test Libraries) and automotive software applications.

SMs are enabled in the design with the help of additional defines (for example, +define+MEMECC for ECC on memories). The user can create a new configuration based on a combination of SMs. Functional strobes, consistent across all configurations, include signals for instruction and data buses, as well as Special Purpose Register accesses to external units like cache and Memory Management Unit. The checker strobes vary depending on the SMs enabled. For instance, if ECC is enabled, the ECC error detection from individual modules is configured as checker strobes. Fault targets and exclusions are user-dependent and adjustable to meet specific module-level injection needs.

We identify potential areas of improvement based on an initial analysis of results obtained from the baseline verification flow (XFS) used in AutoSoC. Thus, we implement ECC on all internal memories, along with the duplication of fetch, control, and load-store unit module duplication with temporal redundancy. This is the configuration on which the proposed verification flow is validated. SA and transient faults are enabled for fault simulation.

We introduce a new metric called "Total relative effort", which highlights the increase in the number of faults to be injected as compared to the individual tools. This provides an insight into the trade-offs required to get an accurate estimation of DC.

B. Results

Table IV presents a comparison of the results of fault simulation campaigns run with the individual tools and the proposed verification flow. As seen from the results, the proposed verification flow provides a more accurate DC estimation between the bounds of individual tool coverages. Our work provides a better representation of the fault space to be actually considered for FuSa verification purposes using the combination of two tools. Of course, we have to inject more faults compared to individual tool flows, thereby signifying more fault simulation effort. Nonetheless, for the largest design, AutoSoC, the overhead is 31% and 22% for XFS and VC Z01X, respectively. This trade-off results in a more accurate DC calculation which in turn leads to the correct assignment of ASIL level. For example, the initial results obtained from the proposed verification flow on the AutoSoC design results

979-8-3315-3968-9/24 $31.00 © 2024 IEEE   288

TABLE IV
COMPARISON OF DC, NUMBER OF FAULTS AND TOTAL RELATIVE EFFORT

		XFS	VC Z01X	Proposed methodology	Total relative effort (XFS)	Total relative effort (VC Z01X)
Adder	DC	87.50%	82.14%	83.42%		
	Fault count	32	44	62	1.94	1.4
Counter	DC	60.00%	58.62%	59.70%		
	Fault count	40	38	48	1.20	1.26
FIFO	DC	68.96%	80.47%	70.36%		
	Fault count	5797	5412	6982	1.20	1.29
AutoSoC	DC	98.36%	96.12%	97.79%		
	Fault count	676213	726097	885839	1.31	1.22

in a DC of 95.88% (ASIL B) as compared to the DC from the baseline verification flow (98.36% specifying an ASIL C). With the update classification feature of the flow, we are able to elevate certain fault classifications, resulting in the final DC of 97.79%. However, such varying classifications could have dire implications on the ASIL, if the fault space is not correctly evaluated. Our work addresses this concern by considering the maximum possible fault space covered by the tools to provide an accurate estimation of DC.

VI. DISCUSSION

We summarize the main takeaways from the results in the following points:

1) **Fault space coverage**: In order to correctly determine the ASIL of an automotive component, an accurate estimation of DC is important. We make sure that the FI simulation space is completely covered by a combination of the tools.

2) **Simulation run-time**: By basing our solution on VC Z01X concurrent engine, we are able to minimize the simulation overhead. Nonetheless, it is also possible to base the framework on XFS.

3) **Transient fault space complexity**: The transient fault space is quite extensive due to the possibility of multiple injection and hold times during a simulation. Fault space pruning needs to be considered to effectively choose transient faults that will propagate to the strobing points, thus preventing the simulation of safe faults.

4) **Future work**: Future methodologies could involve implementing automated test generation for undetected faults. This will reduce manual verification efforts further and expedite fault simulation campaigns, particularly at higher abstraction levels like RTL. Additionally, it remains to be seen if other FuSa EDA tools, such as Siemens'® Kaleidoscope™, can provide better results based on the defined metrics.

VII. CONCLUSION

The increasing demand for safety-critical electronic components in automobiles necessitates a thorough research of FuSa EDA tools used for fault simulation purposes. In this paper, we compare two FuSa EDA tools from Cadence® and Synopsys® for FI simulation to identify discrepancies in results and evaluate their effectiveness in covering the fault space. Neither tool covers the fault space entirely, and therefore, they do not contribute to an accurate DC estimation. Therefore, we introduce a novel verification methodology that combines the tools along with additional features and utilities. This proposed flow is tested on an automotive SoC, achieving a DC of 97.79%. Compared to the baseline verification flow, which yields a DC of 98.36%, the methodology provides a more comprehensive estimation by considering the maximum possible fault space, with an increased fault simulation effort of 1.31x. Our automated approach provides an end-to-end verification framework to conduct FuSa verification and obtain a highly accurate DC.

REFERENCES

[1] J. P. Trovao, "Trends in automotive electronics [automotive electronics]," *IEEE Vehicular Technology Magazine*, vol. 14, no. 4, pp. 100–109, 2019.

[2] "ISO 26262 Road Vehicles - Functional Safety," International Organization for Standardization, Dec. 2018.

[3] A. Cagri Bagbaba *et al.*, "An automated formal-based approach for reducing undetected faults in ISO 26262 hardware compliant designs," in *2021 IEEE International Test Conference (ITC)*, 2021, pp. 329–333.

[4] A. Nardi *et al.*, "Functional safety methodologies for automotive applications," in *2017 IEEE/ACM International Conference on Computer-Aided Design (ICCAD)*, 2017, pp. 970–975.

[5] A. C. Bagbaba *et al.*, "Combining fault analysis technologies for iso26262 functional safety verification," in *2019 IEEE 28th Asian Test Symposium (ATS)*, 2019, pp. 129–1295.

[6] F. A. d. Silva *et al.*, "Efficient methodology for ISO26262 functional safety verification," in *2019 IEEE 25th International Symposium on On-Line Testing and Robust System Design (IOLTS)*, 2019, pp. 255–256.

[7] K.-L. Lu *et al.*, "FMEDA-based fault injection and data analysis in compliance with ISO-26262," in *2018 48th Annual IEEE/IFIP International Conference on Dependable Systems and Networks Workshops (DSN-W)*, 2018, pp. 275–278.

[8] D. Alexandrescu *et al.*, "EDA support for functional safety — how static and dynamic failure analysis can improve productivity in the assessment of functional safety," in *2017 IEEE 23rd International Symposium on On-Line Testing and Robust System Design (IOLTS)*, 2017, pp. 145–150.

[9] F. Ferlini *et al.*, "Enabling ISO 26262 compliance with accelerated diagnostic coverage assessment," *Electronics*, vol. 9, no. 5, 2020.

[10] C. Lopez-Ongil *et al.*, "Autonomous fault emulation: A new FPGA-Based acceleration system for hardness evaluation," *IEEE Transactions on Nuclear Science*, vol. 54, no. 1, pp. 252–261, 2007.

[11] O. Ballan *et al.*, "Verification of soft error detection mechanism through fault injection on hardware emulation platform," in *2010 International Conference on Dependable System and Networks Workshops (DSN-W)*, Jun, 2010.

[12] A. Sherer *et al.*, "Ensuring functional safety compliance for ISO 26262," in *2015 52nd ACM/EDAC/IEEE Design Automation Conference (DAC)*, 2015, pp. 1–3.

[13] F. A. da Silva *et al.*, "Special session: AutoSoC - a suite of open-source automotive SoC benchmarks," in *2020 IEEE 38th VLSI Test Symposium (VTS)*, 2020, pp. 1–9.

[14] "Autosoc benchmark suite." (2020), [Online]. Available: https://www.autosoc.org/home. (accessed: Jan 13, 2023).

3D VNWFET-Based Standard Cell Library Design Flow: From Circuit and Physical Design to Logic Synthesis

Sara Mannaa*, Cédric Marchand*, Damien Deleruyelle*, Bastien Deveautour*, Alberto Bosio*, Christoph Lenz[†]
Oskar Baumgartner[†], Ian O'Connor*

*Ecole Centrale de Lyon, INSA Lyon, CNRS, Universite Claude Bernard Lyon 1, CPE Lyon, INL,
UMR5270, 69130 Ecully, France
[†]Global TCAD Solutions GmbH – Vienna, Austria
email {firstname.lastname}@ec-lyon.fr, bastien.deveautour@cpe.fr, damien.deleruyelle@insa-lyon.fr,
c.lenz@globaltcad.com, o.baumgartner@globaltcad.com

Abstract—The vertical nanowire field effect transistor (VN-WFET) is an emerging technology that promises to improve the sustainability of future transistor scaling beyond the limitations of conventional lateral devices. With its 3D gate-all-around (GAA) architecture, such a technology enables designs with improved energy-efficiency as well as reduced footprint and thus interconnect capacitance. In this work, and based on the compact model of a real VNWFET device, we present the design flow for the generation of a standard cell library starting from the circuit and physical design of logic cells to logic synthesis based on the VNWFET technology. The results on the synthesized benchmark cells, as compared against 45nm and 65nm CMOS libraries, demonstrate a significant decrease in the average dynamic power consumption and delay values up to 71X and 34X respectively, with an average area gain of up to 5X. However, an increase in leakage power consumption (up to 2X on average) was also observed.

Index Terms—Emerging Technology, Vertical Nanowires, Gate-all-around field effect transistor, Standard Cell Library, Physical Design, Logic Synthesis

I. INTRODUCTION

The massive increase in the number of smart devices deployed at the edge, from cellphone light sensors to advanced driver assistance systems, has dramatically increased the quantity of generated raw data to be processed. This translates to a need for ever-higher edge computing performance within stringent energy, delay, security and bandwidth constraints. Thus, the design of hardware processors capable of performing complex computational operations is facing many challenges. Improving such processors starts by improving the basic building blocks and circuits at the heart of key operations. During Moore's law, it was possible to rely on scaling to increase the number of transistors integrated within processors in order to achieve the required performance improvements. However, this approach is now limited by the physical properties of CMOS devices, reflected mainly through the increase in power consumption per chip as devices scale down. Hence, new emerging technologies are under active investigation to overcome these limitations while maintaining the required performance and constraints. One promising alternative consists of moving toward 3D technologies such as the Vertical Nanowire Field Effect Transistor (VNWFET) [1];

an emerging technology that can be considered as "More Moore" technology. While its operation is based on the same principle as that of the field effect transistor, the semiconductor channel is a nanowire sandwiched between two electrodes. This technology relies on a Gate All Around (GAA) architecture which will give improved electrostatic control, reduced short-channel effect and better scalability opportunities. In this work, we present a complete design flow enabling the successful generation of a standard cell library based on VNWFET technology. This includes the characterization of basic logic cells under different drive strengths as well as the generation and verification of the physical layout of the standard cells. We finally tested our library by synthesizing some logic circuits.

The remainder of this paper is organised as follows: Section II describes the main aspects of the VNWFET technology. Section III details the work carried out toward the generation of the standard cell library while Section IV explains our approach targeting the generation of the physical layouts of the cells present in the library, as well as the verification of our design. Then, Section V shows the results of the logic synthesis. Finally, Section VI concludes our work and summarizes possible future work.

II. VERTICAL JUNCTIONLESS TECHNOLOGY

In this work, we use a VNWFET with a junctionless (JL) GAA architecture as shown in Figure 1. It is composed of a homogeneous highly doped nanowire channel patterned into a silicon substrate, which itself is highly doped with boron [1]. In vertical GAA designs, the gate material completely surrounds the nanowire channel, forming a cylindrical control region around the vertical silicon nanowires which improves both power consumption and immunity toward short channel effects. In addition, JL devices [2] can overcome the complex manufacturing process of the p-n junctions (present in traditional transistors) with the continuous shrinking of the devices. These aspects, along with the device's vertical orientation help to push the limits of scaled transistors while maintaining the desired performance and reducing footprint. Throughout this work, we consider a device compact model [3] which is based

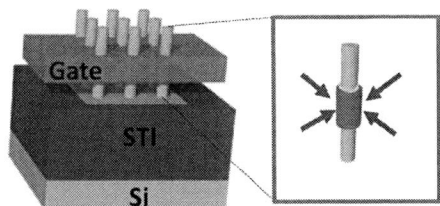

Fig. 1. Vertical gate-all-around junctionless nanowire transistor (JLNT) [4]

on the actual fabricated device described above and the physics of carrier transport in the junctionless architecture.

III. STANDARD CELL LIBRARY CHARACTERIZATION

For the generated library, we adopt the Liberty (.lib) file format, which includes the timing and physical information relative to the standard cells present in the library while also adopting the nonlinear delay model (NLDP). In this sense, Lookup tables (LUTs) are defined to store timing and power values or constraints. For the generation of the VNWFET logic cell library [5], we designed and characterized four elementary logic cells that should be available in such a library: INV1X1_CStatic_JL1, NAND2X1_CStatic_JL1, NOR2X1_CStatic_JL1 and XOR2X1_CStatic_JL1 where the formalism *OPnXk_Style_Technology* indicates the Boolean operation OP, the number of inputs n and the number of outputs k, as well as the logic design style (i.e., Complementary static logic obtained by using n- and p-type VNWFET) and the technology variant used to implement the cells (i.e. JL nanowire transistor with single gate: JL1). In this context, we carried out SPICE simulations where we used the compact model described in Section II implemented as an executable Verilog-A model for the VNWFET. Throughout this work, and based on the fabricated device and associated model, the gate physical length l_g and the nanowire (NW) diameter d_{nw} are set to 18 nm and 22 nm respectively. However, and in order to study the cells' behavior under different drive strength, we designed different instances of each cell by varying the number of NWs. For simple cells such as the inverter, we found that a ratio of 1 between the n-type and p-type NW values will give well-matched rise and fall times of the cells. For the other cells, and in order to achieve drive strengths equivalent to that of the inverter, we redefined the ratio by doubling the number of NWs in the case of two series transistors. After defining the sets of NW parameters, and verifying the logical functionality of the designed cells (i.e. assuring that all output transitions respect the cells' truth table), we performed a detailed study of their static and dynamic behavior by evaluating the energy consumption and delay while varying the number of NWs used per transistor.

We observed that with the increase of the number of NWs the static and dynamic power (calculated by measuring the supply current values during all static combinations of inputs and output transitions respectively) increase whereas the delay (i.e. the time difference between the output voltage and input voltage to reach half V_{dd})

TABLE I
LOGIC CELLS PRESENTED IN OUR VNWFET GENERATED LIBRARY AND THEIR AREA VALUES

Logic Cells	n-type #NW	p-type #NW	Area(um^2)
Combinational Cells			
BUF1X1_Static_JL1 Instance	4	4	0.116
	24	24	0.403
	44	44	0.694
	64	64	0.982
INV1X1_Static_JL1 Instance	4	4	0.188
	24	24	0.764
	44	44	1.344
	64	64	1.921
NAND2X1_Static_JL1 Instance	8	4	0.314
	48	24	1.470
	88	44	2.625
NOR2X1_Static_JL1 Instance	4	8	0.401
	24	48	1.441
	44	88	2.712
XOR2X1_Static_JL1 Instance	8	8	0.902
	48	48	3.791
	88	88	6.681
Sequential Cell			
DFFSR	24NW based logic cells & two 4NW INV		4.544

decreases as shown in (Figure 2. In addition to the above combinational logic cells, we also designed a sequential gate, the D-type Flip Flop (DFF), with synchronous set/reset control signals. Unlike other cells, we only designed one variant, due to convergence limitations in the device compact model. The designed DFF uses mainly 24-NWs logic gates, with two 4-NW INV1X1_CStatic_JL1 gates used as weak inverters in the bistable circuit elements. The current version of the logic cell library is generated by using a commercial tool. The tech library is named JLNT and its cells are summarized in Table I.

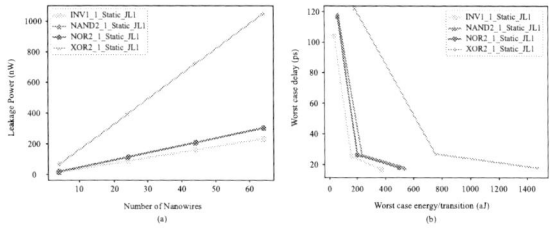

Fig. 2. Plots showing (a) the leakage power and (b) the Pareto Fronts for different logic cells while varying the number of nanowires per VNWFET

IV. PHYSICAL DESIGN AND LAYOUT

Physical design is an important step in the characterization flow of logic cells. It enables an actual estimation of area values and more accurate timing and power consumption values that can be obtained after the parasitic extraction from such designs. We developed an automated physical design flow for logic cells. We based this work on the use of an existing open-source tool [6] which is available as a satisfiability modulo theories (SMT)-based many-tier VFET standard cell (SDC) synthesis framework that provides minimum-sized cell layouts by performing concurrent Place and Route (P&R) of FETs. The SMT tool generates a .conv file, which is essentially a text file describing the unsized layout template of the desired logic cell. In order to generate a full layout in GDSII file format, sized according to the design rules defined by the technology, we extended the flow of the tool. The first step was to extract from the generated .conv file the necessary

979-8-3315-3968-9/24 $31.00 © 2024 IEEE

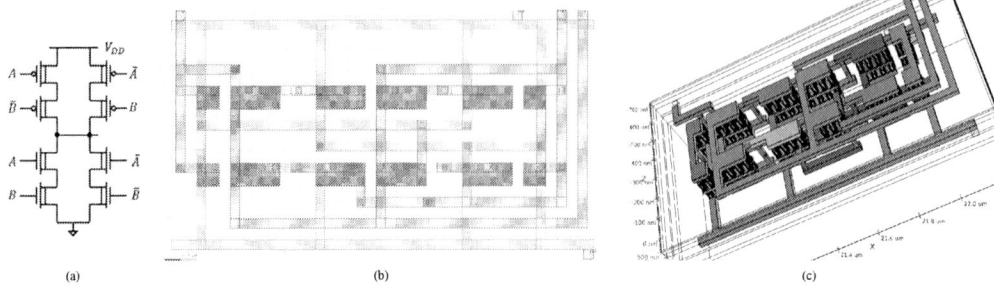

Fig. 3. (a) Schematic (b) layout, and (c) 3D view of a 2-input XOR2 cell

information relative to metal layers, transistors, and pins to be transformed into an actual layout. To this end, we developed a Python script which then adapts the layout template to user-defined technology rules and generates the full layout using the GDSII tool kit (Gdstk [7]; C++ library available as a Python module). We then generated a preliminary physical layout for the cells in the standard cell library as described above (in GDSII file format). This allowed us to obtain the height and width of the cells and thus their area values as summarized in Table I. We set all cells to the same height (0.688 um) to enable conventional row-placement, while width varies according to the complexity of the cell as well as the number of nanowires used per cell. We were then able to generate a Library Exchange File (.lef) containing the physical data for the library cells. In order to verify that the generated layout has no issues and will guarantee the functional logic behavior of the standard cells, we completed a verification test using Global TCAD Solutions (GTS) tools. We chose to carry out the verification on a two-input XOR gate, one of the more complex logic cells. As shown in Figure 3b, the cell under test, in addition to the XOR logic functionality (8 NWs per transistor), also has two inverters (4 NWs per transistor) that are used for inverting the two inputs. After completion of the parasitic extraction, a transient analysis was carried out using the generated netlist. This analysis shows no degradation of the XOR logic function-ality. We then carry out a comparative study in order to quantify the impact of parasitic interconnect networks on key performance metrics. Comparing delay and energy per transition figures, we quantify the average increase in delay as 1.72X and energy/transition as 1.13X when considering the parasitic interconnect network (Figure 4).

V. LOGIC SYNTHESIS

A typical Logic Synthesis flow takes as input a circuit model description, usually expressed in a hardware descrip-tion language (e.g., VHDL) and produces as output the gate level netlist by using building blocks (i.e. logic cells) from the provided standard cell library. To test our generated library, we first targeted the synthesis of some of the ISCAS-85 benchmarks [8] and the subsequent comparison of the obtained results with conventional CMOS logic cell libraries (i.e. 45 nm free PDK [9] and a 65 nm industrial logic cell library PDK) Table II summarizes the main reports (i.e. the number of cells used to synthesize the target benchmarks, the corresponding physical area values

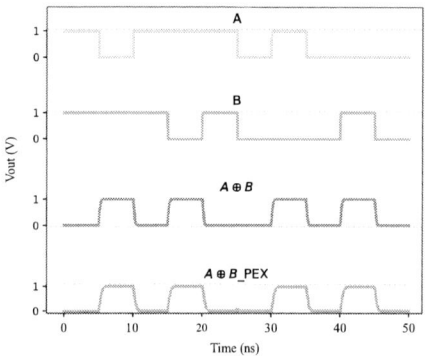

Fig. 4. Transient simulation results with and without parasitic annotation of the XOR gate

as well as the expected delays) obtained from benchmarks synthesis. We can see that the number of cells used by the 45 nm and 65 nm technologies to synthesize the circuits is smaller than that used by our VNWFET library (i.e. JLNT lib). This is due to the fact that these libraries are full libraries containing many more logic cells than those in our library. However, we can still achieve an average area gain using the synthesized cells up to 5.57X and 4.53X as compared to the area values of the 45 nm and 65 nm respectively. Also, the VNWFET library was able to improve the delay values up to 3.33X and 2.81X on average as compared to the full 45 nm and 65 nm libraries respectively. Comparing the power values, we can see that the circuits synthesized with the VNWFET library have

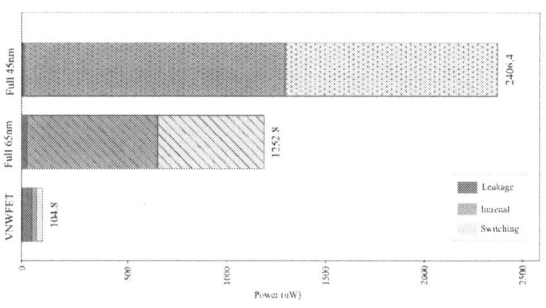

Fig. 5. Detailed power values of benchmark circuit C6288

979-8-3315-3968-9/24 $31.00 © 2024 IEEE

TABLE II
CELL COUNT SCALABILITY, AREA VALUES AND DELAYS OF SYNTHESIZED CIRCUITS

Benchmark Circuits	Cell Count			Area Values (um²)			Delay (ns)		
	VNWFET	Full 45 nm	Full 65 nm	VNWFET	Full 45 nm	Full 65 nm	VNWFET	Full 45 nm	Full 65 nm
C432	172	127	78	47.58	293.78	219.44	0.41	1.24	0.83
C499	229	254	142	128.53	726.95	588.12	0.27	0.90	0.87
C1355	193	248	135	120.22	706.77	562.12	0.23	0.81	0.76
C1908	247	268	175	115.65	711.46	556.92	0.38	1.37	1.12
C6288	1503	625	591	712.76	2857.57	2840.76	0.99	3.12	2.52
Processing Element	VNWFET	Rest. 45nm	Rest. 65nm	VNWFET	Rest. 45nm	Rest. 65nm	VNWFET	Rest. 45nm	Rest. 65nm
N2C2 4_bits	229	530	234	152.85	1055.46	624.52	0.087	0.156	0.064
N2C2 8_bits	769	1674	761	432.43	3288.39	1985.36	0.114	0.185	0.071
N2C2 16_bits	2699	5697	2648	1503.02	11074.54	6764.16	0.187	0.235	0.087
N2C2 32_bits	10020	20608	9884	4597.06	39733.28	24717.68	0.335	0.340	0.122

very low dynamic power values (up to an average decrease of 71X and 34X as compared to the 45 nm and 65 nm technologies). The library also showed a significant improvement on how the dynamic power is distributed among the internal and switching power values. The internal power occupies an average of 32% of the dynamic power instead of 56% and 47% for the 45 nm and 65 nm technologies respectively. However, the leakage power dominates the dynamic power for the cells synthesized by our library (an average of 53% of the total power) which is 2X higher as compared to other technologies. Still, the VNWFET gave the lowest total power i.e. an average of 26X and 13X decrease in total power consumption as compared to the 45 nm and 65 nm libraries respectively. For simplicity, only the detailed power values of one benchmark circuit is shown in Figure 5. We then targeted the synthesis of a Neural Network Compute Cube (N^2C^2) [10], a processing element capable of performing multiplication-accumulation (MAC) operations and can be used for neural networks. We synthesized 4 different instances of the N^2C^2 block with data sizes of 4,8,16 and 32 bits, aiming to assess the scalability of VNWFET in comparison to traditional CMOS technology. Due to the fact that our library contains only basic cells, and in order to make a fair comparison, we restricted the 45 nm library to only the logic cells present in the VNWFET library, while we only restricted some of the 65 nm cells. The obtained reports after the synthesis of the N^2C^2 benchmarks are also shown in Table II. Comparing the cell count, we can see that the synthesized N^2C^2 circuits by our library outperforms the restricted 45nm library, but uses almost the same hardware resources as that of the restricted 65 nm library. However, taking into account the physical cost of the synthesized benchmark circuits, we can clearly see that the VNWFET library provides the lowest area values among the other libraries. Based on the latency reports, we can observe that the VNWFET performs better than the restricted 45 nm however it performs poorly as compared to the 65nm restricted library. This observation contradicts what we observed for the above benchmark circuit; this could be due to the fact the N^2C^2 cells are composed of both sequential and combinational cells and still the restricted 65 nm library contains much more cells than our library. For the power consumption values we observe the same trend; i.e. both versions of 45nm as well as the restricted 65nm always perform worse than that of the VNWFET version in terms of the total power consumption. When looking at the details however, we observe the same behavior as for

the synthesized benchmarks, i.e. the leakage power of the VNWFET library represents its highest consumption.

VI. CONCLUSION AND FUTURE WORK

In this work, we presented a complete flow targeting the generation of a standard cell library based on a real VNWFET device. We detailed our approach starting from the logic cells characterization to the physical design and verification methodology as well as the generation of the library .lef file. We were able to use our generated library to synthesize well-known benchmark circuits and our compute cube then to compare the obtained results with well-known CMOS technologies. Our library generally showed improvements in terms of delay, area, dynamic power and short-channel effect values, but it also showed an increase in the leakage power values. In fact, the increase in leakage power values is expected with more advanced technologies due to the lower threshold voltage and shrinked devices. Further investigations will be carried out on the obtained leakage power values.

ACKNOWLEDGMENT

This work has been funded by the European Union's Horizon 2020 research and innovation programme under grant agreement No 101016776 (FVLLMONTI).

REFERENCES

[1] G. Larrieu *et al.*, "Vertical nanowire array-based field effect transistors for ultimate scaling," *Nanoscale*, vol. 5, no. 6, pp. 2437–2441, 2013.

[2] J.-P. Colinge *et al.*, "Nanowire transistors without junctions," *Nature nanotechnology*, vol. 5, no. 3, pp. 225–229, 2010.

[3] C. Mukherjee *et al.*, "Compact modeling of 3d vertical junction-less gate-all-around silicon nanowire transistors towards 3d logic design," *Solid-State Electronics*, vol. 183, p. 108125, 2021.

[4] I. O'Connor *et al.*, "Fvllmonti: The 3d neural network compute cube (n^2c^2) concept for efficient transformer architectures towards speech-to-speech translation," in *2024 Design, Automation Test in Europe Conference Exhibition (DATE)*, 2024, pp. 1–6.

[5] S. Mannaa *et al.*, "Vnwfet-based technology: From device modelling to standard cell library," in *2023 IEEE 23rd International Conference on Nanotechnology (NANO)*, 2023, pp. 576–581.

[6] D. Lee *et al.*, "Many-tier vertical gate-all-around nanowire fet standard cell synthesis for advanced technology nodes," *IEEE Journal on Exploratory Solid-State Computational Devices and Circuits*, vol. 7, no. 1, pp. 52–60, 2021.

[7] "Gdsii tool kit (gdstk) github repository," Available at = https://github.com/heitzmann/gdstk.

[8] F. Brglez *et al.*, "A neutral netlist of 10 combinational benchmark circuits and a targeted translator in fortran," 06 1985.

[9] "45 nm free pdk liberty file," Available at = https://github.com/The-OpenROAD-Project-Attic/OpenROAD-Utilities/blob/master/TimerCalibration/Free45PDK/gscl45nm.lib.

[10] R. Bishnoi *et al.*, "Energy-efficient computation-in-memory architecture using emerging technologies," in *2023 International Conference on Microelectronics (ICM)*. IEEE, 2023, pp. 325–334.

979-8-3315-3968-9/24 $31.00 © 2024 IEEE

AUTHOR INDEX

Aarella, Seema G.	272, 278
Aftabjahani, Sohrab	13
Agarwal, Ayushi	25
Ahmad, Isaar	25
Ahmadilivani, Mohammad Hasan	176
Ahmed, Bulbul	13
Aldea, C.	156
Alrahis, Lilas	260
Amraoui, Sami El	7
Arrassi, Asmae El	103
Ayache, Mouadh	192
Aznar, F.	156
Babukutty, Subin	284
Baloch, Sajid	172
Bangad, Nikhil Jagdish	278
Barbareschi, Mario	164
Barone, Salvatore	164
Basak, Debajit	248
Bathalapalli, Venkata K. V. V.	266
Baumgartner, Oskar	290
Bende, Ankit	115
Berekovic, Mladen	192
Bhattacharjee, Debjyoti	79
Bhatti, M. Kamran	160
Bhowmik, Abhiroop	284
Bidkar, Darshan Mohan	278
Bilal, Mujahid	160
Biyani, Yashvardhan	103
Blanton, Shawn	152
Bolderik, Bram Van	204
Bosio, Alberto	176, 290
Bossuet, Lilian	196
Castillo, Ernesto Cristopher Villegas	67
Catthoor, Francky	79
Celma, S.	156
Chakraborty, Supriya	138
Chattopadhyay, Anupam	43, 248
Chen, Junchao	220
Cherezova, Natalia	176
Ciesielski, Maciej	73
Condia, Josie E. Rodriguez	220
Copetti, Thiago Santos	138
Corporaal, Henk	148
Corradi, Federico	148
Dasari, Jiteshri	73
Deb, Suman	248
Deleruyelle, Damien	290
Deveautour, Bastien	176, 290

Dharwadkar, Radhika	25
Drechslert, Rolf	115
Dube, Ayushi	85
Dworzak, Thorsten	184
Elaraby, Nahla	208
Elfadel, Ibrahim M.	19
El-Hadbi, Assia	216
Elia, Rafaella	31
Elissati, Oussama	216
Emmanuele, Antonio	164
Eraso, U. Esteban	156
Esmaeilpour, Mohammadreza	127
Esposito, Giuseppe	220
Ewert, Christian	254
Farahmandi, Farimah	13, 168
Fellah-Touta, Anis	196
Feng, Jerrie	43
Fesquet, Laurent	144, 216
Fieback, Moritz	284
Gaillardon, Pierre-Emmanuel	132
Gamil, Homer	200
Ganti, Arun	37
Gaydadjiev, Georgi	121
Gebregiorgis, Anteneh	103
Gergely, Istvan Andras	208
Ghosh, Prokash	25
Giraud, Bastien	1
Glaß, Michael	67
Gogoi, Ankur	37
Gomony, Manil Dev	103, 148, 204
Grenouillet, Laurent	1
Grosso, Vincent	196
Guerrero-Balaguera, Juan-David	220
Guizzetti, Roberto	91
Hamdioui, Said	103, 121
Heemstra, Sonia	204
Heinkel, Ulrich	184
Hroub, Ayman	212
Iskandar, Paulette	55
Ismael, Mohammad	212
Jantsch, Axel	208
Jayaram, Vivekananda	278
Jenihhin, Maksim	176
Jha, Chandan Kumar	115
Joseph, P. J.	25
Joshi, Rajiv	103
Kapoor, Hemangee K.	109
Kathiriya, Manoj Jayantilal	278

Khan, Esrat	19
Knechtel, Johann	200, 260
Kolluru, Sumanth	61
Korb, Matthias	192
Kougianos, Elias	230, 266, 272
Krishnappa, Manjunatha Sughaturu	278
Kritikakou, Angeliki	220
Krstic, Milos	220
Kumar, Aakarshan	266
Kumar, Krishna	25
Kunz, Wolfgang	55
Laguerre, Julie	1
Lappas, Jan	127
Lara-Nino, Carlos Andres	196
Laubeuf, Nathan	79
Lauga-Larroze, Estelle	144
Lenz, Christoph	290
Lettnin, Djones	55, 184
Leveugle, Régis	7
Li, Xiufan	43
Lim, Eugene	43
Lin, Tsung-Han	152
Mahanta, Rishabh	109
Maistri, Paolo	7
Makryniotis, Thomas	121
Maniatakos, Michail	200
Manna, Kanchan	37
Mannaa, Sara	290
Marchand, Cédric	1, 290
Mazzocca, Nicola	164
Mellenthin, Knut	180
Menkovski, Vlado	204
Merchant, Farhad	115
Minhas, Muhammad Kahsif	160
Minhas, Muhammad Kashif	172
Mir, Fouwad	103
Mohammad, Sajeed	168
Mohanty, Saraju P.	230, 266, 272
Moussa, Hasan	144
Mulhem, Saleh	192, 254
Murthy, Nitish Satya	79
Musale, Tejas	37
Muzaffar, Shahzad	19
Nabeel, Mohammed Thari	260
Nabeel, Mohammed	200
Nagar, Jaimini	184
Nagpal, Akshay	278
Nair, Harideep	152
Naser, Nasib	212
Nešković, Andrija	254
Noel, Jean-Philippe	1
O'Connor, Ian	290
O'Connor, Ian	1
Olmos, Bryan	55
Panda, Preeti Ranjan	25
Pappalardo, Salvatore	176
Patkar, Sachin	115
Poehls, Letícia Maria Bolzani	138
Poudel, Amrit Sharma	254
Prieto, Arturo	188
Qassem, Lamees M. Al	19
Rachakonda, Laavanya	230, 236, 242
Raik, Jaan	176
Rajendran, Gokulnath	248
Rama, Enkele	192
Rana, Vikas	115
Rausch, Sebastian	208
Ravinarayanan, Balajiraja	180
Reis, Ricardo	49
Reorda, Matteo Sonza	220
Rhetat, Lucas	1
Rodrigues, Joachim	188
Roy, Sourav	25
Saha, Sujan Kumar	13
Sánchez-Azqueta, C.	156
Santos, Fernando Fernandes Dos	220
Shadmehri, Seyed Hossein Hashemi	138
Shahroodi, Taha	103
Shamsa, Mahdi	230
Shen, John Paul	152
Shisha, Ali	180
Sierra, Robert Limas	220
Silva, Felipe Augusto Da	67
Simon, Sebastian	184
Sinanoglu, Ozgur	260
Singh, Gian	85
Singh, Simranjeet	115
Snelgrove, Ashton	132
Stasiewicz, Samuel	236
Stevens, Kenneth S.	61
Stockham, Skylar	132
Taheri, Mahdi	176
Taly, Emilien	91
Tao, Xingjian	103
Taouil, Mottaqiallah	121, 284
Tehranipoor, Mark	13
Theocharides, Theocharis	31
Traiola, Marcello	220
Urard, Pascal	91
Vargas, Fabian Luis	138
Vatajelu, Elena-Ioana	91
Vellaisamy, Prabhu	152
Verhelst, Marian	79
Veronesi, Alessandro	220

Vizcardo, Luis Enrique Murillo 49
Vrudhula, Sarma 85
Wang, Perry 152
Wang, Siyi.. 43, 248
Waris, Haroon 160, 172
Wehn, Norbert.. 127
Weis, Christian... 127
Yaldagard, Mohammad Amin 103
Yanambaka, Venkata P. 266, 272
Yu, Cunxi 73
Zhang, Yicheng 148
Zhou, Jingbo 13